T0138552

PROCEEDINGS OF

THE FIRST SINO-AMERICAN WORKSHOP ON

MOUNTAIN METEOROLOGY

PROCEEDINGS OF
THE FIRST SINO-AMERICAN WORKSHOP ON
MOUNTAIN METEOROLOGY

BEIJING, 18-23 MAY, 1982

CHAIRMEN: TAO SHIYEN, ELMAR R. REITER, GAO YOUXI

SPONSORED BY ACADEMIA SINICA
AND BY THE U.S. NATIONAL ACADEMY OF SCIENCES,
COMMITTEE ON SCHOLARLY COMMUNICATIONS
WITH THE PEOPLE'S REPUBLIC OF CHINA

EDITORS:

ELMAR R. REITER
COLORADO STATE UNIVERSITY, FT. COLLINS, COLORADO

ZHU BAOZHEN
ACADEMIA SINICA

QIAN YONGFU
ACADEMIA SINICA

SCIENCE PRESS, BEIJING
AMERICAN METEOROLOGICAL SOCIETY
BOSTON, MASSACHUSETTS

1983

Responsible Editor Zhang Lizheng

Published by Science Press, Beijing

Distribution rights throughout the world, excluding the People's Republic of China, granted to the American Meteorological Society, Boston, Massachusetts

Printed by C & C Joint Printing Co., (H.K.) Ltd.

First published 1984

ISBN 0-933876-58-0

Science Press Book No. 3447-44

PREFACE

This volume contains the proceedings of the First Sino-American Workshop on Mountain Meteorology, held in Beijing from 18 to 23 May, 1982. Thanks to the gracious hospitality extended by the Chinese hosts the meeting was a pleasant and memorable experience for all participants. Our memories will be enhanced by the group photograph which was taken outside the conference hall on the premises of the Friendship Hotel. Unfortunately, the bearded countenance of Rick Anthes is missing from the picture, because on that occasion he had strayed beyond the gates of the compound on a photographic foray of his own and could not be located in time. Dr. James Reardon-Anderson, also shown in the photograph, was the official interpreter for the U.S. delegation which owes him a debt of gratitude for his excellent services.

The tedious work in preparing the copy-ready contributions to this volume deserves a few words of recognition: Mr. Ding Yihui collected the manuscripts from the Chinese authors. He also was instrumental in organizing the workshop and in negotiating with the publisher. Many hours of his time were devoted to making this workshop a success. The papers went through several editorial stages which included the help of the U.S. participants. The typing on an IBM System 6 word processing system was left in the capable hands of Mrs. Bonnie Grantham. The many equations in some of the papers turned typing and page preparation into difficult and time-consuming tasks. Mrs. Juanita Veen painstakingly coordinated the reference lists. She also shared with me the job of making spelling and grammar conform to American usage.

I have done my best to present the names of Chinese authors and of Chinese geographic locations in Pinyin, wherever possible. Such transliterations were not carried through consistently with the names of Chinese authors who are U.S. nationals, nor with Chinese names in the lists of references, because older versions of spelling had become traditional there, including the use of hyphens in two-syllable given names. In keeping with Chinese custom, Chinese authors are listed <u>first</u> with their family name, <u>then</u> with their given name. Exceptions to this rule, again, were granted to U.S. nationals of Chinese extraction. This explanatory note, hopefully, will reduce, if not prevent (who reads a Preface?), the confusion to be expected in citations of the papers contained in this volume.

The printing of this book was expertly done by Science Press of Beijing. Financial contributions for manuscript preparation from the U.S. National Academy of Sciences' Committee on Scholarly Communications with the People's Republic of China, and from Colorado State University are gratefully acknowledged.

The photograph on the dust jacket of this book shows a view from Drepung Monastery near Lhasa, Tibet, and conveys a sense of orography and atmopshere interacting over the world's highest plateau.

Fort Collins, Colorado, January 1983 Elmar R. Reiter

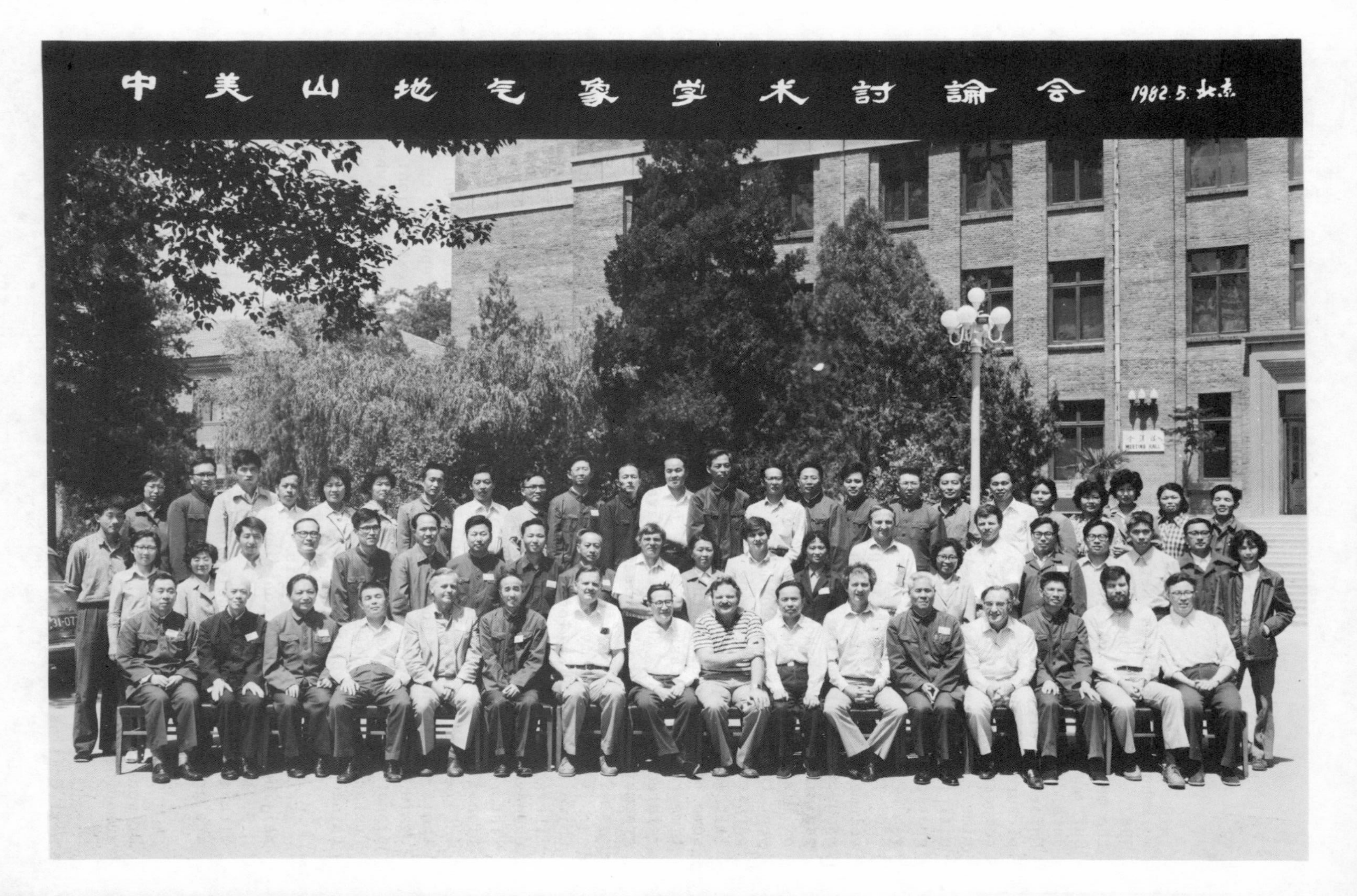
中美山地气象学术討論会 1982.5.北京

PARTICIPANTS

(0 Richard A. Anthes), 1. Chen Longxun, 2. Luo Siwei, 3. Fu Baopu, 4. Sun Chao, 5. W. Lawrence Gates, 6. Zhu Baozhen, 7. Douglas K. Lilly, 8. Ye Duzheng, 9. Elmar R. Reiter, 10. Gao Youxi, 11. John C. Wyngaard, 12. Tao Shiyen, 13. Isidoro Orlanski, 14. Zeng Qingcun, 15. Ronald B. Smith, 16. Zhang Guangkun, 17. Zhou Fengxian, 19. Lu Peisheng, 20. Lin Zhiguang, 21. Yu Zhihao, 22. Wei Tongjian, 23. Lin Benda, 24. Shou Shaowen, 25. Shen Zhibao, 26. Wang Tan, 27. William R. Cotton, 28. Zhang Yan, 29. Bruce A. Albrecht, 30. Cai Zeyi, 31. John B. Hovermale, 32. Lu Yufang, 33. James Reardon-Anderson, 34. Qian Yongfu, 35. Yan Hong, 36. Shen Rujin, 37. Ding Yihui, 39. Li Rongfeng, 40. Dong Shuanglin, 42. Huang Ronghui, 43. Wang Jiaxiu, 44. Li Ci, 46. Wang Anyu, 47. Ji Zhongzhen, 48. Sung Zhengshan, 49. Ge Zhengmo, 50. Sang Jianguo, 52. Zhou Xiaoping, 53. Fang Zongyi, 56. Xu Youfeng, 57. Gao Dengyi, 58. Luo Meixia, 59. Chen Yuxiang, 60. Wang Xiaobai, 61. Zheng Aiying, 62. Qian Chengan.

CONTENTS

OPENING CEREMONY

Ye Duzheng
and
Elmar R. Reiter

0.1 OPENING ADDRESS

by

Ye Duzheng
Vice-President, Academia Sinica
The People's Republic of China

Ladies, Gentlemen, and Comrades:

First of all, please allow me on behalf of the Chinese Academy of Sciences to extend a warm welcome to the American scientists coming from the other side of the Pacific Ocean.

Today, I am very happy to be present here to inaugurate formally this important Workshop on Mountain Meteorology, sponsored and organized jointly by the Chinese Academy of Sciences and the U.S. National Academy of Sciences, with much support by the Committee on Scholarly Communication with the People's Republic of China.

Eleven eminent American scientists and twenty Chinese scientists have gathered together here to present their latest research work and to discuss further the problems of common interest to us. The workshop will also give us an excellent opportunity to explore the areas towards which our major efforts should be directed in the future. I am much pleased at convening this meeting because it is the first scientific meeting in the field of atmospheric science, held by the Academies of Sciences of our two countries, and it marks a forward step towards the scientific exchange on the basis of cooperative studies between the Chinese and American meteorologists during the past few years. I believe that, through helpful discussions and extensive exchange of ideas, the mutual understanding and friendship will be strengthened, and this definitely will promote further development of the research on atmospheric science in our two countries.

The main topic of this workshop will deal with the laws of motion of synoptic systems on various scales and with their modification under the influence of topography. We all feel the need for special studies to tackle this problem because of its theoretical and applied importance. Over the years, we Chinese meteorologists have learned more about planetary and large-scale systems than about small and mesoscale systems. Although much effort has been devoted to the studies of small and mesoscale synoptic systems in recent years, our knowledge about them is far too little. In China many kinds of damaging weather events, for example rainstorms and hailstorms, are closely related to systems on that scale. In the United States the major weather disasters are also caused by small and mesoscale systems. It is an accepted fact that very outstanding and praiseworthy work in this area has been done by our American colleagues. Hence, it is our common hope to discuss in this workshop the formation and development of small and mesoscale synoptic systems, and the effects of topography on them, as one of the main topics in order to promote research work in this field. In addition, the application-oriented research of the problem, with the aim of providing better forecasting techniques, also should be emphasized and encouraged.

In the western part of our country there is the huge and high Tibetan Plateau with the most complex geographical features in the world. This plateau covers a quarter of the land area of our country. Many complex and unique features of synoptic processes in East Asia may be attributed to its existence. The plateau is not only the geographical origin for some synoptic systems and the place where the synoptic systems moving over the plateau from the outside of this region are modified, it can also exert an important effect on synoptic climatology in East Asia and the general circulation over even more extensive areas. In recent years, Chinese meteorologists have done a lot of work on this matter and achieved quite a number of interesting results. We must say, however, that the problem is still with us. As we have gained more and more knowledge about this problem we have also become increasingly aware of the complexities involved.

The Rocky Mountains in North America are the longest mountain range in the world. Its existence as a topographic barrier also makes the synoptic processes there attain prominent features. It has been observed that there are also some unique synoptic systems there and the configurations of many of these systems may be modified over and around the Rocky Mountains. As a result, the Rocky Mountains can not only affect the regional synoptic climatology in North America, but also can have a great effect on the general circulation on a much larger-scale basis.

The American meteorologists have been engaged in the study on the effect of the Rocky Mountains for many years and have obtained quite a few widely recognized results. It is very significant to discuss also the effects of large-scale topography on the circulation systems as another main topic at this workshop.

In short, we Chinese and American scientists have done a lot of work so far and obtained many achievements in the above-mentioned aspects. There is, however, much more and even more difficult work ahead of us. Many scientific problems still remain to be explored further in the future. I believe this workshop will give a great impetus to future research on these problems.

I earnestly hope this meeting can be an excellent beginning for further exchange of scientific research results between the Chinese and American meteorologists in the future. We are willing to work together with more scientists to explore further the laws relevant to atmospheric science and to continue our efforts to strive for making a greater contribution to mankind.

I wish the workshop all success.

May the American colleagues have a nice time during their stay in our country.

Thank you very much.

0.2 INTRODUCTORY REMARKS

by

Elmar R. Reiter
Department of Atmospheric Science
Colorado State University
Fort Collins, CO 80523 USA

The idea for this first Sino-American Workshop on Mountain Meteorology, sponsored by Academia Sinica and by the U.S. National Academy of Sciences, was born almost exactly two years ago when I was given the opportunity to participate in the First International Symposium on the Qinghai-Xizang (Tibet) Plateau in Beijing. Subsequent to that symposium a caravan of about 80 foreign guests and 120 Chinese scientists had the unique opportunity to travel by car across the most fascinating province that we call "The Roof of the World".

There one is reminded of the Chinese folk song

"Skull Mountain up above
Treasure Mountain down below,
The sky is only three foot three away,
Bend your head if you go by foot,
Dismount if you go by horse".

The same sense of closeness to space, but also of solitude that provokes thinking, is expressed in a poem by Han-Shan[1], whose famous writings date from the Tang Dynasty:

I climb the road to Cold Mountain,
The road to Cold Mountain that never ends.
The valleys are long and strewn with stones;
The streams broad and banked with thick grass.
Moss is slippery, though no rain has fallen;
Pines sigh, but it isn't the wind.
Who can break from the snares of the world
And sit with me among the white clouds?

Against this poetic backdrop our workshop will view the mountains of Eurasia and North America not so much as the supports of heaven, but as sources of perturbation for the atmosphere. These sources are manyfold and complex. They act in the form of large obstacles to the flow of the atmosphere that has to pass over or around them; they express themselves in terms of ever varying friction or drag, coupling and decoupling the free atmosphere from the planetary boundary layer according to criteria of thermal stability and wind shear, exciting at times interwoven trains

[1] "Cold Mountain, 100 Poems by the Tang Poet Han-Shan", (translated by Burton Watson, Columbia Univ. Press, 1970).

of gravity waves that reach into the lofty realm of the stratosphere, that plunge foehn winds into valleys and river gorges, and that send squall lines, like messengers, rolling out into the foothills and plains. These perturbations also act in seasonal and diurnal cycles of solar radiation turned into heat and convective motion at the earth's surface which is only "three foot three" from the sky.

In our lectures and discussions we will sit together "among the white clouds", deliberate on our observations and on the theories by which large mountain ranges impact on the atmosphere on all scales from large to small. But what is the purpose of these deliberations? Certainly they should not be conducted only for the sake of scientific argument, nor should they stop at the threshold between basic and applied science. If nothing more came of our joint discussions in this workshop, we would have to share the pessimistic skepticism expressed by the final poem of Wen Yiduo, written in 1930 at the request of Xu Zhimo[2]:

"Who doesn't know
How little these things are worth:
A tree full of singing cicadas, a pot of common wine.
Even the mention of misty mountains, valleys at dawn, or
glittering starry skies
Is no less commonplace, most worthlessly commonplace.
They do not deserve
Our ecstatic surprise, our effort to call them in touching terms,
Our anxiety to coin golden phrases to cast them in song.
I, too, would say that to burst into tears because of an oriole's
song
It too futile, too impertinent, too wasteful."

So, perhaps, we should not become ecstatic about a numerical model which runs successfully on a computer, or about a set of data collected under difficult circumstances on the slopes of Qomolangma Feng (Mt. Everest), unless there is a vision in our discussions of things to come, of ideas blossoming, and of enterprises planned to benefit our two countries and strengthening the bonds of friendship and understanding.

This thought leads me to the second purpose of our joint workshop: It is not only the exchange of scientific ideas and results that brings us together in this assembly. It is of equal importance to build personal relationships, to promote friendship where there was mistrust, to conquer with understanding our differences of opinion, and to learn to respect our national sensitivities. Our two countries have different histories and cultures to look back upon, we have different economic systems that sustain us and different political systems that govern us. Let these be the differences that we will respect: China is at her best when she is Chinese, and America is at her best when she is American.

Let us not, however, emphasize our differences -- let us focus on those aspects which unite our two great nations with common bonds that will strengthen and interweave our commitments for the future:

[2] In: Jonathan D. Spence, 1981: The Gate of Heavenly Peace. The Chinese and Their Revolution, 1885-1980. The Viking Press, New York.

Two nations with great pride, committed to peace at home and abroad;
Two countries with enormous resources waiting to be developed;
Two peoples who hold in high esteem the ideals of diligence, hard work, and endurance in the face of adversities.

Both our nations have struggled through periods of upheaval, through revolutions and through indescribable hardships endured for the sake of freedom. A poem written in 1949 by Hu Feng[2)], a close friend and ally of the famous Lu Xun, summarizes the Chinese revolutionary experience -- but might as well serve as a description of America and the people who sought refuge on her shores:

My comrades at arms
My brothers
I have seen you
Dying in a dank and stench-filled prison,
Starving and freezing in a deserted village.
You - you and the peasants - have fed lice with your flesh,
Have drunk bloody water on the battlefield with your friends.
You have endured repeated hammerings, repeated trials.
You have conquered pain and death.
During these
Many, many years,
Your hope stayed alive
And your will stayed alive.
And today,
At this very moment that stirs you,
Forget all the past
Except that the past
Has purified you, like a newborn child
Lying in a warm cradle
His untainted heart overflowing with the blessing of new life.

But enough now of the description of common experiences that bind us together. There is much work waiting for us: much to be said, more to be done, and more yet to be learned from each other. May this joint workshop be the first step of a long journey on which we will embark together.

SESSION I

PLANETARY AND LARGE-SCALE EFFECTS OF MOUNTAINS ON GENERAL CIRCULATION AND CLIMATE: OBSERVATIONAL ASPECTS

1.1 THERMAL EFFECTS OF THE TIBET PLATEAU ON ATMOSPHERIC CIRCULATION SYSTEMS

Elmar R. Reiter
Colorado State University
Fort Collins, Colorado 80523 USA

ABSTRACT

Detailed case studies are presented for April 1979, which show the relative importance of heat fluxes over the Plateau of Tibet during periods of major anticyclogenesis. It is shown that large-scale planetary-wave adjustments, as well as local adiabatic subsidence in foehn winds played a major role in tropospheric warming. Topographic details, usually smoothed away in the lower boundary description used by numerical models, appear to be important in the generation of foehn effects, but also in generating cyclonic and anticyclonic perturbations.

The orographic blocking and thermal effects of elevated land masses should not be viewed by themselves, but in their interaction with temperature perturbations over the ocean. Data are presented to show that different phases of the southern oscillation appear to have either an amplifying or a damping effect on those planetary wave patterns which are usually ascribed to orographic blocking. Large differences in the seasonal "memories" of midlatitude - tropical teleconnections are observed with different phases of the southern oscillation.

I. PLATEAU EFFECTS DURING SPRING

In a recent study Reiter and Westhoff [12] pointed out that the ultralong planetary wave numbers 1 and 2 show a large interannual variability during spring in a latitude range that contains the Plateau of Tibet and the Himalaya Mountains. Figure 1 shows the ratio between the amplitudes of these planetary waves, once computed from daily 500-mb height fields and averaged by calendar date irrespective of the phase angles of the waves, and secondly computed from 500-mb heights which were first averaged according to calendar date and then harmonically analyzed. (Some time-smoothing has been applied to the data shown in Fig. 1.) Ratio values of 1.0 would indicate that the planetary wave is expected to recur in exactly the same phase position year after year, whereas a value of 0.0 would mean that the phase angle of the wave has a random interannual variability on that particular calendar date.

According to Fig. 1 the interannual variability of wave number 1 is particularly large (ratio values < 0.2) during the transition season March-May and in the latitude region 25 to 40°N. A recent study by Ye [17] indicates that in this latitude region the atmosphere over the

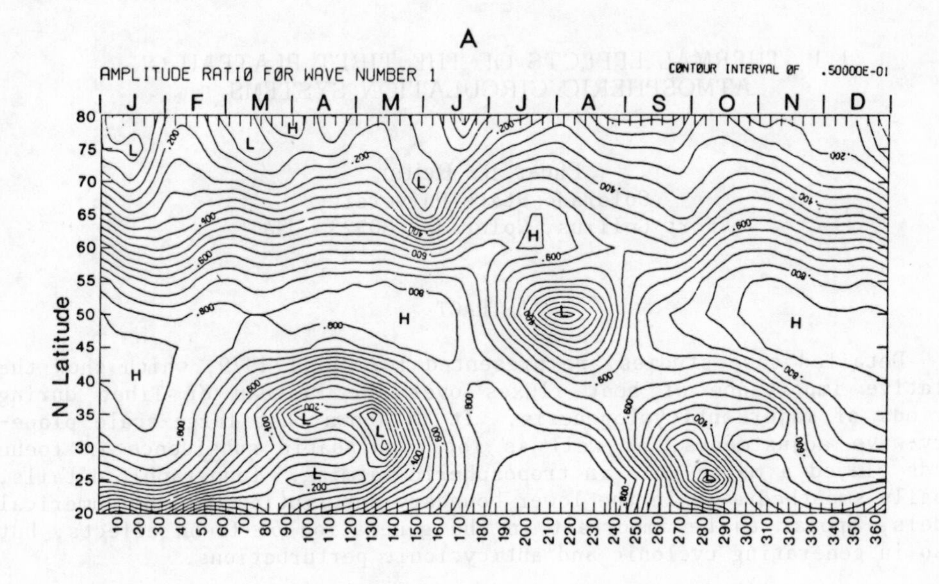

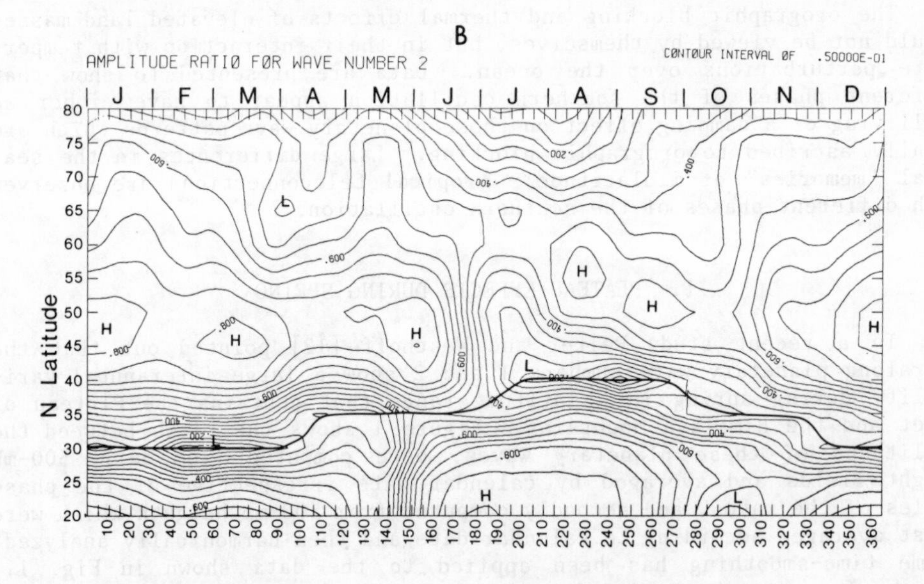

Fig. 1 Amplitudes of (a) planetary wave number 1 and (b) wave number 2, computed from daily 500-mb height fields and averaged by calendar date irrespective of phase angles of waves, divided by the amplitudes computed from 500-mb heights that were first averaged by calendar date and then harmonically analyzed [12].

Plateau of Tibet changes from being a cold source into the role of a heat source during March-April (curves labelled "E" in Fig. 2). This change, which ultimately produces the transition from winter- to summer-monsoon circulation patterns, must have a strong reflection in planetary wave-pattern adjustments, but does not occur on the same date each year -- hence the strong interannual variability of ultralong planetary waves shown in Fig. 1.

Are the planetary wave adjustments during the transition season caused entirely by the sensible and latent heat input of the plateau, or are there other factors at work as well?

To answer this question we analyzed in detail two cases of anticyclogenesis over Tibet [13]. The first occurred during the middle of April 1979 and marked the season's first major breakdown suffered by the subtropical jet stream (STJ) south of the Himalayas. From daily total ozone measurements carried out by TOMS (Total Ozone Mapping Spectrophotometer) on NIMBUS 7 it became evident that prior to April 15, 1979, all cyclonic and anticyclonic wave disturbances that travelled across the Mediterranean were "funnelled" into the STJ south of the Himalayas, where they lost much of their amplitude. On April 15 an anticyclonic wave with relatively low total ozone content was found over Pakistan, and on the next two days developed into a quasi-stationary, upper-troposheric anticyclone over Tibet, disrupting the westerlies of the STJ over northern India. As an example, the analysis of TOMS data for April 17, 1979, is presented in Fig. 3a. This anticyclogenesis was accompanied by a drastic readjustment in the planetary long wave pattern, leading to an intensification of the trough over eastern China, and to the eastward displacement of a trough, upstream of the plateau, from western Europe into the region of the Caspian Sea. This long wave pattern began to collapse on April 21, but a similar development (the second case study in this report) took place, beginning on August 23. Figure 3b shows the ozone data on April 28 near the peak of anticyclogenesis.

The establishment of these anticyclonic systems is clearly evident from time sections at Tibetan radiosonde stations of the altimeter correction, D, [1] defined as

$$D = Z_p - Z \tag{1}$$

where Z_p is the geopotential height of a constant-pressure surface in the standard atmosphere, and Z is the observed height of the same isobaric surface. As an example, the time section at Dingri (No. 55664, 28° 38'N, 87° 05'E) is given in Fig. 4. The long-lasting anticyclonic developments in the upper troposphere during the middle and end of April are quite apparent from this diagram. The two anticyclonic waves during the first half of April are of much shorter duration and are associated with transient disturbances in the polar-front jet (PFJ), which passed to the north of the plateau. These earlier wave passages had little effect on the STJ south of the Himalayas.

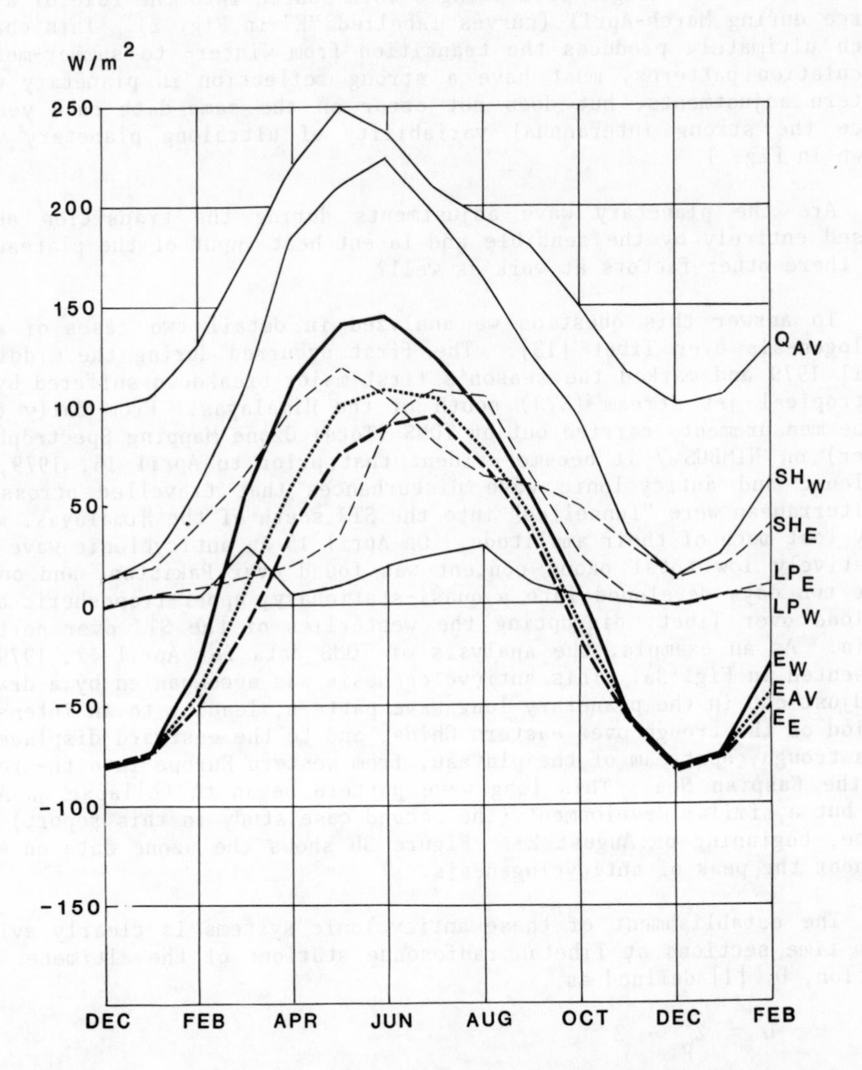

Fig. 2 Annual variation of various terms of the atmospheric heat budget over the Plateau of Tibet (in W/m^2). Q_{AV} = total heat transfer from plateau to atmosphere (turbulent fluxes of sensible and latent heat from ground and effective long-wave radiation). E = atmospheric heat source (turbulent fluxes of latent and sensible heat from ground, absorption of short-wave radiation, effective long-wave radiation, latent heat of precipitation, outgoing long-wave radiation through the atmosphere). SH = turbulent flux of sensible heat, LP = latent heat of precipitation. Subscripts W, E and AV indicate western plateau region, eastern plateau region and average over whole plateau. (Data from Ye [17]; after Reiter and Gao, [13].)

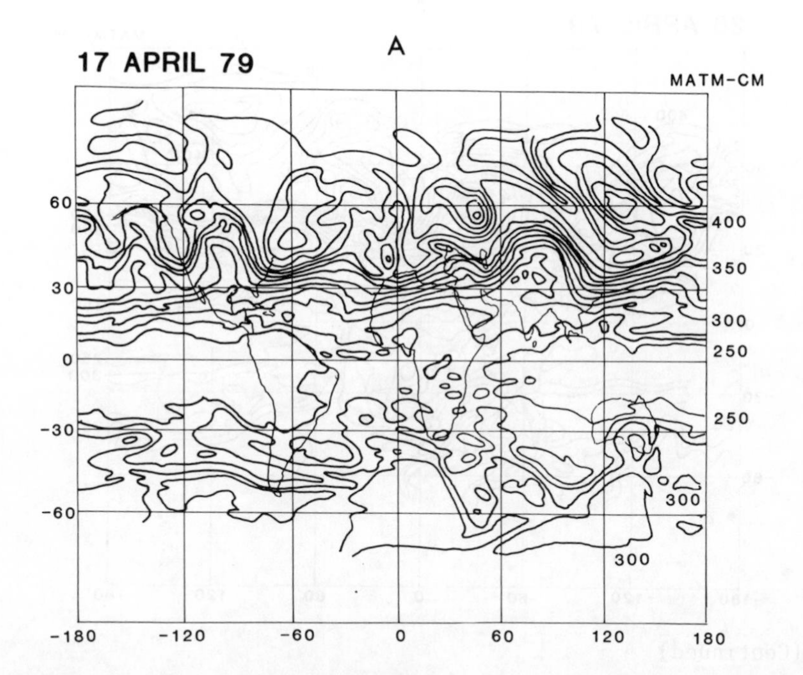

Fig. 3 Total ozone in milli-atmospheric centimeters (matm-cm), as observed by TOMS (Total Ozone Mapping Spectrophotometer) for dates as indicated. Isolines, some of which are labelled along the right margin of the diagram, were transcribed from original color photographs of CRT output (data courtesy of Dr. Arlin Krueger, NASA, Goddard Space Flight Center). The intervals between isolines correspond roughly to 16.5 matm-cm. High values of O_3 usually correspond to large-scale cyclonic features in the planetary circulation, low values O_3 to anti-cyclonic features [13].

Analyses of specific virtual temperature anomalies [1]

$$S^* = \frac{T_p - T^*}{T_p} \qquad (2)$$

(T_p is the temperature of an isobaric surface in the standard atmosphere and T^* is the observed virtual temperature anomaly on the same surface) reveal that the case centered on April 17 was associated with warming mainly in the lower stratosphere and upper troposphere, whereas the case towards the end of April showed strongest warming in the lower troposphere and in the planetary boundary layer (PBL). Figure 5 presents the example of temperature anomalies at Dingri.

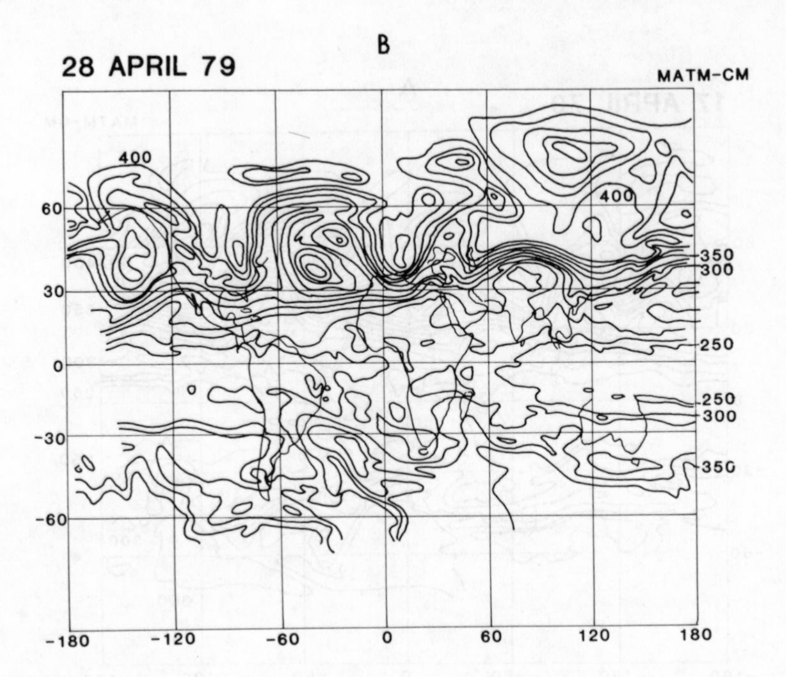

Fig. 3 (Continued)

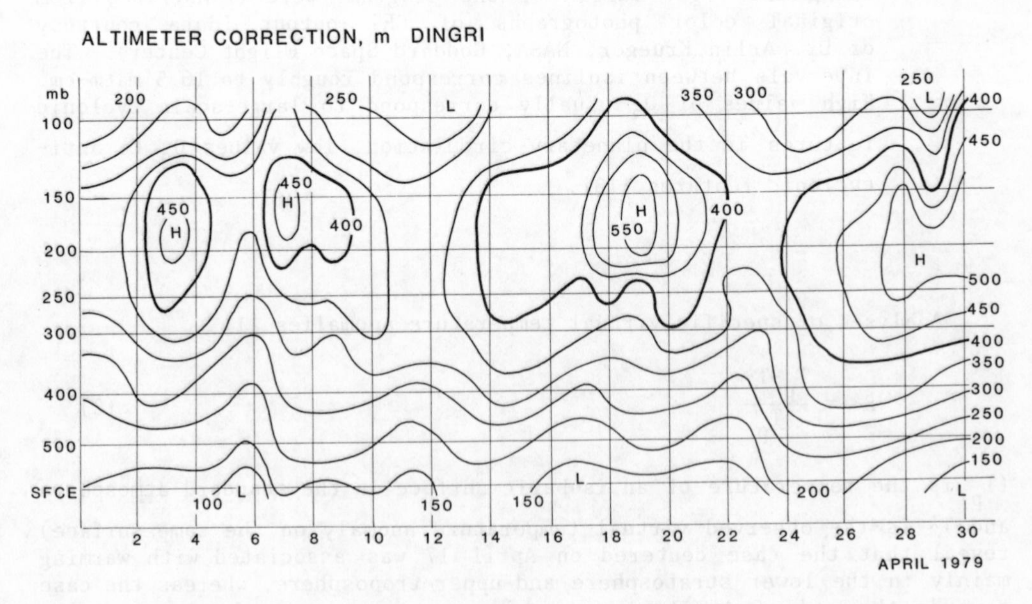

Fig. 4 Time section of altimeter correlation, D (in geopotential meters) for Dingri, April 1979. Isoline of 400 m is emphasized for easier reference [13].

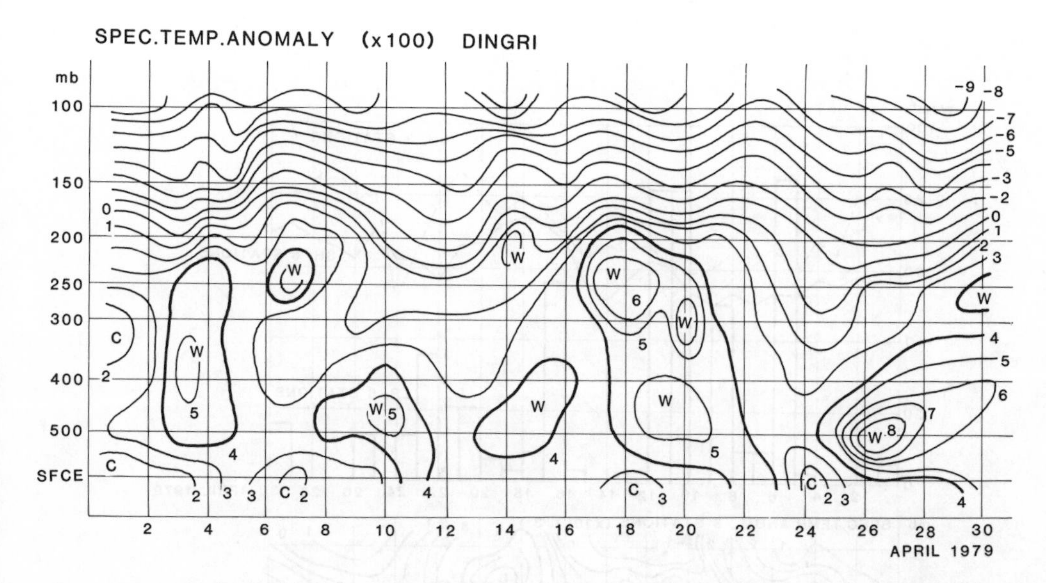

Fig. 5 Specific virtual temperature anomaly, S* (x 100, nondimensional) for Dingri, April 1979. Isoline +4 is emphasized for easier reference [13].

To estimate the heating effects of the plateau the sensible heat transfer

$$SH = C_D \, c_p \, \rho (T_s - T_a) \, U \tag{3}$$

was computed at Tibetan stations. $\rho \cong 0.75$ kg/m^3 is the air density, T_s and T_a are the measured ground surface and air temperatures in °C, $c_p = 10^3$ J/kg °K is the specific heat at constant pressure, $C_D \cong 5 \times 10^3$ (nondimensional) is the drag coefficient and U is the surface wind speed in m/s. In the upper portion of Fig. 6, SH in W/m^2 is shown for three (Lhasa, Qamdo and Lingzhi) and six Tibetan stations (the previous three plus Yushu, Germu and Chayu). Above-normal sensitive heat fluxes prevailed during both cases of anticyclogenesis, but not enough to explain the intensity of the development. Latent heat releases are also shown in Fig. 6. Significant values of these releases appear to occur before, and not during the anticyclogenesis, suggesting that the observed rainfall was caused by cyclonic disturbances and was not yet a major contributor to upper-tropospheric "warm-core" systems as appears to be more often the case during the summer season.

The disruption of the STJ, associated with a westward displacement of the subtropical high-pressure cell from the Persian Gulf to the Bay of Bengal, is illustrated by the 100-mb charts shown in Fig. 7.

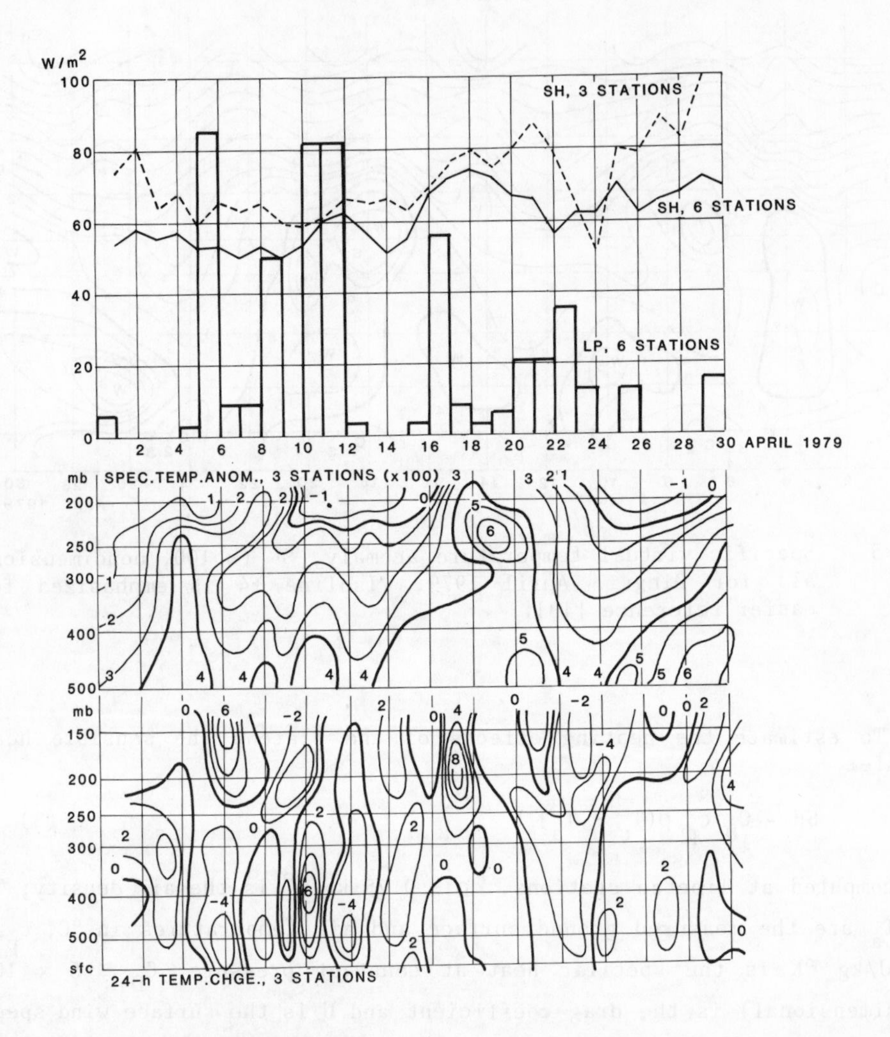

Fig. 6 Upper part: Surface heat flux, in W/m^2 for three (dashed line) and six (solid line) Tibetan stations (see text for station identification). Histogram gives latent-heat release by precipitation (W/m^2). Middle part: Specific virtual temperature anomalies (x 100, nondimensional) averaged over three Tibetan stations. Lower part: 24-hour temperature changes (°C) averaged over three Tibetan stations [13].

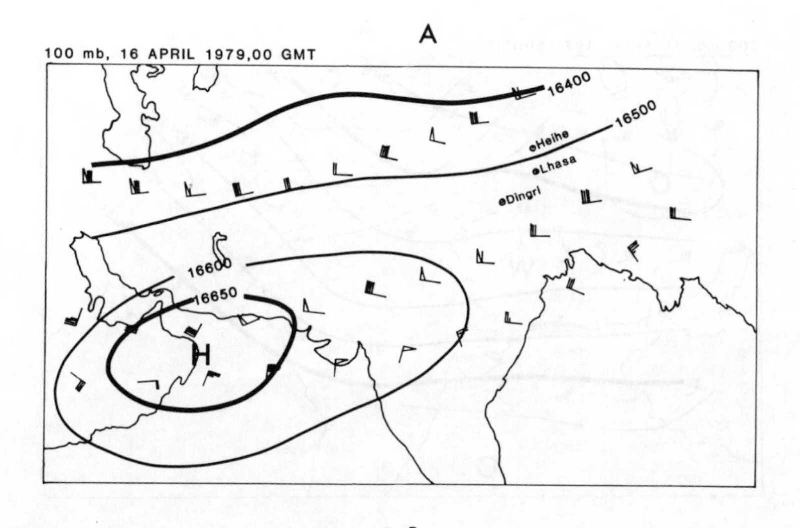

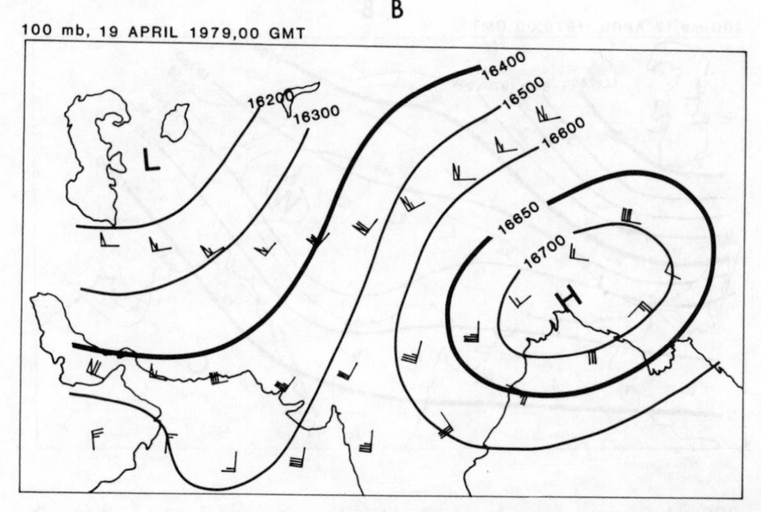

Fig. 7 100-mb contours (geopotential meters) and winds (short barb = 2 m/s, long barb = 4 m/s, triangle = 20 m/s) from NMC gridded data, augmented by Tibetan radiosonde reports, for (a) 16 April, 1979, 0000 GMT and (b) 19 April, 1979, 0000 GMT [13].

That the strong warming between April 16 and 17 near tropopause level, evident from Fig. 5, was not merely due to a horizontal advective process can be demonstrated with Fig. 8. The warm center of -44°C that is found on April 17 north of the Himalaya range at the geographic longitude of the Bay of Bengal remains in the same position also during the next day (map not shown here). On the previous day (April 16) warmest

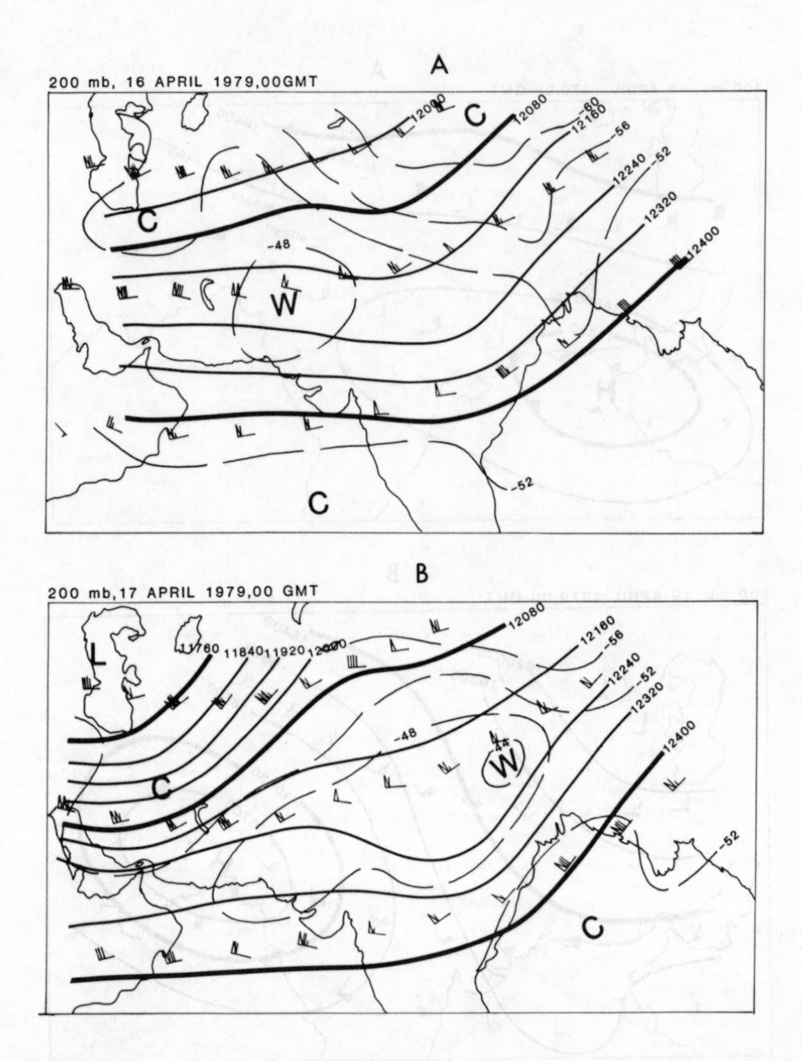

Fig. 8 200-mb contours (geopotential meters, solid lines) and isotherms (°C, dashed lines) for dates as indicated. Wind symbols are the same as explained in the legend of Fig. 7 [13].

temperatures in evidence at 200 mb were -48° over Pakistan. This warming, mainly at stratospheric levels, can be illustrated by a sequence of soundings taken at Lhasa. From Fig. 9 we gather that the major warming event took place between 16 April, 1200 GMT and 17 April, 0000 GMT. It was mainly due to adiabatic sinking motions above the tropopause (Fig. 10). In the planetary boundary layer, as characterized by the 600-mb surface we find a weak low-pressure vortex, also with relatively warm temperatures, situated over eastern Tibet on 18 April, 0000 GMT (Fig. 11).

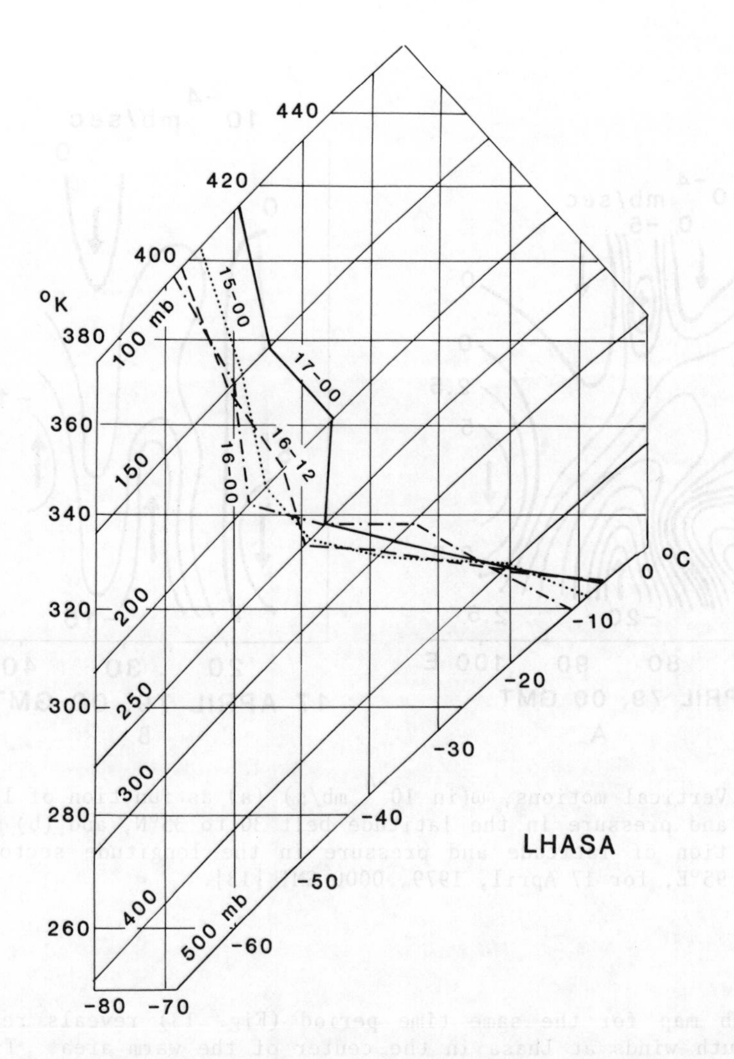

Fig. 9 Lhasa radiosonde data, reported on standard isobaric surfaces (no "significant point" data available) between 15 April, 0000 GMT and 17 April, 0000 GMT, 1979, and plotted on a tephigram. Abscissa shows temperature (°C) and ordinate contains potential temperature (°K) [13].

The case of anticyclogenesis towards the end of April 1979 also was characterized by strong distortions in the isohypses of the 600-mb surface and by a strong warming trend in the PBL (Fig. 12, see also Fig. 5). The shape of the contour lines shown in Fig. 12 are reminiscent of patterns encountered over the Alps during foehn wind conditions. Indeed,

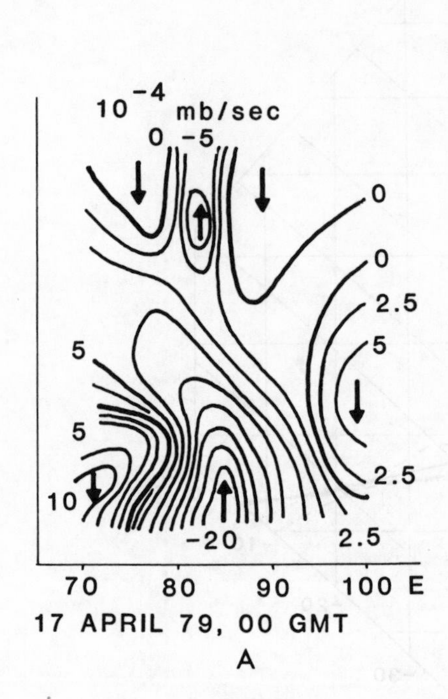

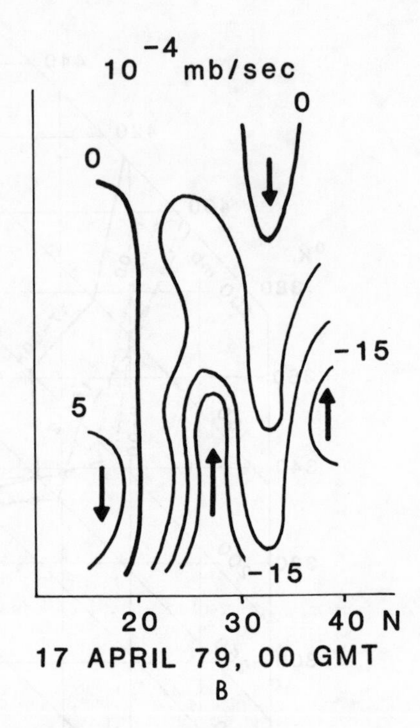

Fig. 10 Vertical motions, ω(in 10^{-4} mb/s) (a) as function of longitude and pressure in the latitude belt 30 to 35°N, and (b) as function of latitude and pressure in the longitude sector 90 to 95°E, for 17 April, 1979, 0000 GMT [13].

the 500-mb map for the same time period (Fig. 13) reveals relatively strong south winds at Lhasa in the center of the warm area. Time sections of winds at three observation stations in that region (Fig. 14), indicate that foehn under southerly wind conditions may have contributed most significantly to the strong warming of the PBL. The adiabatic subsidence associated with such foehn winds also is revealed in the vertical-motion cross sections of Fig. 15.

We, thus, arrive at the conclusion that the two cases of strong anticyclogenesis over Tibet in the second half of April were the result of a collaboration of several factors. Most important were the advection of a high-pressure ridge and a sudden adjustment in the planetary long-wave pattern. This adjustment, most likely, was not only caused by the contribution of sensible heat from the Plateau of Tibet, but also by the general warming of the continental regions during this time of the year. Adiabatic sinking motions, in part associated with foehn winds, cannot be ignored as a major factor in the local warming of the troposphere.

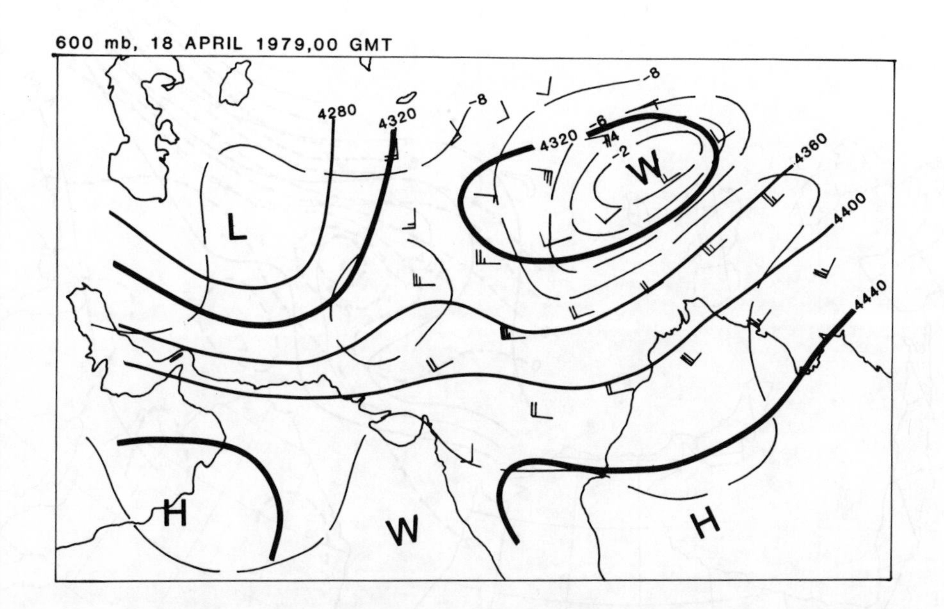

Fig. 11 600-mb contours (geopotential meters, solid lines) and isotherms (°C, dashed lines) from NMC gridded data, augmented by adjusted surface temperature reports from Tibetan radiosonde stations on 18 April, 1979, 0000 GMT. Wind symbols are the same as explained in the legend of Fig. 7 [13].

Numerical models, such as the one reported by Kuo and Qian [9], usually have to contend with a strongly smoothed topography. Tucker [15] used a modified Kuo-Qian model to study the same anticyclonic developments in April 1979 as described above. Because of the smoothed topography, which virtually ignored the existence of the formidable mountain barriers of the Himalaya, Transhimalaya (Nyainqentanglha) and Tien Shan, too much moisture was allowed to flow into the plateau region, producing much more precipitation there than actually observed. The discrepancy could be corrected by prohibiting moisture convergence in the lowest two levels of the model (approximately corresponding to the PBL top near 500 mb). Such a prohibition effectively mimicked the orographic blocking effect of the Himalayas.

Tucker's model experiments produced anticyclonic developments over the plateau, but the sharp convolutions of the contour and isotherm patterns, as for instance evident in Fig. 12, were not revealed by the model output. This discrepancy has to be blamed on the relatively coarse grid used by the model, and on the smoothed model topography. Both factors tend to inhibit the development of local foehn winds, which we have identified as major contributors to tropospheric warming and to the formation of local pressure and temperature anomalies.

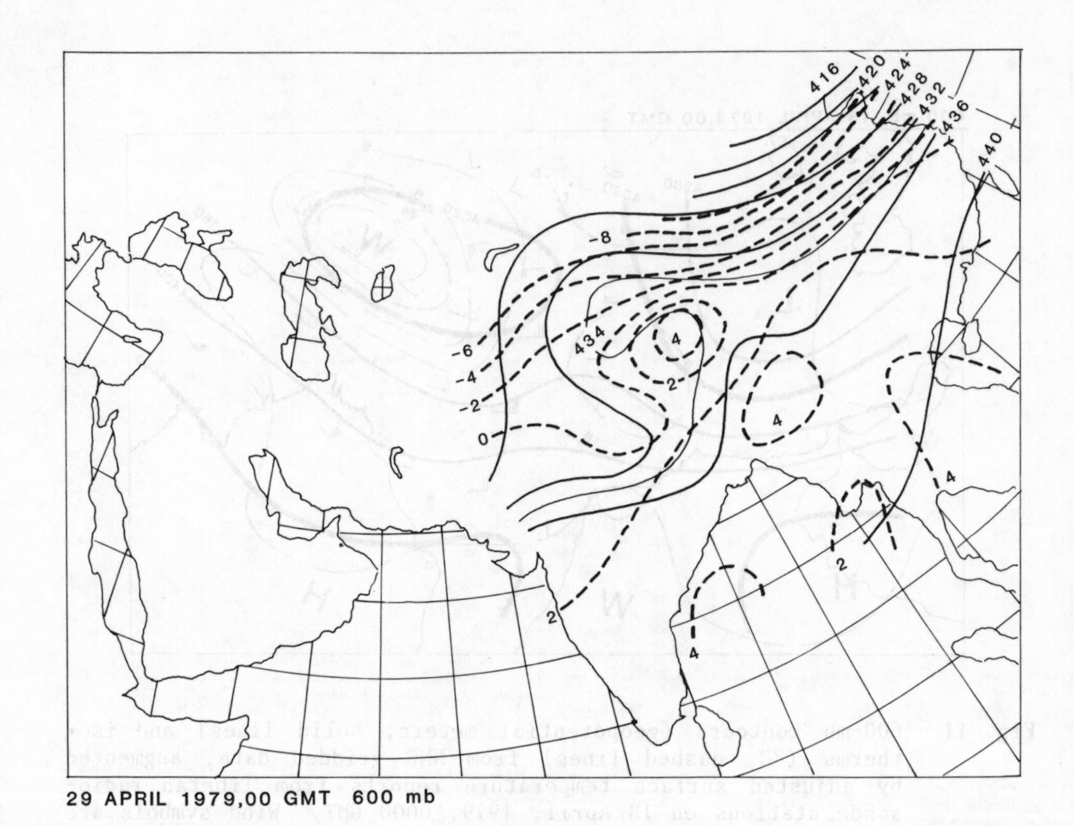

Fig. 12 600-mb contours (geopotential decameters, solid lines) and isotherms (°C, dashed lines) from NMC data, augmented by adjusted Tibetan radiosonde surface temperature reports on 29 April, 1979, 0000 GMT [13].

II. MESOSCALE TOPOGRAPHIC EFFECTS

In the foregoing section we have commented on flow effects, such as foehn winds, that are caused by orographic details usually suppressed in large-scale numerical model specifications of the lower boundary surface. Godev [3, 4] pointed out that the shape of orographic obstacles exercises a certain influence on cyclonic and anticyclonic activity over corrugated terrain. The vertical velocity at the top of the PBL receives a contribution from a term which depends on wind speed (irrespective of wind direction) and on the Laplacian of terrain height, Z_o. Because of the independence of this term from wind direction, its effects should be revealed in a climatological display of frequencies of cyclogenesis and anticyclogenesis comprising a variety of surface wind conditions. Such has been demonstrated by Godev for the Alpine regions of Europe.

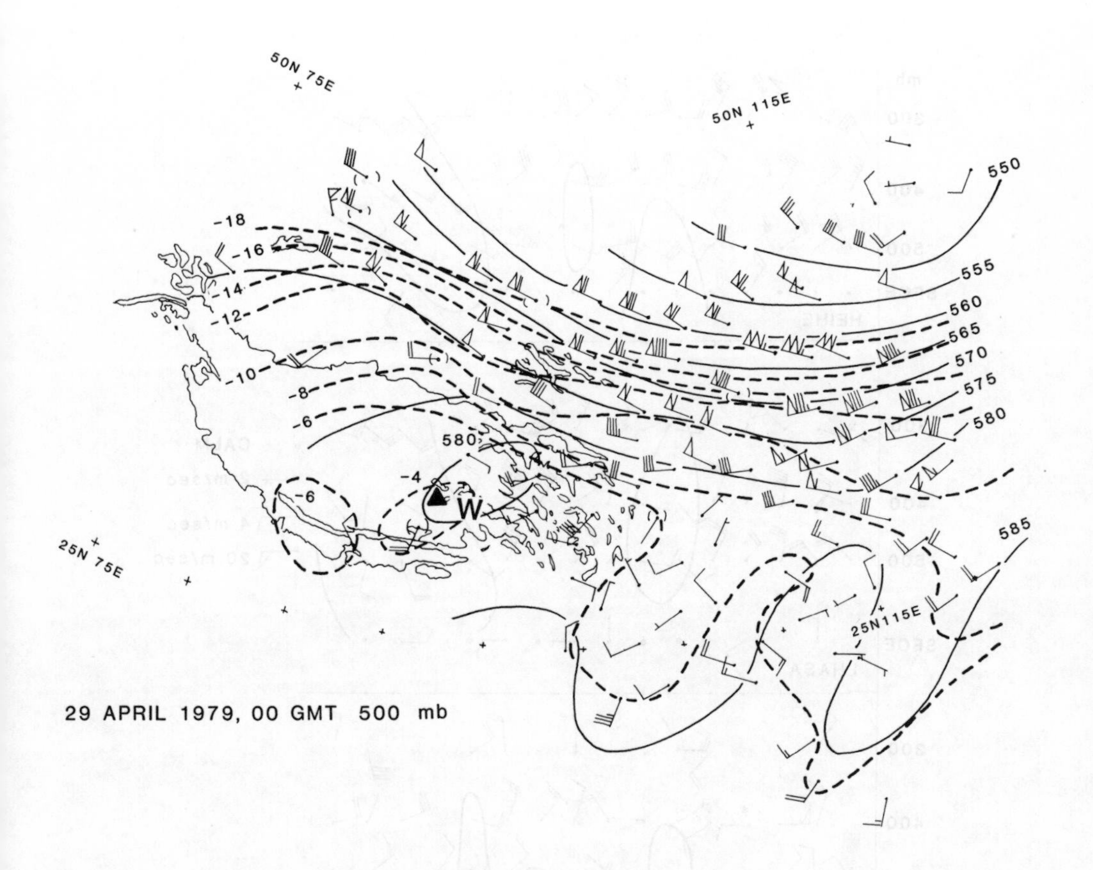

Fig. 13 500-mb contours (geopotential decameters, solid lines) and isotherms (°C, dashed lines) over the People's Republic of China on 29 April, 1979, 0000 GMT. Thin lines indicate the 15,000 ft terrain contour. Wind symbols are: short barb = 2 m/s, long barb - 4 m/s, triangle = 20 m/s. A triangle and "W" mark the location of the 600-mb warm center derived from Fig. 12 [13].

Gao et al. [2] and Ye et al. [18] showed the existence of several low-pressure centers on the mean 600-mb maps for July, using data from the period 1961-1970 (Fig. 16). Apparently there is a nonrandom, preferential distribution of pressure systems in the plateau region that should, somehow, be linked to topographic details.

The Laplacian of Z_o was computed from terrain heights extracted from the Times Atlas [14] on a 1° lat. x 1° long. grid. The values obtained for

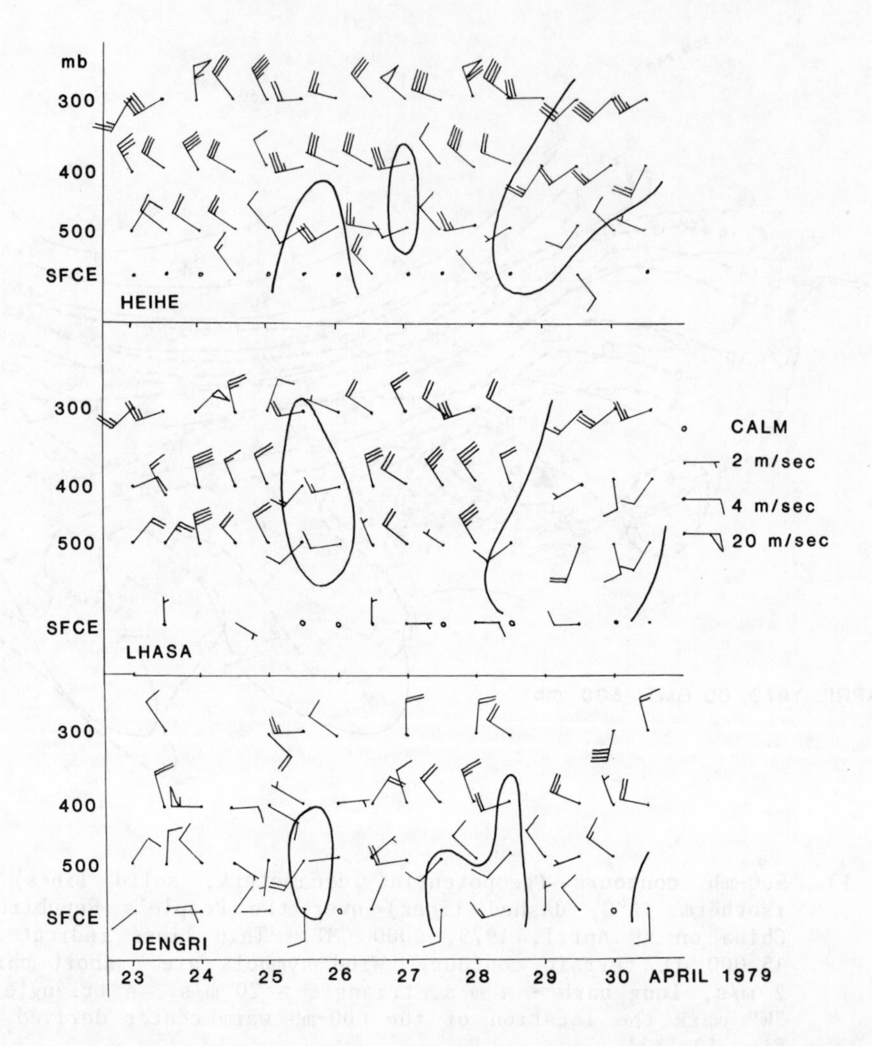

Fig. 14 Time sections of radiosonde winds at Heihe, Lhasa and Dengri. Solid lines indicate wind shift to a southerly direction. Wind symbols are the same as explained in the legend of Fig. 13 [13].

$$\nabla^2 Z_o = \frac{\partial^2 Z_o}{\partial X^2} + \frac{\partial^2 Z_o}{\partial Y^2} \tag{4}$$

were corrected for map scale and convergence of meridians and smoothed in the same diamond-shaped grid pattern used in the Laplacian calculations [11]. Results are shown in Fig. 17. The positions of high and low

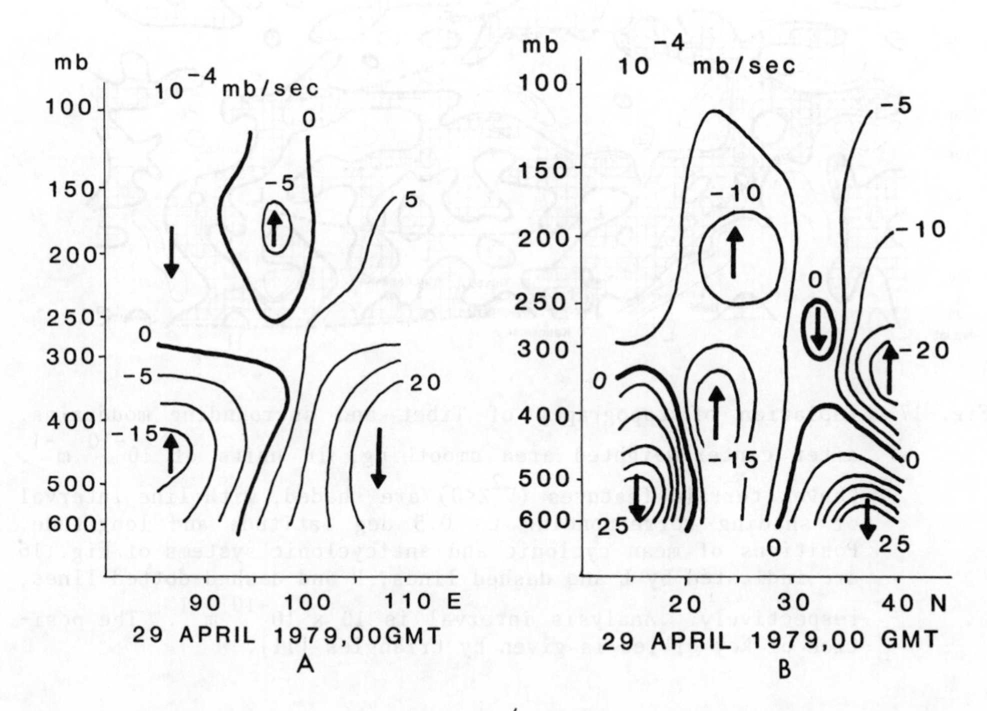

Fig. 15 Vertical motions, ω(in 10^{-4} mb/s), (a) as function of longitude and pressure in the latitude belt 30 to 35°N, and (b) as function of latitude and pressure in the longitude sector 90 to 95°E, for 29 April, 1979, 0000 GMT [13].

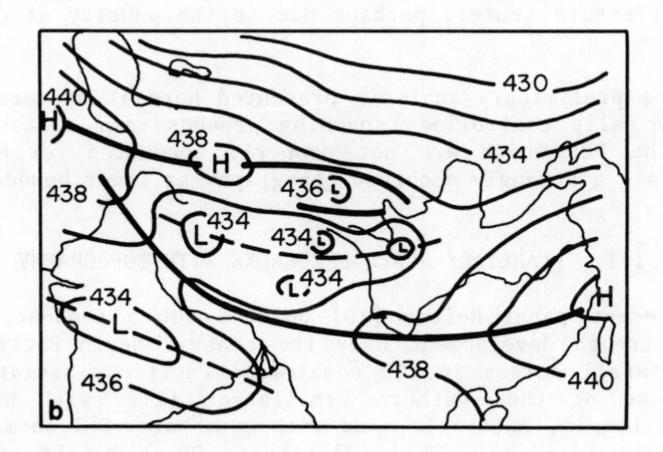

Fig. 16 Mean 600-mb charts in geopotential decameters for July 1961-1970. Dashed: trough lines; solid: ridge lines. (After Ye et al. [18].)

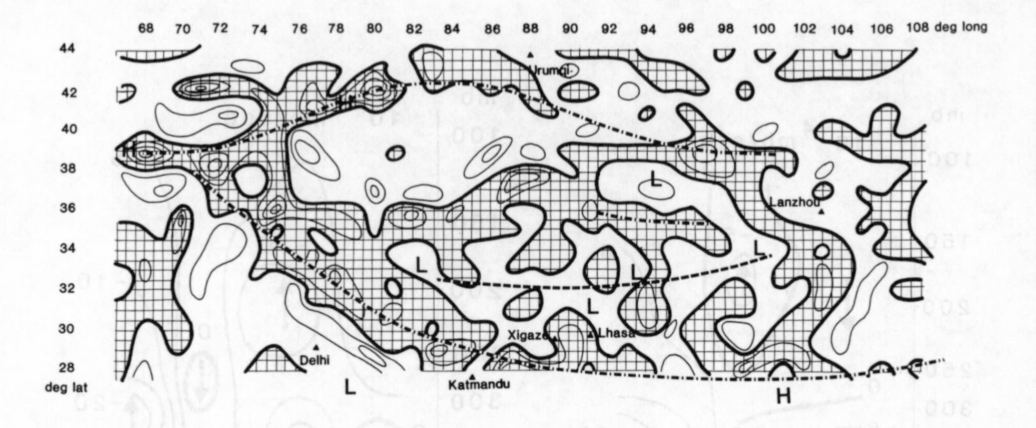

Fig. 17 Laplacian of topography of Tibet and surrounding mountains, after center-weighted area smoothing, in units of 10^{-10} m^{-1}. Convex terrain features ($\nabla^2 Z<0$) are shaded, with line interval of shading corresponding to 0.5 deg latitude and longitude. Positions of mean cyclonic and anticyclonic systems of Fig. 16 are indicated by L and dashed lines, H and dashed-dotted lines, respectively. Analysis interval is 10×10^{-10} m^{-1}. The position of key cities is given by triangles [11].

pressure anomalies shown in this diagram are the same as those given in Fig. 16. There is a good correspondence between concave terrain features and cyclonic developments, whereas convex terrain appears to attract anticyclonic development. The "bowl" of the Taklamakan desert does not show a low-pressure center, perhaps due to the paucity of data in that region.

From the preliminary analysis presented here it appears that there are orographically controlled "spawning grounds" for mesoscale disturbances in the PBL which are not properly accounted for by numerical models that use a strongly smoothed topography as lower boundary.

III. PLANETARY-SCALE FEEDBACKS WITH TOPOGRAPHY

In a recent paper Reiter [10] pointed out a tendency for strong topospheric trough development over the central North Pacific to occur prior to rainfall surges in the equatorial Pacific associated with the negative phase of the southern oscillation (i.e. with high-pressure anomalies at Darwin, Australia, and with warm-water SST anomalies in the equatorial East Pacific). The midlatitude 500-mb height anomalies obtained by contrasting excessively wet and dry months in the equatorial Pacific (Fig. 18) were similar to those shown by Wallace and Gutzler

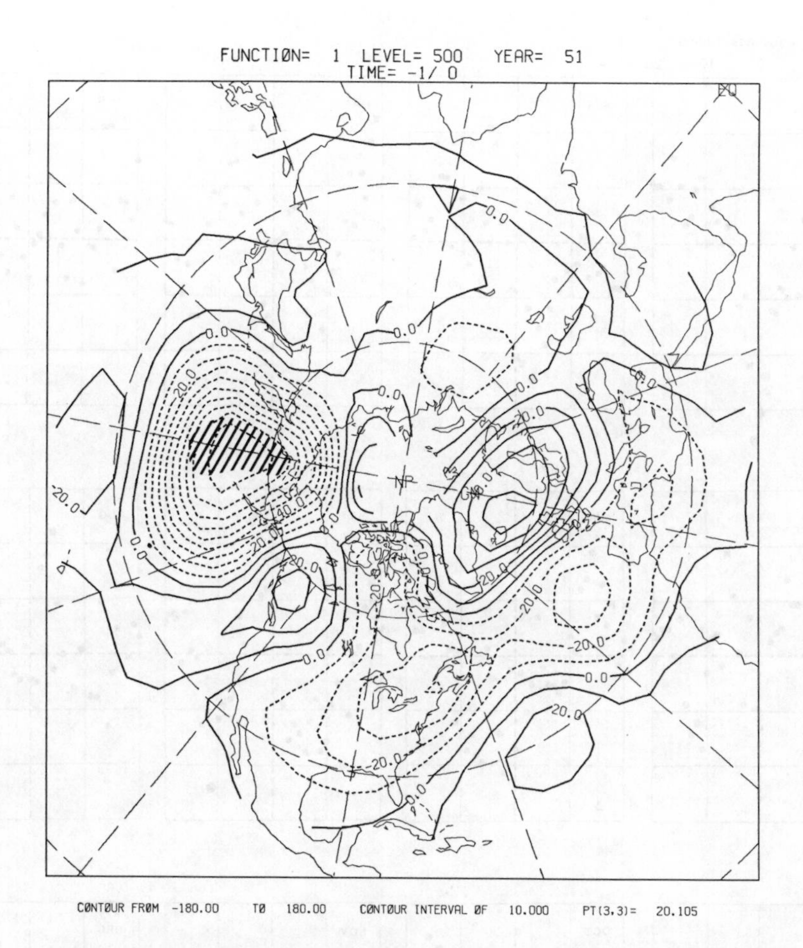

Fig. 18 Mean 500-mb height anomalies averaged with respect to key months 8/51, 11/53, 4/56, 12/57, 4/59, 2/61, 2/64, 9/65, 11/68, 11/72 (precipitation maxima at Line Islands) minus average with respect to key months 10/52, 5/54, 2/56, 2/57, 11/58, 3/61, 6/64, 4/67, 5/68, 7/70, 7/73, 8/75 (precipitation minima at Line Islands). 500-mb difference patterns are shown with -1 month lag against the aforementioned key months. Analysis interval: 10 gpm. The hatched region indicates confidence > 99 percent that two different ensembles were differenced [10].

[16] and also to those derived from a linearized hemispheric numerical model by Hoskins and Karoly [6]. Our own analysis results showed that the coupling between middle and low latitudes was significant only during December, January and February (Fig. 19) and that the lag-correlations tended to be most significant as the midlatitude 500-mb patterns in the Pacific <u>preceded</u> the tropical precipitation extremes by one month, whereas Hoskins and Karoly suggest that the perturbation <u>originates</u> in the

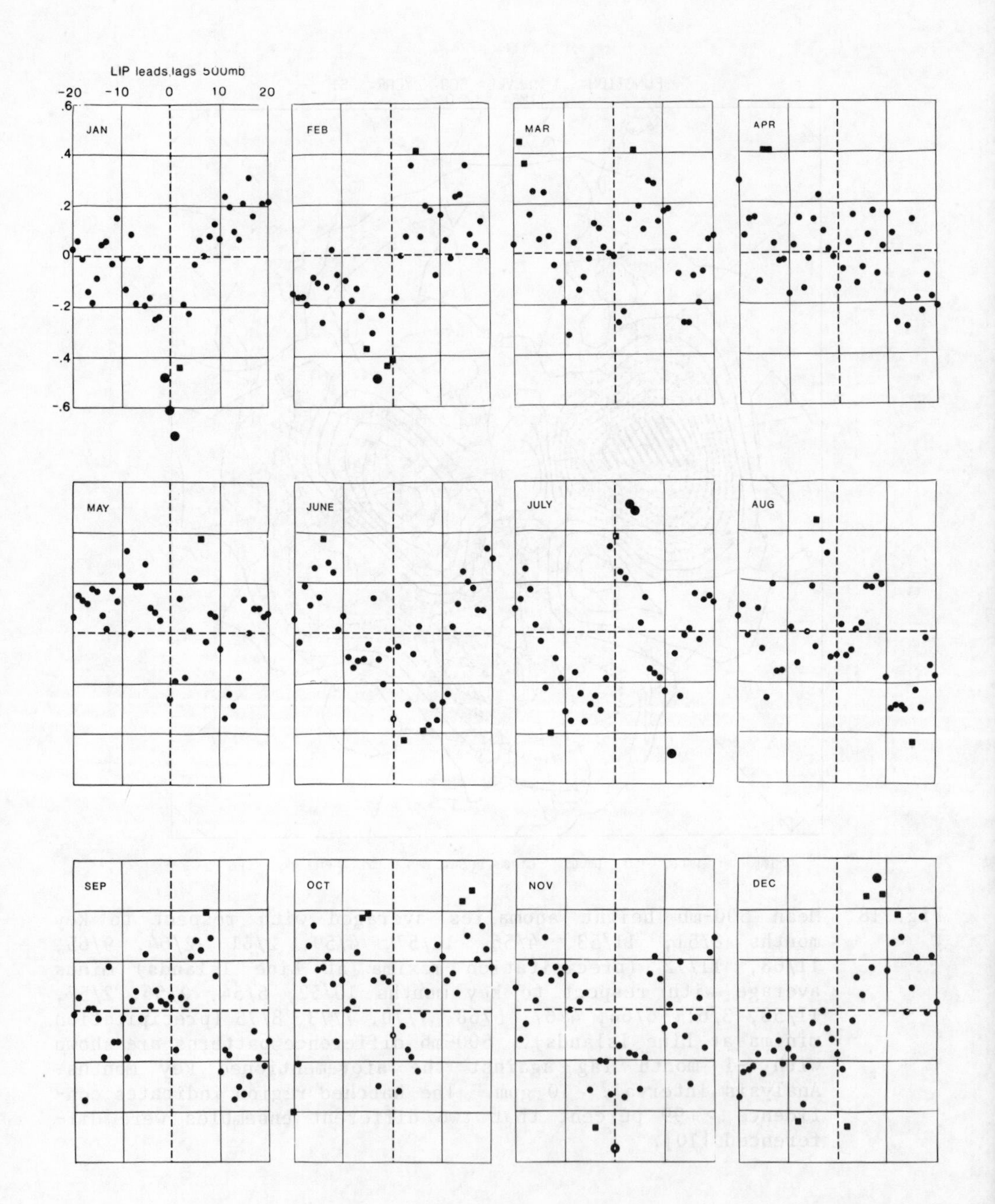

Fig. 19 Monthly lag correlations between Line Islands precipitation index and 500-mb height anomalies at 50 to 60°N, 180 to 160°W. Time lags (in months) along abscissas, correlation coefficients along ordinates. Key months are taken from 500-mb data and precipitation is lagged against these months. Heavy dots indicate 99 percent, squares indicate 95 percent significance of correlations [10].

tropical heat source region. Hanna [5] demonstrated with a nonlinear, global two-layer spectral model initialized with January climatic conditions that preceding trough development in midlatitudes, indeed, can foster an intensification of the intertropical convergence zone (ITCZ). He also showed that planetary wave patterns underlying the anomalies depicted in Fig. 18 form as a result of simultaneous cold SST anomalies in the central North Pacific and warm anomalies in the equatorial East Pacific, both of realistic and not exaggerated magnitudes. If only one of these SST anomalies is introduced into the model, the model has to be over-forced in order to obtain a planetary wave response, similar to the experiments by Julian and Chervin [8]. Reiter's [10] analyses and Hanna's [5] global model results suggest that midlatitude-tropical teleconnections during the negative phase (positive pressure anomaly at Darwin) of the southern oscillation proceed through a narrow seasonal "gate", are strongly coupled to an appropriate distribution of positive and negative SST anomalies, and allow perturbation (wave) energy to pass from middle to low latitudes. The last of the conclusions is in agreement with statements by Huang [7], who found a waveguide for wave number 1 which would allow such an equatorward perturbation transfer.

There is another interesting aspect to Fig. 18: During warm water-related equatorial Pacific precipitation surges the quasi-stationary midlatitude Rossby wave pattern appears to amplify those planetary waves which are normally associated with orographic blocking, namely a trough over the central North Pacific, a ridge over the Rocky Mountains, a trough over the eastern United States, etc. During the opposite phase of the southern oscillation (cool SST's and drought conditions over the equatorial East Pacific) the midlatitude planetary wave pattern anomalies tend to weaken the orographic effects.

The latter conclusion is confirmed by analyzing 500-mb anomalies for "good" and "bad" monsoon years. To define the quality of the Indian summer monsoon we considered the first eigenvector of precipitation for July, August and September and for the period 1949-1978. The eigenvector pattern, together with the locations of stations used for the analysis, is shown in Fig. 20. The time series of this eigenvector is given in Fig. 21. Differences between the 500-mb monthly mean charts for August of the "wet" years 1956, 1959, 1961, 1962, 1964, and 1975 and of the "dry" years 1952, 1957, 1968, 1972, and 1974 are composited in Fig. 22a. Wet Indian monsoon summers appear to favor positive tropospheric height anomalies over the central North Pacific. Contrary to the narrow seasonal "gate" of teleconnections between North Pacific troughs and equatorial Pacific precipitation surges, centered on January, the Indian monsoon teleconnection with the central North Pacific is already apparent during March prior to the monsoon season (Fig. 22b). (In this region the difference patterns between wet and dry summer monsoons are significant at better than the 95 percent level.)

As stated before, experiments with our two-layer spectral GCM gave good results with SST distributions corresponding to the negative phase of the southern oscillation. Model experiments with a heat source placed over the Bay of Bengal and India confirm the positive height anomaly pattern over the North Pacific shown in Fig. 22.

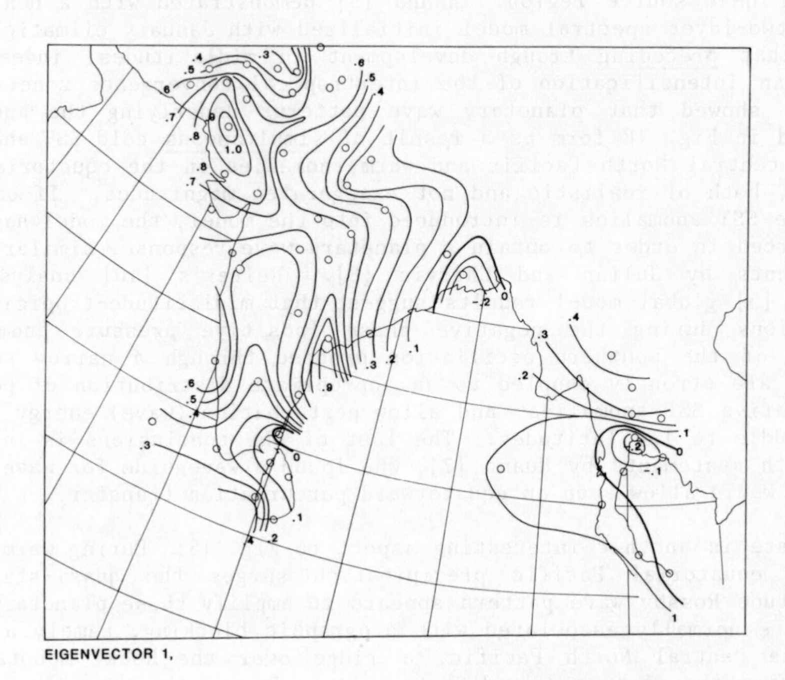

Fig. 20 Pattern of eigenvector 1 of Indian summer precipitation (July, August, September), using data from 1947 to 1978.

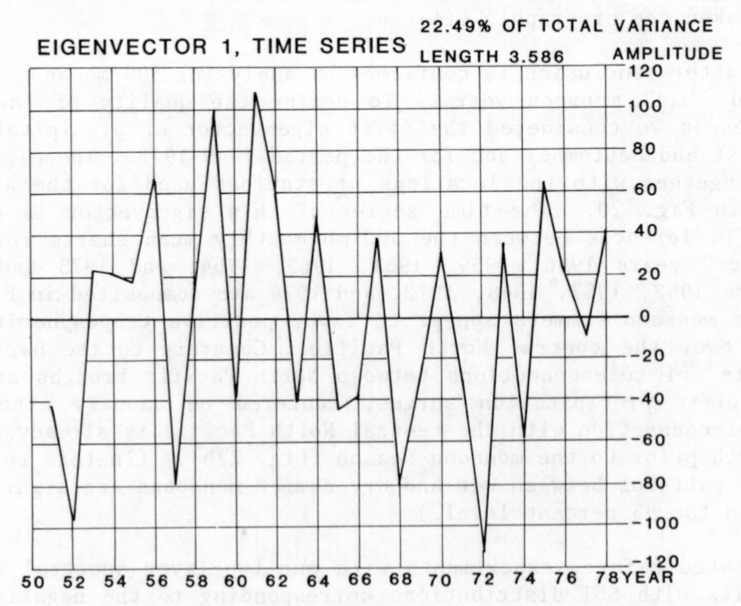

Fig. 21 Time series of eigenvector 1 of Indian summer precipitation. The contribution of this eigenvector to precipitation (in mm) in each year is given by the pattern value (Fig. 20) times the amplitude for that year, divided by the length of the eigenvector (3.586).

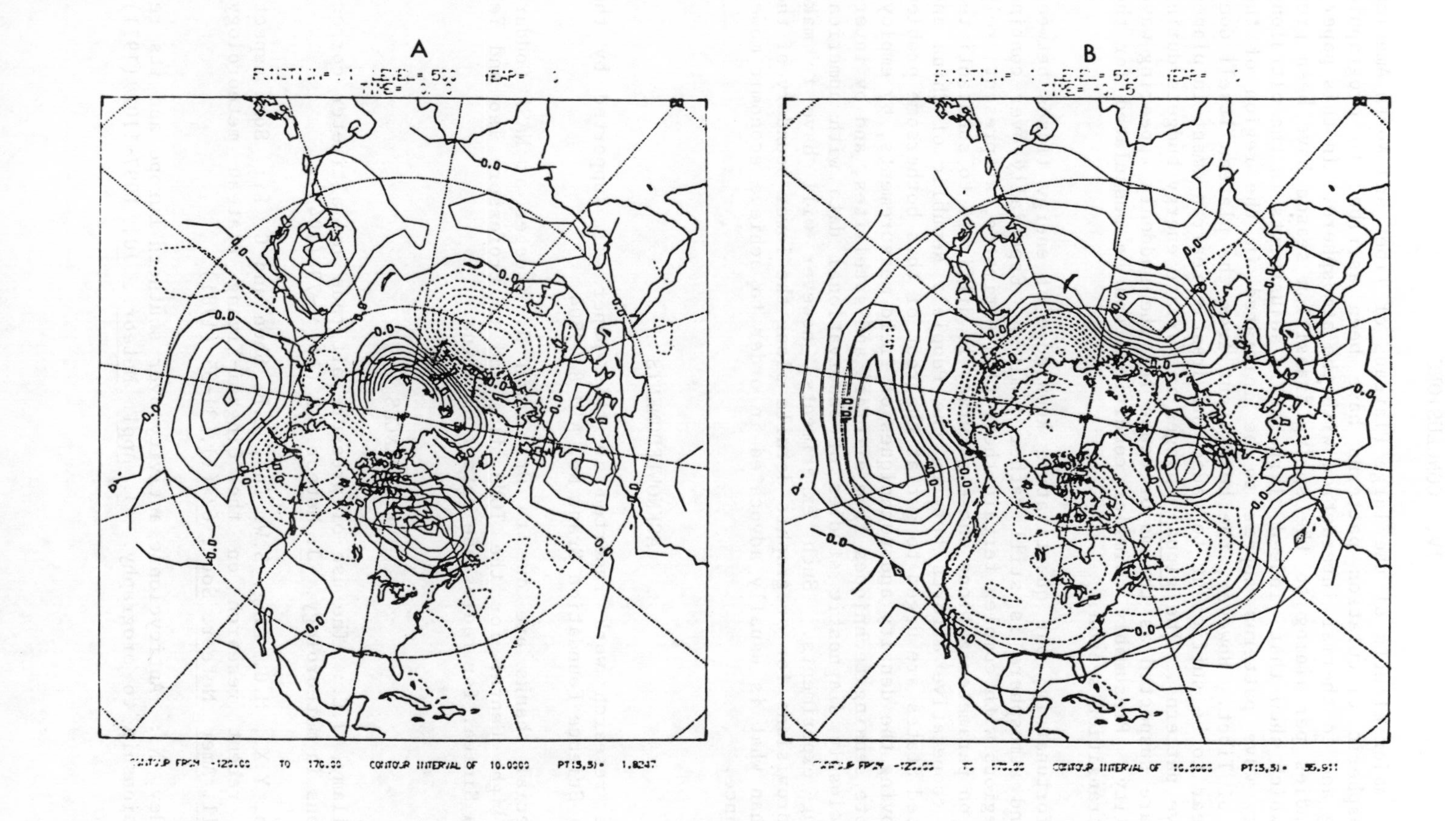

Fig. 22 500-mb pattern differences (in gpm) between years with "good" and "bad" Indian summer monsoon. Time lags are with reference to the month of August of the monsoon years. A = 0 lag, B = -5 months lag.

IV. CONCLUSIONS

The major effects of the large Plateaus of Tibet and North America on atmospheric circulation patterns have been ascribed to orographic blocking and to thermal interaction with the atmosphere. In this paper, case studies pertaining to the spring transition season have been presented which show that, indeed, large-scale adjustments in the ultralong planetary wave patterns appear to be initiated in the region of the Plateau of Tibet. However, the heat input from the plateau itself does not appear to be sufficient to cause the drastic rearrangement in planetary wave patterns. The global changes in surface energy budgets during spring are suspected as the major cause for the suddenly appearing wave instability. Presumably, similar conditions can be postulated for the autumn transition season.

Unfortunately, our quantitative knowledge of energy fluxes between earth and atmosphere is still rather limited. Especially over continental regions with complex terrain these flux estimates, at present, rely heavily on parameterizations whose accuracy is difficult to substantiate. Massive cooperative efforts between the People's Republic of China and the United States are urged to get at the root of this bothersome problem by improving the density and techniques of field measurements, by employing remote sensing techniques from aircraft and satellites, and by interfacing closely diagnostic studies of observational data with numerical modelling experiments. Such experiments, however will have to make better provisions for orographic details along the lower boundary of the model than what is usually advocated in order to achieve economic model performance.

ACKNOWLEDGMENTS

The research work reported in this paper was supported by the National Science Foundation Grant No. ATM 8109504.

Special thanks are due to Dr. Arlin J. Krueger at NASA Goddard Space Flight Center for the TOMS data and to Professors Tao and Ye, Academia Sinica, for radiosonde data from China.

REFERENCES

1. Bellamy, J.C.: The use of pressure altitude and altimeter corrections in meteorology. J. Meteor., 2 (1), 1-79 (1945).

2. Gao, Y.X., M.C. Tang, S.W. Luo, Z.B. Shen and C. Li: Some aspects of recent research on the Qinghai-Xizang Plateau meteorology. Bull. Amer. Meteor. Soc., 62 (1), 31-35 (1981).

3. Godev, N.: Anticyclonic activity over southern Europe and its relationship to orography. J. Appl. Meteor., 10, 1097-1102 (1971).

4. Godev, N.: The dynamical influence of orography and friction on atmospheric processes. Ph.D. Thesis, Bulgarian Academy of Sciences, Geophysical Institute, 27 pp (1972).

5. Hanna, A.F.: Short-term climatic fluctuations forced by thermal anomalies. Ph.D. Dissertation, Department of Atmospheric Science, Colorado State University, Ft. Collins, CO (1982).

6. Hoskins, B.J. and D. Karoly: The steady, linear response of a spherical atmopshere to thermal and orographic forcing. J. Atmos. Sci., 38 (6), 1179-1196 (1981).

7. Huang, R.-H.: Comparison between winter and summer response of the northern hemisphere model atmosphere to forcing by topography and stationary heat sources. Proceedings, Workshop on Mountain Meteorology, Beijing, May 18-23, 1982 (1983).

8. Julian, P.R. and R.M. Chervin: A study of southern oscillation and Walker circulation phenomenon. Mon. Wea. Rev., 106, 1433-1451 (1978).

9. Kuo, H.L. and Y.F. Qian: Influence of the Tibetan Plateau on cumulative and diurnal changes of weather and climate in summer. Mon. Wea. Rev., 109 (11), 2337-2356 (1981).

10. Reiter, E.R.: Surges of tropical Pacific rainfall and teleconnections with extratropical circulation patterns. Proceedings, Symposium on The Global Water Budget, Oxford, England, August 1981 (in print) (1982).

11. Reiter, E.R.: Typical low-tropospheric pressure distributions over and around the Plateau of Tibet. Arch. Meteoro. Geoph. Bioclim. (in press) (1982).

12. Reiter, E.R. and D.R. Westhoff: A planetary wave climatology. J. Atmos. Sci., 38 (4), 732-750 (1981).

13. Reiter, E.R. and Deng-yi Gao: Heating of the Tibet Plateau and movements of the South Asian high during spring. Mon. Wea. Rev., accepted for publication (1982).

14. Times Books London: Atlas of the World, Comprehensive Edition. In collaboration with John Bartholomew and Son Limited, Printed in Great Britain, ISBN 0 7230 0235 5, 227 pp (1980).

15. Tucker, Donna R.: April circulation over the Tibetan Plateau: Investigations with a primitive equation model. Environmental Research Paper No. 36, Colorado State University, Fort Collins, CO (1982).

16. Wallace, J.M. and D.S. Gutzler: Teleconnections in the geopotential height field during the northern hemisphere winter. Mon. Wea. Rev., 109 (4), 784-812 (1981).

17. Ye, D.Zh.: Some aspects of the thermal influences of the Qinghai-Tibetan Plateau on the atmospheric circulation. Archives Meteor. Geoph. Bioclim., Ser. A, 31 (3), 205-220 (1982).

18. Ye, D.Zh., Y.X. Gao, M.C. Tang, S.W. Lo, C.B. Shen, D.Y. Gao, Z.S. Song, Y.F. Qian, F.M. Yuan, G.Q. Li, Y.H. Ding, Z.T. Chen, M.Y. Zhou, K.J. Yang, and Q.Q. Wang: Meteorology of Qinghai-Xizang (Tibet) Plateau. Science Press, Beijing (in Chinese) (1979).

1.2 INFLUENCES OF LAND-SEA TOPOGRAPHY DISTRIBUTIONS ON THE TEMPERATURE FIELD IN JULY

Gao Youxi, Li Ci, Yuan Fumao,
Yang Xiuhua and Yong Xiaochun
Lanzhou Institute of Plateau Atmospheric Physics
Academia Sinica
The People's Republic of China

ABSTRACT

Using global data, temperature anomalies between the surface and 100 mb have been studied for the northern hemisphere summer and southern hemisphere winter. Continental and oceanic anomalies are defined for both hemispheres. The vertical and horizontal extent of large-scale land-sea distributions, but also of smaller-scale features, such as the Mediterranean Sea and the Australian continent, are described. The influence of the Plateau of Tibet in terms of sensible and latent heating is separated from that of the overall land-sea distribution.

I. INTRODUCTION

The planetary atmospheric circulation is mainly affected by the following three factors: the seasonal variation of solar radiation, the land-sea distribution and large-scale orographic barriers, such as the Tibetan Plateau, the Rocky Mountains, the Greenland Plateau and the Antarctic Plateau. Those three factors, acting together, form various stationary circulation patterns, as seen on monthly mean upper-air maps. Even so, each of these factors has its own special influence on the formation of the land-sea monsoon, the plateau monsoon, the interhemispheric monsoon, the large-scale monsoon in the upper troposphere [8] and the stratospheric monsoon as defined by one of the authors [2], and revealed in the nearly opposite directions of prevailing winds in January and July. It is well known now that the dynamical and thermal effects of land-sea and topography distributions play a most important role in disturbing the planetary atmospheric circulation. However, the physical processes, mechanisms and characteristics of these disturbing effects are not yet clear. During the past hundred years, the direct effect of the land-sea distribution has been regarded only as a boundary layer phenomenon. In the 1940's, some authors have shown that direct effects of the land-sea distributions can reach to the higher atmosphere [5, 6]. In the 1970's, some authors believed that effects may reach as high as 10 km [4]. We now believe that the effects of the large-scale land-sea distribution, exercised by the Eurasian continent, the Pacific and Atlantic Oceans and America can extend at least to the tropopause, if not through the whole atmosphere. Recently, Zhang [9] discussed the effect of the meridional land-sea distribution on the temperature field and on atmospheric circulation. However, due to the limitation of data, so far no one has discussed the horizontal extent, vertical height and intensity of perturbations in these fields which are caused by the land-sea distribution.

Using the global data published by Van de Boogaard [7], the authors have calculated temperature anomalies around latitude circles from a 5° latitude by 5° longitude grid at various levels (ground surface, 850, 700, 500, 300, 200, 150 and 100 mb) and then have analyzed the anomaly extent, height and intensity caused by various scale land-sea distributions over the summer and winter hemispheres. Also, the formation mechanisms of temperature anomaly fields will be discussed.

II. ON THE EFFECTS OF LAND-SEA AND TOPOGRAPHY DISTRIBUTIONS

1. Hemispheric Scales

According to the extent of land area over any hemisphere, the authors [1, 3] classified the western half of the northern hemisphere (II)[1], the eastern (III) and western (IV) halves of the southern hemisphere, as water hemispheres. The eastern half of the northern hemisphere (Zone I) should be regarded as a land hemisphere.

As seen from latitudinal temperature anomaly distribution charts, (Figs. 1a, b, c) the most evident positive anomaly is observed in Zone I -- the warmest half of the hemisphere, and the most remarkably negative anomaly in Zone II -- the coldest half of the hemsiphere. Similar anomaly patterns are maintained up to 150 mb. The patterns are reversed at 100 mb and above.

These distributions can also be seen in Table 1a. Zone I is the warmest half-hemisphere. The eastern part of the hemisphere is warmer than the western part. Because the land area in Zone III is somewhat larger than that in Zone IV, Zone III is a bit colder than Zone IV.

Table 1b shows that the differences in hemispheric mean temperature anomalies are rather large, hence, in order to have a better understanding of the monsoon evolution, notably the evolution of the Asian monsoon, consideration of only the land-sea temperature differences between the Asian continent and its surrounding oceans would not be enough. Consideration of the differences between half-hemispheres would be more reasonable.

2. Actual Land-Sea Distribution

The temperature anomalies of actual land and sea areas (Fig. 2) are presented in Table 2a. Based on that table, we also calculated the differences of temperature anomalies between land areas and their adjacent seas to the west and east. Results are presented in Table 2b.

[1] Zones I, II, III and IV are shown in Fig. 2.

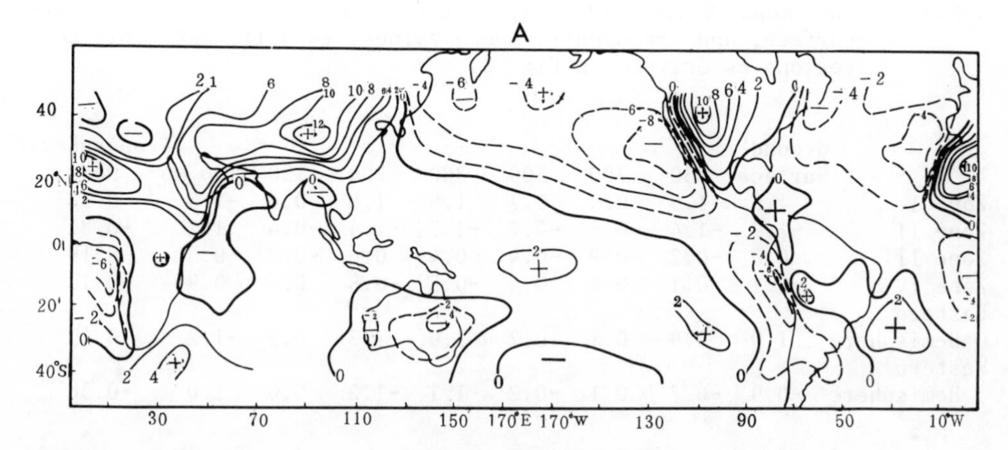

Fig. 1a The temperature anomaly distribution at ground surface, °C

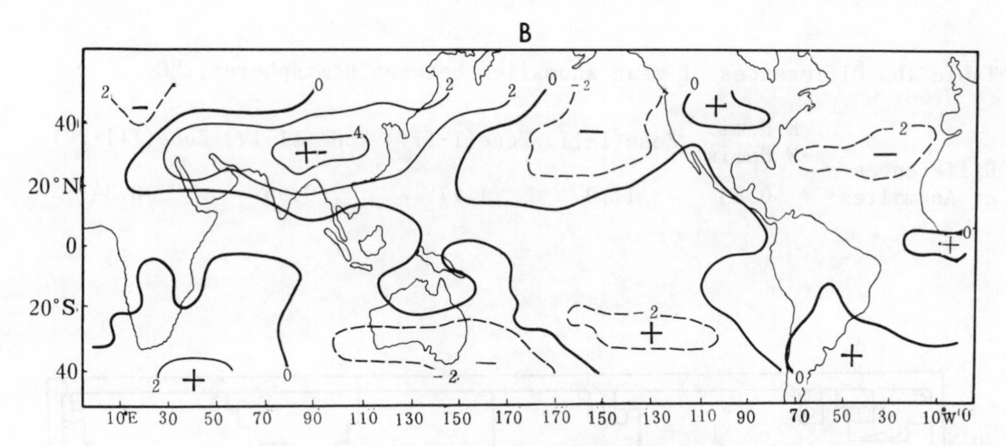

Fig. 1b The temperature anomaly distribution at 500 mb, °C.

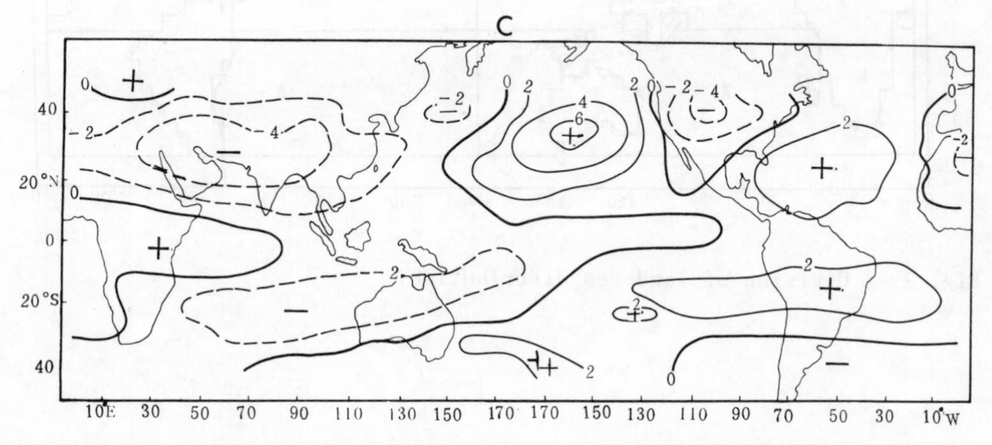

Fig. 1c The temperature anomaly distribution at 100 mb, °C.

Table 1a The mean values of temperature anomalies at each isobaric surface, and tropospheric mean values, in different hemispheric regions as defined in Fig. 2.

	Ground Surface	850	700	500	300	200	150	100 mb	Tropospheric Mean
Zone I	1.8	1.9	0.9	0.7	1.5	1.8	0.6	-1.5	0.98
Zone II	-1.7	-1.4	-0.5	-0.7	-1.5	-1.4	-0.4	1.1	-0.81
Zone III	0.0	-0.2	-0.9	-0.4	0.4	0.6	-0.2	-0.8	-0.19
Zone IV	-0.1	0.1	0.8	0.4	-0.5	-0.8	0.4	0.9	0.15
Eastern Hemisphere	1.0	0.9	0.1	0.2	1.0	1.3	0.3	-1.2	0.45
Western Hemisphere	-0.9	-0.7	0.1	-0.2	-1.1	-1.2	0.0	1.0	-0.38

Table 1b Differences of mean anomalies between hemispheres, °C.

	E Hemis. -W Hemis.	Zone(I-II)	Zone(I-III)	Zone(II-IV)	Zone(III-IV)
Differences of Anomalies	0.83	1.79	1.17	-0.94	-0.34

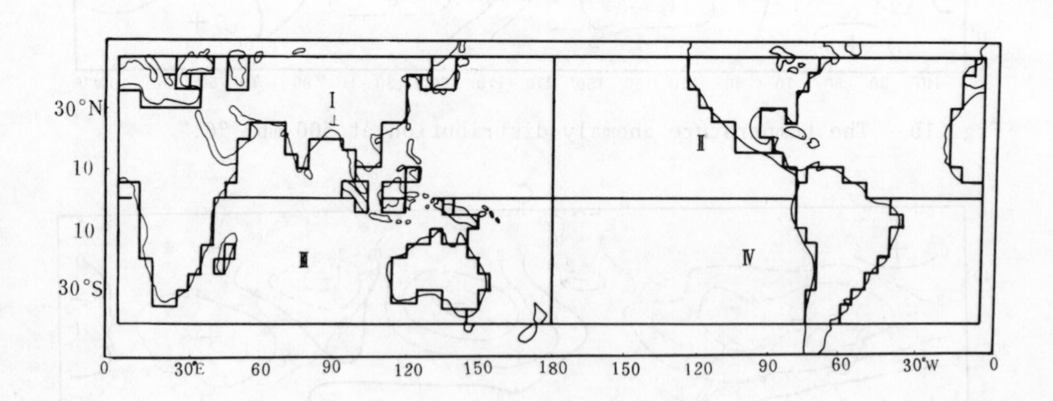

Fig. 2 Division of land-sea distribution.

Table 2a Mean values of temperature anomalies over actual land or sea.

	Surface	850	700	500	300	200	150	100mb	Mean
Eurasian Continent	3.9	3.3	1.9	1.6	3.1	4.1	2.1	-2.8	2.15
N. Hemis. Continents	2.7	2.8	1.2	0.6	1.4	1.8	0.8	-1.6	1.21
N. African Continent	2.6	3.3	0.9	-0.5	0.3	0.5	0.8	-0.9	0.88
Australian continent	-0.6	-1.9	-1.7	-1.3	0.8	1.1	-0.4	-1.2	-0.65
S. African Continent	-0.2	1.6	-1.0	0.0	-0.2	0.6	1.3	0.0	0.26
N. American Continent	0.6	1.2	0.3	-0.2	-0.5	-1.1	-1.4	-0.3	-0.16
S. Hemis. Continent	-0.7	0.4	-0.5	-0.4	-0.2	0.0	0.3	0.1	-0.13
S. American Continent	-0.7	1.2	0.6	-0.1	-1.1	-1.2	0.2	1.4	0.14
N. Atlantic Ocean	-2.3	-3.2	-1.2	-1.1	-2.0	-2.1	-0.2	1.7	-1.30
N. Pacific Ocean	-2.4	-2.3	-0.6	-0.2	-0.9	-0.9	-0.6	0.8	-0.89
N. Hemis. Ocean	-2.0	-2.2	-0.7	-0.4	-1.1	-1.0	-0.4	0.9	-0.86
S. Pacific Ocean	0.4	0.1	0.9	0.3	0.1	-0.1	0.6	0.5	0.35
S. Atlantic Ocean	0.6	-0.7	0.1	0.0	-1.0	-1.4	0.2	0.6	-0.20
S. Hemis. Ocean	0.5	-0.2	0.2	0.1	0.0	-0.1	0.2	0.7	-0.20
N. Indian Ocean	0.6	1.5	0.3	0.3	0.0	0.3	-0.7	-1.3	0.13
S. Indian Ocean	0.7	-0.4	-0.9	-0.1	0.7	0.5	-0.5	1.1	0.10

Table 2b Differences of mean anomalies between land and adjacent ocean areas.

	Ground Surface	850	700	500	300	200	150	100mb	Mean
Eurasian Land & Sea	6.3	6.1	2.8	2.3	4.4	5.6	2.5	-4.1	3.25
N. Hemis. Land & Sea	4.7	5.0	1.9	1.0	2.5	2.8	1.2	-2.5	2.08
N. African Land & Sea	3.6	4.2	1.5	-0.1	1.3	1.4	1.2	-1.1	1.50
N. American Land & Sea	3.0	4.0	1.2	0.5	1.0	0.4	-1.0	-1.6	0.94
Australian Land & Sea	-1.1	-1.8	-1.7	-1.4	0.4	0.9	-0.5	-2.0	-0.90
S. Hemis. Land & Sea	-1.2	0.6	-0.7	-0.5	-0.2	0.1	0.1	-0.6	-0.30
S. African Land & Sea	-0.9	2.2	-0.6	0.7	-2.4	1.1	1.5	-0.9	-0.09
S. American Land & Sea	-1.7	1.5	-0.4	-0.3	-0.6	-0.5	0.6	0.9	0.06

Among the absolute mean anomaly values for the whole air column (ground surface to 100 mb), as shown by the last row in Table 2a, the value over the Eurasian continent (+2.15°C)[2] is the largest, followed by the continents of the northern hemisphere (+1.21°C) and by the North African continent (+0.88°C). Then the mean values decrease in the following sequence: the Australian continent (-0.65°C), South African continent (+0.27°C), North American continent (-0.18°C), the continents of the southern hemisphere (-0.13°C), and the South American continent (+0.14°C) which is the smallest. Among oceans, the North Atlantic (-1.30°C) has the largest value, followed by the North Pacific (-0.89°C),

[2] Hereafter, the order is arranged by absolute values and +, - signs are also added to avoid misunderstanding.

the oceans of the northern hemisphere (-0.86°C), and so on, as shown in Table 2a. The South Indian Ocean (0.10°C) has the smallest value.

The mean difference values between land and sea temperature anomalies for the whole air column (Table 2b) are largest for the Eurasian continent and adjacent seas (+3.25°C), followed by the continents and oceans of the northern hemisphere (+2.08°C), and so on, as shown in the last row of Table 2b. The South American continent and oceans (0.06°C) have the smallest difference.

Even though the Eurasian and the North American continents are situated at nearly the same latitudinal belts, the thermal effects south of 45°N are quite different in summer. The heating effect of the Eurasian continent is very dominant (+2.15°C), the effect reaching as high as 150 mb. A transitional layer is observed between 150 and 100 mb. On the other hand, the heating effect of the American continent (0.73°C) is rather small reaching only 700 to 500 mb. Therefore, the differences between these two continents are remarkable, the layer affected by heating is 10 km higher over Eurasia than over North America. The heating intensity over the former continent is more than twice as strong as that over the latter.

The North Pacific and the Atlantic Oceans are both quite large in extent, and their cooling effects in terms of intensity and influence height are also strong. Transitional layers are observed over both oceans at 150-100 mb.

The South African and the South American continents are similar in latitudinal position and in shape, but their heating or cooling effects (see Table 2a) are quite different, because the former has a large land area to the north.

The heating (cooling) effect of the land or water hemispheres is only one half of that of continents and adjacent oceans (see Table 2a, b). Therefore, it would be improper to emphasize the heating or cooling effect of either continents or oceans only. It is more reasonable to consider land-sea effects together in order to obtain a better understanding of the evolution of general atmospheric circulation patterns.

3. The Extent of the Influence of Land-Sea Distributions.

Because land-sea distributions are different in extent, shape and elevation, the extents of their influence are remarkably different in the horizontal as well as vertical directions.

Horizontal extent: The horizontal influence extent of the land-sea distribution in the northern hemisphere is largest at sea level (Fig. 1a), and decreases with height, reaching a minimum at about 500 mb (Fig. 1b). It increases again and reaches another maximum at 150 mb (Fig. 1c). The North American continent is smaller than the Eurasian continent (Fig. 1a), however, its influence extent is largest at sea level. It decreases faster upward and nearly disappears at 500 mb (Fig. 1b). The situation over the southern hemisphere is more irregular than that over the northern hemisphere, except over the Australian continent. There the pattern is somewhat similar to that over North America, except that Australia is a heat sink in July.

Vertical influence extent: As seen from Tables 1a and 2a, the vertical variation of the intensities of the effects of land-sea distributions shows an opposite phase between land and water hemispheres (Table 1a) and land and ocean (Table 2a) as well. In general, the values of temperature anomalies (positive or negative) usually have a maximum in the surface layer, then weaken upward, reaching a minimum value somewhere near 500 mb. Beyond that level the anomalies increase again and reach the next maximum at 200 to 300 mb. Hereafter, they weaken again and finally become reversed at 150 to 100 mb.

Figure 3 shows the vertical variation of temperature anomaly values over continents and oceans in the northern and southern hemispheres. In the northern hemisphere, temperature anomalies over the continents are positive and over the oceans they are negative (Fig. 3a). The temperature anomalies of the whole troposphere over the Eurasian continent are positive and change into negative ones at 150 mb. The positive temperature anomalies of the North American and North African continents only reach to about 500 mb. The negative temperature anomalies over both the North Pacific and the North Atlantic are rather large and change into positive values at 100 to 150 mb, which are a little bit smaller in absolute value than those over the Eurasian continent. The anomalies over the North Indian Ocean are quite particular, in as much as either positive or negative anomaly values are rather small. The influence of the Eurasian continent and its tropical position might be the cause of such an anomaly behavior over that region.

Figure 3b shows the situation over the southern hemisphere. The direct cooling effect of the South African, South American and Australian continents reaches to 900 mb, 900 mb and 400 mb, respectively. The direct heating effects of the South Atlantic, South Indian and South Pacific Oceans are also restricted to the troposphere and reach only to 900 mb, 900 mb and 300 mb, respectively.

It can be seen that the influences of the land-sea distribution are completely different in the summer and winter hemispheres: not only are the heating or cooling effects of continents and oceans larger in the summer hemisphere than in the winter hemisphere, but also the influence heights in the summer hemisphere are much higher than those in the winter hemisphere. The vertical profiles of the temperature anomaly values in the summer hemisphere are simpler than those in the winter hemisphere (Fig. 3a, b). In addition, the continents and oceans of different sizes and shapes, located in the same hemisphere, have different effects, and the continents or oceans with similar size and shape, but located in different hemispheres, have different effects, too.

4. Effects of Some Special Land-Sea Distributions

The Mediterranean Sea is an inland sea and the Australia continent is a large "island". They are of similar shape and elongation but have different extents and geographical positions. The former is smaller than

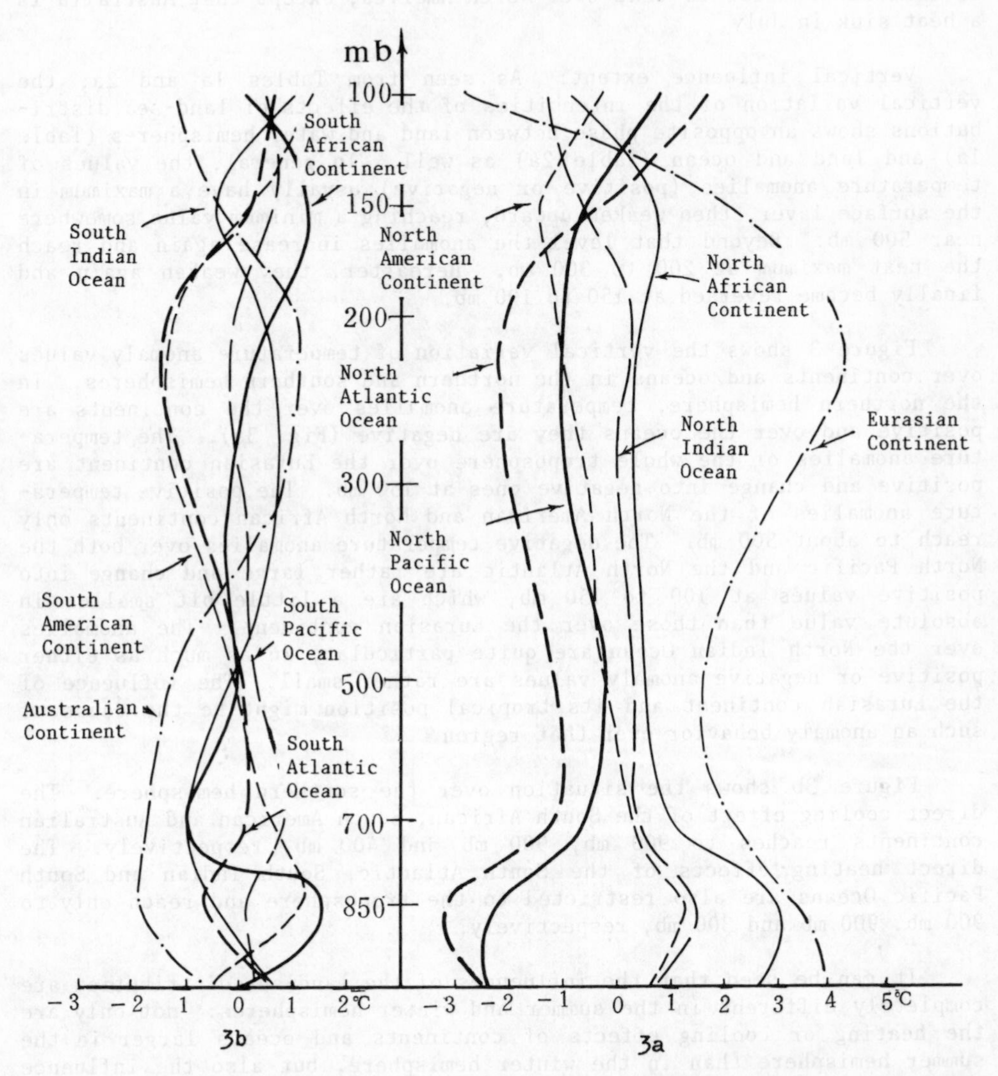

Fig. 3 Vertical profiles of temperature anomalies for different land and sea areas.

the latter and is located farther from the equator. They also belong to different hemispheres. Theoretically, both regions should show the same cooling effects, one characteristic for summer, the other for winter. However, they exert quite different influences.

Table 3 shows the area-mean temperature anomaly of the two regions at various levels. Both regions have, on the average, different thermal effects. It can be seen from Fig. 1a that there are two subareas with

Table 3 The mean temperature anomalies of the Australian continent and the Mediterranean Sea at various levels.

	Ground Surface	850	700	500	300	200	150	100mb	Mean
Australian Continent	-0.6	-1.9	-1.7	-1.3	0.8	1.1	-0.4	-1.2	-0.65
Mediterranean Sea	-0.5	1.2	0.5	-1.1	-1.0	1.7	1.6	-1.6	0.10

negative anomalies which indicate the cooling effect of the Mediterranean Sea in summer at 1000 mb. However, this cooling effect completely disappears at 850 mb and the area-mean anomaly becomes positive at 1000 to 700 mb. The Australian continent has a cooling effect which, according to our analysis, can reach to somewhere near 400 mb.

It seems to us that because the Mediterranean Sea is a relatively small inland sea, the heating effect of the Eurasian continent shows its controlling influence over the whole Eurasian area, and overrides the feeble cooling effect of the Mediterranean. However, the positive anomalies over the Mediterranean area are all weaker than over the surrounding land areas, thus indicating a certain cooling effect.

Because the Indian and Indochina peninsulas are similar in extent, shape, geographical position and land-sea distribution, they should show similar land-sea effects in the lower troposphere (lower than 850 mb) (see Fig. 1a). However, these land-sea effects disappear completely at 850 mb. Above this level, the temperature anomalies become positive all over this region, no matter whether over land or over sea. This fact also clearly indicates the controlling heating effect of the Eurasian continent.

The North and South African continents are attached to each other and to the Eurasian continents, hence the northern continents should extend their influences to the south. Theoretically, the cooling effect of the South African continent in the southern hemisphere winter should strengthen the easterly jet located over the southern part of North Africa. However, in reality the South African continent shows a slight heating effect on the average (see Table 2a). Therefore, the easterlies over North Africa are somewhat diminished from what is expected theoretically.

III. THE EFFECTS OF THE TIBETAN PLATEAU

1. The Tibetan Plateau Enhances the Influence of the Land-Sea Distribution

Table 4 presents the temperature anomaly values at various levels over the geometrical centers of the Eurasian continent and of the Tibetan Plateau.

Table 4 The temperature anomaly values over the center of the Eurasian continent (60°E, 45°N) and over the Tibetan Plateau center (85°E, 35°N), and the maximum global anomaly values and their geographical positions at various levels.

	Ground Surface	850	700	500	300	200	150 mb
Eurasian Continental Center (60°E, 45°N)	4.8	0.5	0.0	-1.2	2.0	7.0	3.7
Tibetan Plateau Center (85°E, 25°N)	12.0			3.5	7.3	7.5	3.8
Actual Maximum Center	13.8			5.0	7.5	7.5	4.4
Geographical Position of Maximum Value	35°N 90°E			30°N 100°E	35°N 90°E	35°N 85°E	40°N 90°E

The anomalies of the maximum positive centers and their geographical positions over Eurasia are given as well. There are several interesting facts to be pointed out: (I) The maximum anomaly centers are neither observed over the Eurasian continental center nor over the Tibetan Plateau center but somewhere else over the plateau. (II) The anomaly values of the maximum center are very close to those over the plateau center, but much greater than those of the geographical center of Eurasia. (III) The maximum anomaly centers at various levels are quite close to the plateau's center but are far removed from the Eurasian continental center. These facts indicate that the existence of the Tibetan Plateau not only forces the maximum heating center (which would appear near the Eurasian geographical center if there were no plateau) to shift roughly 1000 km southward and 3000 km eastward, and also makes the intensity of the temperature anomaly 2 to 4 times larger than that over the Eurasian geographical center.

We conclude, therefore, that the heating effect of the huge plateau is to strengthen the warming effects produced by the Eurasian continent. Even so, because the plateau's extent is much smaller than that of Eurasia (1/20), the heating effect of the plateau can be regarded as "a small perturbation" upon the much larger heating disturbance of the Eurasian continent.

2. Quantitative Estimate of the Influences of Large-Scale Orography on the Temperature Field

Let us assume that the temperature anomaly from the zonal mean, $\Delta T_{\lambda\phi}$, at any grid point is the result of the influence of land-sea distribution and can be written as $\Delta T_{\lambda\phi} = T_{\lambda\phi} - [\bar{T}_{\phi}]$, where $T_{\lambda\phi}$ is the monthly mean temperature at a certain grid point with longitude λ and latitude ϕ, and $[\bar{T}_{\phi}]$ is the zonal-mean temperature along latitude ϕ. We then let $[T_L]$ be the area-mean temperature for a certain continent, such as the

Eurasian or the North American continents. This temperature can be expressed as

$$[T_L] = \frac{\sum_{n=1}^{N} (\Delta T_{\lambda\phi})_n}{N}$$

Here, N is the total number of grid points on the continent. Then, the influence of the large-scale orography on $\Delta T_{\lambda\phi}$ could be considered by using $T_{o\lambda\phi}$ which is calculated from $T_{o\lambda\phi} = \Delta T_{\lambda\phi} - [T_L]$.

Figures 4a, b, c show the values of $T_{o\lambda\phi}$ at the 1000, 500, and 100 mb levels over the Eurasian continent. In Fig. 4a, there are two remarkable, positive anomaly centers. One is situated over the Tibetan Plateau and the other over the Iran Plateau. The intensity of the positive anomalies is strongest at ground surface (Fig. 4a) then decreases upward and reaches a minimum at 500 mb (Fig. 4b). Thereafter it increases again and reaches a second maximum at 200 to 300 mb, then decreases again and finally changes into negative values at 100 mb (Fig. 4c). The horizontal extent of the positive anomaly over the plateau region exceeds by far the actual extent of the plateau.

The situation over the Rocky Mountains (not shown here) is somewhat similar to that over the Tibetan Plateau region, but the horizontal extent of the anomaly decreases very rapidly with height. The area of the positive anomaly is largest near the ground. The anomaly disappears completely at 500 mb, and negative anomalies prevail above that level.

3. Further Estimates of the Effect of the Tibetan Plateau on the Temperature Field

Besides the influence of the large-scale continental elevation, the values $\Delta T_{o\lambda\phi}$ also include the influences of distance from the sea and of the free atmosphere. Suppose the Tibetan Plateau were a west-east elongated rectangle, and $T'_{o\lambda\phi}$ could be written as: $T'_{o\lambda\phi} = T_{o\lambda\phi} - (\bar{T}_{o\lambda\phi})_{90°E}$, where $(\bar{T}_{o\lambda\phi})_{90°E}$ is the mean value of $\Delta T_{o\lambda\phi}$ along longitude 90°E within the rectangle. Then we can consider $T'_{o\lambda\phi}$ as only due to the effect of the orographic elevation, and the effects resulting from the distance from both the sea and the free atmosphere could be considered as eliminated. The calculated results are presented in Table 5.

From this table it can be seen that the values of $T_{o\lambda\phi}$ decrease with both increasing distance from the plateau and height above ground level. The ratio values of $(T_{o\lambda\phi}/\Delta T_{\lambda\phi}) \cdot 100$ reach 60 to 70 percent in the layer at and below 500 mb, and 20 to 50 percent above 500 mb. The ratio values of $(T'_{o\lambda\phi}/\Delta T_{\lambda\phi}) \cdot 100$ are 20 to 50 percent. Those numbers mean that about 20 to 50 percent of the influence of land-sea and topography distribution are induced by orographic elevation, and about 20 to 40 percent

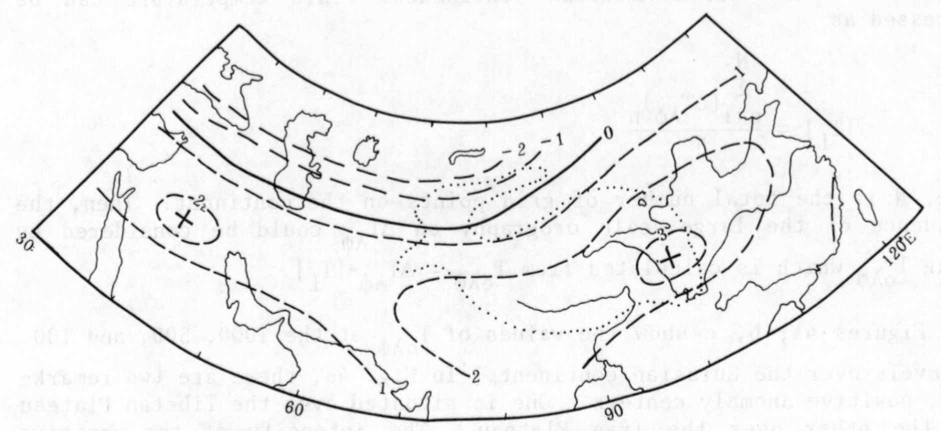

Fig. 4a The distribution of temperature anomalies at the earth's surface due to the effects of the Tibetan Plateau.

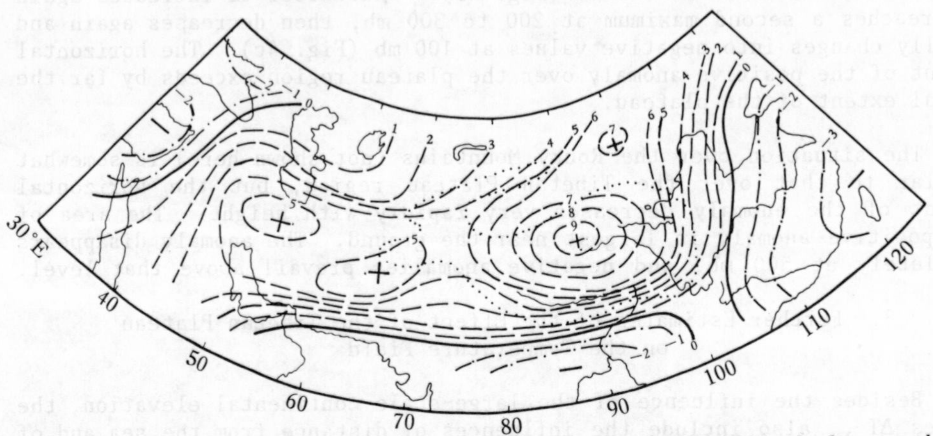

Fig. 4b The distribution of temperature anomalies at 500 mb due to the effects of the Tibetan Plateau.

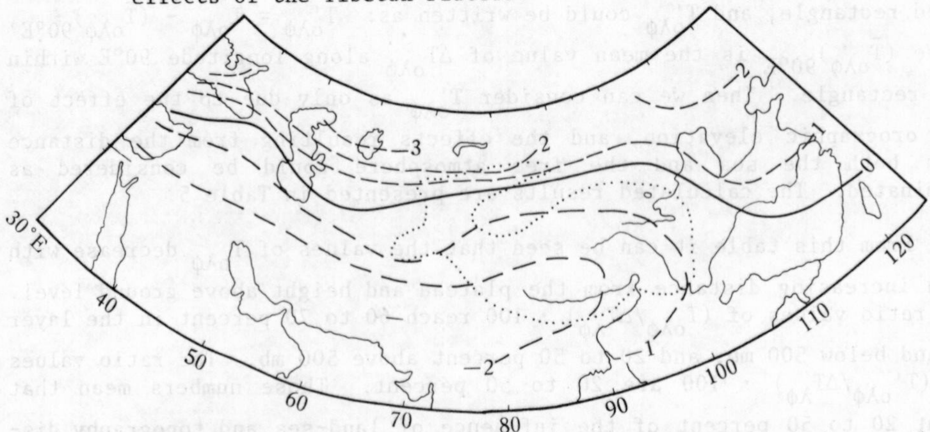

Fig. 4c The distribution of temperature anomalies at 100 mb due to the effects of the Tibetan Plateau.

Table 5 The temperature anomalies along 90°E at various levels due to land-sea effect (first line), Tibetan Plateau effect (second) and to plateau elevation effect (fourth). The third and the fifth lines are the ratios of the second to the first and the fourth to the first lines.

		45°N	40°	35°	30°	25°	20°	15°	Mean
Ground Surface	$\Delta T_{\lambda\phi}$	8.0	10.4	13.8	10.5	1.8	-0.5	-0.2	6.3
	$T_{o\lambda\phi}$	4.1	6.5	9.9	6.6	-2.1	-4.4	-4.1	2.4
	$T_{o\lambda\phi}/\Delta T_{\lambda\phi}$ (%)	51	63	72	63				
	$T'_{o\lambda\phi}$	1.7	4.1	7.5	4.2	-6.5	-6.8	-6.5	
	$T'_{o\lambda\phi}/\Delta T_{\lambda\phi}$ (%)	16	39	54	40				
500 mb	$\Delta T_{\lambda\phi}$	0.8	2.3	2.5	5.1	3.6	2.7	1.4	2.6
	$T_{o\lambda\phi}$	-0.8	0.7	1.9	3.5	2.0	1.1	-0.2	1.5
	$T_{o\lambda\phi}/\Delta T_{\lambda\phi}$ (%)		30	76	69	56	41		
	$T'_{o\lambda\phi}$	-2.3	-0.8	0.4	2.0	0.5	-0.4	-1.3	
	$T'_{o\lambda\phi}/\Delta T_{\lambda\phi}$ (%)			16	39	14			
300 mb	$\Delta T_{\lambda\phi}$	2.7	5.1	7.5	7.3	5.3	3.4	1.6	3.9
	$T_{o\lambda\phi}$	-0.4	2.6	4.4	4.2	2.2	0.3	-1.5	1.5
	$T_{o\lambda\phi}/\Delta T_{\lambda\phi}$ (%)		49	58	57	41	9		
	$T'_{o\lambda\phi}$	-1.9	1.1	2.9	2.7	0.7	-1.2	-3.0	
	$T'_{o\lambda\phi}/\Delta T_{\lambda\phi}$ (%)		22	39	37	11			
200 mb	$\Delta T_{\lambda\phi}$	5.6	6.9	7.4	6.8	5.6	4.3	2.7	4.6
	$T_{o\lambda\phi}$	1.5	2.8	3.3	2.7	1.5	0.2	-1.4	1.3
	$T_{o\lambda\phi}/\Delta T_{\lambda\phi}$ (%)	27	41	45	40	27	5		
	$T'_{o\lambda\phi}$	0.2	1.5	1.8	1.4	0.2	-1.1	-2.7	
	$T'_{o\lambda\phi}/\Delta T_{\lambda\phi}$ (%)	4	22	24	21	4			

are produced by the distance from the sea and from the free atmosphere. Only about 30 to 40 percent of the thermal effects are caused by the land-sea distribution.

Table 6 shows values of the gradient of $\Delta T_{\lambda\phi}$ and $\Delta T_{o\lambda\phi}$ along a certain latitude or longitude line, measured in different directions

Table 6 The magnitude of temperature anomalies, their distance from the zero line (in number of grid intervals) and the gradient for land-sea, plateau and free atmosphere at various levels.

	Land-Sea	Plateau-Free Atm.	Land-Sea	Plateau-Free Atm.	Land-Sea	Plateau-Free Atm.	Land-Sea	Plateau-Free Atm.
W-E Direction	13.8	9.9	5.1	3.4	7.5	4.4	7.5	3.4
Number of Grid Intervals	7	4	12	7	13	9	13	6
Gradient (°C/5 Long.)	2.0	2.5	0.4	0.5	0.6	0.5	0.6	0.6
S Direction	13.8	9.9	5.1	3.4	7.5	4.4	7.5	3.4
Number of Grid Intervals	3	2	5	2	5	2	6	4
Gradient (°C/5 Lat.)	4.6	5.0	1.0	1.7	1.5	2.2	1.3	0.9
N Direction	13.8	9.9	5.1	3.4	7.5	4.4	7.5	3.4
Number of Grid Intervals	2	2	3	2	3	2	2	2
Gradient (°C/5 Lat.)	6.9	5.0	1.7	1.7	2.5	2.2	3.8	1.7

(east-west and northward or southward) at various levels. $\Delta T_{\lambda\phi}$ (or $\Delta T_{o\lambda\phi}$) is the maximum value at the center of the anomaly distribution analyzed on charts and distances are measured along the same longitude (or latitude) line northward, southward (or east-westward) from the center to the intersection point of the 0°C line. It can be seen from Table 6 that the temperature gradients which are produced either by the land-sea distribution or by the difference between the plateau and the free atmosphere are nearly the same in magnitude.

From this rough analysis we conclude that the influence of the Tibetan Plateau on the temperature field is very important, not only due to the fact that its horizontal influence extent exceeds by far its actual size and its vertical influence region can reach very high, at least through the whole troposphere, but also because its influence intensities are larger than those of the land-sea distribution. Thus, if the influence of the Eurasian continent and adjacent oceans (Pacific, Indian or Atlantic Oceans) can produce land-sea monsoons, the effect of the contrasts between the plateau and the free atmosphere will generate the so-called "plateau monsoon".

IV. HEATING MECHANISM DUE TO THE LAND-SEA INFLUENCE

The heating mechanism due to land-sea and topography distribution effects can be divided into three kinds: i.e. sensible heating from nonelevated underlying surfaces, latent heating from condensation or sublimation and sensible heating from elevated plateau surfaces.

Figure 5 indicates that the heating effect of the Eurasian continent contains the first two kinds of mechanisms: the sensible heating in the surface boundary layer (1000 to 850 mb), and the latent heating in upper levels (400 to 250 mb). Because sensible heating is uniform in the boundary layer due to the mixing process, the positive temperature anomalies are also uniform in the 1000 to 850 mb layer. Even above this layer, the sensible heating due to turbulent transfer and thermal convection is still active in the free atmosphere; however, it is much weaker than that in the boundary layer. In addition, it is very easy for the stronger winds in the free atmosphere to carry the heat away. Therefore, the values of positive temperature anomalies decrease with height more remarkably in the layer 850 to 500 mb (or 850 to 700 mb). After reaching minimum values at 500 mb (or 700 mb) the anomaly values increase again and reach another maximum, which is formed mainly by the latent heat release at 200 to 300 mb. From 200 to 300 mb upward, due to decreased condensation and sublimation, the positive anomalies decrease again. Of course, large sensible heating effects of the Tibetan Plateau would naturally have an important contribution to the warming process of the air in the midtroposphere.

Let us consider the relative importance of sensible and latent heating at various levels as shown in Fig. 5. In this diagram the vertical temperature anomaly profiles for the following gridpoints are shown: the plateau (35°N, 90°E), Nanjing (30°N, 115°E) and Assam (25°N, 90°E).

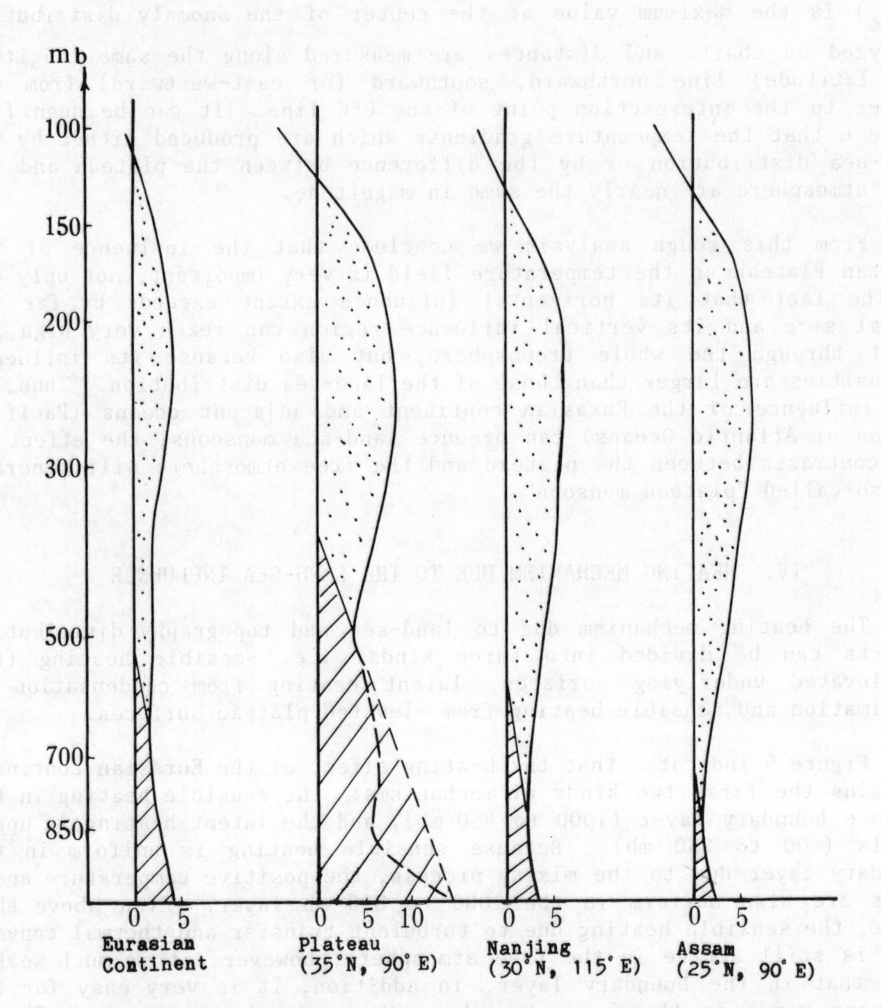

Fig. 5 Vertical profiles of the temperature anomalies over the Eurasian continent, over the Plateau (35°N, 90°E), near Nanjing (30°N, 115°E) and near Assam (25°N, 90°E). Striped: sensible heating; dotted: latent heating. Abscissa: temperature anomalies in °C.

Suppose that the heating effect in the layer from the earth's surface to 500 mb (or 700 mb, etc.) were influenced only by sensible heat which has a linear distribution with height, then the striped area would present the heating effect of sensible heat, and the dotted area would represent the heating effect of latent heat. Therefore, the monthly mean cloud base heights over the plateau, Nanjing and Assam are 500 mb, 850 mb and 850 mb, respectively. The layer below cloud base is heated only by sensible heat. The layer between cloud base and the intersecting point of

Table 7 The relative importance of sensible and latent heating for three regions at various levels.

	Ground Surface	850	700	600	500	300	200	150	sum	Total Latent Heat
Plateau Region $\Delta T_{\lambda\phi}$ (35°N, 90°E)				6.0	3.5	7.5	7.4	3.8	28.2	18.7
Relative Importance ($\Delta T_{\lambda\phi}/28.2$)				(21)	(12)	27	26	13		
Relative Importance of Latent Heat (%) $\frac{\Delta T_{\lambda\phi}}{18.7}$						40	40	20		
Nanjing Region $\Delta T_{\lambda\phi}$ (30°N, 115°E)	3.0	2.3	2.6		4.0	5.3	5.9	2.0	25.1	19.8
Relative Importance ($\Delta T_{\lambda\phi}/28.2$)	(12)	(9)	10		16	21	24	8		
Relative Importance of Latent Heat (%) $\frac{\Delta T_{\lambda\phi}}{18.7}$			13		20	27	30	10		
Assam Region $\Delta T_{\lambda\phi}$ (25°N, 90°E)	1.8	1.0	1.3		3.6	5.3	5.6	2.1	20.7	17.9
Relative Importance ($\Delta T_{\lambda\phi}/28.2$)	(9)	(5)	6		17	26	27	10		
Relative Importance of Latent Heat (%) $\frac{\Delta T_{\lambda\phi}}{18.7}$			7		20	30	31	12		

the 0°C isoline and the dashed line is heated by both latent and sensible heat. The layer above that intersection point is heated mainly by latent heat (including precipitation and sublimation heat).

The relative importance of different heating over these three places at various levels can be seen from Table 7. It seems that the sensible heating (see numbers in parentheses) over the plateau is more important than that over both Nanjing and Assam, and the importance of latent heating is different at various places and levels.

Maximum latent heating is observed everywhere near 200 to 300 mb but the values over the plateau are generally largest, especially in the high troposphere.

V. CONCLUSIONS

1. Because the eastern half of the northern hemisphere is the warmest of all global regions, it is necessary to consider the interaction between this part of the hemisphere and other regions.

2. The influence of large-scale land-sea distributions can extend throughout the whole troposphere, but small-scale effects only reach into the lower troposphere.

3. The extent and intensity of temperature anomalies due to the land-sea influence are largest and strongest near the ground, decrease with height and reach a minimum at 500 mb. Thereafter, they expand and increase again upward and reach a second maximum at 100 mb or 200 to 300 mb. Above these levels they decrease rapidly and finally assume an opposite sign at 100 mb.

4. The role of the Tibetan Plateau strengthens the effect of the land-sea distribution. About 60 percent of the observed temperature anomalies are produced by the effects of the Tibetan Plateau.

5. The gradients of temperature anomalies produced by the land-sea effect and by the Tibetan Plateau free-atmosphere effect are almost of the same magnitude. We conclude, then, that the land-sea effect can produce land-sea monsoons, the effect of the Tibetan Plateau free-atmosphere differences produces the plateau monsoon.

REFERENCES

1. Gao, You-xi: The influence of land-sea distribution and Tibetan Plateau on the climate of China. Proceeding of Tibetan Plateau Meteorology, 1975-1976, 34-46 (in Chinese) (1976).

2. Gao, You-xi: Some problems of monsoons. Proceeding of Meteorological Society Annual Meeting of Yunnan Province: Monsoon Monograph (in Chinese) (1979).

3. Gao, You-xi and Li Ci: The interhemispheric monsoons -- an important mechanism of the interaction between northern and southern hemisphere atmosphere. Plateau Meteo., 1 (1) (in Chinese) (1982).

4. ICSU/WMO: The monsoon experiment. GARP Publication Series No. 18, W.M.O., Geneva, Switzerland (1976).

5. Rossby, C.G. et al.: Relation between the intensity of the zonal circulation of the atmosphere and the displacements of the semi-permanent centers of action. J. Mar. Res., 2, 38-55 (1939).

6. Smagorinsky, J.: The dynamic influence of the large-scale heat sources and sinks on the quasi-stationary mean motions of the atmosphere. Quart. J. Roy. Met. Soc., 75, 417-428 (1953).

7. Van de Boogaard, Henry: The mean circulation of the tropical and subtropical atmosphere -- July. NCAR Tech. Note, Sept. (1977).

8. Xie Yi-bing, Chen Sou-jun, et al.: The seasonal variation of general atmospheric circulation and long-range forecasting in summer half-year rainfall over the Changjiang valley (preliminary report). Papers of Lanzhou Conference on Medium and Long Range Forecasting (in Chinese) (1958).

9. Zhang, Jia-cheng: The thermal effect of meridional sea-land distribution on the general atmospheric circulation in Eurasia and its contiguous areas . Acta Meteorologica Sinica, 38 (3) (in Chinese) (1980).

1.3 THE EFFECTS OF THE LAND-SEA DISTRIBUTION AND OF THE PLATEAU ON THE MEAN MERIDIONAL CIRCULATION AT LOW LATITUDES IN JULY

Luo Siwei, Yao Lanchang and Lü Shihua
Lanzhou Institute of Plateau Atmospheric Physics
Academia Sinica
The People's Republic of China

ABSTRACT

Global and regional mean meridional circulation patterns are presented for July, using data from the period 1961-1974 from both hemispheres. The circulation patterns in different longitude sectors clearly reveal oceanic and continental influences. The "chimney" role of the Qinghai-Xizang Plateau in exporting energy in the upper troposphere is evident from the analyses. Heat sources and sinks within these circulation patterns are estimated quantitatively using the thermodynamic equation.

I. INTRODUCTION

The mean meridional circulation is a main component [7, 3] of the general circulation of the atmosphere. It is not yet well understood because past observation data are not enough. Vuorela and Tuominen [5] estimated the global mean circulation of mass flux only with the south-north wind component, and others [8, 1] discussed the mean meridional circulation limited to the northern hemisphere and to a single year or month. Recently Ye [9] calculated the global and regional mean meridional circulations averaged over about 10 years of data, but these patterns were also limited to the northern hemisphere.

In this paper the mean meridional circulations and physical parameters, such as the vertical velocity component, heating fields and energy transports are calculated at grid points with a resolution 5°x5° on the 850 mb, 700 mb, 500 mb, 300 mb, 200 mb and 100 mb surfaces in the global latitude zone between 45°N-40°S for the July mean conditions of 1961-1974 [4].

The vertical velocity ω is calculated by integrating the equation of continuity $D + \partial\omega/\partial P = 0$. Its lower boundary condition is defined by the equation $\omega_s = \vec{V}_s \cdot \nabla P_s$ where ω_s is the vertical velocity at the surface, P_s is the surface pressure and $\vec{V}_s$ is the surface wind. The heating field is calculated from the thermodynamic equation

$$Q = \partial T/\partial t + \vec{V} \cdot \nabla T + \omega(\partial T/\partial P - gT/RP)_o$$

The total energy transported by the monsoon circulation in the north-south direction is calculated by the equation

$$E_y = \frac{1}{g}\int_0^{P_o} (C_p \bar{T} + \bar{\Phi} + L\bar{q})\, \bar{V}\, dP_o$$

The results obtained show that the effects of the large-scale land-sea distribution and of the plateau on the meridional circulation are strongly evident. Because the underlying surfaces in the eastern Pacific Ocean, the Atlantic Ocean and the European and African continents are similar in both hemispheres, the Hadley cells in these regions are, too. Over eastern Asia and the western Pacific Ocean, on the other hand, there are typical monsoon circulations. The Qinghai-Xizang Plateau plays a "chimney" role in the export of energy in the upper topopsphere.

II. THE MEAN MERIDIONAL CIRCULATION

1. Global Distribution

The global mean meridional circulation (Fig. 1) shows that the Hadley cell over the northern hemisphere is much weaker than over the southern hemisphere. The results shown here are similar to the ones obtained by Ye [7, 8, 9], but our Hadley cell is weaker than the one obtained by Vuorela and Tuominen [5], and is shifted somewhat southward. Vuorela estimated the meridional circulation of mass flux from the global

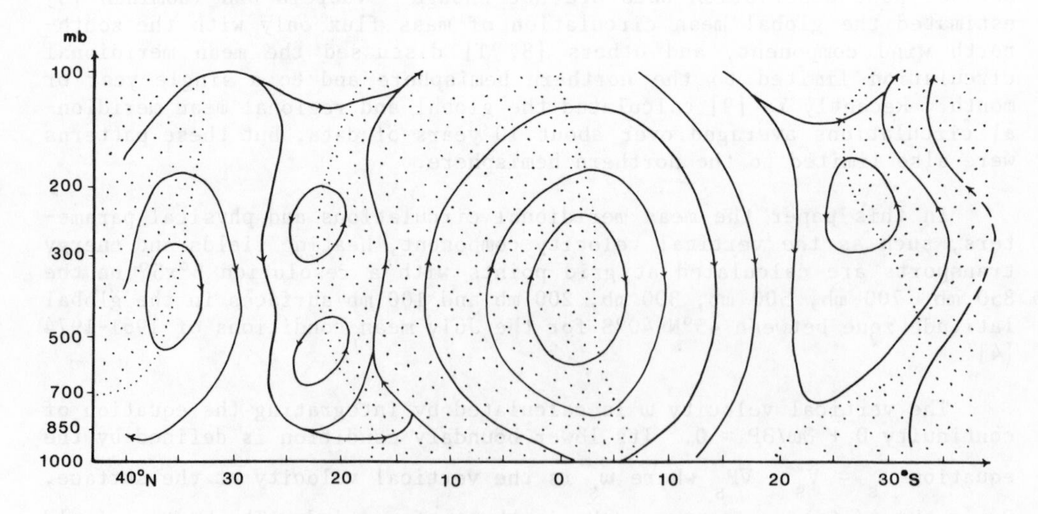

Fig. 1 The global mean meridional circulation for July with vertical velocity magnified 200 times. The ordinate is labelled in mb and the abscissa in degrees latitude. The dotted areas denote heat sources and the blank areas heat sinks.

mean north-south wind component and Ye [7, 8, 9] calculated the meridional circulation from 10 years of data. In this paper we calculated the fluxes with data from more than 10 years so the results should be more reliable.

It is worthwhile to note that the Hadley cells in both hemispheres are asymmetric. This fact will be discussed later.

2. Eastern Pacific, Atlantic, Europe and Africa

Figures 2 to 4 show the meridional circulations over the eastern Pacific (175°-120°W), over the Atlantic (0°-55°W) and over Europe and Africa (5°-60°E). The meridional circulation over the eastern Pacific (Fig. 2) is rather similar to the global one. The major differences are that the Hadley cells in both hemispheres are almost symmetric and that the indirect cell over the northern hemisphere is shifted northward and is located to the north of 35°N.

The meridional circulation over the Atlantic (Fig. 3) is similar to the one over the eastern Pacific Ocean. So is the meridional circulation over Europe and Africa, except that the indirect cell of the southern hemisphere is shifted to the south of 40°S. All these three meridional circulations over the northern hemisphere are nearly similar to that obtained by Ye [7, 8, 9] and the major differences between them appear to depend upon the adopted longitude zone of averaging.

3. Asia, Western Pacific and America

Figure 5 shows the meridional circulation over Asia. The above mentioned Hadley cell of the northern hemisphere disappears and is replaced by a huge monsoon circulation over the latitude zone 40°W-25°S.

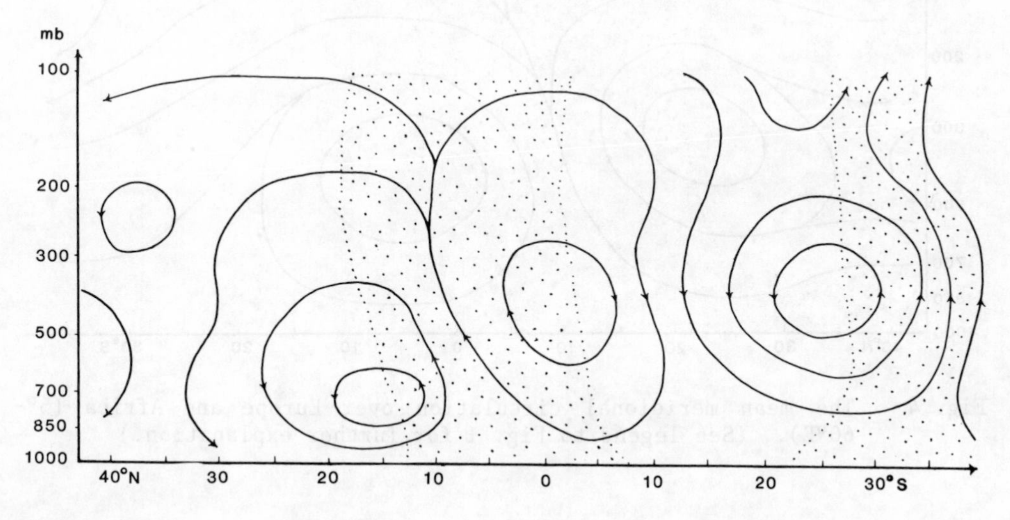

Fig. 2 The mean meridional circulation over the eastern Pacific (175°-120°W). (See legend to Fig. 1 for further explanation.)

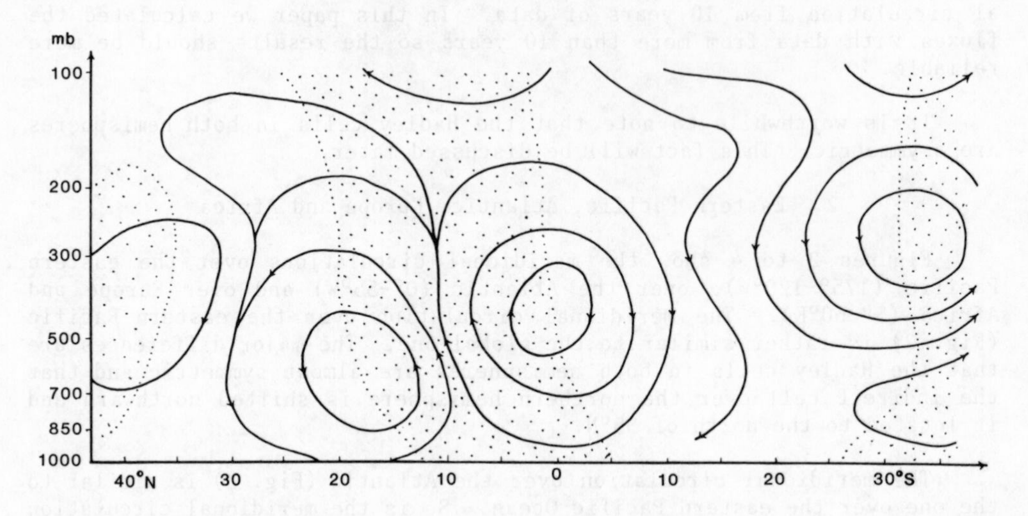

Fig. 3 The mean meridional circulation over the Atlantic (0°-55°W). (See legend to Fig. 1 for further explanation.)

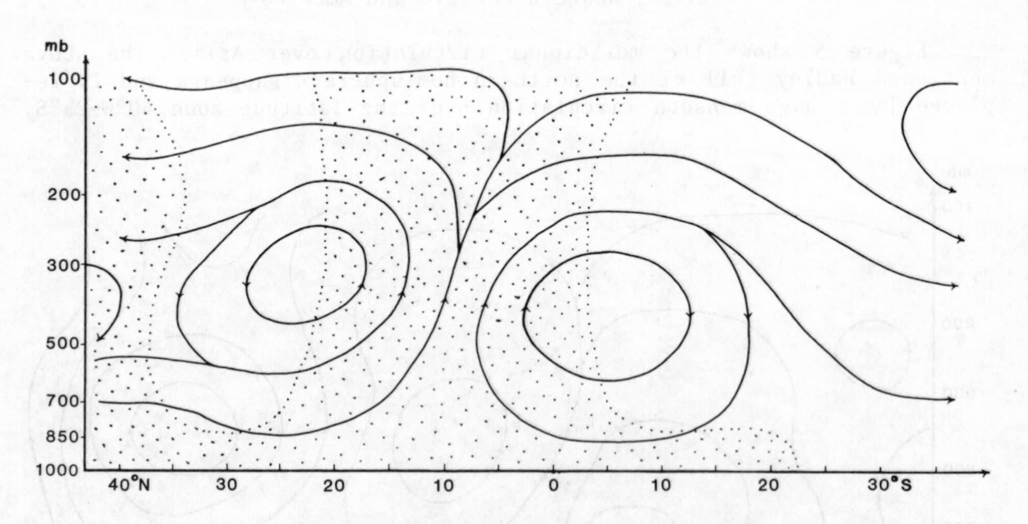

Fig. 4 The mean meridional circulation over Europe and Africa (5°-60°E). (See legend to Fig. 1 for further explanation.)

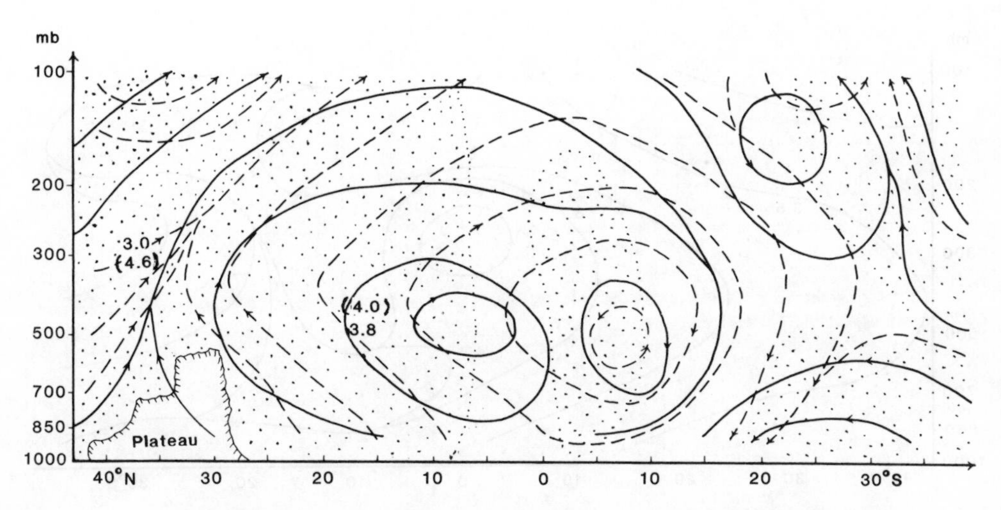

Fig. 5 The mean meridional circulation over Asia (full lines) and along 90°E (dashed lines). The numbers denote heat source centers over Asia and over the plateau (in parentheses) and are given in °C/day.

This circulation has two centers located at 9°W and 7°S, respectively. Over the plateau (75°-105°E) there is an even better defined monsoon circulation than over the whole of Asia. The air ascends in the latitude zone 17°S-40°N, moves northward in the low troposphere, then turns southward in the upper troposphere and descends in the subtropical belt of the southern hemisphere. The center of the monsoon cell is located at 7°S on the 600-mb surface and there is a convergent zone over the plateau below 300 mb. In Fig. 5 we only give the meridional circulation along 90°E, because it is the same as the one averaged over the plateau longitude sector.

The meridional circulation over the western Pacific is similar to the one over Asia (see Fig. 6). The air ascends north of 5°S, moves northward, and turns southward in upper troposphere, but it does not descend to sea level. A part of the airflow turns westward and the other part first descends to 500 mb, then turns eastward and ascends again. The meridional circulation along 160°E is the same as the one shown in Fig. 6, but dissimilar to the one obtained by Ye [9].

The meridional circulation over America is a transitional pattern (Fig. 7). Although there exists a weak Hadley cell in the lower troposphere over the northern hemisphere, the indirect cell of the northern hemisphere is almost connected with the southern Hadley cell and becomes a monsoon circulation. The meridional circulation along 100°W, as indicated by the dashed lines in Fig. 7, is similar to that over America, but somewhat different from the one obtained by Ye [7, 8, 9].

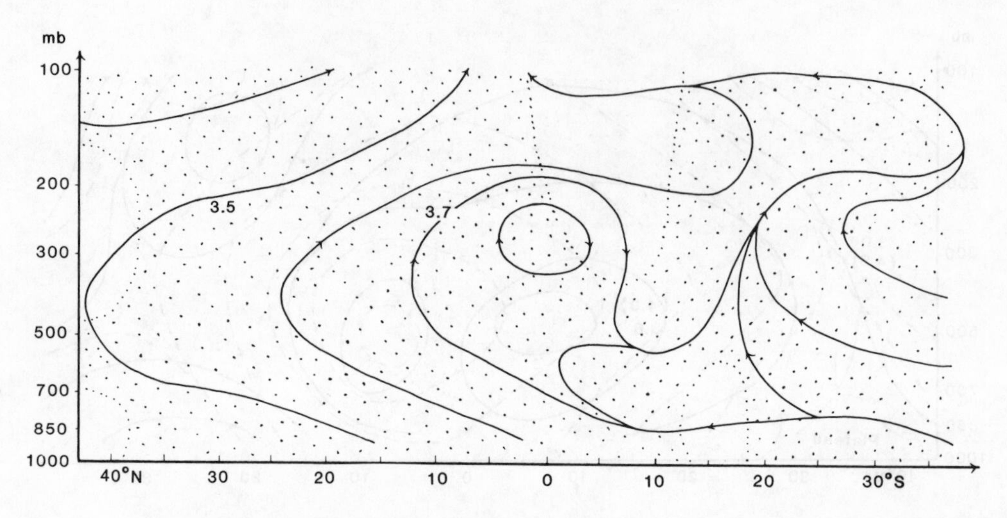

Fig. 6 The mean meridional circulation over the western Pacific.

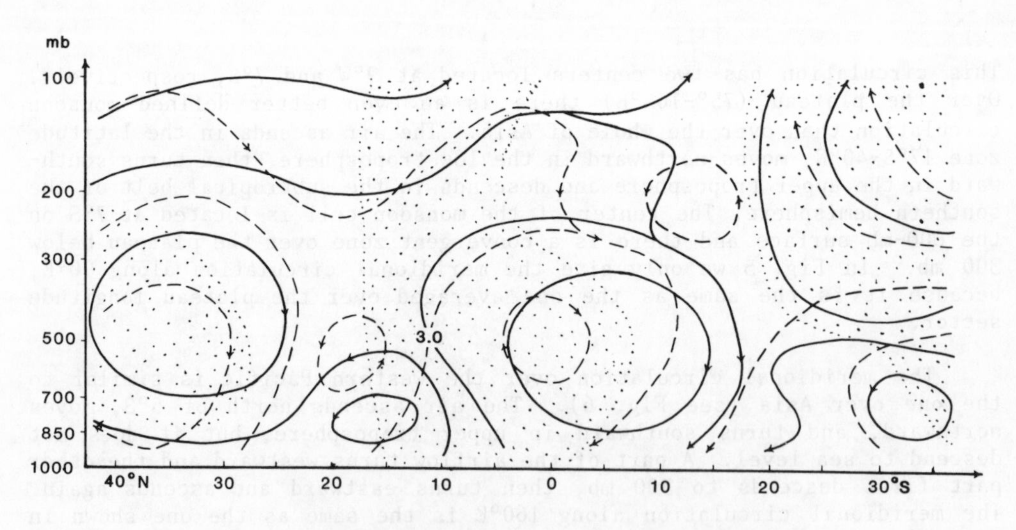

Fig. 7 The mean meridional circulation over America (full lines) and along 100°W (dashed lines).

III. HEATING FIELD

1. Global Distribution

Almost everywhere there are ascending motions over heat sources and descending motions over the heat sinks in the meridional circulations shown in the diagrams above. In Fig. 1 there exists a heat source corresponding to the thermal equator in the latitude zone 0°-20°N. The intensity of the heat source center is about 2.0°C/day, situated at 5°-15°N between 300-200 mb. This location corresponds to the position of the ascending branch of the Hadley cell. On both sides of this heat source there are heat sinks in the subtropical belts of both hemispheres, but the northern one (its center being -0.9°C/day), is weaker than the southern one (its center being -3.1°C/day). These sinks correspond to the positions of the descending branches of the Hadley cells. Certainly these two heat sinks not only depend upon the surface cooling, but also on dynamic effects.

2. Eastern Pacific, Atlantic, Europe and Africa

Comparing Fig. 1 with Figs. 2 to 4, we find that the distribution of heating fields over the eastern Pacific, Atlantic, Europe and Africa is similar to the global one, but the heat sinks in Figs. 2 to 4 are more symmetric in both hemispheres, so are the Hadley cells. In the northern hemisphere the heat sink over the Atlantic is weaker than over the eastern Pacific and it is the weakest over Europe and Africa, but the heat source over Europe and Africa is the largest. This observation may be related to the land-sea distribution. The larger the land mass (Europe and Africa) the stronger will be the heat source. Over a large ocean (east Pacific) the heat sink would be more intense.

The underlying surfaces in these three regions have a mutual characteristic feature, i.e. the land-sea distributions are more or less uniform and symmetric in both hemispheres. In the eastern Pacific the surface is all water in both hemispheres. In the Atlantic sector the surface is also mainly water, although there is some land mass, but it is nearly symmetric in both hemispheres. In Europe and Africa the surface is mainly land in low and middle latitudes, with a nearly symmetric distribution in both hemispheres. These observations lead to the following conclusion: The Hadley cell is symmetric in both hemispheres if the underlying surface is uniform and symmetric on large scales in both hemispheres, no matter whether the surface consists of water or land.

3. Asia, Western Pacific and America

Over Asia (Fig. 5) the whole region north of 5°S is a heat source, except for a small area north of the equator. The above mentioned subtropical heat sink of the northern hemisphere (about 20°-35°N in Fig. 4) disappears. In this heat source region there are two centers: one is located at 10°-20°N between 500 and 300 mb with an intensity of 3.8°C/day, corresponding to the thermal equator in summer which is shifted 5° latitude farther northward than the global mean. The other heat source center is situated at 35°-40°N near the 300-mb surface. Its intensity is about 3.0°C/day and is due to the thermal effect of the land mass in Asia. In the meridional cross section over the plateau (75°-105°E), there are also two heat source centers, the northern one with 4.6°C/day is stronger than the southern one which is about 4.0°C/day. Along 90°E there are also two heat source centers of more than 6.0°C/day. These values confirm the very significant heating role of the plateau.

In the meridional cross section over the western Pacific the heating field is similar to that over Asia (Fig. 6). There are also two heat source centers in the northern hemisphere: one is situated at 5°-10°N near 300 mb with an intensity of 3.7°C/day and the other one is located at 30°N between 300 and 200 mb with an intensity of 3.5°C/day. The heat sink of the southern hemisphere is weaker than in the Asian sector, probably because in the latitude zone 0°-20°S the land mass in the western Pacific sector is much larger than in the Asian sector. The subtropical heat sink of the northern hemisphere in this region disappears most likely because of the following reasons: (1) The sensible heat flux at the sea surface is affected by warm currents and by the south-west monsoon (-10 cal/cm^2 day).[1] (2) A larger net radiation is induced by small cloud amounts in the subtropical belt (about 100 cal/cm^2 day). (3) Larger evaporation is induced by dry air coming from the continent (about 200 cal/cm^2 day). The total heat source situated near 30°N is about 200-300 cal/cm^2 day [2] which coincides with the result obtained by Yao [6].

In the meridional cross section of America (Fig. 7) in the northern hemisphere in low latitudes there exists a heat source situated at 10°N between 700-300 mb. This source corresponds to the thermal equator and its intensity is about 3.0°C/day. The heat source at latitude 30°-40°W is probably induced by land masses and by the Rocky Mountains, because the area ratio of the sea-to-land mass is 3:8. There is a heat sink at 20°-30° N which is different from Asia. This sink is probably due to the difference of land-sea distribution between these two regions because the area ratio of sea-to-land mass in North America is 8:2, but in Asia it is 1:9.

In the meridional cross section of 90°E (Fig. 5, dashed lines) and 100°W (Fig. 7, dashed lines) the meridional circulations are different although the underlying surfaces are very similar. Along 90°E there is a typical monsoon circulation and along 100°W the circulation is similar to that over America. Along 160°E the heating field distribution and circulation are similar to those over the western Pacific and contrary to those over the eastern Pacific, although the underlying surface there is similar to that of the eastern Pacific. This shows that the meridional circulation along a narrow longitude zone is not directly determined by its underlying surface but by the large-scale land-sea distribution.

[1] $1\text{Wm}^{-2} = 2.388 \times 10^{-5}$ cal cm^{-2} s^{-1}.

IV. THE MAINTENANCE AND ENERGY TRANSPORT OF THE MONSOON CIRCULATION

As mentioned above, the monsoon circulation over Asia is maintained by the heat source of the northern hemisphere. In general, in the subtropical belt, one finds a heat sink and descending motions, but over Asia this pattern is replaced by ascending motions and a heat source induced by the thermal effects of land masses and of the plateau. The Hadley cell disappears and the Ferrel cell of the northern hemisphere is connected with the Hadley cell of the southern hemisphere and forms a huge monsoon circulation which weakens the global mean Hadley cell of the northern hemsiphere and strengthens that of the southern hemisphere.

In the following we shall discuss the contribution of energy transport to the monsoon circulation. Table 1 gives the values of the vertical transport of total energy in the monsoon circulation over Asia (unbracketed values) and over the longitude sector of the plateau (bracketed values). The negative sign denotes upward transport and the positive sign denotes downward transport. The ordinate is in mb and the abscissa is in degrees latitude. There are two peak values of the monsoon circulation over Asia, one is situated at about 35°N near 100 mb (-0.083 cal/cm^2 s), which corresponds to the thermal effect of the land mass and plateau and the other is situated at about 15°N near the sea surface (-0.075 cal/cm^2 s) and between 500 and 300 mb (-0.062 cal/cm^2 s) which

Table 1 The vertical transport of total energy in the monsoon circulation over Asia (unbracketed values) and over the Tibetan Plateau (bracketed values). Units: 10^{-1} cal/cm^2 s. The negative sign denotes upward transport and the positive sign denotes downward transport.

mb \ Latitude	40	35	30	25	20	15	10	5°N
100	-0.06	-0.83	-0.77	-0.30	-0.05	-0.18	0.34	0.53
100	[0.17	-0.98	-0.76	-0.20	-0.04	-0.18	-0.02	-0.15]
200	-0.12	-0.75	-0.70	-0.33	-0.19	-0.37	0.09	0.30
200	[-0.07	-0.89	-0.67	-0.17	-0.30	-0.24	-0.14	-0.00]
300	-0.13	-0.57	-0.53	-0.30	-0.33	-0.61	-0.16	0.10
300	[-0.49	-0.69	-0.55	-0.13	-0.50	-0.77	-0.43	-0.46]
500	-0.09	-0.28	-0.29	-0.31	-0.47	-0.63	-0.16	0.05
500	[-0.29	-0.35	-0.35	-0.23	-0.67	-0.80	-0.50	-0.55]
700	0.03	-0.08	-0.08	-0.32	-0.39	-0.41	-0.02	0.10
700				[-0.38	-0.52	-0.48	-0.29	-0.34]
850	0.06	-0.06	-0.06	0.24	-0.29	-0.25	-0.00	0.15
850				[-0.31	-0.39	-0.29	-0.15	-0.09]
1000	0.28	0.23	-0.24	-0.72	-0.94	-0.75	0.28	1.89
1000				[-1.06	-1.41	-0.95	-0.17	1.82]

corresponds to the effect of the thermal equator in low latitudes. There are also two peak values of the monsoon circulation over the plateau at the same latitude, one (0.098 cal/ cm^2 s) situated at 35°N near 100 mb and the other (-0.08 cal/cm^2 s) situated at 15°N near 500-300 mb. These values are larger than the respective values over Asia, and the peak value at 35°N is larger than that at 15°N, i.e. the thermal effect of the plateau is stronger than that of the thermal equator.

The intensity of vertical energy transports near 500 mb over the plateau (2.10 cal/cm^2 min) is about two orders of magnitude larger than that of the global mean Hadley cell (0.01 cal/cm^2 min) and one order of magnitude larger than the total vertical transport of sensible and latent heat in the global latitude zone 30°-40°N (0.01 cal/cm^2 min). Such large differences perhaps depend on the different calculation methods and available data, but we still think that the vertical energy transport of the monsoon circulation over the plateau is very large.

Where does the energy transport go? What influence does it produce? We shall discuss these questions briefly.

Table 2a gives the north-south energy transport over Asia (5°-40°N). Below 500 mb and to the south of 35°N this transport is directed northward and it is stronger in the lower layer (850 mb). Above 500 mb the transport is southward and its intensity is about three times larger than the low-level northward transport. Near the 200-mb level the energy transport is largest and there are two centers, one situated at 35°N and the other at 10°N.

Table 2a The north-south transport of total energy in the monsoon circulation over Asia. Positive values denote northward transport and negative southward transport.

Unit: 10^5 cal/mb cm sec

mb \ °N	40	35	30	25	20	15	10	5°N
100	-0.2	-0.3	-0.3	-0.3	-0.3	-0.3	-0.2	-0.1
200	-0.4	-0.3	-0.2	-0.2	-0.2	-0.3	-0.4	-0.5
300	-0.3	-0.3	-0.1	-0.0	-0.1	-0.1	-0.2	-0.2
500	-0.1	-0.0	0.0	0.1	0.0	0.1	0.1	0.1
700	-0.1	-0.0	0.1	0.1	0.1	0.1	0.0	0.0
850	-0.1	0.0	0.1	0.2	0.1	0.2	0.1	0.1

Table 2b gives the west-east energy transport. Eastward transport prevails except for a region to the south of 30°N and above 500 mb where the transport is westward. The center of westward energy transport is located at 15°N on 100 mb and is twice as large as the global mean. The eastward transport center of energy is situated at 40°N near 200 mb and is one-third times larger than the global mean. The north-south energy transport is one order of magnitude smaller than the west-east transport. From these data we infer that in the area of the South-East Asian monsoon the export of energy mainly occurs in the upper troposphere.

Table 2b The west-east transport of total energy in the monsoon circulation over Asia. The positive values denote eastward transport and the negative values westward transport.

Unit: 10^5 cal/mb cm sec

mb \ °N	40	35	30	25	20	15	10	5
100	1.5	0.8	-0.5	-1.6	-2.3	-2.5	-2.0	-1.3
200	2.4	1.4	0.1	-0.6	-1.2	-1.3	-1.5	-1.6
300	1.5	1.2	0.3	-0.3	-0.6	-0.5	-0.6	-0.8
500	0.6	0.3	0.2	-0.1	-0.0	0.2	0.2	0.1
700	0.2	0.1	0.0	0.0	0.3	0.5	0.6	0.5
850	0.1	-0.0	0.1	0.2	0.3	0.6	0.8	0.5

A small part of this energy is transported to the southern hemisphere and is equal to the total energy transport over the other parts of the equator. The larger part of this energy is transported westward and eastward. The intensity of this transport is about one order of magnitude larger than the global mean. The southward energy transport over the southern side of the plateau in the upper troposphere is equal to the total northward energy transport in all of the Hadley cell over the same latitude.

V. CONCLUSIONS

According to the above discussion the effects of large-scale land-sea distribution and of the plateau on the mean meridional circulation are very evident. Where the land-sea distribution is nearly uniform and symmetric in both hemispheres (such as in the eastern Pacific, Atlantic, European and African sectors), so is the Hadley cell. In Asia there exists a huge monsoon circulation induced by the thermal effects of the land mass and of the plateau in the northern hemisphere. These effects generate heat sources and ascending motion instead of the heat sink and descending motion in the subtropical belts of the other parts of the globe.

The monsoon circulation over the western Pacific is related to the subtropical heat source in the northern hemisphere. Because the land area in the subtropical belt is small over North America, there exists only a weak monsoon circulation.

Since the sense of the monsoon circulation is opposite to that of the Hadley cell in the northern hemisphere and coincident with the Hadley cell in the southern hemisphere, the two Hadley cells are asymmetric, the northern one being much weaker than the southern one.

The export of energy by the monsoon circulation is very large. Furthermore, the export of energy over the plateau is largest in strength and most extensive in both vertical and horizontal extent. We think that the Qinghai-Xizang Plateau plays a "chimney" role in the export of energy in the upper troposphere.

ACKNOWLEDGMENTS

The authors are indebted to comrades Zhang Mingjuan, Pan Ruinian, and Chu Zhenshan for their assistance in data processing and final ediţing of the original manuscript.

REFERENCES

1. Chen, Qui-Shi et al.: The mean stream field and meridional circulation in area of Pacific trade winds and southwest monsoon in the southeast part of Asia in July of 1958. Q. J. Meteor. Soc. China, 34, 28-350 (1982).

2. The Oceanic and Meteorological Group, Institute of Oceanography, Academia Sinica: Atlas of sea surface heat budget in Northwest Pacific. Science Press (1979).

3. Palmén, E. and C.W. Newton: Atmospheric circulation systems. New York, Academic Press (1969).

4. Van de Boogaard, H.: The mean circulation of the tropical and subtropical atmosphere - July in 1961-1974. Technical Note, NCAR/TN - 118 + STR (1977).

5. Vuorela, L.A. and I. Tuominen: On the mean zonal and meridional circulations and the flux of moisture in the northern hemisphere during the summer season. Pure Appl. Geophys., 57, 167-180 (1964).

6. Yao, Lanchang et al.: The monthly mean heating fields of atmosphere and their annual variation over Asia. Q. J. Plateau Meteor. China, No. 3 (1982).

7. Ye, T.-C. and P.-C. Chu: Some fundamental problems of the general circulation of the atmosphere. Science Press (in Chinese) (1958).

8. Ye, T.-C. et al.: The mean meridional circulation and the angular momentum budget in 1950. Q. J. Meteor. Soc. China, 57, 307-322 (1956).

9. Ye, T.-C. et al.: The average summer vertical circulation to the south of 45°N of northern hemsiphere and its relation to the distribution of heat sources and sinks in the atmosphere. Q. J. Meteor. Soc. China, 39, 28-35 (1981).

1.4 OROGRAPHIC RAIN ON THE WESTERN GHATS

Ronald B. Smith and Yuhlang Lin
Department of Geology and Geophysics
Yale University
New Haven, Connecticut 06511 USA

ABSTRACT

The heavy monsoon rainfall along the Malabar Coast is clearly associated with influence of the Western Ghats yet, the primary source of the rain is deep convection rather than orographically forced ascent. To investigate this problem we combine an analytical treatment of thermally and orographically induced mesoscale circulations, with a steady state model of turbulent buoyant plumes. The results indicate that there are two internally consistent modes of airflow over the mountains. With no precipitation, the orographic disturbance is insufficient to destabilize the air column with regard to deep cumulus development. On the other hand, the disturbance produced by the heating computed from the observed rainfall, does alter the environment in such a way as to allow cumulonimbus development. This seems to offer a partial explanation for the rainy and dry spells on the coast observed during the summer monsoon. The existence of a heat-induced off-shore trough is also predicted by the model.

I. INTRODUCTION

A good example of the local enhancement of precipitation by orography is the large annual rainfall recorded along the Malabar Coast and on the windward slopes of the Western Ghats in India (Fig. 1). This rainfall occurs almost entirely during the 3-4 month summer period when the coast lies in the path of the west-southwest monsoon winds crossing the Arabian Sea. For the purpose of discussion, it is possible to identify five existing theories which might account for the observed rainfall.

1) Smooth uplift [25], [26], [27] in which the vertical motion leading to condensation and precipitation is directly forced by the upsloping terrain below.

2) Diurnal convection (e.g. [18]) in which the temperature difference between land and sea, or heated mountain slopes produce circulations which trigger the onset of convection each afternoon.

3) Coastal trough [6], [9], [14] which, like a synoptic scale trough in midlatitudes, is thought to be associated with rainy weather.

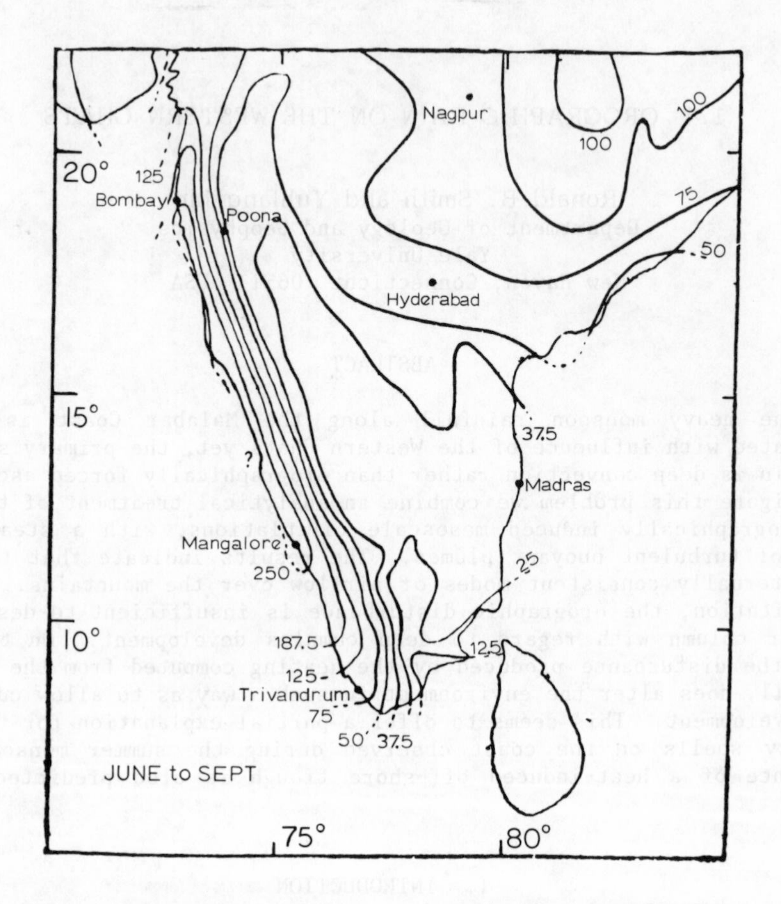

Fig. 1 The areal distribution of rainfall over India during the summer monsoon months of June to September (from [20]). The rainfall is concentrated just upstream of the Western Ghats. The seaward extent of the high rainfall region is not known.

4) Lifting instability (e.g. [5]) in which the orographic lifting triggers deep convection.

5) Low-level feeder clouds [1], [2] which redistribute the rainfall reaching the ground while not influencing the main hydrometeor production aloft.

While none of these theories can be said to be irrelevant to the problem at hand, there is contradictory evidence for each. The showery nature of the rainfall and the satellite observation that the rainfall is caused by deep cumulonimbus clouds seems to speak against (1) in its pure form but yet low-level forced orographic lifting along the lines of (5) is helping to enhance rainfall on local hills and to smooth out the showery fluctuations there. There is a measurable diurnal modulation to the rainfall, supporting (2), but this is a minor effect in this particular region. The empirical association (3) between rainfall and a coastal trough is supported by the analysis in this paper, but the trough is seen to be a result of the rainfall rather than a cause. The widely discussed mechanism of destabilization by orographic lifting (4) seems to be most relevant here, yet the suggestion of this paper is that it needs a substantial modification. The mesoscale lifting that triggers the convection is largely produced by the latent heating itself, rather than the mountain. The rainfall system is then nearly self-sustaining, with the mountain acting only to stabilize the system and hold it in place.

In this paper we will construct a more detailed, but still semi-empirical, model of the Malabar Coast rainfall. From this model we conclude that of the four existing conceptual models, no. 4 is closest to the truth but that it is oversimplified. In fact the latent heating in the rainclouds generates instability much the way a propagating squall line might do. The somewhat smaller influence of the mountain serves to anchor the system and to maintain its organization. This behavior gives the system a discrete on-off character leading to rainy and dry spells.

II. THE MONSOON WIND

The structure of the summer winds approaching India has been studied for a number of years [10], [5], [11], [21], [23], [3]. Progress in this research has recently accelerated due to the observations from the MONEX project.

Throughout the summer monsoon period, the low-level winds across the Arabian Sea approach India with a speed of about 15 m/s (at 850 mb), and a direction more or less perpendicular to the coast (Fig. 2). Embedded in the flow are minor disturbances which account for variations in cloudiness over the sea and for a time-varying streakiness to the wind strength. In the upper troposphere, the winds are reversed, blowing from the east. Disturbances exist in these easterlies as well, and these may also play an important role in modulating rainfall in the region.

III. THE NATURE OF THE RAINFALL ALONG THE MALABAR COAST

As shown in Fig. 3, the land immediately along the coast receives considerable annual rainfall while the western slopes of the Ghats receive even more. Immediately beyond the crest of the mountains the rainfall decreases to a low value and this amount is mostly associated with tropical depressions approaching from the east. Based on estimates of rainfall far out over the sea, the coastal rainfall is at least thrice the upstream value. The lack of rainfall data just off the coast makes it difficult to estimate how far offshore the enhanced region extends. Rainfall estimates from satellite [22] may not be accurate enough to resolve this question. From limited radar data showing cumulus cloud development [31], it is likely that the enhancement extends at least 30 km offshore.

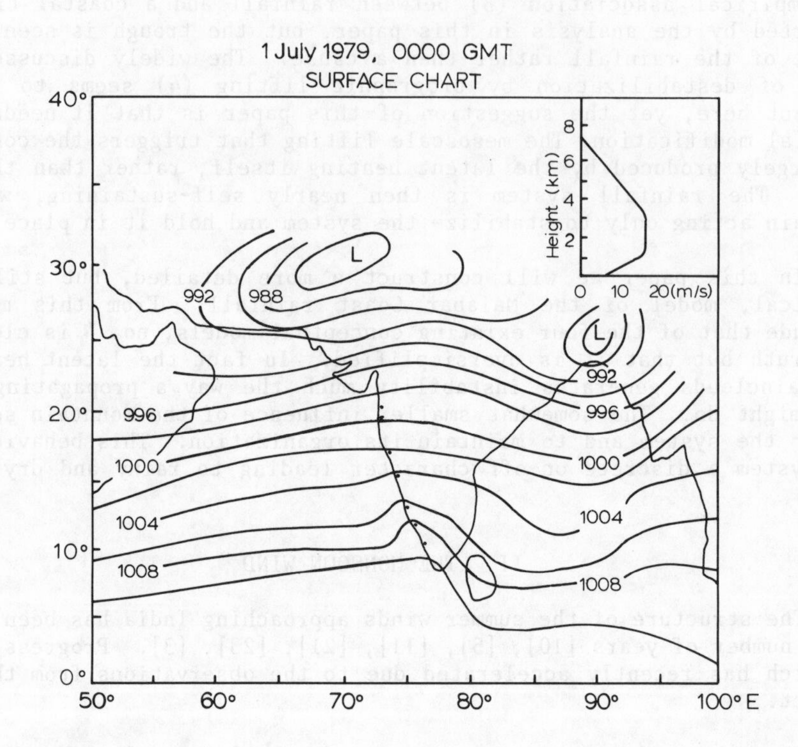

Fig. 2 A typical surface chart for the Indian Ocean during the summer monsoon. This particular chart is for 1 July 1979 at 0000 GMT. Also shown is a vertical sounding taken from the western coast of India showing the component of wind speed perpendicular to the coast. Above 7 km the winds are easterly.

Part (d) of Fig. 3 shows the distribution of rainfall across a section of the coast where mountains are broken by a gap (the Palghat Gap). The marked difference between this profile and the others is, in our view, that the gap is so narrow that it doesn't influence the enhanced shower development along the shoreline, but the local enhancement due to orographically forced low-level feeder clouds is missing.

Evidence that the origin of the rainfall is not smooth orographic lifting is quite clear. Local observations indicate that the rainfall is very irregular with adjacent 3-hour averages being almost uncorrelated, while the onshore winds continue unabated. Thus, in spite of the obvious importance of low-level feeder clouds, it seems that the ultimate cause of the rain is the development of deep convection.

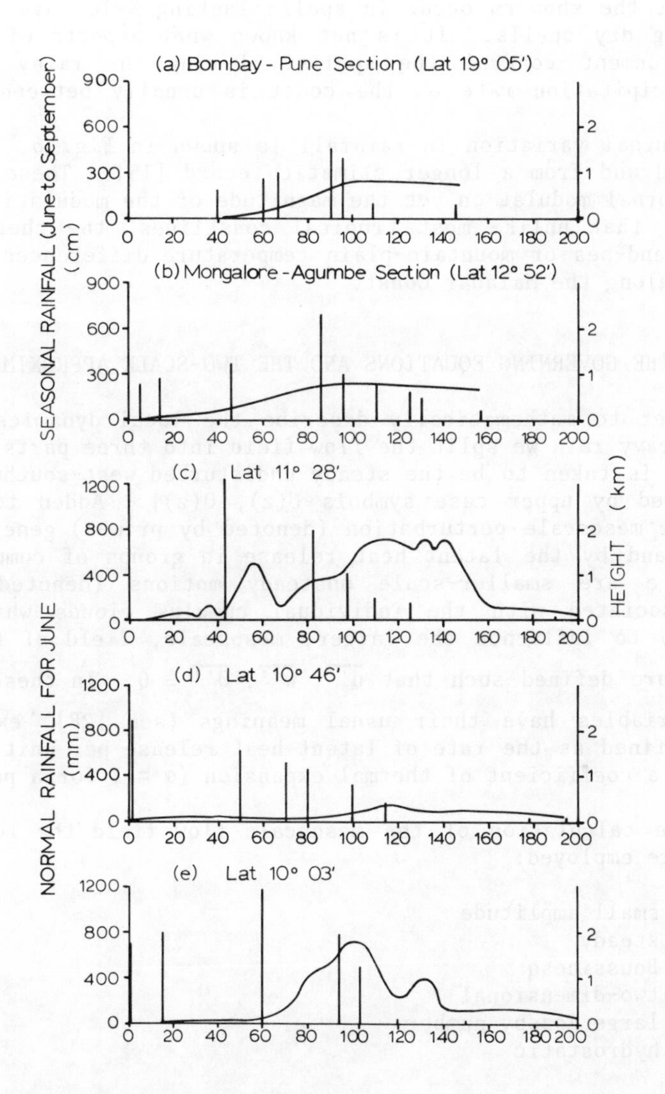

Fig. 3 The distributions of rainfall for several sections across the Western Ghats. (a) Bombay-Pune section with latitude about 19°05'. The seasonal rainfall of June to September (in cm) and mountain profile are plotted. (b) Same as (a), except for Mongalore-Agumbe section with latitude about 12°52'. (From [27].) (c) Cross section at latitude 11°28'. The normal rainfall for June (in mm) and mountain profile are plotted. (d) Cross section at latitude 10°46', through the Palghat gap. There is no increase of rainfall from the coast toward the eastern side. (e) Same as (c), except at latitude 10°03'. (From [17].).

From Fig. 4, and the more exhaustive survey shown in Fig. 5, it is evident that the showers occur in spells lasting 5-10 days separated by equally long dry spells. It is not known what aspects of the larger-scale environment control the spells. During the rainy spells, the average precipitation rate on the coast is usually between 2-4 mm/hr.

The diurnal variation in rainfall is shown in Fig. 6, both for the MONEX period and from a longer climatic record [15]. These data show a definite diurnal modulation yet the magnitude of the modulation is small. We conclude that unlike most tropical coastlines, the thermal forcing caused by land-sea or mountain-plain temperature differences is of minor importance along the Malabar Coast.

IV. THE GOVERNING EQUATIONS AND THE TWO-SCALE APPROXIMATION

In order to mathematically describe the local dynamics during the spells of heavy rain we split the flow field into three parts. First, the basic state is taken to be the steady undisturbed west-southwest monsoon flow [denoted by upper case symbols $U(z)$, $\theta(z)$]. Added to this is a steady-state mesoscale perturbation (denoted by primes) generated by the orography, and by the latent heat release in groups of cumulus clouds. Third, there are smaller-scale unsteady motions (denoted by double primes) associated with the individual cumulus clouds which can act collectively to influence the larger, mesoscale, field of flow. These quantities are defined such that $\overline{u''}$, $\overline{w''}$, $\overline{\theta''} \equiv 0$. In these equations, all the variables have their usual meanings (see [28]) except $\overset{\bar{o}}{L}(x,z)$ which is defined as the rate of latent heat release per unit volume, and α, which is a coefficient of thermal expansion ($\alpha = 1$ for a perfect gas).

For the calculation of the mesoscale flow field the following assumptions are employed:

- small amplitude
- steady
- Boussinesq
- two-dimensional
- large Rossby number
- hydrostatic

within these assumptions the governing equations are

$$\rho_o U \frac{\partial u'}{\partial x} + \rho_o w' \frac{\partial U}{\partial z} = - p'_x - \rho_o \left\{ \frac{\partial \overline{u''^2}}{\partial x} - \frac{\partial \overline{u''w''}}{\partial z} \right\} \qquad (1)$$

$$0 = -p'_z - \rho' g - \rho_o \left\{ \frac{\partial \overline{u''w''}}{\partial x} - \frac{\partial \overline{w''^2}}{\partial z} \right\} \qquad (2)$$

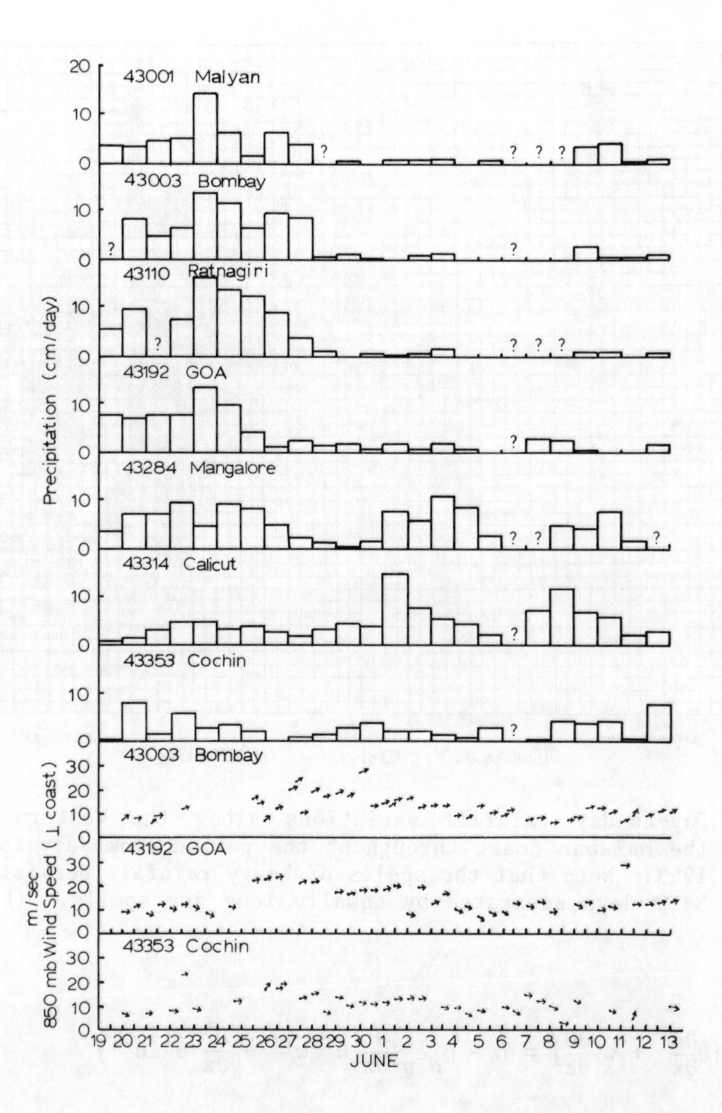

Fig. 4 A time series of daily rainfall amounts from seven stations along the Malabar Coast during the period 19 June to 13 July 1979. The stations are presented in order - north to south as shown in Fig. 2. Although the rainfall is convective in nature, the daily rainfall totals tend to be more continuous with rainy spells and dry spells extending over a substantial fraction of the coast. The bottom part of the figure shows the strength of the onshore winds from three coastal stations during the same period.

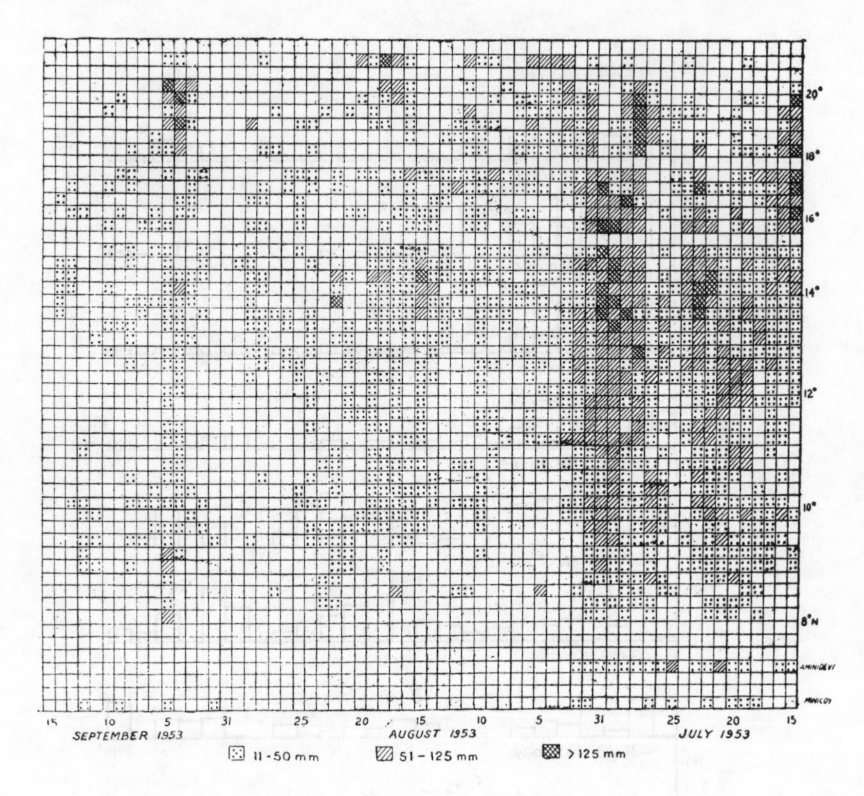

Fig. 5 Day-to-day rainfall variations along the Western Ghats and the Malabar Coast throughout the period from July to September 1953. Note that the spells of heavy rainfall persist for about 5-10 days separated by equally long dry spells. (From [20].)

$$\rho_o c_p \{U\frac{\partial\theta'}{\partial x} + w'\frac{\partial\theta}{\partial z}\} = \bar{L} - \rho_o c_p \{\frac{\partial}{\partial z}\overline{\theta''w''} - \frac{\partial}{\partial x}\overline{\theta''u''}\} \quad (3)$$

$$\rho'/\rho_o = \alpha\theta'/\theta_o \quad (4)$$

$$\frac{\partial u'}{\partial x} + \frac{\partial w'}{\partial z} = 0 \quad (5)$$

The averages denoted by the overbar are, in principle, taken with respect to time as the mesoscale flow is independent of this variable. In practice however, additional averaging over small regions of space (e.g. $\Delta x \sim 10$ km, $\Delta z \sim 300$ m) would be necessary to obtain statistical significance. This in turn requires that the steady mesoscale flow and the unsteady cumulus dynamics are really occurring on two different scales. This is a crucial assumption and one that is difficult to defend.

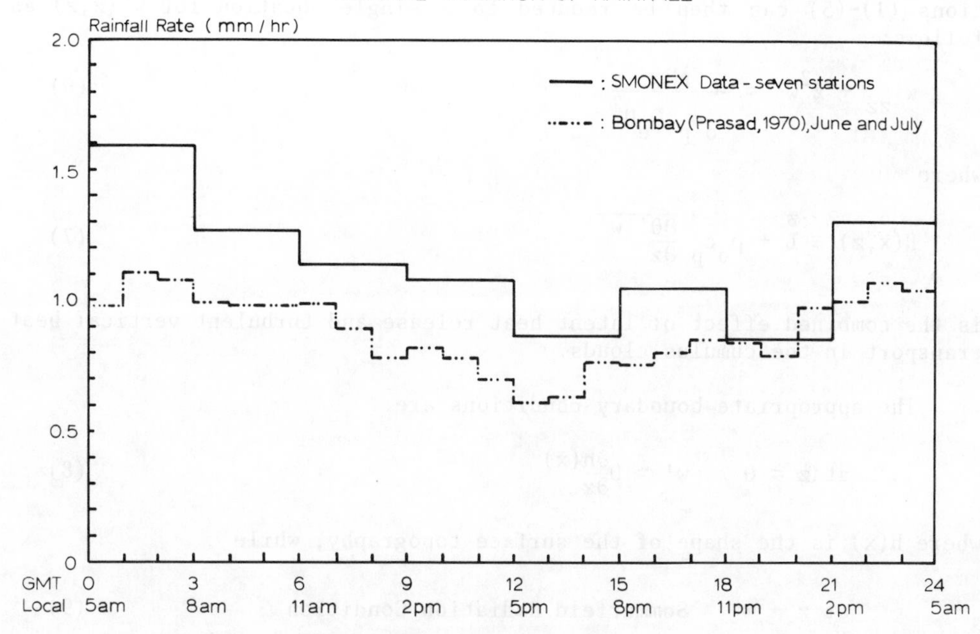

Fig. 6 Average rainfall rate (mm/hr) as a function of time of day. The solid curve is from 3-hourly data from the SMONEX data set. It includes data from the seven surface stations shown in Figs. 2 and 3, for the periods June 21 to July 6 and July 9 to July 13, 1979. The dashed line was calculated from hourly data from Bombay for the months of June and July, 1948-1965 by Prasad [15]. Both curves show a minor early morning maximum.

That this scale split is possible is suggested by the nature of local rainfall statistics. From a single station, a long record, perhaps longer than a "spell", is required to obtain a representive value for average rainfall rate. On the other hand, if several stations are included, especially similarly situated stations along the coast, then a representative rainfall rate can be obtained from a short time average (i.e. $\sim$ a few hours). This suggests that the effect of cumulus clouds can be considered as a statistically stationary forcing function over the period of a rainy spell.

The problem can be reduced to its most basic physical elements if we make the following further assumptions:

- neglect vertical shear in the background flow (Eq. 1)
- neglect the Reynolds stresses due to the cumulus clouds (Eqs. 1, 2)
- neglect the horizontal transport of sensible heat by the cumulus clouds (Eq. 3).

These assumptions are not required for mathematical tractability. Equations (1)-(5) can then be reduced to a single equation for $w'(x,z)$ as follows:

$$w'_{zz} + \ell^2 w' = \frac{gH\alpha}{\rho_o c_p \Theta_o U^2} \tag{6}$$

where

$$H(x,z) \equiv \bar{\bar{L}} + \rho_o c_p \frac{\partial \overline{\theta'' w''}}{\partial z} \tag{7}$$

is the combined effect of latent heat release and turbulent vertical heat transport in the cumulus clouds.

The appropriate boundary conditions are:

$$\text{at } z = 0 \qquad w' = U\frac{\partial h(x)}{\partial x} \tag{8}$$

where h(x) is the shape of the surface topography, while

$$\text{at } z \to \infty \qquad \text{Sommerfeld Radiation Condition} \tag{9}$$

The solutions to Eqs. (6), (8), and (9) with various choices for H(x,z) have been discussed by Smith and Lin [29], including the questions about vertical momentum flux, the phase relation between H and w', decay of the disturbance downstream, and the possibility of negative mountain drag. In this paper we will simply apply these results to the Western Ghats problem.

The most convenient closed-form solution arises from a heating function given by $H(x,z) = Q(x)/2d$ for $z_H + d \partial z > z_H - d$

$$\text{where} \quad Q(x) = Q\left(\frac{b_1^2}{(x+c)^2+b_1^2} - \frac{b_1 b_2}{(x+c)^2+b_2^2}\right) \tag{10}$$

and $H(x,z) = 0$ for $z > z_H + d$ and $0 > z > z_H - d$

and topography given by

$$h(x) = \frac{ha^2}{x^2+a^2} \tag{11}$$

In these formulae

Q is a measure of the strength of the vertically integrated heating ($Q = \max\{Q(x)\}$ if $b_2 >> b_1$)

z_H is the central altitude of the heating

d is the height above and below z_H to which the heating extends

b_1 is the half-width of the heating distribution

b_2 is the half-width of a compensating cooling required to keep the solutions bounded at infinity (normally $b_2 >> b_1$)

c is the horizontal distance between the center of the heating and the peak of the mountain

h is the height of the mountain

a is the half-width of the mountain.

The solutions for vertical displacement, pressure, and momentum flux associated with Eqs. (10) and (11) are given in the appendix. These expressions are rather lengthy because of all the geometric parameters that they include, but they can be easily plotted to investigate various combinations of orographic and thermal forcing.

V. THE DISTURBANCE TO THE MONSOON WIND BY THE COMBINED EFFECT OF OROGRAPHY AND CUMULUS HEATING

We first examine the disturbance caused by orography alone by plotting (in Fig. 7) the streamlines given by A1 with Q = 0. For the Western Ghats we choose h = 800 meter, a = 40 km, U = 10 m/s, $\ell = N/U = 0.001\ m^{-1}$. The result is the familiar pattern of vertically propagating mountain waves first presented by Queney [16]. Note that the lifting of the low-level air begins well upstream but we will see in a later section that this lifting is insufficient to trigger the growth of cumulus clouds. Between 3 and 4 kilometers there is orographically induced descent.

In order to estimate the value of H(x,z) we can integrate Eq. (7) vertically. If $\overline{\theta''w''}$ vanish as at z = 0 and ∞ then

$$Q(x) \equiv \int_0^\infty H(x,z)dz = \int_0^\infty \overline{\overset{o}{L}}(x,z)dz \quad (12)$$

If we further assume that all the condensed water falls immediately to the ground, then from Eq. (12) and from the conservation of water mass

$$Q(x) = \lambda \overset{o}{R}(x) \quad (13)$$

where λ is the latent heat of condensation per unit mass (2.5×10^6 J/kg) and $\overset{o}{R}(x)$ is the rainfall at the ground (kg/sec-m^2). Raingauge measurements of $\overset{o}{R}(x)$ allow us to estimate Q(x), but not the way in which the heat input is distributed in the vertical.

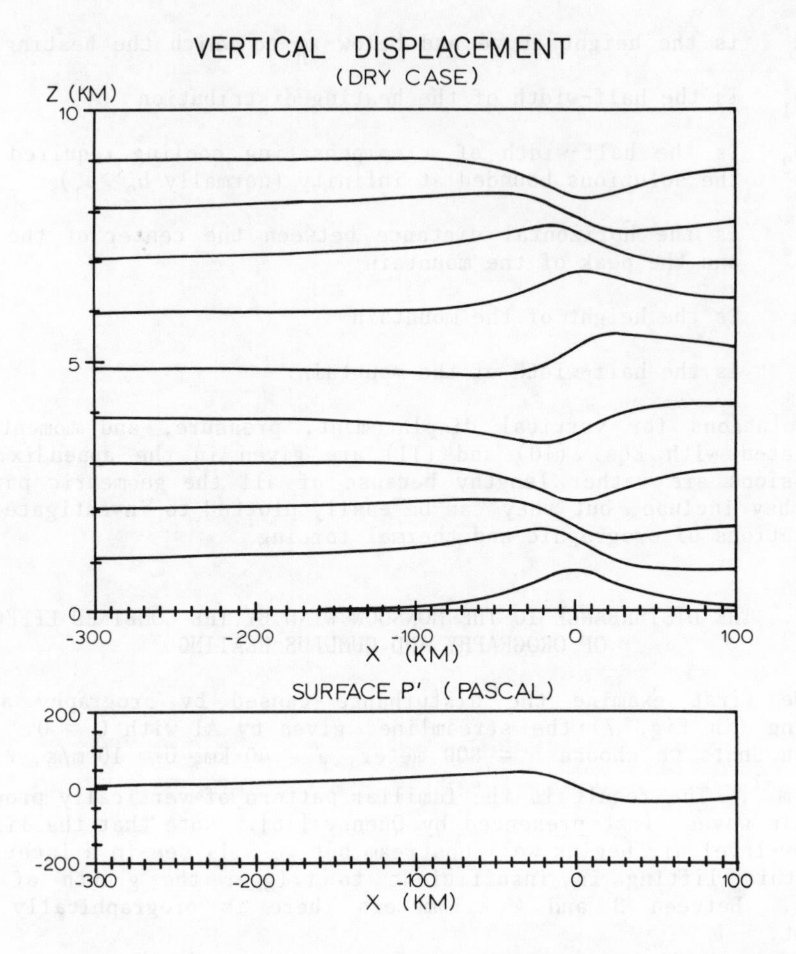

Fig. 7 Hydrostatic adiabatic flow over a bell-shaped mountain. The surface perturbation pressure is shown at bottom.

For the purpose of illustration we choose a maximum rainfall rate 1.7 mm/hr ($\sim$4.8 x 10^{-4} kg/sec-m^2) which from Eq. (13) gives Q = 1200 W/m^2. Rather arbitrarily we choose z_H = 3 km and d = 1.5 km. This gives a maximum value of the local heating rate of H(x,z) = Q/2d = 0.4 W/m^3. From the horizontal distribution of rainfall we select b_1 = 40 km, b_2 = 200 km and c = 100 km. The results are shown in Fig. 8.

The comparison of Figs. 7 and 8 indicates that the heating is producing a disturbance which is at least as large as that of the mountain.

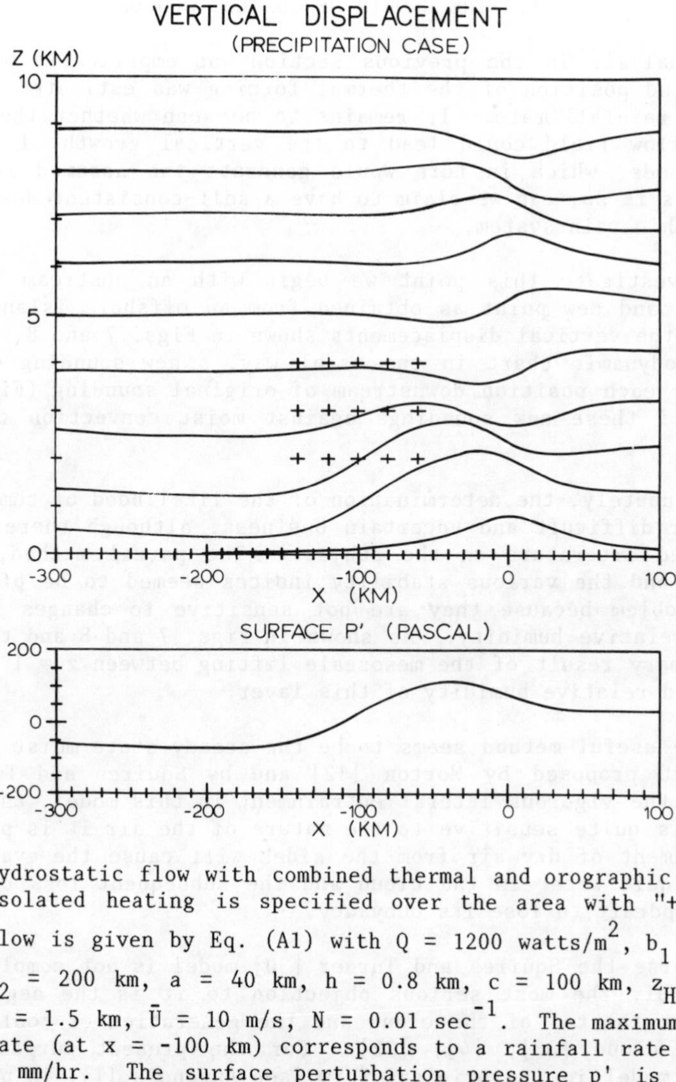

VERTICAL DISPLACEMENT
(PRECIPITATION CASE)

SURFACE P' (PASCAL)

Fig. 8 Hydrostatic flow with combined thermal and orographic forcing. Isolated heating is specified over the area with "+". This flow is given by Eq. (A1) with $Q = 1200$ watts/m^2, $b_1 = 40$ km, $b_2 = 200$ km, $a = 40$ km, $h = 0.8$ km, $c = 100$ km, $z_H = 3$ km, $d = 1.5$ km, $\bar{U} = 10$ m/s, $N = 0.01$ sec^{-1}. The maximum heating rate (at x = -100 km) corresponds to a rainfall rate of about 1 mm/hr. The surface perturbation pressure p' is drawn at the bottom. Note that there is a wide pressure trough produced by the heating, which is absent in the adiabatic case.

The nature of the thermally induced displacements are moderately sensitive to the choice of z_H (see [29] for a discussion of this point). For this choice ($z_H = 3$ km) there is a region of strong low-level convergence and ascent which may be able to trigger cumulus growth. This choice is reasonable as waves generated in the upper troposphere would be absorbed by the critical level in midtroposphere. Also, the condensation and sensible heat flux convergence in the upper troposphere may be largely offset by the evaporation of detrained liquid water.

VI. THE GROWTH OF CUMULUS CLOUDS

The analysis in the previous section was empirically based as the magnitude and position of the thermal forcing was estimated from observations of rainfall rate. It remains to be seen whether the calculated mesoscale flow field could lead to the vertical growth of small-scale cumulus clouds, which in turn would generate the assumed latent heat. Only if this is so, can we claim to have a self-consistent description of the orographic rain system.

To investigate this point we begin with an upstream sounding of temperature and dew point as obtained from an offshore island (see Fig. 9). Using the vertical displacements shown in Figs. 7 and 8, and working on a thermodynamic chart in the usual way, a new sounding can be calculated for each position downstream of original sounding (Fig. 8). The stability of these new soundings against moist convection can then be examined.

Unfortunately, the determination of the likelihood of cumulus growth is a rather difficult and uncertain business, although there is no lack of published literature on the subject. The parcel method, the slice method [8] and the various stability indices seemed to be of little use in this problem because they are not sensitive to changes in low- and mid-level relative humidity. As shown in Figs. 7 and 8 and then in Fig. 9, the primary result of the mesoscale lifting between $z = 1$ and 3 km is an increased relative humidity of this layer.

A more useful method seems to be the steady state moist plume model of the sort proposed by Morton [12] and by Squires and Turner [30]. Because of the vigorous lateral entrainment in this model, the growth of the plume is quite sensitive to the nature of the air it is penetrating. The entrainment of dry air from the sides will cause the evaporation of existing liquid water in the cloud and the subsequent loss of heat will cause the updraft to lose its buoyancy.

Of course the Squires and Turner [30] model is not completely reliable. Probably the most serious objection to it is the neglect of entrainment at the top of the cloud and the generation of cool downdrafts within the cloud [33], [4], [32]. For the present purposes we will accept the model in its original form (see Appendix II), in part because its behavior and shortcomings are well documented (e.g. its overestimation of vertical velocity) and in part because we think it may be accurate enough to give a reasonable comparison of stability between soundings which are not too different. Before applying the Squires-Turner model to the present problem, their test cases were redone to check both their computational methods and ours. Our calculations matched theirs to within a few percent.

The result of a steady state plume calculation on three environmental soundings is shown in Fig. 10. A moist plume starting at cloud base ($z = 1$ km) in the original sounding "A" will grow upwards at first due to latent heat release but will soon lose buoyancy as it enters the stable, drier air near 3 km. For this particular sounding, the plume is just barely able to penetrate into the less stable air near 4.5 km, where

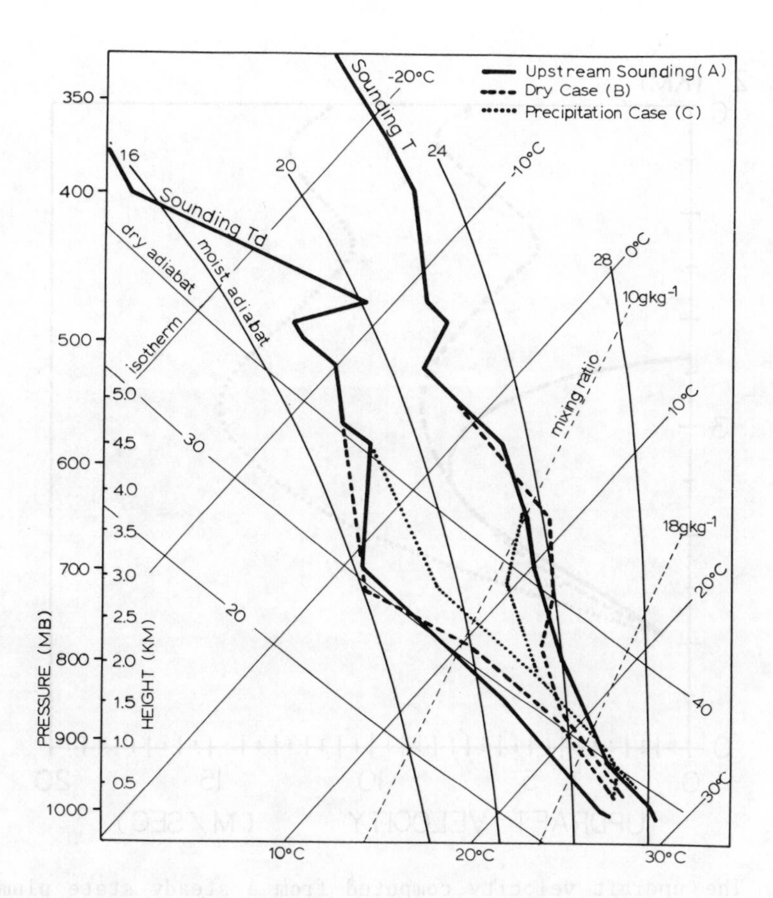

Fig. 9 A thermodynamic chart of the original upstream sounding, and the soundings modified by the vertical displacements shown in Figs. 7 and 8, at x = -100 km. The solid curves are the upstream sounding of temperature (T) and dew point (Td) at 0000 GMT, 1 July 1979 at Amihi which is about 300 km upstream. The dashed curves are the upstream sounding and modified by the mountain alone. The dotted curves are the upstream sounding and modified by the vertical displacements caused by the specified heating and mountain. The background lines are isotherms, dry and moist adiabats, and saturation mixing ratio.

it can reaccelerate. This seems consistent with the observation of frequent small cumulus clouds over the sea, which occasionally grew into cumulonimbus.

A similar calculation was done for sounding "B", representing the original sounding "A" modified by the dry flow field (Fig. 7). The result just above cloud base was much the same. At an altitude of 3 to 4

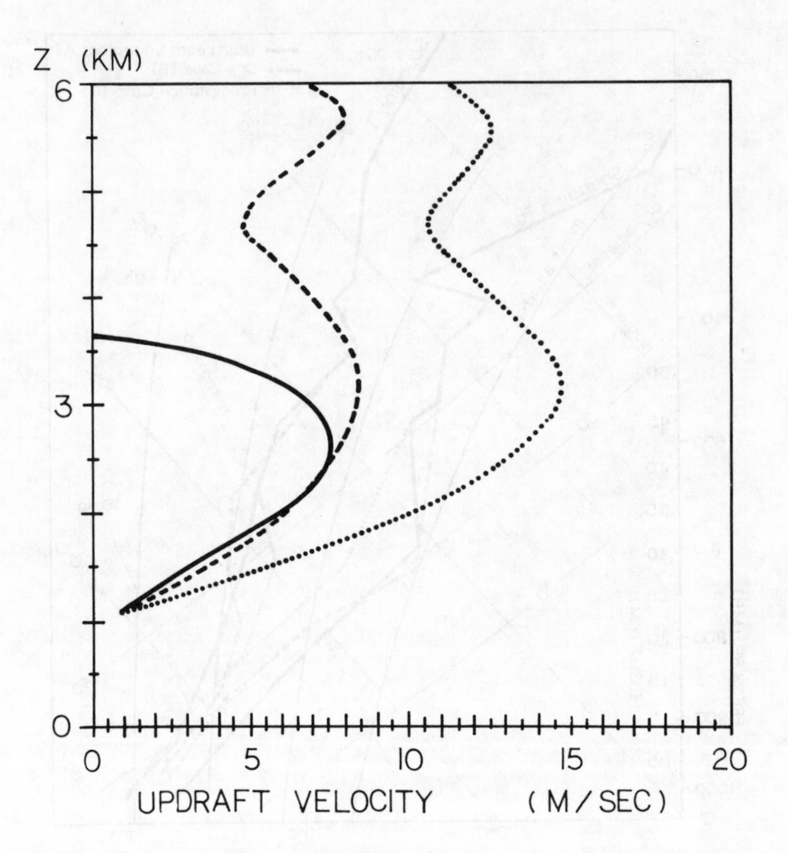

Fig. 10 The updraft velocity computed from a steady state plume model ([30]) using three environmental soundings (Fig. 8). The cloud base, initial updraft velocity and cloud base radius are assumed to be 1 km, 1 m/sec, and 1 km, respectively. The cloud base temperatures depend on the sounding and modified soundings. (a) The result of the calculation done using the original sounding (solid curve). (b) Adiabatic case (dashed curve), the sounding is modified by the orography as shown in Fig. 9. (c) Precipitation case (dotted curve), the sounding is modified by the orography and specified heating as shown in Fig. 9.

km, orographically forced descent has occurred and the increased stability prevents the vertical development of cumulus clouds.

The implication of this result is that the perturbation to the monsoon wind by the mountain alone may not be sufficient to destabilize the air column, it may even act to stabilize it against deep cloud growth. This is consistent with the occurrence of long dry spells during which the monsoon winds continued to blow against the coast and the mountains.

Finally, the plume model was run on sounding "C" - the original sounding modified by the flow field generated by combined heating and orography. In this case the environmental air above cloud base has been so strongly modified by mesoscale lifting that the plume can maintain its buoyancy and grow quickly to a significantly greater altitude. This result gives the self-consistency we mentioned earlier. The mesoscale perturbation computed from the observed heating rates, and with observed topography, is such as to destabilize the air column and allow cumulus to grow which could, in turn, produce the observed heat. There is also the suggestion that the perturbation from the heating alone - without the mountain - may be large enough to do this. Note in this regard that the plume model together with Eq. A1 (with h = 0) could describe a self-sustaining propagating squall line moving at a speed U over flat terrain.

VII. THE SURFACE PRESSURE FIELD

The trough of low pressure just upstream of the heating, illustrated in Fig. 8, is associated hydrostatically with a region of abnormally warm air aloft. This warm region is generated (according to Eq. 3) by direct heating and by heating induced subsidence. This trough is of interest for two reasons. First, something like it has been observed and reported by several authors and second, it raises the possibility (first suggested by Smith and Lin [29]) of a negative mountain drag.

Several authors [6], [9], [19], [14], [13] have discussed the observed association between heavy coastal rain and the existence of a mesoscale coastal trough as indicated by the low pressure at certain coastal stations. The structure of this "trough" offshore is not known. In these papers, the trough has been considered as an atmospheric disturbance which would produce heavy rain in much the same way as might a synoptic scale trough in midlatitudes. The possibility of using it as a forecasting tool has been discussed although it has not been possible to predict the occurrence of the trough.

Against this background, the current results offer a different interpretation. We find that the trough could arise as a result of the heating aloft rather than vice versa. Of course in the full model there is no cause-effect relationship because the downstream side of the trough is responsible for the low-level velocity convergence which triggers the cumulus growth. This theoretical result is consistent with the observation of Hoxit et al. [7].

It is clear that if the thermally induced surface low were strong enough and located over the windward slopes of the mountain, then a negative drag (or mountain thrust) would occur. This does not violate any fundamental principle as the latent heating represents an external energy source which allows the mountain to accelerate, rather than decelerate, the atmosphere passing over it.

We have looked at the predictions of the theory applied to the Western Ghats and at direct measurements of surface pressure in the region. Our results are somewhat inconclusive but seem to suggest that during heavy rain the thermal low is strong enough to reverse the drag but that it is usually located too far upstream. Thus, while negative drag remains a real possibility, we can as of yet, offer no marked example.

VIII. DISCUSSION

In the preceding sections we have constructed a conceptual model of orographic rain. The number of assumptions involved in the analysis, and the lack of detailed data from the region, make it impossible to verify the model. It must remain, for the time being at least, a working hypothesis rather than a verified theory.

The most intriguing idea to come out of the model is the idea of multiple steady state solutions to the problem of moist airflow approaching a mountain range. We find internally consistent solutions both with and without rain. This seems to agree with the occurrence of wet and dry spells during the summer monsoon period. This implies that the system may have a bit of hysteresis, and that the question of how the system chooses between the two states might be a difficult one. We have not yet looked into the various subtle and not-so-subtle changes in the synoptic scale environment which could influence the system.

The discussion is also limited by the fact that we have not yet mathematically coupled the cumulus development with a time dependent mesoscale model. Only steady state solutions were considered herein. Until this is done we can say nothing about the stability of the system or how it might flip between the rain and no-rain states. Because the heating seemed to produce a slightly stronger mesoscale response than the orography, we tend toward a description of the orographic rain as a "squall line anchored to the mountain". A description of the "anchoring" process also requires a time dependent model with cumulus parameterization.

It may be that the mechanism described herein will find application on other tropical coastlines during periods of fresh onshore winds. In particular we find that the careful study of Sakakibara [24] concerning the cumulus development in the Kyushu region of Japan during southeast winds, to be very much in accord with the results reported herein for the Western Ghats.

ACKNOWLEDGMENTS

The discussions with C. Emanuel, R. Bleck, J. Klemp, C. Lord, D. Fitzjarrald, A.K. Mukherjee, U.S. De, D.R. Sikka, D. Durran, J. McGinley were helpful in the course of this research. Special thanks to D.R. Sikka for acting as host during a visit to the Indian Institute of Tropical Meteorology and to J. Fein of the National Science Foundation for making the visit possible. This research was supported in part by a grant (ATM 80-23348) from the Atmospheric Sciences Division of the National Science Foundation.

APPENDIX I

STEADY STATE PERTURBATIONS ON THE MEAN WIND

The vertical displacement produced by combined thermal and orographic forcing (Eqs. (10) and (11)) can be derived as (see [29])

$$\eta(x,z) = \eta_1(x,z) - A \sin \ell z \cdot [(\tan^{-1}\frac{x+c}{b_1} - \tan^{-1}\frac{x+c}{b_2}) \cdot$$

$$(\sin \ell(z_H+d) - \sin \ell(z_H-d)) - \frac{1}{2}\ell n(\frac{(x+c)^2+b_2^2}{(x+c)^2+b_1^2}) \cdot$$

$$(\cos \ell(z_H+d) - \cos \ell(z_H-d))],$$

$$\text{for } z < z_H-d$$

$$\eta(x,z) = \eta_1(x,z) + A[\cos \ell z \cdot (\tan^{-1}\frac{x+c}{b_1} - \tan^{-1}\frac{x+c}{b_2}) +$$

$$\frac{1}{2}\sin \ell z \cdot \ell n(\frac{(x+c)^2+b_2^2}{(x+c)^2+b_1^2})](\cos \ell z - \cos \ell(z_H-d)) - A \sin \ell z$$

$$[(\tan^{-1}\frac{x+c}{b_1} - \tan^{-1}\frac{x+c}{b_2}) \cdot (\sin \ell(z_H+d) - \sin \ell z) -$$

$$\frac{1}{2}\ell n(\frac{(x+c)^2+b_2^2}{(x+c)^2+b_1^2})(\cos \ell(z_H+d) - \cos \ell z)],$$

$$\text{for } z_H-d < z < z_H + d$$

$$\eta(x,z) = \eta_1(x,z) + A[\cos \ell z \cdot (\tan^{-1}\frac{x+c}{b_1} - \tan^{-1}\frac{x+c}{b_2}) +$$

$$\frac{1}{2}\sin \ell z \cdot \ell n(\frac{(x+c)^2+b_2^2}{(x+c)^2+b_1^2})](\cos \ell(z_H+d) - \cos \ell\,(z_H-d))$$

$$\text{for } z > z_H + d \qquad \text{(A1)}$$

where $\eta_1(x,z)$ and A are defined as

$$\eta_1(x,z) = \frac{ha(a \cos \ell z - x \sin \ell z)}{x^2 + a^2}$$

$$A = \frac{gQb_1}{2d\rho_o c_p \bar{T} U^3 \ell^2}$$

The surface perturbation pressure at the ground can be computed from Eq. (A1) and using Bernoulli's equation

$$P'(x,o) = -\rho_o U u'(x,o) = \rho_o U^2 \frac{\partial\eta}{\partial z}\bigg|_{z=0}$$

The result is

$$P'(x,o) = -\frac{\rho_o U^2 ha\ell x}{x^2+a^2} - A\ell \cdot [(\tan^{-1}\frac{x+c}{b_1} - \tan^{-1}\frac{x+c}{b_2}).$$

$$(\sin \ell(z_H+d) - \sin \ell(z_H-d)) - \frac{1}{2}\ell n(\frac{(x+c)^2+b_2^2}{(x+c)^2+b_1^2}) \cdot$$

$$(\cos \ell(z_H+d) - \cos \ell(z_H-d))] \qquad \text{(A2)}$$

The momentum flux can be computed from

$$F = \rho_o \int_{-\infty}^{\infty} u'w' \, dx$$

The solution associated with Eq. (A1) is

$$F = -\frac{\pi}{4}\rho_o h^2 NU - \pi A\ell\rho_o U^2 ha[(\frac{a+b_1}{(a+b_1)^2+c^2} - \frac{a+b_2}{(a+b_2)^2+c^2}) \cdot$$

$$(\sin \ell(z_H+d) - \sin \ell(z_H-d)) + (\frac{c}{(a+b_1)^2+c^2} - \frac{c}{(a+b_2)^2+c^2}) \cdot$$

$$(\cos \ell(z_H+d) - \cos \ell(z_H-d))]$$

$$\text{for } z < z_H-d$$

$$F = -\frac{\pi}{4}\rho_o h^2 NU - \pi A^2 \ell\rho_o U^2 (\cos \ell(z_H+d) - \cos \ell(z_H-d) \cdot$$

$$(\cos \ell z - \cos \ell(z_H-d))\, \ell n(\frac{(b_1+b_2)^2}{4\, b_1 b_2}) - \pi A\ell\rho_o U^2 ha \cdot$$

$$[(\frac{a+b_1}{(a+b_1)^2+c^2} - \frac{a+b_2}{(a+b_2)^2+c^2}) (\sin \ell(z_H+d) - \sin \ell z) +$$

$$\left(\frac{c}{(a+b_1)^2+c^2} - \frac{c}{(a+b_2)^2+c^2}\right) (\cos \ell(z_H+d) - 2 \cos \ell(z_H-d) + \cos \ell z)]$$

$$\text{for } z_H-d < z < z_H + d$$

$$F = -\frac{\pi}{4} \rho_o h^2 NU - \pi A^2 \ell \rho_o U^2 (\cos \ell(z_H+d) - \cos \ell(z_H-d))^2 \cdot$$

$$\ell n\left(\frac{(b_1+b_2)^2}{4\, b_1 b_2}\right) - 2\pi A \ell \rho_o U^2 ha(\cos \ell(z_H+d) - \cos \ell(z_H-d)) \cdot$$

$$\left(\frac{c}{(a+b_1)^2+c^2} - \frac{c}{(a+b_2)^2+c^2}\right)$$

$$\text{for } z > z_H + d \qquad \text{(A3)}$$

APPENDIX II
AN ENTRAINING JET MODEL

The method described by Squires and Turner [30] uses the following equations to describe a steady state entraining jet.

$M = b^2 u\rho$	mass flux definition
$\frac{d}{dz}M = 2b\alpha u\rho_o$	conservation of mass
$\frac{d}{dz}(Mu) = gb^2(\rho_o - \rho - \rho\sigma)$	conservation of momentum
$\frac{d}{dz}(M\sigma) + \frac{d}{dz}(Mq) = q_o \frac{dM}{dz}$	conservation of total water.

In these equations the quantities

b	plume radius
u	updraft velocity
ρ	plume density
α	entrainment coefficient ($\alpha = 0.10$)
ρ_o	environment density
g	acceleration of gravity
σ	specific liquid water
q	specific humidity of plume
q_o	specific humidity of environment

are functions of height (z). In addition, the perfect gas law, an energy equation, and the Clausius-Clapeyron equation are needed to complete the system. For each calculation it is necessary to specify the environmental $\rho_o(z)$, $q_o(z)$ or equivalent, and the initial values for b, z, u at cloud base. In most cases it is sufficient to take the plume virtual temperature as equal to that of the environment at cloud base.

REFERENCES

1. Bergeron, T.: Studies of the orogenic effect on the areal fine structure of rainfall distribution. Met. Inst., Uppsala University, Rep. No. 6 (1968).

2. Browing, K.A.: Structure, mechanism and prediction of orographically enhanced rain in Britain. In Orographic Effects in Planetary Flows, GARP Publication Series #23, WMO, Geneve (1980).

3. Cadet, D. and H. Ovarlez: Low-level airflow circulation over the Arabian Sea during the summer monsoon as deduced from satellite-tracked superpressure balloons. Part I: Balloon trajectories. Quart. J. Roy. Met. Soc., 102, 805-816 (1976) .

4. Cotton, W.R.: On parameterization of turbulent transport in cumulus clouds. J. Atmos. Sci., 32, 548-564 (1975).

5. Das, P.K.: The monsoon. St. Martin's Press Inc., New York, 162 pp (1972).

6. George, P.A.: Effects of off-shore vortices on rainfall along the west coast of India. Indian J. Met. Geophys., 7, 225-240 (1956).

7. Hoxit, L.R., C.F. Chappell and J.M. Fritsch: Formation of mesoslows or pressure troughs in advance of cumulonimbus clouds. Mon. Wea. Rev., 104, 1419-1428 (1976).

8. Iribarne, J.V. and W.L. Godson: Atmospheric thermodynamics. D. Reidel Pub. Co., Holland, 222 pp (1973).

9. Jayaram, M.: A preliminary study of an objective method of forecasting heavy rainfall over Bombay and neighborhood during the month of July. Indian J. Met. Geophys., 16, 557-564 (1965).

10. Krishnamurti, T.N. and Vince Wong: A planetary boundary-layer model for the Somali jet. J. Atmos. Sci., 36, 1895-1907 (1979).

11. Krishnamurti, T.N., Philip Ardanuy, Y. Ramanathan and Richard Pasch: On the onset vortex of the summer monsoon. Mon. Wea. Rev., 109, 344-363 (1981).

12. Morton, B.R.: Buoyant plumes in a moist atmosphere. J. Fluid Mech., 2, 127-144 (1957).

13. Mukherjee, A.K.: Dimension of an "off-shore vortex" in East Arabian Sea as deduced from observations during MONEX 1979. In Results of Summer MONEX Field Phase Research (Part A), FGGE Op. Rep. 9, WMO, Geneva, G. Grassman (ed.), 176-183 (1980).

14. Mukherjee, A.K., M.K. Rao and K.C. Shah: Vortices embedded in the trough of low pressure off Maharashtra-Goa coasts during the month of July. Indian J. Met. Hydrol. Geophys., 61-65 (1978).

15. Prasad, B.: Diurnal variation of rainfall in India. Indian J. Met. Geophys., 21, 443-450 (1970).

16. Queney, P.: The problem of airflow over mountains. A summary of theoretical studies. Bull. Amer. Met. Soc., 29, 16-26 (1948).

17. Ramachandran, G.: The role of orography on wind and rainfall distribution in and around a mountain gap: Observational study. Indian J. Met. Geophys., 23 (1), 41-44 (1972).

18. Ramage, C.S.: Monsoon meteorology. Academic Press, New York, 296 pp. (1971).

19. Ramakrishnan, A.R.: On the fluctuations of the west coast rainfall during the southwest monsoon of 1969. Indian J. Met. Geophys., 23, 231-234 (1972).

20. Ramakrishnan, K.P. and B. Gopinatha Rao: Some aspects of the non-depressional rain in peninsular India during the southwest monsoon. In Symposium on the Monsoon World, Indian Meteorological Department (1958).

21. Rao, G.V., W.R. Schaub, Jr. and J. Puetz: Evaporation and precipitation over the Arabian Sea during several monsoon seasons. Mon. Wea. Rev., 109, 364-370 (1981).

22. Rao, M.S.V. and J.S. Theon: New features of global climatology revealed by satellite derived oceanic rainfall maps. Bull. Amer. Met. Soc., 58, 1285-1288 (1977).

23. Saha, K.R. and S.N. Bavadekar: Moisture flux across the west coast of India and rainfall during the southwest monsoon. Quart. J. R. Met. Soc., 103, 370-374 (1977).

24. Sakakibara, H.: Cumulus development on the windward side of a mountain range in convectively unstable air mass. J. Met. Soc. Japan, 57, 341-348 (1979).

25. Sarker, R.P.: A dynamical model of orographic rainfall. Mon. Wea. Rev., 94, 555-572 (1966).

26. Sarker, R.P.: Some modifications in a dynamical model of orographic rainfall. Mon. Wea. Rev., 95, 673-684 (1967).

27. Sarker, R.P., K.C. Sinha Ray and U.S. De: Dynamics of orographic rainfall. Indian J. Met. Hydrol. Geophys., 29, 335-348 (1978).

28. Smith, R.B: The influence of mountains on the atmosphere. Advances in Geophysics, 21, 87-230 (1979).

29. Smith, R.B. and Y.L. Lin: The addition of heat to a stratified airstream with application to the dynamics of orographic rain. Quart. J. R. Met. Soc., 108, 353-378 (1982).

30. Squires, P. and J.S. Turner: An entraining jet model for cumulonimbus updrafts. Tellus, 14, 422-434 (1962).

31. Srivastava, G.P., B.B. Huddar and V. Srinivasan: Radar observations of monsoon precipitation. Indian J. Met. Geophys., 17, 249-252 (1966).

32. Telford, J.W.: Turbulence, entrainment, and mixing in cloud dynamics. Pageoph., 113, 1067-1084 (1975).

33. Warner, J.: On steady-state one-dimensional model of cumulus convection. J. Atmos. Sci., 22, 1035-1040 (1970).

REPORT ON ALPEX AS OF MAY 1982

Ronald B. Smith
Department of Geology and Geophysics
Yale University
New Haven, Connecticut 06511 USA

ABSTRACT

The Alpine Experiment (ALPEX) is an international cooperative experiment involving more than a dozen countries, intended to study the influence of mountains on the atmosphere. The project was organized as part of the Global Atmospheric Research Program under the World Meteorological Organization. Early in the planning, the Alps were chosen as the site for the study because of (1) the variety of mountain-induced phenomena which occur there, (2) the history of meteorological research in that area, and (3) the enthusiastic interest of the European weather services which suggested that an observing system of unprecedented quality could be constructed there.

The 13-month ALPEX Observing Period began on 1 September 1981 and continues until 31 September 1982. The more intensive Special Observing Period (SOP) occurred from 1 March until 30 April 1982. The enhanced observational system, including the high density rawinsonde network and three long-range research aircraft, were available only during the two-month SOP. Because of this, it is likely that the major scientific results of ALPEX will come from the study of cases which occurred during the SOP (see Fig. 1).

As of this writing, it is far too early to describe the scientific results of ALPEX. The final data set will not be finished for several months and it will be long after that before the scientists have completely analyzed the data. For now, it is only possible to list some of the phenomena which were encountered during the SOP.

Lee cyclogenesis: At least five cases of lee cyclogenesis occurred during the SOP. Several aspects of the storm development were studied by balloon soundings and research aircraft, including the upstream jet structure, the exit region of the jet during cyclogenesis, the blocking caused by the Alps, and the structure of the mature cyclone over the Mediterranean.

Airflow deflection and blocking by Alps: About four good cases of strong northwest winds against the Alps were observed, in which the low-level air was observed to slow and become deflected to the east and west of the Alps. The flow in midtroposphere was observed to cross over the Alps, sometimes descending on the south side to form a foehn, and sometimes continuing at constant level.

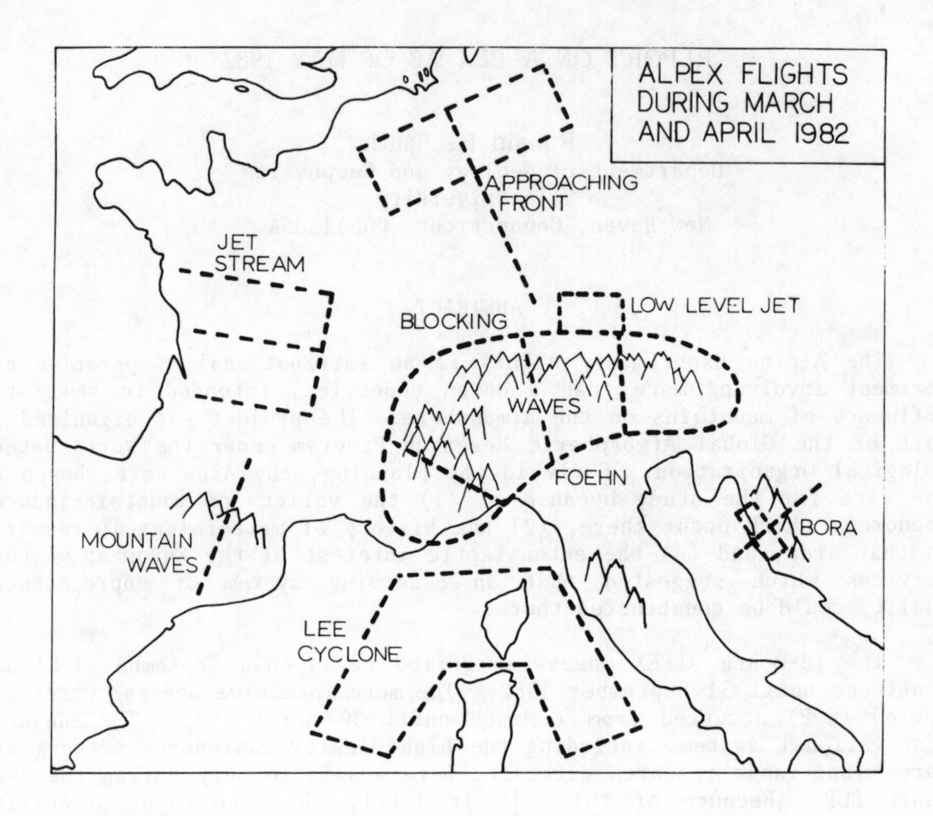

Fig. 1 This diagram shows, in schematic form, the types of research flights which were made by the three research aircraft (NCAR Electra, NOAA P-3, DFVLR Falcon) during the Special Observing Period of ALPEX.

Frontal passage: Many cases of frontal passage over the Alps occurred during the SOP but because of the difficulty in predicting the exact time of passage, only about four cases were studied with the research aircraft.

Thermally-induced low-level jet: One case of strong low-level jet flow was observed on the plains just north of the Alps.

Bora: Five cases of strong bora winds on the Yugoslavian coast were studied with the research aircraft. Strong airflow acceleration upstream of the coast, and strong turbulence in the lee, were encountered.

Mountain waves: Several cases of moderate strength vertically propagating waves were investigated over the Alps and over the Pyrenees.

Mistral: Several occurrences of strong mistral wind occurred in the lower Rhone valley during the SOP. These cases were primarily investigated using light aircraft from the French station at St. Raphael. On one occasion, the large aircraft operating from Geneva were used for such a study.

Over the next few months and years the scientific results of ALPEX will appear in the literature. These results will provide new understanding of the influence of mountains on the atmosphere, and may lead to the development of similar projects in other mountainous regions of the world.

ALPEX BIBLIOGRAPHY

Report of the first session of the intergovernmental planning meeting on ALPEX, GARP Special Report No. 36, WMO, Geneva.

Report of the second session of the intergovernmental planning meeting on ALPEX, GARP Special Report No. 39, WMO, Geneva.

ALPEX Experiment Design, GARP-ALPEX No. 1, WMO, Geneva.

ALPEX Flight Programme, GARP-ALPEX No. 2, WMO, Geneva.

ALPEX Operational Centre Plan, GARP-ALPEX No. 3, WMO, Geneva.

ALPEX Data Management Plan, GARP-ALPEX No. 4, WMO, Geneva.

MED-ALPEX, Oceanographic contribution to ALPEX Experiment Design, GARP-ALPEX No. 5, WMO, Geneva.

Oceanographic effects of planetary flows, GARP Publication Series No. 23, WMO, Geneva.

Kuettner, J.P. and T.H.R. O'Neill: ALPEX-the GARP mountain subprogram. Bull. Am. Meteorol. Soc., 62, 793-805 (1981).

Smith, R.B.: ALPEX update. Bull. Am. Meteorol. Soc., 63, 186-188 (1982).

1.5 WIND AND TEMPERATURE CHANGES OVER EURASIA DURING THE ONSET AND WITHDRAWAL OF THE SUMMER MONSOON OF 1979

Ding Yihui
Institute of Atmospheric Physics
Academia Sinica
The People's Republic of China

T. Murakami and T. Iwashima
University of Hawaii
Honolulu, Hawaii 96822 USA

ABSTRACT

During early summer of 1979, notable circulation changes occurred in the wind and temperature fields over extensive areas of the Eurasian continent north of 15°N. Among these areas, the most significant one is the Afghanistan-western Tibetan Plateau region which was characterized by an abrupt increase of 300-mb temperatures and intensification of a 300-mb anticyclone around 4 June, i.e. about two weeks prior to the monsoon onset over India. At the same time, a strong upper jet stream in the westerlies at 300 mb exhibited a distinct northward shift from about 30°N to 35°-40°N, and the easterlies at 700 mb exhibited rapid establishment and intensification over extensive areas near 25°N. A similar northward shift of a 300-mb jet stream also took place very far upstream (55°E) of the Tibetan Plateau on about 3 June. But in contrast to Yin's [7] early findings, this northward shift of the 300 mb jet stream was not observed over the eastern Tibetan Plateau. Concurrently, rapid development of an upper tropospheric anticyclone and an increase of temperature also occurred in association with the establishment of intense monsoon rains over the East China Sea-Japan region. Computations have shown that the increased temperature in these regions may be primarily caused by the diabatic heating over eastern China-Japan and Afghanistan-western Tibetan Plateau which precedes the Indian monsoon onset by more than two weeks.

Massive changes in wind and temperature also occurred over Eurasia north of 15°N during the late summer of 1979 when the summer monsoon retreated. An elongated west-southwest to east-northeast band of negative ΔT extended from northeast Africa through northern Tibet, northern China and beyond, with the maximum temperature decrease (exceeding -12°C) occurring over the northeast Tibetan Plateau. It is also demonstrated that an abrupt decrease in 300-mb zonal-mean temperatures occurred over an extensive midlatitude zone (40° to 55°N) around 18 August, i.e. about five days prior to the monsoon withdrawal over South Asia.

The foregoing results clearly suggest that the drastic wind and temperature changes during the onset and withdrawal phase of the summer monsoon are not confined to the immediate vicinity of the Tibetan Plateau, but also occur zonally over an extensive region across the entire Eurasian continent. Therefore, the thermal and dynamical effects on the seasonal transition and monsoon activity exerted by the Tibetan Plateau might not be so important as thought before. Its major effect appears to show up in the regional enhancement of the zonal belt of the planetary-scale circulation. The impact coming from the middle and high latitudes on the activities of the summer monsoon should be stressed and further elaborated.

I. INTRODUCTION

Limited research indicated a possible link between circulation changes over northern Eurasia during late spring and the onset of summer monsoon over South Asia. Winston and Krueger [6] found a large albedo difference between the early spring season of 1975 and 1976 over Soviet central Asia. They postulated that this large interannual albedo difference may influence the onset and intensity of the following summer monsoon circulations.

The onset of the 1979 summer monsoon over central India was declared to be 19 June [4]. Before and during the monsoon onset massive changes occurred in the wind and temperature fields over extensive areas of the Eurasian continent north of 15°N [3]. They pointed out that the increase in upper-tropospheric temperature and heating over the midlatitude regions could be a prerequisite to the establishment of the summer monsoon near India.

Chinese meteorologists [5] made an extensive study concerning principal weather systems over Eurasia. During the transition period the tropospheric temperature decreases rather sharply over and around the Tibetan Plateau. It is highly probable that this rapid temperature decrease, on the one hand, is related in some way to the monsoon withdrawal from the Indian region and, on the other hand, is related to the rapid cooling and the activities of cold air at middle and high latitudes.

The present study examines the possible link between the onset and withdrawal of the 1979 summer monsoon, and wind and temperature changes over Eurasia north of 15°N, with emphasis on the interaction of the circulation features between low and middle latitudes, and on the role of the Tibetan Plateau in causing these changes.

II. DATA AND COMPUTATION PROCEDURES

This study utilized twice-daily u, v and T data at eight levels (100, 200, 300, 400, 500, 700, 850 and 1000 mb) for the monsoon season of 1979 (1 May to 30 September). These data were extracted from the operational objective analyses of the National Meteorological Center (NMC), Washington, D.C. at 2.5 latitude-longitude resolution over Eurasia from 0° to 150°E and 15° to 70°N. The computational procedures used here are identical to those described by Murakami and Ding [3]. The reader is directed to their paper for a complete description of computational procedures.

III. WIND AND TEMPERATURE CHANGES DURING THE EARLY SUMMER OF 1979

Krishnamurti and Ramanathan [2] showed that the zonally averaged 850 mb winds over the Arabian Sea (5° to 10°N, 50° to 70°E) increased substantially between 5 and 15 June -- from less than 5 m/s before 5 June to greater than 15 m/s after 15 June. During this transition period, notable circulation changes also occurred over the entire Eurasian continent. For example, the time-latitude section of 300 mb U along 75°E (Fig. 1, middle) exhibits a distinct northward shift of strong (20 m/s) westerlies from about 30°N during May to 35° to 40°N after mid-June. A similar northward shift of the 300-mb jet stream also takes place very far upstream (55°E) of the Tibetan Plateau on about 3 June (Fig. 1, left). Yin

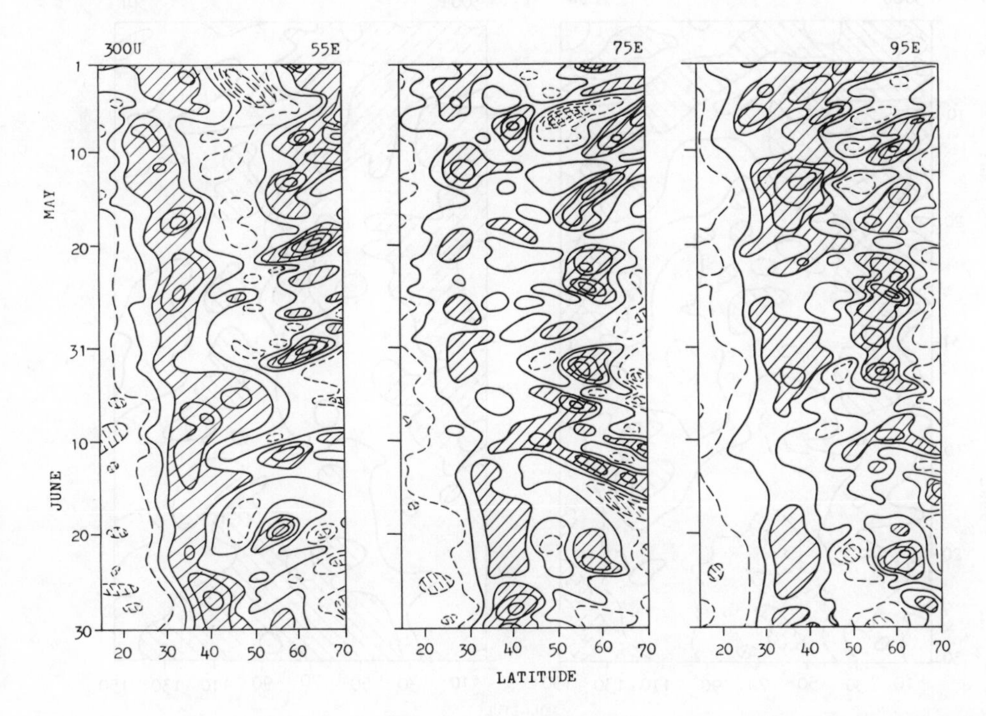

Fig. 1 Time-latitude sections of 300-mb U, averaged over a region extending 10 degrees of longitude east and west of 55°E (left), 75°E (middle), and 95°E (right), respectively, during the early summer (1 May to 30 June) of 1979. Intervals are 10 ms^{-1} with full and dashed lines representing positive, zero and negative intervals, respectively. Hatching denotes region of greater than 20 ms^{-1} westerlies. Dashed hatching denotes regions of greater than 10 ms^{-1} easterlies.

[7] postulated that the onset of the Indian monsoon occurs almost simultaneously with a sudden shift in the location of the upper-tropospheric subtropical jet stream from the southern to the northern periphery of the Tibetan Plateau. However, this is not evident in our study as the time-latitude section along 95°E (eastern Tibetan Plateau) does not exhibit a similar northward shift of the 300-mb u-component jet stream (Fig. 1, right).

The salient feature in Fig. 2 (left) is the establishment of 300-mb easterlies between about 50° and 90°E along 25°N after 7 June. Similarly, increases in 300-mb easterlies also occur east of about 130°E at 25°N. In Fig. 2 (right), the 300-mb temperatures at 30°N are greater than 245°K near 100°E and less than 245°K near 70°E between 20 and 31 May. In comparison, 300-mb temperatures exceed 245°K over an extensive region from 50°E to about 110°E after mid-June. Thus, a substantial 300-mb temperature increase occurs near 70°E, 30°N on about 10 June.

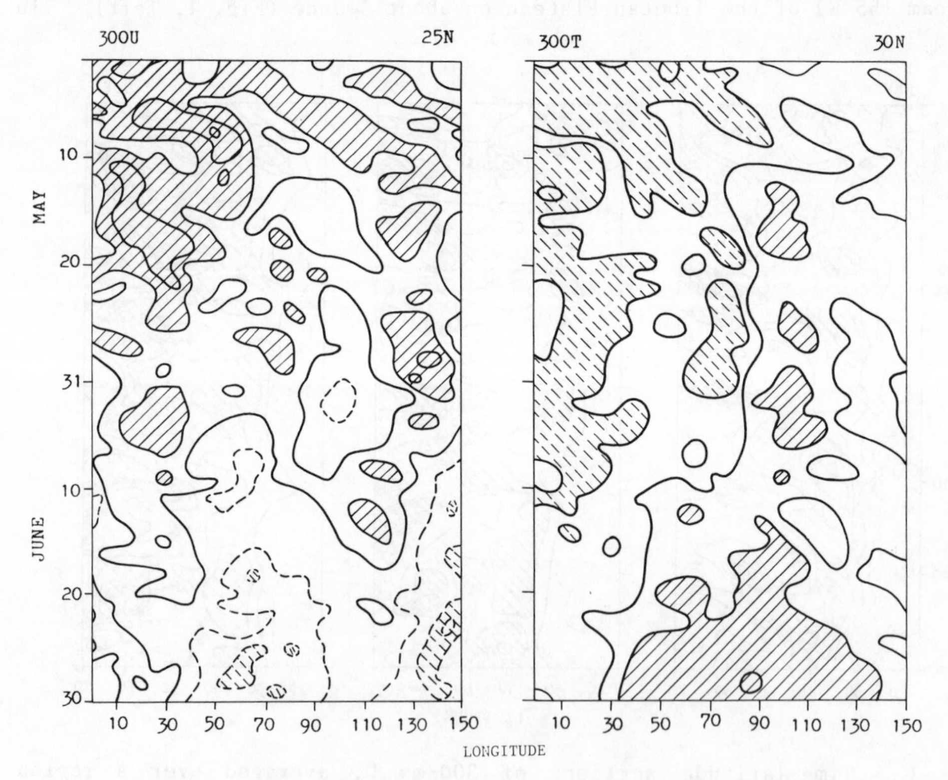

Fig. 2 Left: Time-latitude section of 300-mb U at 25°N. Intervals are 10 ms^{-1} with full and dashed lines representing positive, zero, and negative intervals, respectively. Hatching denotes regions of greater than 20 ms^{-1} westerlies. Dashed hatching denotes regions of greater than 10 ms^{-1} easterlies. Right: 300-mb T (5° intervals) at 30°N. Hatching denotes regions of greater than 245°K. Dashed hatching denotes regions of less than 235°K.

At 700 mb, the temperature near the western end of the Tibetan Plateau appears to increase on around 4 June (Fig. 3, right), about six days earlier than the corresponding temperature increase at 300 mb. This 700-mb temperature increase nearly coincides with the intensification of 700-mb easterlies near 25°N at around 130° to 150°E, 65° to 95°E and 10° to 30°E, respectively (Fig. 3, left). Concurrently, 700-mb monsoon westerlies become established over India and Indochina (not shown).

We have shown evidence that significant changes in wind and temperature occur over Eurasia during the first half of June. To further investigate these changes, we selected two contrasting periods, i.e. one from 15 to 30 May and the other from 20 to 30 June. For brevity, these two periods are hitherto referred to as "pre-" and "post-" onset phases, respectively. In spite of marked regional differences in the timing of the onset over southeast and East Asia, the term "onset" in this study refers to that for the Indian summer monsoon.

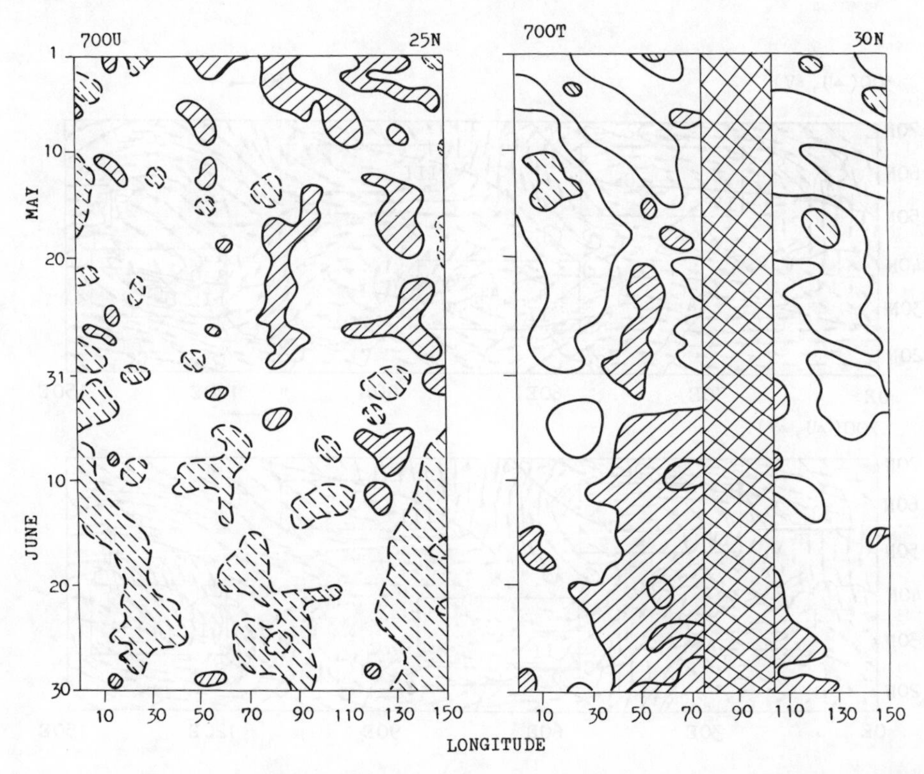

Fig. 3 Left: Time-longitude section of 700-mb U at 25°N. Hatching indicates regions of greater than 10 ms westerlies. Dashed hatching denotes regions of easterlies. Right: 700-mb T (5° intervals) at 30°N. Hatching denotes regions of greater than 285°K. Dashed hatching denotes regions of less than 285°K. Shading are for land mass projecting above 700 mb.

The circulation changes between the pre- and post-onset phases, (Δu, Δv)-vectors, were computed by:

$$U = U(20\text{-}30/6) - U(15\text{-}30/5),$$
$$V = V(20\text{-}30/6) - V(15\text{-}30/5),$$

where U(15-30/5) and U(20-30/6) represent pre- and post-onset mean zonal winds, respectively. Similarly, V(15-30/5) and V(20-30/6) denote the mean meridional winds for the two phases. The results of these computations at 300 mb and 700 mb are shown in Fig. 4 with the symbols "A" and "C" denoting anticyclonic and cyclonic circulation changes, respectively. Similarly, the differences in temperature at 300 and 700 mb were also computed in this manner (refer to Fig. 5).

At 300 mb (Fig. 4, top), the most significant feature is the existence of a well-organized, east-west oriented axis of anticyclonic circulation changes at approximately 25° to 35°N, with three distinct "A"

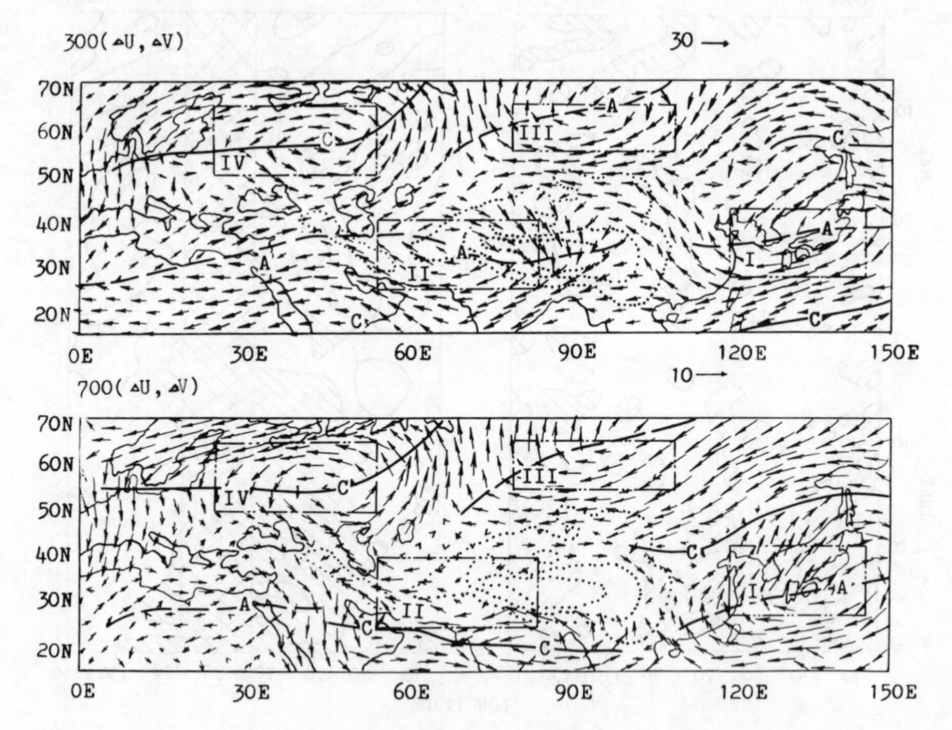

Fig. 4 Differences in the mean wind vectors (arrows) between the pre- and post-onset phases. The symbol "A" denotes anticyclonic changes and the symbol "C" denotes cyclonic changes. Top: At 300 mb with unit wind vector of 30 ms. Bottom: At 700 mb ($10\ ms^{-1}$ unit wind vector). Also shown are Regions 1 to 4 which underwent a notable change in the wind field. The thermal equation was applied to Regions 1 and 2.

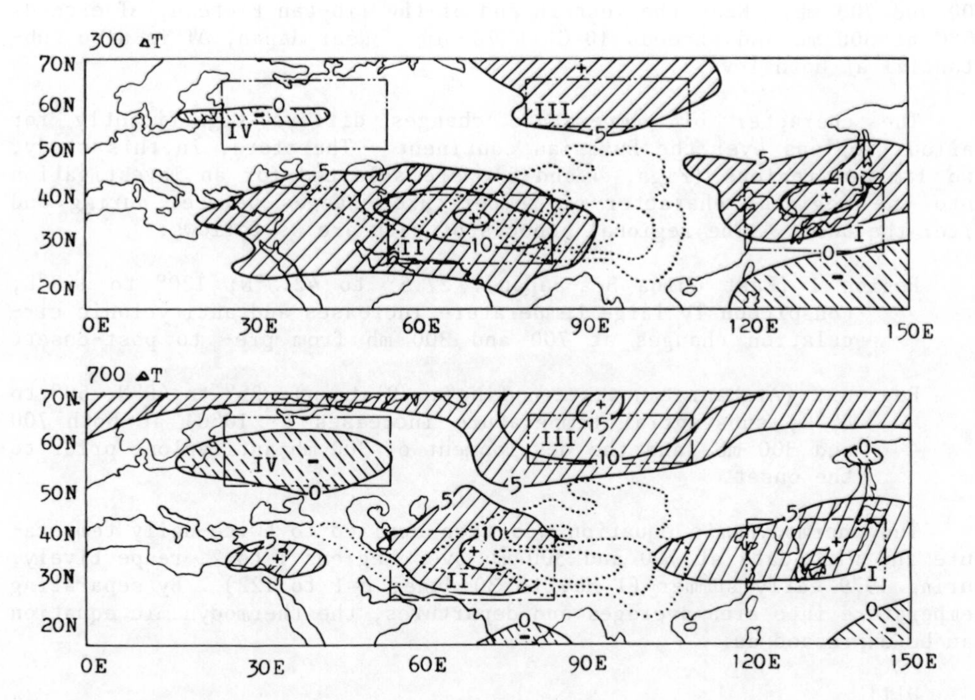

Fig. 5 Differences in mean temperature fields between pre- and post-phases, at 300 mb (top) and 700 mb (bottom), respectively, with intervals of 5°C. "Plus" and "minus" signs denote positive and negative T centers. Also shown are Regions 1 to 4 which underwent a notable change in the temperature field. The thermal equation was applied to Regions 1 and 2.

cells at (37°N, 140°E), (35°N, 70°E), and (35°N, 35°E). Of particular interest is the presence of a pronounced "A" circulation cell near the western end of the Tibetan Plateau. As mentioned earlier, this cell is associated with the dissipation of the upper trough prior to the onset and the development of the monsoon high following the onset. However, this upper-tropospheric monsoon high seems to develop as part of the anomalous anticyclonic circulation, changes occurring at about 35°N across the entire Eurasian continent during the onset phase.

At 700 mb (Fig. 4, bottom), pronounced "C" cells dominate the Arabian Peninsula, the northern Arabian Sea, central India, and the northern Bay of Bengal. These cells are associated with the post-onset development of the low-level monsoon trough. Interestingly, these "C" circulation changes over the monsoon region are contrasted with marked 700-mb "A" circulation changes to the east (southern China and Japan) and west (northern Africa).

A comparison between Figs. 4 and 5 indicates that "A" cells are generally associated with positive ΔT centers at both 300 and 700 mb, while "C" cells are generally associated with negative ΔT centers at both

300 and 700 mb. Near the western end of the Tibetan Plateau, ΔT exceeds 15°C at 300 mb and exceeds 10°C at 700 mb. Near Japan, ΔT is also substantial at both levels.

The character of temperature changes differs significantly for various regions over the Eurasian continent. Therefore, in this study, two limited regions (Figs. 4 and 5) were selected for an investigation into the regional character of temperature changes before, during and after the onset. The regional aspects of each are as follows:

Region 1 (East China Sea-Japan): 27.5° to 42.5°N; 120° to 145°E; conspicuously large temperature increases and anticyclonic circulation changes at 700 and 300 mb from pre- to post-onset.

Region 2 (Afghanistan-western Tibetan Plateau): 25° to 40°N, 50° to 85°E; substantial temperature increases (~ 10°C) at both 700 and 300 mb, and the development of 300 mb anticyclone prior to the onset.

The thermodynamic equation was then applied to twice-daily temperature and wind data at 700 and 300 mb over Regions 1 to 2, respectively, during 1979 early summer (1 May to 30 June; n=1 to 122). By separating temperature into area averages and departures, the thermodynamic equation can be expressed as:

$$\frac{\partial [T]}{\partial t} = A + B + C + D \qquad (1)$$

where

$$A = -[u\frac{\partial T''}{\partial t} + v\frac{\partial T''}{\partial y}],$$

$$B = [\omega](\frac{K}{p} - \frac{\partial}{\partial p})\ [T],$$

$$C = [\omega''(\frac{K}{p} - \frac{\partial}{\partial p})\ T''],$$

$$D = [Q]/c_p.$$

Here, the A term is the temperature advection. The B term represents adiabatic temperature change due to the area-mean vertical motion. The C term represents adiabatic temperature change due to vertical motion of spatial eddies and the D term, which is related to diabatic heating, was estimated as a residual in Eq. (1). It may include the effects of sensible heat, latent heat and radiation.

Figure 6 (top) reveals that [T] at 300 mb over Region 1 is nearly constant (~ 236 K) through May. This period is followed by a monotonic [T] increase after 4 June (n=70). Prior to this date, the diabatic effect, D (Fig. 6, bottom) is generally negative (cooling). Conversely, D becomes predominantly positive (heating) after that date. Therefore, the diabatic heating, D, is primarily responsible for the increased [T] after 4 June. Here, the most important finding is that heating D over

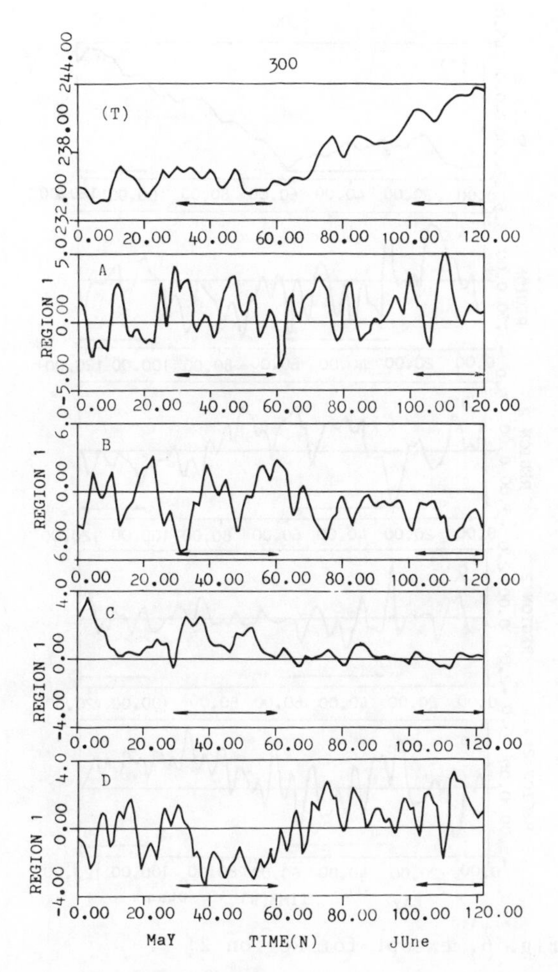

Fig. 6 Time series (twice daily; n=1 to 122) of [T] (°C) and the computed results of A, B, C and D terms (10^{-5}°C s^{-1}) in the thermal Eq. (1), obtained at 300 mb over Region 1. Also shown are the pre- and post-onset periods defined in this study.

Region 1 begins approximately two weeks prior to the onset of the summer monsoon over India (~ 19 June).

Interestingly, the 300-mb [T] over Region 2 also tends to increase after the beginning of June (Fig. 7, top). This increase is accompanied by positive (heating) D after about 1 June (Fig. 7, bottom). Thus, as in Region 1, the heating D over the Afghanistan-western Tibetan Plateau region also precedes the Indian monsoon onset by more than two weeks.

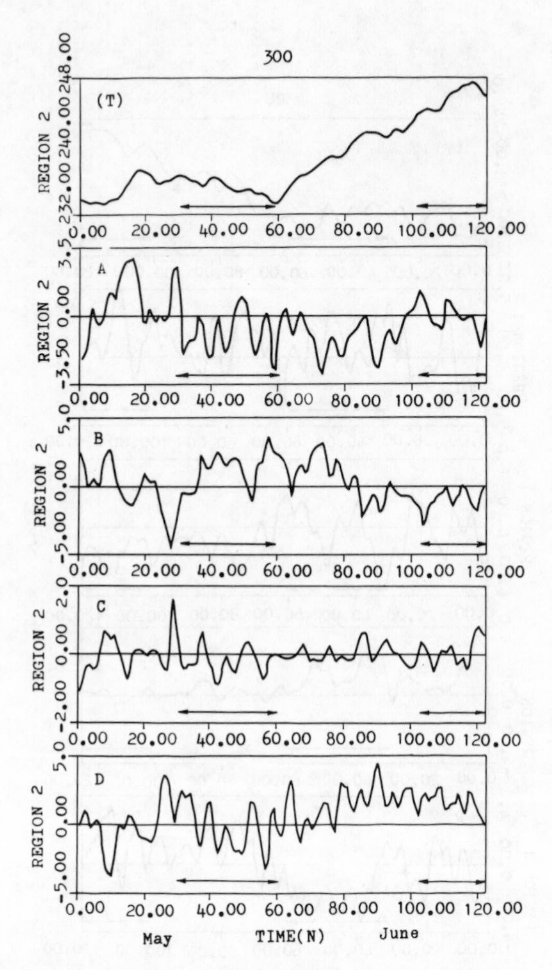

Fig. 7 As in Fig. 6, except for Region 2.

IV. WIND AND TEMPERATURE CHANGES DURING THE LATE SUMMER OF 1979

After the 1979 summer monsoon over central India commenced on about 19 June, the Indian monsoon was quite active until the middle of July (Fig. 8, left). During this active monsoon period, 700-mb westerlies over the central India-Indochina region (15°N, 70° to 100°E) were substantially strong and occasionally exceeded 10 m/s. During the following two weeks, the 700-mb westerlies were very weak. This period of break monsoon was characterized by below normal rainfall over central India [1]. On about 29 July the summer monsoon revived and lasted for approximately three weeks. The second break monsoon with weak 700-mb westerlies began about 20 August. The summer monsoon never revived again over most areas in South and southeast Asia. Thus, it was difficult to determine an exact withdrawal date of the 1979 summer monsoon [4]. The prolonged break was believed to be one important cause for the deficit of monsoon rains over many areas in India.

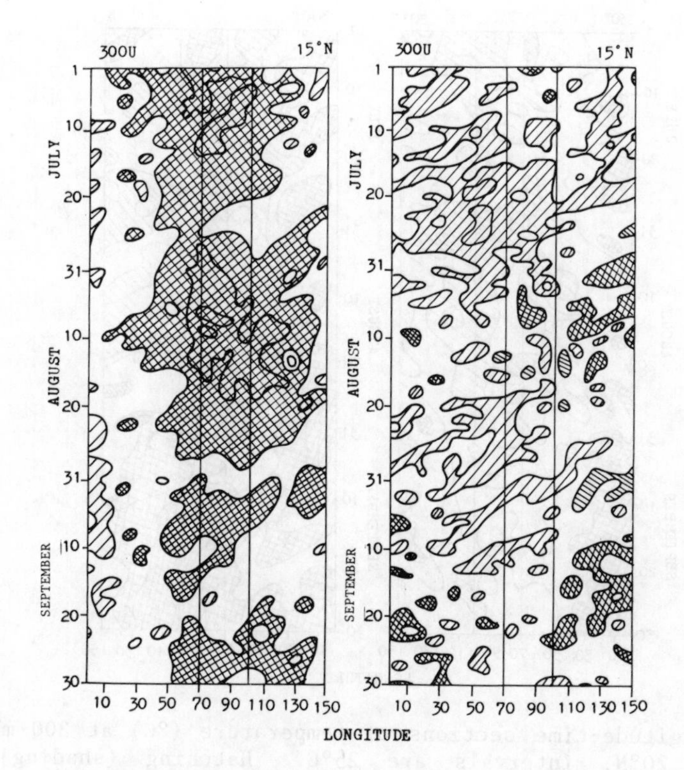

Fig. 8 Longitude-time sections of zonal winds with 10 units (unit: m s^{-1}) per interval. Shading indicates region of westerly zonal winds. Hatching denotes region of greater than 10 ms^{-1} easterly winds. Left: for 700-mb U at 15°N. Right: for 300-mb U at 15°N.

The equally obvious intraseasonal changes are also seen in the time-longitude section for 300-mb u-component along 15°N (Fig. 8, right). It reveals that the active monsoons were characterized by weak upper easterlies or even upper westerlies, whereas the break monsoons were associated with stronger 300-mb easterlies. Around 20 August, when the second active phase ended, the 300-mb easterlies became substantially stronger. They remained strong until the end of August, then they gradually diminished.

In Fig. 9 (left), 300-mb temperatures along 20°N indicate significant fluctuations over the entire Asian continent (0° to 150°E), with fluctuations most prominent over the monsoon region (70° to 100°E). Over this region, a sharp 300-mb temperature decrease occurred around 23 August. Therefore, in this study the date of the summer monsoon withdrawal was defined as 23 August.

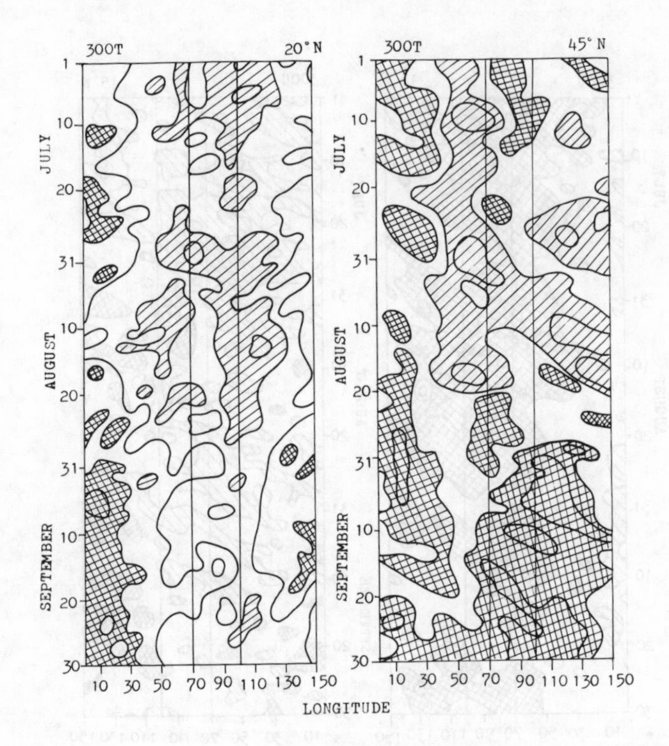

Fig. 9 Longitude-time sections of temperature (°C) at 300 mb. Left: at 20°N, intervals are 25°C. Hatching (shading) denotes regions of higher (lower) than -27.5° (-32.5°)C. Right: at 45°N, intervals are 5°C. Hatching (shading) denotes regions of higher (lower) than -35° (-40°)C.

The 300-mb temperature along 45°N (Fig. 9, right) exhibits an even more drastic drop than along 20°N on 20 August (approximately three days earlier than the monsoon withdrawal). Temperature changes are most pronounced near Kazakh (60° to 70°E), where temperatures decreased from above -35°C prior to 20 August to below -40°C after that date.

An attempt was made to more clearly identify the spatial character of temperature decreases by computing differences between T averaged over 20 days (29 July to 17 August) of active monsoon and 20 days (11 to 30 September) after monsoon withdrawal. In Fig. 10, immediately evident is an elongated west-southwest to east-northeast band of negative ΔT extending from northeast Africa through northern Tibet to northeast China and beyond. Hence, these temperature decreases represent a very large-scale phenomenon. The largest temperature decrease (exceeding -12°C) occurs over the northwest Tibetan Plateau.

The distribution of ΔT in Fig. 10 indicates a zonal character in temperature changes. Therefore, we will examine the processes by which

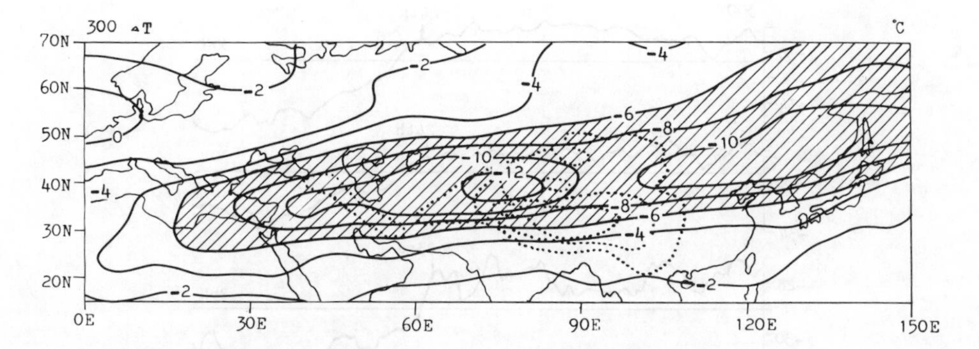

Fig. 10 Differences in the 20-day mean 300-mb temperature between 29 July to 17 August and 11 to 30 September, 1979. 2°C intervals with hatching representing regions of less than -6°C temperature differences. Dotted lines represent smoothed topographic heights at 1 km intervals.

the zonal mean temperatures over Eurasia change during late summer. The time-series of twice-daily, 300-mb zonal mean T, averaged between 0° and 150°E at 5° latitudinal intervals from 15°N to 60°N, are shown in Figs. 11 and 12, respectively. In Fig. 11, note that there is a sharp decrease in 300-mb temperature at 20° and 25°N around 23 August. The monsoon withdrawal from south and southeast Asia (50° to 100°E), as confirmed in Fig. 9 (left), contributes most to this sharp decrease. Farther to the north, 300-mb $\bar{T}$ at 30° and 35°N decreases gradually after 24 August. The most interesting feature in Fig. 12 is the near simultaneous occurrence of a sharp decrease in 300-mb $\bar{T}$ over extensive latitudinal zones around 18 August. This decrease is most clearly defined near 45° to 50°N. Thus, 300-mb temperature at higher latitudes decreases sharply about five days prior to the monsoon withdrawal. These features may be an indication of some form of lateral coupling between the high and low latitudes during the monsoon withdrawal phase.

V. CONCLUDING REMARKS

This study has presented evidence that the upper-tropospheric circulation over the Eurasian continent underwent significant changes during the early summer (1 May to 30 June) and late summer (1 August to 30 September) of 1979. These changes are best exemplified by the rapid development of the anticyclonic system above 300 mb by and an abrupt increase of 300-mb temperatures near the western end of the Tibetan Plateau approximately two weeks prior to the monsoon onset over India. At the same time, the strong upper westerly jet stream at 300 mb exhibited a distinct northward shift from about 30°N to 35°-40°N and the easterlies at 700 mb exhibited a rapid establishment and intensification over extensive areas near 25°N. These changes led to the rapid disappearance of the upper troughs south of the Himalayas. Of particular interest in late summer is the sudden decrease in 300-mb zonal mean $\bar{T}$, averaged over

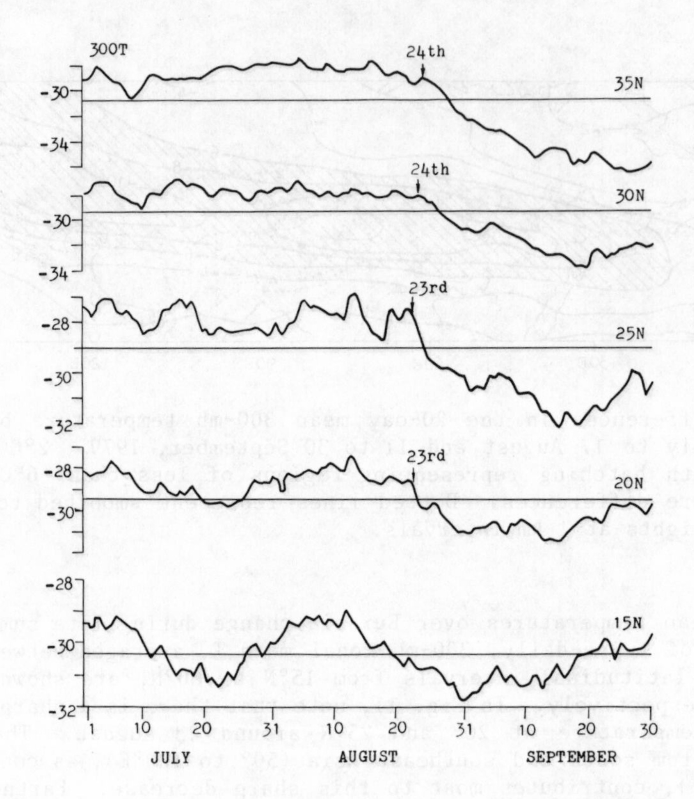

Fig. 11 Time series of 300-mb zonal mean temperature averaged between 0° and 150°E at 15°, 20°, 25°, 30°, and 35°N, respectively. Arrows indicate the initial date of sharp $\bar{T}$ decreases.

Eurasia (0° to 150°E) along 20° to 25°N, around 23 August 1979. The monsoon's withdrawal from South Asia contributed the most to this sharp temperature decrease. A similar decrease in 300-mb temperature also occurred over extensive midlatitude zones (40° to 55°N) around 18 August, i.e. about five days prior to the monsoon withdrawal. Owing to the fact that the wind and temperature changes for the early summer and late summer of 1979 both preceded the onset and withdrawal of the summer monsoon, it seems that the activities of the blocking highs at middle and high latitudes may exert an important influence on the timing of onset and withdrawal of the summer monsoon.

In addition, although the Tibetan Plateau may play an important role in the establishment of an upper-tropospheric anticyclone in the early summer and in causing the rapid change in temperature in this area, it should be noted that similar rapid changes also occurred over an extensive midlatitude (~ 35°N) region of the Eurasian continent. In particular, a nearly simultaneous intensification of the upper-tropospheric anticyclone over the East China Sea-Japan is conspicuous. This region is also characterized by a sharp increase in upper-tropospheric temperature

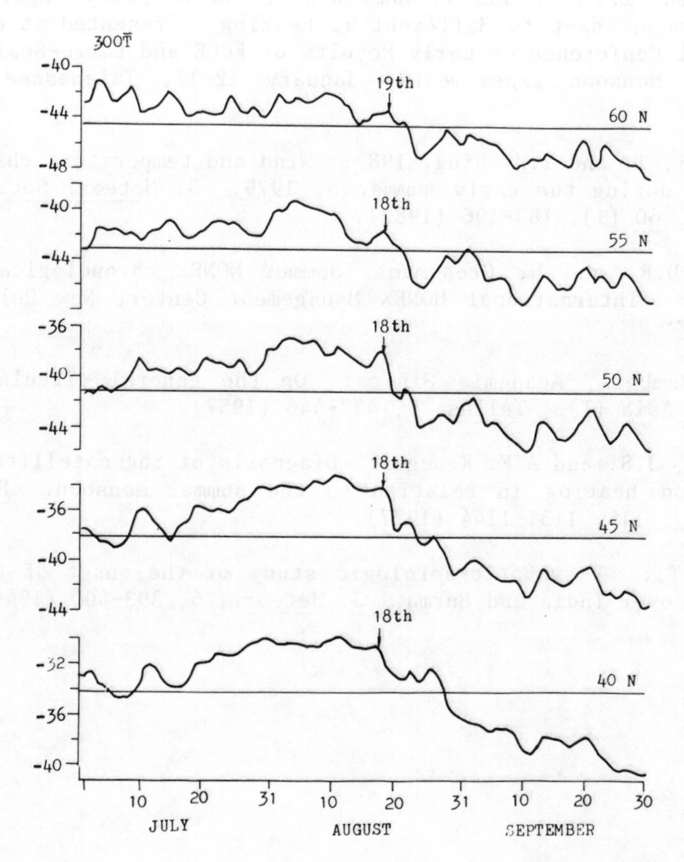

Fig. 12 As in Fig. 10, except at 40°, 45°, 50°, 55°, and 60°N.

and diabatic heating during early June. Over the Afghanistan-western Tibetan Plateau region, the upper tropospheric heating exceeded 1.5°C per day after 1 June. Thus, the increase in upper-tropospheric heating over these midlatitude regions may be a prerequisite for the establishment of the summer monsoon near India. Thermal and dynamical effects on the seasonal transition and monsoon activity exerted by the Tibetan Plateau might not be as important as thought before. The plateau's major effect appears to show up in the regional enhancement of the zonal belt of the planetary-scale circulation.

REFERENCES

1. Chang, C.C.: A contrasting study of the rainfall anomalies between central Tibet and central India during the summer monsoon season of 1979. Bull. Amer. Meteor. Soc., 62 (1), 20-22 (1981).

2. Krishnamurti, T.N. and Y. Ramanathan: Sensitivity experiments on the monsoon onset to differential heating. Presented at the International Conference on Early Results of FGGE and Large-Scale Aspects of its Monsoon Experiments, January 12-17, Talahassee, Florida (1981).

3. Murakami, T. and Y.H. Ding, 1982: Wind and temperature changes over Eurasia during the early summer of 1979. J. Meteor. Soc. of Japan Ser. II, 60 (1), 183-196 (1982).

4. Sikka, D.R. and R. Grossman: Summer MONEX chronological weather summary. International MONEX Management Center, New Delhi, India (1980).

5. Staff Members, Academia Sinica: On the general circulation over eastern Asia (I). Tellus, 9, 432-446 (1957).

6. Winston, J.S. and A.F. Krueger: Diagnosis of the satellite-observed radiation heating in relation to the summer monsoon. Pure Appl. Geophys., 115, 1131-1144 (1977).

7. Yin, M.T.: A synoptic-aerologic study of the onset of the summer monsoon over India and Burma. J. Meteor., 6, 393-400 (1949).

1.6 ANNULUS SIMULATION OF THE NORTHERN SUMMER MEAN GENERAL CIRCULATION AND COMPARATIVE STUDY OF FLOW PATTERN CHARACTERISTICS OVER EASTERN ASIA

Sung Chengshan, Wang Yunkuan, Wang Guifang
Institute of Atmospheric Physics
Academia Sinica
The People's Republic of Chian

ABSTRACT

In this paper we simulate the northern summer mean general circulation in a rotating annulus. The results show that the thermal effect is a main factor for the formation of the summer circulation, especially the heating effect of large-scale topography.

We have explained the similarity and difference of summer flow patterns between East Asia and North America in such a way that the highs are formed by the heating effect of topographies and they must interact with the surrounding flow pattern at the same time.

The experiment shows that the local cooling effect of the troposphere over the Indian Ocean and South Asia may be one reason for the formation of the southwest monsoon and the upper tropospheric anticyclone.

The experimental results are further verified by synoptic facts.

I. INTRODUCTION

Great progress of experimental simulation of the atmospheric phenomenon have been made. Yeh and Chang [7] have simulated the heating effect of the Qinghai-Xizang Plateau. Since then, work has been done on structure and movement of the Qinghai-Xizang high and the subtropical high over the Pacific Ocean [2] [6] [8]. However, the objective of these simulations is mainly confined to a particular circulation system. For a better understanding of the general circulation as a whole, we try to model the northern hemisphere circulation in a fluid experiment.

Once we have obtained a hemispheric model, we can explore the relationships between the different regional circulation systems and determine the relative importance of various factors. For example, in summer there are some similarities (and some differences) between the circulation patterns of North America and East Asia. The main reason is the heating effect of the Rocky Mountains and the Qinghai-Xizang Plateau - a point we will discuss in detail. Finally, we shall discuss the formation of the monsoon circulation.

II. EXPERIMENT DESIGN

Our experimental setup is fundamentally the same as in Yeh et al. [7]. Figure 1 shows a sketch of the experiment design. The diameter of the outer cylinder is 38.3 cm. The working medium is a mixture of glycerin and water with a specific gravity of 1.043. The tracers are small plastic balls of milk-white color with a diameter about 0.5 to 1.0 mm. Their specific gravity is the same as the working medium. The depth of the working medium is about 6 cm.

1. Topography

In the annulus (Fig. 1) we put scale models of Iran, the Qinghai-Xizang Plateau, and the Rocky Mountains in their proper position. These are made of plastic with shapes similar to the smoothed actual topography. The model of the Qinghai-Xizang Plateau has been connected with the Iran Plateau. The highest point is 3.0 cm, while the model of the Rocky Mountains is 1.1 cm high.

2. Distribution of Heat or Cold Sources

For simplicity we only simulate two main kinds of diabatic sources: sensible heat and latent heat with proper intensity. In the annulus, all

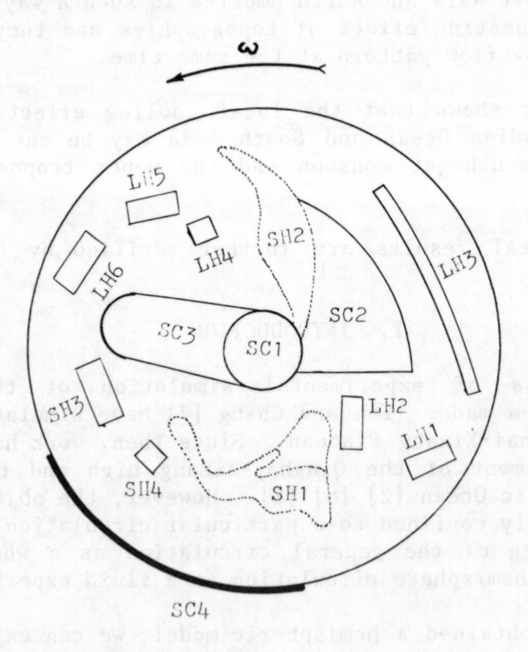

Fig. 1 Schematic diagram of the topography and diabatic sources in the annulus. SH1 to SH4: sensible heat sources, SC1 to SC4: cold heat sources; LH1 to LH6: latent heat sources. The boundaries of the topography are indicated by a dashed line in all the charts in this paper.

of the heat sources are made of electric resistance wire. We can control the intensity of electric current to obtain a suitable heating power rate.

The sensible heat flux from the surface of the earth in June [1] and the budget of the troposphere in July [4] are used to design the distribution of heat or cold sources. To form sensible heat sources, the SH1 and SH2 (Fig. 1), electric resistance wires are set mainly on the surface of the central parts of the Rocky Mountains, the Iranian Plateau and the Qinghai-Xizang Plateau in the form of spirals. Other heat sources (SH3, SH4, LH1 to LH6) are made of electric resistance wire wrapped on a square or rectangular framework. SH3 and SH4 are placed on the surfaces of North Africa and the Arabian Peninsula, respectively.

Latent heat sources, LH1 to LH6, are arranged at lower latitudes and on the east coast of continents at a height of 2.5 to 3.0 cm from the bottoms to model the effect of latent heat.

The cold sources, SC1 to SC3, are made from small copper boxes. SC1 is connected with SC2, whose area is about 160 cm^2. The area of SC3 is about 126 cm^2. Cold water may enter the box through a tube over the polar area to produce cold sources. In the experiment, a cylinder with a diameter of 4.6 cm was put in the polar region. The circulating cold water can be diverted to the inner cylinder to form a cold source, SC1. To investigate the role of the temperature field over the area of the Indian Ocean and South Asia, we made the external cold source, SC4, by cooling the water bath in a limited area.

The intensities of all cold sources are estimated by the temperature of cold water forming them.

III. EXPERIMENT RESULTS

We shall present simulated flow patterns in summer and then discuss the effects of various factors.

1. Summer Flow Pattern

Plate 1a is a simulated upper-layer circulation. Plate 1b is a lower-layer circulation. The test conditions are shown in Table 1. It gives the intensity of various heat sources. In the first experiment to be described, the cold sources over the Pacific and Atlantic Oceans (SC2, SC3) were not used. Compared to the real atmosphere (Fig. 2), we can see that the main features of the summer general circulation have been simulated very well.

In the upper layer (Plate 1a), the simulated circulation systems are as follows:

The westerlies are located at middle and high latitudes around the polar region with a flow intensity of about 0.21 m/s, corresponding to 24

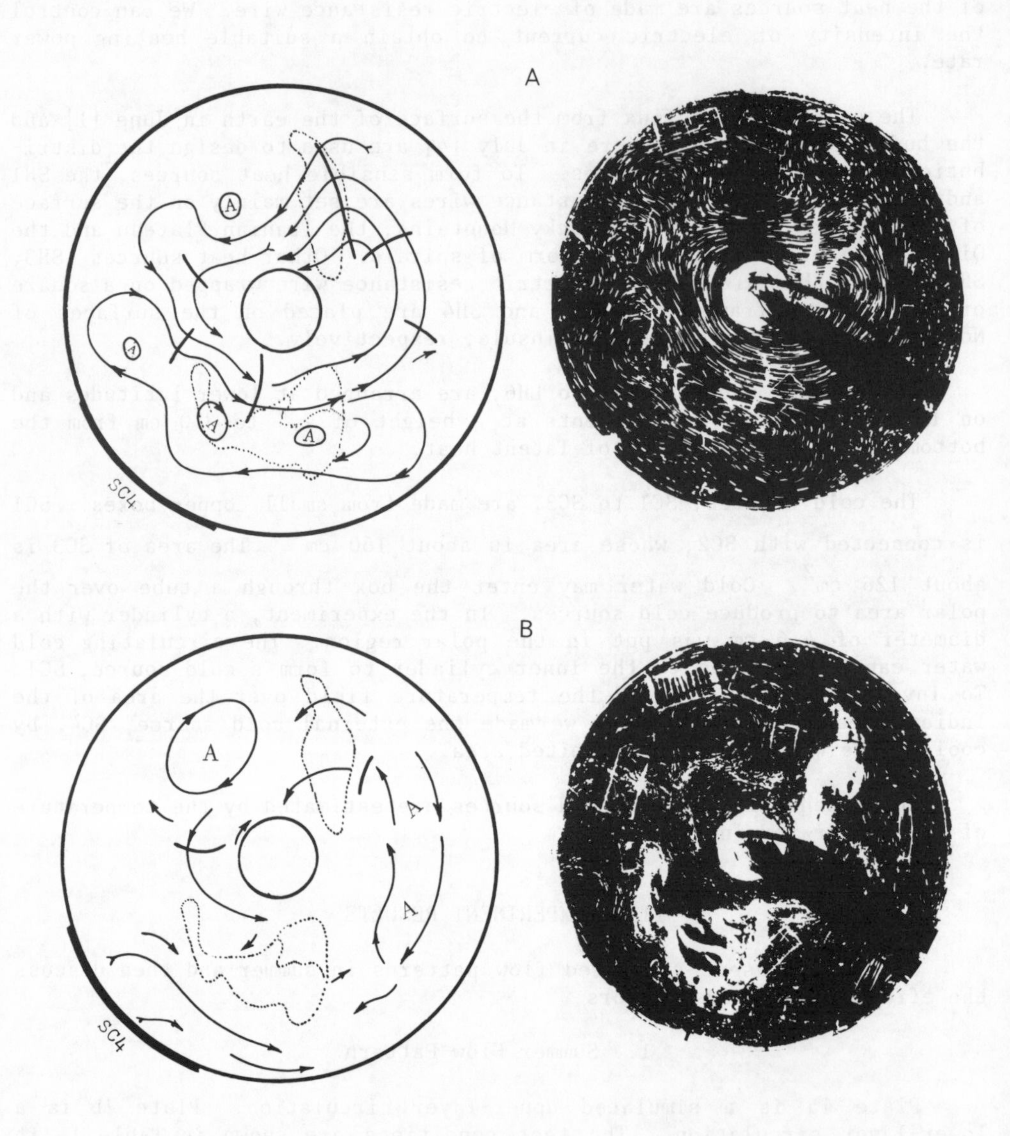

Plates 1a,b Simulated northern hemisphere summer flow field. Plate 1a: upper layer, Plate 1b: lower layer. The period of rotation of the annulus (T) is 31.3 sec. The times of exposure of photography are 9 sec (upper layer) and 36 sec (lower layer). Short-white lines show the tracks of tracers in all plates in this paper. The difference in temperature between the inner cylinder and the experiment medium is -0.8°C. The temperature of the water bath (SC4) drops by 2°C. Experimental steps: 1. Cold water was diverted to inner cylinder, 2. add SH1 to SH4 (31 annulus days after 1.), 3. add LH1 to LH6 (43 annulus days), 4. forming SC4 (58 annulus days).

Table 1 Heating area and heating power rate of various heat sources in the annulus.

Heat Source	Heating Area (cm^2)	Mean Heating Power Rate ($watt/cm^2$)
SH1	68	0.0086
SH2	33	0.0050
SH3	3x6	0.0110
SH4	2.5x2.5	0.0087
LH1	2x5	0.0210
LH2	2x5	0.0210
LH3	1x20	0.0170
LH4	2.5x2.5	0.0088
LH5	2x5	0.0120
LH6	3x6	0.0063

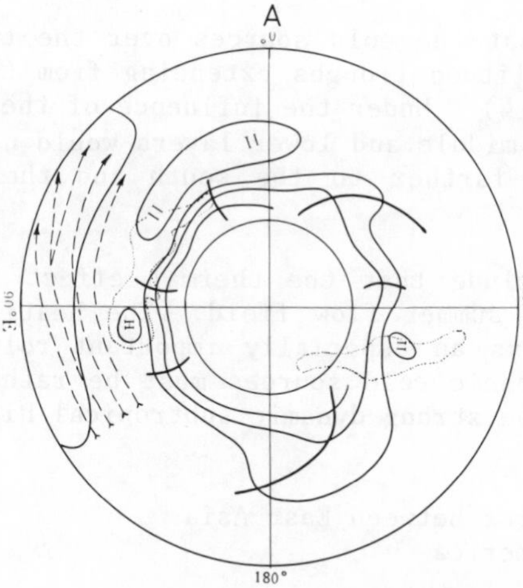

Fig. 2 Observed summer circulation (reproduced from [3]). Upper: 300 mb; lower: 700 mb. Solid lines indicate geopotential height contours, dashed lines indicate streamlines.

m/s in the real atmosphere according to kinematic similarity. A high-pressure zone is seen over the Eurasian continent. The centers of highs are located over North Africa, the Iranian Plateau and the Qinghai-Xizang Plateau, respectively. Easterlies are located over the south side of these highs. Intensity of the easterlies is about 0.094 cm/s corresponding to 11 m/s in the real atmosphere. Two large-amplitude troughs are present over the Pacific and Atlantic Oceans, respectively, which are the upper tropospheric midoceanic troughs induced by a thermal disturbance effect of topographies in the westerlies. There are other main troughs located in the region between the highs over the Eurasian continent.

In the lower layer (Plate 1b), we can see that the subtropical high over the two large oceans (Pacific and Atlantic) and the phenomena associated with the southwest monsoon over the Indian Ocean and South Asia have been simulated. The intensity of the southwest flow is about 0.048 cm/s corresponding to 6 m/s in actual observations. The simulated vertical structure is analogous to that observed. For example, right above the southwest monsoon are easterlies, while the Pacific subtropical high is located under the upper subtropical west wind forming to the south of the upper tropospheric midoceanic trough.

But some failures must be noticed. Because the Rocky Mountain high could not remain over the mountains, the wave pattern over North America shifts eastward by about 90° longitude. This phenomenon is associated with the stability of the Rocky Mountain high. We shall discuss the question later. Furthermore, the troughs downstream of the Rocky Mountains and of the Qinghai-Xizang Plateau have not be simulated.

In the above experiment, cold sources over two large oceans are not incorporated. What role do they play?

Further experiments have proven that the cold sources over the two large oceans can produce two large-amplitude troughs extending from the lower to the upper layers (as in Plate 4). Under the influence of these troughs, the subtropical highs over the middle and lower layers would not be obvious. Their position would be farther to the south and their intensity would be very weak.

From the discussion above we conclude that the thermal effect is very important for the formation of a summer flow field. The heating effect of large-scale topographies plays an especially important role, while the direct effects of summer oceanic cold sources must be rather weak, otherwise we could not observe the strong dynamic subtropical high over the oceans.

2. Contrast of Flow Patterns Between East Asia and North America

There are some similarities and differences between flow patterns of East Asia and North America. The similarities are presumably due to the heating effect of large-scale topography. The Rocky Mountains and the Qinghai-Xizang Plateau are each located over the western part of these two continents.

In the following, we shall primarily deal with the Rocky Mountain high and make some comparisons with the Qinghai-Xizang high in order to better understand the summer flow field features over East Asia and North America.

(1) Structure of the Rocky Mountain high produced by the heating effect of mountains. In summer, the maximum intensity of the sensible heat source reaches 8,000 cal/cm^2/month (125 watts/m^2) over the Rocky Mountains. Plates 2a and b show the circulation produced by the heating effect of the Rocky Mountains. The upper layer (Plate 2a) is an anticyclone called the Rocky Mountain high in our experiment, while in the

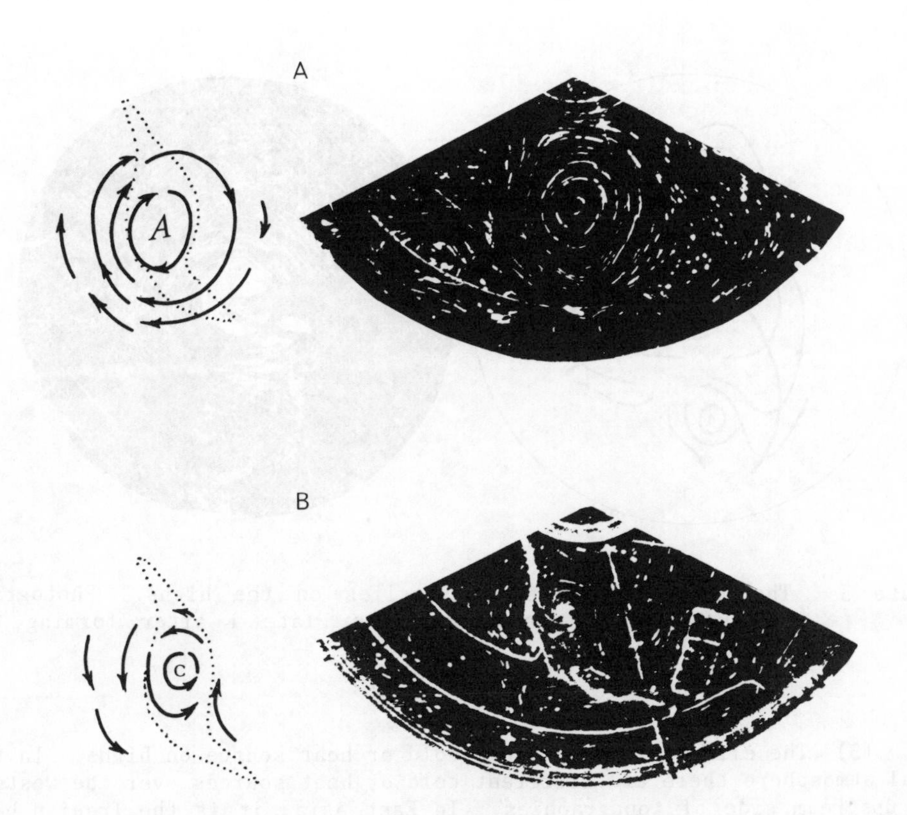

Plates 2a,b Circulation systems formed by the heating effect of the Rocky Mountains. Plate 2a: upper layer, Plate 2b: lower layer. T = 34.2 sec, heating intensity of SH2: 0.0096 $watt/cm^2$.

lower layer (Plate 2b) it is a cyclonic circulation. This structure is similar to the Qinghai-Xizang high. Because the height of the heat source over the Rocky Mountains is lower than the Qinghai-Xizang Plateau, the height of the transition layer from low to high is likewise lower than over the plateau in Asia.

(2) The effect of westerlies on the movement of highs. Plate 3 gives the flow field showing the interaction between highs and westerlies. We can see that under the influence of westerlies, the Rocky Mountain high does not remain over the mountains. A trough over the western side of the mountains is formed by a heating effect of topography on the westerlies. This situation forces the Rocky Mountain high to shift farther eastward. In contrast, the Qinghai-Xizang high remains over the plateau. This fact shows that under similar conditions of westerlies, the Rocky Mountain high is more unstable than the Qinghai-Xizang high.

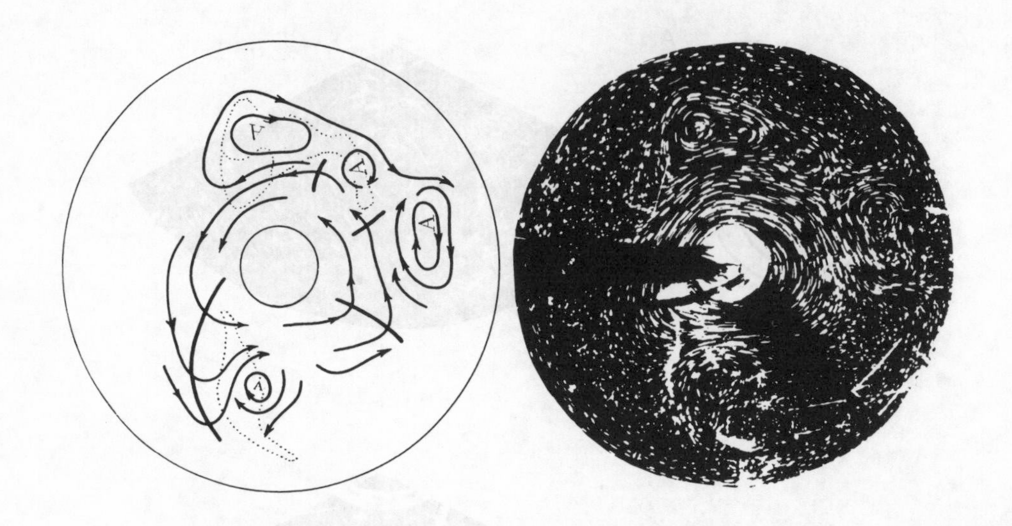

Plate 3 The influence of the westerlies on the highs. Photograph is taken in same experiment as Plate 1 after forming the westerlies and highs.

(3) The effect of an upstream cold or heat source on highs. In the real atmosphere there are different cold or heat sources over the western or upstream side of topographies. In East Asia, it is the Iranian heat source, while near the western periphery of the Rocky Mountains the California current produces a cold source with a cooling rate of about 200 $cal/cm^2/day$ (97 $watts/m^2$) in the atmosphere [4].

Experiments show that the Iranian high, induced by the Iranian heat source, is favorable to the maintanence of the Qinghai-Xizang high. A trough may be seen in the area between the Iranian high and the Qinghai-Xizang high, but it does not cause the Qinghai-Xizang high to move eastward.

To examine the effect of the Pacific cold source on the position of the Rocky Mountain high, we made the oceanic cold sources (SC2, SC3) with a temperature of 2.5°C lower than the experiment medium. From Plate 4 we can see that two large-amplitude troughs are formed over the oceans. The Pacific trough invades the western part of the Rocky Mountains. Ahead of this trough is a strong south-southwest flow forcing the Rocky Mountain high to move eastward.

(4) Comparison of summer flow patterns between East Asia and North America. From the above experiment results we can see that the highs formed by the heating effect of topographies can be influenced by the surrounding flow field. When the westerlies are weak, or when a trough is far away from the mountains, these highs would remain over the mountains. Otherwise, they moved eastward. From this viewpoint, we can divide the summer flow field into two basic types over East Asia and

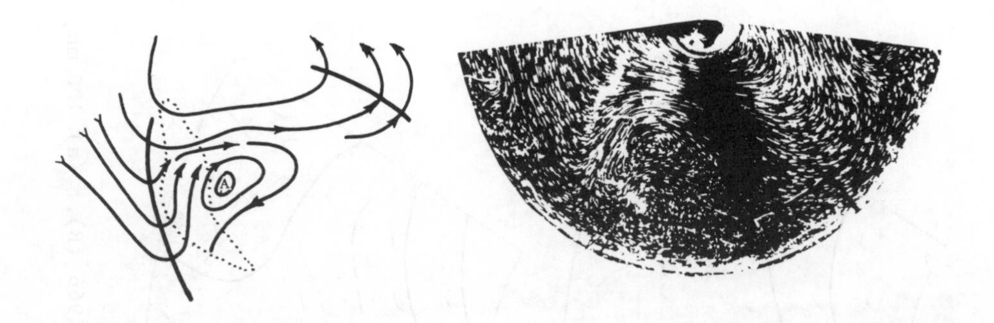

Plate 4 Influence of the Pacific cold source on the Rocky Mountain high. T = 31.2 sec. Temperature of cold water forming the cold sources is 2.5°C lower than the experiment medium. Heating intensity of SH2: 0.0097 watt/cm.

North America. One of them is the condition where the high center is located over the mountains, with weak interaction between the highs and the surrounding flow field (westerlies or troughs, etc.). Another type shows strong interaction forcing the highs to move eastward.

Over North America the circulation falls into the type of weak interaction. The mean circulation of August, 1967 is an example (Fig. 3A). From this figure, we can see that an anticyclone is over the Rocky Mountains at 100 mb. A weak trough is located over the central part of the Pacific Ocean far away from the Rocky Mountains. The maximum intensity of the westerlies is about 18 m/s at 200 mb. In the lower layer (850 mb), a warm low-pressure system is located over the Rocky Mountains. Comparing this situation with the simulated structure of the Rocky Mountain high (see Plates 2a, b), we can see that the two have similar features. In Fig. 3A it also has been shown that the subtropical high is located over the Atlantic Ocean.

In the case of strong interaction, the circulation is similar to the mean circulation of July, 1966 (Fig. 3B). A deep trough is located near the west part of the Rocky Mountains, while the intensity of westerlies to the northeast of the mountains is about 28 m/s at 200 mb, a larger speed than August, 1967. The North American continent is occupied by large-scale anticyclone systems in the whole troposphere over the eastern side of the Rocky Mountains.

Over East Asia, the flow pattern is similar to that over North America. In the period of active monsoon, the Qinghai-Xizang high is over the plateau. But in the period of break monsoon, a strong trough invades the western part of the plateau, and the Qinghai-Xizang high moves eastward. To compare these situations, Fig. 4 shows the basic flow pattern over East Asia in summer. We note the similarity with the North American pattern.

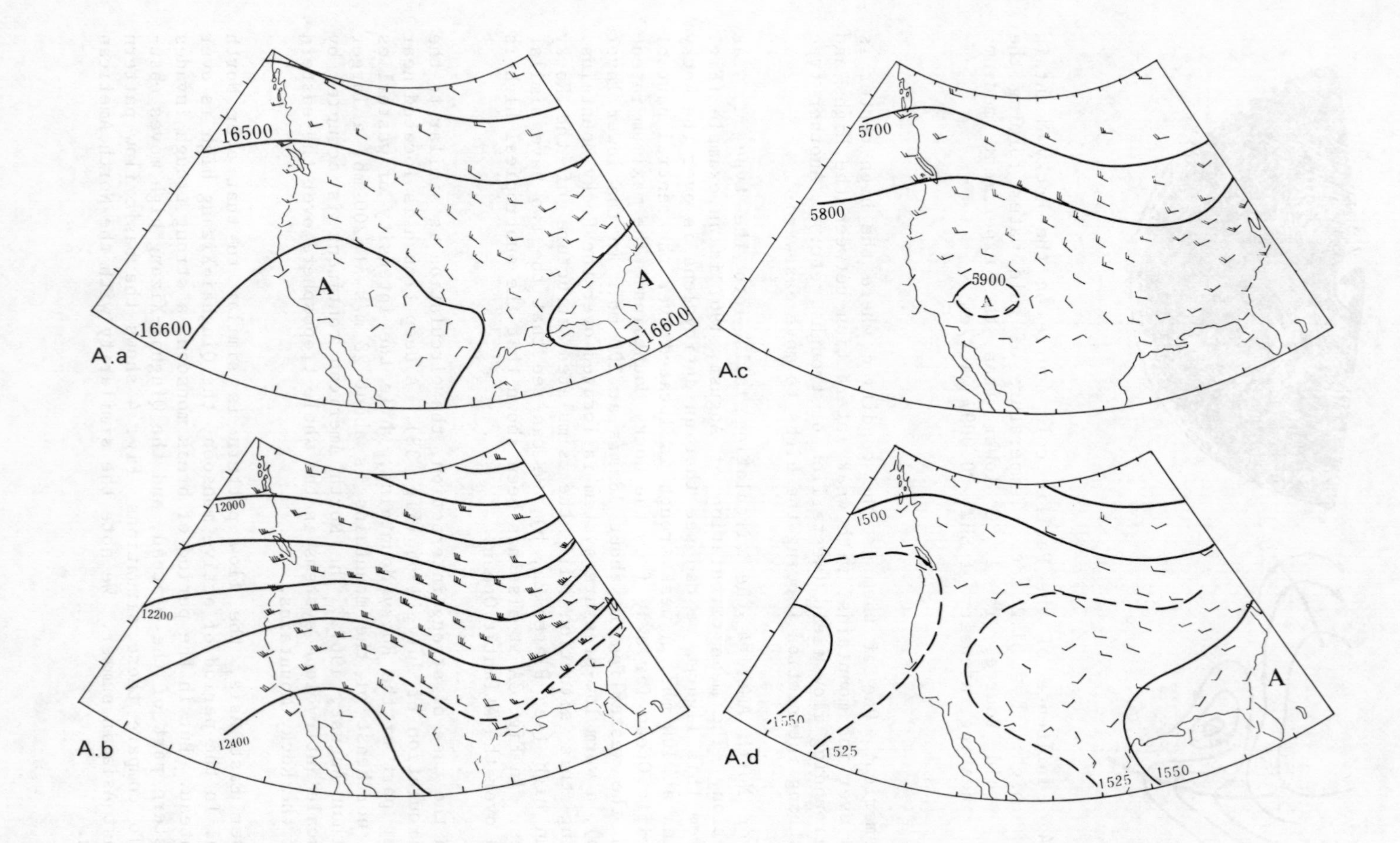

Fig. 3 Monthly mean wind and height contours in August 1977 (A), and July 1966 (B). (a) 100 mb, (b) 200 mb, (c) 500 mb, (d) 850 mb.

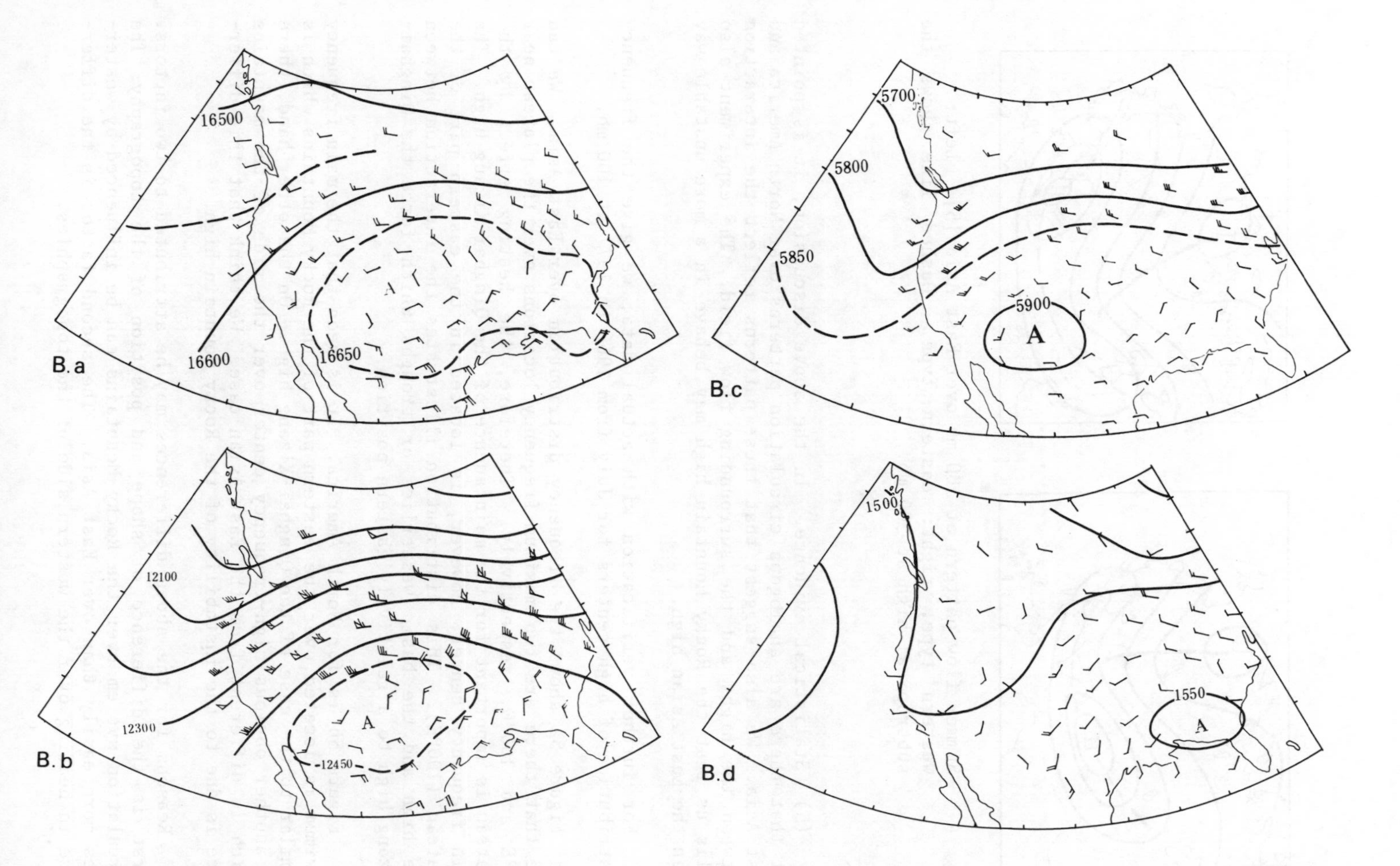

Fig. 3 (Continued)

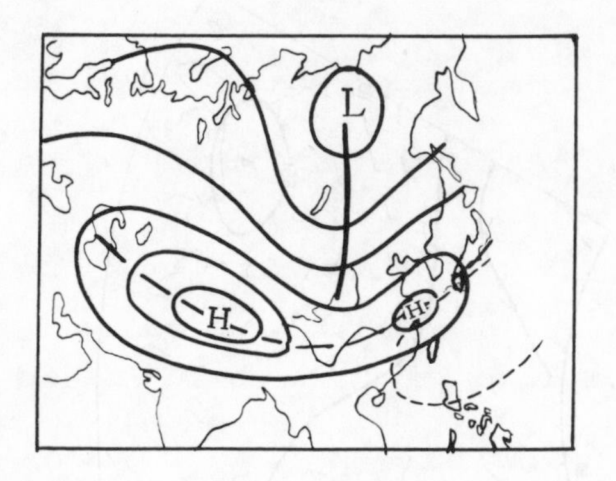

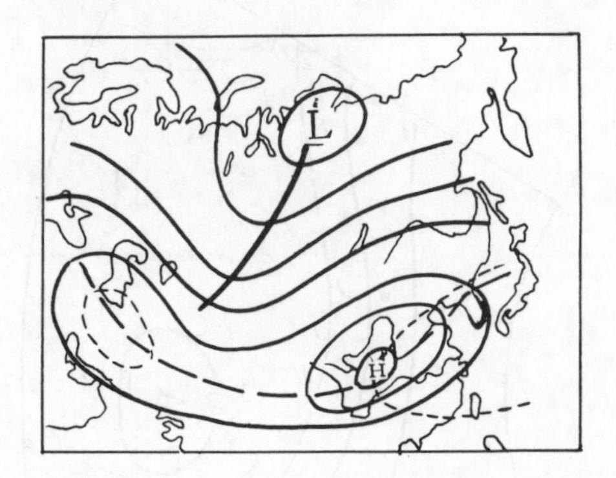

Fig. 4 Summer flow pattern at 100 mb over East Asia [5]. Left: western type, right: eastern type. Dashed line shows the subtropical high at 500 mb.

(5) Statistical evidence. In the above discussion, it is pointed out that there are analogous circulation patterns over North America and East Asia. We also suggest that these patterns reflect the interaction between the highs and the surrounding flow field. The experiment also tells us that the Rocky Mountain high may behave in a more unstable way than the East Asian high.

For further verification with actual data, we give the frequency distribution of high centers for July from 1966 to 1977 at 100 mb.

Figure 5a shows the frequency distribution over East Asia. We can see that there are two maximum frequency locations over the plateau near 80°E and 100°E, respectively. Therefore, the heating effect by the plateau is important for the maintanence of the Qinghai-Xizang high. The main frequency center, however, is located in the eastern part of the plateau (100°E). This distribution illustrates the interaction between the high and the basic westerlies or troughs which force the Qinghai-Xizang high to a relatively eastern position.

Figure 5b is for North America. It is seen that the main frequency maximum is located over the eastern part of the Rocky Mountains which is similar to the case of the Qinghai-Xizang high. On the other hand, there is another obvious high-frequency center over the side of the mountains which is different from the East Asian case. We think that this difference is due to the instability of the Rocky Mountain high.

Reasons for the above differences may be attributed to two factors. First is the difference of shape and position of the topography. The circulation system over the Rocky Mountains can be influenced by westerlies more easily than over East Asia. The second factor is the difference in heating over the western side of the topographies.

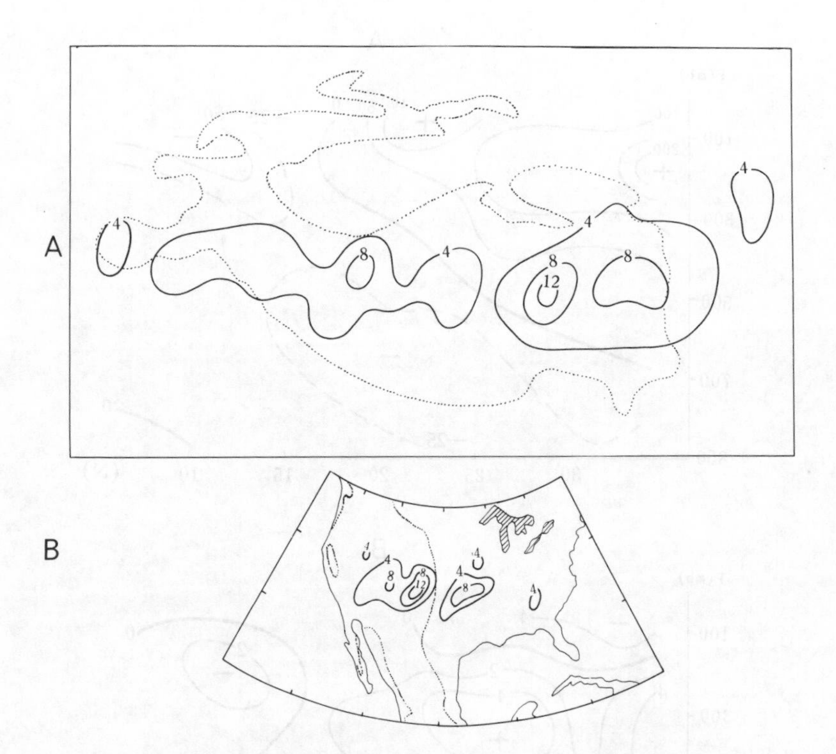

Fig. 5 Frequency distribution of high centers from 1966 to 1977 at 100 mb. Upper: East Asia, lower: North America. (The frequency has been calculated in each 2° square. The areas with values ≥ 4 are shown.)

3. Simulation of the Monsoon Circulation

In summer, the southwest monsoon and the upper troposphere anticyclone are very important circulation systems. The warming effect over the Eurasian continent, especially the heating effect of the Qinghai-Xizang Plateau, plays an important role. We think that an opposite change of the tropospheric temperature field at low latitudes over the Indian Ocean and South Asia may be another reason for their formation.

First, let us examine the observational facts. In July, the temperature in the area of the Indian Ocean and South Asia is about 4°C lower than the eastern part of North Africa and is 6°C lower than the northern part of India at 850 mb [3]. This low latitude cooling can be illustrated by a seasonal change of temperature. Figure 6 shows the difference of temperature and geopotential height from May to June along 80°E. Note that the temperature in the troposphere over the northern part of India increases by about 4°C. The geopotential height also increased. These increases are due to the heating effect of the Asian continent (including the plateau). At the same time, the temperature in the troposphere over

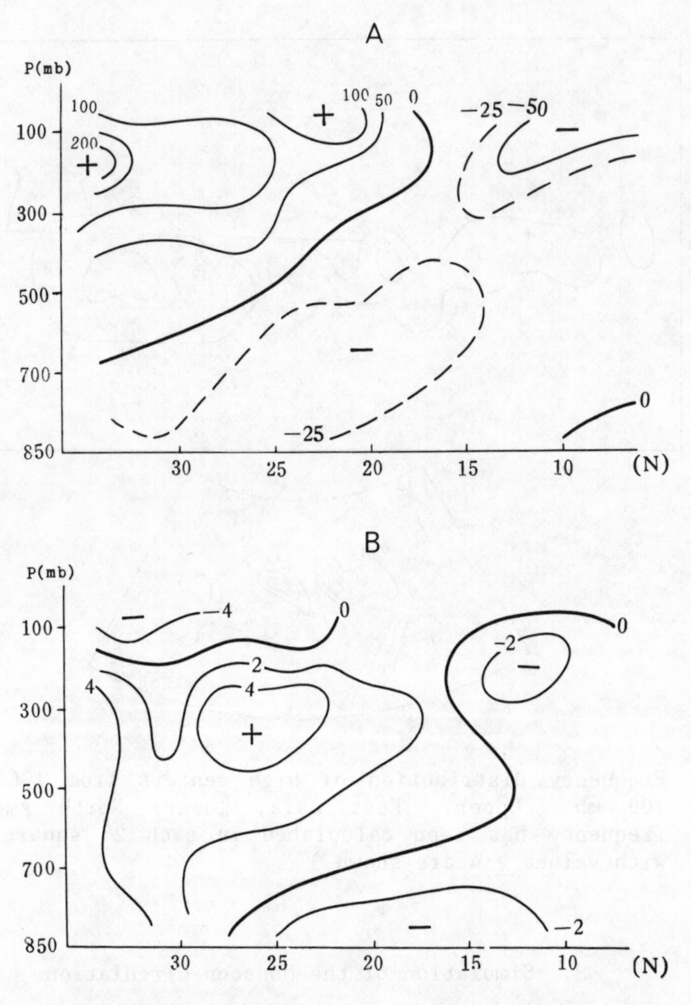

Fig. 6 The difference of temperature and geopotential height from May to June along 80°E. (a) Geopotential height (gpm), (b) temperature (°C).

the southern tip of the Indian subcontinent (south of 10°N) drops by about 2°C, which is equal to half of the temperature increase in the northern part of India.

To examine the role of the temperature field at low latitudes, we conducted the experiment with only a cold source put over the tropical area. When the basic flow field reaches a quasi-stationary state, we decrease the temperature in the area of the water bath (SC4), as shown in Fig. 1. If the temperature of the water bath is 2°C lower than the working medium, the resulting flow field is as shown in Plates 5a and b.

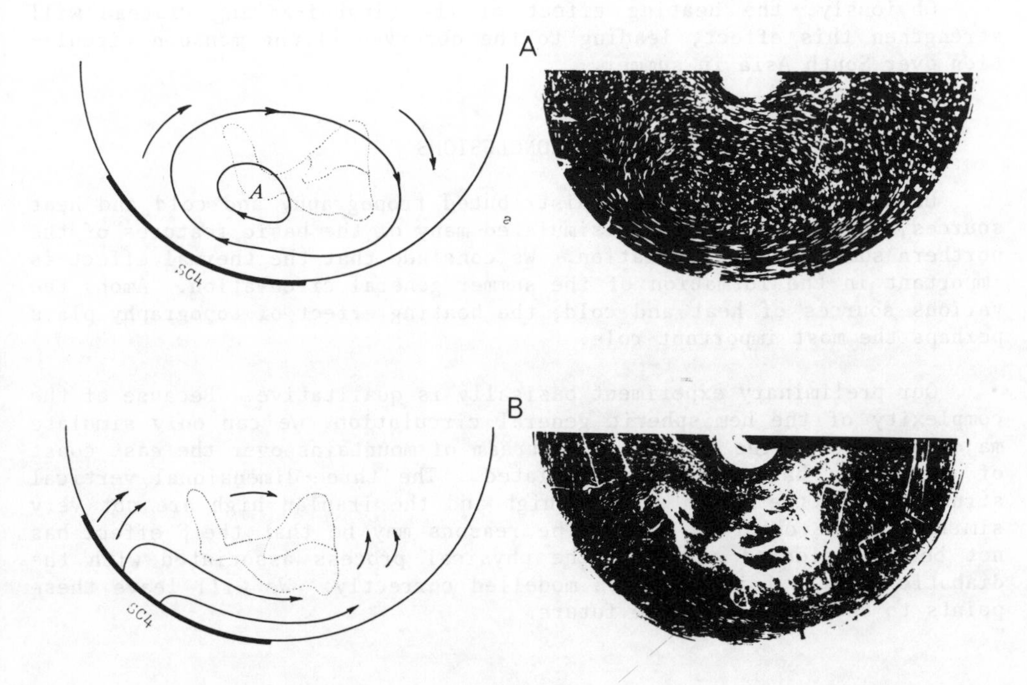

Plates 5a,b The horizontal circulation induced by the cooling effect over the Indian Ocean and South Asia. Plate 5a: upper layer, Plate 5b: lower layer. T = 31.2 sec.

Plate 5a is the upper-layer circulation. Note the large-scale anticyclone over the plateau. To the south of the anticyclone are the easterlies. The speed of the flow reaches about 0.12 cm/s, corresponding to 14 m/s in the real atmosphere.

Plate 5b shows the lower-layer circulation. We can see that a broad-scale southwest flow exists over the Indian Ocean and South Asia. The intensity of the flow near the boundary is 0.04 cm/s, corresponding to 5 m/s in the real atmosphere. It also can be seen that near the western part of the cold water bath, the direction of flow is almost perpendicular to the wall of the outer cylinder which is similar to a cross-equatorial flow.

The above experiment shows that the local cooling effect over the area at low latitudes over the Indian Ocean may be one of the main factors for the formation of the southwest monsoon and the upper troposphere anticyclone. The meridional monsoon circulation seems to be produced by a cooling effect in the troposphere at low latitudes, resulting in the horizontal flow pattern described above. Therefore, the southwest monsoon and the upper tropospheric anticyclone are interrelated.

Obviously, the heating effect of the Qinghai-Xizang Plateau will strengthen this effect, leading to the observed strong monsoon circulation over South Asia in summer.

IV. CONCLUSIONS

Using a flow model with distributed tropography and cold and heat sources, we have successfully simulated many of the basic features of the northern summer mean circulation. We conclude that the thermal effect is important in the formation of the summer general circulation. Among the various sources of heat and cold, the heating effect of topography plays perhaps the most important role.

Our preliminary experiment basically is qualitative. Because of the complexity of the hemispheric general circulation, we can only simulate major features. The troughs downstream of mountains over the east coast of continents have not been simulated. The three-dimensional vertical structures of the North African high and the Iranian high are not very similar to the observed ones. The reasons may be that the β effect has not been considered and that the physical process associated with the diabatic effects have not been modelled correctly. We will leave these points to be improved in the future.

REFERENCES

1. Budyko, M.I.: Climate and life. International Geophysics Series, 18, 140-260 (1974).

2. Chang, Ke-su, et al.: The annulus simulation of movement of the Tsinghai-Tibetan high and its application to the forecast of summer flow patterns of the high troposphere. Scientia Sinica, 20 (5), 632-650 (1977).

3. Van de Boogaard, Henry: The mean circulation of the tropical and subtropical atmosphere - July. NCAR/TN-118+STR, National Center for Atmospheric Research, Boulder, Colorado (1977).

4. Kubota, I.: Seasonal variation of energy sources in the earth's surface layer and in the atmosphere over the northern hemisphere. J. Met. Soc. Japan, 48 (1), 30-46 (1970).

5. Lou, Si-wei: Qinghai-Xizang high in summer at 100 mb. Meteorology of Qinghai-Xizang Plateau, edited by T.C. Yeh, Y.X. Gao, et al., Chap. 10. Science Press, Beijing (in Chinese), 127-139 (1979).

6. Research Group on Experimental Simulation: An experimental simulation of the three-dimensional structure of airflow fields over the Qinghai-Xizang Plateau in summer. Scientia Atmospherica Sinica, 4, 247-255 (1977).

7. Yeh, Tu-cheng, Chang Chieh-chien, et al.: A preliminary experimental simulation of the heating effect of the Tibetan Plateau in the general circulation over eastern Asia in summer. Scientia Sinica, 17 (3), 397-420 (1974).

8. Zhou, Ming-yu, et al.: The experimental simulation of the western Pacific subtropical high in summer. Scientia Atmospherica Sinica, 4 (1), 12-20 (1980).

SESSION II

PLANETARY AND LARGE-SCALE EFFECTS OF MOUNTAINS ON GENERAL CIRCULATION AND CLIMATE: THEORETICAL ASPECTS

2.1 THE EFFECTS OF MOUNTAINS ON THE ATMOSPHERIC GENERAL CIRCULATION AND CLIMATE

W. Lawrence Gates
Department of Atmospheric Sciences
and
Climatic Research Institute
Oregon State University
Corvallis, Oregon 97331 USA

ABSTRACT

The physical basis of the effects of large-scale mountains on the atmospheric general circulation is seen to consist of closely coupled dynamical and thermal processes. The more or less separate treatment of these effects in theoretical studies and simplified models is briefly reviewed in terms of the resolution, smoothing, and properties of various coordinate systems and conservation principles. The basic dynamical result of orography's barrier to the flow is the formation of a planetary-scale lee wave, while the basic thermal effect is the provision of an increased sensible and latent heat source over the elevated terrain.

A combined treatment of orography's dynamical and thermal effects is afforded by atmospheric general circulation models, and the results of previous simulations with the NCAR and GFDL models both with and without mountains are briefly reviewed. The major effects of the mountains in winter is an intensification and downstream displacement of the quasi-stationary long-wave troughs over the major continents, and the enhancement of southerly flow and precipitation southeast of large-scale mountains in summer. From a phenomenological viewpoint the mountains in the northern hemisphere serve to localize cyclogenesis to their lee and to focus the subsequent storm tracks northeastward.

The results of new GCM simulations made with the Oregon State University (OSU) model with and without mountains are reported for both winter and summer. These integrations (which have been carried to statistical equilibrium) generally confirm earlier results, and clearly show the role of the mountains in the maintenance of both midlatitude and tropical circulations and the associated climatic distributions of precipitation. In a special examination of orographic effects in East Asia, the Tibetan Plateau is found to have a profound effect on the regional summer climate through its influence on the southeast Asian monsoon.

The outstanding problems in the study of orography's effects on the general circulation and climate which require further research are seen to be the improved resolution and/or parameterization of subgrid-scale processes over mountainous terrain, the development of improved techniques for the estimation of the statistical significance of simulated climate changes, and the application of models of the coupled atmosphere-ocean system.

I. INTRODUCTION

It is commonly accepted by meteorologists and climatologists that mountains significantly affect the weather in many parts of the world, and that the large-scale distribution of orography has important effects on the atmospheric general circulation and hence on the regional and global climate. When searched for details, however, this concensus contains surprisingly little dynamical information, and the basic question of the relative importance of orographic and thermal effects has not yet been answered satisfactorily. There is, therefore, a clear need for further theoretical and numerical modelling research on this question, as well as new observational and diagnostic studies.

After noting the general physical effects of mountains on the large-scale atmospheric circulation, I shall briefly comment on representational aspects of the problem related to resolution, coordinates and the parameterization of subgrid-scale effects, and then briefly review the treatment of the dynamical and thermal effects of large-scale mountains in atmospheric models. On the basis of this background, a review of the sensitivity of the climate to orography in the various numerical simulations which have been performed with atmospheric general circulation models will be given. The preliminary results from a new set of simulations recently completed at Oregon State University (OSU) will then be presented, in which the global climatic effects of large-scale orography will be systematically examined for both the winter and summer seasons. Finally, I shall comment on a number of outstanding problems in this area and indicate the directions of needed future research.

1. General Physical Effects: Dynamical vs. Thermal

Because of the kinematic boundary condition's requirement that the air flow tangentially to the earth's surface, a mountain presents an impenetrable barrier to the wind. Depending upon the atmosphere's structure and motion, as well as on the size, shape and location of the mountains, the air is therefore forced to go over and/or around the mountain. Because of the nonlinearity of the atmosphere's dynamics, however, this alteration of the flow occurs over a wide spectrum of spatial scales, and is seen in both transient and quasi-stationary components. The dynamical effect of a mountain barrier also depends upon its orientation relative to the incident airflow, and in this respect the effects of predominantly meridional mountain chains (such as the Rockies and the Andes) may be expected to be different from those of more block-like mountains (such as the Himalayas and the Antarctic massif). For this same reason a mountain's dynamical effect may be different in different seasons.

A mountain barrier may also change the way in which the atmosphere is heated in several ways. First, the air over the mountain may be heated by the upward flux of sensible and/or latent heat from the elevated surface of the mountain at a different rate than that which occurs at the same elevation in the absence of the mountain. Depending upon the season and the surface heat and water budgets, a mountain may therefore serve as an elevated heat source for the atmosphere. Such heating may in turn influence the circulation over and near the mountain, and result in a local reinforcement of dynamical and thermal effects. A similar interaction of the orographically induced heating and the airflow may also occur on larger scales of motion in which the atmospheric heating rate is critically affected by the large-scale flow. For example, an orographically-induced Rossby or planetary wave in the lee of a mountain barrier may be partially maintained by the sensible and latent heating which may occur preferentially in association with motions in the trough. This close (and essentially nonlinear) coupling between the heating and the motion is characteristic of the atmosphere, and is responsible for the difficulty which has been experienced in attempts to separate the dynamical and thermal effects caused by mountains.

The long-term effects of orography may be viewed in terms of the quasi-stationary or steady response of the atmosphere, as portrayed either in the solutions of appropriate models or in the averages of the observed data as shown in Figs. 1 and 2. These effects, however, are in reality the statistical average of a series of transient synoptic states. While some of this variation is accounted for by an essentially random sequence of weather events, the fixed geography of the mountains gives rise to several large-scale phenomena which are important determinants of the overall orographic effect. Chief among these are the lee trough noted earlier and the associated occurrence of cyclogenesis in the lee of a large-scale mountain barrier, and the characteristic occurrence of the summer monsoon in southeast Asia. Other events such as tropospheric blocking and the occurrence of sudden stratospheric warming have also been related to the effects of mountains in the planetary circulation.

2. Representational Problems

Mountains present the most complex topography on the earth's surface, with many local variations of elevation which cannot be resolved in a large-scale model. The mountain heights are therefore usually averaged in some way to represent the mean elevation on the scale of the resolution of the model. This results in a general smoothing of the terrain height and the suppression of the higher mountain peaks. Even if this area averaging is done from a high-resolution tabulation of the local elevation, the smoothed mountains as seen by models of different horizontal (and vertical) grids will in general not be the same in either height or extent. Systematic tests of this resolution effect have not yet been carried out, although from recent work of Arakawa and Lamb [1] it appears to be especially important in the case of relatively steep mountain slopes. As noted below, there are also difficulties in the representation of the pressure gradient force near steep terrain in the vertical coordinate systems commonly in use.

Aside from the resolution of the mountains themselves, there are a variety of physical processes caused by the mountains but which occur on scales too small to be resolved. These include the flux of energy by gravity waves which may result in localized mountain waves, the turbulent fluxes of momentum and energy caused by irregular topography, the systems of valley and mountain breezes resulting from the diurnal differential heating of the mountain surfaces, and the small-scale convection, cloudiness and precipitation over mountains which may result from localized vertical fluxes of sensible (and latent) heat. The parameterization of these effects in terms of the large-scale flow is generally inadequate,

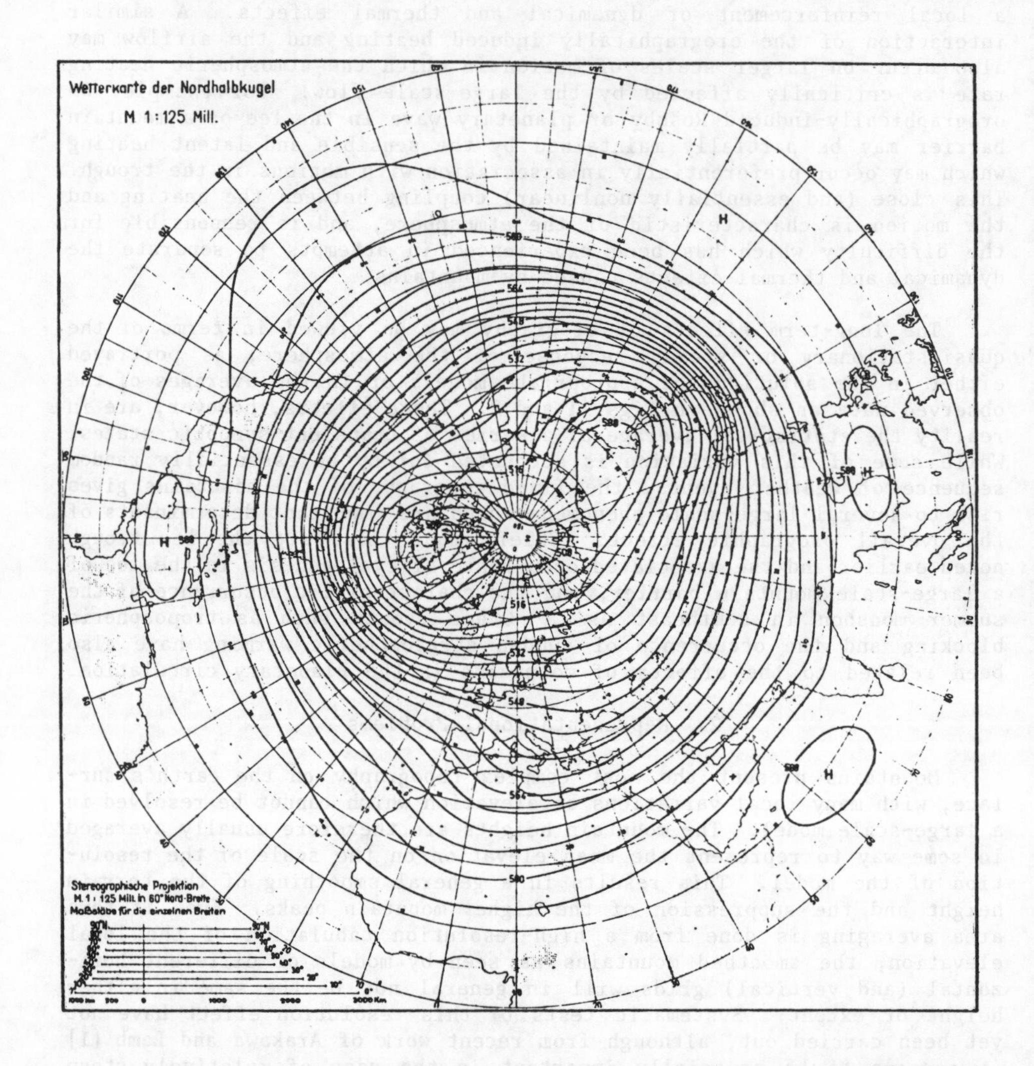

Fig. 1 The observed mean distribution of 500-mb geopotential height (dm) in the northern hemisphere for January. (From [16].)

and may account for at least a portion of the difficulties which models experience in portraying mountain effects. Other representational problems are the analysis of observational data in the vicinity of mountains and the assembly (or initialization) of data for use in numerical integrations; while these latter problems are of primary concern in weather analysis and prediction, they are relatively unimportant in studies of the general circulation and climate.

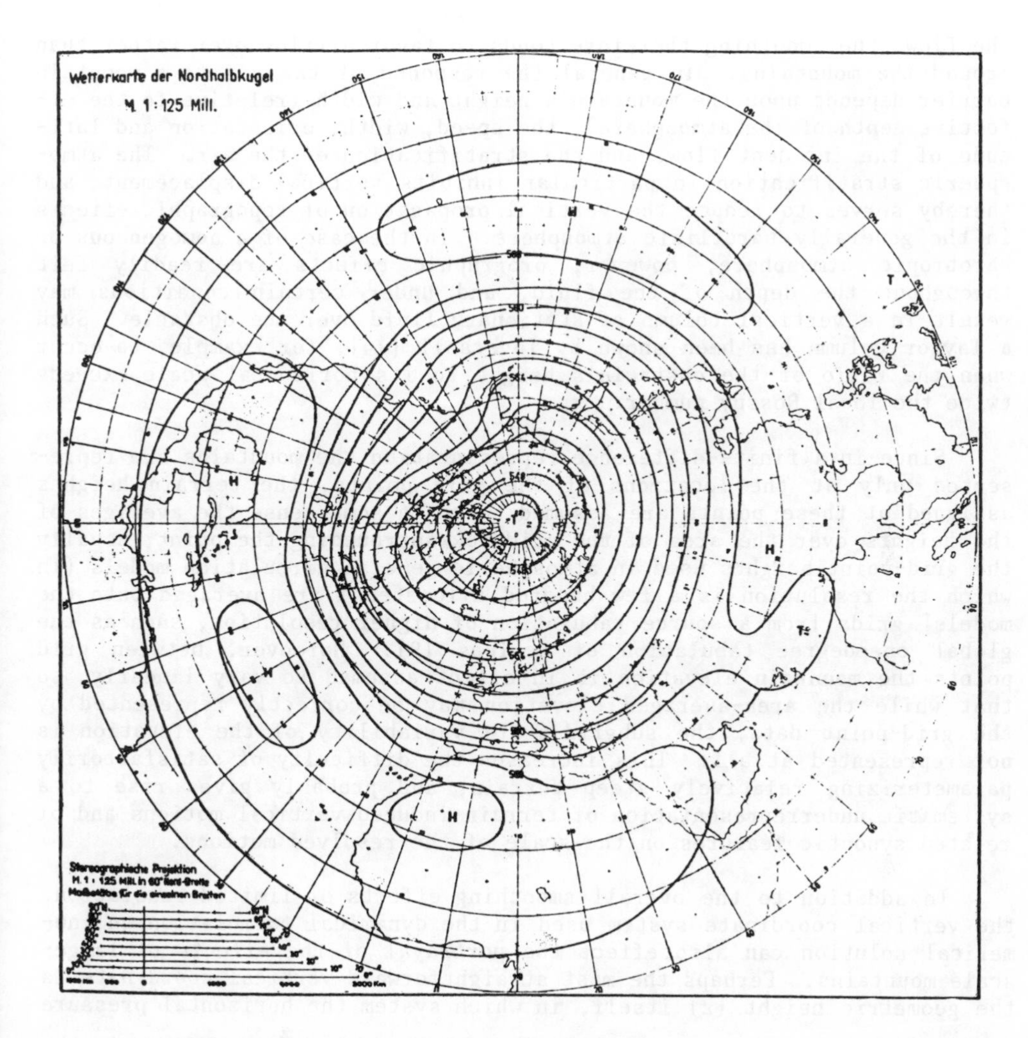

Fig. 2 Same as Fig. 1, except for July.

II. TREATMENT OF THE EFFECTS OF LARGE-SCALE MOUNTAINS

Although it is difficult to separate the dynamical and thermal effects of mountains on the large-scale behavior of the atmosphere, this distinction is a useful guide to the various studies of mountain effects which have been made with simplified models.

1. Basic Dynamical Effect: Flow Barrier

(1) Resolution, smoothing and coordinates. As noted earlier, the basic result of a mountain's presence is the fact that the air cannot flow through it. When averaged onto a model's grid, the smoothed mountain is lower than the actual mountain, and presents a solid barrier to

the flow; the smoothing therefore tends to favor airflow over rather than around the mountains. In general the response of the air to a mountain barrier depends upon the mountain's height and width (relative to the effective depth of the atmosphere), the speed, width, orientation and latitude of the incident flow, and the stratification of the air. The atmospheric stratification in particular inhibits vertical displacement, and thereby serves to reduce the vertical propagation of topographic effects in the generally baroclinic atmosphere. In the case of a homogeneous or barotropic atmosphere, however, orographic effects are readily felt throughout the depth of the fluid, and under certain conditions may result in a vertical column of stationary fluid over the obstacle. Such a Taylor column has been shown by Ingersoll [14], for example, to occur when the ratio of the obstacle's height to its horizontal scale exceeds twice the local Rossby number.

Since in a finite-difference representation the mountains are represented only at the locations of the grid points, the terrain heights assigned at these points are assumed to be in some sense the averages of the heights over the area of the grid box surrounding the point; usually the grid-point heights used in atmospheric general circulation models (in which the resolution is a few hundred kilometers) are averaged onto the models' grids from a source tabulation of higher resolution, such as the global one-degree tabulation of Scripps [9]. Moreover, between grid points the mountain elevation is in effect assumed to vary linearly, so that while the area-averaged elevation may be correctly represented by the grid-point data, the subgrid-scale variability of the elevation is not represented at all. This increases the difficulty of satisfactorily parameterizing relatively steep terrain, and probably gives rise to a systematic underrepresentation of terrain-induced vertical motions and of related synoptic features on the scale of the resolved motions.

In addition to the overall smoothing effects of limited resolution, the vertical coordinate system used in the dynamical formulation and numerical solution can also effect the portrayal of the effects of large-scale mountains. Perhaps the most straightforward vertical coordinate is the geometric height (z) itself, in which system the horizontal pressure gradient force in the equations of motion is simply $\alpha\vec{\nabla}_H p$, where α is the specific volume, p is the pressure, and $\vec{\nabla}_H$ is the horizontal gradient operator. In this system the volume occupied by mountains is removed from the computational domain, and the pressure force is calculated as usual on the surrounding level coordinate surfaces. When applied in general circulation models, as by Kasahara and Washington [19] for example, such a "blocking" method requires lateral boundary conditions on the vertical surfaces which now represent the sides of the mountains on the grid, or the use of the surface boundary condition $w = \vec{v}_H \cdot \vec{\nabla} h$ between grid points, where $w = \dot{z}$ is the vertical velocity, $\vec{v}_H$ the horizontal velocity and h the local surface elevation. The variability of the density as a coefficient of the pressure force term in this system may contribute to numerical inaccuracy in a finite-difference calculation unless special precautions are taken. A similar blocking technique may be employed when pressure is used as a vertical coordinate, in which case the pressure force becomes $\vec{\nabla}_p\phi$, where ϕ is the geopotential. Now, however, the surface of the earth is no longer at a fixed coordinate level (as it is in geometric or Cartesian coordinates) since the surface pressure is generally variable; special procedures are therefore required in the vicinity of mountains in this system also.

This difficulty of isobaric coordinates is alleviated to some extent by the use of a dimensionless coordinate (σ) in which the pressure is divided by the local surface pressure, i.e., $\sigma = p/p_s$, as first introduced by Phillips [25]. In such a sigma-coordinate system the surface of the earth is always at $\sigma = 1$ with the surface boundary condition $\dot{\sigma} = 0$, while the mass continuity equation becomes a prognostic tendency equation for the surface pressure p_s. In this system the pressure gradient force, however, becomes $\alpha\vec{\nabla}_\sigma p + \vec{\nabla}_\sigma\phi$, rather than the single term expressions as in the Cartesian and isobaric systems. In the vicinity of steep mountains the pressure force is characteristically a small difference between these terms which are individually large and of opposite signs; this may lead to considerable error even when special differencing procedures are used. Other errors may be introduced if the dependent variables on the sigma surfaces are found by interpolation from isobaric or geometric surfaces. Such errors can be reduced by the use of a reference atmosphere and enforcement of the conservation of potential temperature and its square in the vertical, or the use of other special procedures.

We may also use the potential temperature (θ) itself as a vertical coordinate, in which case the pressure gradient force becomes $\vec{\nabla}_\theta(c_pT + \phi)$ where T is the temperature. Although adiabatic motion is here represented by $\dot{\theta} = 0$, this system has not been widely used in global general circulation modelling since near the earth's surface (even in the absence of mountains) the diabatic heating is large and the θ-surfaces may intersect the ground at a steep angle or become folded.

(2) Conservation principles. In addition to the problem of accurately representing the pressure gradient force in the presence of mountains, other computational problems may arise in the treatment of the advection terms in a model when the dynamical effect of mountains is included, as reviewed by Kasahara [15]. Unless appropriate conservation principles are observed in the numerical solution, the large-scale effects of mountains may be misrepresented even with relatively high resolution. In the shallow-water equations, for example, the potential vorticity $\eta(H-h)^{-1}$ should be conserved, where η is the absolute vorticity, H is the height of the (free) fluid surface and h is the height of the surface (mountain) over which the fluid is flowing. As discussed below, this conservation principle profoundly affects the behavior of both transient and stationary flow patterns, and is not automatically met in many finite-difference schemes. Arakawa has proposed a differencing scheme for the shallow-water equations based upon the conservation of potential enstrophy $\eta^2(H-h)^{-1}$, and has subsequently shown that its use can give a substantial improvement in simulation accuracy over steep mountains, even with relatively coarse horizontal resolution [1].

(3) Large-scale flow alteration. In spite of the limitations noted above with respect to resolution, coordinates and conservation principles, the basic large-scale dynamical effect of mountains is an alteration of the synoptic-scale flow pattern. In contrast to the small-scale gravity waves which may occur in the lee of mountains under certain conditions of vertical stability and wind shear, the effect of mountains on the large- or planetary-scale motion in middle latitudes is basically governed by vorticity considerations. The strong tendency of the air to conserve its absolute potential vorticity (or enstrophy) favors the formation of a Rossby wave trough in the lee of a mountain barrier for the case of strong westerly flow in the midtroposphere, while such an effect is absent in the case of an easterly current.

Studies of the effects of idealized barriers to the flow in channels show that the planetary vorticity gradient (or β-effect) may result in a downstream series of stationary waves, depending upon the relative depth and horizontal scale of the flow as well as on its speed and stratification. For relatively wide channels and little stratification, the mountains serve to excite the stationary planetary Rossby wave whose length is given by $2\pi(U/\beta)^{\frac{1}{2}}$ where U is the zonal (westerly) speed and $\beta = \partial f/\partial y$ is the meridional gradient of the Coriolis parameter. The dynamics of such quasi-stationary forced planetary waves also involve the effects of friction in the planetary boundary layer, as first shown by Charney and Eliassen [4]. The earth's spherical shape has also recently been shown by Grose and Hoskins [12] to be important in the generation of wave trains on a planetary scale.

Although a quasi-stationary state is seldom approached in the atmosphere, these considerations would indicate that the orographic effect is generally stronger in winter than in summer due to the stronger wintertime westerly flow. In most cases there appears to be an interaction between the forced planetary-scale waves and the transient free waves, with lee cyclogenesis being the most commonly observed result downstream of the Rockies, Alps and southern Andes; resonant interaction between forced and free waves has also been suggested as a cause of the phenomenon of "blocking". In the case of the Tibetan Plateau, lee cyclogenesis is most common in winter when the westerly jet is south of the mountains; the movement of the jet to the north of the plateau is closely related to the onset of the Mei-yü or East Asian summer monsoon rainfall regime as described by Kasahara [16].

The excitation of planetary waves by large-scale mountains is also closely related to the vertical propagation of energy into the stratosphere, and thereby possibly responsible for the phenomenon of "sudden stratospheric warming". Depending upon the vertical distribution of the zonal wind speed and temperature in the stratosphere, the wave energy of the longer tropospheric planetary waves may propagate upward and produce wave amplification in the stratosphere and thus increase the heat transport, which in turn results in anomalous warming in the polar latitudes.

2. Thermal Effect: Elevated Heat Source

(1) Sensible heating. In addition to the purely dynamical or barrier effect, large-scale mountains also exert a strong thermal effect on the circulation by virtue of their role as an elevated source of heat. It is well-known that the warming (cooling) of the atmosphere by the upward (downward) flux of sensible heat from the surface is primarily responsible for the maintenance of relatively low (high) sea-level pressure over the continents in summer (winter). As suggested by Bolin [3] and Staff Members, Academia Sinica [29, 30], and as shown by Smagorinsky [28] using a simplified atmospheric model, the presence of mountains tends to increase this monsoonal effect by virtue of the fact that the surface heating or cooling now takes place at a higher elevation. In fact, it has been shown by Stern and Malkus [31] that the effect of surface heating on the large-scale airflow is in many ways analogous to that of an "equivalent mountain". Therefore, when a mountain is also present the thermal effect augments that of the barrier itself, and may result in the eastward displacement of the surface low pressure. Kuo and Qian [20] have shown that diurnal heating over the Tibetan Plateau has an analogous effect by maintaining an average upward vertical motion and lower surface pressure as a result of the mountain-valley breeze circulation induced by the surface heating.

(2) Latent heating. Even more important than the direct heating over mountains, however, is the latent heating which may result from either the surface sensible heating itself or from the orographically-induced air flow. In the former case the latent heating tends to deepen the layer of convection over the mountains and to increase the resulting precipitation. In the case of an orographically-induced planetary wave or lee cyclogenesis, the production of cyclonic vorticity to the east of the trough tends to produce large-scale vertical motion, which may in turn result in synoptic-scale condensation. Such latent heating may be particularly intense when the orographically-induced trough is near the eastern coast of a continent. This interaction between the dynamical and thermal effects of mountains makes it difficult to separate their roles; most modelling studies in which both are included, however, have concluded that their effects on the large-scale flow are generally comparable (see, for example, [5], [21], [13], [2]).

III. SENSITIVITY OF GCM-SIMULATED CLIMATE TO MOUNTAINS

1. Preliminary Estimates

The first systematic estimates of the dynamical and thermal effects of large-scale mountains on the atmospheric general circulation were made in the 1950's with simplified models, some of which have already been cited. The barotropic model was first applied to this problem by Charney and Eliassen [4], who showed that the behavior of both the transient free waves and the quasi-stationary forced waves of the midtroposphere could be successfully accounted for with the one-dimensional quasi-geostrophic barotropic vorticity equation. By including mountains (via their effect on the near-surface vertical motion) they showed that the otherwise rapid retrogression of the longest planetary waves could be effectively reduced, and by also including the effect of surface friction (as given by the near-surface geostrophic vorticity) they were able to reproduce satisfactorily both the phase and amplitude of the stationary long waves in middle latitudes. This approach was extended by Bolin [3], who showed that further improvement in the results could be obtained by taking into account the north-south extent of both the westerlies and of the mountain barriers over which they flow. Further refinements in the analysis of orography's effects on the general circulation in simplified models were made by Gambo [6] and Murakami [24], while Smagorinsky [28] was the first to comprehensively examine the corresponding thermal effects using a linearized quasi-geostrophic model.

2. Previous GCM Integrations

In spite of their limitations, the successes of these earlier studies served to underscore the importance of orographic effects on the general circulation, and have served as a basis for subsequent studies using general circulation models (GCMs). Such models provide the opportunity to examine the dynamical and thermal effects of orography at the same time, and allow consideration of transient and nonlinear effects over the global domain.

While most GCM simulations of orographic effects have been made at the National Center for Atmospheric Research (NCAR) or at the Geophysical Fluid Dynamics Laboratory (GFDL), an early study was made by Mintz [23] using a two-level global model. In an extended integration for northern hemisphere wintertime conditions (but without hydrology), he speculated that the Tibetan Plateau helped to maintain the Siberian high in winter, but that the inclusion of orography had little effect on the total atmospheric kinetic energy. Although Mintz did not make an extensive analysis of the mountains' effects on the simulated climate, his general results have been supported by later simulations using the NCAR two- and six-level GCMs [18], [19] and the GFDL nine- and eleven-level GCMs [21], [13], [22].

(1) Circulation and energetic effects. In their first experiments both with and without orography using the NCAR two-level GCM, Kasahara and Washington [18] found that the inclusion of mountains made an overall improvement in the simulation of the global distribution of January sea-level pressure as shown in Fig. 3, but made little difference in the circulation at midtroposphere. It has subsequently been recognized, however, that their 15-day integration was not long enough to permit a statistical equilibrium to be established, and that the model had not been satisfactorily calibrated against observation. In subsequent simulations of January climate with the NCAR six-layer GCM, Kasahara and Washington [19] found that the inclusion of large-scale mountains had little effect on either the distribution of sea-level pressure or precipitation, and that the dynamical effect of orography in general appeared to be less than its thermal effect. When this model was extended to include the stratosphere, Kasahara et al. [17] found that the inclusion of orography enhanced the amplitude of the planetary-scale waves in the upper atmosphere as expected from linear theory.

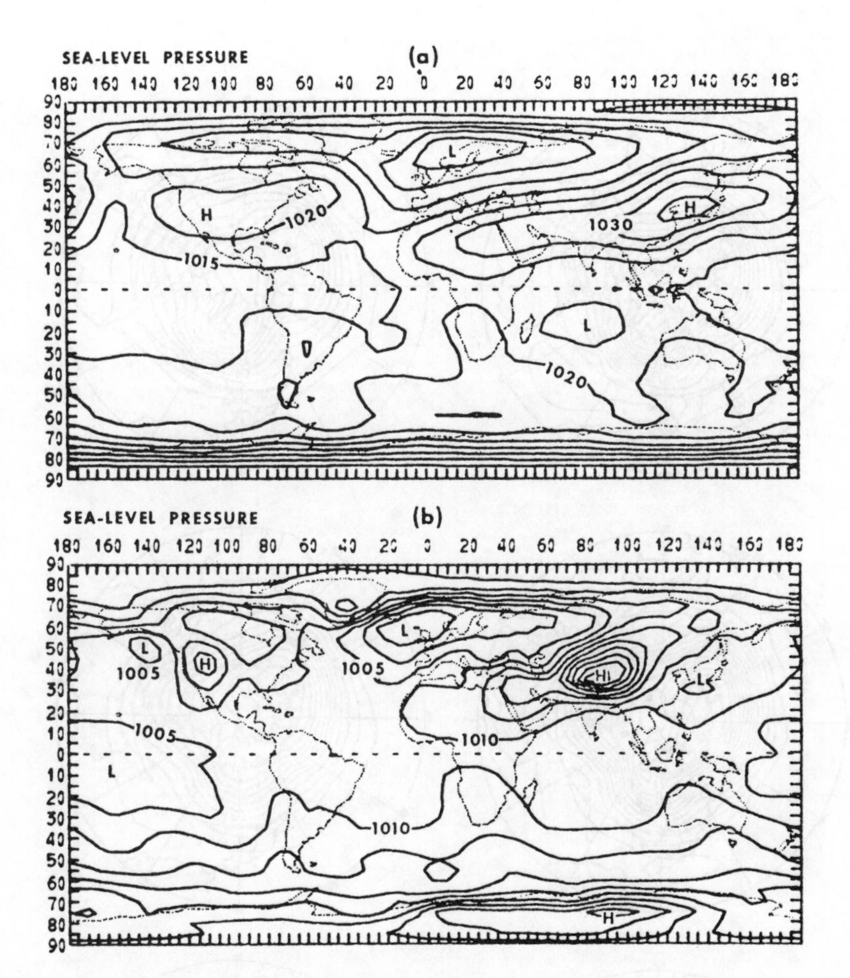

Fig. 3 The mean distribution of January sea-level pressure (mb) simulated by the NCAR two-level GCM with mountains (above) and without mountains (below). (From [18].)

Using the GFDL nine-layer GCM in experiments both with and without mountains, Manabe and Terpstra [21] found that the inclusion of orography tended to increase the amplitude of the quasi-stationary long waves, as shown in Fig. 4, but otherwise had little effect on the large-scale pressure distribution as shown in Fig. 5. Although their model omitted diurnal and seasonal variations (and used prescribed distributions of sea-surface temperature, cloudiness and humidity), Manabe and Terpstra found that the meridional transport of heat and momentum by the stationary or standing eddies was increased relative to that by the transient eddies upon the inclusion of mountains in the model, especially in the northern hemisphere. They concluded that the January Siberian high is maintained primarily by the Tibetan Plateau (in agreement with Mintz [23]), while the January Aleutian low is maintained primarily by thermal

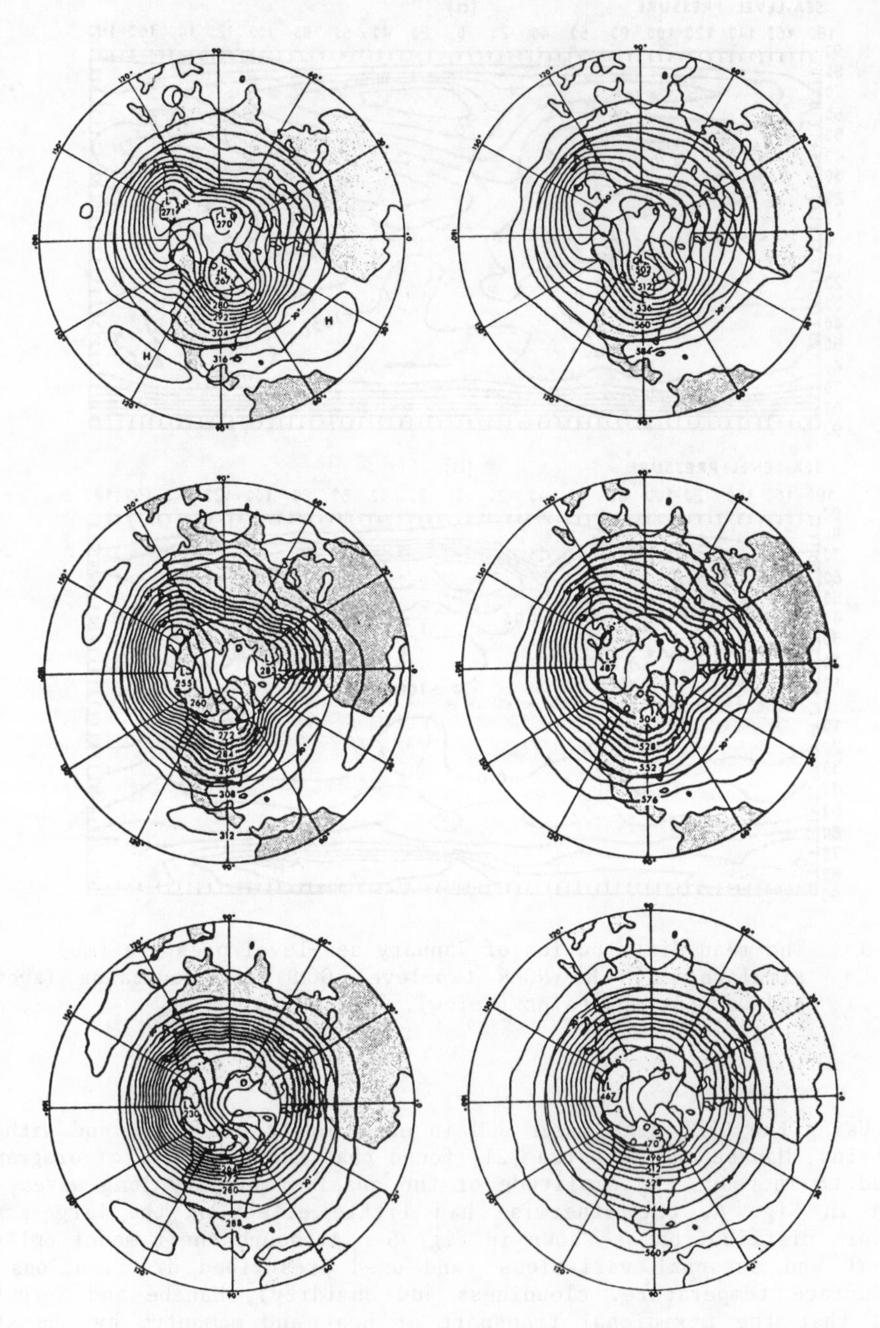

Fig. 4 The mean distribution of January geopotential height (dm) at 700 mb (left column) and 500 mb (right column) as observed (top), and as simulated by the GFDL 9-level GCM with mountains (middle) and without mountains (bottom). (From [21].)

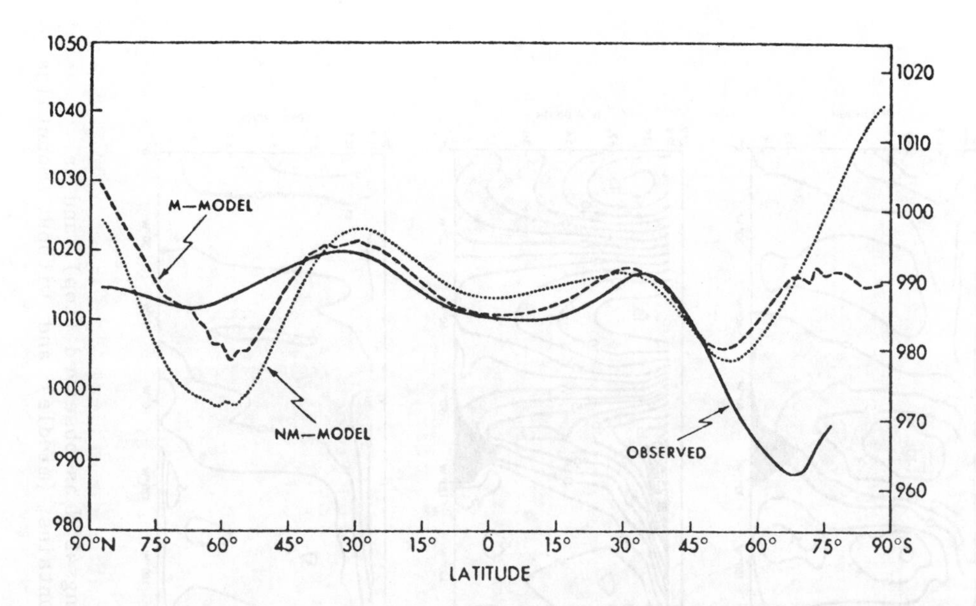

Fig. 5 The distribution of zonally-averaged January sea-level pressure (mb) as observed [23] and as simulated by the GFDL 9-level GCM both with mountains (M) and with no mountains (NM). The left-hand scale is for the observed and mountain cases, and the right-hand scale is for the no-mountain case. (From [21].)

effects (in disagreement with Kasahara and Washington [18]). They also found that the inclusion of orography increased the kinetic energy of the stationary waves and decreased that of the transient waves, while leaving the total kinetic energy virtually unchanged; their solutions for the mean meridional wind both with and without mountains are shown in Fig. 6.

In a study of the role of mountains in the Asian monsoon circulation with the GFDL 11-level GCM, Hahn and Manabe [13] found that the mountains helped to maintain relatively low surface pressure and relatively high surface temperature over the Tibetan Plateau in July, and thereby permitted the penetration of tropical moist air farther into the Asian continent than would otherwise be the case. The effects of orography in the southern hemisphere were also examined in a July simulation by Mechoso [22], who found that removal of the Antarctic continent effectively removed the surface easterlies in high southern latitudes and increased midtropospheric temperatures.

(2) Phenomenological consequences. Numerical experiments with GCMs have also shown systematic orographic effects on the phenomena of cyclogenesis, storm tracks, and the onset of the Asian monsoon. For example, Manabe and Terpstra [21] found that the orography in their model increased cyclogenesis in the lee of the Rockies and in the lee of the Tibetan Plateau as shown in Fig. 7, while Washington et al. [32] using

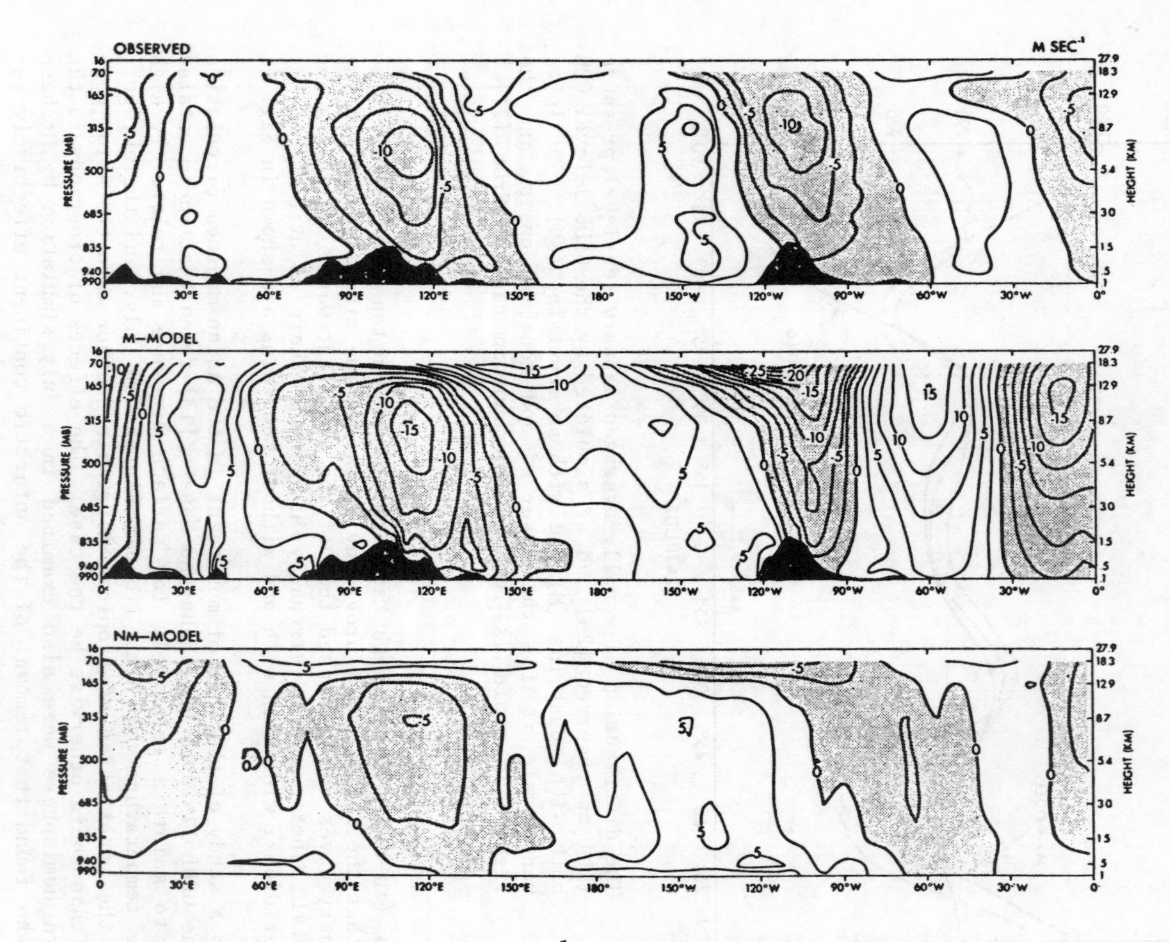

Fig. 6 The mean January meridional wind (m sec^{-1}) along 45°N as observed (top), and as simulated by the GFDL 9-level GCM with mountains (middle) and without mountains (bottom). (From [21].)

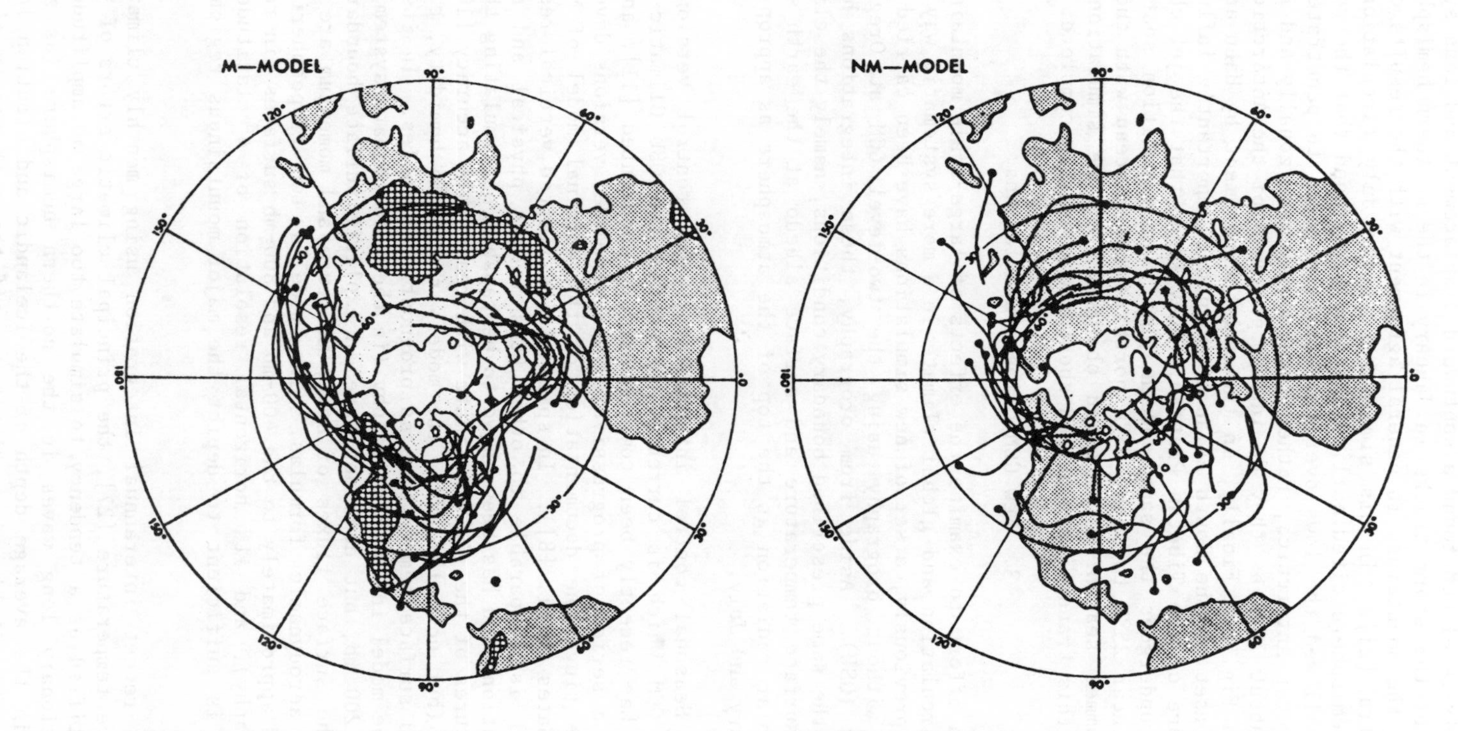

Fig. 7 Storm tracks simulated in the northern hemisphere during January by the GFDL 9-level GCM with mountains (left) and without mountains (right). (From [21].)

the NCAR six-level GCM found a southward displacement and less systematic structure of the storm tracks in January in the northern hemisphere upon removal of the mountains, in general agreement with the results of Manabe and Terpstra [21]. In his simulation of the July circulation in the southern hemisphere cited earlier, Mechoso [22] found that the removal of mountains allowed the lows over the southern ocean to penetrate to high latitudes over Antarctica, rather than moving more zonally and remaining north of about 70°S as they do in the presence of the Antarctic Plateau as seen in Fig. 8. Finally, in the model experiment by Hahn and Manabe [13] the onset of the South Asian monsoon was importantly influenced by the presence of the Tibetan Plateau. They found that the jet characteristically undergoes the same sudden jump from a position south of the mountains near 25°N to about 45°N north of the plateau with the advance of the summer season as is often observed; in the simulation without mountains this transition required about two months to complete.

3. New GCM Climate Simulations

In an effort to examine the effects of large-scale mountains on the general circulation and global climate in a more systematic way than has been done previously, a set of new simulations have been carried out both with and without orography using the two-level GCM at Oregon State University (OSU). Aside from orography, these integrations have been made with the same prescribed boundary conditions, namely the climatological sea-surface temperature and surface albedo at the earth's surface and the solar radiation at the top of the atmosphere as appropriate for both January and July.

(1) Seasonal control integrations. The control version of the two-level GCM which is currently in use at the OSU Climatic Research Institute has recently been comprehensively documented [11], and is the latest in a series of progressively improved model versions developed in the decade since the documentation of the original model of Mintz and Arakawa (Gates et al. [8]). In spite of its limited vertical resolution, this model is comparable to other GCMs in its physical and numerical sophistication, and has been shown to be capable of simulating the large-scale features of the global climate with reasonable accuracy [10], [26]. Like many (but not all) GCMs, the model treats the humidity, cloudiness and ground surface temperature as prognostic variables. In its present version the model includes orography (in the σ-coordinate system) and has a top at 200 mb, but does not have an explicit surface boundary layer; instead the surface fluxes of heat, moisture and momentum are given by the bulk aerodynamic formulas. The model's two tropospheric levels correspond approximately to the 400-mb and 800-mb surfaces (in regions of low orography), and its horizontal resolution of 4° latitude and 5° longitude is sufficient to depict the major mountainous regions of the world.

In a recent interannual integration using monthly climatological sea-surface temperature [27], the principal climatic errors of the model were identified as a tendency to simulate too large an amplitude for the quasi-stationary long waves in the northern hemisphere (as seen, for example, in the average depth of the Icelandic and Aleutian lows), and excessive precipitation in midlatitudes off the east coasts of Asia and

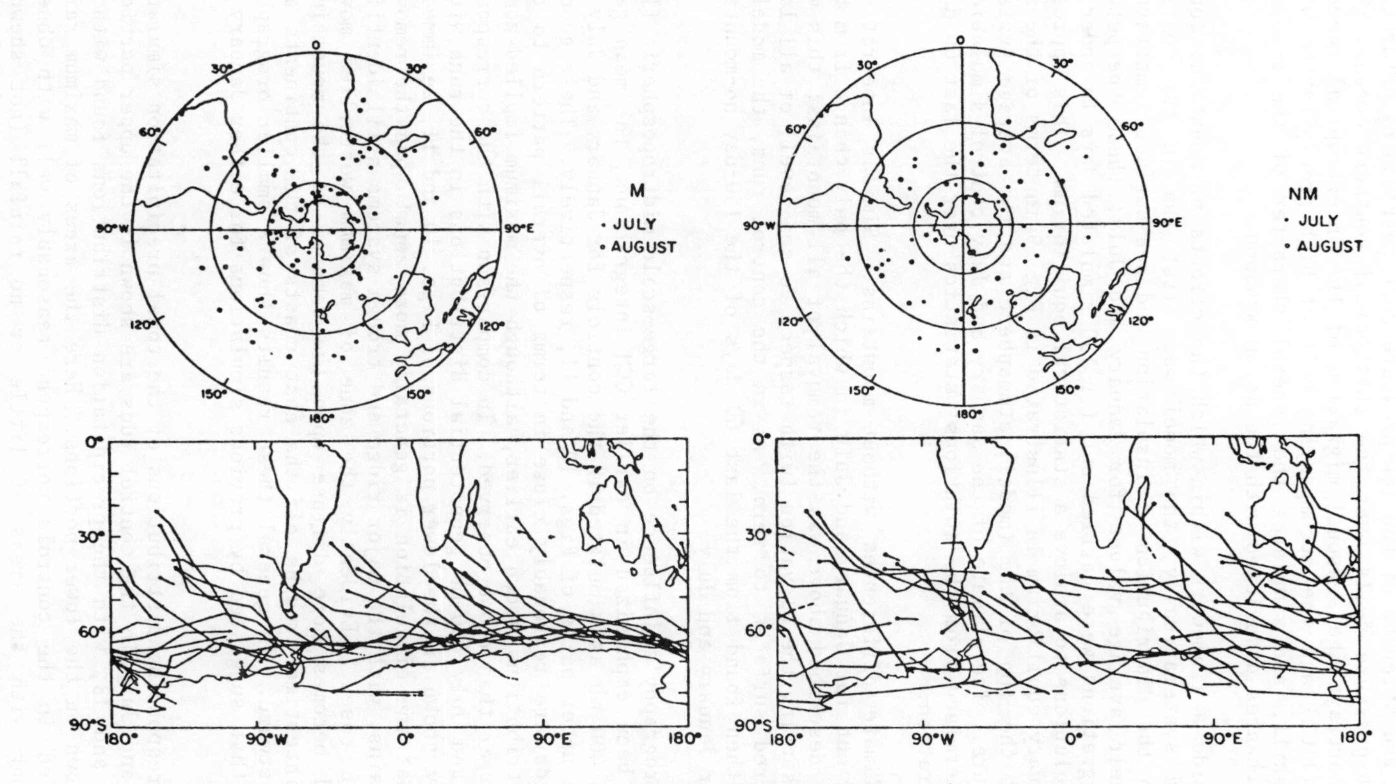

Fig. 8 Locations of cyclogenesis (above) and storm tracks (below) simulated in the southern hemisphere during July and August by the GFDL 11-level GCM with mountains (left) and without mountains (right). (From [22]).

North America. On the other hand, the model successfully simulates both the position and strength of the low-pressure zone surrounding Antarctica and the high-pressure cells over the subtropical southern oceans. The model also portrays the seasonal migration of the intertropical convergence zone (ITCZ) and the accompanying belt of high precipitation with reasonable skill, and simulates the general character of the seasonal circulation changes associated with the Asian monsoon.

To provide a framework within which the effects of mountains could be evaluated systematically, the model was first run in its control version with the distribution of insolation and sea-surface temperature fixed at their average values for January and July; these "perpetual season" integrations were carried out for 150 simulated days in order to permit the solutions to achieve a statistical equilibrium. This approach toward a January equilibrium is illustrated in Fig. 9 in terms of the net radiation at the top of the (model) atmosphere and the average surface air temperature. The climate of the January and July control simulations was then determined from the solutions' statistics over the last 60 days of the integration.

(2) Climate differences without mountains. Similar integrations were carried out for January and July in which the only change from the control runs described above was the removal of all mountains; this was achieved by setting the elevation (with respect to sea level) of all land and ice-covered surfaces to zero. As in the control runs, the model's climate was then found from the last 60 days of the 150-day no-mountain solutions for January and July.

Since orography's influence on the large-scale midtropospheric flow pattern has been emphasized in earlier GCM integrations, the mean geopotential at 400 mb as simulated in the controls for January and July is shown in the upper parts of Figs. 10 and 11, respectively. These geopotential fields are reasonably close in terms of overall pattern to the observed distributions shown earlier, although the maximum implied zonal winds are larger than those observed. In comparison with the corresponding January and July 400-mb geopotential distributions in the runs without orography shown in the lower portions of Figs. 10 and 11, we immediately see that the circulation is generally more zonal upon the removal of the mountains, with the major ridge and trough systems still identifiable in both cases. Evidently the zone of maximum westerlies moves poleward and becomes more diffuse upon the removal of the mountains, while the wintertime troughs off the east coasts of the continents are displaced westward. In general these results show a smaller orographic effect than that suggested by previous simulations for either January or July.

The corresponding distributions of the total precipitation simulated for January and July in the control runs are shown in the upper portions of Figs. 12 and 13, with the precipitation distributions found without mountains shown in the lower portions. Here the areas of maximum rainfall simulated in the controls correspond reasonably well with observations, along with the areas of little or no rainfall (not shown); especially notable is the intense precipitation in the equatorial Indian

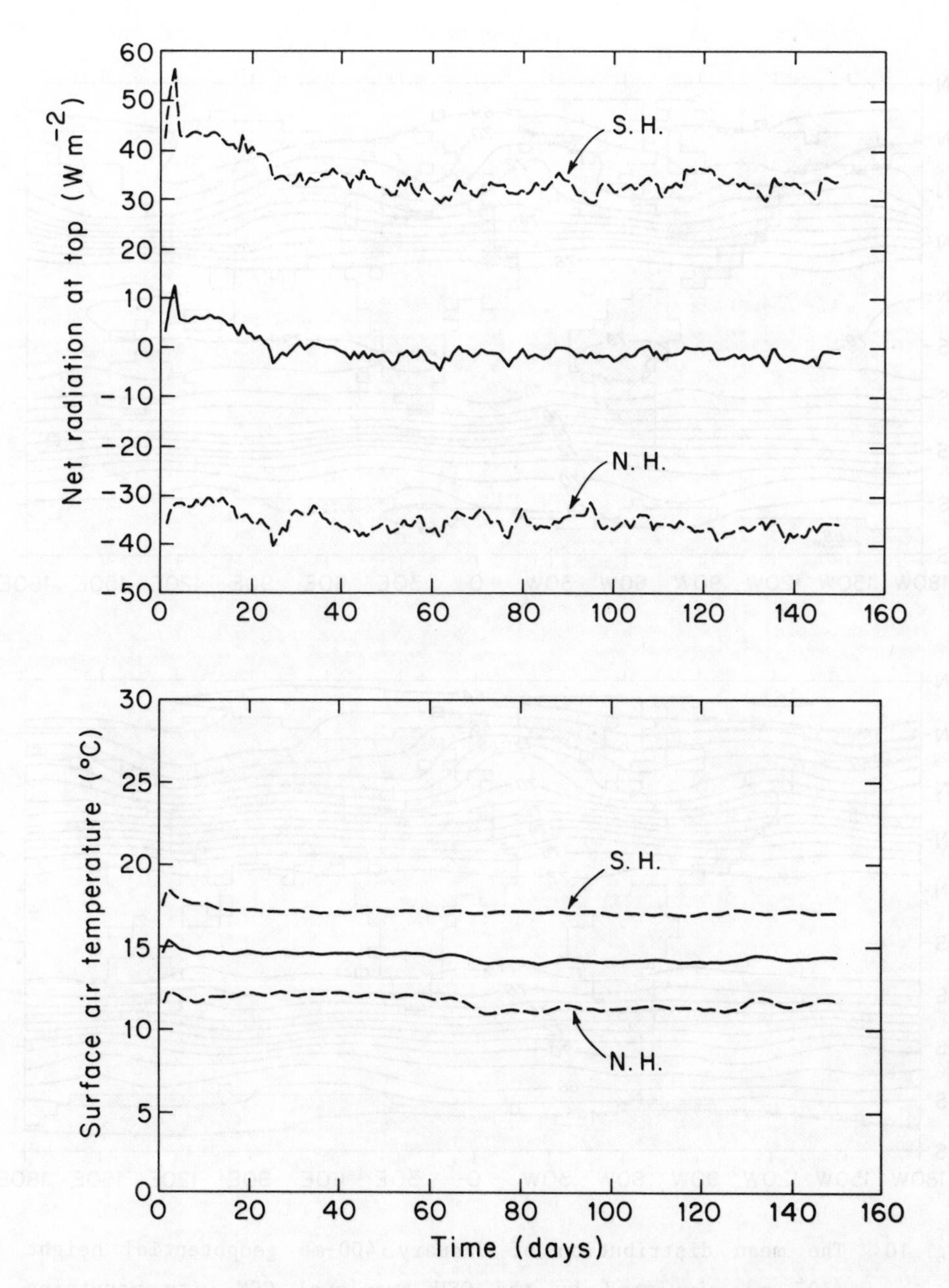

Fig. 9 The spin-up and transient behavior of the July simulation without mountains with the OSU two-level GCM in terms of the net radiation at the top of the (model) atmosphere (above) and the mean surface air temperature (below). The global average is shown by the full line, with the dashed lines showing the averages over the northern (N.H.) and southern (S.H.) hemispheres.

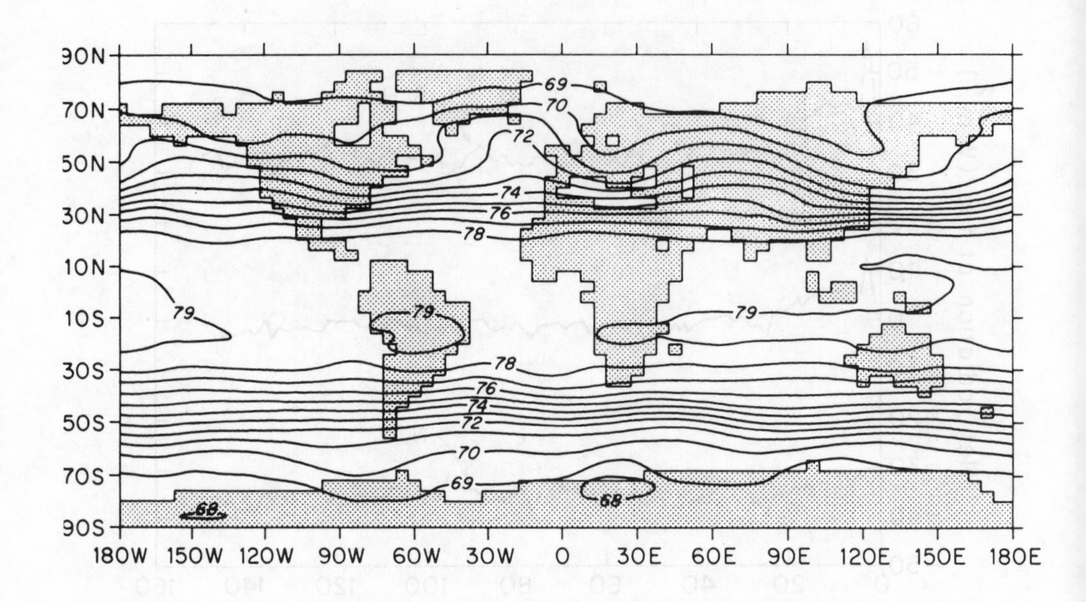

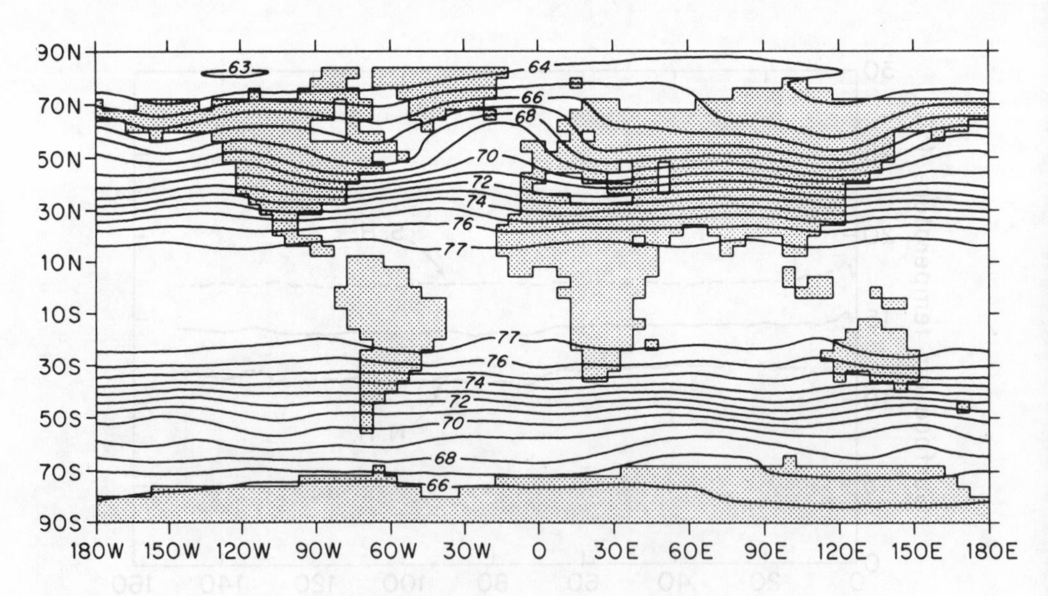

Fig. 10 The mean distribution of January 400-mb geopotential height (10^2 m) simulated by the OSU two-level GCM with mountains (above) and without mountains (below).

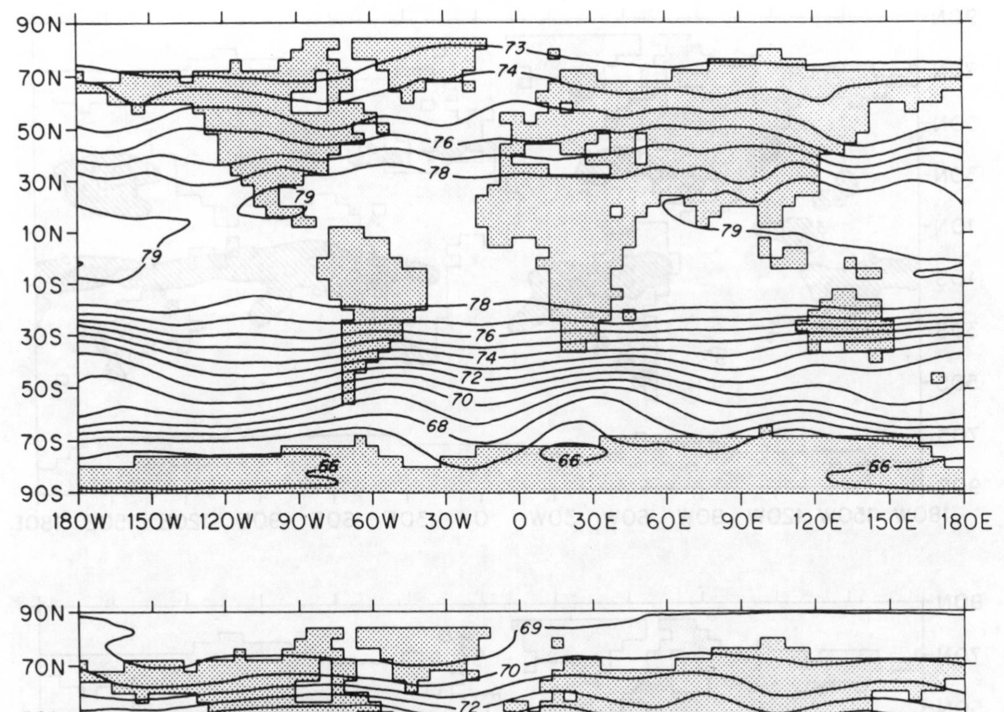

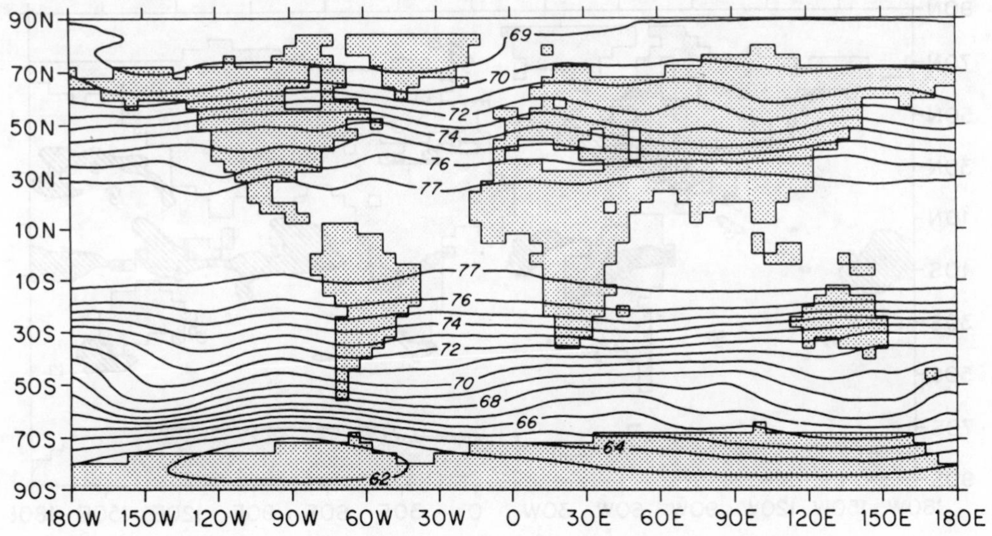

Fig. 11 Same as Fig. 10, except for July.

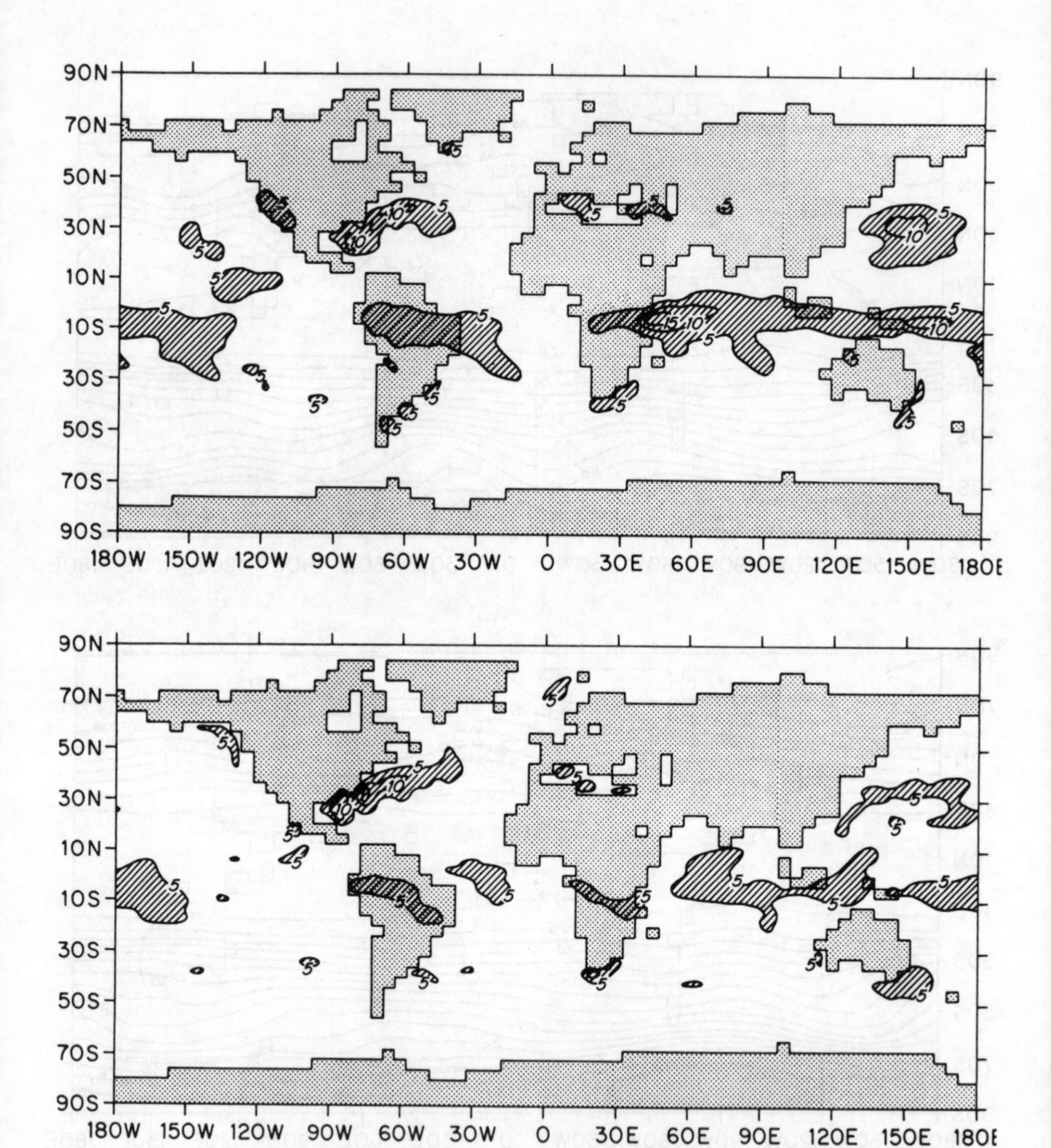

Fig. 12 The distribution of total January precipitation (mm day^{-1}) simulated by the OSU two-level GCM with mountains (above) and without mountains (below).

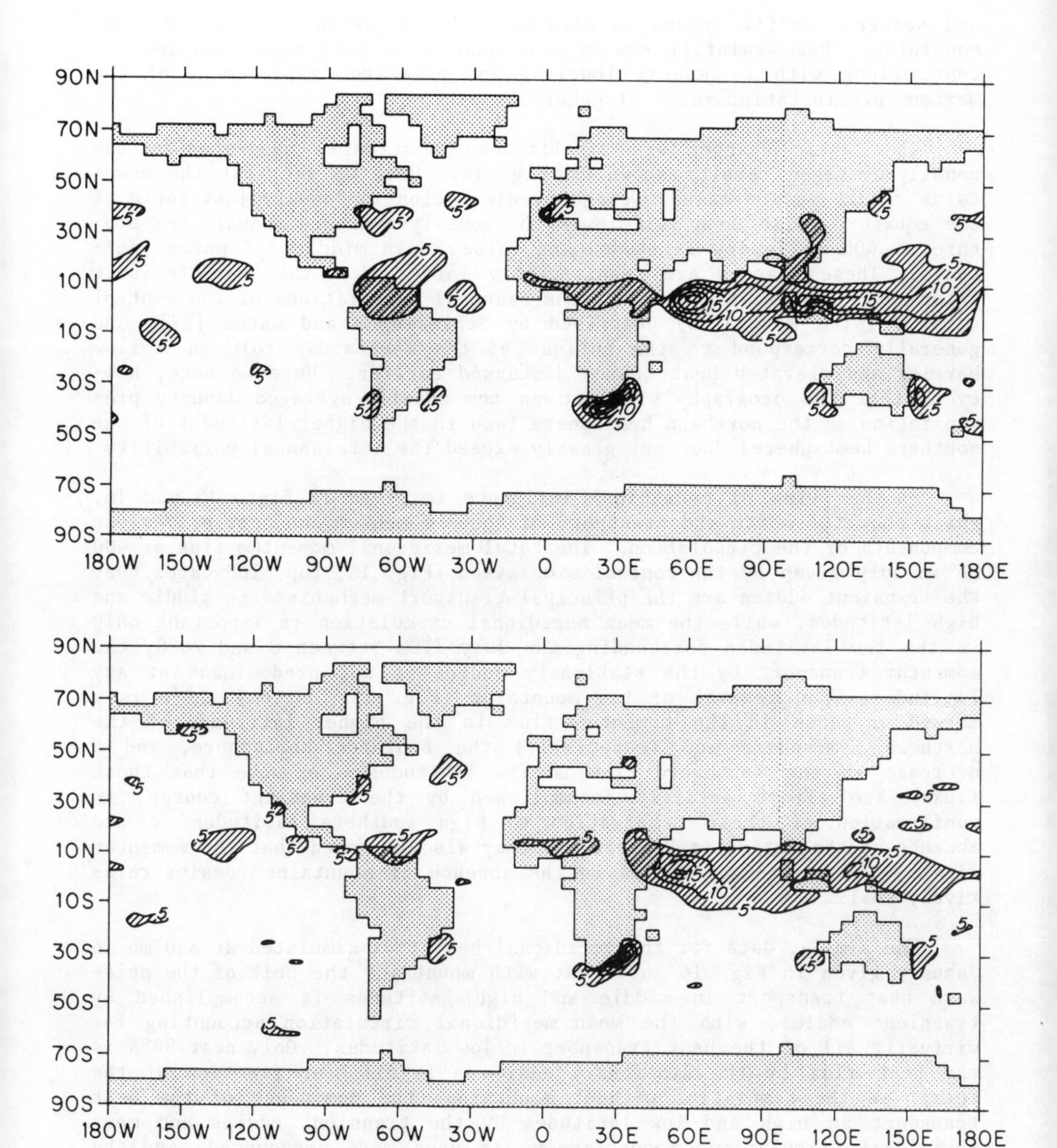

Fig. 13 Same as Fig. 12, except for July.

and western Pacific Oceans associated with the Asian monsoon. Without mountains, these rainfall maxima are reduced in both magnitude and extent, along with a general lowering and poleward displacement of the maximum precipitation rates elsewhere.

That these changes are significant is strongly suggested by the zonally-averaged results shown in Fig. 14. Here we see that the mountains result in increased average precipitation in January just south of the equator, while mountains lower the zonally-averaged January temperature at 400 mb in low latitudes and raise it in middle and polar latitudes. These changes are significantly larger than the characteristic interannual variabiltiy found in the seasonal integrations of the control version (with mountains) described by Schlesinger and Gates [27], and generally correspond to the notions of the mountains' role as a flow barrier and elevated heat source discussed earlier. Here we note, however, that the orography's effect on the zonally-averaged January precipitation in the northern hemisphere (and in the higher latitudes of the southern hemisphere) does not greatly exceed the interannual variability.

Another view of mountains' influence is given in Figs. 15 and 16, which show the northward transport of zonal momentum and heat by various components of the circulation. The total meridional momentum flux at 400 mb in July given by the control simulation (Fig. 15, top) indicates that the transient eddies are the principal transport mechanisms in middle and high latitudes, while the mean meridional circulation is important only in the low latitudes surrounding the July ITCZ between 0 and 20°S; the momentum transport by the stationary eddies is not predominant at any latitude. Upon removal of the mountains (Fig. 15, bottom) there is a marked increase in the momentum flux in the higher latitudes of the northern hemisphere and (especially) the southern hemisphere, and a decrease in the southward flux in low latitudes. We note that these fluxes are almost entirely accomplished by the transient eddies, in confirmation of their penetration to high southern latitudes in the absence of the Antarctic Plateau. It may also be noted that the momentum flux by the stationary eddies in the absence of mountains remains relatively small.

The similar data for the meridional heat flux simulated at 800 mb in January given in Fig. 16 show that with mountains the bulk of the poleward heat transport in middle and high latitudes is accomplished by transient eddies, with the mean meridional circulation accounting for virtually all of the heat transport in low latitudes. Only near 50°N is the heat flux by the stationary eddies a significant fraction of the total. In the simulation without mountains, the dominance of the heat transport at high and low latitudes by the transient eddies and mean meridional circulation, respectively, is even more pronounced, and the transport by the stationary eddies, which are presumably directly maintained by the mountains, is reduced to near insignificance.

To provide a more detailed view of the influence of mountains on the circulation and climate over the East Asian region, the average 400-mb geopotential simulated for January and July is shown in Figs. 17 and 18, respectively. Here the broadening and westward shift of the trough near 140°E which occurs upon the removal of orography in both summer and winter is more clearly seen than in the global distributions from which

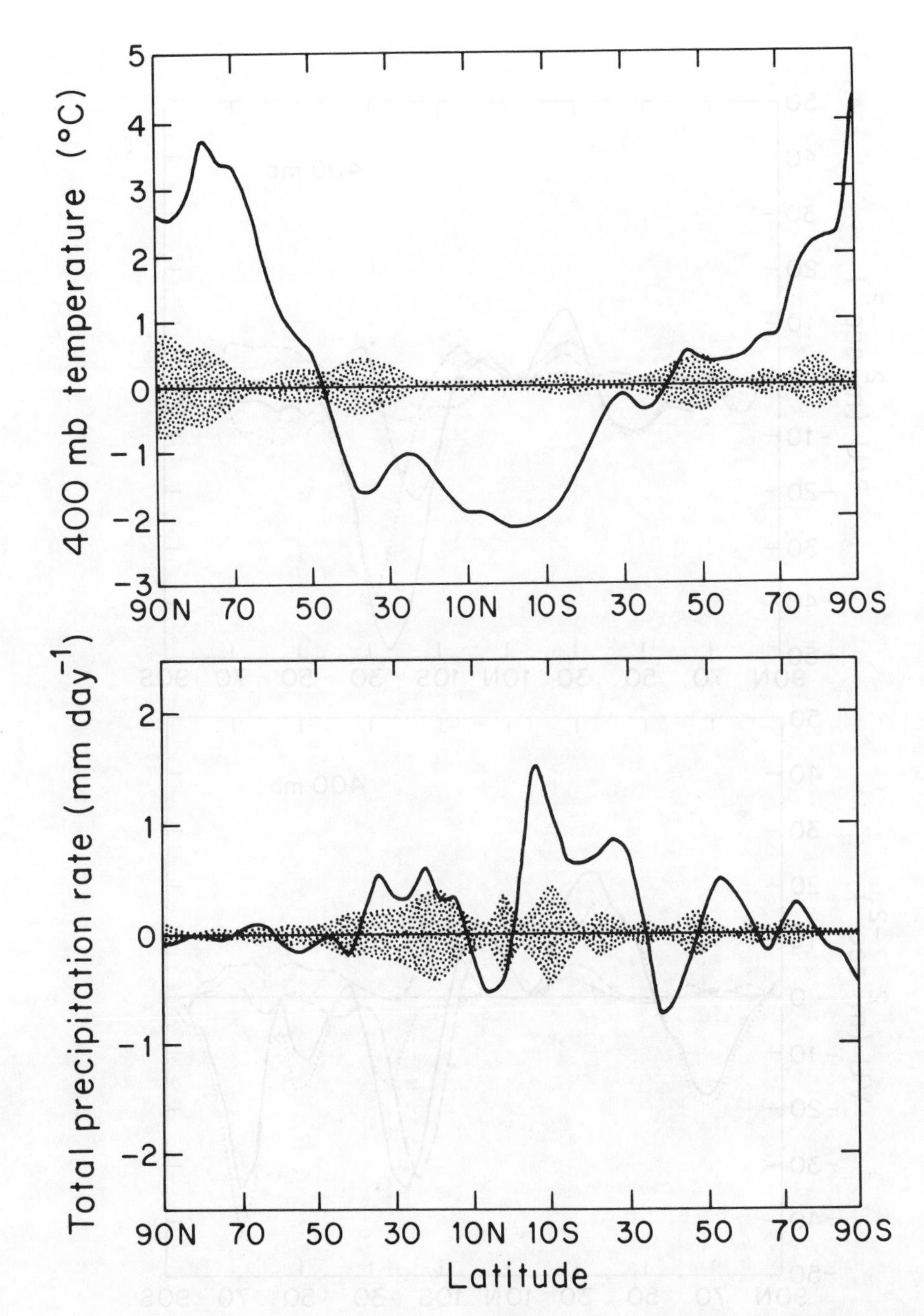

Fig. 14 The distribution of the effect of orography on the zonally-averaged January 400-mb temperature (above) and on the total precipitation (below) as simulated by the OSU two-level GCM. The full line shows the mountain <u>minus</u> no-mountain differences, and the shaded area (symmetric about zero) shows the root-mean-square variability of an ensemble of January means about their average found in an interannual control integration.

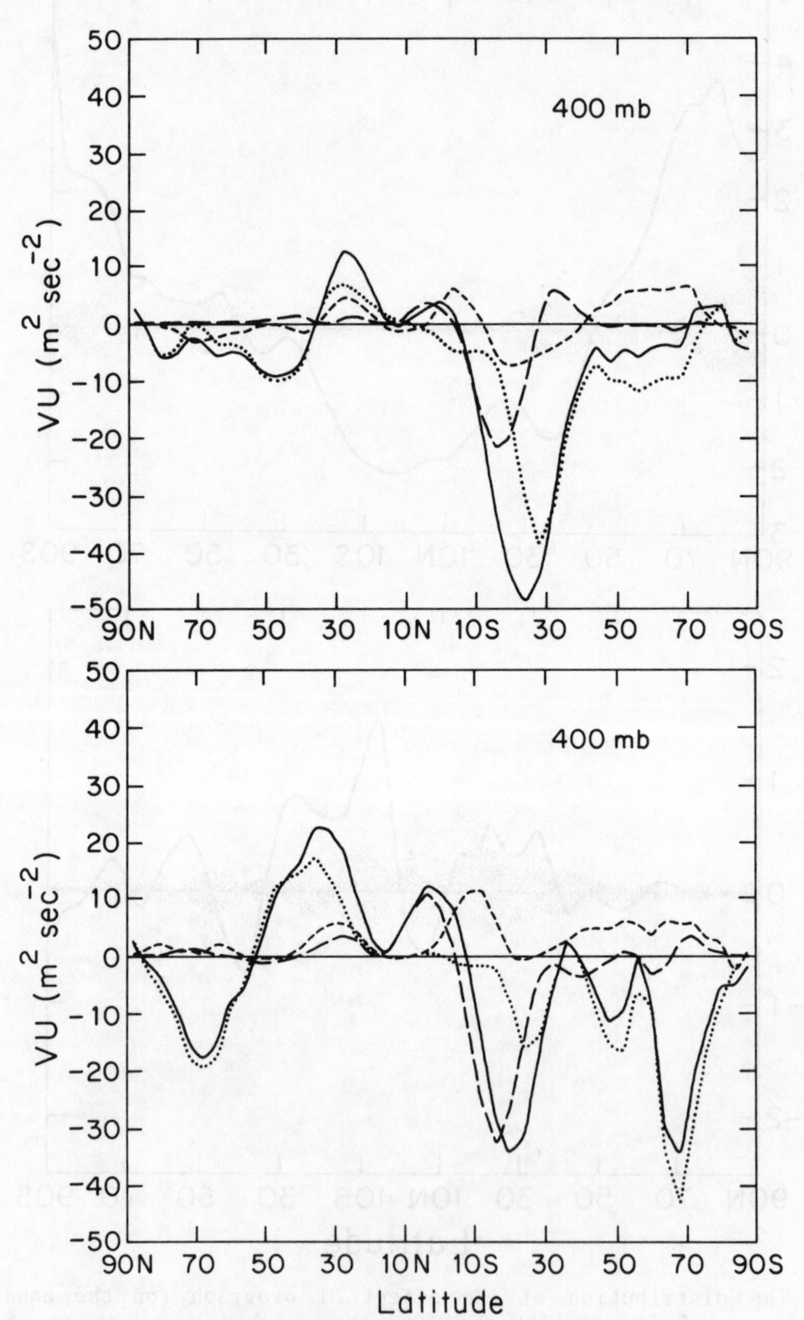

Fig. 15 The northward transport of zonal momentum at 400 mb as simulated by the OSU two-level GCM with mountains (above) and without mountains (below). Shown here is the total flux (full line), the flux by transient eddies (dotted line), the flux by stationary eddies (short-dashed line), and the flux by the mean meridional circulation (long-dashed lines).

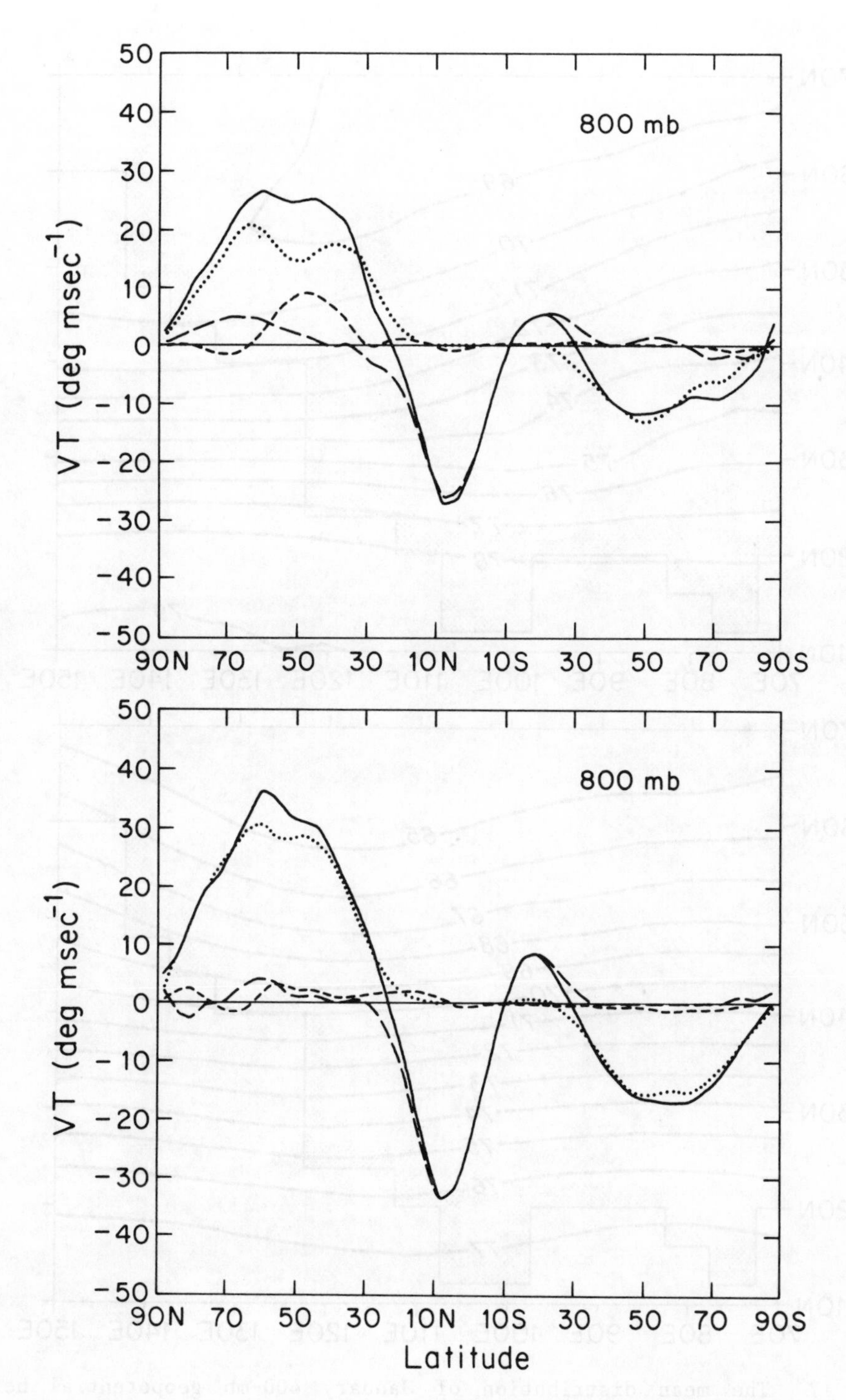

Fig. 16 Same as Fig. 15, except for the heat flux at 800 mb in January.

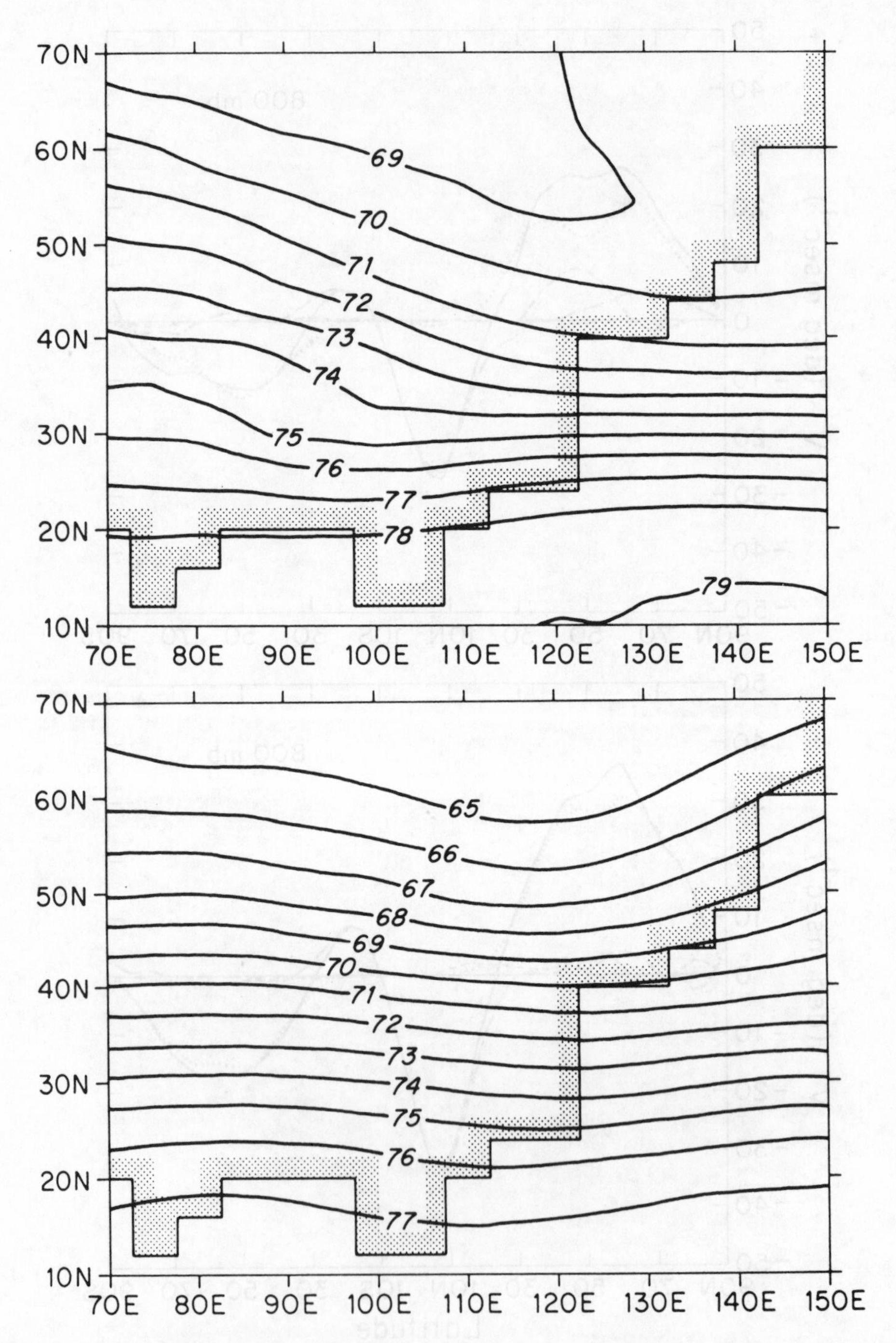

Fig. 17 The mean distribution of January 400-mb geopotential height (10^2 m) simulated over East Asia by the OSU two-level GCM with mountains (above) and without mountains (below).

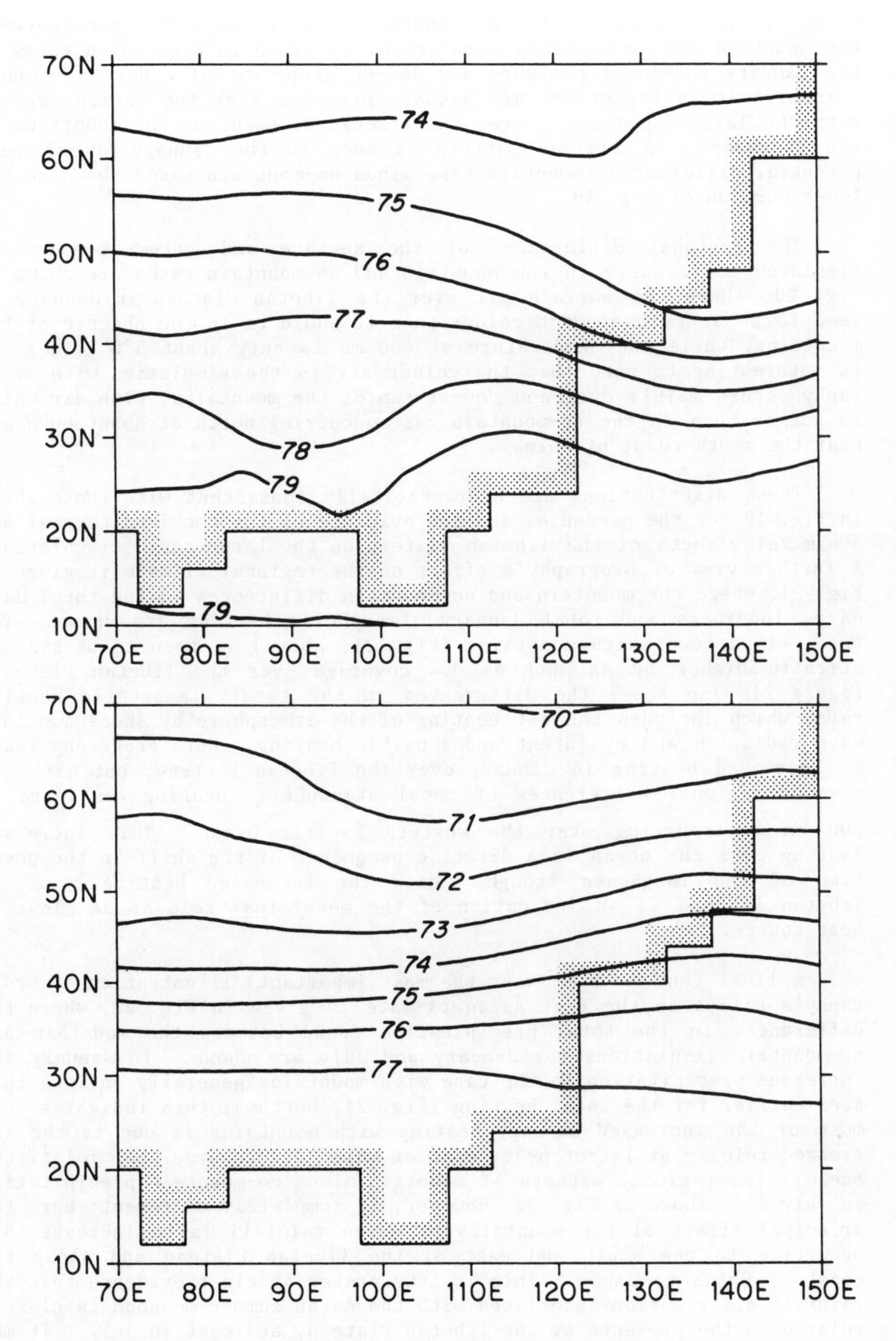

Fig. 18 Same as Fig. 17, except for July.

these close-up views are taken. Another view of the differences between the mountain and no-mountain simulations is given in Fig. 19 in terms of the January sea-level pressure and 400-mb geopotential. Here the mountains result in higher average January pressure over the entire region, with the largest increases over the Tibetan Plateau and the mountains to the northeast. A similar pattern is seen in the January 400-mb geopotential differences (mountain case minus no-mountain case) shown in the lower portion of Fig. 19.

The regional differences of the surface and 400-mb temperature simulated for January in the mountain and no-mountain cases are shown in Fig. 20. Here the surface air over the Tibetan Plateau in January is seen to be as much as 30°C colder than it would be in the absence of the mountains, while the temperature at 400 mb is only about 5°C colder. It is interesting to note that the colder air in the simulation with orography occurs mainly over and downstream of the mountains, with air which is warmer than in the no-mountain case occurring north of about 60°N and near the south coast of China.

These distributions are hydrostatically consistent with those shown in Fig. 19 for the pressure, and are evidence of the combined thermal and dynamical effects of the Tibetan Plateau on the large-scale circulation. A further view of orography's effect on the regional climate is given in Fig. 21, where the mountain and no-mountain differences in the total January cloudiness and total January atmospheric heating are shown. The total cloud cover with mountains (Fig. 21, above) is seen to be systematically higher by as much as 0.4 coverage over the Tibetan Plateau. Figure 21 also shows the differences in the total atmospheric heating rate, which includes the net heating of the atmosphere by short and long wave radiation and by latent and sensible heating. Here orography leads to increased heating in January over the Tibetan Plateau, but with the maximum regional differences of total atmospheric heating (as large as 300 Wm^{-2}) occurring over the western Pacific Ocean. This increased heating over the ocean is a direct consequence of the shift in the position of the long-wave trough, while the increased heating over the Tibetan Plateau is an indication of the mountains' role as an elevated heat source.

A final (but in some ways the most important) illustration of orography's effect on the East Asian climate is given in Fig. 22, where the differences in the total precipitation found between the mountain and no-mountain simulations for January and July are shown. In January the increased precipitation in the case with mountains generally follows that seen earlier for the total heating (Fig. 21, bottom); this indicates that most of the increased January heating with mountains is due to the increased release of latent heat, most of which occurs over the midlatitude ocean. The regional pattern of mountain minus no-mountain precipitation in July also shown in Fig. 22, however, is completely different; here the principal effect of the mountains on Asian rainfall is to increase that occurring to the south and east of the Tibetan Plateau and along the coast of China to Japan. This is interpreted as clear evidence that the rainfall distribution associated with the Asian summer monsoon is closely related to the presence of the Tibetan Plateau, at least in July. It may

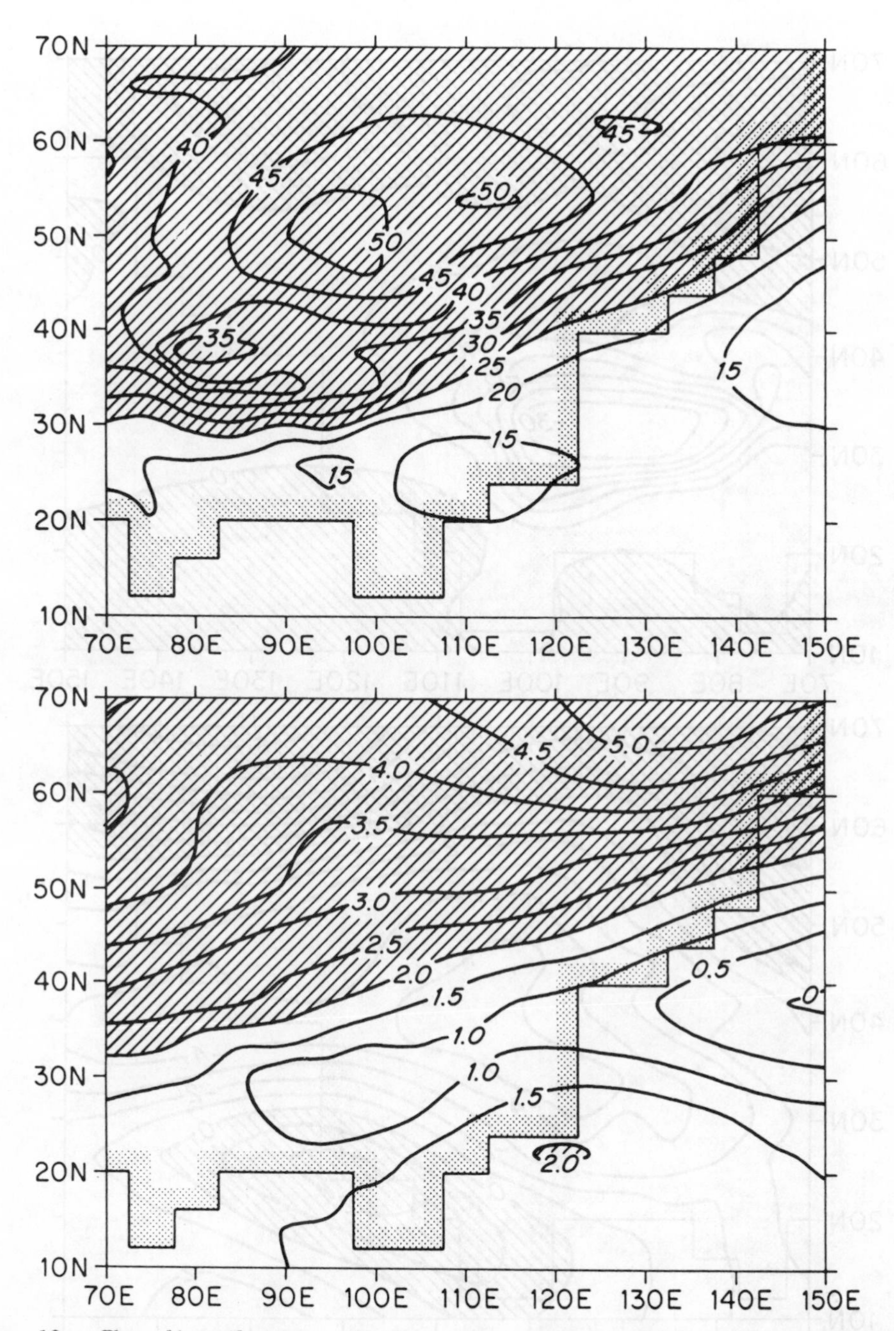

Fig. 19 The distribution of the effect of orography on the January sea-level pressure (mb) (above) and on the 400-mb geopotential height (10^2 m) (below) as simulated over East Asia by the OSU two-level GCM. The isolines are the mean mountain minus no-mountain differences, with the higher values shaded.

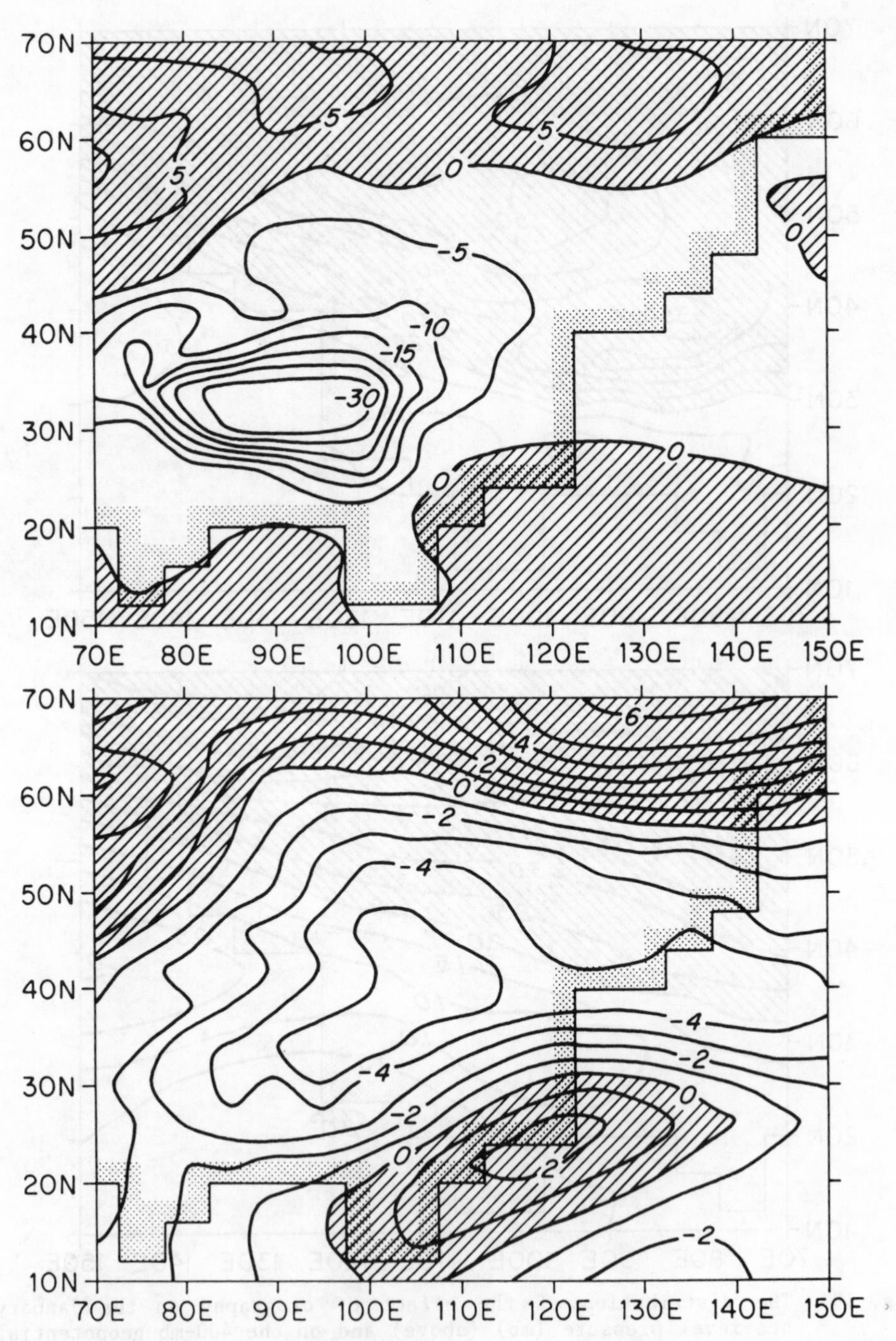

Fig. 20 The distribution of the effect of orography on the January surface air temperature (°C) (above) and on the 400-mb temperature (°C) (below) as simulated over East Asia by the OSU two-level GCM. The isolines are the mean mountain minus no-mountain differences, with the positive values shaded.

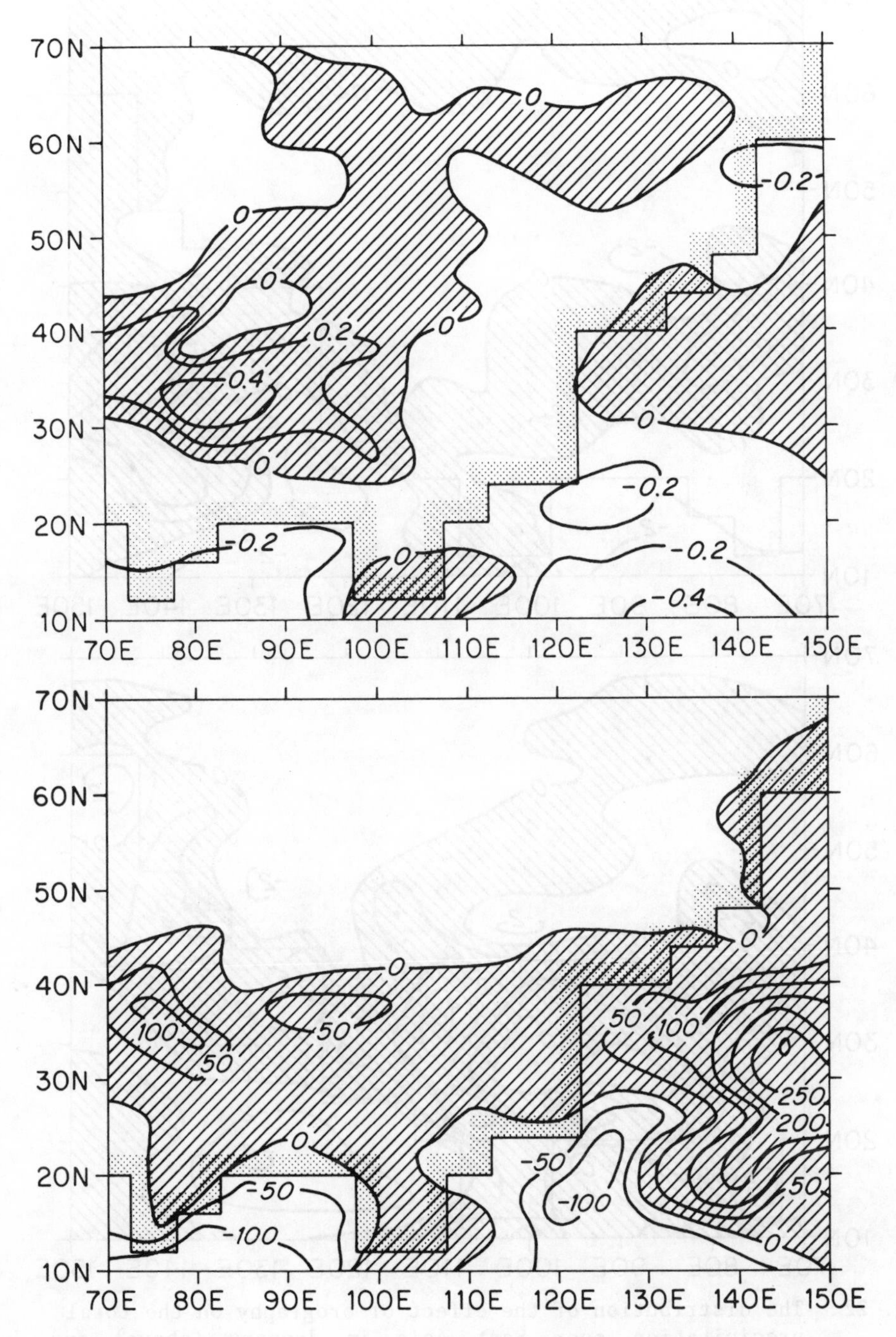

Fig. 21 Same as Fig. 20, except for the January total fractional cloudiness (above) and net atmospheric heating (Wm^{-2}) (below).

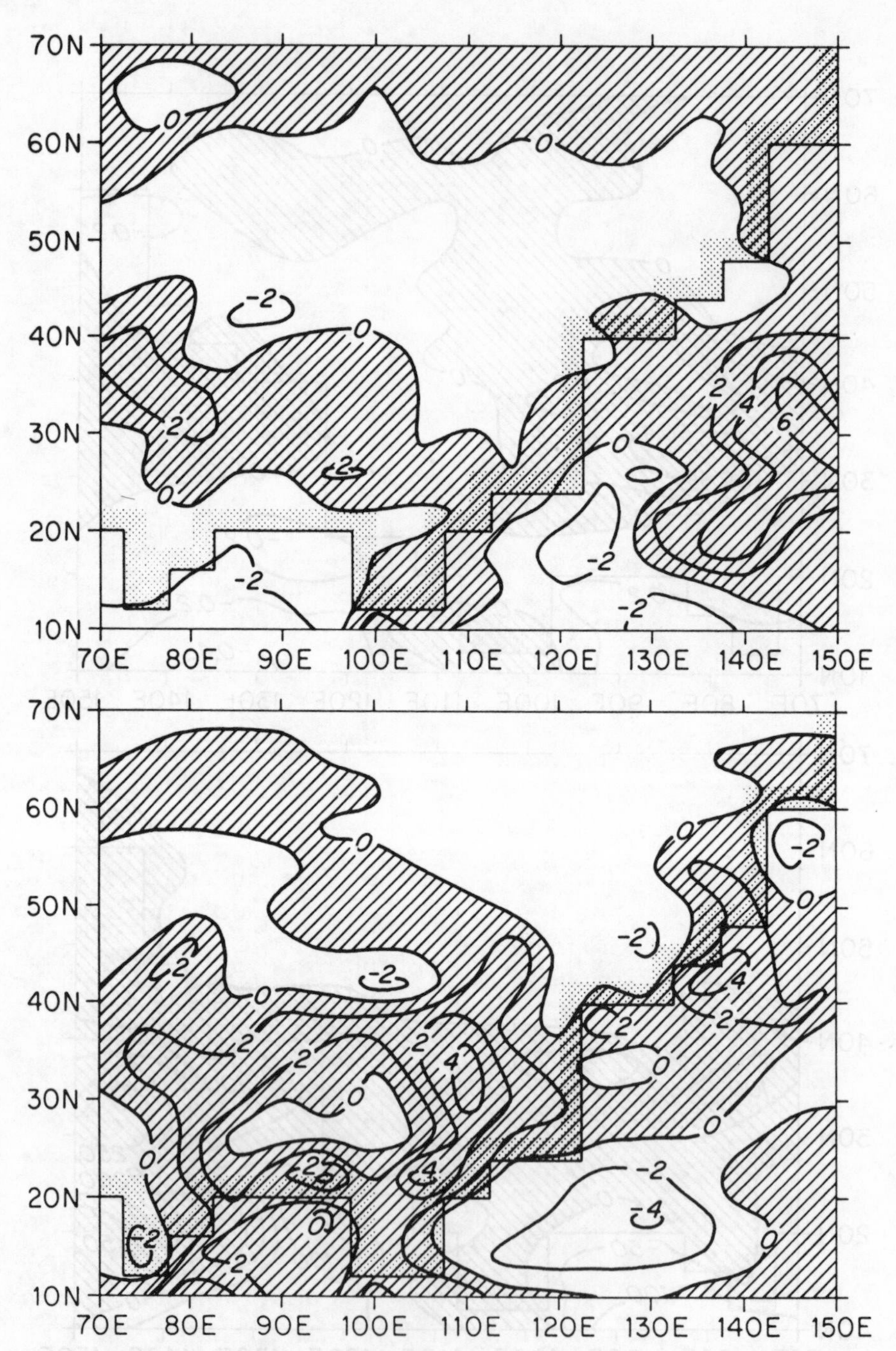

Fig. 22 The distribution of the effect of orography on the total precipitation over East Asia in January (above) and July (below) as simulated by the OSU two-level GCM. Here the isolines (mm day^{-1}) are the mean mountain minus no-mountain differences, with the positive values shaded.

be speculated that the Tibetan Plateau similarly influences the precipitation distribution throughout the warm season of the year, through its apparent control of the northward shift of the Asian jet as noted earlier.

The above examples of orography's influence on the general circulation and climate as simulated in "perpetual season" integrations with the OSU two-level GCM are drawn from a more comprehensive analysis which is currently underway [7]. Here orography's role in the global energetics is being studied, along with the effects of mountains on a wide variety of climatic variables in both summer and winter. An interannual integration of the OSU model without mountains is now being planned, and its analysis in comparison with the interannual control run already completed should provide further insight into the mountains' effect on the time-dependent mechanics of the general circulation and their role in such regional seasonal phenomena as the monsoon and the associated seasonal shifts in surface climate.

IV. OUTSTANDING PROBLEMS AND FUTURE RESEARCH

Aside from the straightforward application of GCMs in interannual simulations both with and without mountains as noted above, there are several outstanding problems connected with the design of models and the analysis of their results which require further research. Chief among these are the problems of model resolution and parameterization, the problem of estimating the statistical significance of model-simulated results, and the general problem of satisfactorily coupling the ocean and atmosphere into a single interactive model. A few comments on each of these questions are given below.

1. Resolution and Parameterization

The effects of the horizontal resolution on the statistics of the solution of a model (and of GCMs in particular) are not well known, even for standard or control model versions. In those few GCM simulations which have been made with different resolution, usually only a halving in the horizontal grid size is examined, and then only in a limited experimental manner in order to select an operational grid. These studies have indicated, however, that increased resolution alone does not always yield an improved simulation, which in turn suggests that parameterizations which are developed and tuned for one scale of resolution may not be applicable to another scale, especially in the presence of orography. Extended simulations with several different models both with and without mountains, preferably with more than two different resolutions, are needed to examine this question, along with further sensitivity tests on the effects of various representations of the flow over large-scale mountains, such as the use of potential enstrophy conserving schemes. Further research is also needed on the effects of the smoothing which is inherent in mountains' representation on a particular grid. It seems likely, for example, that the large-scale air flow effectively responds to the envelope of the mountains' height as represented by the elevation of the small-scale orography, rather than to their smoothed (and therefore necessarily lowered) height; if this is the case, then all current

GCMs may be systematically underestimating (as well as mispresenting) orography's effect.

A related problem is the successful parameterization of the various subgrid-scale processes which are of importance in mountainous terrain, especially those which serve to modify the large-scale vertical fluxes of heat, moisture and momentum through the planetary boundary layer. The parameterization of moist convective adjustment which has been used with reasonable success in tropical and oceanic areas (see [33]) may require modification in the presence of strong vertical motions associated with forced upslope flow over mountains, along with appropriate provision for the presence of orographic cloudiness. The parameterization of the frictional drag over mountain terrain also needs further attention, both with respect to the incorporation of the effects of the large variations in effective surface roughness over short distances and the representation of the drag effects of mesoscale mountain waves. Additional research should also be given to the calculation of the pressure torque exerted by mountains, as this often appears to be of the same sign as the mountain-induced frictional torque and may, therefore, play an important role in the atmosphere's balance of zonal momentum.

2. Significance Estimates

Once solutions of a model both with and without mountains have been generated and the mountains' effects formally separated, there remains the problem of determining how much of the apparent differences due to orography is actually due to the mountains and how much is due to the inevitable differences in the statistics of the model's solutions caused by the essentially unpredictable synoptic-scale motions. Existing techniques for the separation of such a climatic signal from the background natural fluctuations (or climatic noise) do not adequately take into account the solutions' correlations in space and time or the nonnormality of many simulated climatic statistics. Longer simulations and/or the generation of ensembles of solutions, along with improved statistical estimation techniques, will be necessary for definitive delineation of the seasonal and geographical distribution of orographic effects.

3. Ocean-Atmosphere Coupling

Perhaps the most profound problem requiring attention in the further study of orography's effect on the atmospheric general circulation and climate is the successful coupling of the atmosphere and ocean into a single model, as a prelude to the subsequent analysis of the properties of an extended integration of the coupled system. All GCM studies of the effects of mountains which have been made so far have used the same (usually climatological) distribution of sea-surface temperature in the no-mountain case as is used in the control case with mountains. This means that however much the removal of the mountains may effect the low-level airflow and temperature, for example, the sea-surface temperature is not free to respond to these changes; this compounds the already difficult problem of separating the purely dynamical and thermal effects of orography. When a coupled ocean-atmosphere model (including as a minimum an interactive oceanic mixed layer) is used in future mountain and no-mountain experiments, this problem of separation will remain, but at least the total effect of orography will be more completely simulated than heretofore.

ACKNOWLEDGMENTS

This research was supported by the National Science Foundation under Grant ATM-8001702 and by the OSU Climatic Research Institute. I would like to thank Robert L. Mobley for his assistance in carrying out the integrations of the OSU two-level GCM reported here, and Cindy Beck for her assistance in preparing the manuscript.

REFERENCES

1. Arakawa, A. and V.R. Lamb: A potential enstrophy and energy conserving scheme for the shallow water equations. Mon. Wea. Rev., 109, 18-36 (1981).

2. Ashe, S.: A nonlinear model of the time-average axially asymmetric flow induced by topography and diabatic heating. J. Atmos. Sci., 36, 109-126 (1979).

3. Bolin, B.: On the influence of the earth's orography on the general character of the westerlies. Tellus, 2, 184-195 (1950).

4. Charney, J.G. and A. Eliassen: A numerical method for predicting the perturbations of the middle latitude westerlies. Tellus, 1 (2), 38-54 (1949).

5. Derome, J. and A. Wiin-Nielsen: The response of a middle-latitude model atmosphere to forcing by topography and stationary heat sources. Mon. Wea. Rev., 99, 564-576 (1971).

6. Gambo, K.: The topographical effect upon the jet stream in the westerlies. J. Meteor. Soc. Japan, 34, 24-38 (1956).

7. Gates, W.L.: The climatic effects of large-scale mountains as simulated for January and July with the OSU two-level GCM. Report of the Climatic Research Institute, Oregon State University, Corvallis (in preparation) (1982).

8. Gates, W.L., E.S. Batten, A.B. Kahle and A.B. Nelson: A documentation of the Mintz-Arakawa two-level atmospheric general circulation model. R-877-ARPA, The Rand Corporation, Santa Monica, California, 408 pp. (1971).

9. Gates, W.L. and A.B. Nelson: A new (revised) tabulation of the Scripps topography on a 1° global grid. Part I: Terrain heights. R-1276-1-ARPA, The Rand Corporation, Santa Monica, California, 132 pp. (1973).

10. Gates, W.L. and M.E. Schlesinger: Numerical simulation of the January and July global climate with a two-level atmospheric model. J. Atmos. Sci., 34, 36-76 (1977).

11. Ghan, S.J., J.W. Lingaas, M.E. Schlesinger, R.L. Mobley and W.L. Gates: A documentation of the OSU two-level atmospheric general circulation model. Report No. 35, Climatic Research Institute, Oregon State University, Corvallis, 395 pp. (1982).

12. Grose, W.L. and B.J. Hoskins: On the influence of orography on large-scale atmospheric flow. J. Atmos. Sci., 36, 223-234 (1979).

13. Hahn, D.G. and S. Manabe: The role of mountains in the South Asian monsoon circulation. J. Atmos. Sci., 32, 1515-1541 (1975).

14. Ingersoll, A.P.: Inertial Taylor columns and Jupiter's great red spot. J. Atmos. Sci., 26, 744-752 (1969).

15. Kasahara, A.: Computational aspects of numerical models for weather prediction and climate simulation. In Methods in Computational Physics, Vol. 17, General Circulation Models of the Atmosphere, Academic Press, New York, pp. 1-66 (1977).

16. Kasahara, A.: Influence of orography on the atmospheric general circulation. In Orographic Effects in Planetary Flows, GARP Publications Series, No. 23, WMO, Geneva, pp. 1-49 (1980).

17. Kasahara, A., T. Sasamori and W.M. Washington: Simulation experiments with a 12-layer stratospheric global circulation model. I. Dynamical effect of the earth's orography and thermal influence of continentality. J. Atmos. Sci., 30, 1229-1251 (1973).

18. Kasahara, A. and W.M. Washington: Thermal and dynamical effects of orography on the general circulation of the atmosphere. In Proc. WMO/IUGG Symp. Num. Wea. Pred. in Tokyo (26 Nov.-4 Dec. 1968), Meteorological Society of Japan, Tokyo, pp. IV 47-IV 56 (1969).

19. Kasahara, A. and W.M. Washington: General circulation experiments with a six-layer NCAR model, including orography, cloudiness and surface temperature calculations. J. Atmos. Sci., 28, 657-701 (1971).

20. Kuo, H.L. and Y.F. Qian: Influence of the Tibetan Plateau on cumulative and diurnal changes of weather and climate in summer. Mon. Wea. Rev., 109, 2337-2356 (1981).

21. Manabe, S. and T.B. Terpstra: The effects of mountains on the general circulation of the atmosphere as identified by numerical experiments. J. Atmos. Sci., 31, 3-42 (1974).

22. Mechoso, C.R.: Topographic influences on the general circulation of the southern hemisphere: A numerical experiment. Mon. Wea. Rev., 109, 2131-2139 (1981).

23. Mintz, Y.: Very long-term global integration of the primitive equations of atmospheric motion. In WMO-IUGG Symposium on Research and Development Aspects of Long-Range forecasting (Boulder, Colorado, 1964), WMO Tech. Note No. 66 (WMO-No. 162. TP. 79), WMO, Geneva, pp. 141-167 (1965).

24. Murakami, T.: The topographical effect upon the stationary upper flow patterns. Papers Meteor. Geophys., 7, 69-89 (1956).

25. Phillips, N.A.: A coordinate system having some special advantages for numerical forecasting. J. Meteor., 14, 184-185 (1957).

26. Schlesinger, M.E. and W.L. Gates: The January and July performance of the OSU two-level atmospheric general circulation model. J. Atmos. Sci., 37, 1914-1943 (1980).

27. Schlesinger, M.E. and W.L. Gates: Preliminary analysis of the mean annual cycle and interannual variability simulated by the OSU two-level atmospheric general circulation model. Report No. 23, Climatic Research Institute, Oregon State University, Corvallis, 47 pp. (1981).

28. Smagorinsky, J.: The dynamical influence of large-scale heat sources and sinks on the quasi-stationary mean motions of the atmosphere. Q. J. Roy. Meteor. Soc., 79, 342-366 (1953).

29. Staff Members, Academia Sinica: On the general circulation over eastern Asia, I. Tellus, 9, 432-446 (1957).

30. Staff Members, Academia Sinica: On the general circulation over eastern Asia, II. Tellus, 10, 58-75 (1958).

31. Stern, M.E. and J.S. Malkus: The flow of a stable atmosphere over a heated island. Part I. J. Meteor., 10, 30-41 (1953).

32. Washington, W.M., B. Otto-Bliesner and G. Williamson: January and July simulation experiments with the 2.5° latitude-longitude version of the NCAR general circulation model. NCAR Tech. Note NCAR/TN-123+STR, National Center for Atmospheric Research, Boulder, Colorado, 39 pp. (Vol. 1), 61 pp. (Vol. 2) (1977).

33. World Meteorological Organization: Parameterization of subgrid-scale processes. GARP Publications Series No. 8 (Report of the JOC Study Conference in Leningrad, 20-27 March 1972), WMO, Geneva, 101 pp. (1972).

2.2 COMPARISON BETWEEN THE WINTER AND SUMMER RESPONSE OF THE NORTHERN HEMISPHERIC MODEL ATMOSPHERE TO FORCING BY TOPOGRAPHY AND STATIONARY HEAT SOURCES

Huang Ronghui
Institute of Atmospheric Physics
Academia Sinica
The People's Republic of China

ABSTRACT

Stationary planetary waves responding to forcing by the northern hemispheric topography and stationary heat sources are investigated by means of a quasi-geostrophic, steady state, 34-level model, with Rayleigh friction, the effect of Newtonian cooling and the horizontal kinematic thermal diffusivity included in a spherical coordinate system.

It is discovered from the computation of the refractive index square of stationary planetary waves that another waveguide pointing from the lower troposphere in middle latitudes toward the upper troposphere in low latitudes is present in the propagation of stationary planetary waves, in addition to the polar waveguide in winter. However, a waveguide also points from the lower troposphere in middle latitudes toward the upper troposphere in middle-low latitudes near 30°N, in addition to the waveguide propagating vertically to the upper troposphere in high latitudes in summer.

The computed results show that the vertical distributions of amplitude and phase, and stationary disturbance patterns at constant height responding to forcing by both the hemispheric topography and stationary heat sources, are in good agreement with the observations either in winter or in summer. In addition, the forced stationary planetary waves in winter differ appreciably from those in summer. This difference shows that the formation and propagation of stationary planetary waves occur in connection with the zonal mean wind.

I. INTRODUCTION

The response of a model atmosphere in middle latitude to forcing by topography and stationary heat sources was investigated by Huang and Gambo [7] with a β-plane approximation multilevel model. The computed results are in good agreement with observed results. In that paper, however, it was assumed that the motion takes place on a β-plane centered at 45°N and the mean zonal wind is constant. Dickinson [4] emphasized the role of horizontal wind shear in the vertical propagation of stationary planetary waves. Thus, the response of a hemispheric model atmosphere to forcing by topography and stationary heat sources must be discussed in a model where the vertical and horizontal wind shears are considered.

The response of a northern hemispheric model atmosphere to forcing by topography has been studied by many authors. For example, Staff Members, Academia Sinica [15] studied the effects of the forcing by topography and stationary heat sources on the formation of standing troughs and ridges with a two-level quasi-geostrophic model. Egger [5] investigated the linear response of a hemispheric model atmosphere to forcing by topography by means of a two-level primitive equation model. Ashe [1] showed with a two-level model that the nonlinear terms in the mean state equation are important in the formation of averaged asymmetric flow.

However, in these simplified models, such as the two-level model mentioned above, the effect of the vertical propagation of stationary planetary waves excited in the troposphere cannot be treated correctly in the stratosphere. Due to the poor vertical resolution in the simplified models, the vertical distributions of the amplitude and phase of stationary planetary waves responding to forcing by the northern hemispheric topography and stationary heat sources cannot be obtained correctly.

Matsuno [12] showed that the planetary scale, stationary disturbances in the winter stratosphere are considered to be upward propagating internal Rossby waves forced from below. In this computation, a multilevel model in the spherical coordinate system was used and the observed value of height at the 500 mb level was used as the lower boundary condition. The stationary planetary waves responding to forcing by the northern hemispheric topography and stationary heat sources in winter have been investigated by Huang and Gambo [8] using a 34-level spherical coordinate model. The considerable differences between the winter and summer response to forcing by topography and stationary heat sources are investigated in this paper.

II. THE MODEL AND PARAMETERS

The steady state, quasi-geostrophic vorticity and thermodynamic equations in which Rayleigh friction, the effect of Newtonian cooling and the horizontal kinematic thermal diffusivity are included, may be expressed as

$$\bar{u}\frac{\partial}{a\cos\phi\,\partial\lambda}(\zeta') + v'\frac{\partial}{a\,\partial\phi}(\bar{\zeta} + f) = f\frac{\partial\omega}{\partial p} - R_f\zeta', \tag{1}$$

$$\bar{u}\frac{a}{a\cos\phi\,\partial\lambda}\left(\frac{\partial\phi'}{\partial p}\right) - 2\Omega_o\sin\phi\frac{\partial\bar{u}}{\partial p}v' + \sigma\omega = -\frac{RH}{c_p p} - \alpha_R\frac{\partial\phi'}{\partial p} + K_T\nabla^2\left(\frac{\partial\phi'}{\partial p}\right), \tag{2}$$

respectively. The notations used in the above equations are as follows:

a: radius of the earth

$\bar{u}$: the basic zonal wind speed

v': meridional component of perturbation motion

ϕ': geopotential of perturbation

ζ': vertical component of relative perturbation vorticity

$\bar{\zeta}$: vertical component of relative vorticity of the basic state

$\sigma = -\alpha \frac{\partial \ell n\theta}{\partial p}$: static stability parameter (α: specific volume, θ: potential temperature)

H: diabatic heating per unit time and unit mass

R: gas constant (0.287 kJ kg^{-1} deg^{-1})

c_p: specific heat at constant pressure (1.004 kJ kg^{-1} deg^{-1})

α_R: Newtonian cooling coefficient

K_T: horizontal kinematic thermal diffusivity

R_f: Rayleigh friction coefficient of perturbation

f: Coriolis parameter

ω: vertical p-velocity (dp/dt)

If we include the second approximation to v' in the planetary vorticity advective term, we divide the planetary vorticity advection term into $2\Omega_o \cos\phi v'/a$ and remainder. Here $2\Omega_o \cos\phi v'/a$ expresses the advection of planetary vorticity by the north-south wind and has been shown to be dominant in the vorticity equation of planetary scale disturbances. If we divide the meridional component of the perturbation wind into the components of geostrophic wind and nongeostrophic wind (second approximation), we can obtain the second approximation to v' from the equation of motion. Thus

$$v' = \frac{1}{2\Omega_o \sin\phi}\left(\frac{1}{a\cos\phi}\frac{\partial\phi'}{\partial\lambda} - \hat{\Omega}\frac{1}{2\Omega_o \sin\phi}\frac{\partial^2\phi'}{\partial\phi\partial\lambda}\right) \qquad (3)$$

$$\hat{\Omega} = \frac{\bar{u}}{a\cos\phi}$$

Thus, the model equations are as follows:

$$\hat{\Omega}_{n-\frac{1}{2}} \frac{\partial}{\partial\lambda}\left\{\frac{1}{2\Omega_o \sin\phi}\frac{1}{a^2}\left[\frac{\sin^2\phi}{\cos\phi}\frac{\partial}{\partial\phi}\left(\frac{\cos\phi}{\sin^2\phi}\frac{\partial\phi'}{\partial\phi}\right) + \frac{1}{\cos^2\phi}\frac{\partial^2\phi'}{\partial\lambda^2}\right]\right\}_{n-\frac{1}{2}}$$

$$+ \frac{1}{a} q_{n-\frac{1}{2}} \frac{1}{2\Omega_o \sin\phi}\frac{1}{a\cos\phi}\frac{\partial\phi'_{n-\frac{1}{2}}}{\partial\lambda} = f\left(\frac{\partial\omega}{\partial p}\right)_{n-\frac{1}{2}} - (R_f)_{n-\frac{1}{2}} \times$$

$$\frac{1}{2\Omega_o \sin\phi}\frac{1}{a^2}\left[\frac{\sin\phi}{\cos\phi}\frac{\partial}{\partial\phi}\left(\frac{\cos\phi}{\sin\phi}\frac{\partial\phi'}{\partial\phi}\right) + \frac{1}{\cos^2\phi}\frac{\partial^2\phi'}{\partial\lambda^2}\right]_{n-\frac{1}{2}},$$

$$n = 1, 2, \ldots, 35. \qquad (4)$$

$$\hat{\Omega}_n \frac{\partial}{\partial\lambda}\left(\frac{\partial\phi'}{\partial p}\right)_n - \left(\frac{\partial\hat{\Omega}}{\partial p}\right)_n \frac{\partial\phi'_n}{\partial\lambda} + \sigma_n\omega_n = -\left(\frac{RH}{c_p p}\right)_n - (\alpha_R)_n \left(\frac{\partial\phi'}{\partial p}\right)_n$$

$$+ (K_T)_n \times \frac{1}{a^2}\left[\frac{\partial^2}{\partial\phi^2} - \tan\phi\frac{\partial}{\partial\phi} + \frac{1}{\cos^2\phi}\frac{\partial^2}{\partial\phi^2}\right]\left(\frac{\partial\phi'}{\partial p}\right)_n$$

$$n = 1, 2, \ldots, 34. \qquad (5)$$

Here q is expressed as

$$q = \left[2(\Omega_o + \hat{\Omega}) - \frac{\partial^2\hat{\Omega}}{\partial\phi^2} + 3\tan\phi\frac{\partial\hat{\Omega}}{\partial\phi}\right]\cos\phi$$

For the upper boundary conditions, we assume that the vertical p-velocity vanishes at the top of the model, i.e.

$$\omega = 0, \quad \text{at } p = p_t \quad (\text{or } Z = Z_t) \qquad (6)$$

For the lower boundary condition at $p = p_s$ (p_s: surface pressure), we assume that the vertical p-velocity is caused by surface topography where the standard pressure is p_G, and also by Ekman pumping resulting from the viscosity in the Ekman layer. Thus, the vertical p-velocity at $p = p_s$ is given as

$$\omega_s = \vec{V}_s \cdot \nabla p_G - \frac{p_s \cdot F}{2f}\zeta'_s \qquad \text{at } p = p_s \qquad (7)$$

where $\vec{V}_s$ is the horizontal velocity vector at $p = p_s$, and $p_s = 1000$ mb for simplicity. F is the friction coefficient and will be treated as a constant ($4 \times 10^{-6}\ s^{-1}$), and ζ'_s is the vorticity perturbation at the surface.

1. The Vertical and Meridional Difference Scheme

The vertical difference scheme used in this model is the same as that discussed in the paper of the β-plane approximation model. That is, the vertical grid increments (ΔZ) used in this model are as follows:

Z:		
	0-12 km	ΔZ = 1.5 km
	12-30 km	ΔZ = 2.0 km
	30-60 km	ΔZ = 3.0 km
	60-92 km	ΔZ = 4.0 km

The top of this model's atmosphere is defined as Z = 92 km (i.e. $p_t = 1.140 \times 10^{-3}$ mb in winter, $p_t = 8.459 \times 10^{-4}$ mb in summer). We divide the atmosphere into 34 layers from the earth's surface to 92 km.

Substituting Eqs. (6) and (7) into Eqs. (4) and (5), we can eliminate ω_n (n = 1, 2, ... 35). Thus, we obtain 35 linear differential equations in regard to ϕ'

We shall assume that the perturbation equations have a zonal structure given by $e^{ik\lambda}$, i.e. any solution to the model Eqs. (4) and (5) can be expressed as

$$\phi'(\lambda,\phi,p) = R_e \sum_{k=1}^{K} \Phi_k(\phi,p)e^{ik\lambda} \tag{8}$$

where k is the wave number in the longitudinal direction.

When Eq. (8) is substituted into the linear differential equations obtained above, we can get 35 linear differential equations in regard to $\Phi_k(\phi,p)$. In order to solve the linear algebraic equations in regard to $\Phi_k(\phi,p)$, the finite-difference scheme with the grid interval of $\Delta\phi = 5°$ is used in the latitudinal direction. The finite-difference analogue thus formulated makes a system of linear equations for $\Phi_{j,n}$ at 19 x 35 points (19 and 35 points in the ϕ and p directions, respectively). In order to solve these linear algebraic equations, the lateral boundary conditions are necessary. We require that $\Phi_k(\phi,p)$ should vanish at the pole and equator, i.e.

$$\Phi_k(\phi,p) = 0, \qquad \phi = \frac{\pi}{2}$$

$$\Phi_k(\phi,p) = 0 \qquad \phi = 0 \tag{9}$$

Since the relaxation methods are not generally applicable for the linear algebraic equations obtained above, the method proposed by Lindzen and Kuo [10] is used to solve such a system of linear equations.

Therefore, if the forcing functions of topography and diabatic heating are considered as known from observations, the response of a model atmosphere to forcing by the hemispheric topography and stationary heat sources can be computed from these linear algebraic equations. The vertical distributions of the amplitude and phase for stationary planetary waves in the northern hemisphere also can be computed.

2. Parameters

(1) Static stability σ_n: The static stability parameter, σ_n, is calculated in this hemispheric model from the mean temperature and density at 45°N in January and July obtained for the U.S. Standard Atmosphere. For simplicity, we will assume that the static stability parameter does not change with latitude.

(2) The vertical profile of the basic zonal mean wind: The observed zonal mean wind in the global atmosphere up to the mesopause level is

presented by Murgatroyd [13]. But these results cannot be used directly in this study because the data include small-scale features which may have significant influence on the calculation of the refractive index square of stationary planetary waves. For this reason, we shall use the vertical distribution of the zonal mean wind constructed by Matsuno [12] up to height 60 km in winter. In addition, the vertical distribution from height 60 km to 92 km is extrapolated referring to the vertical distribution presented by Holton [6]; the vertical distribution thus obtained is similar to that constructed by Lordi et al. [11] and is shown in Fig. 1.

The vertical distribution of the zonal mean wind, computed by Murgatroyd [13], is used. In order to eliminate small-scale features, we have smoothed the results computed by Murgatroyd and the results are shown in Fig. 2.

(3) The coefficient of Rayleigh friction R_f: The value of R_f up to 30 km height is assumed to be $0.1 \times 10^{-6}\ s^{-1}$ [6]. We shall assume plausible values above 30 km, while large values of R_f are incorporated above 40 km to eliminate the influence of upward propagating waves reflected at the top of the model.

(4) The horizontal kinematic thermal diffusivity coefficient and the coefficient of Newtonian cooling shall be used.

III. COMPUTATION OF THE REFRACTIVE INDEX OF STATIONARY PLANETARY WAVES

In order to discuss the characteristics of the vertical and horizontal propagation of stationary planetary waves, and in a sense to better understand the results of numerical computations of stationary planetary waves induced by the topographical and thermal effect, we shall compute the refractive index square of stationary planetary waves as proposed by Charney and Drazin [3] in winter and summer.

If there is no heat source and Rayleigh friction, the effect of Newtonian cooling and the horizontal kinematic thermal diffusivity are not considered, and if we assume that the atmosphere is nearly isothermal, we can obtain the potential vorticity equation in the Z-coordinate system from Eqs. (1) and (2):

$$\frac{\sin^2\phi}{\cos\phi}\frac{\partial}{\partial\phi}\left(\frac{\cos\phi}{\sin^2\phi}\frac{\partial\psi_k}{\partial\phi}\right) + \ell^2\sin^2\phi\,\frac{\partial^2\psi_k}{\partial z^2} + Q_k\psi_k = 0 \qquad (10)$$

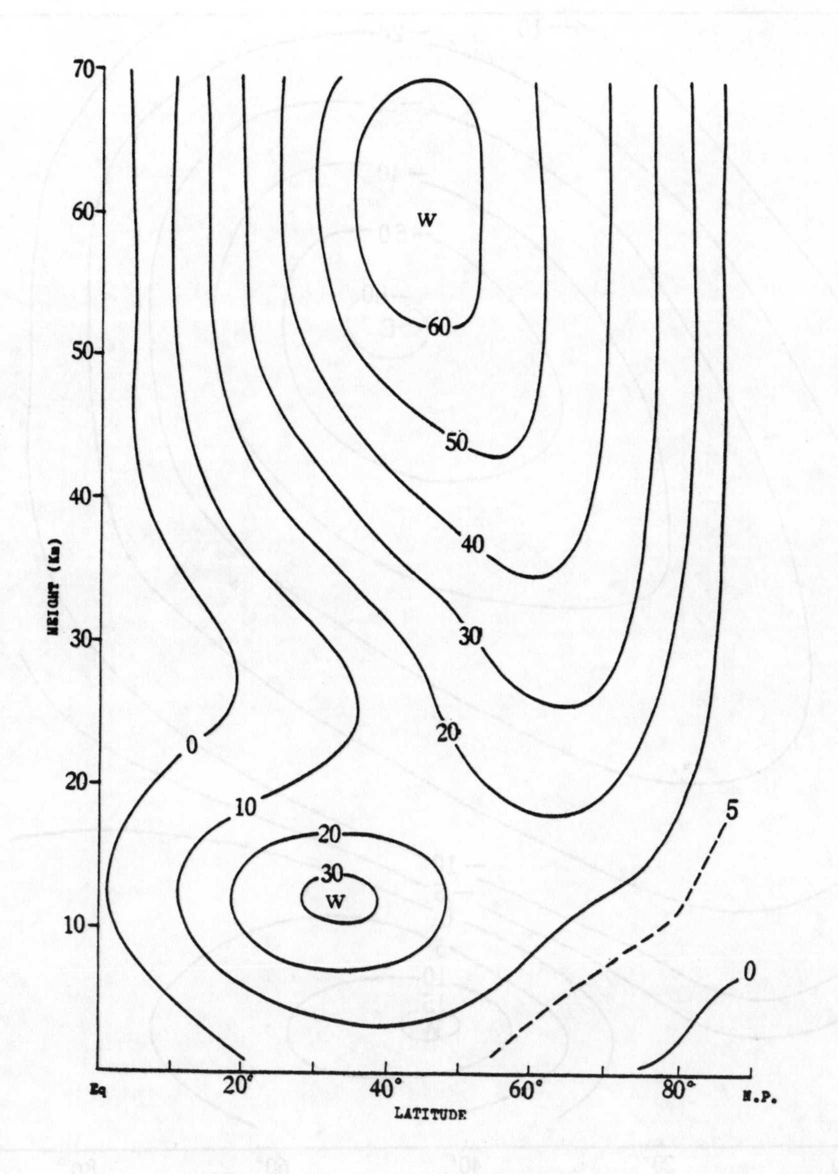

Fig. 1 The model basic state zonal mean wind distribution (ms^{-1}) in the winter.

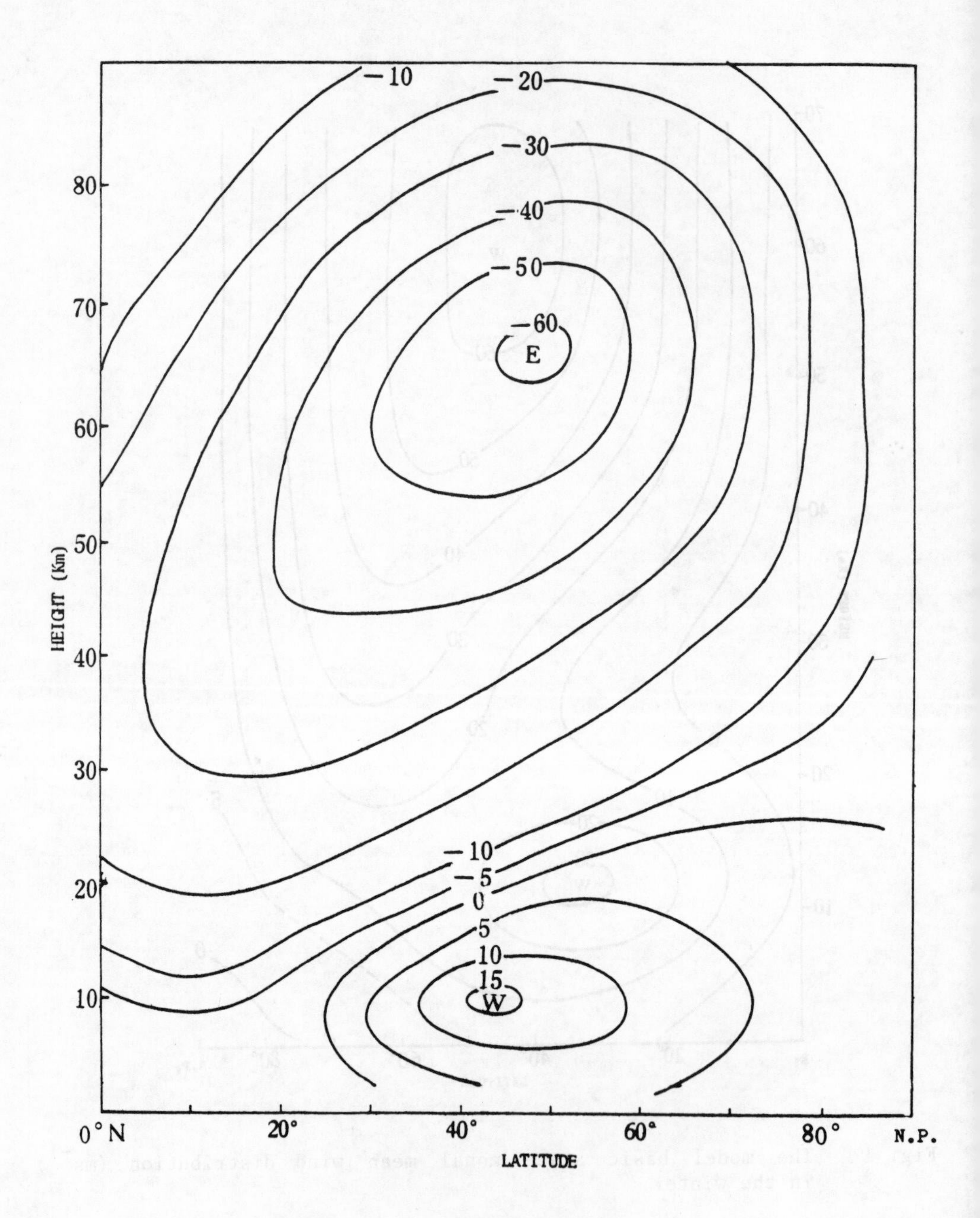

Fig. 2 As in Fig. 2, except for summer.

$$Q_k = Q_o - \frac{k^2}{\cos^2\phi} \tag{11}$$

$$Q_o = [2(\Omega_o+\hat{\Omega}) - \frac{\partial 2\hat{\Omega}}{\partial\phi^2} + 3\tan\phi \frac{\partial\hat{\Omega}}{\partial\phi} - \ell^2\sin^2\phi(\frac{\partial^2\hat{\Omega}}{\partial Z^2} - \frac{1}{H_o}\frac{\partial\hat{\Omega}}{\partial Z})]/\hat{\Omega}$$

$$- \ell^3\sin^2\phi \frac{1}{4H_o^2}$$

$$\ell = 2\Omega_o a/\tilde{N} \quad .$$

Here H_o is 7 km and $\tilde{N}$ is 2×10^{-2} s^{-1} in the isothermal atmosphere. Equation (10) is the two-dimensional wave equation; it describes the wave propagation in the ϕ and Z directions in an isothermal atmosphere. Here Q_k is regarded as the refractive index square of waves for zonal wave number k.

(1) Winter: The distribution of Q_k in winter is computed from the zonal mean wind shown in Fig. 1 by Eqs. (11) and (12); the dashed curves in Fig. 3 show the refractive index square Q_1 for the k = 1 wave.

We note that a minimum of Q_1 is located in the stratosphere and the mesosphere in high latitudes, and that another minimum of Q_1 is found in the lower layer of the stratosphere in middle latitudes. Because the Q_k level may be considered as the wave front of planetary waves for wave number k, the stationary planetary waves are propagated through the belt of larger values of Q_1 and blocked by the region of small values of Q_1. Therefore, it may easily be understood that there are two waveguides in the vertical and lateral propagation of stationary planetary waves for wave number k = 1, i.e. a waveguide extending from the troposphere toward the upper stratosphere is the polar waveguide suggested by Dickinson [4]. It is discovered in the present investigation that another waveguide extends from the lower troposphere in middle latitudes to the upper stratosphere in low latitudes. For the sake of better understanding the characteristics of the waveguide mentioned above, we computed the response of a model atmosphere to forcing by idealized topography and by idealized stationary heat sources at 40°N and 80°N, respectively. The computed results have proved that the two waveguides discussed above for the vertical and meridional propagations of stationary waves responding to forcing by idealized topography, or by idealized stationary heat sources do exist. The schematic waveguides for the stationary planetary wave k = 1, responding to forcing by idealized topography at 40°N, are denoted by arrows in Fig. 3.

(2) Summer: Similarly, the distribution of Q_k in summer is computed from the zonal mean wind shown in Fig. 2 by Eqs. (11) and (12). The dashed curves in Fig. 4 show the refractive index square Q_1 for wave

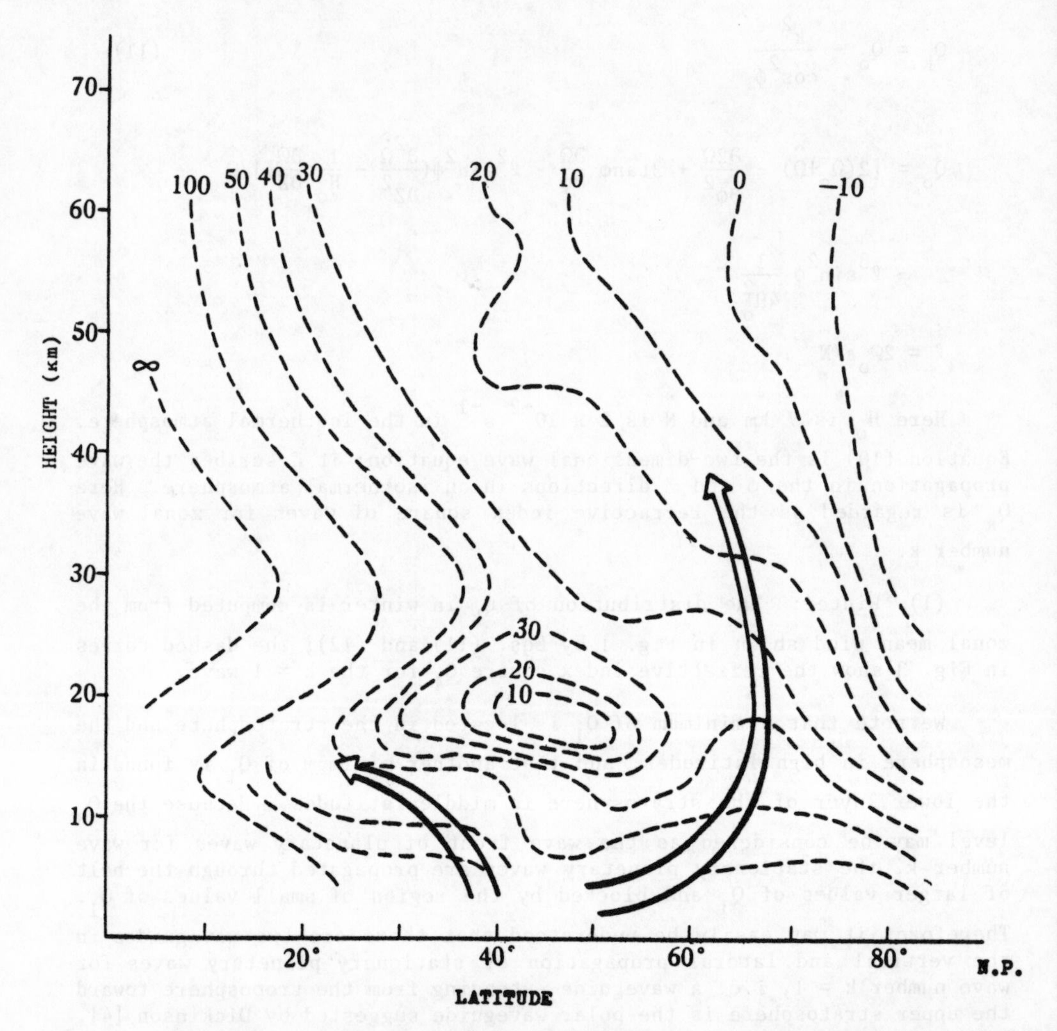

Fig. 3 Schematic diagram of the waveguide of a stationary planetary wave for wave number k = 1, responding to forcing by idealized topography at 40°N. Dashed curves show the refractive index square, Q_1 for wave number k = 1.

number k = 1. We note that a minimum of Q_1 is located in the lower stratosphere at middle to high latitudes. The values of Q_1 in the troposphere and stratosphere at low latitudes, and in the stratosphere at middle and high latitudes, are negative. Therefore, the stationary planetary waves in summer cannot be propagated into the stratosphere. This differs from the characteristics of the distribution of Q_1 in winter because of the differences in the zonal mean wind, according to the

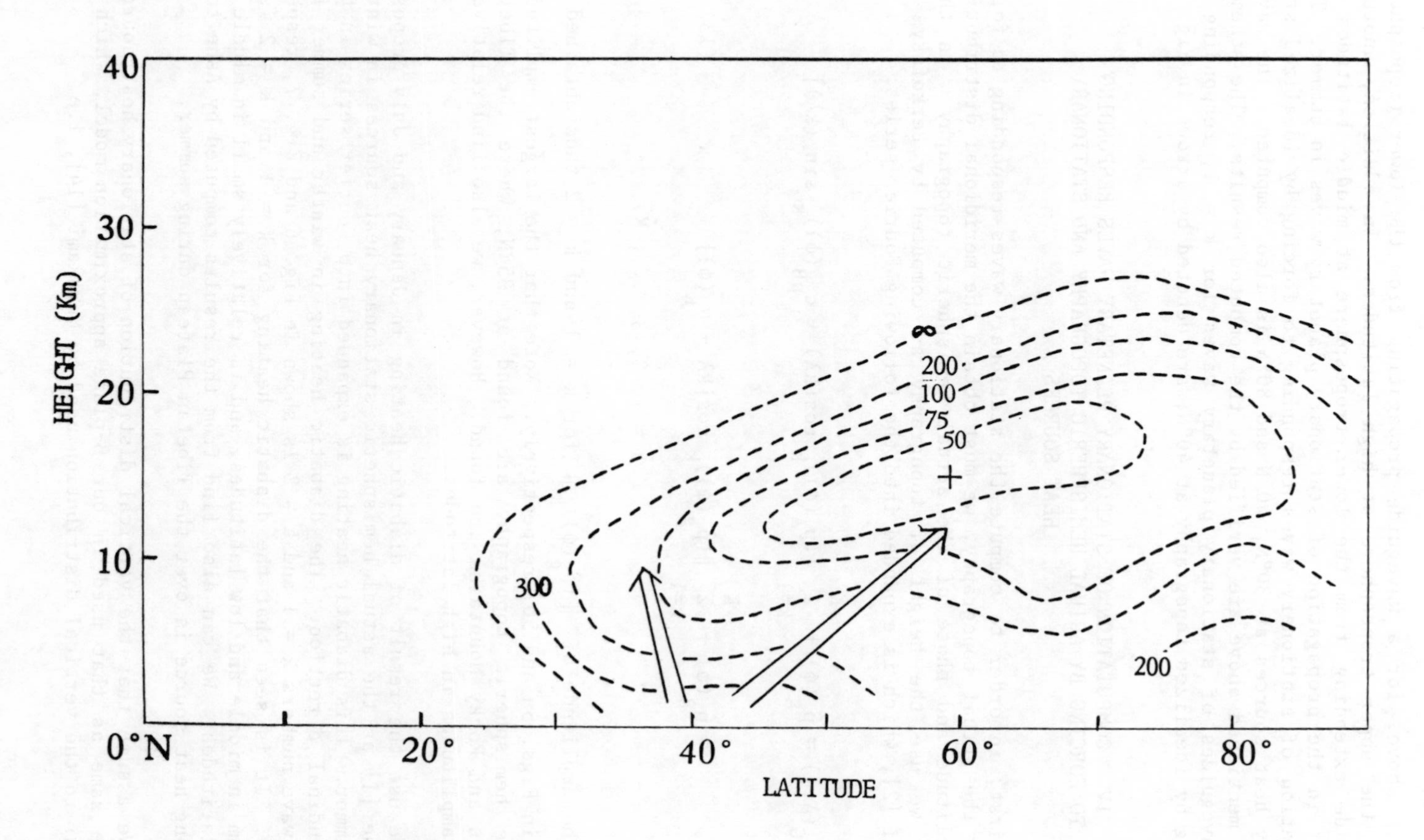

Fig. 4 As in Fig. 3, except for summer.

analysis above, for a waveguide propagating from the lower troposphere toward the upper troposphere at high latitudes. In addition, another waveguide extending from the lower troposphere at middle latitudes is present in the propagation of stationary planetary waves in summer. The propagation of stationary waves responding to forcing by idealized stationary heat sources at 30°N, 40°N and 80°N is also computed. The waveguides mentioned above are verified by the computed results. The schematic waveguides of stationary planetary waves for $k = 1$, responding to forcing by idealized topography at 40°N, are denoted by arrows in Fig. 4.

IV. COMPUTATION OF STATIONARY PLANETARY WAVES RESPONDING TO FORCING BY ACTUAL HEMISPHERIC TOPOGRAPHY AND STATIONARY HEAT SOURCES

First, in order to compute the stationary waves responding to forcing by the actual topography, we must obtain the meridional distribution of amplitude and phase of the actual hemispheric topography. In this paper, we use the height of topography as computed by Berkofsky and Bertoni [2], which is expanded into the following Fourier series:

$$p_G(\lambda,\phi) = \hat{p}_g(\phi) + \sum_{k=1}^{K} [(p_A(\phi))_k \cos(k\lambda) + c_{pB}(\phi))_k \sin(k\lambda)]$$

$$= \hat{p}_g(\phi) + \sum_{k=1}^{K} \|(\hat{p}_g(\phi))_k \cos[k\lambda - \alpha_k(\phi)] \qquad (13)$$

$$k = 1, 2, 3, \ldots, K.$$

The amplitude of $\|(\hat{p}_g(\phi))_k\|$ for $k = 1$ and $k = 2$ thus obtained is shown in Figs. 5a and 5b, respectively. Note that the largest amplitudes of the hemispheric topography are found at 35°N, where the Tibetan Plateau and Rocky Mountains are found. However, we also find relatively large amplitudes in high latitudes.

We use the result of diabatic heating in January and July proposed by Ashe [1] as the actual hemispheric stationary heat sources in winter and summer. This diabatic heating is expanded into Fourier series in the longitudinal direction. The diabatic heating in winter and summer for zonal wave numbers $k = 1$ and $k = 2$ is shown in Fig. 6 and Fig. 7, respectively. It is seen that the diabatic heating for $k = 1$ and $k = 2$ is a maximum in middle and low latitudes, and is relatively small in middle to high latitudes. We can also find from the results computed by Ashe that a strong heat source is over the Tibetan Plateau during summer.

We assume that the vertical distribution of stationary heat sources is the same as that used in our β-plane approximation model, which is similar to the vertical distribution used by Murakami [14], i.e.

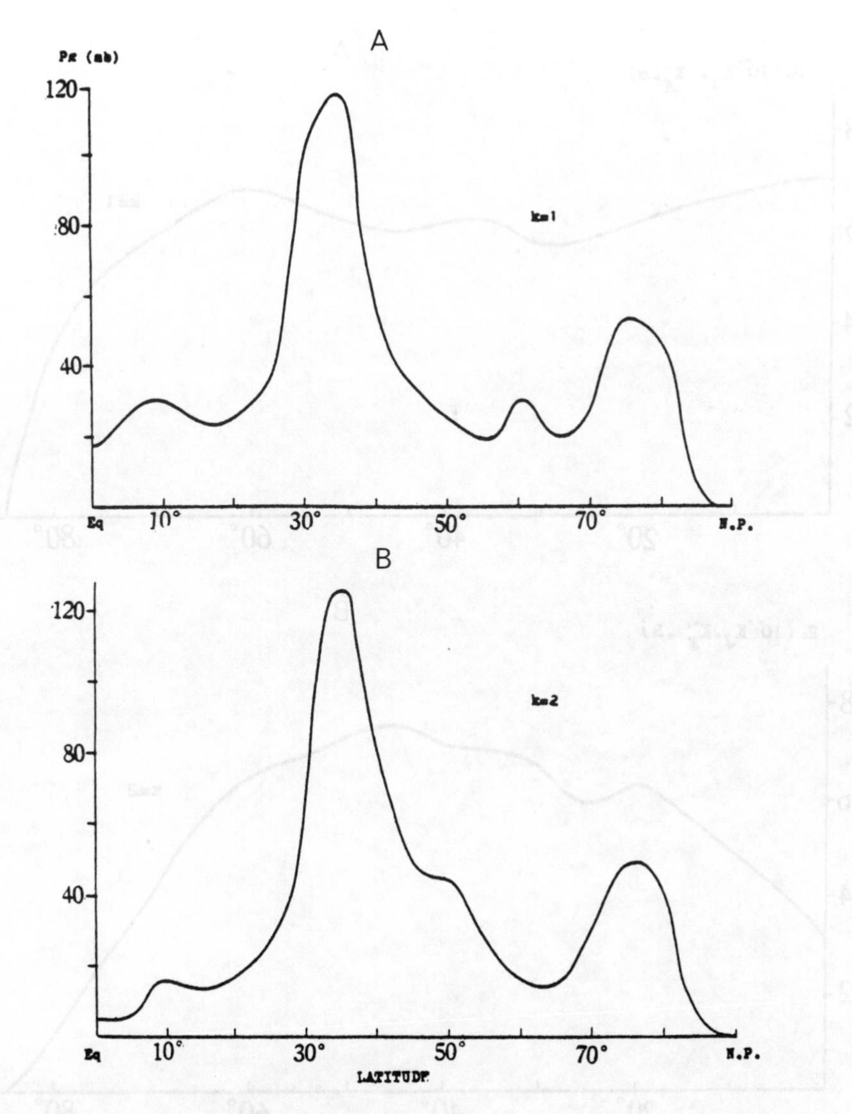

Fig. 5 The amplitude of zonal wave component 1 (a) and 2 (b) for standard pressure at ground.

$$(\hat{H}_o(\phi,p))_k = (\hat{H}_o(\phi))_k \exp(-(\frac{p-\bar{p}}{d})^2) \qquad (14)$$

Here, we assume d = 300 mb and $\bar{p}$ = 500 mb. This means that the distribution of diabatic heating has a maximum at 500 mb and decreases exponentially above and below 500 mb. Since strong heat sources and cooling sources are mainly in middle and low latitudes, according to the actual vertical distribution of diabatic heating in summer, the height of maximum diabatic heating at low latitudes is higher than that in middle and high latitudes. Thus, we assume $\bar{p}$ = 400 mb south of 40°N in summer.

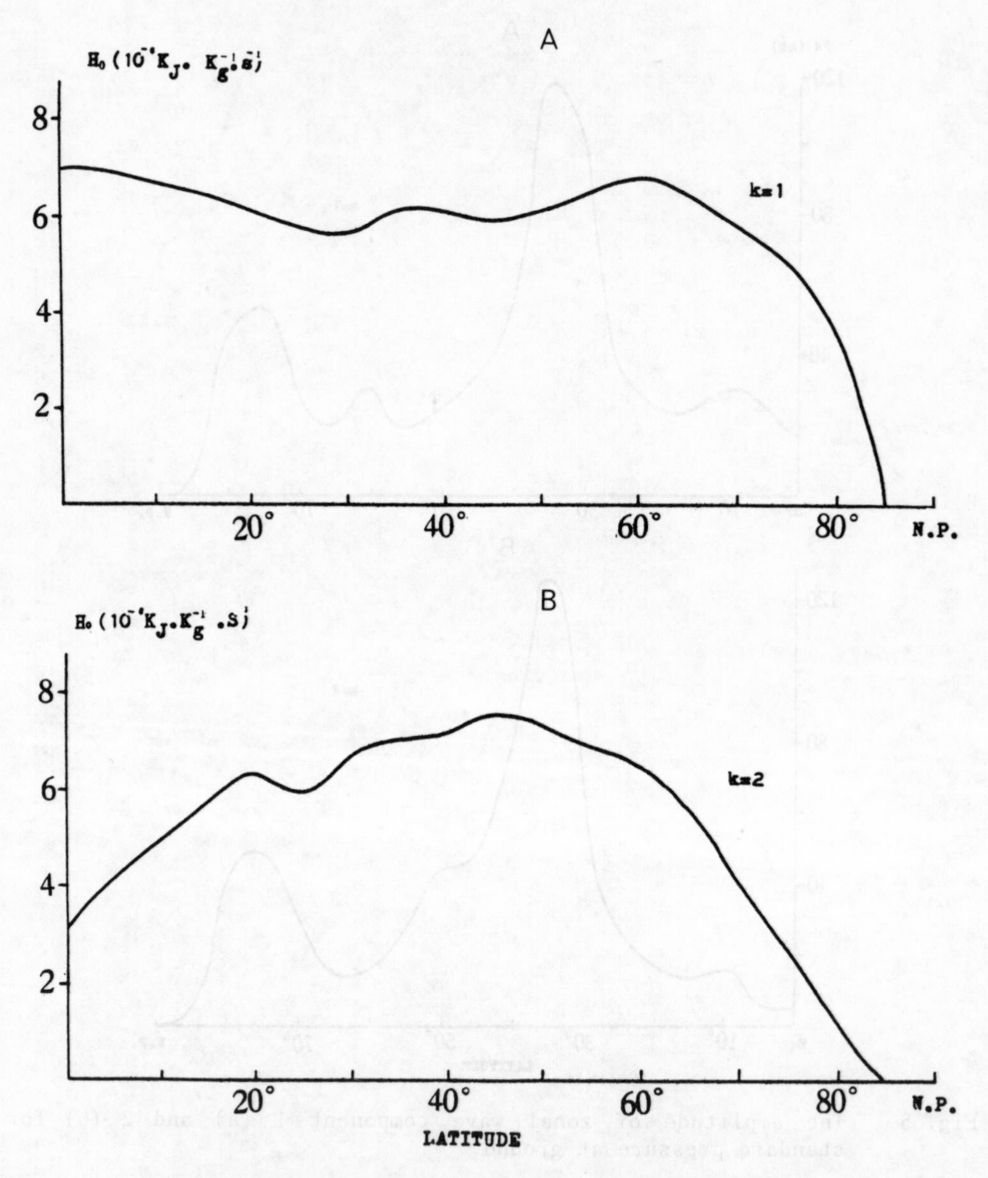

Fig. 6 The diabatic heating at 500 mb for zonal wave number 1 (a) and 2 (b) in winter.

When we substitute the forcing function of both actual topography and diabatic heating into the linear algebraic Eqs. (4) and (5), we obtain the vertical distributions of amplitude and phase for stationary planetary waves responding to these actual hemispheric forcing mechanisms.

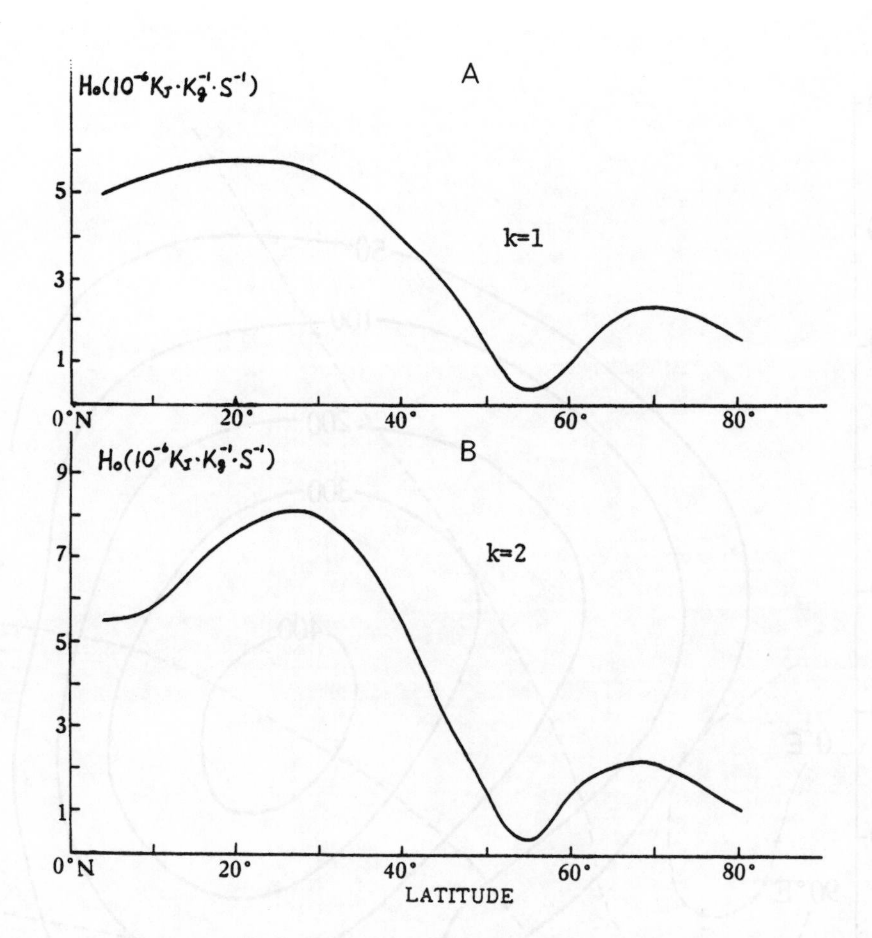

Fig. 7 As in Fig. 6, except for summer.

1. Winter

The vertical distributions of the amplitude and phase in response to forcing by both actual hemispheric topography and stationary heat sources in winter for zonal wave numbers k = 1 and k = 2 are shown in Figs. 8 and 9, respectively. For the sake of comparison, the latitude-height sections of the observed amplitude and phase for k = 1 in January, averaged over the years 1964 to 1970 in the troposphere and over 1965 to 1969 in the stratosphere, are reproduced from the paper of van Loon et al. [16] in Figs. 10 and 11.

Here, we find the following results from Figs. 8 through 11:

1. The amplitude of stationary planetary waves responding to forcing by both actual topography and stationary heat sources has a maximum at 38 km height and 60°N for k = 1 and at 27 km height and 60°N for k =

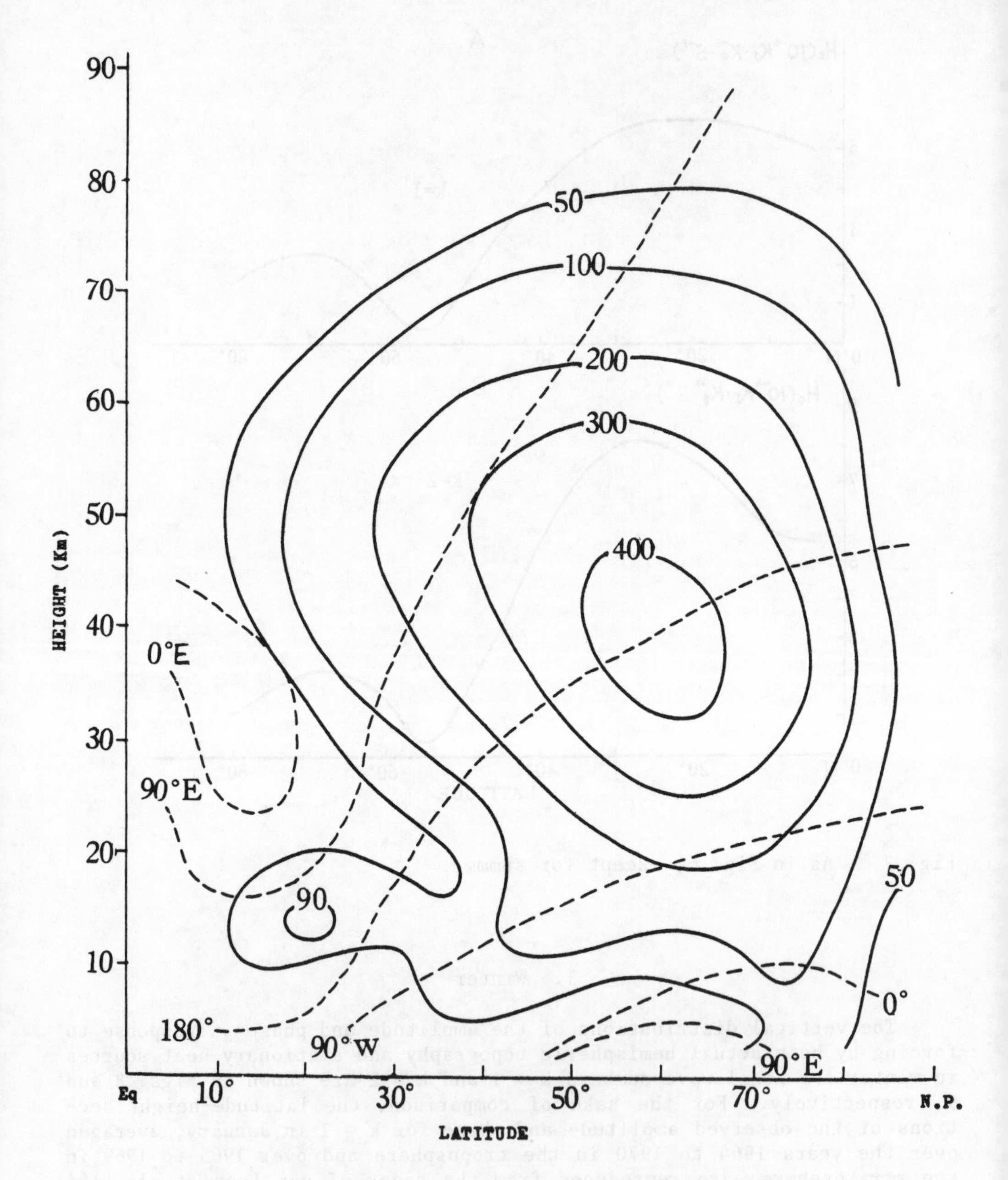

Fig. 8 The vertical distribution of amplitude (m) (solid curves) and phase (dashed curves) of stationary planetary waves, responding to forcing by both the northern hemispheric topography and stationary heat sources for wave number k = 1. Position of ridges is shown on the equal phase line.

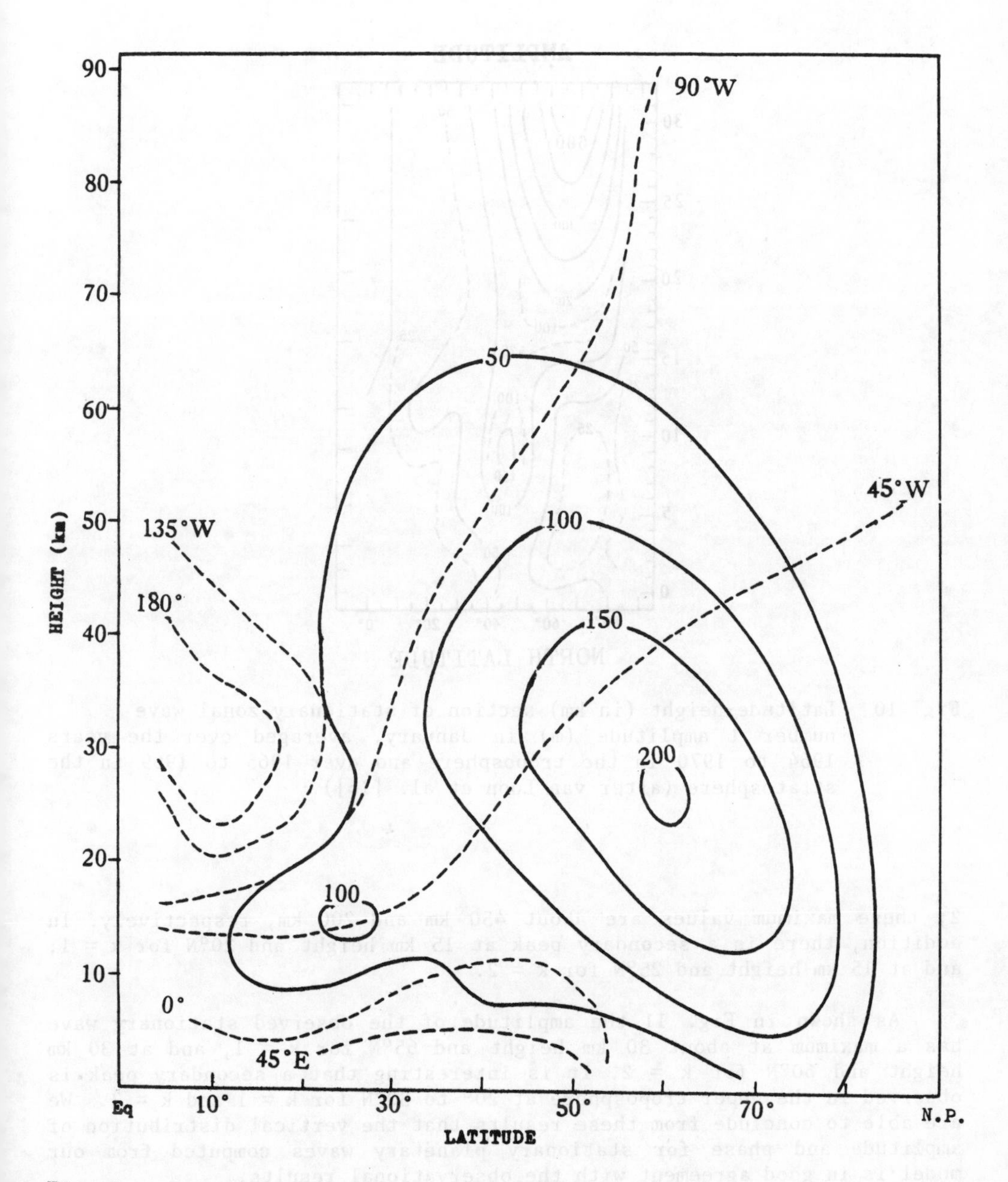

Fig. 9   As in Fig. 8, except for wave number k = 2.

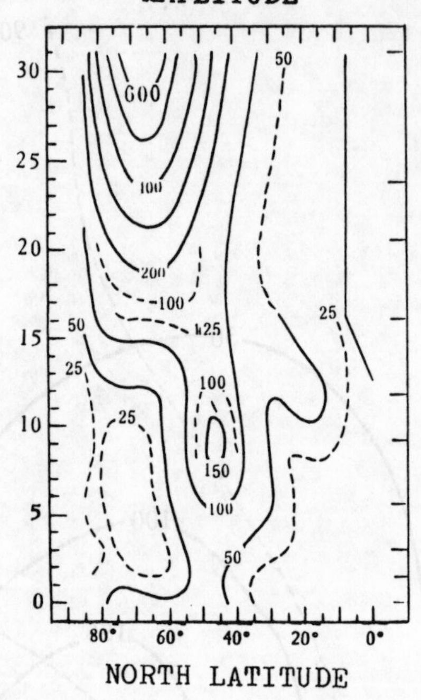

Fig. 10 Latitude-height (in km) section of stationary zonal wave number 1 amplitude (m) in January, averaged over the years 1964 to 1970 in the troposphere and over 1965 to 1969 in the stratosphere (after van Loon et al. [16]).

2; these maximum values are about 450 km and 200 km, respectively. In addition, there is a secondary peak at 15 km height and 20°N for k = 1, and at 15 km height and 25°N for k = 2.

As shown in Fig. 11 the amplitude of the observed stationary wave has a maximum at about 30 km height and 65°N for k = 1, and at 30 km height and 60°N for k = 2. It is interesting that a secondary peak is observed in the upper troposphere at 20° to 30°N for k = 1 and k = 2. We are able to conclude from these results that the vertical distribution of amplitude and phase for stationary planetary waves computed from our model is in good agreement with the observational results.

2. A remarkable feature in the amplitude distributions, computed from our model, is that disturbances are mainly confined to high latitudes, centered at 60°N. This is one of the well-known characteristics of stratospheric circulation. Moreover, the secondary peaks of stationary planetary waves are obtained in the upper troposphere at low latitudes. This secondary peak may be considered as a reason for the

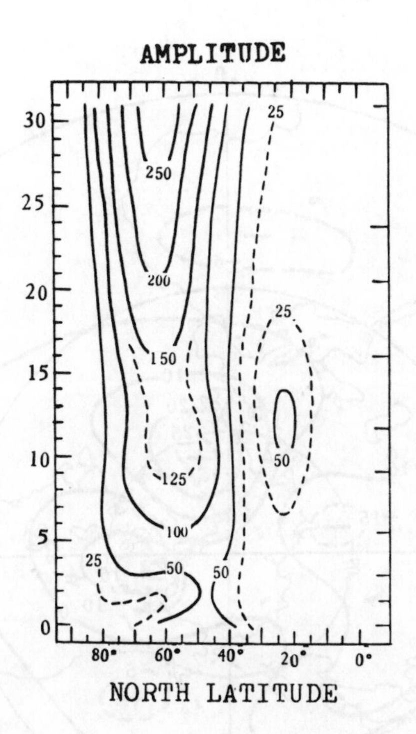

Fig. 11 As in Fig. 10, except for zonal wave number 2.

formation of stationary planetary waves at low latitudes. Webster [17] suggested that most of the time-independent circulation in low latitudes is forced by heating and orography within the tropics and subtropics. However, we consider that the forcing effects from higher latitudes are more important in the formation of stationary planetary waves in the lower stratosphere at low latitudes.

Next, we compute the disturbance pattern on an isobaric surface (or at constant height) responding to forcing by actual topography and stationary heat sources in winter by means of Eq. (8), by combining zonal wave components k = 1 to 3. Figure 12 shows the computed disturbance pattern at the 12 km level. For the sake of comparison, we calculated the disturbance pattern at 200 mb in January, averaged over the years 1972 to 1977 from observed data as shown in Fig. 13. We see that the computed result is in good agreement with the observed one. In both computed and observed patterns negative anomalies are found over the eastern coast of Asia and North America, while positive anomalies are found near the west side of the Rocky Mountains and the Atlantic Ocean. It is interesting to note that there are three subtropical standing disturbance patterns over the region at about 20°N in Fig. 13, i.e., the positive anomaly is found over the western coast of the Pacific Ocean and North Africa. We remark that the computed position of the subtropical high over the West Pacific Ocean is in good agreement with the observed one.

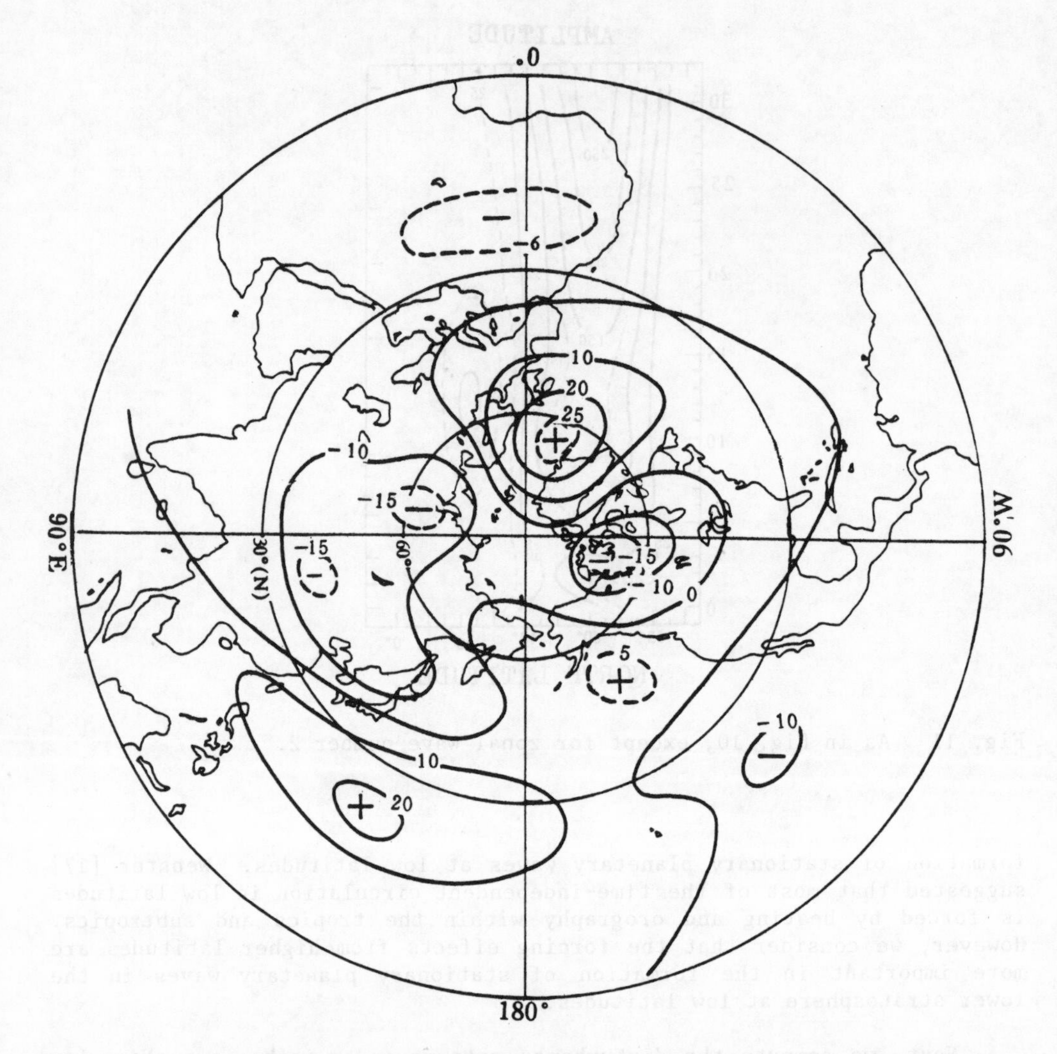

Fig. 12 The disturbance pattern (in dm) at the 12 km level, responding to forcing by both the northern hemispheric toporaphy and stationary heat sources.

2. Summer

The vertical distributions of amplitude and phase responding to forcing by both actual hemispheric topography and stationary heat sources in summer for zonal wave numbers $k = 1$ and $k = 2$ are shown in Figs. 14 and 15, respectively. Iwashima [9] calculated latitude-height sections of the observed amplitude and phase for $k = 1$ and $k = 2$ in July, averaged over the years 1965 to 1977.

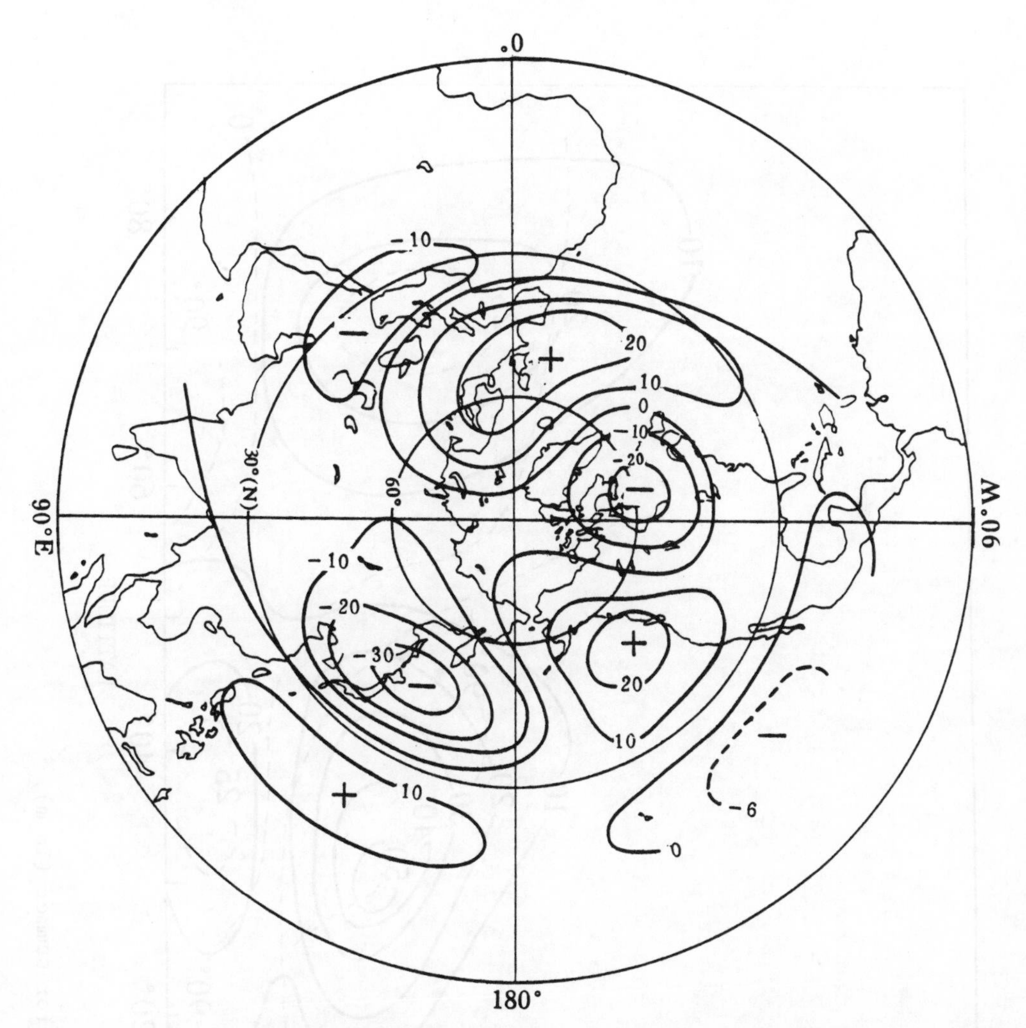

Fig. 13 The observed January planetary wave height pattern (in dm) due to zonal wave numbers 1 to 3 at 200 mb.

We find the following results from Figs. 14 and 15:

1. The amplitudes of stationary planetary waves responding to forcing by both actual topography and stationary heat sources have a maximum at 13 km height and 30°N for $k = 1$ and $k = 2$; this maximum value is about 60 m for $k = 1$ and $k = 2$. Moreover, there is a secondary peak at 13 km height and 70°N, while the amplitude for $k = 2$ is small at high latitudes. However, the computed amplitudes for wave numbers 1 and 2 are relatively small compared with the observed ones.

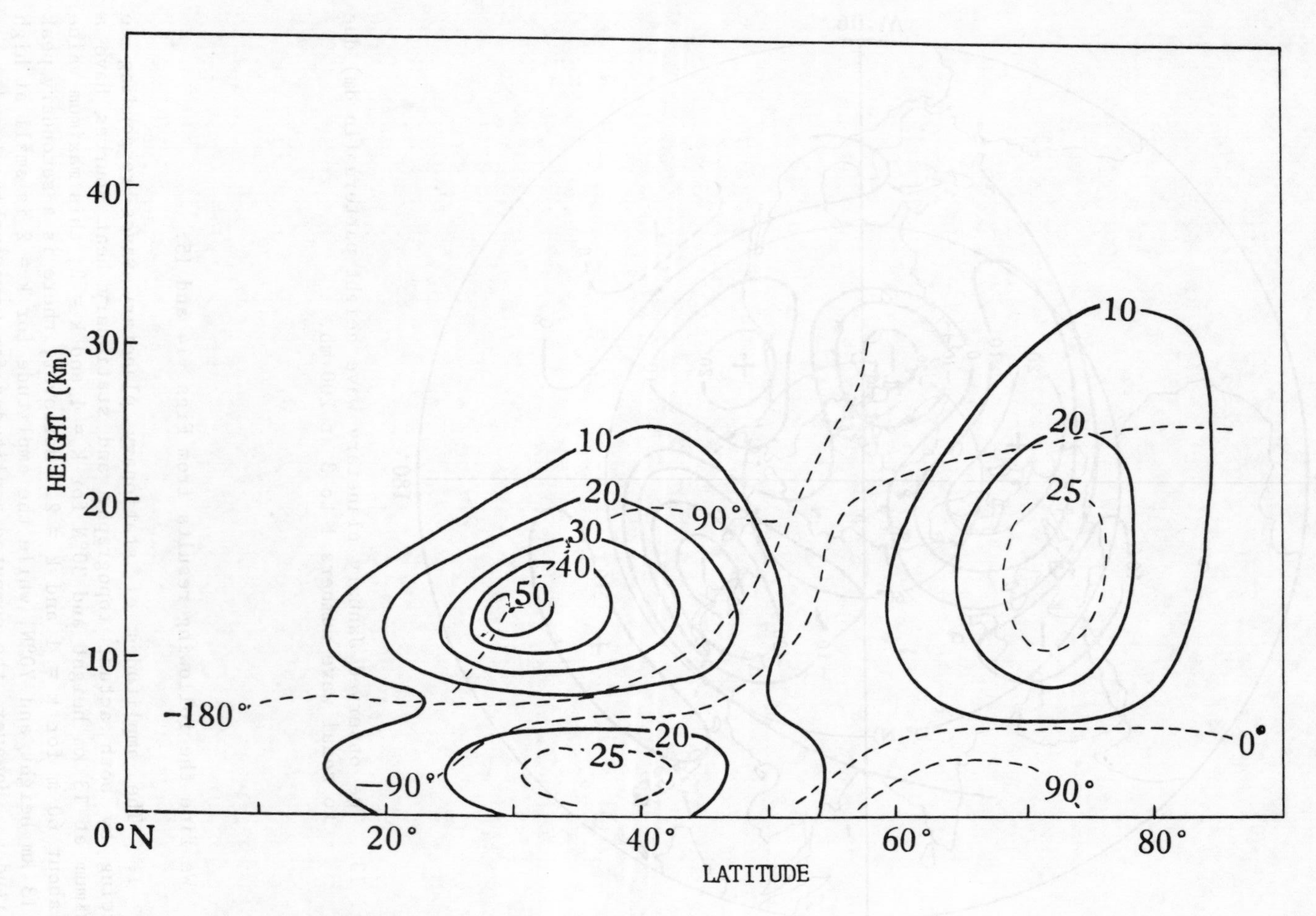

Fig. 14 As in Fig. 8, except for summer (in m).

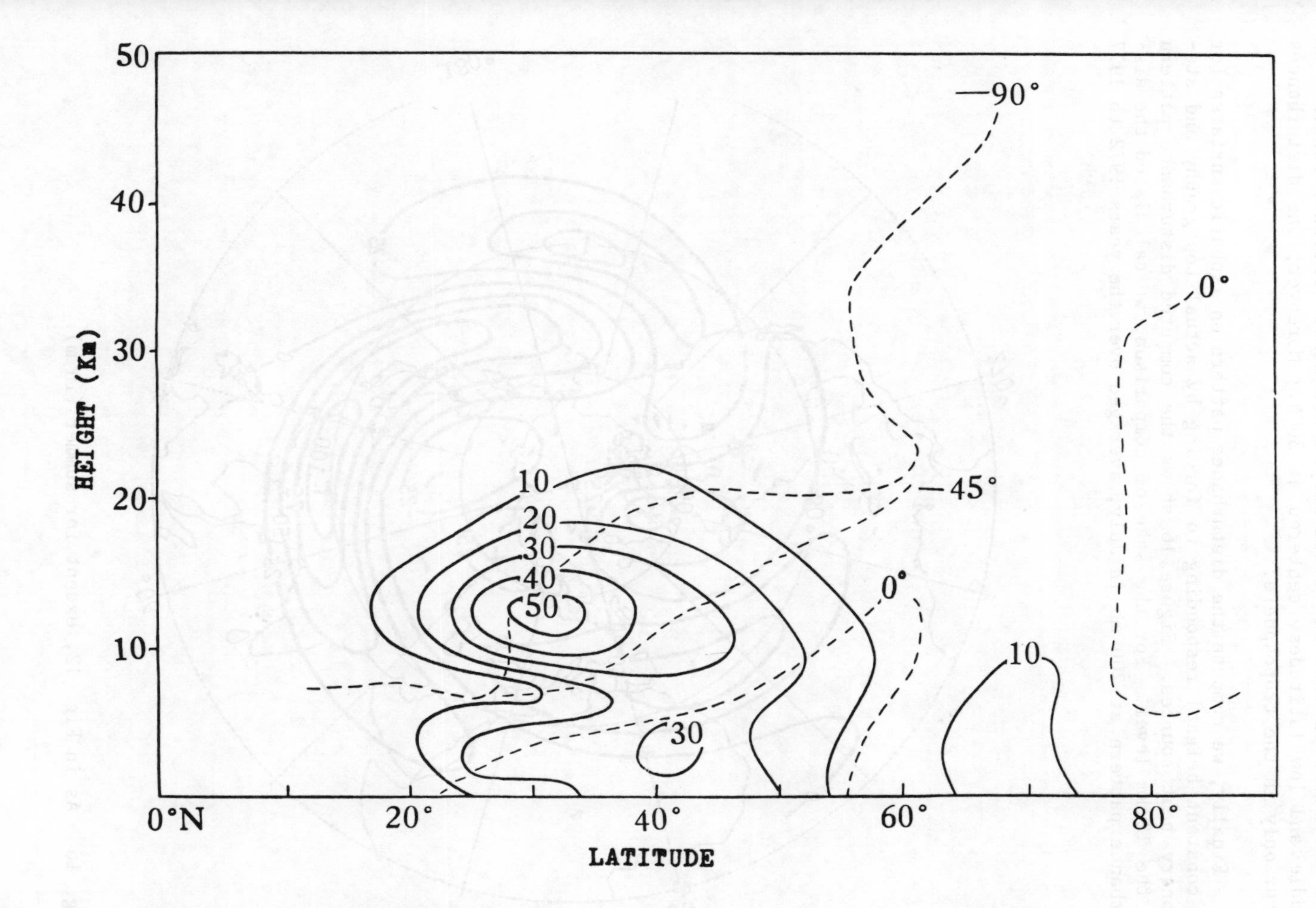

Fig. 15 As in Fig. 14, except for wave number 2.

2. A remarkable feature in the amplitude distributions computed from our model is that the disturbances are almost entirely confined to middle and low latitudes, centered at 30°N. Moreover, the disturbances occur only in the troposphere.

Finally, we compute the disturbance pattern on isobaric surfaces (or at constant height) responding to forcing by actual topography and stationary heat sources. Figure 16 shows the computed disturbance pattern at the 12 km level. For the sake of comparison, we calculated the disturbance pattern at 200 mb in July, averaged over the years 1972 to 1977

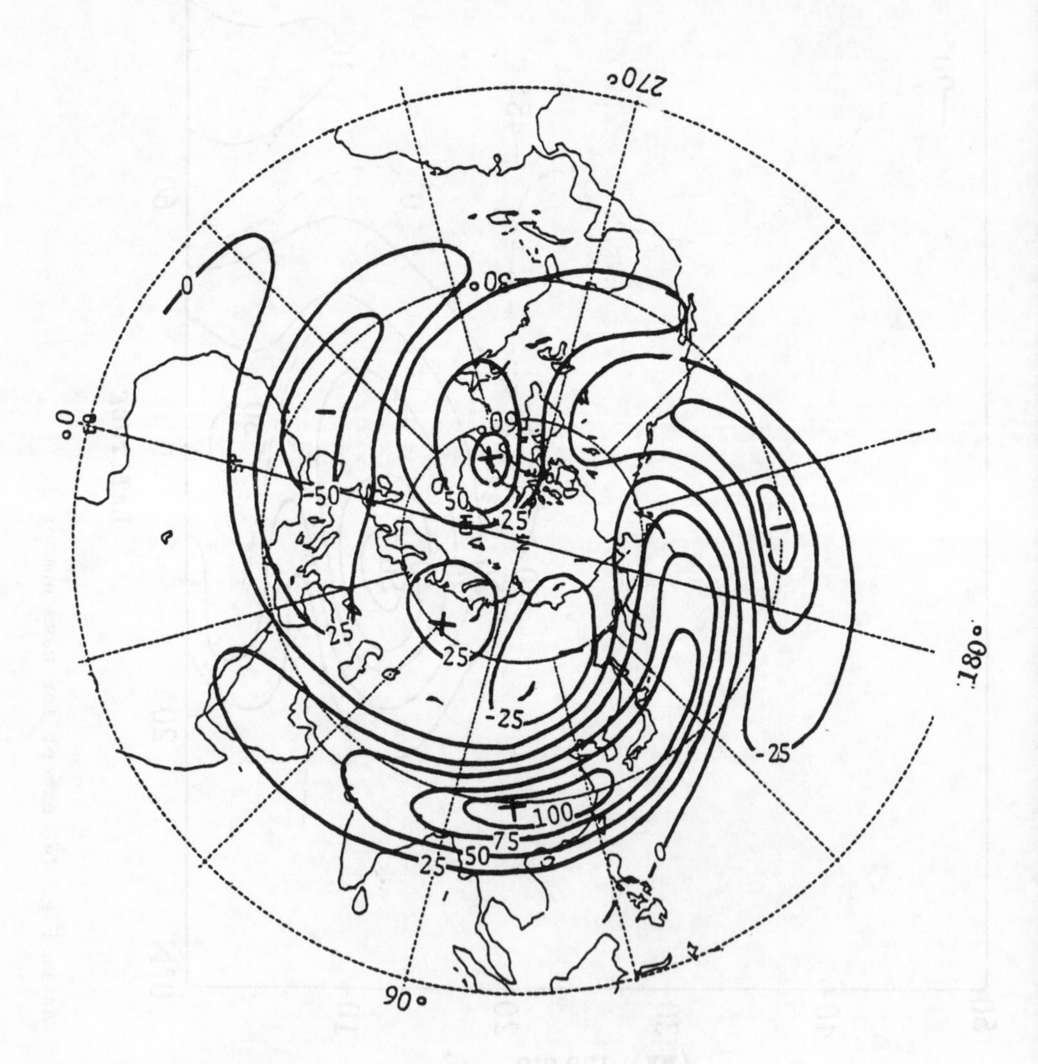

Fig. 16 As in Fig. 12, except for summer (in m).

from the observed data and it is shown in Fig. 17. We find that the computed result is in good agreement with the observed one. In both computed and observed patterns, the major positive anomaly is found over the Tibetan high. Other positive anomalies are found over the north coasts of America and Europe.

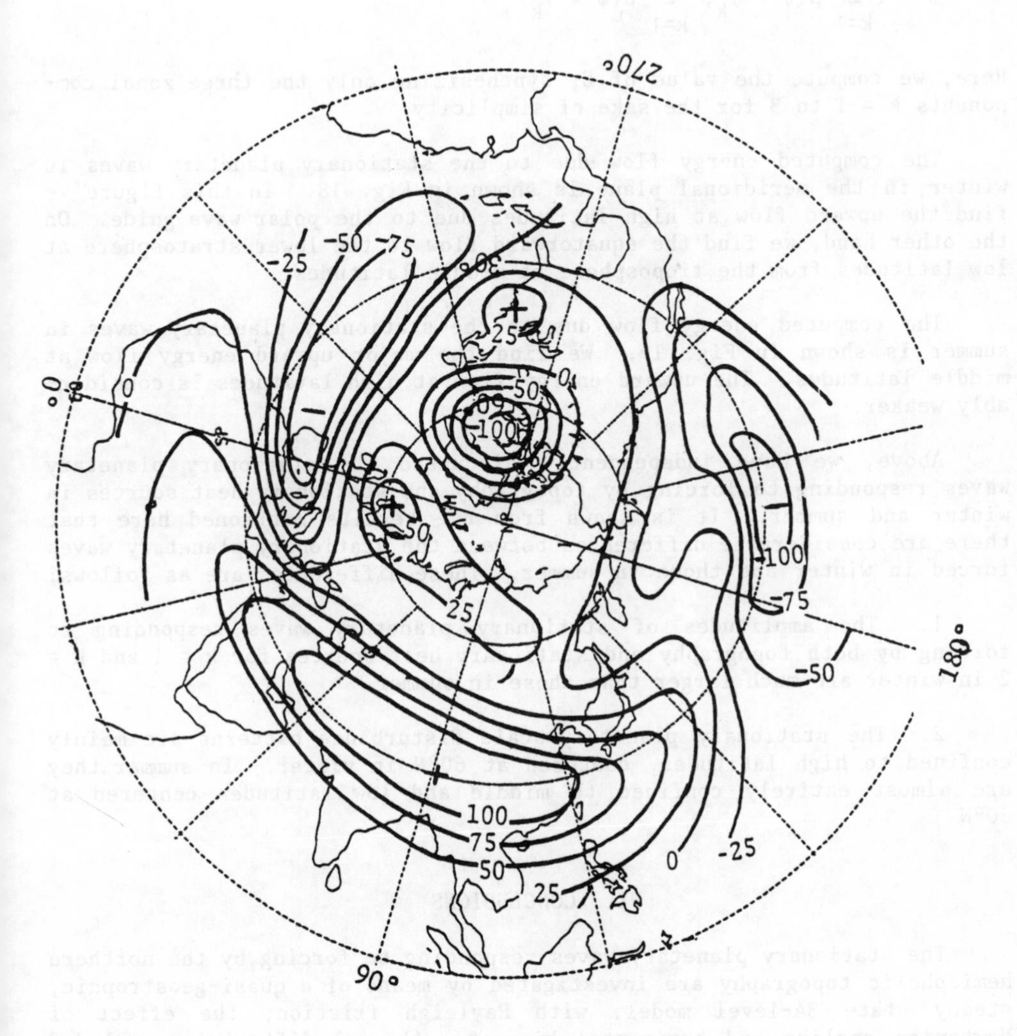

Fig. 17 As in Fig. 13, except for summer (in m).

V. ENERGY FLOWS DUE TO THE STATIONARY PLANETARY WAVES

The eddy momentum and heat fluxes can also be directly used to compute the energy flux by means of the following relation:

$$E = \left(\sum_{k=1}^{3} \bar{\rho}(\overline{\phi' v'})_k, \quad \sum_{k=1}^{3} \bar{\rho}(\overline{\phi' w'})_k \right)$$

Here, we compute the value of E, synthesizing only the three zonal components k = 1 to 3 for the sake of simplicity.

The computed energy flow due to the stationary planetary waves in winter in the meridional plane is shown in Fig. 18. In this figure we find the upward flow at high latitudes due to the polar wave guide. On the other hand, we find the equatorward flow to the lower stratosphere at low latitudes from the troposphere at middle latitudes.

The computed energy flow due to the stationary planetary waves in summer is shown in Fig. 19. We find the major upward energy flow at middle latitudes. The upward energy flow at high latitudes is considerably weaker.

Above, we have independently discussed the stationary planetary waves responding to forcing by topography and stationary heat sources in winter and summer. It is shown from the results mentioned here that there are considerable differences between the stationary planetary waves forced in winter and those in summer. These differences are as follows:

1. The amplitudes of stationary planetary waves responding to forcing by both topography and stationary heat sources for k = 1 and k = 2 in winter are much larger than those in summer.

2. The stationary planetary-scale disturbance patterns are mainly confined to high latitudes, centered at 60°N in winter. In summer they are almost entirely confined to middle and low latitudes centered at 30°N.

VI. CONCLUSIONS

The stationary planetary waves responding to forcing by the northern hemispheric topography are investigated by means of a quasi-geostrophic, steady state 34-level model, with Rayleigh friction, the effect of Newtonian cooling and horizontal kinematic thermal diffusivity included in a spherical coordinate system. The computed results are in good agreement with the observed ones. Moreover, there are considerable differences between the stationary planetary waves in winter and those in summer; these differences are mainly due to the waveguides which occur in close connection with the zonal mean wind. Thus, the anomaly of stationary planetary waves occurs in connection with the zonal mean wind.

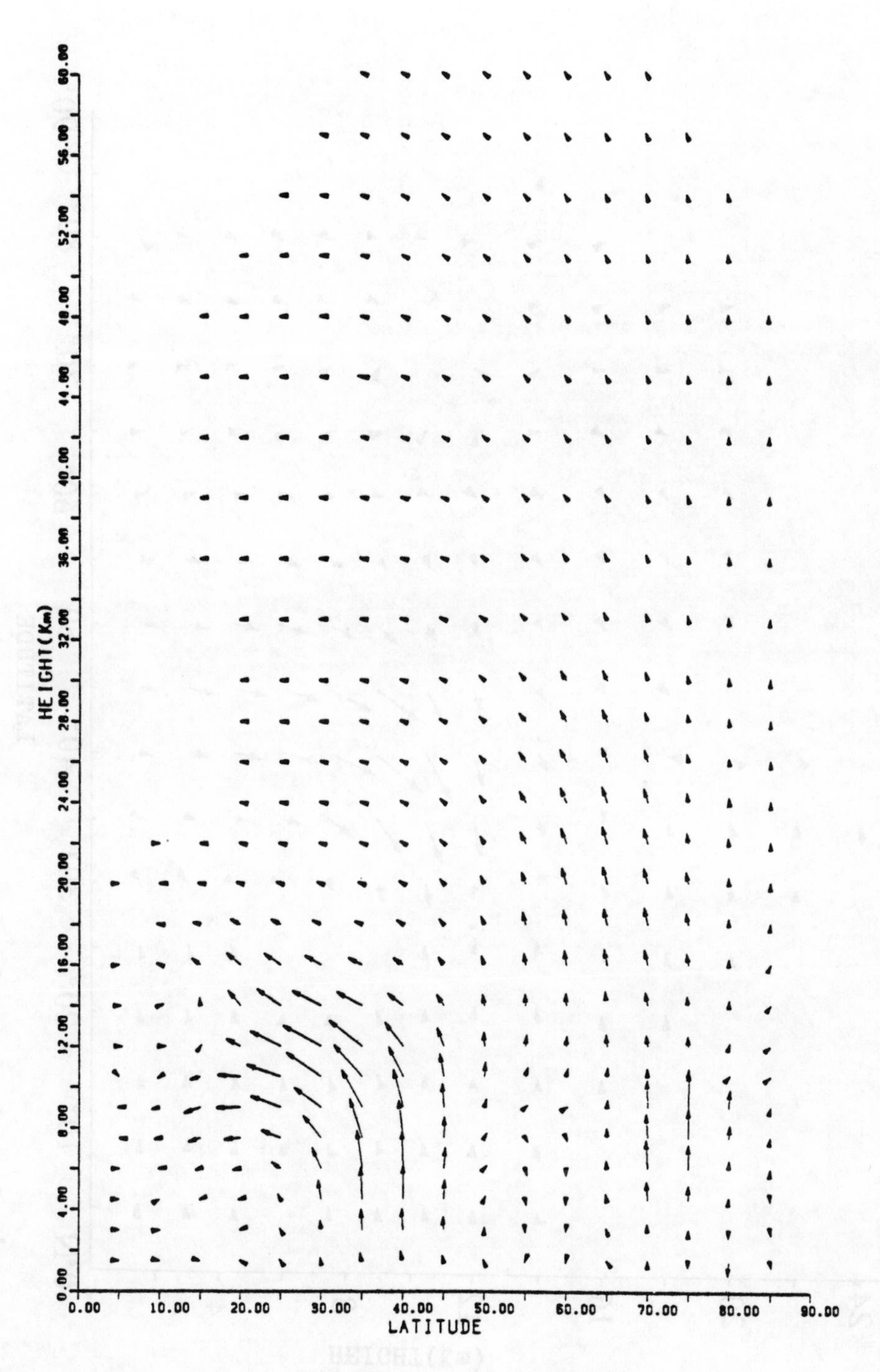

Fig. 18 The vertical distribution of energy flux due to stationary waves 1 to 3, responding to forcing by both the northern hemispheric topography and stationary heat sources.

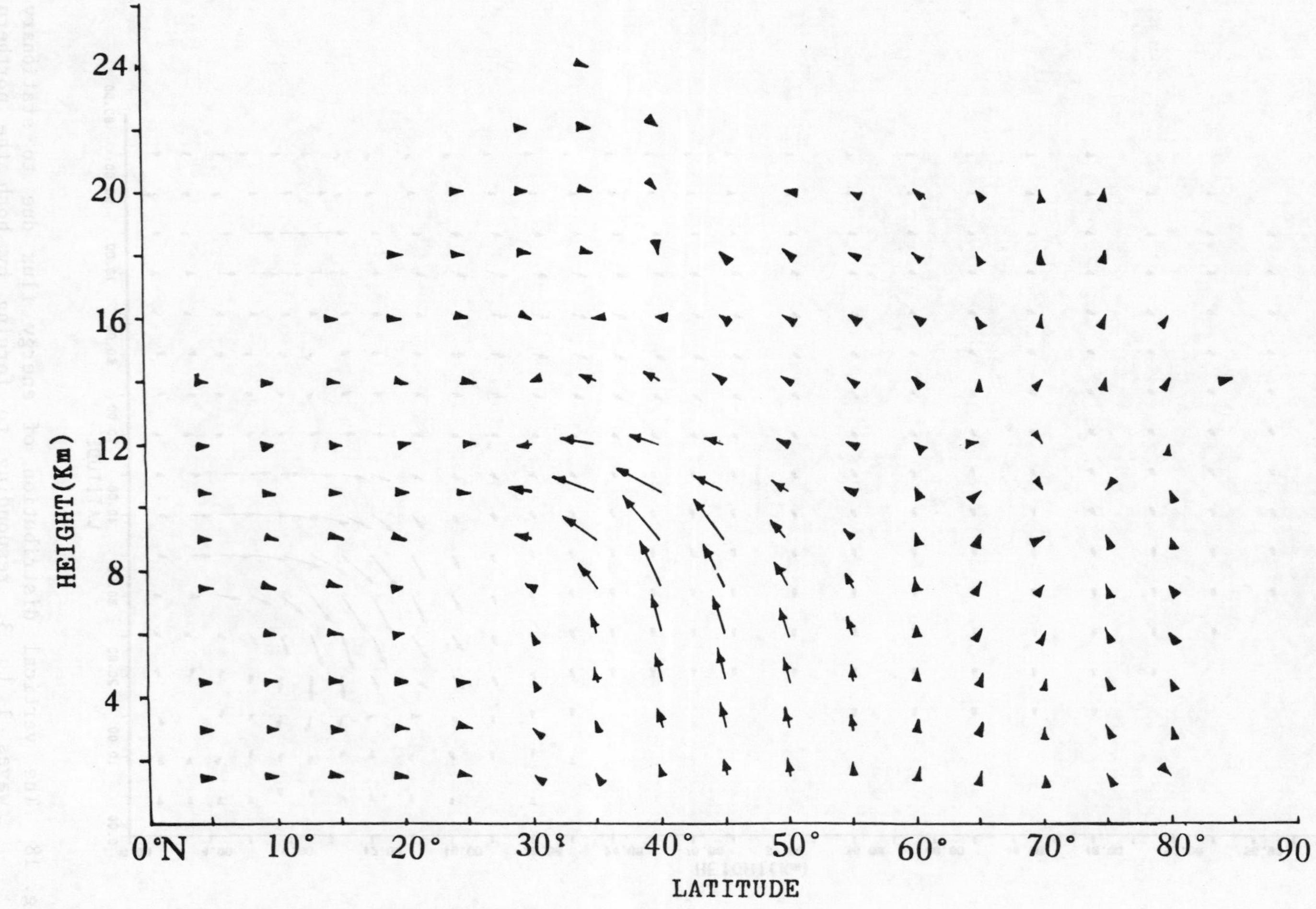

Fig. 19 As in Fig. 18, except for summer.

In the results computed above the effects of summer are less than the observed ones, due to the nonlinear terms which are not incorporated in this model. Thus, by doing so we may obtain a better understanding of the nature of planetary-scale waves.

ACKNOWLEDGMENTS

The author is grateful to Prof. K. Gambo (University of Tokyo) for his valuable comments and suggestions during the course of this study. Thanks are due to Prof. Zhu Baozhen for discussion of this work.

REFERENCES

1. Ashe, S.: Nonlinear model of the time-average axially asymmetric flow induced by topography and diabatic heating. J. Atmos. Sci., 36, 109-126 (1979).

2. Berkofsky, L. and E.A. Bertoni: Mean topographic charts for the entire earth. Bull. Amer. Meteor. Soc., 36, 350-354 (1955).

3. Charney, J.G. and P.G. Drazin: Propagation of planetary-scale disturbances from the lower into the upper atmosphere. J. Geophys. Res., 66 (1), 83-110 (1961).

4. Dickinson, R.E.: Planetary Rossby waves propagating vertically through weak westerly wind waveguides. J. Atmos. Sci., 25, 984-1002 (1968).

5. Egger, J.: The linear response of a hemispheric two-level primitive equation model to forcing by topography. Mon. Wea. Rev., 104, 351-363 (1976).

6. Holton, J.R.: A semispectral numerical model for wave-mean flow interactions in the stratosphere: Application to sudden stratospheric warmings. J. Atmos. Sci., 33, 1639-1647 (1976).

7. Huang, Rong-hui and K. Gambo: The response of a model atmosphere in middle latitudes to forcing by topography and stationary heat sources. J. Meteor. Soc. of Japan, 59, 220-237 (1981).

8. Huang, Rong-hui and K. Gambo: The response a hemispheric multi-level model atmosphere to forcing by topography and stationary heat sources. Part I: Forcing by topography; Part II: Forcing by stationary heat sources and forcing by topography and stationary heat sources. Special Issue Commemoration of the Centennial of the Meteor. Soc. of Japan, 60, 78-108 (1982).

9. Iwashima, T.: Analysis of standing wave in the atmosphere. Presented at the Annual Meeting of the Meteorological Society of Japan, March (1981).

10. Lindzen, R.S. and H.L. Kuo: A reliable method for the numerical integration of a large class of ordinary and partial differential equations. Mon. Wea. Rev., 97, 732-734 (1969).

11. Lordi, N.J., A. Kasahara and S.K. Kao: Numerical simulation of stratospheric sudden warmings with a primitive equation spectral model. J. Atmos. Sci., 37, 2746-2767 (1980).

12. Matsuno, T.: Vertical propagation of stationary waves in the winter northern hemisphere. J. Atmos. Sci., 27, 871-883 (1970).

13. Murgatroyd, R.J.: The structure and dynamics of the stratosphere. The Global Circulation of the Atmosphere. Royal Meteoro. Soc., London, 155-195 (1969).

14. Murakami, T.: Equatorial tropospheric waves induced by diabatic heat sources. J. Atmos. Sci., 29, 827-836 (1972).

15. Staff Members, Academia Sinica: On the general circulation over eastern Asia. Tellus, 10, 299-312 (1958).

16. van Loon, H., R.L. Jenne and K. Labitzke: Zonal harmonic standing waves. J. Geoph. Res., 78, 4463-4471 (1973).

17. Webster, P.J.: Response of the tropical atmosphere to local, steady forcing. Mon. Wea. Rev., 100, 518-541 (1972).

2.3 THE EFFECTS OF TOPOGRAPHY ON THE STATIONARY PLANETARY WAVES IN WINTER AND SUMMER

Lin Benda
Department of Geophysics
Beijing University
The People's Republic of China

ABSTRACT

In this paper a primitive equation linear wave model is developed to examine the effects of topography and diabatic heating on the formation and maintenance of stationary planetary waves in winter and summer. It is found that in winter the topographic forcing plays an important role in the formation and maintenance of stationary waves both in the stratosphere and troposphere. The vertical structure and horizontal wave pattern of the stationary waves forced by topography are very similar to those observed. In summer, on the other hand, the diabatic heating, especially the latent heating connected with the active convection in low latitudes, is a main contribution to the maintenance of the summer monsoon circulation in the subtropics; the topographic effect cannot reproduce a realistic wave structure and pattern of the summer monsoon circulation although it makes some contribution to the stationary waves in high latitudes.

I. INTRODUCTION

In recent years, both observational and theoretical studies have led to considerable progress in understanding the role of planetary waves in the dynamics of the atmosphere. It was found that the behavior of stationary planetary waves is quite different in winter and summer. In winter, the stationary waves can propagate upwards into the stratosphere. However, in summer stationary waves with significant amplitude are hardly observed in the stratosphere: They are trapped in the troposphere with a strength much weaker than in winter. One of the significant features of the winter stratospheric circulation over the northern hemisphere is the prevalence of ultralong planetary waves which are quasi-stationary in nature and consist mainly of zonal wave numbers 1 and 2.

It is generally believed that the stratospheric planetary waves are forced in the troposphere by topography and by diabatic heating associated with land and sea thermal contrasts, and propagate vertically from the troposphere. Many numerical studies have been performed to examine the role of topography and diabatic heating in the formation and maintenance of winter stationary waves. Bolin [2], Zhu [26, 27], Smagorinsky [22], Murakami [16], and Zhu and Lei [28] gave some examples. Many authors agree that the relative importance of topographic forcing increases with altitude, and hence the stratospheric stationary waves are primarily excited by topography. However, in regard to the relative importance of topography and diabatic heating for the planetary waves in the troposphere, there is no general agreement. For example, Bolin [2] and Smagorinsky [22] concluded that thermal effects can account for the essential features of the observed distribution of sea level pressure. Sanka-Rao and Saltzman [20] indicated that a good deal of the longitudinal variance of the mean tropospheric state in the middle latitudes is explainable by the effects of asymmetric surface heating. But Kasahara et al. [8] and Manabe and Terpstra [13] reported that their general circulation models with mountains not only reproduced quite realistic stratospheric circulations but also reproduced tropospheric flow fields which were in good qualitative agreement with the observed fields. They suggested that the large-scale mountains are also chiefly responsible for maintaining the quasi-stationary waves in the troposphere. It appears that, although a number of numerical experiments have been done to examine and compare the effects of topography and diabatic heating on the stationary waves, the question of the relative importance of these forcing effects is still outstanding.

Compared to the situation for winter, the summer stationary waves have received relatively little attention, although there have been studies concerned with this topic. For example, Zhu [26] used a two-layer baroclinic model to examine the effects of topography and diabatic heating on summer steady westerly disturbances. He found that in summer the effect of topography is much weaker than that of diabatic heating in Asia, but they have the same order in America. Sanka-Rao [19] used a quasi-geostrophic model with a simplified mean zonal wind independent of latitude to simulate the summer wave pattern. He was able to reproduce the basic features of the Asian monsoon circulation with reasonable phase reversal in the vertical. Webster [24] used a linear steady-state two-layer primitive equation model to examine the response of the tropical atmosphere to local steady forcing. He found that the dominant factor in low latitudes is the latent heating, whereas in high latitudes advective process becomes more important. Egger [5] used a similar two-layer model but extended the domain of the model to cover the whole hemisphere in order to examine the horizontal pattern of stationary waves in summer. He successfully simulated the summer monsoon circulation in the subtropics, but his model failed to simulate the observed wave pattern in high latitudes. Defant et al. [4] did an observational study of the stationary waves and provided the structure of stationary waves in summer for different wave numbers. He found that the height variation in the subtropical belt in July is thermally forced by the heating of both continents, thus wave numbers 1 and 2 play dominant roles.

Although there have been some observational theoretical studies on summer stationary waves, they all have some limitations. In some models the basic state was too simplified; in others the vertical resolution was too coarse (for example, using a two-layer model). Some experiments were mainly concerned with the responses in the tropics or in the westerlies.

In the present paper, we used a primitive equation linear wave model in spherical coordinates with realistic basic state and forcings and a high vertical resolution to simulate the behavior of stationary waves in winter and summer. The role of different forcing mechanisms for the formation and maintenance of stationary waves will be compared.

II. THE MODEL EQUATIONS AND BOUNDARY CONDITIONS

Basically, two kinds of numerical models have been used to examine the behavior of stationary waves. One is the general circulation model such as described by Kasahara et al. [8] and Manabe and Terpstra [13], in which the variables are integrated under certain boundary and initial conditions and then their long-term (for example, 30 days) averages are calculated to separate the stationary characteristics. Another is the so-called linear wave model, for example those of Matsuno [14] and Schoeberl and Geller [21], in which the variables are divided into zonal mean and perturbation, and the steady-state perturbation solutions are directly found for a single harmonic component for certain atmospheric basic states and boundary conditions.

The characteristics of stationary waves forced by topography and diabatic heating have been examined by Kasahara et al. [8] and Manabe and Terpstra [13] with GCM. It is of interest to use a linear wave model for the same purpose. These experiments should provide more insight into the behavior of stationary waves. Besides, from the comparison of the two kinds of model experiments, one can test the ability of the linear wave model in reproducing the observed behavior of stationary waves.

The basic equations of the model are the steady, linearized primitive perturbation equations in the spherical coordinate system. The forcings include topography and diabatic heating, and Rayleigh friction as well as Newtonian cooling are incorporated as damping processes. Let λ, θ, z represent the longitude, latitude and vertical coordinates, respectively, where $z \equiv - H\ln(p/p_s)$ and H is a constant scale height. The horizontal momentum equations, hydrostatic equation, continuity equation and the first law of thermodynamics may then be written as follows:

$$\frac{\bar{u}}{a\cos\theta}\frac{\partial u'}{\partial\lambda} + \left[\frac{1}{a\cos\theta}\frac{\partial}{\partial\theta}(\bar{u}\cos\theta) - 2\Omega\sin\theta\right]v' + \frac{\partial\bar{u}}{\partial z}w' = - \frac{1}{a\cos\theta}\frac{\partial\phi'}{\partial\lambda} - \kappa u' \tag{1}$$

$$\frac{\bar{u}}{a\cos\theta}\frac{\partial v'}{\partial\lambda} + \left[2\Omega\sin\theta + \frac{2\bar{u}}{a}\tan\theta\right]u' = - \frac{1}{a}\frac{\partial\phi'}{\partial\theta} - \kappa v' \tag{2}$$

$$\frac{\partial\phi'}{\partial z} = \frac{RT'}{H} \tag{3}$$

$$\frac{1}{a\cos\theta}\frac{\partial u'}{\partial\lambda} + \frac{1}{a\cos\theta}\frac{\partial}{\partial\theta}(v'\cos\theta) + \frac{\partial w'}{\partial z} - \frac{w'}{H} = 0 \tag{4}$$

$$\frac{\bar{u}}{a\cos\theta}\frac{\partial T'}{\partial\lambda} + \frac{\partial\bar{T}}{a\partial\theta}v' + \left[\frac{R\bar{T}}{c_p H} + \frac{\partial\bar{T}}{\partial z}\right]w' = \frac{Q'}{c_p} - \kappa_T T' \tag{5}$$

Here the prime designates the perturbation variables and the overbar zonally averaged variables; u, v, and w are zonal, meridional and vertical velocity components, ϕ the geopotential, T the temperature, Q the rate of diabatic heating per unit mass, κ the Rayleigh drag coefficient and κ_T the Newtonian cooling coefficient.

Following Holton [6], we introduce nondimensional variables as follows:

$$u^* = u' \cos\theta/2\Omega a, \quad v^* = v' \cos\theta/2\Omega a, \quad \bar{u}^* = \bar{u}\cos\theta/2\Omega a$$
$$w^* = w'/2\Omega H, \quad \phi^* = \phi'/(2\Omega a)^2, \quad T^* = RT'(2\Omega a)^2$$
$$\bar{T}^* = R\bar{T}/(2\Omega a)^2, \quad Q^* = RQ'/(2\Omega)^3 a^2 c_p, \quad z^* = z/H$$

For the nondimensional perturbation variables, we assume zonal harmonic solutions

$$\begin{bmatrix} u^* \\ v^* \\ w^* \\ \phi^* \\ Q^* \end{bmatrix} = \begin{bmatrix} u \\ v \\ w \\ \phi \\ Q \end{bmatrix} e^{ik\lambda + z^*/2}$$

where u, v, w, ϕ, and Q are complex amplitudes of the perturbation which are functions of latitude and height. Substituting the above expressions into Eqs. (1) to (5) and eliminating T between Eqs. (3) and (5), we get the following relationships among the complex amplitudes:

$$i\bar{\omega}_D u + \bar{a}v + \frac{\partial \bar{u}^*}{\partial z^*} w = -ik\phi$$

$$i\bar{\omega}_D v + \bar{\beta}u = -\cos\theta \frac{\partial\phi}{\partial\theta} \tag{6}$$

$$\frac{ik}{\cos^2\theta} u + \frac{i}{\cos\theta}\frac{\partial v}{\partial\theta} + \frac{\partial w}{\partial z^*} - \frac{w}{2} = 0$$

$$i\bar{\omega}_T \left[\frac{\partial}{\partial z^*} + \frac{1}{2}\right]\phi - \bar{\gamma}v + S_o w = Q$$

where

$$\bar{\omega}_D = \frac{k\bar{u}^*}{\cos^2\theta} - \frac{i\kappa}{2\Omega}, \qquad \bar{\omega}_T = \frac{k\bar{u}^*}{\cos^2\theta} - \frac{i\kappa_T}{2\Omega}$$

$$\bar{a} = \frac{1}{\cos\theta}\frac{\partial\bar{u}^*}{\partial\theta} - \sin\theta, \qquad \bar{\beta} = \sin\theta + \frac{2\sin\theta}{\cos^2\theta}\bar{u}^*$$

$$\bar{\gamma} = \frac{\sin\theta}{\cos^2\theta}\frac{\partial\bar{u}^*}{\partial z^*}, \qquad S_o = \frac{R\bar{T}^*}{c_p} + \frac{d\bar{T}^*}{dz^*} = \frac{N^2H^2}{(2\Omega a)^2}$$

Here N^2 is the Brunt-Väisälä frequency. From the first two equations of Eq. (6), u and v can be expressed in terms of ϕ and w:

$$u = \bar{\sigma}\,[-\bar{a}\cos\theta\,\frac{\partial\phi}{\partial\theta} - k\bar{\omega}_D\phi + i\bar{\omega}_D\,\frac{\partial\bar{u}^*}{\partial z^*}\,w]$$

$$v = \bar{\sigma}\,[i\bar{\omega}_D\cos\theta\,\frac{\partial\phi}{\partial\theta} - ik\bar{\beta}\phi - \bar{\beta}\,\frac{\partial\bar{u}^*}{\partial z^*}\,w] \tag{7}$$

where $\bar{\sigma} = (\bar{a}\,\bar{\beta} + \bar{\omega}_D^2)^{-1}$. Substituting the expression for v into the fourth equation of Eq. (6), w can be expressed in terms of ϕ and Q:

$$w = \frac{Q}{S} + \frac{i\bar{\varepsilon}}{S}\frac{\partial\phi}{\partial\theta} - \frac{i\bar{\omega}_T}{S}\frac{\partial\phi}{\partial z^*} - \frac{i\bar{\delta}}{S}\,\phi \tag{8}$$

where

$$\bar{\varepsilon} = \bar{\gamma}\,\bar{\sigma}\,\bar{\omega}_D\cos\theta, \quad \bar{\delta} = \frac{\bar{\omega}_T}{2} + \kappa\,\bar{\beta}\,\bar{\gamma}\,\bar{\sigma}, \quad S = S_o + \bar{\gamma}\,\bar{\beta}\,\bar{\sigma}\,\frac{\partial\bar{u}^*}{\partial z^*}$$

Then, eliminating w between Eqs. (7) and (8), we can express u and v in terms of ϕ and Q:

$$u = -(\bar{\sigma}\,\bar{a}\cos\theta + \bar{\omega}_D\,\bar{\varepsilon}\,\bar{\lambda})\frac{\partial\phi}{\partial\theta} + \bar{\omega}_T\bar{\omega}_D\bar{\lambda}\frac{\partial\phi}{\partial z^*} + \bar{\omega}_D(\bar{\delta}\,\bar{\lambda} - k\bar{\sigma})\phi + i\bar{\omega}_D\bar{\lambda}\,Q$$

$$v = i(\bar{\sigma}\,\bar{\omega}_D\cos\theta - \bar{\beta}\,\bar{\varepsilon}\,\bar{\lambda})\frac{\partial\phi}{\partial\theta} + i\bar{\beta}\,\bar{\omega}_T\bar{\lambda}\frac{\partial\phi}{\partial z^*} + i\bar{\beta}(\bar{\delta}\,\bar{\lambda} - k\bar{\sigma})\phi - \bar{\beta}\,\bar{\lambda}\,Q \tag{9}$$

where

$$\bar{\lambda} = \frac{\bar{\sigma}}{S}\frac{\partial\bar{u}^*}{\partial z^*}$$

Finally, substituting Eqs. (8) and (9) into the third equation of Eq. (6), we can get a single second order partial differential equation for ϕ as follows:

$$A\frac{\partial^2\phi}{\partial\theta^2} + B\frac{\partial^2\phi}{\partial\theta\partial z^*} + C\frac{\partial^2\phi}{\partial z^{*2}} + D\frac{\partial\phi}{\partial\theta} + E\frac{\partial\phi}{\partial z^*} + F\phi = G \tag{10}$$

where

$$A = \frac{\bar{\mu}}{\cos\theta}$$

$$B = \frac{\bar{\nu}}{\cos\theta} + \frac{\bar{\varepsilon}}{S}$$

$$C = -\frac{\bar{\omega}_T}{S}$$

$$D = \frac{1}{\cos\theta}\left[\frac{\partial\bar{\mu}}{\partial\theta} + \bar{\beta}\bar{\zeta}\right] + \frac{\partial}{\partial z^*}\left(\frac{\bar{\varepsilon}}{S}\right) - \frac{k}{\cos^2\theta}(\bar{\sigma}\ \bar{a}\cos\theta + \bar{\omega}_D\ \bar{\varepsilon}\ \bar{\lambda}) - \frac{\bar{\varepsilon}}{2S} \quad (11)$$

$$E = \frac{k}{\cos^2\theta}\bar{\omega}_D\ \bar{\omega}_T\ \bar{\lambda} + \frac{1}{\cos\theta}\frac{\partial\bar{\nu}}{\partial\theta} - \frac{\partial}{\partial z^*}\left[\frac{\bar{\omega}_T}{S}\right] + \frac{\bar{\omega}_T}{2S} - \frac{\bar{\delta}}{S}$$

$$F = \frac{k}{\cos^2\theta}\bar{\omega}_D\ \bar{\zeta} + \frac{1}{\cos\theta}\frac{\partial}{\partial\theta}(\bar{\beta}\ \bar{\zeta}) - \frac{\partial}{\partial z^*}\left(\frac{\bar{\delta}}{S}\right) + \frac{\bar{\delta}}{2S}$$

$$G = i\frac{\partial}{\partial z^*}\left(\frac{Q}{S}\right) - \frac{i\ \bar{\beta}\ \bar{\lambda}}{\cos\theta}\frac{\partial Q}{\partial\theta} - i\left[\frac{1}{\cos\theta}\frac{\partial}{\partial\theta}(\bar{\beta}\ \bar{\lambda}) + \frac{k\ \bar{\omega}_D\ \bar{\lambda}}{\cos^2\theta} + \frac{1}{2S}\right]Q$$

Here G is the forcing term caused by diabatic heating and

$$\bar{\mu} = \bar{\delta}\ \bar{\omega}_D\cos\theta - \bar{\beta}\ \bar{\varepsilon}\ \bar{\lambda}, \qquad \bar{\nu} = \bar{\beta}\ \bar{\omega}_T\ \bar{\lambda}, \qquad \bar{\zeta} = \bar{\delta}\ \bar{\lambda} - k\ \bar{\sigma}$$

To solve Eq. (10), two boundary conditions for both the latitudinal and vertical directions must be specified. In our experiments, the numerical integration is performed for the region from the North Pole to the equator and from the ground to 60 km for winter and 40 km for summer. At the pole and the equator, it is assumed that $\phi = 0$, since there is ordinarily a zero wind line in lower and middle levels near the equator which would absorb the wave energy and make the wave amplitude decay significantly. It is reasonable to use this approximation as the lateral boundary condition at the equator.

At the lower boundary, topography and diabatic heating are incorporated as the main forcing sources for the stationary waves. The vertical perturbation velocity forced by topography is

$$w'_\eta = \frac{\bar{u}}{a\cos\theta}\frac{\partial\eta'}{\partial\lambda}$$

where η' denotes the perturbation height of the topography. Letting

$$\eta' = H\eta^*, \qquad \eta^* = \eta e^{ik\lambda}$$

and

$$w'_\eta = 2\Omega H w^*_\eta, \qquad w^*_\eta = w_\eta e^{ik\lambda}$$

we get

$$w_\eta = \frac{i\ k\ \bar{u}^*}{\cos^2\theta}\eta$$

substituting w_η into Eq. (8), the lower boundary condition can be written

$$\bar{\varepsilon}\frac{\partial\phi}{\partial\theta} - \bar{\omega}_T\frac{\partial\phi}{\partial z^*} - \bar{\delta}\phi = \frac{k\,\bar{u}^*\,S}{\cos^2\theta}\eta + iQ \quad \text{at} \quad z = 0 \tag{12}$$

where the terms on the right-hand side correspond to the forcing terms which are induced by topography and diabatic heating. From Eqs. (12) and (10) it can be seen that the topography can only excite planetary waves through the lower boundary, whereas the diabatic heating can force planetary waves both at the lower boundary and in the interior. Additionally, the forcing due to diabatic heating is determined not only by the distribution of the heating rate itself but also by the distribution of its spatial derivatives. These differences between the topographic and the diabatic forcing are likely to have some influence on the behavior of the forced waves.

As the upper boundary condition, it is assumed that above the upper boundary the amplitude of the response vanishes ($\phi = 0$). Since both Rayleigh friction and Newtonian cooling are included in this model, it is appropriate to assume that waves which are propagated from low levels have been damped so strongly by these dissipation processes that there is no significant amplitude response which would be reflected from the upper boundary. Some experiments have been done to test the effect of the upper boundary by setting the boundary at 60 km and 100 km and keeping other parameters unchanged. The results show little difference in the amplitude response, implying that there is no reflection from the upper boundary.

Equation (10) with the above boundary conditions is solved by finite differencing using a numerical method analogous to that of Lindzen and Kuo [11].

III. THE BASIC STATE

In Eqs. (10) and (12) the coefficients such as A, B, C, D, E, F, G, ..., etc. depend upon the basic state of the atmosphere, including mean zonal wind, forcing and dissipation. In order to solve the equation, this basic state must be specified.

1. The Mean Zonal Wind

There are two ways in which the mean wind profile can be specified for this problem:

The first is a finite difference form in which the mean zonal wind $\bar{u}$ is specified for each grid point, as done by Matsuno [14]. The second is an analytic form in which $\bar{u}$ is expressed as an analytic function of latitude and height by properly choosing some parameters to fit the real atmospheric conditions. Tung [23] used this type of mean zonal wind.

Since the wave behavior is very sensitive to the mean wind structure, errors caused by the finite difference scheme in calculating the second order derivatives of the mean wind field could result in large errors in the wave response. In order to avoid this kind of error and to facilitate the calculation of coefficients such as A, B, C, ..., which are complicated differential functions of the mean zonal wind, and to permit the wind structure (latitudinal or vertical shear) to be easily changed, we use analytic forms of mean zonal wind in our experiments, and assume

$$\bar{u}(\theta, z) = 2\Omega a \cos(\theta - \theta_o) V \tag{13}$$

where

$$V = \Sigma v_i \qquad i = 1, 2, \ldots, i_N$$

is a function of latitude and height in which

$$v_1 = u_i \operatorname{sech}[B_i(\theta-\theta_i)] \operatorname{sech}[A_i(z^*-z_i^*)] \tag{14}$$

Figures 1 and 2 are the standard mean wind models for winter and summer to be used in our experiments. For the winter case, $i_N = 4$ and $v_1(\theta_1 = 5°,\ z_1^* = 27.5/H,\ u_1 = -13)$ creates the easterlies in the middle levels of low latitudes, $v_2(\theta_2 = 30°,\ z_2^* = 12/H,\ u_2 = 30)$ corresponds to the subtropical jet in the upper troposphere, and $v_3(\theta_3 = 45°,\ z_3^* = 60/H,\ u_3 = 64)$ and $v_4(\theta_4 = 65°,\ z_4^* = 35/H,\ u_4 = 28)$ are used to reproduce the polar night jet with tilted vertical axis. This structure of mean zonal wind is similar to that used by Matsuno [14]. In the summer case, $i_N = 3$ and $v_1(\theta_1 = 45°,\ z_1^* = 12/H,\ u_1 = 24)$ corresponds to the subtropical jet, $v_2(\theta_2 = 50°,\ z_2^* = 65/H,\ u_2 = -55)$ creates an easterly jet in the upper stratosphere and $v_3(\theta_3 = 15°,\ z_3^* = 30/H,\ u_3 = -38)$ reproduces the tilted structure of the easterlies in low to middle latitudes. This structure is similar to the observed structure [18].

2. The Dissipation Parameters

In this model, both Rayleigh friction and Newtonian cooling are incorporated. The vertical profile of the Newtonian cooling coefficient above the stratosphere is taken from Holton and Wehrbein [7] and that below the stratosphere is taken from Schoeberl and Geller [21]. The Rayleigh function coefficient for winter is also taken from Holton and Wehrbein [7]. But the distribution of the Rayleigh friction coefficient for summer is treated in a different way from winter: In the winter case, κ, is taken to be $5 \times 10^{-7}\ s^{-1}$ below 50 km. Since in winter the critical line is located at a very low latitude (near 5°N) where the main forcing (topography) is very weak, only a weak damping near the lower boundary is needed to prevent a large response near the critical line. However, in summer the main forcing (the latent heating) has its maximum magnitude at almost the same latitude as the zero wind line (near 25°N). It is found that the Rayleigh friction coefficient, $\kappa = 5 \times 10^{-7}\ s^{-1}$, at the lower boundary is too small and produces an unrealistically large

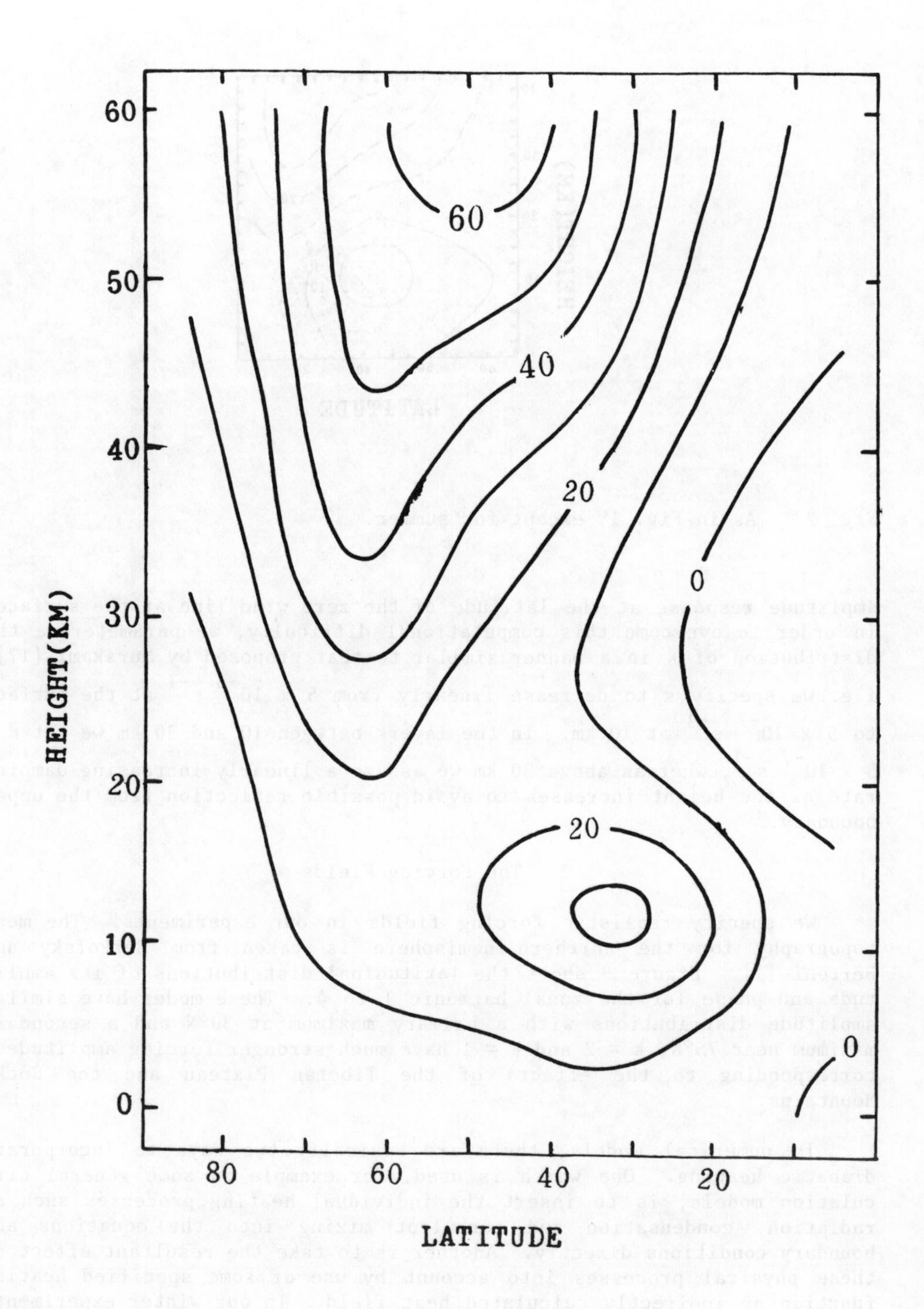

Fig. 1 The mean zonal wind structure for winter.

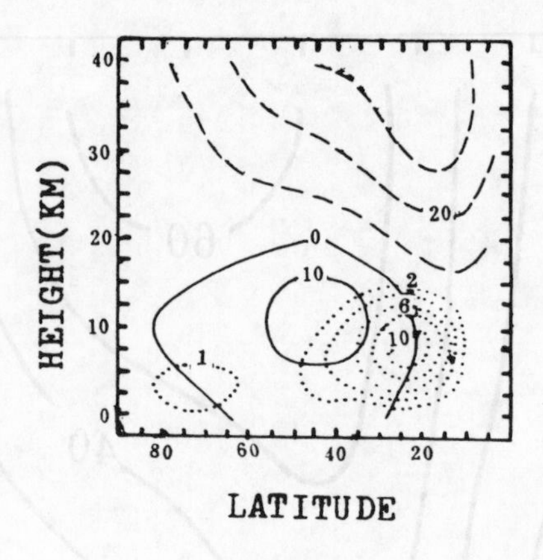

Fig. 2 As in Fig. 1, except for summer.

amplitude response at the latitude of the zero wind line at the surface. In order to overcome this computational difficulty, we parameterize the distribution of κ in a manner similar to that proposed by Murakami [17], i.e. we specify κ to decrease linearly from 5×10^{-6} s^{-1} at the surface to 5×10^{-7} s^{-1} at 10 km. In the layers between 10 and 30 km we set κ = 5×10^{-7} s^{-1}, whereas above 30 km we assume a linearly increasing damping rate as the height increases to avoid possible reflection from the upper boundary.

3. The Forcing Fields

We specify realistic forcing fields in our experiments. The mean topography for the northern hemisphere is taken from Berkofsky and Bertoni [1]. Figure 3 shows the latitudinal distributions of its amplitude and phase for the zonal harmonic 1 to 4. These modes have similar amplitude distributions with a primary maximum at 30°N and a secondary maximum near 75°N; k = 2 and k = 1 have much stronger forcing amplitudes, corresponding to the effects of the Tibetan Plateau and the Rocky Mountains.

In numerical models, there are basically two ways to incorporate diabatic heating. One which is used, for example in some general circulation models, is to insert the individual heating processes such as radiation, condensation and turbulent mixing into the equations and boundary conditions directly. Another is to take the resultant effect of these physical processes into account by use of some specified heating function or indirectly calculated heat field. In our winter experiments the latter method is used, but in the summer experiments the former method is used.

The time averaged diabatic heating fields for certain levels in the troposphere for winter have been calculated by some authors, for example,

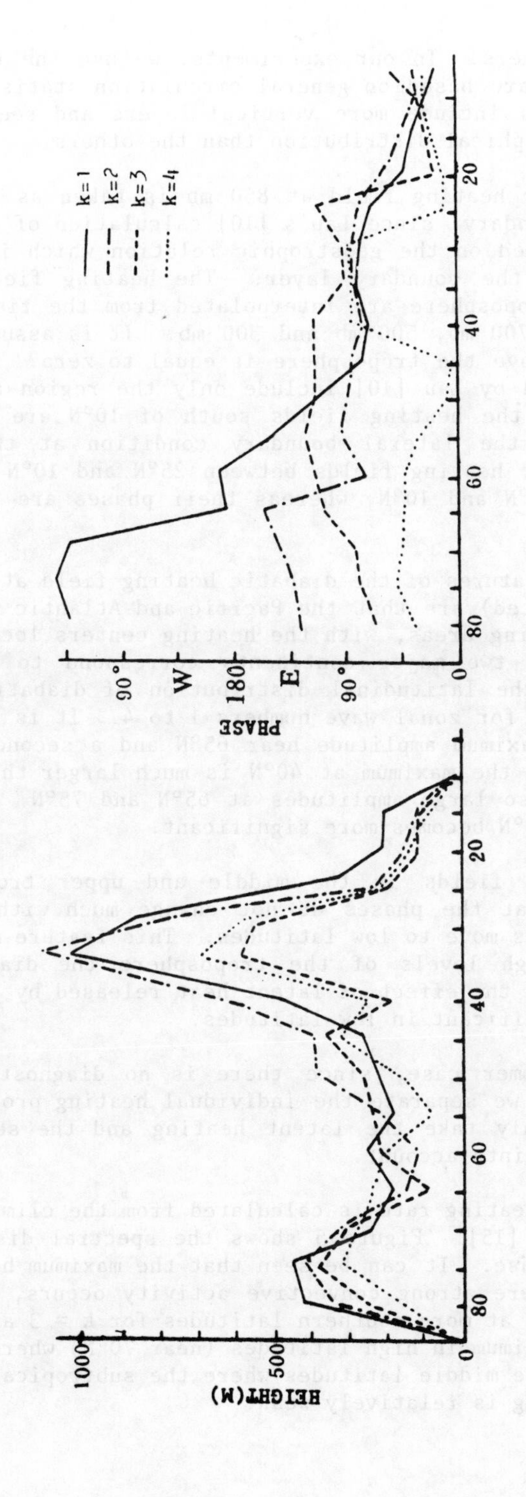

Fig. 3 The spectral distribution of topography Amplitude (left) and phase (right).

Lau [10] and others. In our experiments, we use the recent results of Lau [10], which are based on general circulation statistics, since these calculated fields include more vertical layers and seem to have a more reasonable geographical distribution than the others.

The diabatic heating field at 850 mb is taken as the heating field at the lower boundary, since Lau's [10] calculation of the 1000 mb heating field is based on the geostrophic relation which is not a good approximation for the boundary layer. The heating fields for the grid levels in the troposphere are interpolated from the fields at the lower boundary and at 700 mb, 500 mb and 300 mb. It is assumed that the diabatic heating above the troposphere is equal to zero. Since the heating fields calculated by Lau [10] include only the region north of 25°N, it is assumed that the heating fields south of 10°N are equal to zero in order to match the lateral boundary condition at the equator. The amplitudes of the heating fields between 25°N and 10°N are interpolated from those at 25°N and 10°N, whereas their phases are taken to be equal to that at 25°N.

The basic features of the diabatic heating field at the lower boundary (figure omitted) are that the Pacific and Atlantic Oceans correspond to two main heating areas, with the heating centers located in the western parts. The two major continents correspond to cooling regions. Figure 4 shows the latitudinal distribution of diabatic heating amplitudes and phases for zonal wave numbers 1 to 4. It is seen that for $k = 1$, there is a maximum amplitude near 65°N and a secondary maximum near 40°N. For $k = 2$ the maximum at 40°N is much larger than that of $k = 1$, and there are also large amplitudes at 65°N and 75°N. For $k = 3$ and 4, the maximum at 75°N becomes more significant.

The heating fields in the middle and upper troposphere (figure omitted) show that the phases do not change much with height, but the maximum amplitudes move to low latitudes. This feature may be due to the fact that at high levels of the troposphere the diabatic heating is mainly related to the effect of latent heat released by convective activity which is significant in low latitudes.

For the summer case, since there is no diagnostic total heating field available, we separate the individual heating process and, as done by Egger [5], only take the latent heating and the sensible heat flux from the surface into account.

The latent heating rate is calculated from the climatic distribution of precipitation [15]. Figure 5 shows the spectral distribution of the amplitude and phase. It can be seen that the maximum heating appears in low latitudes where strong convective activity occurs, i.e. at 25°N for $k = 1$ and 2, and at more southern latitudes for $k = 3$ and 4. There is a small heating maximum in high latitudes (near 70°N) where the polar front is active. In the middle latitudes where the subtropical jet is located, the latent heating is relatively weak.

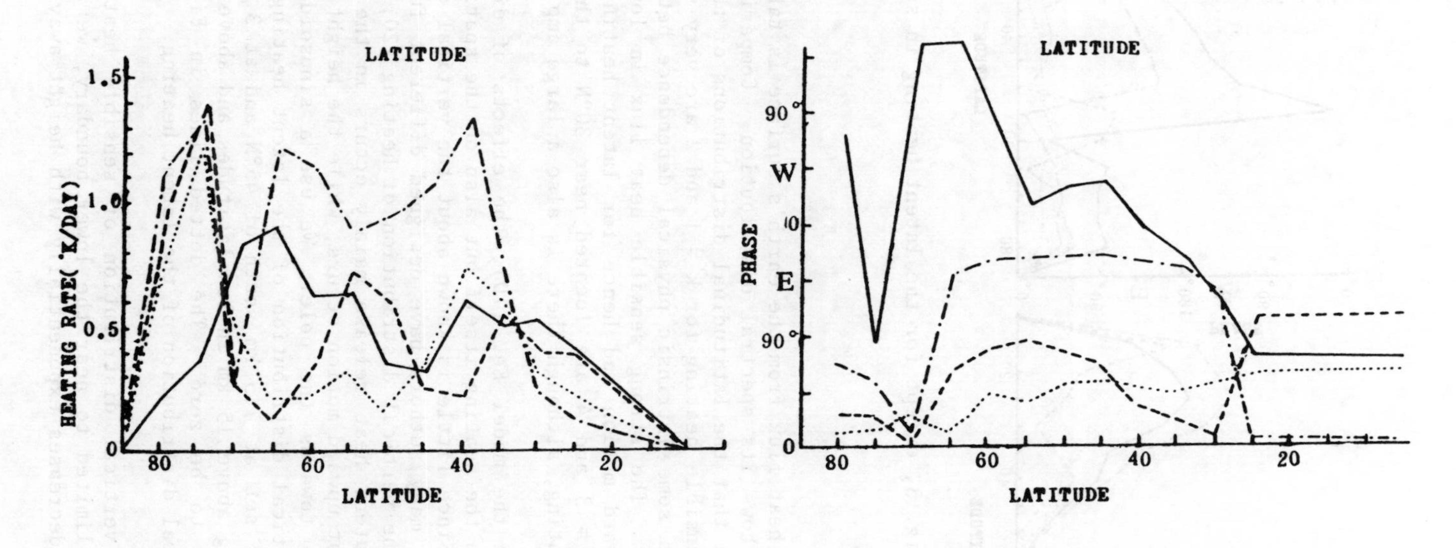

Fig. 4 As in Fig. 3, except for the diabatic heating in winter.

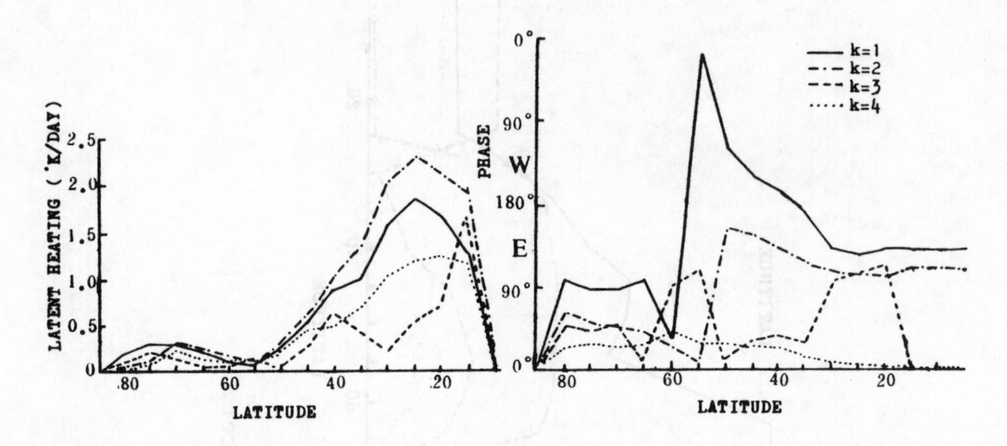

Fig. 5 As in Fig. 3, except for the latent heating in summer.

The sensible heat flux from the earth's surface is taken from Budyko [3], and Fig. 6 shows its spectral distribution. Comparing Fig. 6 with Fig. 5 it is seen that the latitudinal distributions of the latent heating and of the sensible heating for k = 1 and 2 are very similar, implying that there is some intrinsic physical dependence between these two heating processes. The strong sensible heat flux in low latitudes is favorable for upward motion and hence for latent heating. The maximum amplitudes for k = 3 and 4 are located near 30°N to the north of the maximum latent heating, although there is also a large amplitude at 15°N.

As seen from the model Eq. (10), the effects of diabatic heating depend not only on the heating itself but also on the spatial derivatives of the heating. Since little is known about the vertical distribution of diabatic heating, many authors have designed different functional forms to parameterize the vertical distribution of heating [20, 6, 17]. It is known that the latent heat release mainly occurs in the midtroposphere where the strongest upward motion occurs, while the height of the maximum heating decreases towards the pole. We use a sinusoidal function to simulate this vertical distribution of the latent heating. The maximum heating height is set at 7.5 km south of 45°N and at 3.85 km north of 45°N. The heating above 15 km in low latitudes and above 7.5 km in high latitudes is set to be zero. The dotted lines in Fig. 2 show the latitudinal-vertical distribution of the latent heating.

As for the vertical distribution of sensible heating, since its effect is mainly limited to near the lower boundary, we assume that the sensible heating decreases exponentially with height away from the lower boundary.

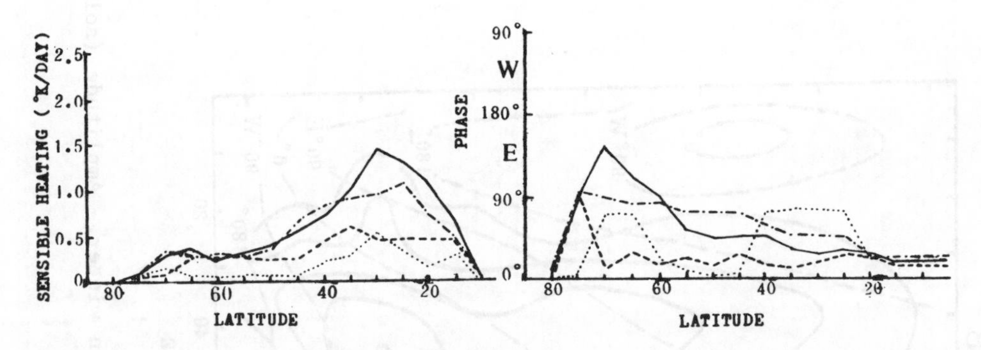

Fig. 6 As in Fig. 3, except for the sensible heat flux from the surface in summer.

IV. THE CHARACTERISTICS OF THE WINTER STATIONARY WAVES FORCED BY TOPOGRAPHY AND DIABATIC HEATING

Now we examine separately the characteristics of the stationary waves excited by realistic topography and diabatic heating. The vertical structure will be examined for each zonal wave number, but the horizontal wave pattern will be investigated only for the composite waves.

First we examine the stationary waves forced by topography. Figure 7 shows the vertical structure of the amplitudes and phases for $k = 1$ to 4. Wave number 1 has the strongest vertical propagation, in which the maximum amplitude is located in the middle stratosphere near 65°N. Both the magnitude and vertical structure are close to the ones observed [25]. It is seen that wave number 2 can still propagate into the stratosphere but with less amplitude; the maximum amplitude is also close to the observed value. It can also be seen that wave numbers 3 and 4 have a behavior different from that of 1 and 2. They are trapped waves whose maximum amplitude responses are limited to the troposphere at 5 km and 2.5 km near 35°N for $k = 3$ and $k = 4$, respectively. All these characteristics of stationary waves are in good agreement with the real atmosphere.

Let us now look at the horizontal wave pattern. Figures 8 and 9 show the computed horizontal wave structures at 5 km and 30 km, respectively, which are composited from $k = 1$ to $k = 4$. For comparison, we show the observed winter mean geopotential height fields of 500 mb and 10 mb in Figs. 10 and 11, respectively. We examine the computed features in the midtroposphere first. From Fig. 8 it is seen that there are two significant stationary troughs off the eastern coast of the Asian and American continents, i.e. in the lee of the Tibetan Plateau and of the Rocky Mountains, respectively, and another weak trough at 30°E. There are three ridges located over the Eurasian continent (70°E), Alaska (140°W) and the eastern Atlantic (10°W). This three-wave pattern is in good agreement with the observed wave structure in the middle troposphere

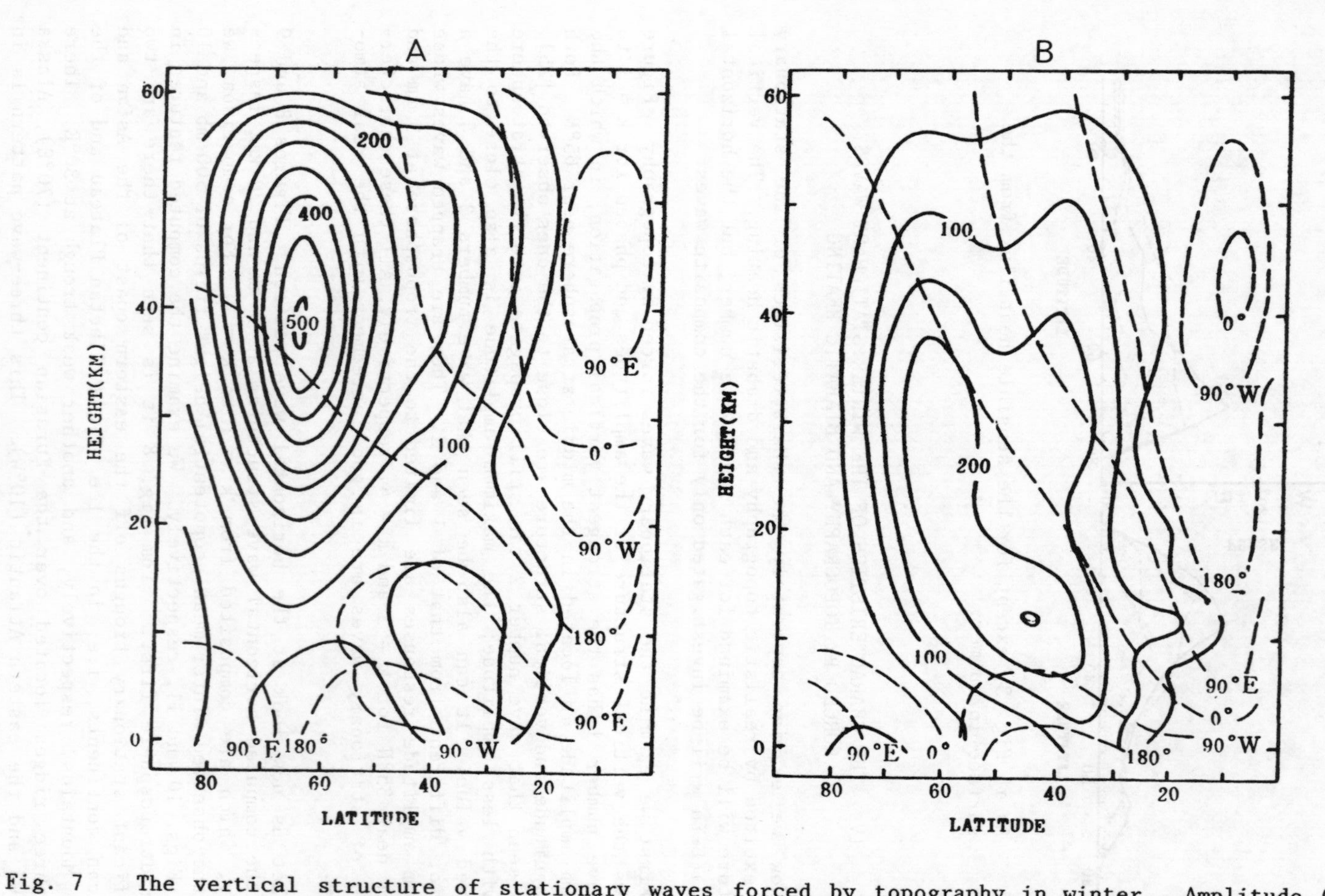

Fig. 7 The vertical structure of stationary waves forced by topography in winter. Amplitude (solid lines) and phase (dashed lines), (a)-(d) for k = 1 to 4.

Fig. 7 (Continued)

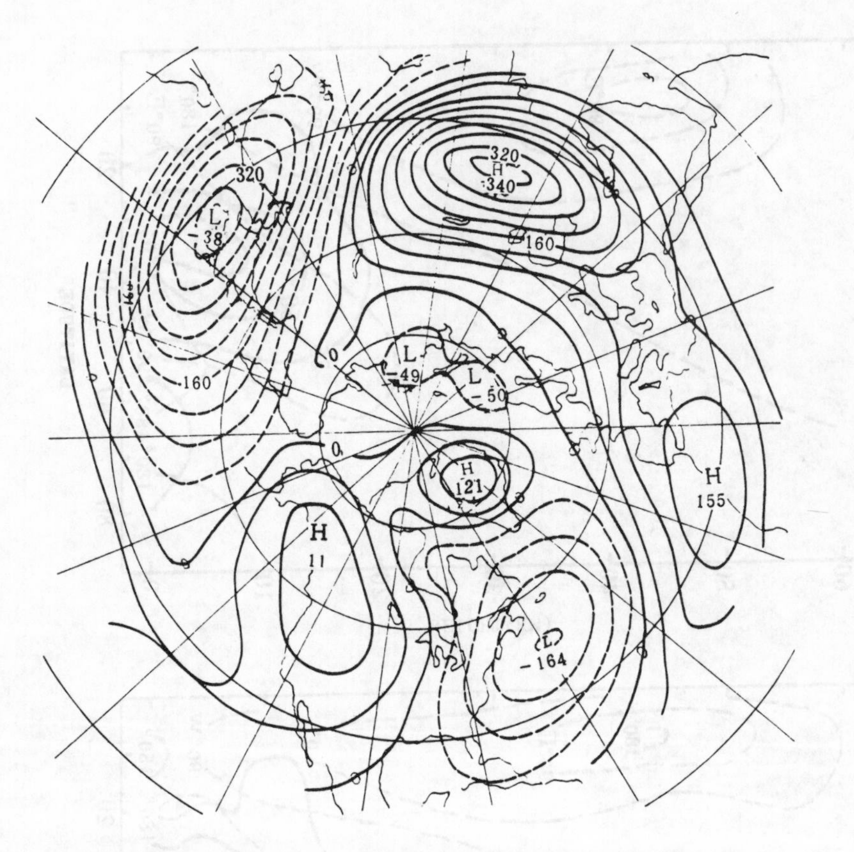

Fig. 8 The horizontal wave pattern of stationary waves forced by topography in winter at 5 km.

in winter (Fig. 10). One main difference between the computed and observed structures is that the computed ridge over Asia is stronger and has a more southerly location than the observed ridge. This difference may be due to the fact that in our model the effects of topographic forcing are taken into account only by creating a forcing vertical velocity through the mean zonal wind advection; another effect of mountains -- the horizontal bifurcating effect -- is omitted in this model. However, for the Tibetan Plateau which is mainly oriented zonally, this horizontal bifurcating effect may also play an important role. If this effect is taken into account, it would reduce the forcing vertical velocity and create a ridge and a trough along the northern and southern sides of the plateau, respectively, which would both reduce the intensity of the stationary ridge and shift it northwards.

Next we look at the wave pattern at 30 km. For the sake of direct comparison, we show the observed disturbed geopotential height field at 10 mb in Fig. 12 [9]. By comparing Fig. 9 with Fig. 12, it can be clearly seen that the main features of the observed stratospheric perturbation

Fig. 9 As in Fig. 8, except at 30 km.

field in winter are reproduced quite well by this topographic forcing model: Wave number one is dominant. The Aleutian high, which is the most striking feature of the midstratospheric flow field in winter, has been reproduced and has reasonable location and strength, and the low centers in northern Europe and the eastern coast of Canada are also reproduced, but with slightly weaker amplitudes.

From the comparisons above it is seen that both realistic tropospheric and stratospheric wave patterns are successfully reproduced by the primitive equation linear wave model with topographic forcing. Manabe and Terpstra [13] found similar results by use of a GCM with mountains. The fact that two different kinds of models lead to very similar results strongly confirms the important role of topographic forcing for the formation and maintenance of stationary waves both in the stratosphere and in the troposphere during winter. It also shows the ability of the primitive equation linear wave model to simulate the behavior of stationary waves.

Fig. 10 The observed winter mean geopotential height field at 500 mb.

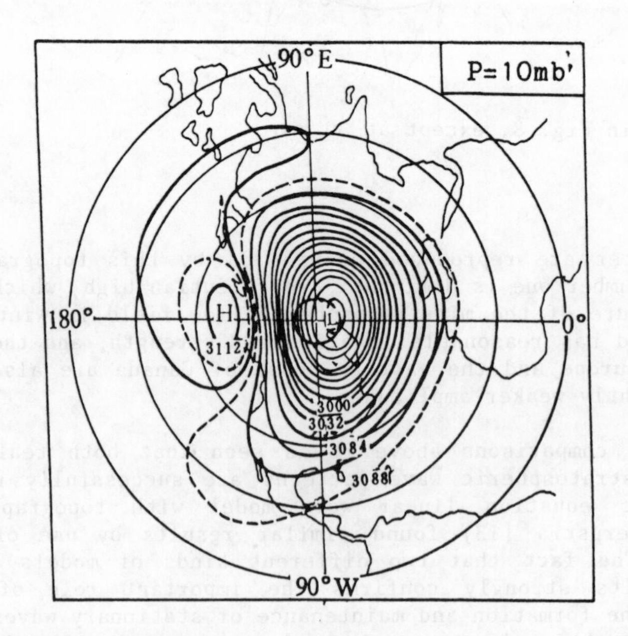

Fig. 11 As in Fig. 10, except for 10 mb.

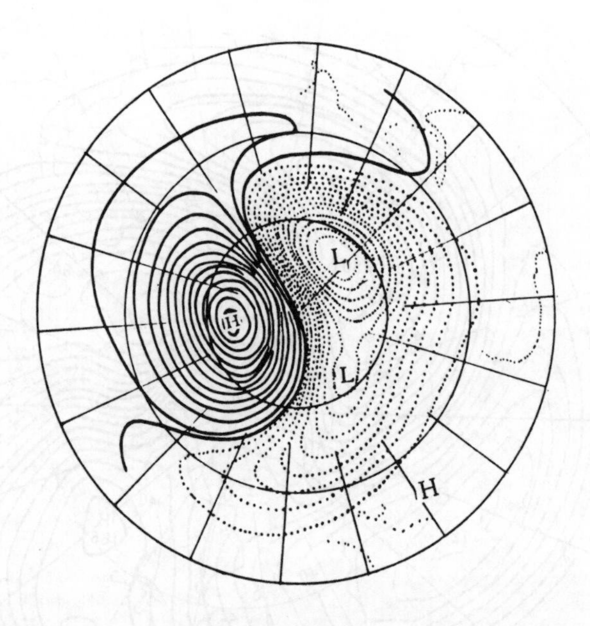

Fig. 12 The observed disturbed geopotential height field for winter at 10 mb.

We now analyze the results from the model in which the waves are forced by diabatic heating. The computed vertical structure of the wave shows that waves number 1 and 2 can also propagate upwards into the stratopshere, as was the case for topographic forcing. But the amplitude responses for k = 1 and 2 in the stratosphere are much weaker than those with topographic forcing. Another difference is that in the case of topographic forcing, waves number 3 and 4, although they cannot propagate into the stratosphere, still have quite large contributions in the troposphere. However, in the case of diabatic heating forcing, these shorter waves have negligible amplitude responses even in the troposphere (figures not shown). Therefore, the stationary waves forced by diabatic heating mainly consist of waves number 1 and 2, which can propagate into the stratosphere but only with weak amplitude responses.

Now let us examine the horizontal wave pattern. Figures 13 and 14 show the wave pattern generated by diabatic heating at 5 km and 30 km. It is seen from Fig. 13 that in the middle troposphere the diabatic heating produces a wave number two pattern in which two ridges are located over the main heating regions in the Pacific and Atlantic Oceans. There are two troughs, one located near 50°E which corresponds to one of the main cooling regions, and another over the Midwest Pacific Ocean, which is about 65° west of the main cooling region over the American continent. Smagorinsky [22] indicated that the phase distribution of stationary waves relative to the heating sources and sinks has something to do with the effect of friction. If friction is not included, the trough (ridge) appears to the east of the heating sources (sinks). If the

Fig. 13 As in Fig. 8, except for the stationary waves forced by diabatic heating.

effect of friction is incorporated in the model, the opposite situation will happen. Zhu [26] also found that the trough (ridge) of stationary waves occurs to the west of the heating sources (sinks). It seems that the results from our experiment are not in agreement with the above conclusions. In our view, the relative phase positions have much to do with the diabatic heating field which is used. In Smagorinsky [22] and Zhu [26], only the vertical mean heating in the lower troposphere has been taken into account, and the heating field in the upper troposphere and the vertical variation of heating were not included. These differences in the heating fields to be used might influence the phase relation between the waves and the heating field. By comparing the computed wave pattern with the observed one (Fig. 10), it is found that over Europe and the middle Atlantic the computed waves have a phase arrangement similar to the observed. However, over Asia, the Pacific and America they are almost out of phase. Thus the diabatic heating is partly responsible for maintaining the trough over Europe and the ridge over the Atlantic, but the stationary waves over Asia, the Pacific and America cannot be explained by the effects of diabatic heating.

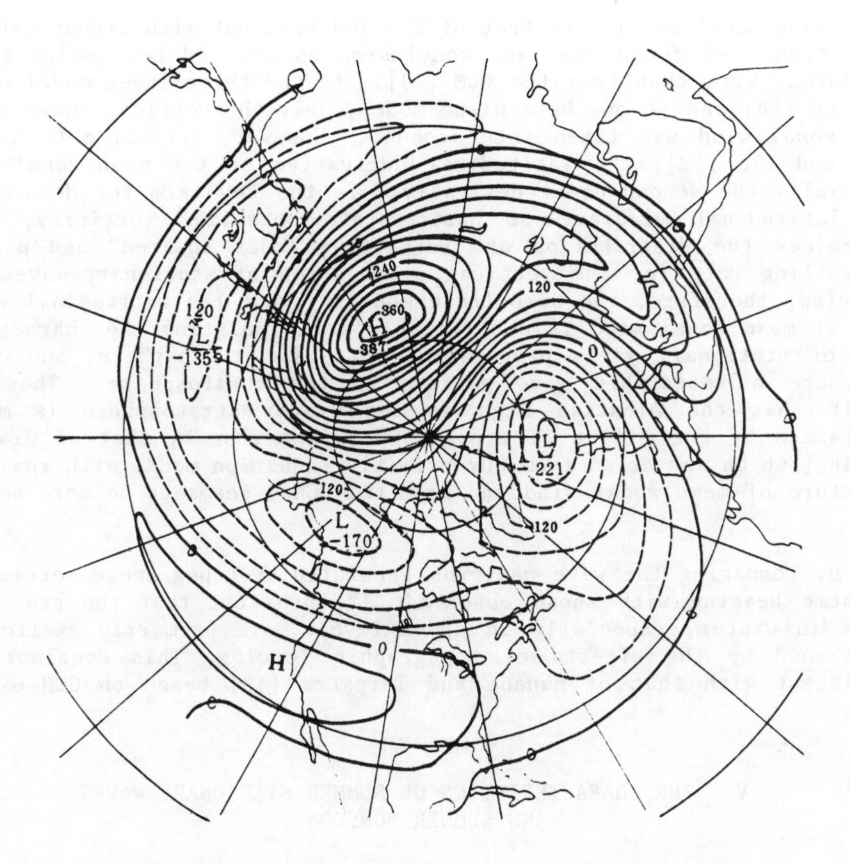

Fig. 14 As in Fig. 9, except for the stationary waves forced by diabatic heating.

We now compare the stratospheric wave patterns. From Fig. 14 it is seen that stationary waves in middle to high latitudes are dominated by wave number 1, as was the case for tropographic forcing. There is a main ridge over the Asian continent, a main trough over the eastern Atlantic, and another weak trough over Alaska. This wave pattern has about 70° phase lag from the observed (Fig. 12). Thus the model with diabatic heating, although it can produce a low center over the Atlantic which has some contributions to maintaining the polar vortex of the middle stratosphere in winter, cannot reproduce a realistic stratospheric wave pattern. Not only is its wave amplitude much weaker than the observed one, but also its phase arrangement is unrealistic on the whole. Also the most striking feature of the middle stratospheric circulation in winter -- the Aleutian high -- cannot be properly reproduced by this model. Zhu and Lei [28] discussed the mechanism for maintaining the Aleutian high by use of a three-layer model and concluded that it was the effect of diabatic heating which mainly contributed to maintaining the Aleutian high. We also did some experiments with a quasi-geostrophic beta plane model similar to that of Zhu and Lei, but with higher vertical resolution, and found the same conclusion as Zhu and Lei, which is not consistent with that from the GCM [13]. In the three-layer model of Zhu and Lei [28] and in our beta plane model, only the vertical shear of the mean zonal wind was taken into account. However, according to Matsuno [14] and Lin [12], the latitudinal derivative of the mean zonal wind, especially the second order derivative, is the key term for determining the latitudinal gradient of basic state potential vorticity, which determines the distribution of "refractive index squared" and plays a controlling role for the vertical propagation of stationary waves. In our view, therefore, the baroclinic model without the latitudinal variation of mean zonal wind is not suitable for simulating the characteristics of stationary waves propagating into the stratosphere, and cannot reproduce a reasonable wave pattern in the stratosphere. Thus, the result that the Aleutian high in the winter stratosphere is mainly maintained by the effect of topography rather than by that of diabatic heating, which was drawn from our primitive equation model with realistic structure of mean zonal wind and from the GCMs, seems to be more believable.

By comparing the wave patterns generated by topographic forcing and diabatic heating with those observed, it turns out that the stationary waves in winter, especially in upper levels, are primarily excited and maintained by the effects of topographic forcing. This conclusion is consistent with that of Manabe and Terpstra [13] based on GCM experiments.

V. THE CHARACTERISTICS OF SUMMER STATIONARY WAVES AND SUMMER MONSOON

In summer, the mean zonal wind has a very different structure from that of winter. It is seen from Fig. 2 that the stratospheric polar night jet disappears and is replaced by an easterly jet at 50°N. The stratosphere of the whole northern hemisphere is occupied by easterlies. In the troposphere, the subtropical jet moves northward to 45°N, with the zero wind line in low latitudes also moving northwards. It can be anticipated that this different mean wind structure would produce a quite different wave structure and pattern.

We first examine the vertical structure. For comparison, we show the observed structure of summer stationary waves in Fig. 15 [4]. It can be seen from this figure that summer stationary waves have two maximum responses in the troposphere, one in low latitudes and one in high latitudes, for all modes. There is a relatively weak response in middle latitudes where the subtropical jet is located. Another characteristic is the decrease of amplitude response with wave numbers; wave numbers 1 and 2 have the major contributions.

Let us look now at the amplitude response forced by topography. Since the structures are similar for different modes, we show the structure for k = 2 in Fig. 16 as a representative example. It is seen that

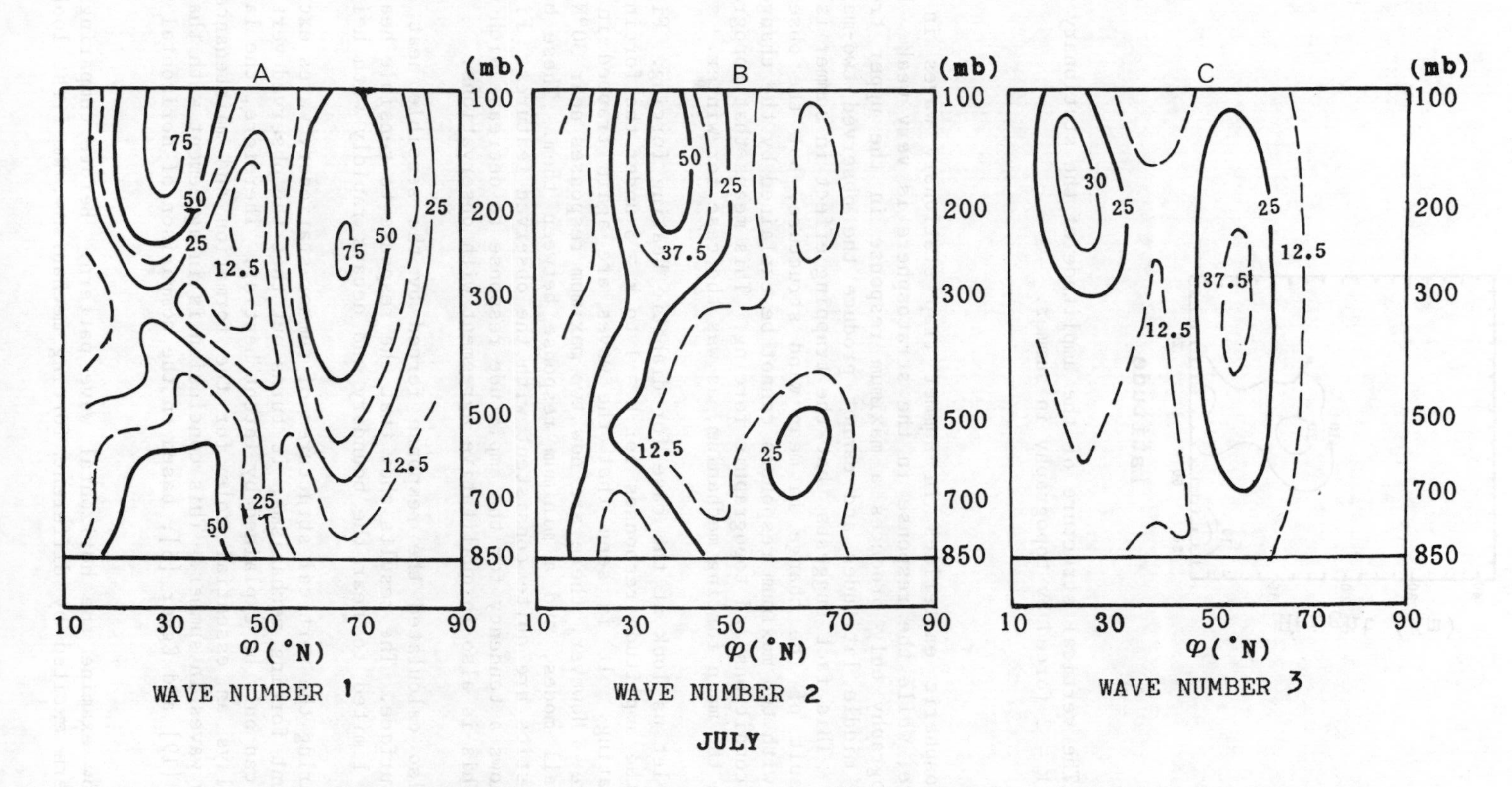

Fig. 15 The observed vertical structure of stationary waves in summer. (a)-(c) for k = 1 to 3.

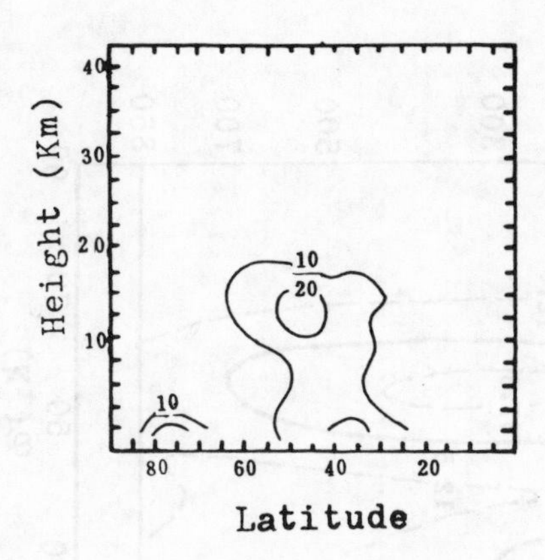

Fig. 16 The vertical structure of the amplitude of the stationary wave k = 2 forced by topography in summer.

the stratospheric easterlies in summer trap stationary waves in the troposphere, while the response in the stratosphere is very weak. However, topography only produces a maximum response in the upper troposphere of middle latitudes and cannot produce the observed two-maxima structure. This fact suggests that the trapping effect in summer is the direct result of the change in mean wind structure, but the observed structure with two maximum responses cannot be explained by the change in mean wind conditions and topographic forcing. This means that topography may not be the main forcing mechanism, as was the case for winter.

Next let us look at the case for diabatic heating forcing. Figure 17 shows the amplitude responses for k = 1 to k = 3 under the forcing of latent heating. It is seen that the waves are again trapped in the troposphere. However, there are now two maximum responses near 30°N and 70°N for all modes, and a minimum response between them. These basic characteristics are quite consistent with the observed features. Figure 17 also shows a tendency for the amplitude response to decrease with wave number. This is also in qualitative agreement with observations.

We also calculated the response forced by the sensible heat flux from the surface. The results show that the response to sensible heating is mainly limited to near the boundary and decays rapidly with height.

Comparing the vertical structures of summer stationary waves excited by different forcing mechanisms, it turns out that the observed vertical structure can only be explained by latent heating. Therefore, the latent heating plays an essential role for the formation and maintenance of stationary waves in summer. This conclusion is in agreement with that of Sanka-Rao [19] and Egger [5], based on the comparison of horizontal wave patterns.

We now examine the horizontal wave pattern. Before comparing the wave pattern excited by different forcing mechanisms, we first look at

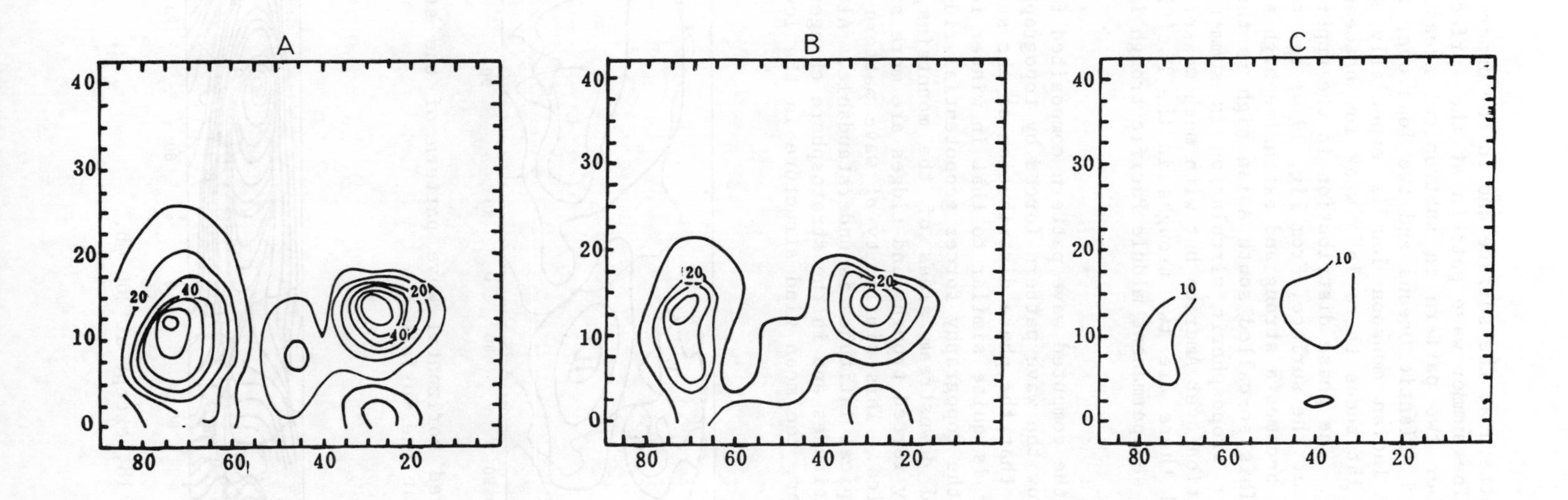

Fig. 17 As in Fig. 16, except for the stationary waves k = 1 to 3 forced by the latent heating in summer.

the observed wave pattern at the surface and in the upper troposphere, as shown in Fig. 18. The summer wave pattern at the surface is characterized by a wave number two pattern in subtropical areas with two highs over the Pacific and Atlantic Oceans and two lows over the two continents, of which the Indian monsoon low is especially significant and extensive. In high latitudes there is a weak low between Greenland and northeastern America. The phase distribution in the upper troposphere is the reverse of that at the surface. From Fig. 19 we see that the surface monsoon low in India becomes a strong and extensive high at upper levels, centered at 90°E. This so-called south Asian high is the most striking feature of the upper tropospheric circulation in summer. There is an anticyclonic circulation over America but with much weaker intensity than that over Asia, and there are two troughs in the Pacific and Atlantic Oceans of which the semipermanent middle Pacific trough is very significant.

We now look at the computed wave pattern composited from k = 1 to 4. Figures 20 and 21 show the wave pattern forced by topography at z = 0 and 15 km. It is found that the phase distribution at the surface forced by topography in summer is quite similar to that in winter in the subtropical latitudes, i.e. the topography forces geopotential ridges and troughs on the upstream and downstream sides of the mountains, respectively. These topographically forced troughs and ridges are more significant over Asia than over America. This similarity of wave pattern for summer and winter in the subtropical latitudes is understandable. Although the mean structure in high latitudes and in the stratosphere changes significantly from summer to winter, the mean wind structure in the lower troposphere

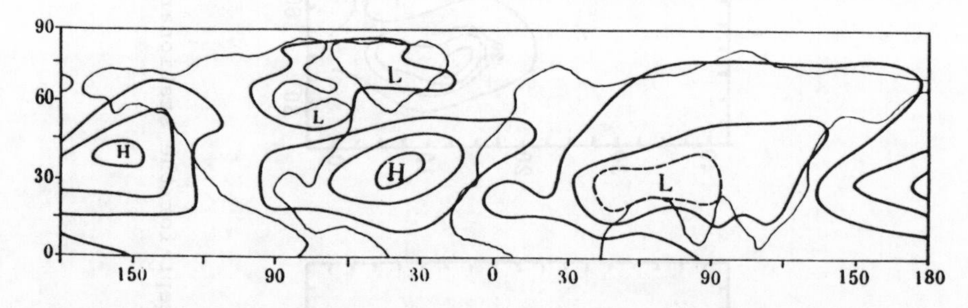

Fig. 18 The observed horizontal wave pattern of the stationary waves in summer at the surface.

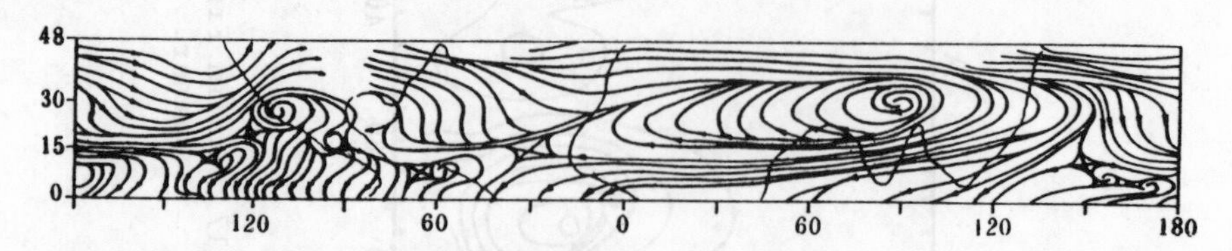

Fig. 19 As in Fig. 18, except at 200 mb.

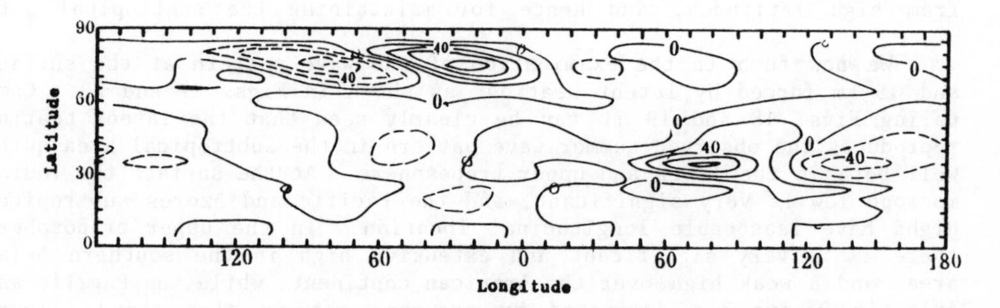

Fig. 20 The horizontal wave pattern of stationary waves forced by topography in summer at z = 0 km.

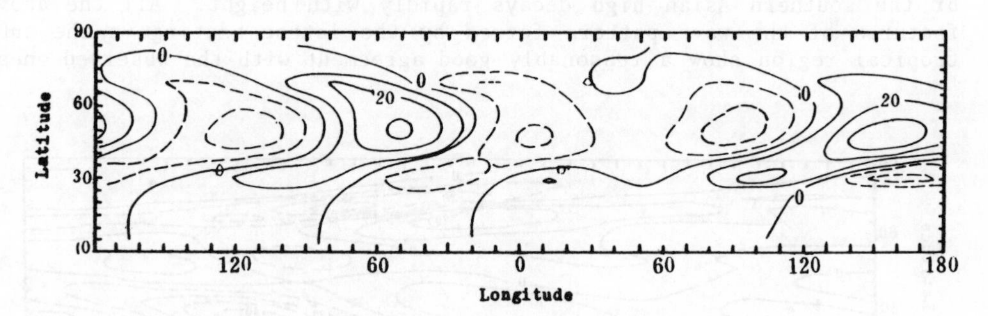

Fig. 21 As in Fig. 20, except at 15 km.

of the subtropical region does not change much, and the subtropics are occupied by westerlies in both summer and winter. The westerlies, according to the principle of potential vorticity conservation, should produce a ridge and a trough on the upstream and downstream sides of the mountains, respectively. However, by comparing Fig. 20 with Fig. 18, it is found that topography cannot reproduce the monsoon circulation in the Asia-Pacific area, but in high latitudes it forces a low between Greenland and northern America which is consistent with that observed.

In the upper troposphere, it is found that topography produces a wave train in middle latitudes, in which there is a high center located in the Aleutian region just like the winter case, but there is only a weak high in southern Asia. The observed strong high in the upper troposphere in the southern Asian area in summer apparently cannot be explained by the topographic effect alone. Also, the wave pattern excited by topography does not clearly show the observed phase reversal in the vertical. Therefore, the topographic forcing in summer cannot produce a reasonable monsoon circulation. However, the wave pattern forced by topography in the upper troposphere shows a northwest-southeast tilt in high latitudes and a northeast-southwest tilt in low latitudes. These orientations of the perturbations are favorable for stationary waves to transport westerly momentum poleward from low latitudes and equatorward from high latitudes, and hence for maintaining the subtropical jet.

We now turn to the examination of the wave pattern at the surface and 15 km forced by latent heating, as shown in Figs. 22 and 23. Comparing Figs. 18 and 19 it can be clearly seen that the latent heating reproduces the observed summer wave pattern in the subtropical area quite well both in the lower and upper troposphere. At the surface the Indian monsoon low is very significant, and the Pacific and Azores subtropical highs have reasonable longitudinal location. In the upper troposphere there is a very significant and extensive high in the southern Asian area, and a weak high over the American continent, while the Pacific and Atlantic Oceans are dominated by troughs. It is also found (figure omitted) that the monsoon low decreases its intensity with height and changes its phase (becoming a high) at about 7.5 km. Above this level the geopotential height increases significantly with height and reaches its maximum magnitude at about 15 km. Above the tropopause the intensity of the southern Asian high decays rapidly with height. All the above features of the wave pattern forced by the latent heating in the subtropical region show a reasonably good agreement with the observed ones.

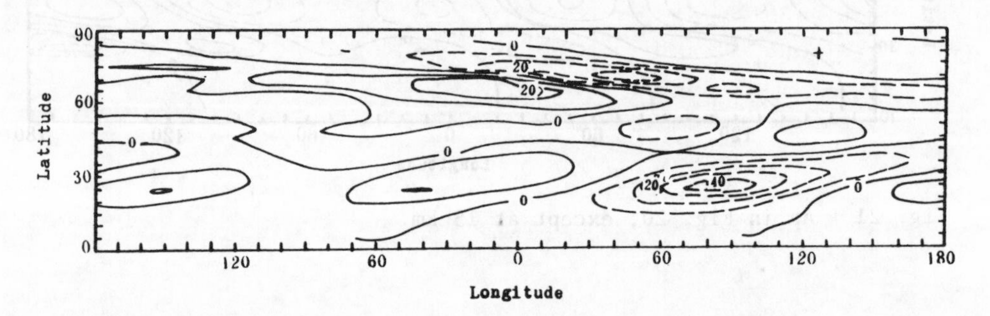

Fig. 22 As in Fig. 20, except for the stationary waves forced by the latent heating in summer.

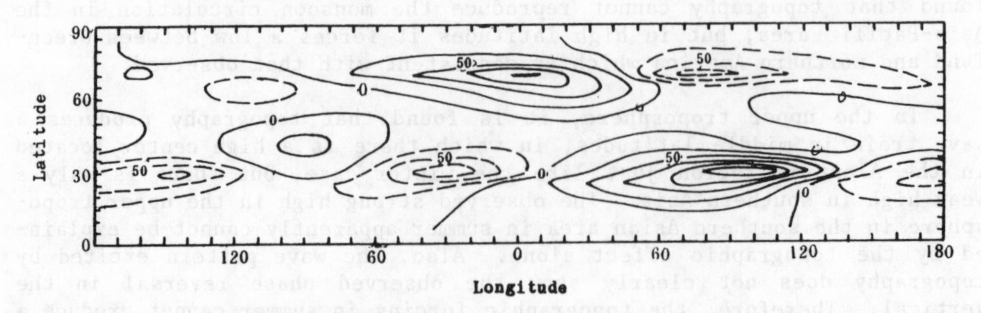

Fig. 23 As in Fig. 21, except for stationary waves forced by latent heating in summer.

However, it is found that the agreement between the computed and observed wave patterns in high latitudes is poor. The computed amplitude response seems to be too strong and the phase distribution is unrealistic. In particular, the low between Greenland and northeastern America at the surface cannot be reproduced by the latent heating. This means that the latent heating is not the main factor for maintaining the stationary waves in high latitudes.

We have also calculated the wave pattern forced by the sensible heat flux from the surface. Figure 24 shows the pattern at the surface (the response at upper levels is negligible). Comparing Fig. 18 and 22, it is found that the sensible heating not only produces a basically realistic wave pattern in the subtropical latitudes, like the latent heating, but also simulates a more realistic pattern in high latitudes. In particular, the low between Greenland and northeastern America has been reproduced. The similarity of the surface responses between the latent heating and the sensible heating in the subtropical region is evidence of the interdependence between the effects of these two heating processes. On the other hand, the fact that the topography and sensible heating produce more realistic surface wave patterns in high latitudes suggests that it is the topography and the sensible heat flux from the surface rather than the latent heating which play a more important role for stationary waves at the surface in high latitudes.

According to the above comparison of both the vertical structures and horizontal wave patterns, it turns out that in summer, unlike in winter, the diabatic heating plays a dominant role for the formation and maintenance of stationary planetary waves. Latent heating takes a crucial part for the monsoon circulation in the subtropical region, but in high latitudes the sensible heat flux from the surface plays a more important role for the surface wave pattern. Topographic forcing, although it cannot produce a realistic vertical wave structure and horizontal wave pattern in the subtropics, has an important contribution to the maintenance of the subtropical jet and the surface wave pattern in high latitudes.

VI. CONCLUSIONS

The present experiments, based on a primitive equation linear wave model, have examined the behavior and the mechanism of maintaining stationary planetary waves in winter and summer. The main results can be summarized as follows:

(1) The vertical propagating ability of stationary waves is mainly controlled by the structure of mean zonal wind. In winter and summer the mean zonal wind has quite different structures, which are responsible for the different behavior of stationary waves in these two seasons. In winter, the middle-high latitudes of the northern hemisphere are occupied by westerlies. There is a significant polar night jet in the stratosphere of high latitudes which plays a wave-guiding role for the vertical propagation of stationary waves and allows the largest scale stationary waves ($k = 1$ and 2) to propagate upwards into the stratosphere. Thus the

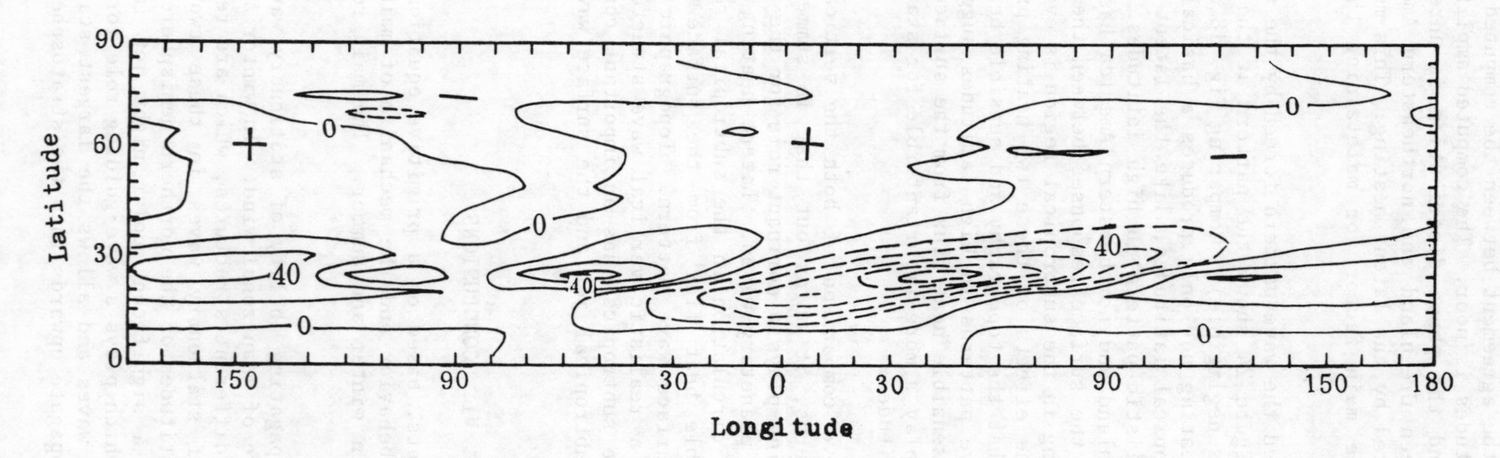

Fig. 24 As in Fig. 20, except for the stationary waves forced by the sensible heat flux from the surface.

stationary waves in the winter stratosphere mainly consist of waves number 1 and 2; the shorter stationary waves are trapped in the troposphere. In summer, the polar night jet disappears and the stratosphere of the northern hemisphere is dominated by easterlies which block the vertical propagation of stationary waves, and hence in summer even the largest scale stationary waves cannot propagate into the stratosphere. The flow field in the stratosphere is nearly oriented zonally and the stationary waves in summer are trapped in the troposphere.

(2) The stationary waves are forced and maintained by topography and diabatic heating. However, there are different mechanisms for maintaining the stationary waves in winter and summer. In winter, the stationary waves forced by topography are quite close to the observed ones, both in the vertical structure for a single mode and in the horizontal wave pattern for composited waves. In contrast, the structure and pattern of waves forced by diabatic heating are quite different from the observed ones. These facts suggest that the stationary waves in winter are mainly forced and maintained by the effects of topography. In summer, however, the topography cannot reproduce a reasonable monsoon circulation in the subtropics, whereas the latent heating reproduces a quite realistic summer monsoon circulation in this region. Thus, in summer it is the diabatic heating rather than the topography which plays the main role in maintaining the monsoon circulation. In high latitudes, the situation is different. The latent heating is no longer the controlling factor, while the advective process plays an important role. The topography reproduces a more realistic surface wave pattern than the latent heating and also takes an important part in maintaining the subtropical westerly jet in middle latitudes.

(3) The stationary waves represent the climatic characteristics of ultralong wave activity. This study on the effects of topography and diabatic heating on the stationary waves is therefore helpful in understanding the climatic effects of topography.

ACKNOWLEDGMENTS

I want to sincerely thank Prof. J.R. Holton for encouraging and helping me in this study. This work was supported by the National Science Foundation, Atmospheric Research Section, NSF Grant ATM 79-24687.

REFERENCES

1. Berkofsky, B. and E.A. Bertoni: Mean topographic charts for the entire earth. Bull. Amer. Meteor. Soc., 36, 350-354 (1955).

2. Bolin, B.: On the influence of the earth's orography on the general character of the westerlies. Tellus, 2, 184-195 (1950).

3. Budyko, M.I.: Climate and life. Academic Press, New York, 508 pp. (1974).

4. Defant, F., A. Osthaus and P. Speth: The global energy budget of the atmosphere, Part II: The ten-year mean structure of the stationary large-scale wave disturbances of temperature and geopotential height for January and July (northern hemisphere). Beitr. Phys. Atmos., 52, 229-246 (1979).

5. Egger, J.: On the theory of planetary standing waves: July. Beitr. Phys. Atmos., 51, 1-14 (1978).

6. Holton, J.R.: A diagnostic model for equatorial wave disturbances: the role of vertical shear of the mean zonal wind. J. Atmos. Sci., 28, 55-64 (1971).

7. Holton, J.R. and W.M. Wehrbein: A semispectral numerical model for the large-scale stratospheric circulation. Rept. No. 1, Middle Atmosphere Project, University of Washington, Seattle (1980).

8. Kasahara, A., I. Sasamori and W.M. Washington: Simulation experiments with a 12-layer stratospheric global circulation model. I. Dynamical effect of the earth's orography and thermal influence of continentality. J. Atmos. Sci., 30, 1229-1251 (1973).

9. Labitzke, K. and Collaborators: Climatology of the stratosphere in the northern hemisphere. Met. Abhand. Band, 100/Heft 4, M/15 (1972).

10. Lau, N.-C.: The observed structure of tropospheric stationary waves and the local balances of vorticity and heat. J. Atmos. Sci., 36, 996-1016 (1979).

11. Lindzen, R.S. and H.L. Kuo: A reliable method for the numerical integration of a large class of ordinary and partial differential equations. Mon. Wea. Rev., 96, 732-734 (1969).

12. Lin, B.-D.: The behavior of winter stationary planetary waves forced by topography and diabatic heating. J. Atmos. Sci., 39, 1206-1226 (1981).

13. Manabe, S. and T.B. Terpstra: The effects of mountains on the general circulation of the atmosphere as identified by numerical experiments. J. Atmos. Sci., 31, 3-42 (1974).

14. Matsuno, J.: Vertical propagation of stationary planetary waves in the winter northern hemisphere. J. Atmos. Sci., 27, 871-883 (1970).

15. Möller, F.: Vierteljahreskarten des Niederschlags für die ganze Erde. Petermanns Geogr. Mitt., 95, 1-7 (1951).

16. Murakami, T.: Vertical transfer of energy due to stationary disturbances induced by topography and diabatic heat sources and sinks. J. Meteor. Soc. Japan, 45, 205-321 (1967).

17. Murakami, T.: Influence of midlatitude planetary waves on the tropics under the existence of critical latitude. J. Met. Soc. of Japan, 52 (3) (1974).

18 Murgatroyd, R.J.: The structure and dynamics of the stratosphere. The Global Circulation of the Atmosphere, G.A. Corby, Ed., London, Roy. Met. Soc., 159-195 (1969).

19. Sanka-Rao, M.: On global monsoon -- further results. Tellus, 22, 648-654 (1970).

20. Sanka-Rao, M. and B. Saltzman: On the steady-state theory of global monsoon. Tellus, 21, 308-329 (1969).

21. Schoeberl, M.R. and M.A. Geller: The propagation of planetary scale waves into the upper atmosphere. Aeronomy Rep. 70, Dept. of Elect. Eng., University of Illinois, 269 pp (1976).

22. Smagorinsky, J.: The dynamical influence of large-scale heating sources and sinks on the quasi-stationary motions of the atmosphere. Quart. J. Roy. Meteor. Soc., 79, 342-366 (1953).

23. Tung, K.K.: A theory of stationary long waves. Part II: Resonant Rossby waves in the presence of realistic vertical shears. Mon. Wea. Rev., 107, 735-750 (1979).

24. Webster, P.: Response of tropical atmosphere to local, steady forcing. Mon. Wea. Rev., 100, 518-541 (1972).

25. van Loon, H., R.L. Jenne and K. Labitzke: Zonal harmonic standing waves. J. Geophys. Res., 78, 4463-4471 (1973).

26. Zhu, Bao-Zhen: The steady-state perturbations of the westerlies by the large-scale heat sources and sinks and the earth's orography (II). Acta Meteor. Sinica, 28, 198-224 (1957).

27. Zhu, Bao-Zhen: A preliminary study on the activities of ultralong waves in relation to the dynamic control of large-scale orography and heat sources. Acta Meteor. Sinica, 34, 285-298 (1964).

28. Zhu, Bao-Zhen and Lei Xiao-En: A study on the formation of main flow patterns in the stratosphere. Acta Meteor. Sinica, 38, 289-299 (1980).

SESSION III

PLANETARY AND LARGE-SCALE EFFECTS OF MOUNTAINS ON GENERAL CIRCULATION AND CLIMATE: NUMERICAL MODELLING

3.1 NUMERICAL EXPERIMENTS WITH THE SICHUAN FLOODING CATASTROPHE (11-15 July 1981)

John B. Hovermale
NOAA, NWS, National Meteorological Center
Washington, DC 20233 USA

ABSTRACT

Experiments with a movable fine-mesh (MFM) model are described whose physical parameterizations are similar to those suggested by Anthes [1] for hurricane simulations. The model is initialized by conditions extracted from the U.S. National Meteorological Center's (NMC) global optimum interpolation (OI) analyses. The model calculations were applied to a major flooding catastrophe in Sichuan Province during the period 11-15 July, 1981, which was characterized by extensive mesoscale convective development. Even though the orographic features of Tibet were entered into the model in a strongly smoothed fashion, the model was capable of predicting mesoscale development.

I. INTRODUCTION

The flow of air in the vicinity of large mountain barriers creates many unique weather anomalies of varied space and time scales. At the climatic end of the spectrum are the windward rain sheds and the associated lee-side steppes. A contrasting example at the very short-range is the local or regional mountain-valley circulation driven by diurnally varying pressure-density solenoids. Some of the most difficult situations to understand (and therefore to predict) are those which involve interactions of topographically forced motions of varied scales which in turn react with traveling, free disturbances in the atmosphere. The rapid growth of disturbances sometimes generated by uncommonly strong physical instabilities add further uncertainty to the ultimate outcome of weather predictions in mountainous regions.

A lee side weather problem of common concern in China and the United States relates to mesoscale heavy precipitation systems that occasionally bring devastation to areas of size between 10^2 km^2 to 10^6 km^2. Meteorological scenarios in these instances are many times similar regardless of which side of the earth we consider. Convective motions are usually closely associated with, and embedded in, a small but synoptically observable low-level disturbance. There is often a sharp contrast in the large-scale environment between warm, very moist, tropical air and dry continental air along a southwest-to-northeast shear line. High elevation solar heating may be a factor in determining the timing and magnitude of the event. To be sure one can find contrasts between

Chinese and American weather systems of this general type (e.g., U.S. environments seem more favorable for tornado genesis) but the similarities remain remarkable.

In this paper, we shall attempt through numerical simulation to better understand the causes behind the development of flash flooding events mentioned above. We shall deal with actual events rather than perform simulations with idealized data to ascertain the degree of realism that can be achieved with current operationally oriented systems. The ability to predict both onset and decay of precipitation will be studied.

Viewed from a mathematical perspective, this experiment is designed to study the speed and vigor with which two large-scale, large amplitude factors (flow and barrier) can interact nonlinearly to produce significant small-scale reactions. Viewed meteorologically, this experiment is designed to study the relative importance of large-scale controls over mesoscale weather in intermountain regions. Viewed in one practical sense, this experiment is designed to give further insight into needs for smaller-scale weather information in mountainous regions.

II. BRIEF DESCRIPTION OF MODEL

The model employed for the study is a limited area on-call operational system of conventional numerical and physical design that has been tested up to the present with 10 equally spaced layers in sigma coordinates. The so-called movable fine-mesh (MFM) weather prediction system has been functioning within the NMC operational framework since the summer of 1975. Whenever hurricanes present potential problems to United States coastal areas, or when significant precipitation events threaten to produce flooding or other hazards, the MFM has been run to determine the degree of increase in operational prediction accuracy with ultrahigh resolution grids.

The considerable reduction in grid increment achieved in this system (it has been applied at resolutions between 50 and 120 km) relative to current operational modes (e.g. Δx = 195 km) is made possible primarily by the relocatable characteristic of the grid. The computer program has been designed to allow the grid not only to be centered over weather systems at the start of forecasts but permits grid shifts following the predicted storms. In this way, ultrahigh resolution calculations in areas not intimately related to the internal behavior of storms can be eliminated with the savings directed toward more precise definition of storm dynamics.

The success of this approach depends critically on the realism with which external information from large-domain, low resolution models can be provided to the system. For experience has shown that, over periods reaching to two days, large-scale atmospheric features exert important controls over the behavior of individual cyclones. To a lesser extent, over the same periods, cyclones influence larger-scale behavior, but for operational expediency this effect is not involved beyond the domain of the limited area grid [area ≅ $(3000)^2 km^2$].

The finite difference structure of the model conforms to the original second-order semimomentum form developed by Shuman [6] for the operational hemispheric model employed at NMC during the 1960's and the 1970's. This was done to help facilitate the exchange of information inward across the lateral boundaries of the MFM. (The MFM accepts, but does not feed back, information to its host, large-scale, model.) Through experimentation it was discovered that conformity of finite difference structure between the host and limited area grids was not a major factor in ensuring a smooth transition at the interface between the two models.

Inconsistencies between solutions on the two grids and the reflected waves they excite are controlled adequately by diffusive-type terms on all predictive variables. Deviations of inner grid variables from outer grid values obtained by linear time interpolation are the factors employed to determine the magnitude of the damping. The same approach has been used by the European Center for Medium Range Forecasting for its limited area experimentation (high frequency waves that escape the lateral boundary zone are damped by a time filter that also helps control high frequency energy introduced by parameterization physics).

The physical parameterizations for the model closely parallel those introduced by Anthes et al. [1] for hurricane simulations. Coefficients for vertical and horizontal eddy diffusion have been adjusted to account for the current grid resolutions of 60 km and 90 mb. Precipitation is released according to conventional large-scale checks for excess of specific humidity beyond some fraction (maximum relative humidity = 90 percent) of the staturation specific humidity. Precipitation also may be released before 90 percent relative humidity is reached if subgrid scale convection is expected to exist. Such decisions are made on the basis of cloud layer moisture convergence and conditional static instability closely following the logic explained by Kuo [3].

Surface exchange of sensible and latent heat are treated in the same manner as by Anthes et al. [1] and radiation influences are ignored.

III. THE LARGE-SCALE SPECIFICATION OF INITIAL STATE

The initial conditions for the experiment have been extracted from the U.S. National Meteorological Center's (NMC) global optimum interpolation (OI) analysis. Economic considerations coupled with relatively sparse data coverage over most of the globe force the maximum resolution of this system to be coarser than desired, especially for small synoptic scales, and certainly for mesoscale phenomena. A form of dynamic (nonstatic) initialization is applied to this analysis to create as much internal balance as possible between the meteorological variables.

The so-called method of nonlinear modal initialization (see [4]) is combined with the NMC operational global, spectral model to achieve large-scale balance. It is worth noting at this point that the method, as utilized at NMC, first establishes the linear balance relationships between meteorological variables of Rossby-type modes. Then the pressure fields are adjusted to account for mountains, assuming a calm atmosphere.

Lastly, all variables are adjusted by the solution of a balanced relationship between nonlinear terms of the Rossby modes and gravity modes (see [8]). This procedure essentially results in a state of balance which the global spectral model would approach if integrated for 12 to 24 hours under adiabatic, frictionless conditions; the latter assumptions must be made to permit a solution by the Machenhauer method.

This solution, obtained with a model whose minimum wave scale is 12° of longitude both in the atmosphere and the orography, is interpolated to the MFM 60 km grid in final condition for time integration.

It is obvious that, due to higher resolution and energy sources and sinks, the initialization will not be in balance with the MFM model, nevertheless those imbalances leading to low frequency oscillations will be reduced substantially below those inherent in the original analysis. Higher frequency oscillations that are generated by imbalances relative to the limited domain model will be dissipated rapidly by the time filter mentioned earlier.

IV. DESCRIPTION OF SYNOPTIC SITUATION

The major catastrophe in Sichuan Province during the period 11-15 July 1981 conforms closely to the general type of synoptic event discussed by Tao and Ding [7] in regard to flooding threats in southwest China. That is to say, several characteristic conditions existed. First, an upper-level midlatitude trough was passing to the north of the Tibetan Plateau. Secondly, there was a zone of strong temperature and moisture gradient along a southwest to northeast line east of the plateau. This zone, logically associated with strong low-level shear, can be readily identified with severe weather events. In reporting on a talk by Mrs. Sun Shu-qing, Professor E. Reiter noted that westward displacement of the Pacific subtropical high was a significant precursor of heavy rain. The 11-15 July case not only had this source of latent fuel, but heavily moisture-laden Bay of Bengal air as well.

Although several times were selected as initial conditions for integrations in the current study, one case will be emphasized in this paper because it exhibits a full cycle from onset to decay of the heavy rainfall.

The large-scale flow as depicted by NMC's global analyses at 0000 GMT, 12 July undoubtedly lacked definition in comparison to the actual atmosphere. A broad trough at 500 mb was portrayed over and to the north of the plateau with axis at approximately 95°E (see Fig. 1). The associated vorticity suggested a rather weak forcing based on quasi-geostrophic principles. The isotachs suggested the typical, but weaker, pattern of jets around the plateau area with strongest winds located at the eastern and southern slopes.

The positionings of a Pacific subtropical high and the monsoon low over India seemed well specified as did their intensities. Such quasi-steady features are more accurately described than transcribed phenomena

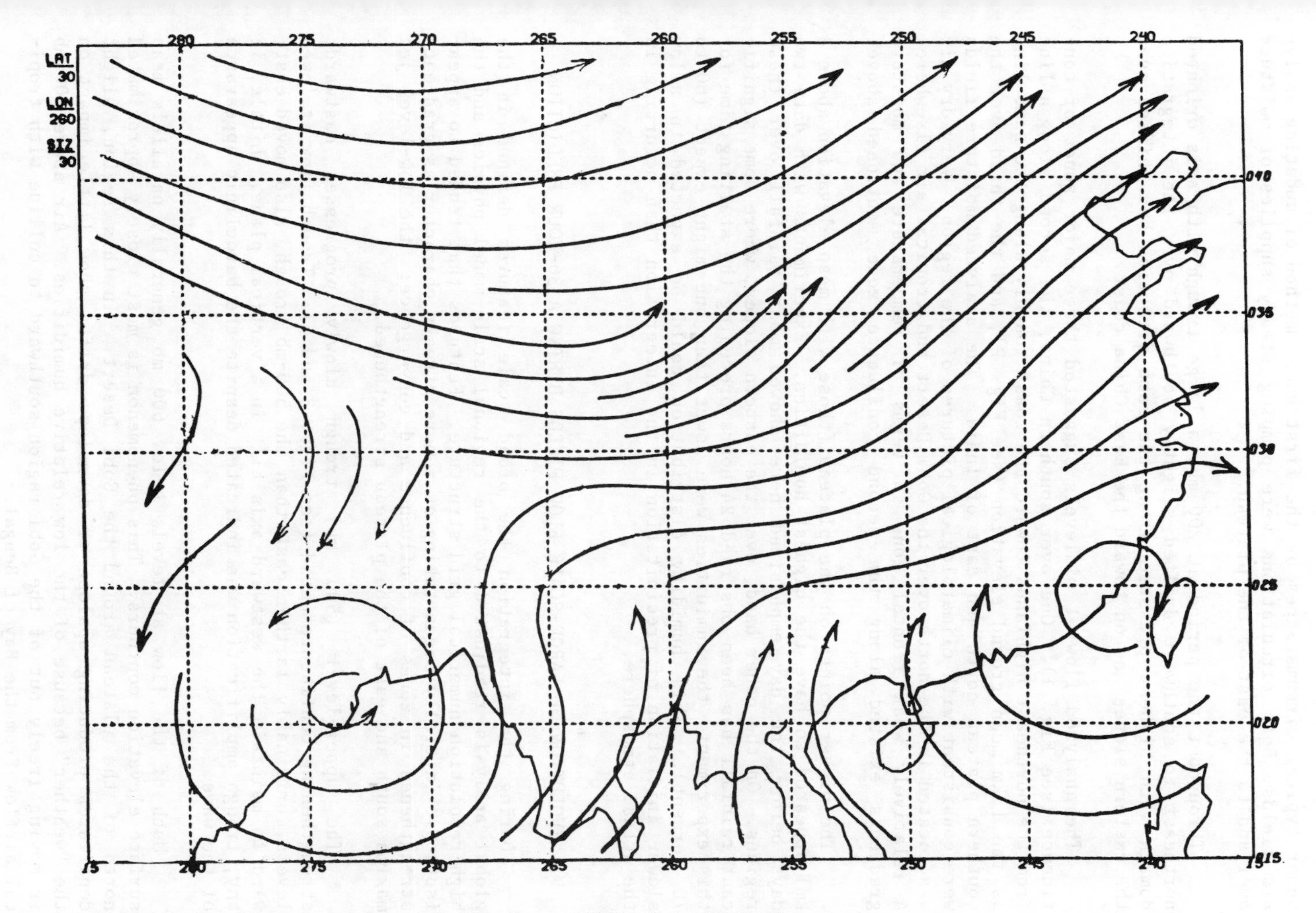

Fig. 1 Streamline analysis at 500 mb from NMC global analysis.

under typical circumstances by the first guess method of updating analysis fields. Both circulations were pumping steady supplies of moisture over and to the east of the plateau.

Throughout the period at 200 mb, a large trough with axis oriented northeast to southwest drifted slightly southward over the integration domain with highest pressure along the same direction but shifted over the eastern slopes and on toward the East China coast.

The analyzed flow at midlevels suggested three major zones of confluence (see Fig. 1). One over southern China, the second along a line from the northeast highlands toward the coast roughly along the same line as the 1 km mean ground elevation (see Fig. 2), and the third over the southern plateau north and east of India. The analyzed moisture fields were consistent with climatological pictures of the region. Very dry air was located in the north over the Gobi Desert and tropical air elsewhere. A relatively weak frontal zone in terms of temperature and moisture gradients existed along the second confluence zone mentioned above.

The higher points on the plateau (those with mean elevation above 3 km) appeared to have the highest humidities. Experiments with data two days before the day emphasized here showed most rainfall over these regions. On the other hand, over the eastern slopes, where some significant rainfall has been observed 24 hours preceding the starting time for this experiment, the humidities were lower than one might expect (60 to 70 percent). This humidity distribution could be expected to allow slower initiation of precipitation in the integration than occurring in the actual atmosphere.

V. EVOLUTION OF LARGE-SCALE FLOW FIELDS DURING A 36-HOUR PREDICTION

During the integration the broad scale features defined in the global analysis readjusted to the regional scale model physics and the high resolution numerical grid structure. Features that tended to appear isotropic initially began taking on a more streaky character as gradients strengthened in zones of confluence and convergence: the low-level jet maxima south and east of the plateau strengthened.

The upper-level (500 mb) trough slowly progresses eastward, strengthening confluence ahead (see Figs. 3 and 4). Trough flow at lower levels, originally farther east than the 500-mb trough, also moved eastward to maintain the westward axis tilt in a vertical plane. Little, if any, trough amplification was indicated despite the baroclinic appearance of the wave.

Much of the flow at levels below 600 mb generally parallels mean surface elevation contours. This phenomenon is most closely approximated north of the plateau around the Gobi Desert. In this region, slight departures producing upslope or downslope drifts show little impact on the "weather" because of the low relative humidities. Air above 600 mb is moving freely out of the Gobi region southward to collide with tropical airflow from the Bay of Bengal.

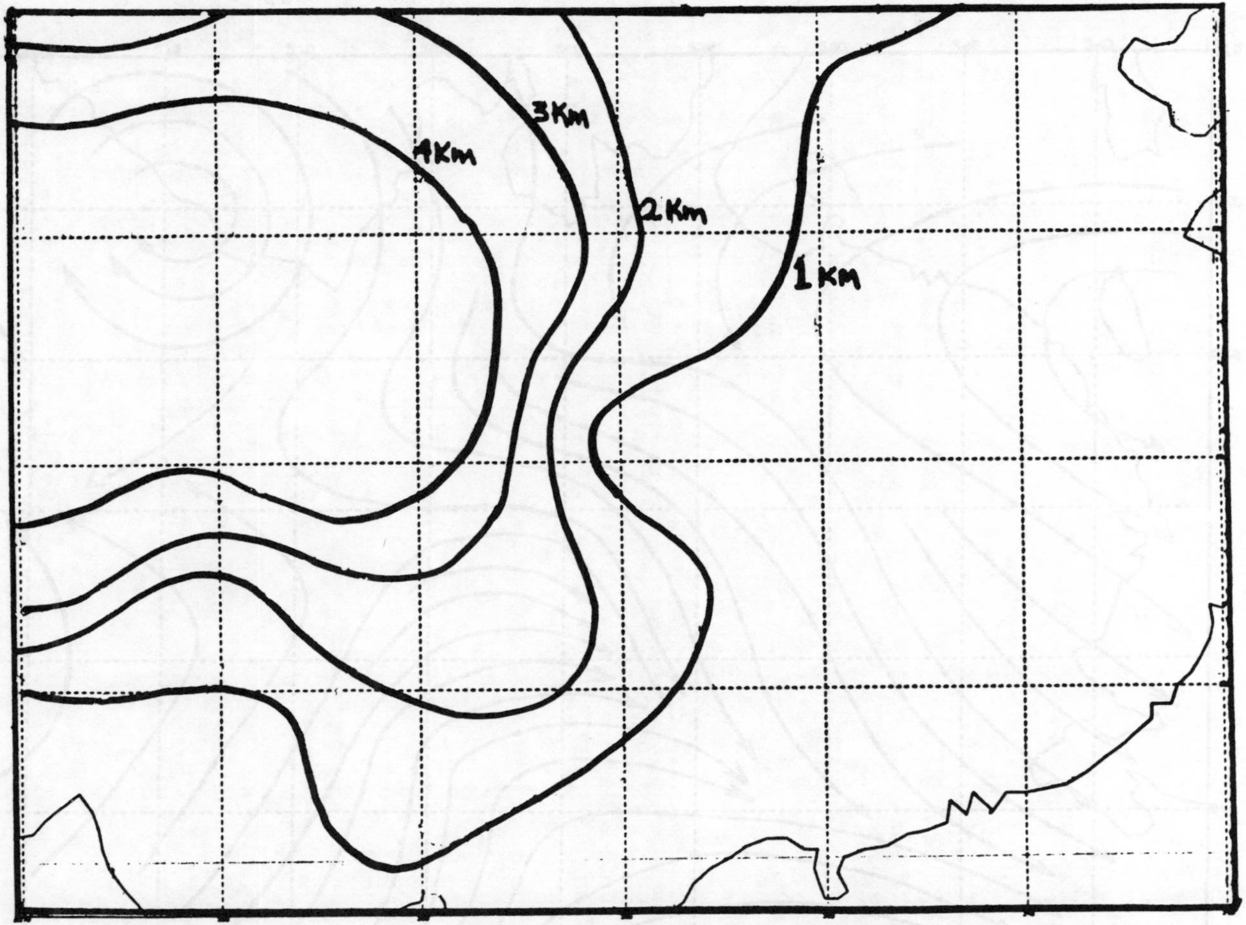

Fig. 2 Smoothed orography pattern over China employed by NMC models.

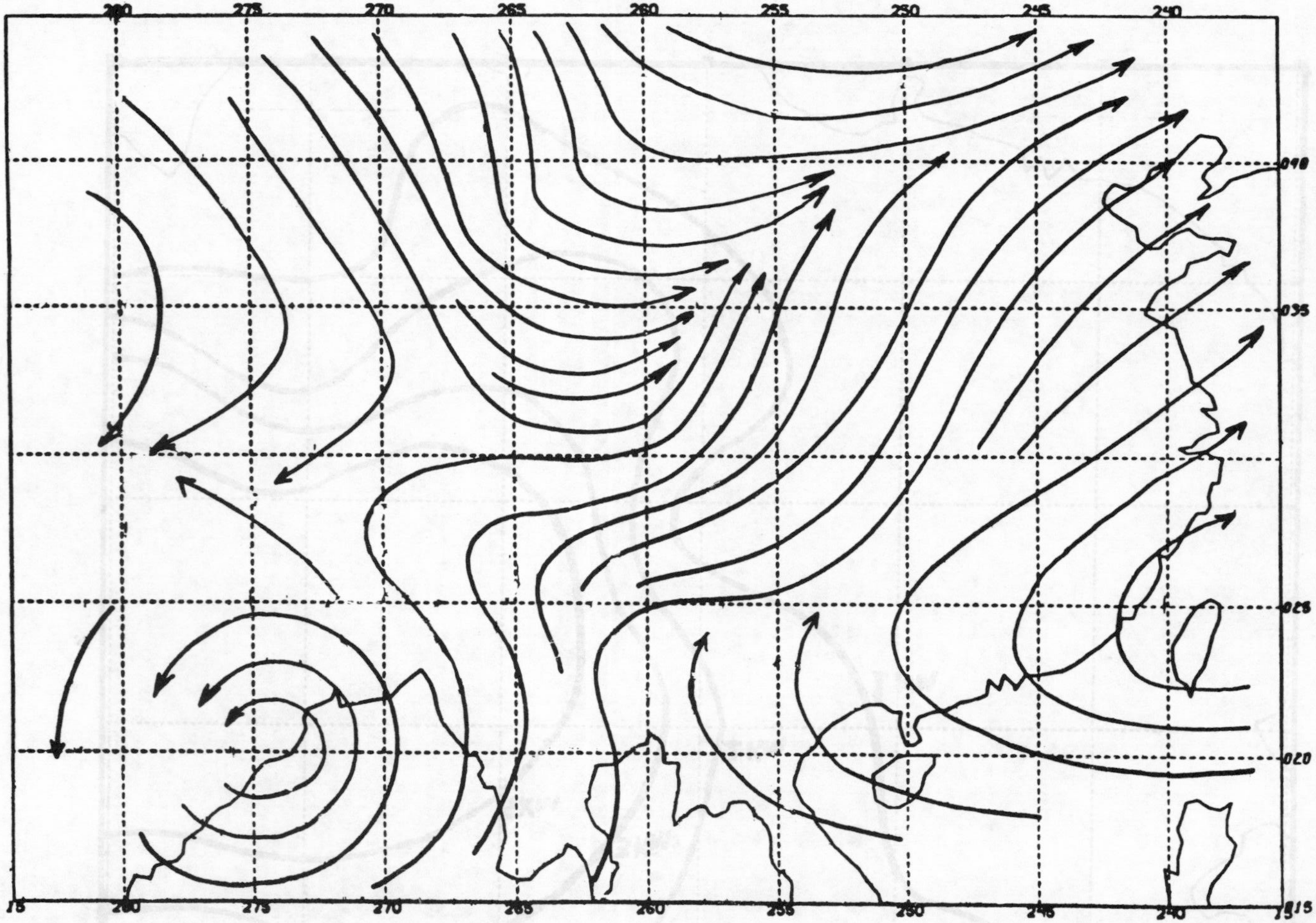

Fig. 3 Streamline analysis at 500 mb, 24 hours of time integration.

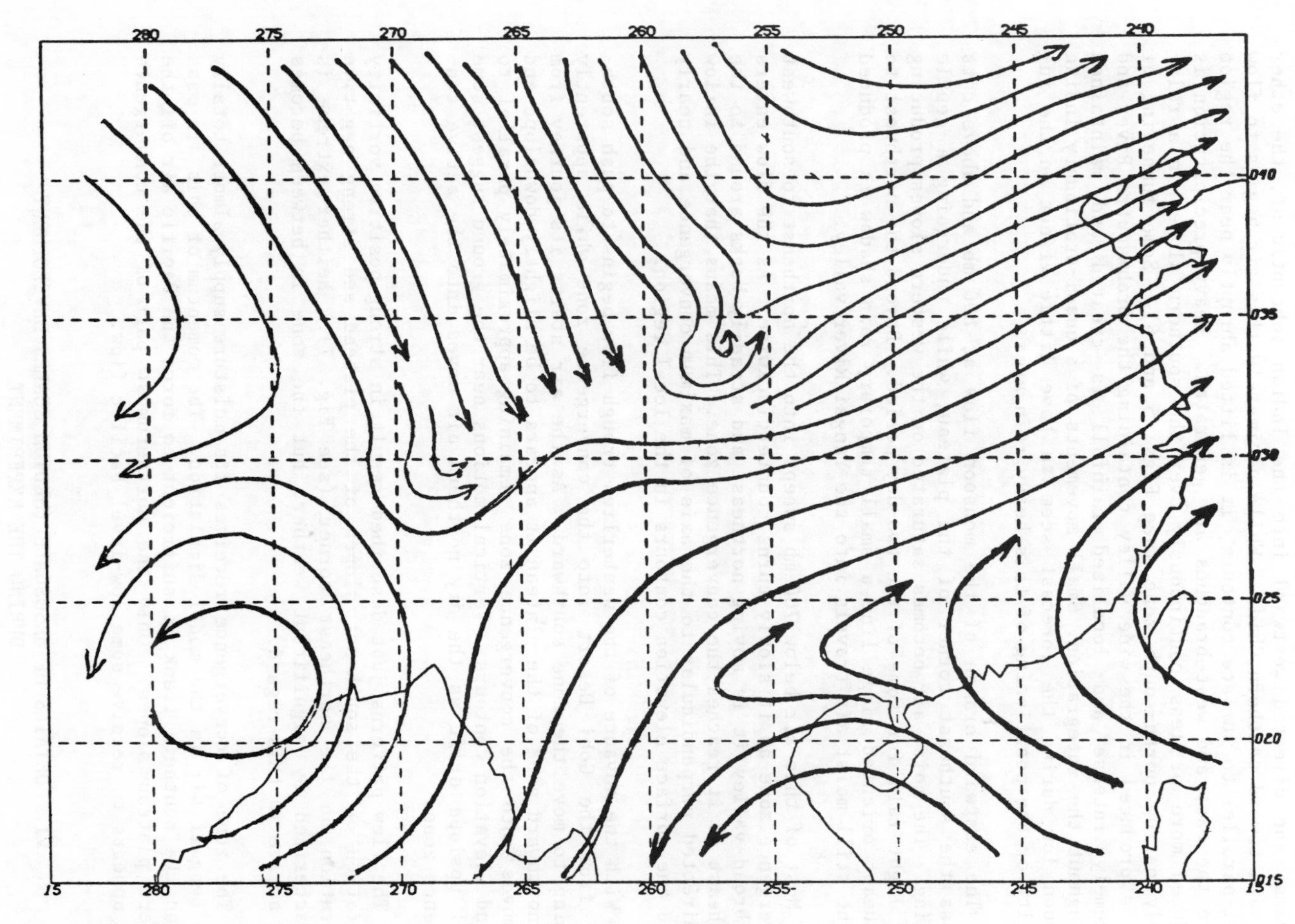

Fig. 4 Streamline analysis at 500 mb, 36 hours of time integration.

The northward moving Bengali air is forced to separate into two streams, one directed westward into the Indian subcontinent, the other eastward toward southern China. While ultimately this air begins to flow quasi-parallel to surface contours, it is lifted abruptly near the region where the east and west branches are generated. Heavy precipitation is produced more or less continuously over the southern slopes in a relatively narrow north-south zone (see Figs. 5 and 6). Some funneling of air is produced in the wide valley containing the Brahmaputra River and extremely intense and localized rainfall is created and maintained throughout the integration. Small movements of a quasi-stationary Indian monsoon low during the forecast seem to have little effect on the intensity or the positioning of this heavy rain area.

The eastward branch of the monsoon flow at 800 mb and above cuts across the southeast corner of the plateau, while undergoing a gentle lifting. The moist air becomes saturated on the western slopes producing only light rain for some of the two-day period. Beyond the northwest to southeast oriented ridge line a small temporary rain shadow is produced as the still moist air travels into the Yangzi River valley.

Most of this air below 700 mb sweeps into the northeast to southwest convergence zone as it slowly turns counterclockwise. As the flow enters the broad valley it is moving northeast and steadily backs around to the southeast as it reaches the convergence zone. This means that the inflow is directed perpendicular to the axis of maximum convergence and nearly so to the surface elevation contours in the local region.

With the advance of the westerlies trough flow begins to push southward from the Gobi Desert into the convergence zone while apparently helping to move the zone southward. As the air starts its journey from the northward side of the plateau it appears to be slightly downslope and it moves into the convergence zone remaining approximately parallel to ground elevation contours. Vertical motions near the ground suggest some small upslope drift as the dry northern air moves into the active convergence zone.

The flow patterns just described result in strong positive vorticity generation on the southeast slopes of the plateau and strong negative generation on the northeast corner (see Fig. 7). Neither extreme is characterized by significant weather, but the zone in between becomes most active in this regard.

The zone of convergence receives its moisture supply almost totally from Bengali air in the model simulation. The momentum of this air mass around the southern flank is sufficient to force the Pacific air off the eastern plateau slopes. Only the northernmost part of the convergence zone appears to receive some low-level Pacific flow.

VI. DETAILS OF MESOSCALE (MEDIUM SCALE) DEVELOPMENT DURING THE EXPERIMENT

A comparison of charts at various levels suggests that the low-level mesoscale flow sometimes appears decoupled from the 500-mb flow. A low-level circulation that forms at 800 mb and below in the Sichuan

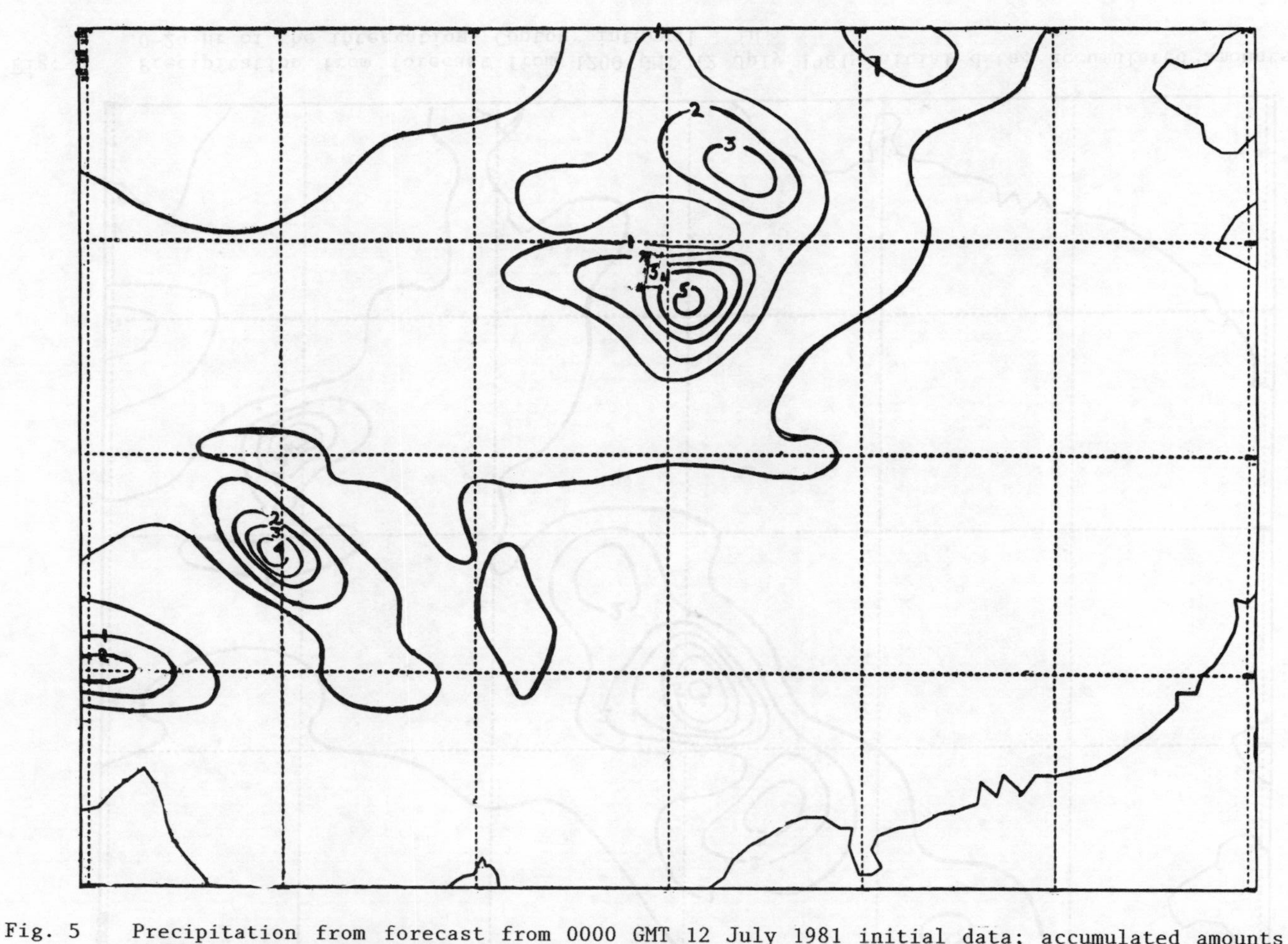

Fig. 5 Precipitation from forecast from 0000 GMT 12 July 1981 initial data; accumulated amounts from 12-36 hours of the integration. Contour interval 1 in.

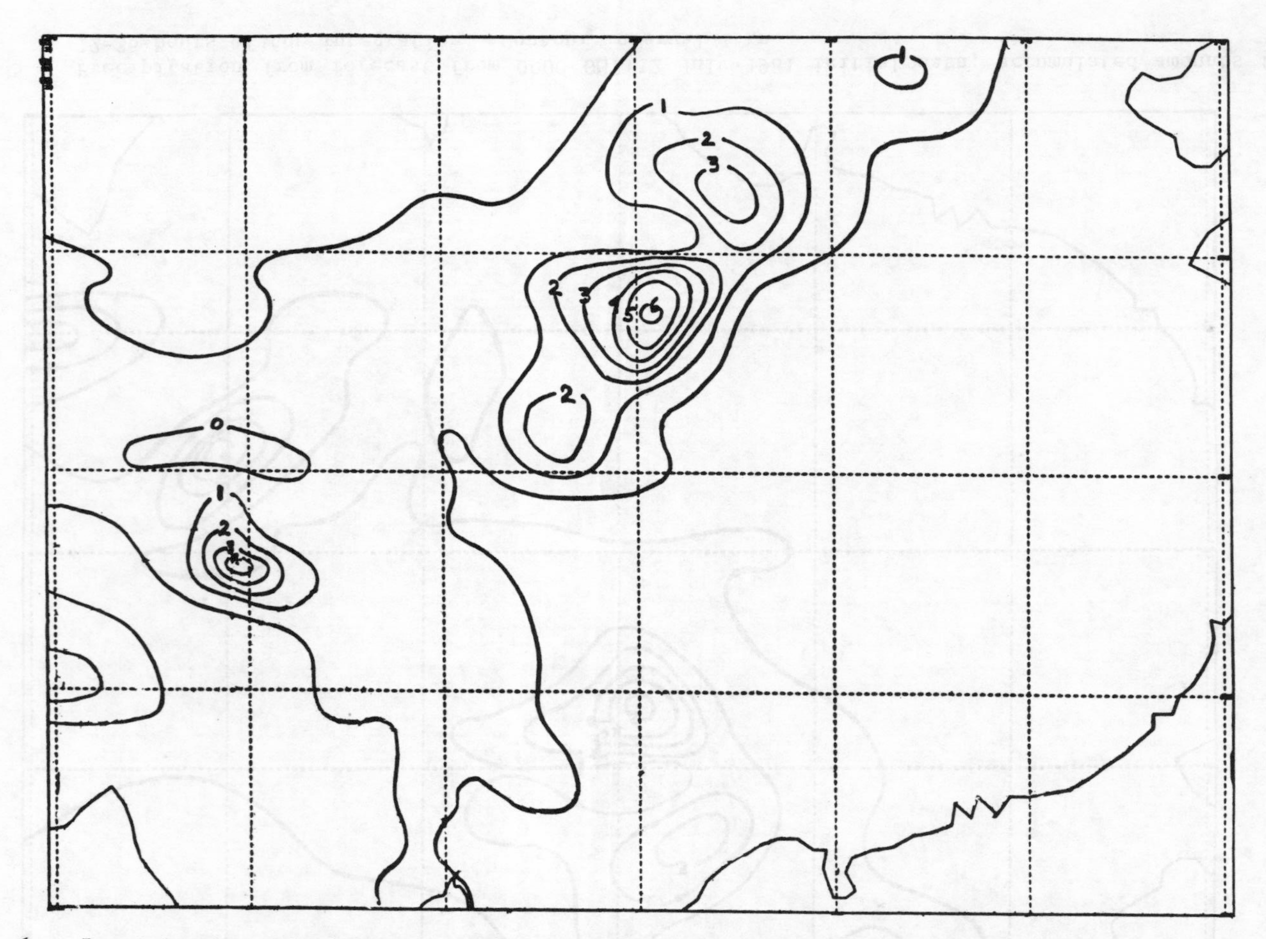

Fig. 6 Precipitation from forecast from 1200 GMT 12 July 1981 initial data; accumulated amounts from 0-24 hr of the integration. Contour interval 1 in.

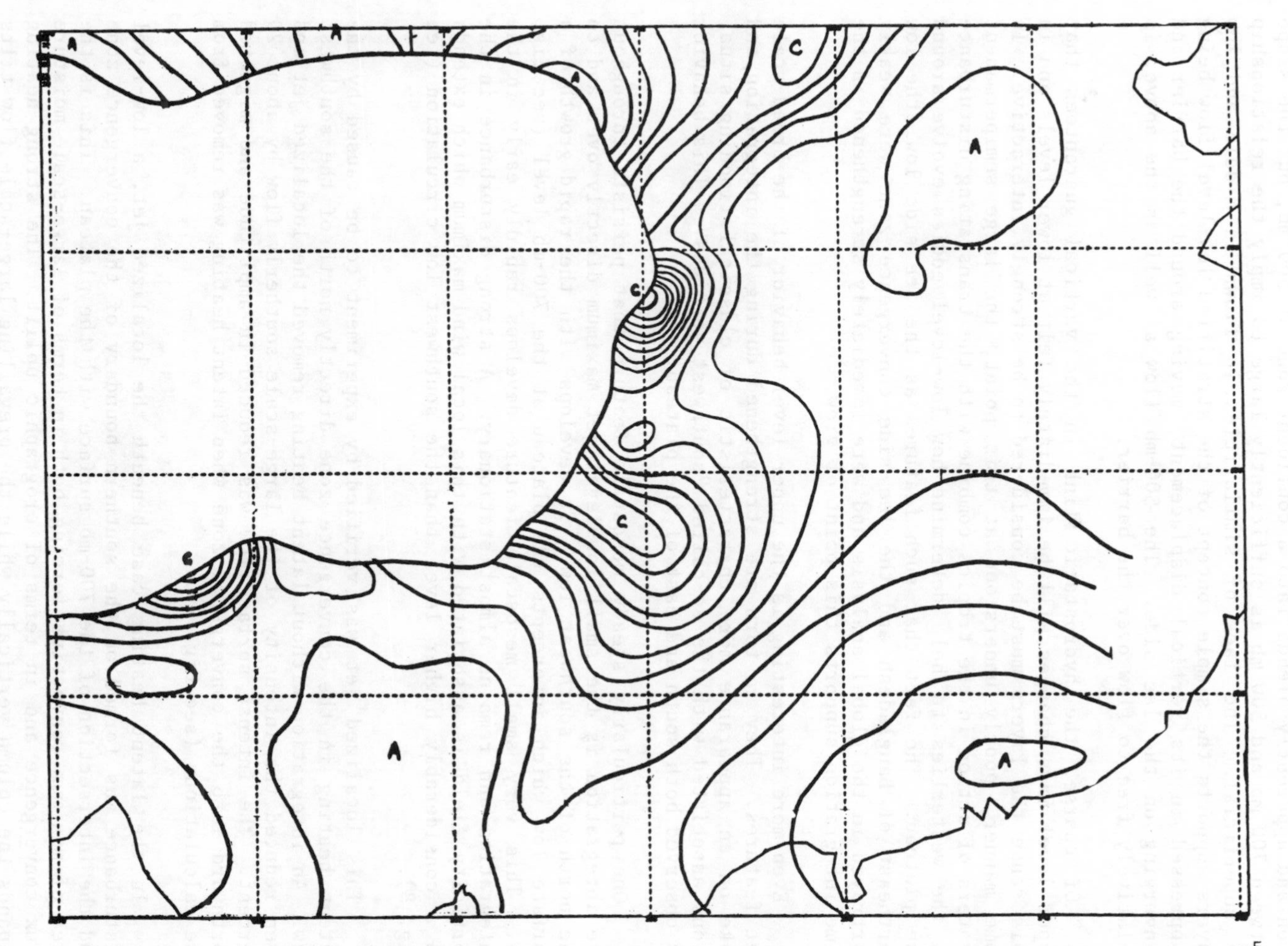

Fig. 7 Vorticity field at 700 mb after 12 hours of time integration. Contour interval 10^{-5} sec^{-1}.

region tilts upward toward the northwest but is lost by 500 mb. The low-level convergence zone northeast of this low tilts in the same direction and apparently blends into a confluent zone at 500 mb, but the slope between 700 mb and 500 mb is sufficiently large to imply the relationship is superficial. The lack of similarities in flows in middle and low layers supports the simple concept of the stratified low-level flow being suppressed in its vertical displacement, moving around the barrier and converging on the lee side. The 500-mb flow as well as the above, is relatively free to flow over the barrier.

Of course, the hydrostatic link in the vertical guarantees that upper level mass changes will be immediately felt at lower levels and in this sense the layers must be considered to be strongly interactive. In some manner, poorly understood at this point, the large semipermanent centers of action in the tropics combine with the translating disturbance in the westerlies to help determine how low-level eddies evolve around the plateau. The fact that such features as the lee side low, the low northeast of Bangladesh and the lee side convergence zone were weakly portrayed in the global analysis and were immediately strengthened in the time integration supports this point of view.

Even more interesting is the upper level behavior of the broad-scale jet features. They in turn are strengthened during the integration and take on an appearance more characteristic of classical flooding situations associated with the so-called "southwest vortices". This behavior is observed both south and east of the plateau.

One particularly steady mesoscale feature that persists throughout the integration is the small low-level jet maximum directly over and to the north of the southwest low. It develops with the rapid growth of a surface low which intercepts the plateau at the 700-mb level (see Fig. 7). This very small mesoscale feature develops rapidly early in the integration and remains almost stationary. A strong disturbance in the vorticity field is associated with this local wind maximum which extends to a considerably higher level than the southwest low circulation (see Fig. 8).

This localized jet was verified by experiment to be caused by the latent heating in the convergence zone directly north of the southwest low. An integration without latent heating removed the localized jet and even reduced the intensity of the large-scale southerly flow by about 20 percent. The intense surface low was reduced in amplitude and migrated northward into the convergence zone when latent heating was removed from the calculations (see Fig. 9).

In the latent heating case beneath the localized jet, a low-level disturbance was formed on the southern boundary of the convergence zone and the intersection of the 700-mb surface with the plateau. This is the area of maximum convective forcing both in terms of large-scale moisture flux convergence and in terms of orographic uplift. The strong heating expands the column vertically while the prevailing large-scale flow tilts the warm column upstream (compare Figs. 7 and 8). The small jet maximum forms on the west side of this column in an attempt to approach quasi-geostrophic balance. The scale of this phenomenon (about 400-500 km) is

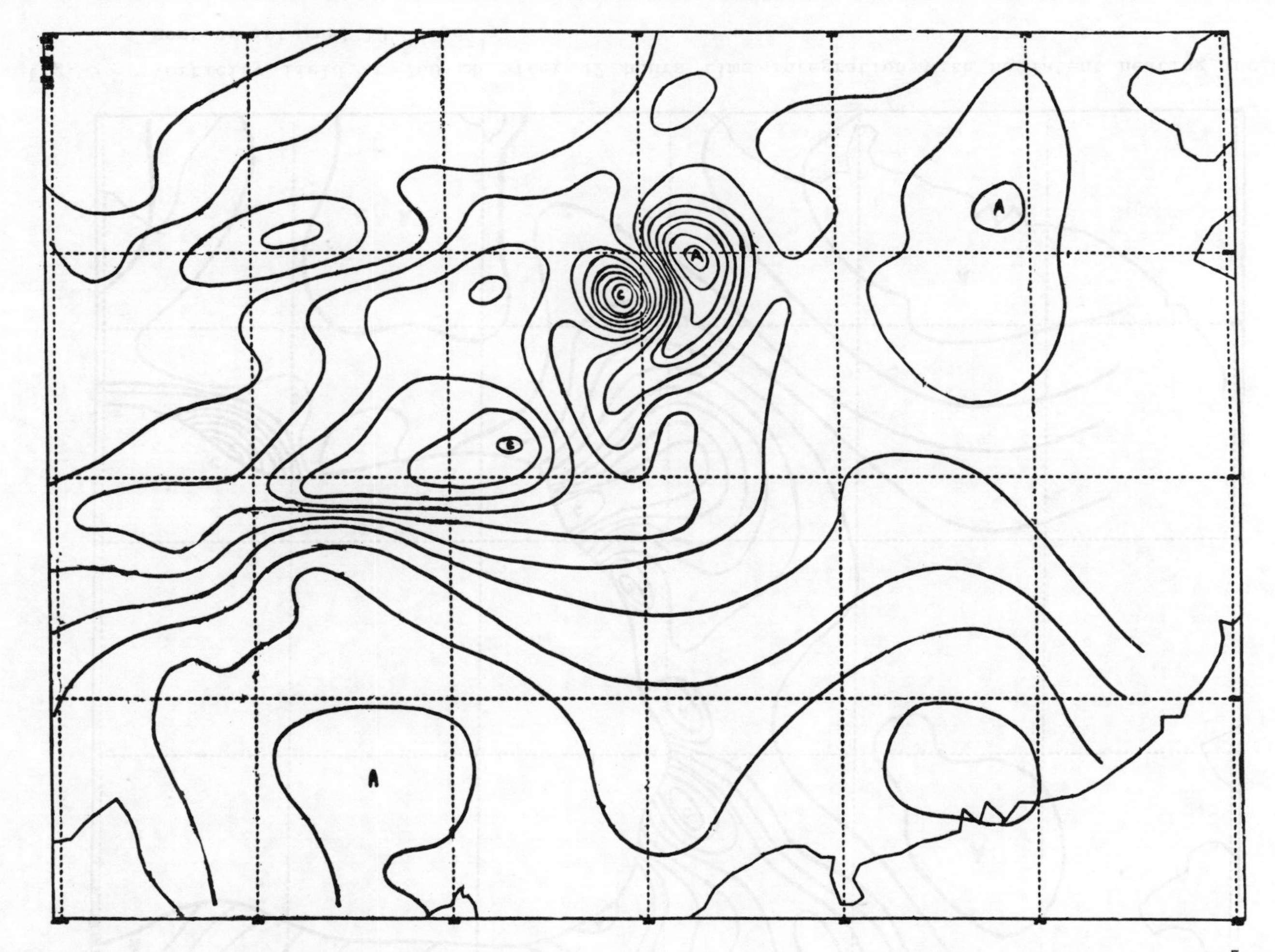

Fig. 8 Vorticity field at 500 mb after 12 hours of time integration. Contour interval 10^{-5} sec^{-1}

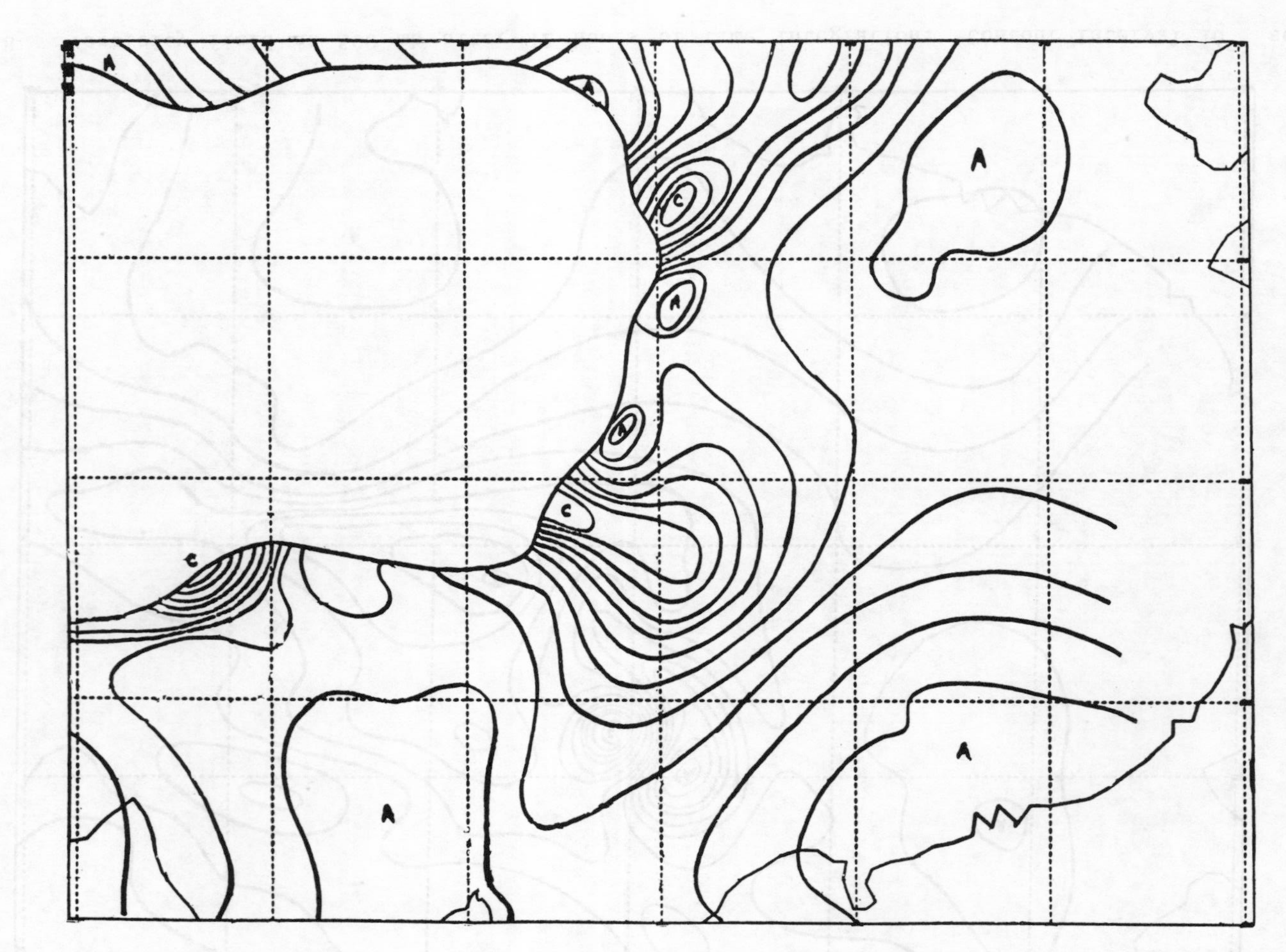

Fig. 9 Vorticity field at 700 mb after 12 hours time integration with no latent heating included; contour interval 10^{-5} sec^{-1}.

only a half to a third of that of the southwest low itself, but is large enough to be treated numerically accurately on the 60 km grid.

The foregoing discussion may cause one to gain the misimpression that the disturbances generated owe their existence to a CISK type of instability. While in their mature states the mesoscale convective eddies are dominated by their latent heat energy source, they formed similarly in their earliest stages whether latent heating was included or not (see Fig. 9). Thus the hydrodynamic forces determined the timing of initial growth and to a major extent the position of eddy generation. But thereafter thermodynamic forcing controls almost completely the amplification and creation of heavy rainfall.

After the generation of the initial convective mesoscale disturbance described above, further developments occurred downstream along the convergnece zone. These had the character of their ancestors but were steadily weaker in each succeeding generation. Thus the development of significant weather downstream was not by wave movement but rather by energy propagation.

The fact that the initial vortex is in a quasi-steady state strongly suggests that mesoscale convective forcing is related to controlling features in the large-scale flow fields. The slightly stronger trough in the mean topography in this region seems also to help to define the location of the steady state situation. A similar situation of a convectively created jet can be seen also in the heavy rain area northwest of Bangladesh.

It is interesting to note that both of the above convective mesoscale phenomena occur near the two main regions of climatological low-level vortex generation analyzed by Tao and Ding [7] (see Fig. 10). The seeming verification of climatological events encourages the construction of improved experiments beyond the one described here to better understand the physical processes leading to storm development. If the intimate relationship between heavy precipitation and these 400-500 km quasi-steady mesoscale convective eddies revealed in this experiment is maintained with further physical numerical refinements, the link between large-scale flow regimes and heavy precipitation may be more precisely defined.

About 24 to 36 hours after the start of the integration from 0000 GMT July 12 data, the main southwest vortex (itself relatively nonconvective) began to move eastward away from the strongest eastern slopes of the plateau. This resulted in a stronger push of northern desert air southward along the eastern plateau slopes especially above the surface boundary layer. The dryness and the low-level vertical stabilization discouraged convection and the strong mesoscale eddies in the convergence zone weakened considerably at low levels. Their remnants in the middle troposphere remained and moved eastward with the ambient flow.

To assume that this was the only event causing the cessation of rainfall would be to ignore a major implication of the experiment. In the region northeast of the moving southwest low, the convergence zone remained well defined. The flow of moist air from the south was even stronger than before. But the missing factor was the strong upslope flow

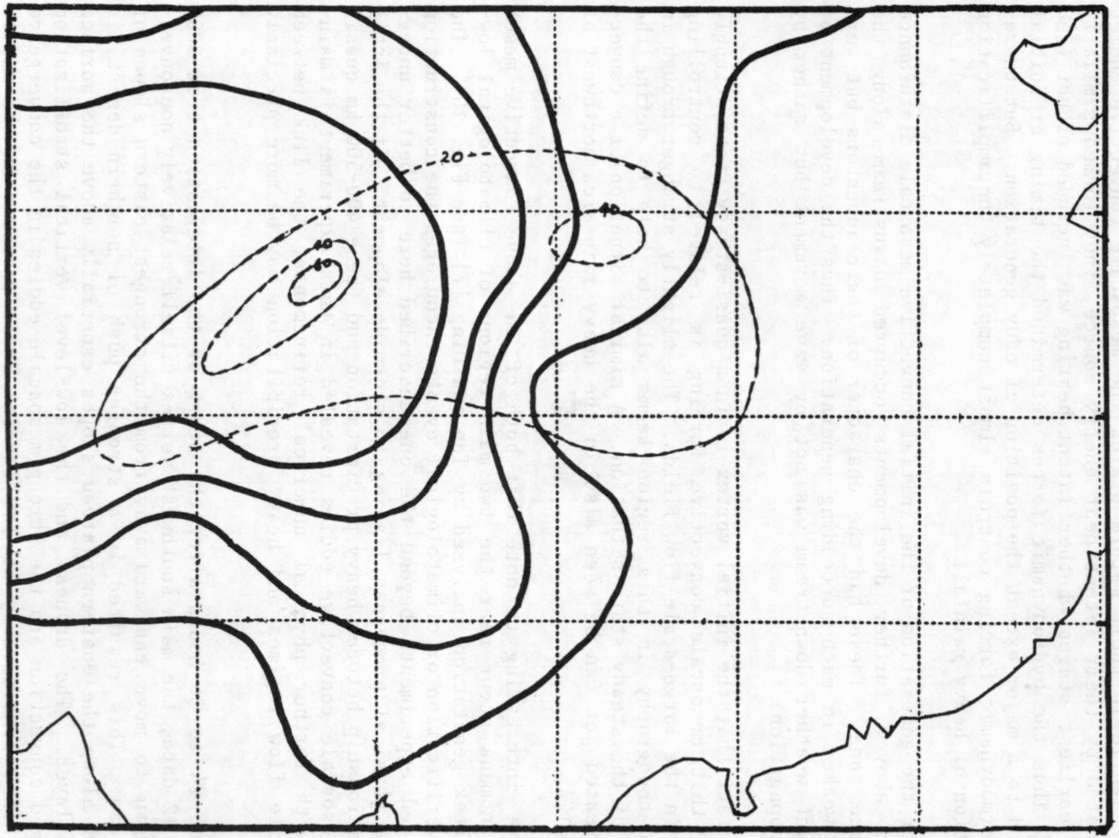

Fig. 10 Frequency of low-level cyclonic vorticies of Tibetan Plateau, May to September 1969-76 as analyzed by Tao and Ding [7]

coinciding with the convergence zone. Apparently in the model simulation at least all three ingredients need to be combined to create the mesoscale convective features that produce the heavy amounts of precipitation. In the model experiment these significant rain producers would have been killed even if warm moist air was ficticiously supplied to the Gobi air mass as it pushed southward.

The preceding scenario briefly stated visualizes first a synoptic flow field with several major features impinging on a strongly smoothed approximation of Tibetan orography. This flow interaction with topography leads directly to several new features or amplification of existing features; both changes could be classified as large mesoscale events. Finally, these mesoscale flow features are positioned in such a way with respect to the mountain that small mesoscale systems grow rapidly. If the final conditions show good agreement with what actually occurred, one might be tempted to see the solution to the problem as primarily one of carefully analyzing the synoptic-scale flow. One might accept the dominance of synoptic forcing for this specific case.

But at least one more important element should be considered. It can be best explained with an example. Several years ago I worked with a student, Oleyar [5], who attempted to understand casual relationships in the development of a severe weather line in Oklahoma 8-9 June 1966. A diagnostic study of the case is given by Fankhauser [2]. The case involved significant synoptic and large mesoscale forcing and an advancing dry line on the high plains east of the Rocky Mountains. However, although the synoptic events could be simulated from initial conditions in model integrations, the small mesoscale, severe weather line did not develop.

Model integrity can always be questioned in such instances and it is possible that the subtle forcing required to start release of the tremendous latent energy in convective instability could not be simulated. But the more likely omission was discovered from hourly surface analyses performed a few hours before the onset of the severe weather. During this period it was noted that a narrow streak of hot air existed just east of the approaching dry line in a southerly flow (see Fig. 11). Midday radiation interacting with water vapor on ground features were the most likely candidates as creators of this anomaly.

Whatever the cause, the warm surface streak was part of the atmosphere and it should be included in a correct simulation. The temperature was raised by allowing stronger heat flux from the ground in this region and in several hours the line storm developed and survived on its own generated latent energy into the night (see Fig. 12). The area of convection and the mesoscale motions developed within it agreed closely with the independent diagnostic study of Fankhauser [2].

The point of contrasting this experience with the experiment described in this paper is to emphasize that the key to correct simulation is the correct specification of small mesoscale forcing. Sometimes orography may provide input, at other times the sun and surface irregularities need to be known, sometimes actual measurements of forcing may be needed.

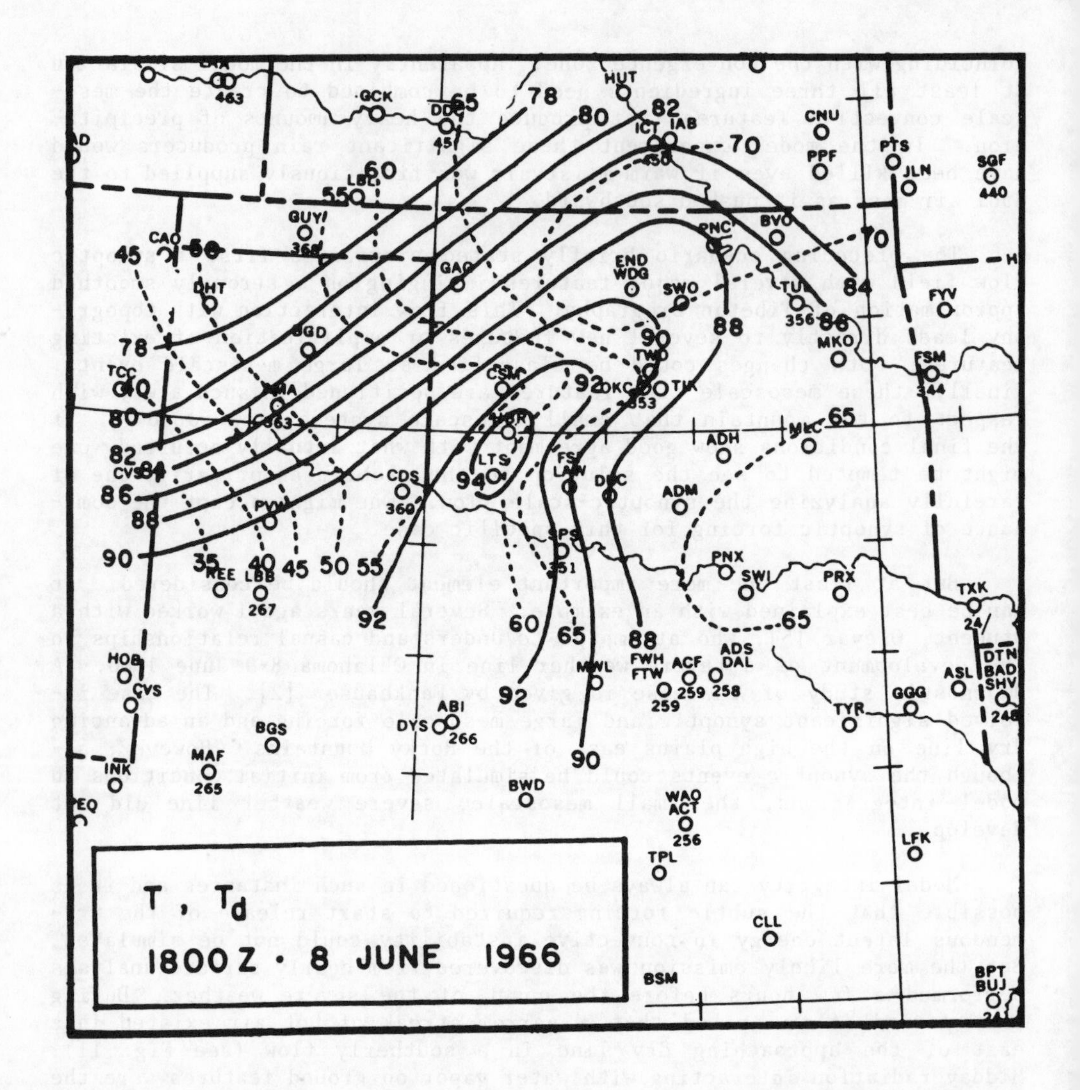

Fig. 11 Mesoscale temperature and dewpoint analysis at 1800 Z, 8 June 1966. Temperature (solid) and dewpoint (dashed) are in °F.

REFERENCES

1. Anthes, R.A., J.W. Trout and S.L. Rosenthal: Comparisons of tropical cyclone simulations with and without the assumption of circular symmetry. Mon. Wea. Rev., 99, 759-766 (1971).

2. Fankhauser, J.C.: The derivation of consistent fields of wind and geopotential height from mesoscale rawinsonde data. J. Appl. Meteor., 13, 637-646 (1974).

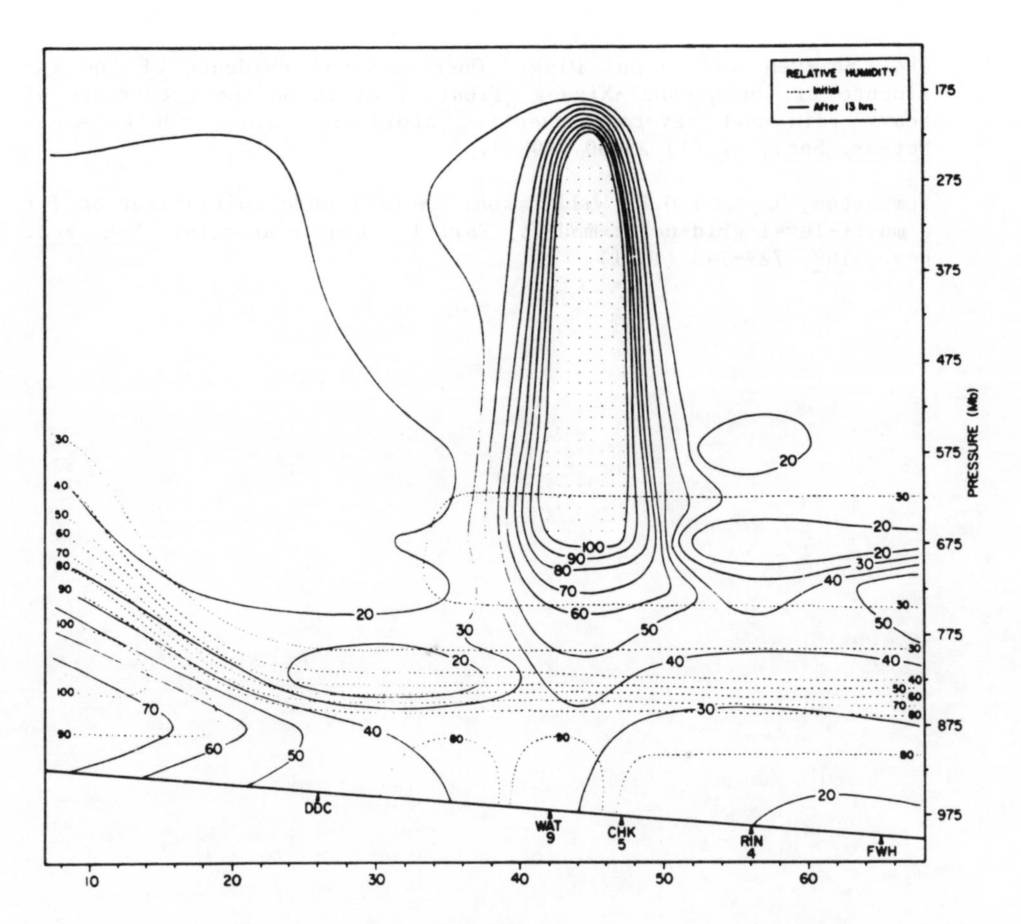

Fig. 12 Cross section of relative humidity before and after generation of a squall line in a numerical simulation [5].

3. Kuo, H.L.: Further studies of the parameterization of the influence of cumulus connection on large-scale flow. J. Atmos. Sci., 31, 1232-1240 (1974).

4. Machenhauer, B.: On the dynamics of gravity oscillations in a shallow water model, with application to nonlinear normal mode initialization. Beiträge zur Physik der Atmosphäre, 50, 253-271 (1977).

5. Oleyar, J.E.: An investigation of the mechanisms of squall line formation. Pennsylvania State University, Masters Thesis, 95 pp. (1972).

6. Shuman, F.G.: Numerical experiments with primitive equations. Proc. Intrn. Symp. Numerical Weather Prediction, Tokyo, Meteorological Society of Japan, 85-107 (1962).

7. Tao, Shi-yen and Yi-hui Ding: Observational evidence of the influence of the Qinghai-Xizang (Tibet) Plateau on the occurrence of heavy rain and severe convective storm in China. Bull. Amer. Meteor. Soc., 62 (1) 23-30 (1981).

8. Temperton, C. and D.L. Williamson: Normal mode initialization for a multi-level grid-point model. Part I: Linear aspects. Mon. Wea. Rev., 109, 729-743 (1981).

3.2 ON THE ATMOSPHERIC HEAT SOURCE OVER ASIA AND ITS RELATION TO THE FORMATION OF THE SUMMER CIRCULATION

Chen Longxun
Institute of Atmospheric Physics
Academia Sinica
The People's Republic of China

and

Li Weiliang and He Jianhua
Academy of Meteorological Science
National Meteorological Bureau
The People's Republic of China

ABSTRACT

Several numerical experiments were conducted to test the effectiveness of the heat source of the Plateau of Tibet in generating the observed monsoon circulation over southeast Asia. It was found that a heat source over the Bay of Bengal can produce circulation patterns which more closely resemble reality, than a heat source over Tibet. As a matter of fact, an abnormally strong heat source over the plateau is unfavorable for the establishment of a normal summer monsoon circulation.

I. INTRODUCTION

During recent years, many meteorologists have worked on the possible role of the Tibetan Plateau in the general circulation. It is commonly believed that in summer the Tibetan Plateau is a vast atmospheric heat source and the thermal effect of the Tibetan Plateau is therefore thought to play an important role in the general circulation. Through numerical experiments, some studies have shown that the upper Asian anticyclone and the tropical upper-tropospheric trough (TUTT) over the Pacific can be simulated if a steady atmospheric heat source is placed only over the Tibetan Plateau. Other studies, using laboratory simulations, also have obtained the same results if the heat source is placed only over the plateau. In these studies, there is the same assumption, i.e. the plateau is the primary heating center in southeast Asia during summer. In the authors' opinion this may not be true.

It is known that the release of latent heat is the most important component in the heat budget during summer, and the rainfall in the monsoon region south of 25°N is much larger than that over the plateau. For example, monthly rainfall at the northeastern coast of the Bay of Bengal in July is over 800 mm compared with an average of 100 mm on the plateau. Taking into consideration air density variations, this means that the latent heating rate over the Bay of Bengal is about five to six times larger than that over the plateau. Thus, it seems that a more realistic position of the primary summer atmospheric heating center over East Asia is over the Bay of Bengal.

In a paper concerning the heat budget on cloudy days in China, Chen et al. [2] pointed out that the maximum rate of heating in China is found in the western part of Yunnan Province to be 2.6°C/day. These results need to be expanded upon with a new series of data. Only then can we perform the numerical experiments of the summer circulation by using a realistic map of the atmospheric heating.

As will be seen in our paper, the realistic position of the heating center in summer is situated over the Bay of Bengal, the northwestern coast of Burma and the South China Sea. Heating over the Tibetan Plateau is relatively weak. In the design of the numerical experiments we placed a realistic heating center over the Bay of Bengal and achieved more realistic circulations than if we placed the heating center over the Tibetan Plateau. The authors opinion is that in the previous studies the thermal effect of the Tibetan Plateau is overestimated. In this paper, a brief summary of our recent studies is given.

II. ATMOSPHERIC HEAT SOURCE IN JULY

We have designed a diagnostic equation for the calculation of the diabatic heating Q/C_p. It has been used to estimate the average values of heating over Asia in July.

The equations for the σ-coordinate primitive-equation model used in our paper are defined as follows:

$$\frac{\partial}{\partial t}(\pi u) = -m^2\left[\frac{\partial}{\partial x}\left(\frac{\pi u}{m}u\right) + \frac{\partial}{\partial y}\left(\frac{\pi v}{m}u\right)\right] - \frac{\partial}{\partial \sigma}(\pi\dot{\sigma}u) + f^*\pi v - m\pi\left(\frac{\partial \Phi}{\partial x} + \frac{RT}{\pi}\frac{\partial \pi}{\partial x}\right) + Fx \tag{1}$$

$$\frac{\partial}{\partial t}(\pi v) = -m^2\left[\frac{\partial}{\partial x}\left(\frac{\pi u}{m}v\right) + \frac{\partial}{\partial y}\left(\frac{\pi v}{m}v\right)\right] - \frac{\partial}{\partial \sigma}(\pi\dot{\sigma}v) - f^*\pi u - m\pi\left(\frac{\partial \Phi}{\partial y} + \frac{RT}{\pi}\frac{\partial \pi}{\partial y}\right) + Fy \tag{2}$$

$$\frac{\partial}{\partial t}(\pi T) = -m^2\left[\frac{\partial}{\partial x}\left(\frac{\pi u}{m}T\right) + \frac{\partial}{\partial y}\left(\frac{\pi v}{m}T\right)\right] - \frac{\partial}{\partial \sigma}(\pi\dot{\sigma}T) + \frac{RT}{C_p}\left(\frac{\partial \pi}{\partial t} + mu\frac{\partial \pi}{\partial x} + mv\frac{\partial \pi}{\partial y}\right) + \frac{RT}{C_p\sigma}\pi\dot{\sigma} + \frac{\pi Q}{C_p} + FT \tag{3}$$

$$\frac{\partial \pi}{\partial t} = -m^2\left[\frac{\partial}{\partial x}\left(\frac{\pi u}{m}\right) + \frac{\partial}{\partial y}\left(\frac{\pi v}{m}\right)\right] - \frac{\partial}{\partial \sigma}(\pi\dot{\sigma}) \tag{4}$$

$$\frac{\partial\Phi}{\partial\sigma} = -\frac{RT}{\sigma} \qquad (5)$$

$$f^* = f + \frac{\tan\phi}{a} u \qquad (6)$$

where a is the earth's radius, Fx and Fy are the components of eddy frictional forces. The term $\dot{\sigma}$ is the vertical velocity in the σ-coordinate and can be expressed as follows:

$$\dot{\sigma} = \{(1-\sigma)\int_1^0 D d\sigma - \int_1^\sigma D d\sigma\} + m\{(1-\sigma)\int_1^0 u d\sigma - \int_1^\sigma u d\sigma\}\frac{\partial \ln\pi}{\partial x}$$

$$+ m\{(1-\sigma)\int_1^0 v d\sigma - \int_1^\sigma v d\sigma\}\frac{\partial \ln\pi}{\partial y} \qquad (7)$$

where D is the divergence, π is the surface pressure and m is the coefficient of Mecator projection of which the central point is at 23.5°N and is equal to $\sec\phi/\sec\phi_o$, ϕ_o = 23.5°N. If the quasi-geostropical relation is only applied to the local term of the vorticity equation, i.e.

$$\frac{\partial}{\partial t}\left(\frac{\partial v}{\partial x} - \frac{\partial u}{\partial y}\right) \doteq \frac{\partial}{\partial t}\left(\frac{1}{f}\nabla\Phi\right) \qquad (8)$$

then we obtain a diagnostic equation of Q/C_p by using Eqs. (1) to (8).

$$\nabla^2 \frac{Q}{C_p} = \frac{f\sigma}{R}\frac{\partial}{\partial\sigma}\{\frac{\partial}{\partial x}(u\frac{\partial v}{\partial x}+v\frac{\partial v}{\partial y}) - \frac{\partial}{\partial y}(u\frac{\partial u}{\partial x}+v\frac{\partial u}{\partial y})\} + \frac{f\sigma}{R}\frac{\partial}{\partial\sigma}\{f^*\frac{\partial}{\partial x}(\frac{u}{m}) + f^*\frac{\partial}{\partial y}(\frac{v}{m})\}$$

$$+ \frac{f\sigma}{mR}\frac{\partial f^*}{\partial y}\frac{\partial v}{\partial\sigma} + f\sigma\frac{\partial}{\partial\sigma}\{\frac{\partial T}{\partial x}\frac{\partial \ln\pi}{\partial y} - \frac{\partial T}{\partial y}\frac{\partial \ln\pi}{\partial x}\} - \frac{f\sigma}{R}\frac{\partial}{\partial\sigma}\{\frac{\partial}{\partial x}(\frac{Fx}{m}) - \frac{\partial}{\partial y}(\frac{Fy}{m})\} \qquad (9)$$

$$+ \nabla^2\{\frac{RT}{C_p}m^2[\frac{\partial}{\partial x}(\frac{u}{m}) + \frac{\partial}{\partial y}(\frac{v}{m})\}+\nabla^2\{\frac{RT}{C_p\pi}\frac{\partial}{\partial\sigma}(\pi\dot{\sigma})\}+\nabla^2\dot{\sigma}\frac{\partial T}{\partial\sigma} - \nabla^2\frac{RT}{C_p\sigma}\dot{\sigma}+\nabla^2 m(u\frac{\partial T}{\partial x}+v\frac{\partial T}{\partial y})$$

This is a diagnostic equation for Q/C_p, and is applied to the data set developed by Van de Boogaard [8]. The time periods from which these data are extracted are from 1968 to 1974 in China and from 1961 to 1974 in the other regions. But for reasons of consistency, we have added to the data set by calculating the mean values of temperature, dew point, and geopotential height of July 1961 to 1974 for all of the stations in China, and the mean wind for the major stations in the same years. These mean values are obtained at the surfaces 850, 700, 500, 300, 200, and 100 mb. Equation (9) is a Poisson equation for Q/C_p and can be solved by iterative methods. In the solution, Q is defined as zero at the boundaries of the computation region and the grid increment is 3.5° longitude. The computation region is extending from 40°E to 165°W and from 25°S to 48°N.

Figure 1 shows the computational result of the diabatic heating in July. The unit used in this figure is °C/day in an entire column of the atmosphere. In this figure, where heat is gained through the diabatic

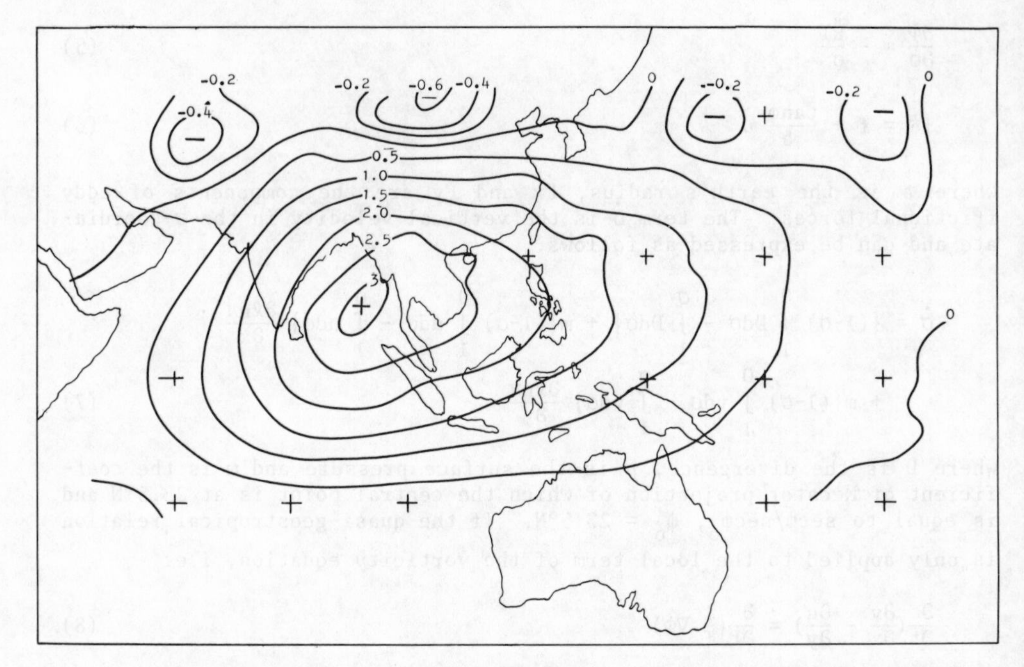

Fig. 1 Distribution of the atmospheric heat source for July based on Eq. (9) (indirect method), (°C/day).

process, Q/C_p is positive. It is shown in Fig. 1 that there exists a massive scale heat source over South Asia and its center is located over the northeastern Bay of Bengal. There is no heating center over the Tibetan Plateau. On the average, the maximum rate of heating is 3°C/day in the Bay of Bengal and is less than 1°C/day over the Tibetan Plateau. We can say that the heating over the plateau is merely an extended part of the Bay of Bengal center. Chen and Li [3] have computed the distribution of Q/C_p in July by use of a direct method and have found that there are two heating centers in the summer monsoon region. The first one is over the Bay of Bengal, but its maximum rate of heating can be as high as 8°C/day. The second one is over the South China Sea and the maximum rate is about 6°C/day. Over the Tibetan Plateau, the average rate and the maximum rate are also less than 1°C/day and 1.2°C/day, respectively, in the direct method. According to these results, the thermal effect of the Tibetan Plateau in summer may have been overemphasized by previous studies.

III. MONTHLY DISTRIBUTION OF THE ATMOSPHERIC HEAT SOURCE

The atmospheric heat source can be estimated by other means. If R_∞ is the net radiation in an earth-atmosphere system observed by the meteorological satellite at the top of the atmosphere, R_o is the radiation balance at the earth's surface calculated by the common climatological method, SC is the exchange of sensible heat between the earth and the atmosphere, and LH is the heating adding to the atmosphere due to the release of latent heat by the condensation process, then the atmospheric heat source HS can be expressed as follows:

$$HS = R_\infty - R_o + SC + LH \tag{10}$$

A positive HS implies a heat source. Winston et al. [9] have calculated the values of R_∞ during the period from February 1974 to June 1979. Their results are used in this paper as a climatological example. The values of R_o and SC have been estimated by Gao et al. [4] over the mainland of China, by the Research Group of Oceanic Meteorology [9] over the Pacific and the South China Sea, and by Hastenrath et al. [5] in the Bay of Bengal. The climatological maps of the rainfall used here have been published by the National Meteorological Bureau [1] for the mainland of China and by Jaeger [6] for other regions. By using the above-mentioned data, a series of monthly maps of the atmospheric heat source HS in the region from 10°N to 45°N and from 70°E to 150°E are obtained and are shown in Figs. 2 to 13. In these maps, the full lines show the heating rate in a column of atmosphere and the units are °C/day. The main features are as follows.

1. The Tibetan Plateau and Southern Asia (West of 105°E)

There is a heat sink over the Tibetan Plateau during the period from October through February and a heat sink from November through April in

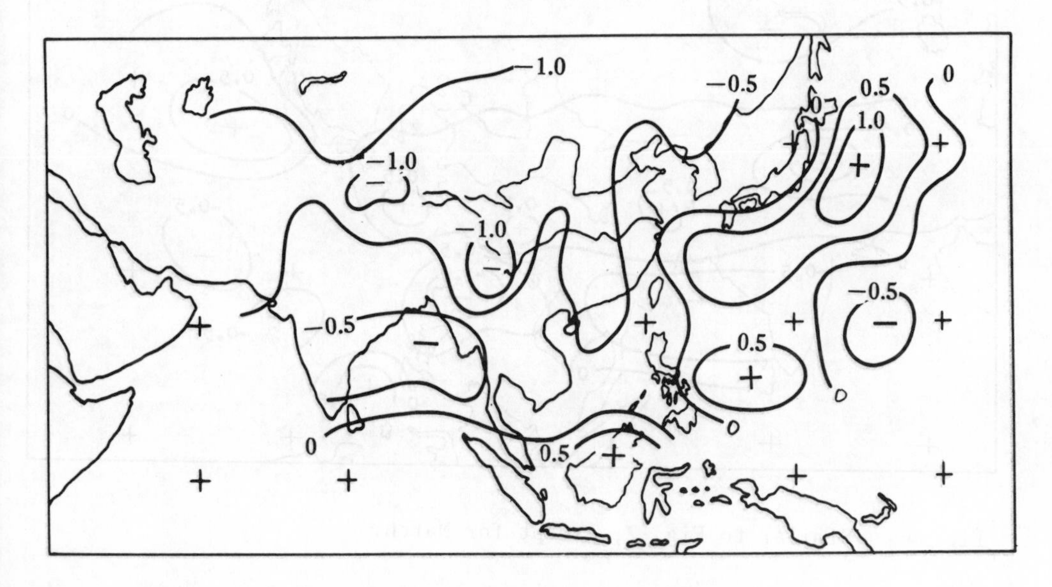

Fig. 2 Distribution of the atmospheric heat source for January based on Eq. (10) (direct method), (°C/day).

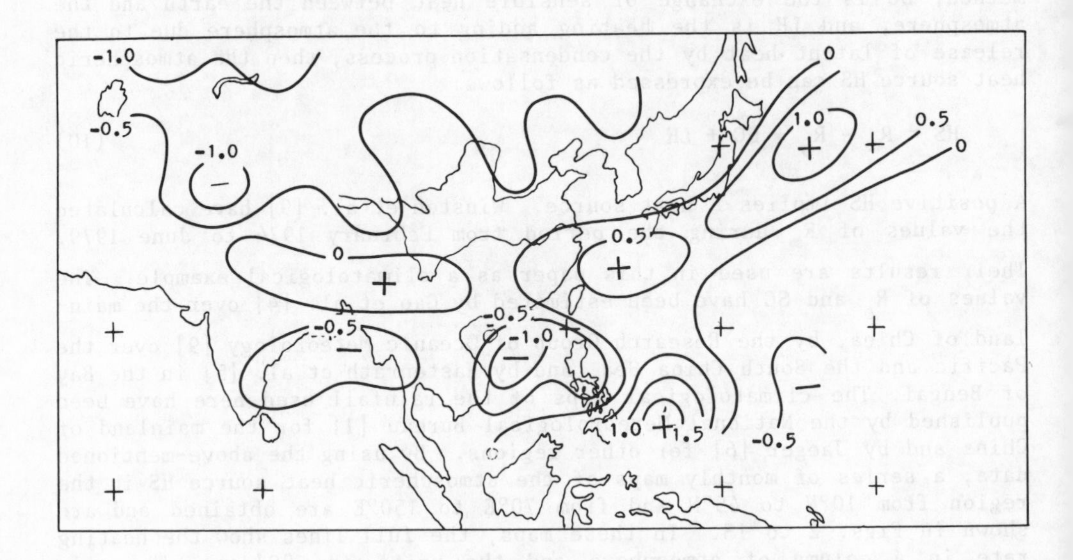

Fig. 3 Similar to Fig. 2, except for February.

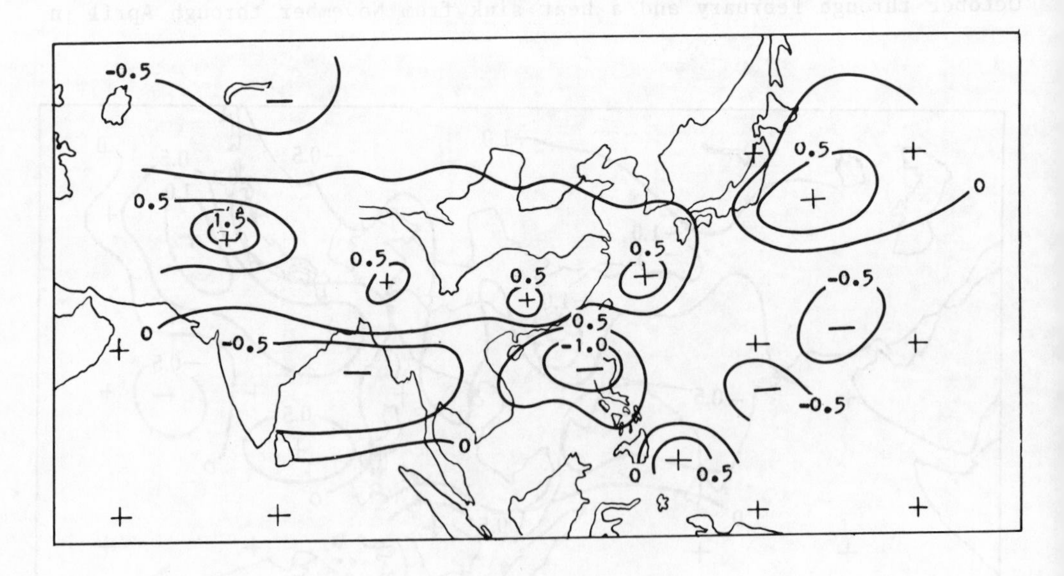

Fig. 4 Similar to Fig. 2, except for March.

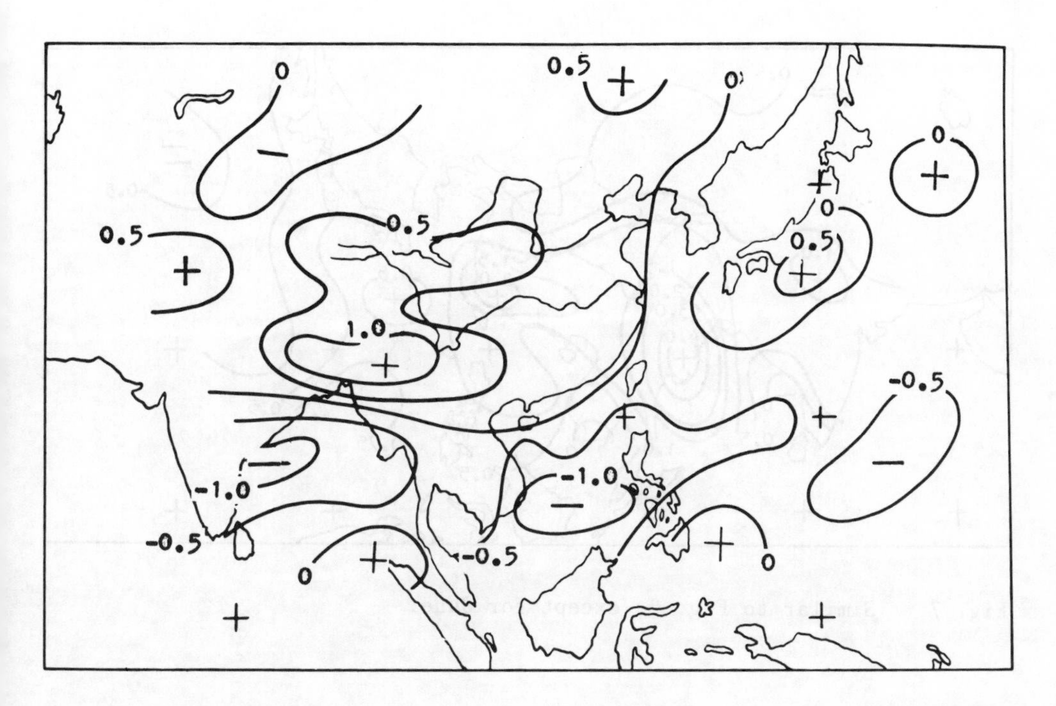

Fig. 5 Similar to Fig. 2, except for April.

Fig. 6 Similar to Fig. 2, except for May.

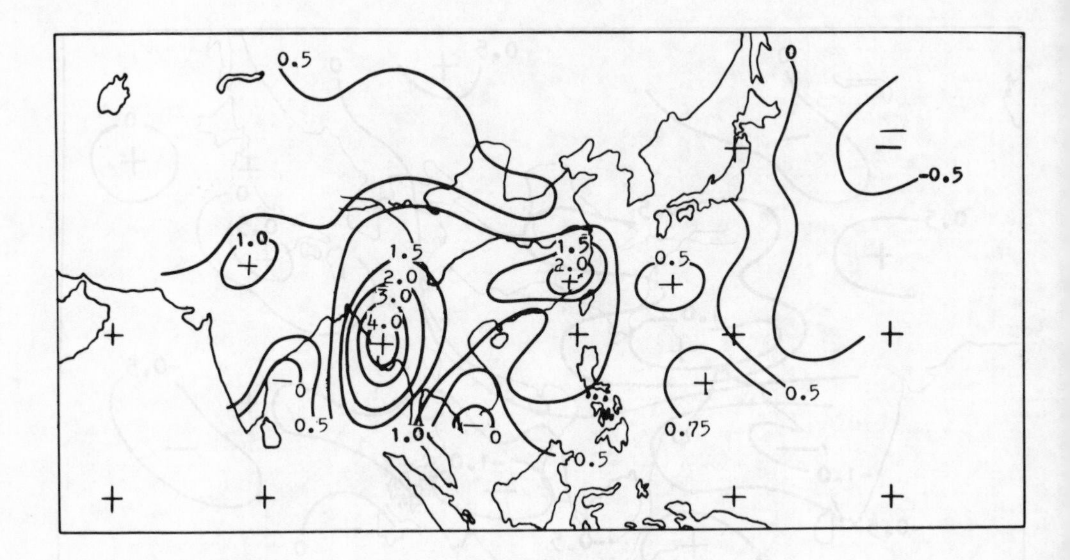

Fig. 7   Similar to Fig. 2, except for June.

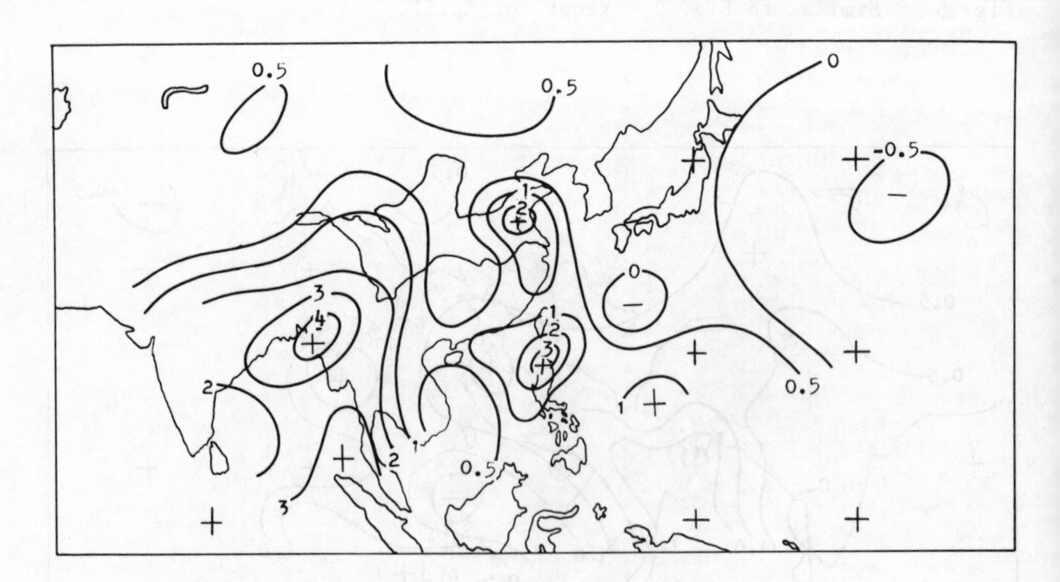

Fig. 8   Similar to Fig. 2, except for July.

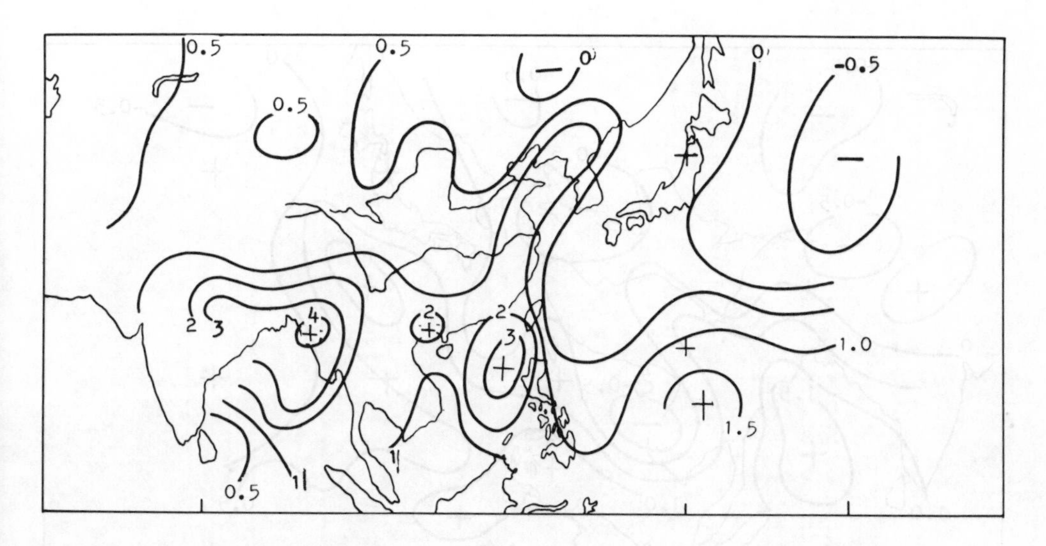

Fig. 9 Similar to Fig. 2, except for August.

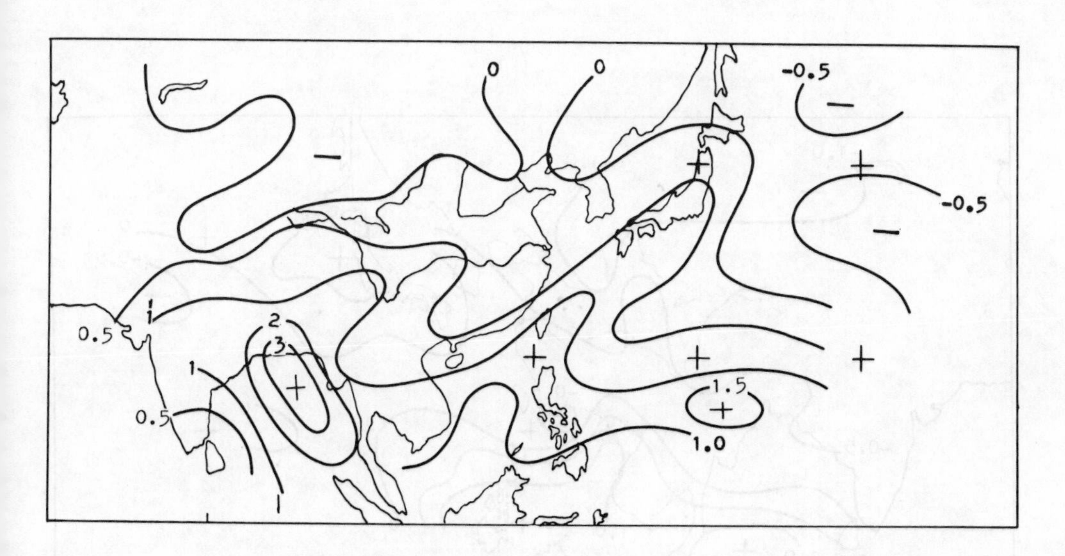

Fig. 10 Similar to Fig. 2, except for September.

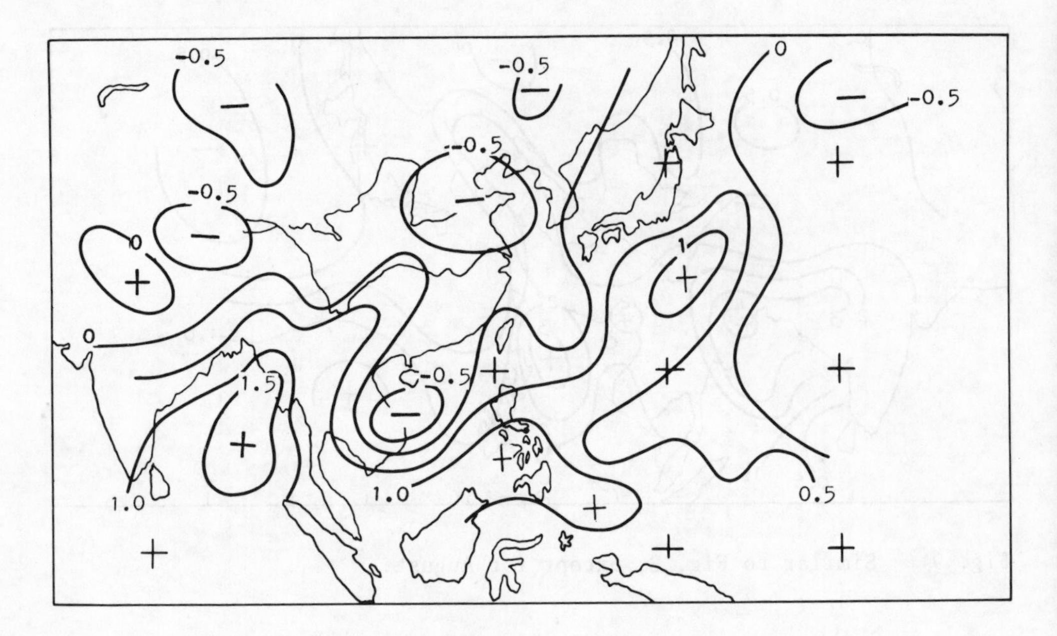

Fig. 11   Similar to Fig. 2, except for October.

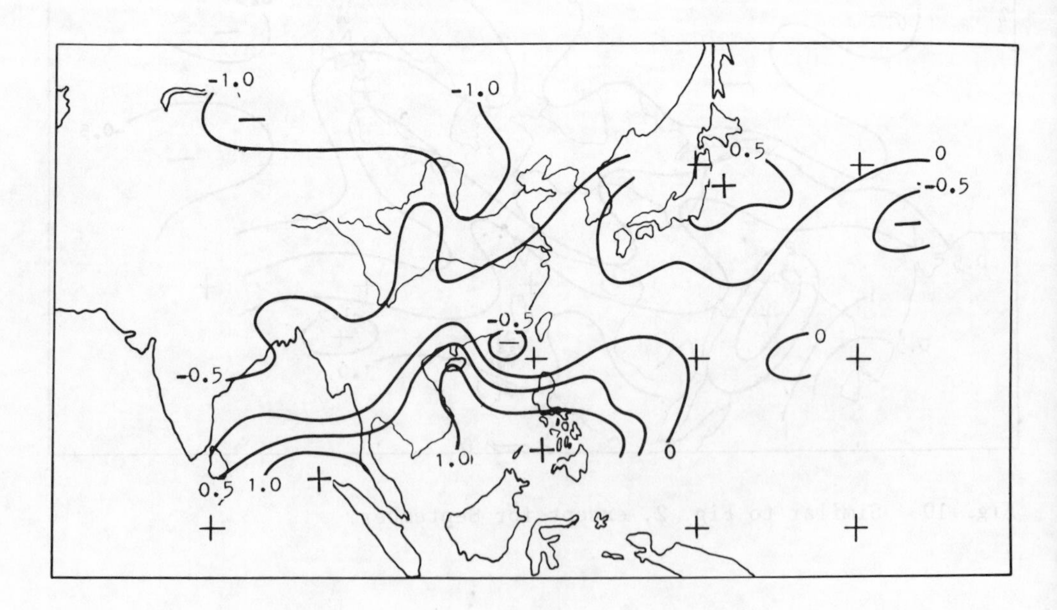

Fig. 12   Similar to Fig. 2, except for November.

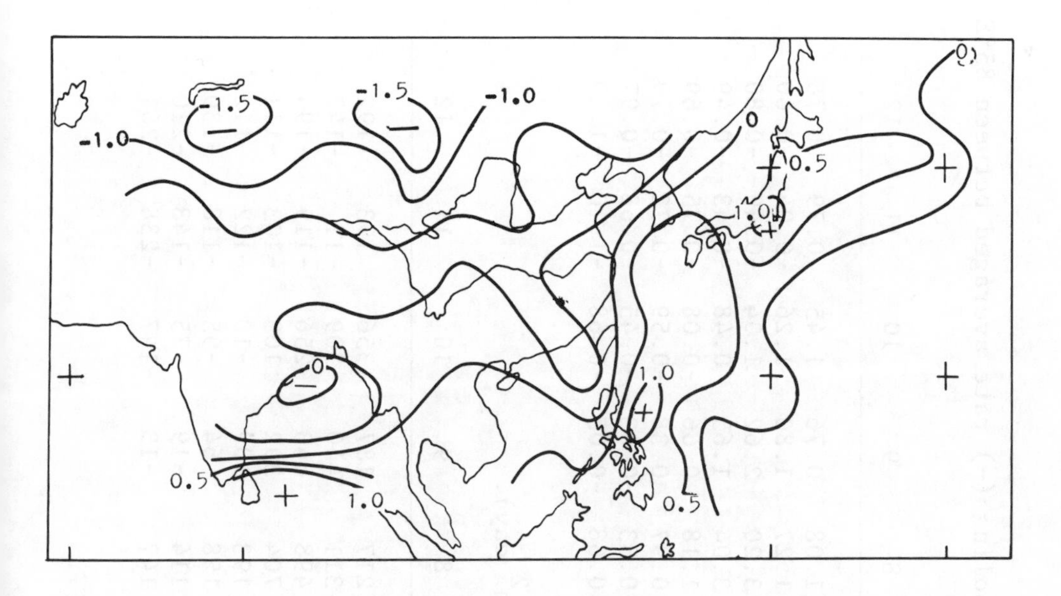

Fig. 13 Similar to Fig. 2, except for December.

the Bay of Bengal. In winter, for example in January (Fig. 2), there are two cooling centers over the plateau. The first one is located over the east part of the plateau (about 100°E, 30°N), and its maximum rate is -1.2°C/day. The latter is located over the northwestern plateau and its maximum cooling rate is about -1°C/day. Table 1 shows the latitudinal distribution of the heat source averaged between 85°E and 95°E (three points) for 12 months and Table 2 shows the seasonal variation of the heat source averaged for some selected regions. In the latter table, the term "plateau" signifies the region where elevation is above 3000 meters (averaged by 13 points), the "northern India" designation is the region between 85°E and 100°E at 25°N, the "Bay of Bengal" signifies the region between 85°E and 95°E, 10°N and 20°N (averaged by 9 points), the "South China Sea" contains the region between 110°E and 120°E, 10°N and 20°N (averaged by 9 points) and the "mainland of East China" represents the region between 25°N and 35°N, 110°E and 120°E (9 points). It can be seen in these tables that the maximum cooling rate appears in December over the plateau (-0.82°C/day) and in March and April in the Bay of Bengal (-0.59°C/day). During the calendar year northern India is the first heat source in South Asia. As we can see in Fig. 3, it becomes a heat source in this region after February and its heating rate is still the strongest in the region west of 105°E in March and April. The plateau becomes a heat source in March, but its heating rate is weaker than that over northern India. There is an abrupt alteration of the heating in the Bay of Bengal from April to May. The Bay of Bengal becomes a heat source in May and the maximum heating region also shifts from northern India to the Bay of Bengal at the same time. The heat source in the Bay of Bengal is strongest in South Asia during the periods from May to October. In July, the average heating rate over the Bay of Bengal is about 3°C/day and the maximum heating rate can be 6 to 8°C/day as suggested in a paper given by

Table 1a Latitudinal distribution of the heating (+) and cooling (-) rate averaged between 85°E and 95°E for 12 months (°C/day).

Month	1	2	3	4	5	6	7	8	9	10	11	12
10°N	0.36	-0.18	-0.12	-0.23	0.78	0.94	2.41	1.08	0.76	1.45	0.79	0.78
15°N	-0.52	-0.55	-0.78	-0.80	0.40	2.10	2.26	1.27	1.80	1.26	-0.04	-0.60
20°N	-0.83	-0.87	-0.75	-0.75	0.42	3.17	3.95	3.20	2.62	1.04	-0.49	-0.80
25°N	-0.25	0.08	0.31	0.96	1.38	2.16	3.21	3.04	1.65	0.48	-0.43	-0.49
30°N	-0.52	-0.27	0.33	0.60	0.77	1.32	2.19	1.18	0.66	-0.08	-0.75	-0.69
35°N	-0.37	-0.50	0.16	0.69	0.72	0.83	0.53	0.97	0.38	-0.59	-0.72	-0.79
40°N	-0.74	-0.64	-0.28	0.17	0.58	0.41	1.09	0.53	-0.03	-0.45	-0.68	-0.97
45°N	-1.10	-0.91	-0.46	-0.07	0.48	0.52	0.45	0.68	-0.05	-0.68	-1.10	-1.20

Table 1b Same as Table 1a, except for the intensity (cal/cm^2 day).

Month	1	2	3	4	5	6	7	8	9	10	11	12
10°N	87	-36	-39	-57	325	231	580	217	187	356	193	192
15°N	-128	-134	-193	-197	231	519	554	311	440	309	-11	-147
20°N	-203	-213	-184	-184	103	778	330	498	643	256	-119	-194
25°N	-61	18	72	223	275	495	738	704	392	109	-103	-111
30°N	-72	-44	27	97	128	218	305	193	109	-10	-122	-112
35°N	-133	-83	26	117	118	137	140	158	64	-95	-118	-130
40°N	-172	-142	-62	36	126	89	79	114	-19	-75	-148	-210
45°N	-248	-196	-97	-15	103	113	96	147	-13	-147	-238	-261

Table 2 Average atmospheric heat source.

Month	1	2	3	4	5	6	7	8	9	10	11	12
A	-0.60	-0.39	0.17	0.47	0.46	0.92	1.04	0.98	0.37	-0.46	-0.69	-0.82
B	-122	-70	29	82	109	165	183	176	94	-60	-119	-161
C	-0.28	0.13	0.20	0.89	1.19	2.10	2.93	2.76	1.35	0.54	-0.48	-0.48
D	-0.34	-0.53	-0.55	-0.59	0.94	2.07	2.80	1.84	1.81	1.25	0.09	-0.21
E	-0.12	-0.63	-0.66	-0.67	-0.01	0.67	1.37	1.55	1.14	0.52	0.38	-0.07
F	-0.60	-0.09	0.01	0.15	0.82	1.27	0.89	1.29	0.41	-0.26	-0.53	-0.56

A. Tibetan Plateau (°C/day). B. Tibetan Plateau (cal/cm^2 day). C. North part of India and Burma (°C/day). D. Bay of Bengal (°C/day). E. South China Sea (°C/day). F. Mainland of East China (°C/day).

Chen and Li [3]. In the same month, the average heating rate is about 1°C/day over the plateau (it is 0.8°C/day in [2]). So we can conclude that the heat source over the Tibetan Plateau in summer is rather weak in comparison with that over the Bay of Bengal.

The heat sink invades the northern plateau in September and extends to the whole plateau after October. In the same month, the Bay of Bengal remains a heat source but becomes a heat sink after November.

2. The South China Sea and the Pacific Ocean

During the period from October to April, the South China Sea is a heat sink and its maximum cooling rate is about 1°C/day. It becomes a heat source after May and appears to be a heating center during July and August. In July, this heating center is located over the east part of the South China Sea and the maximum heating rate is 4°C/day (see Fig. 8), but according to another paper given by Chen and Li [3], it can be as high as 6°C/day. This is one of the two major heating centers in South Asia in July but is weaker than that over the Bay of Bengal.

The western Pacific (west of 140°E) is a heat source from May through February and is a heat sink during March and April. In winter, there appears to be a heating belt situated just over the warm sea current and is caused by the exchange of sensible heat. However, the heat source in the warm season is situated over the Intertropical Convergence Zone (ITCZ) and is caused primarily by latent heat release.

3. The Mainland of East China

Distribution of the heat budget is rather simple in this region. It is a heat source from May to September and a heat sink from October to January. The region near the coast of China is the first to appear as a heat source on the mainland of East China. This zone extends westward in March as a heating belt is formed over the region from the plateau to the Pacific between 25°N and 35°N.

IV. NUMERICAL EXPERIMENT OF THE MONSOON CIRCULATION INFLUENCED BY THE COMBINATION OF HEAT SOURCE AND OROGRAPHIC INFLUENCES

For studies of the influence of the heat source and the Tibetan Plateau on the monsoon circulation in Asia in summer, we have designed a two-dimensional five-layer, primitive equation σ-coordinate, numerical model and performed a series of numerical experiments. The basic equations are as follows:

$$\frac{\partial \pi u}{\partial t} = -\frac{1}{a\cos\phi}\frac{\partial}{\partial\phi}(\pi uv\cos\phi) - \frac{\partial \pi\dot{\sigma}u}{\partial\sigma} + (f + \frac{\tan\phi}{a}u)\pi v - g\frac{\partial \tau_u}{\partial\sigma} + Fu \quad (11)$$

$$\frac{\partial \pi v}{\partial t} = - \frac{1}{a\cos\phi} \frac{\partial}{\partial \phi}(\pi vv \cos\phi) - \frac{\partial \pi \dot{\sigma} v}{\partial \sigma} - (f + \frac{\tan\phi}{a} u)\pi u - g\frac{\partial \tau_v}{\partial \sigma} + Fv$$

$$- (\pi\frac{\partial \Phi}{a\partial \phi} + \pi RT\frac{\partial \ell n P}{a\partial \phi}) \tag{12}$$

$$\frac{\partial \pi T}{\partial t} = - \frac{1}{a\cos\phi} \frac{\partial}{\partial \phi}(\pi vT \cos\phi) - \frac{\partial \pi \dot{\sigma} T}{\partial \sigma} + \frac{\pi \sigma \alpha}{C_p}(\frac{\partial \pi}{\partial t} + v\frac{\partial \pi}{a\partial \phi} + \frac{\pi \dot{\sigma}}{\sigma}) + FT + \frac{\pi Q}{C_p} \tag{13}$$

$$\frac{\partial}{\partial \sigma}(\frac{\partial P}{\partial t} + \pi \dot{\sigma}) + \frac{1}{a\cos\phi} \frac{\partial}{\partial \phi}(\pi v \cos\phi) = 0 \tag{14}$$

$$\frac{\partial \Phi}{\partial \sigma} = -\pi \alpha \tag{15}$$

$$\frac{\partial \pi}{\partial t} + \frac{1}{a\cos\phi} \int_0^1 \frac{\partial}{\partial \phi}(\pi v \cos\phi)\, \alpha\sigma = 0 \tag{16}$$

$$\alpha = \frac{RT}{P} \tag{17}$$

where $\sigma = P-P_t/P_s-P_t$, $\pi = P_s-P_t$, P_t is the pressure at the top of the model and is assumed to be 10 mb, f is the Coriolis parameter, Φ is the geopotential height of the σ surface, Q/C_p is the diabatic heating, τ is the surface friction or eddy vertical mixing of momentum. The simple scheme for τ is expressed as

$$\tau_{\vec{V}} = \frac{g\rho^2}{\pi} K_v \frac{\partial \vec{V}}{\partial \sigma} \tag{18}$$

in the atmosphere, and

$$\tau_{\vec{V}} = C_D \rho_S |\vec{V}_s| \vec{V}_s \tag{19}$$

at the surface, where $\vec{V}_s$ is the velocity vector at the top of the surface boundary layer and is assumed to be equal to the wind velocity at the surface. ρ_s is density, K_v and C_D are the eddy coefficient of the vertical momentum mixing and the surface drag coefficient. The term, $g\rho^2/\pi K_v$, is assigned the value 0.15 $m^2 s^{-1}$. The term, C_D, is assumed to be 0.0015 in the ocean region, 0.003 over the flat land, and 0.008 in the region of the plateau. F_A is the eddy horizontal diffusion of the physical element A. For simplicity, F_A is written in the following form:

$$F_A = k \nabla^2 \pi A \tag{20}$$

where ∇^2 is the Laplacian operator, k is the viscosity coefficient given as $5 \times 10^5 \ m^2 \ s^{-1}$ in this model.

The finite difference scheme adopted here is the horizontal staggered system and in the vertical coordinate the variables are defined at the center of each layer; $\dot{\sigma}$ is defined at the interface between layers and is zero at the upper and lower boundaries. Figure 14 shows the distribution of the variables adopted. The horizontal grid size is 2.5° of latitude and longitude. The variables u and v are defined at 88.75°N, 86.25°N ... 86.25°S, 88.75°S while T, Φ, $\dot{\sigma}$, π are defined at 90°N, 87.5°N, ..., 87.5°S, 90°S; Φ_s is the surface elevation and is presented in Fig. 15. The maximum height adopted here is 4000 m. The average slope of orography on the south side of the plateau is 1 km/275 km. Two types of heating are parameterized. The first one involves the transfer of sensible heat in the planetary boundary layer and is expressed as follows:

$$\frac{\pi Q}{C_p} = \frac{2g}{\Delta\sigma} \rho_s \ C_D \ |\vec{V}|(T_s - T_a) \tag{20}$$

where T_s is the ground temperature and is assumed steady. The T_s data are the climatological values for July and are shown in Fig. 15. The

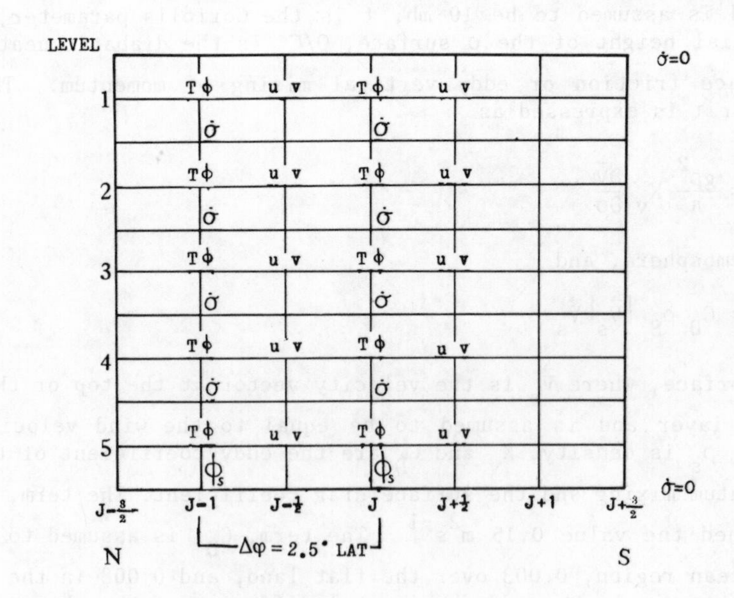

Fig. 14 Grid scheme adopted in the experiments.

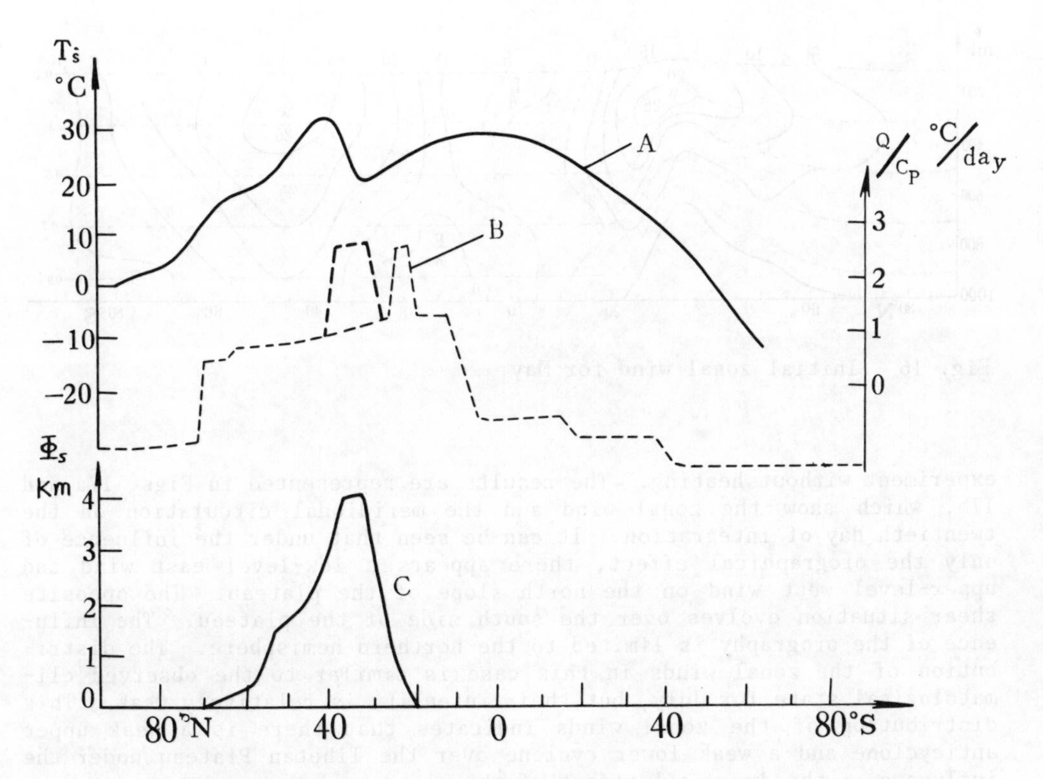

Fig. 15 Distributions of ground temperature (°C) (A), atmospheric heating, (°C/day) (B), and the orographic height (m) (C).

atmospheric temperature, T_a, at the surface is derived from the temperature at $\sigma = 0.9$ and $\sigma = 0.7$ surfaces. The second form of heating is convective in nature; its columnar means are estimated from Fig. 1 and are shown in Fig. 15.

The initial fields of height, temperature and zonal wind are also adopted from the climatological data for May (Fig. 16), but the initial meridional wind, v, and the vertical velocity, $\dot{\sigma}$, are assumed to be zero. Thus the values of v and $\dot{\sigma}$ after time integration are the new values generated by heating and orographical effects. The mutual adaption between the initial values is accomplished by an initial time integration for the first 24 hours without heating and thereafter with heating. We found that the integrated solution arrives at a quasi-steady state after the twentieth day and, the results of that day are presented.

1. The Influence of the Orographic Effect of the Tibetan Plateau on the Monsoon Circulation

In order to better understand the influence of dynamical effects of the Tibetan Plateau on the monsoon circulation, we also performed an

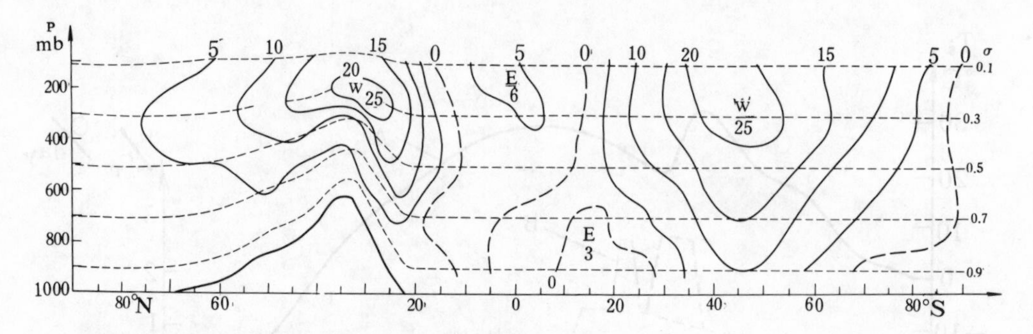

Fig. 16 Initial zonal wind for May.

experiment without heating. The results are represented in Figs. 17a and 17b, which show the zonal wind and the meridional circulation on the twentieth day of integration. It can be seen that under the influence of only the orographical effect, there appears a low-level east wind and upper-level west wind on the north slope of the plateau. The opposite shear situation evolves over the south side of the plateau. The influence of the orography is limited to the northern hemisphere. The distribution of the zonal winds in this case is similar to the observed climatological state for July, but their intensity is relatively weak. This distribution of the zonal winds indicates that there is a weak upper anticyclone and a weak lower cyclone over the Tibetan Plateau under the influence of the dynamical effect of the plateau alone.

The meridional circulation of Fig. 17b shows that there are cells situated over both sides of the plateau with ascending motion over the plateau and descending motions on both sides. The descending region over the southern slopes coincides with the desert of southern Xinjiang in China and the drought belt of the Ganges Valley in India. Other aspects of the simulation do not agree as readily with observed facts: (1) The upper easterly jet and the westerly jet over South Asia and the upper westerly jet in the southern hemisphere are missing and the southwest monsoon near the ground is too weak on the south side of the plateau. (2) In the meridional section there is no indication of a monsoon circulation which rises over the Indian monsoon trough and descends around 20°S.

2. Influence of the Effect of Steady Surface Heating on the Monsoon Circulation

Simulation experiments of the Asian southwest monsoon in relation to several different heating distributions will be presented. The first one is an experiment relating to the effect of the steady exchange of sensible heat between the ground and the air. Figures 18a and 18b represent this simulation on the twentieth day of integration. Figure 18a shows that the most effective area of the steady ground heat source is located over the plateau. The distribution of the zonal wind is very similar to that in Fig. 17a, but there appears a weak east wind near the surface at

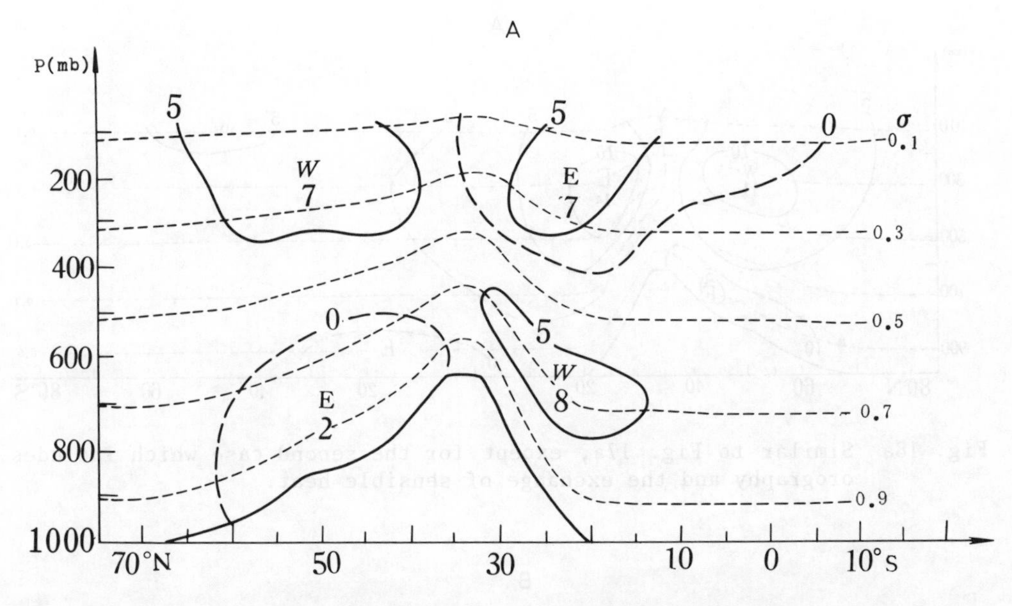

Fig. 17a Computed zonal wind (m/s) for the first case of pure orographic effects.

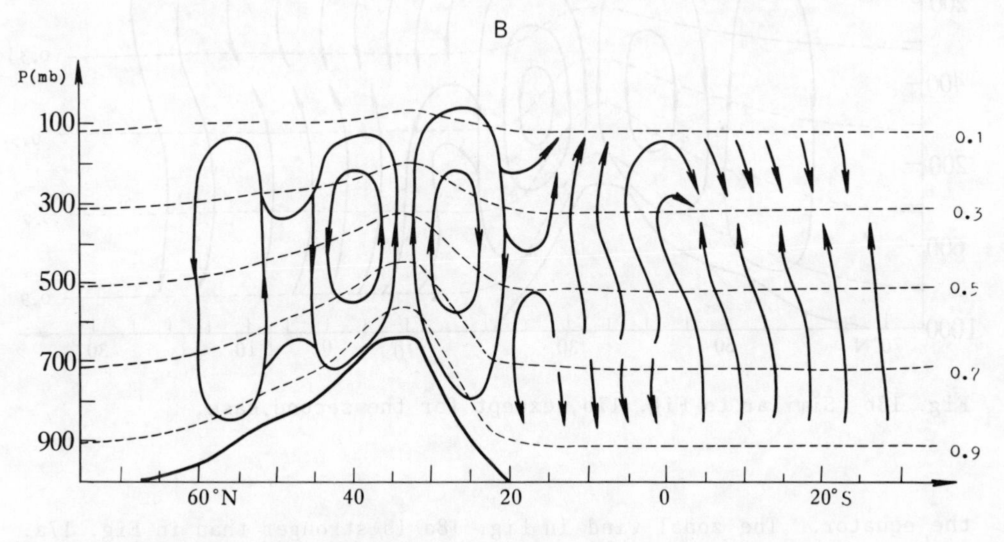

Fig. 17b Computated meridional circulation for the first case of pure orographic effects.

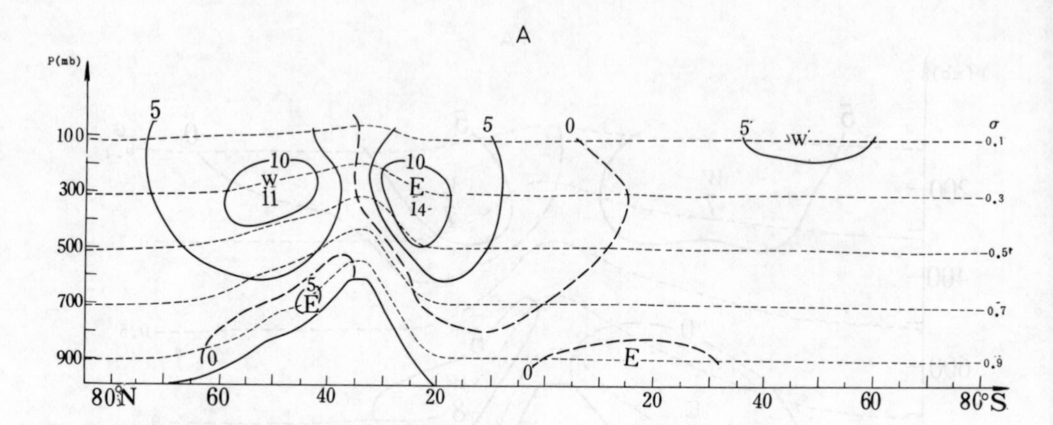

Fig. 18a Similar to Fig. 17a, except for the second case which includes orography and the exchange of sensible heat.

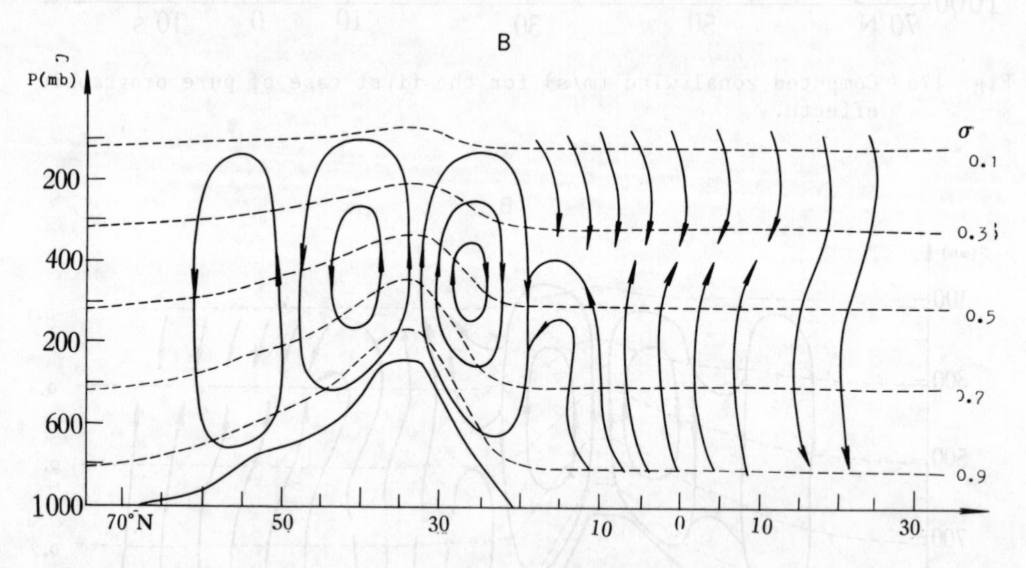

Fig. 18b Similar to Fig. 17b, except for the second case.

the equator. The zonal wind in Fig. 18a is stronger than in Fig. 17a. There are strong easterlies with maximum speeds of 14 ms^{-1} at upper levels ($\sigma = 0.3$) and westerlies as high as 7 ms^{-1} at $\sigma = 0.9$ on the south side of the plateau. The strength of the winds is in closer agreement with actual fact than in the former experiment. However, the strength of the easterlies is still weak and their depth too shallow. The upper westerly jets in both hemispheres do not occur in this case. Figure 18b presents the meridional circulation. The type of cells shown are the same as in Fig. 17b, but their strength has increased to some extent. This result also does not agree with observations.

3. Influence of the Effect of Steady Internal Atmospheric Heating on Monsoon Circulation

In this case, the orography and steady atmospheric heating are included together, but the heating center is placed over the Bay of Bengal. The results are shown in Figs. 19a and 19b. At lower levels, westerlies appear on the south slopes of the plateau and easterlies remain between the equator and 55°S. At upper levels ($\sigma = 0.1$) easterlies are found between the equator and 40°N with a maximum speed of 16 ms^{-1}. On the north side of the plateau, the winds are easterly at lower levels and westerly at upper levels. Though the strength of the winds is somewhat too weak, the distribution approximates well the July climatic state.

The corresponding meridional circulation is shown in Fig. 19b. It is composed of three primary cells. Two of them are over the plateau slopes and their strengths are weaker than the cells presented in Fig. 17b and Fig. 18b. The third cell has a significant vertical circulation. Its ascending branch is around 17.5°N and the area of ascent extends to the vicinity of the equator. Its subsiding branch is in the vicinity of 20°S. The so-called monsoon meridional cell simulated in this case is closer to reality than that developed in the former two experiments. The main defect of this simulation relates to the westerlies and the easterlies on both sides of the plateau which are still weak as compared with the actual state. The reasons causing this defect probably are as follows: (1) The strength of the maximum rate of heating adopted in this experiment was too weak. The realistic maximum rate of heating is about 6 to 8°C/day, but merely 3°C/day were adopted in our simulation. (2) The vertical distribution of the atmospheric heating assumed in our model was not realistic.

4. Influence of Maximum Heating Over the Tibetan Plateau on the Monsoon Circulation

For a study of thermal effects of the Tibetan Plateau the meridional distribution of steady atmospheric heating was enhanced. It was not changed over lower elevations, but over the plateau the heating rate was increased from 1°C/day to 2.76°C/day. The results on the twentieth day of integration are shown in Figs. 20a and 20b. The fundamental features are similar to those in Figs. 19a and 19b, but the upper easterly jet on the south side of the plateau and the westerly jet in the southern hemisphere are weakening. This result implies a further deterioration or damping of the solution relating to the normal monsoon circulation in South Asia if the heating over the plateau is too large.

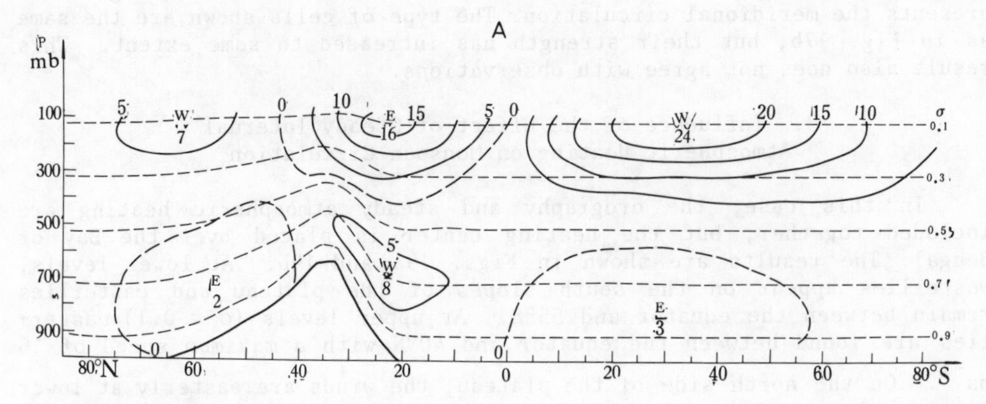

Fig. 19a Similar to Fig. 17a, except for the third case which includes orography and the atmospheric heat source when the heating center is located over the Bay of Bengal.

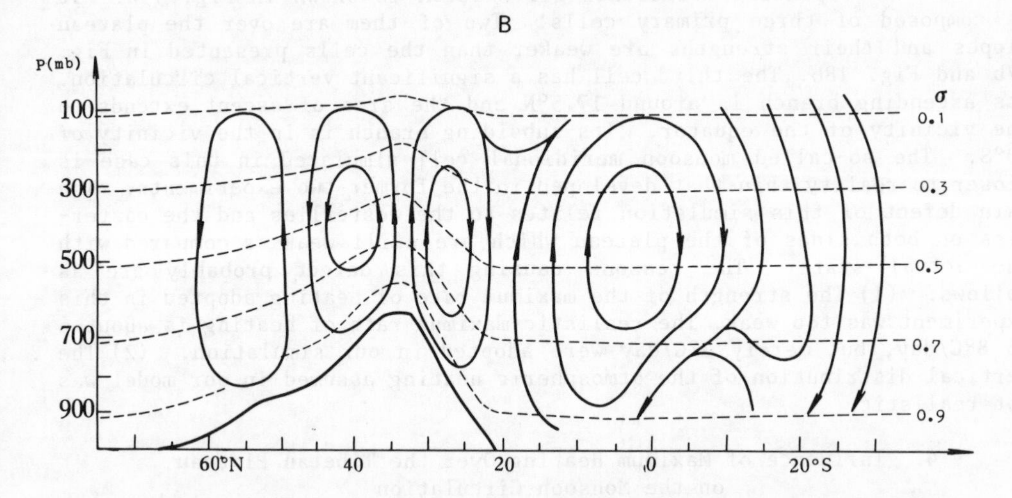

Fig. 19b Similar to Fig. 17b, except for the third case.

V. CONCLUSIONS

From the above discussion, the following conclusions can be drawn:

(1) From October through February the Tibetan Plateau is a heat sink, whereas from March through September it is a heat source. The average rate of heating and cooling is less than 1°C/day.

(2) The Bay of Bengal, northwestern Burma and northeastern India are heat sources after May and change to the strongest centers of heating in Asia during May, June, July and August. The average rate of heating is 3°C/day and the maximum rate of heating can approach 8°C/day.

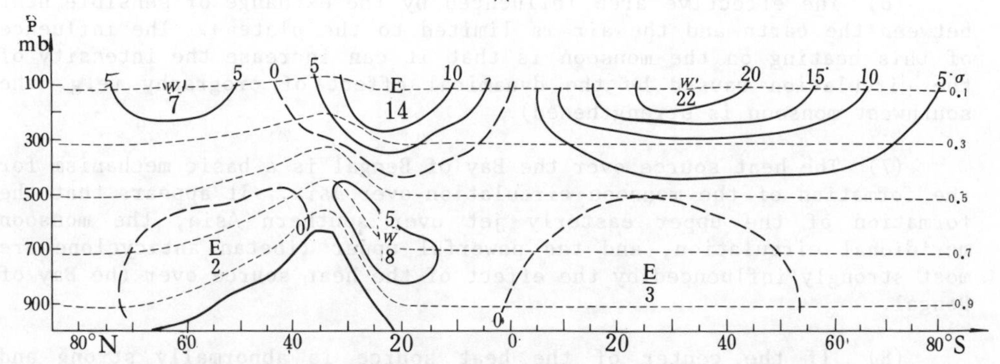

Fig. 20a Similar to Fig. 17a, except for the fourth case which includes orography and the atmospheric heat source when the heating centers are located over the Bay of Bengal and over the plateau.

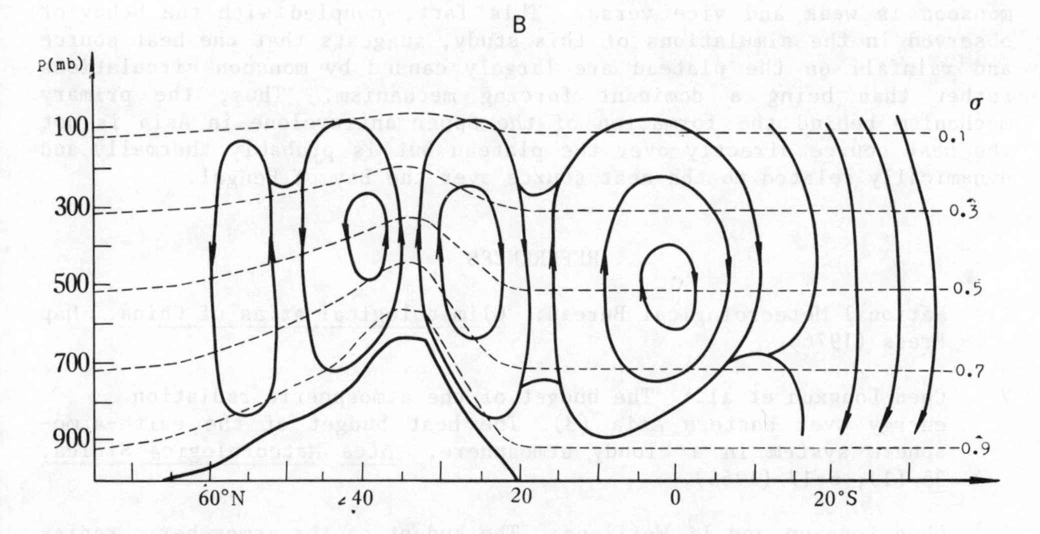

Fig. 20b Similar to Fig. 17b, except for the fourth case.

(3) In Asia, the earliest heat sources occur in North India and the mainland of East China in February and then shift to the Bay of Bengal in May.

(4) In the South China Sea, there is also a heating center in summer and the maximum rate of heating can be 6°C/day, so that it is somewhat weaker than that over the Bay of Bengal.

(5) The dynamical effect of the Tibetan Plateau is an important factor in the formation of the monsoon circulation over South Asia.

(6) The effective area influenced by the exchange of sensible heat between the earth and the air is limited to the plateau. The influence of this heating on the monsoon is that it can increase the intensity of the circulation caused by the dynamical effect of orography (e.g. the southwest monsoon is strengthened).

(7) The heat source over the Bay of Bengal is a basic mechanism for the formation of the monsoon circulation over Asia. It appears that the formation of the upper easterly jet over southern Asia, the monsoon meridional circulation, and the powerful upper Tibetan anticyclone are most strongly influenced by the effect of the heat source over the Bay of Bengal.

(8) If the center of the heat source is abnormally strong and located over the Tibetan Plateau, it will be unfavorable for the establishment of a normal summer monsoon circulation.

In the authors' opinion, the thermal effect of the Tibetan Plateau in summer has been overestimated in previous studies. According to synoptic experience, rainfall on the plateau is high when the Indian monsoon is weak and vice versa. This fact, coupled with the behavior observed in the simulations of this study, suggests that the heat source and rainfall on the plateau are largely caused by monsoon circulations rather than being a dominant forcing mechanism. Thus, the primary mechanism behind the formation of the upper anticyclone in Asia is not the heat source directly over the plateau but is probably thermally and dynamically related to the heat source over the Bay of Bengal.

REFERENCES

1. National Meteorological Bureau: Climatological atlas of China. Map Press (1976).

2. Chen Longxun et al.: The budget of the atmospheric radiation energy over eastern Asia (3). The heat budget of the earth-atmosphere system in a cloudy atmosphere. Atca Meteorologica Sinica, 35 (1), 6-17 (1965).

3. Chen Longxun and Li Weiliang: The budget of the atmospheric radiation energy and the distribution of the atmospheric heat source over Asia in July. Proceedings of Symposium on the Summer Monsoon in China. The People's Press of Yunnan (1981).

4. Gao, Guodong, Lu Yurong et al.: Physical climate atlas of China. Agriculture Press (1981).

5. Hastenrath, S. et al.: Climate atlas of the Indian Ocean, Part I, II. The University of Wisconsin Press (1979).

6. Jaeger, L.: Monatskarten des Niederschlags für die ganze Erde. Berichte des Deutschen Wetterdienstes, 139 (1976).

7. Research Group of Oceanic Meteorology, Institute of Oceanography, Academia Sinica: Atlas of the heat balance at sea surface in the northwest Pacific. Science Press (1979).

8. Van de Boogaard, H.: Mean circulation of the tropical and subtropical atmosphere -- July. NCAR, Boulder, Colorado (1977).

9. Winston, J.S. et al.: Earth-atmosphere radiation budget. Analyses derived from NOAA satellite data, June 1974-February 1978. NOAA National Environmental Satellite Service (1979).

3.3 NUMERICAL EXPERIMENTS SIMULATING THE EFFECTS OF HEAT SOURCES AND OROGRAPHY IN EAST ASIA ON THE SEASONAL CHANGES IN ATMOSPHERIC CIRCULATION

Wang Anyu
Lanzhou Institute of Plateau Atmospheric Physics
Academia Sinica
The People's Republic of China

Hu Qi and Qing Guangyien
Department of Geology and Geography
University of Lanzhou
The People's Republic of China

ABSTRACT

The influences of heat sources and orography in East Asia on the seasonal change of atmospheric circulation are investigated through numerical experimentation with a two-layer primitive equation model. Simulations show basic agreement with the observed annual variation of the average planetary troughs and ridges produced by the orography in East Asia. The simulated results with diabatic heating are more similar to climatology than those without diabatic heating. In addition, experimental results reveal that heating over East Asia has an important contribution to the southward progression and withdrawal of the westerly jet and the formation of the Tibetan summer anticyclone.

I. INTRODUCTION

As a first approximation of forces governing atmospheric circulations, solar radiation can be prescribed as zonally uniform. On the other hand, underlying surface forcing and differential heating related to such conditions can vary significantly along latitude circles. Such zonal variability can have profound effects on the formation of atmospheric activity centers, average planetary troughs and ridges, and their seasonal changes. In this paper we attempt to investigate orographically induced circulations through use of a two-layer model and 12 monthly mean heat source distributions.

II. THE EXPERIMENTAL CONSIDERATIONS

The numerical model is of the primitive equation type and vertically structured in p-coordinates [2]. We present a sketch of the vertical structure of the numerical model in Fig. 1 and of the model domain and topography in Fig. 2. The model domain is 40° to 150°E and 10° to 65°N with a horizontal grid size of 5°x5°. In the model, the highest elevation point is 4700 m above sea level. The maximum orographic gradient is 0.0054. The time step is 15 minutes, and the time integrations are

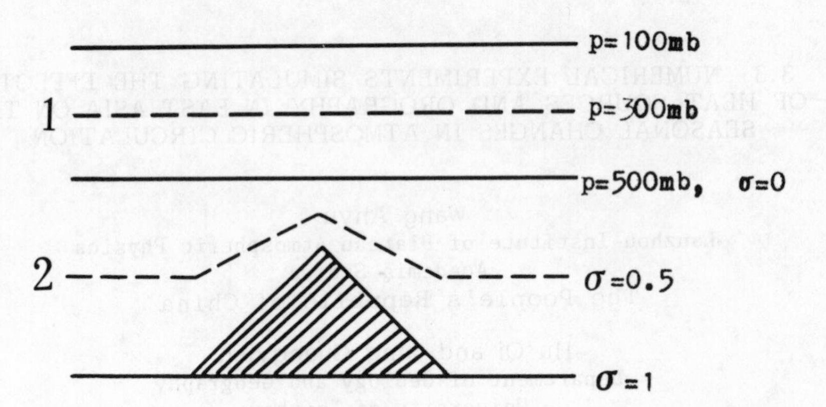

Fig. 1 Sketch of the vertical structure of the numerical model.

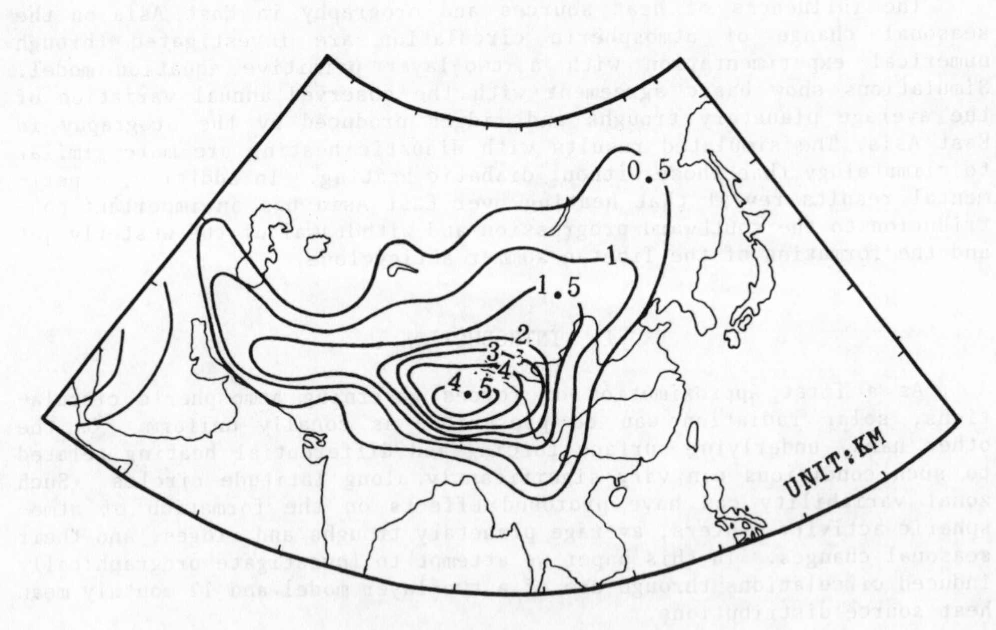

Fig. 2 The model domain and topography.

carried out to five days. Climatological temperatures and geopotential height data for each month are employed as initial conditions at the 300 mb, 700 mb, and 500 mb levels.

Figures 3, 4 and 5 show the initial zonal mean distributions of 300 mb and 700 mb temperatures and 500 mb geopotential heights for every month. The "zonal mean" fields are defined only in terms of averages over the model domain. The 12 monthly mean heat source distributions employed have been completed by Yiao et al. [4]. These data were smoothed to conform to the 5°x5° grid size of the experimental model, and the results are presented in Figs. 6 to 11. It is seen that after winter the Asian continental heat source appears first over the southern Tibetan Plateau and northern India, whereas in summer the most intensive heat sources appear over the Bangladesh region and the South China Sea. In September all Asian continental heat sources are transformed into heat

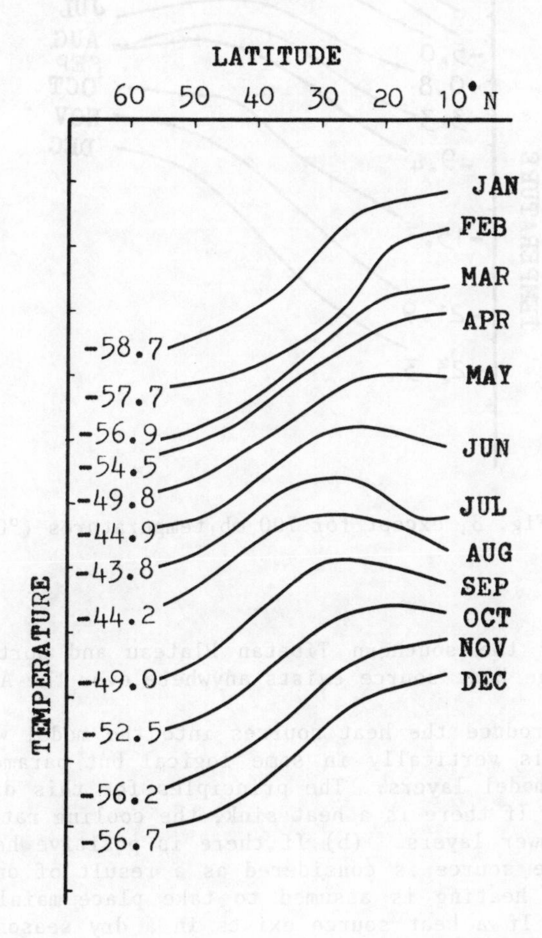

Fig. 3 The initial zonal mean distribution of 300 mb temperatures (°C) for every month. (Marks on ordinate correspond to 10°C.)

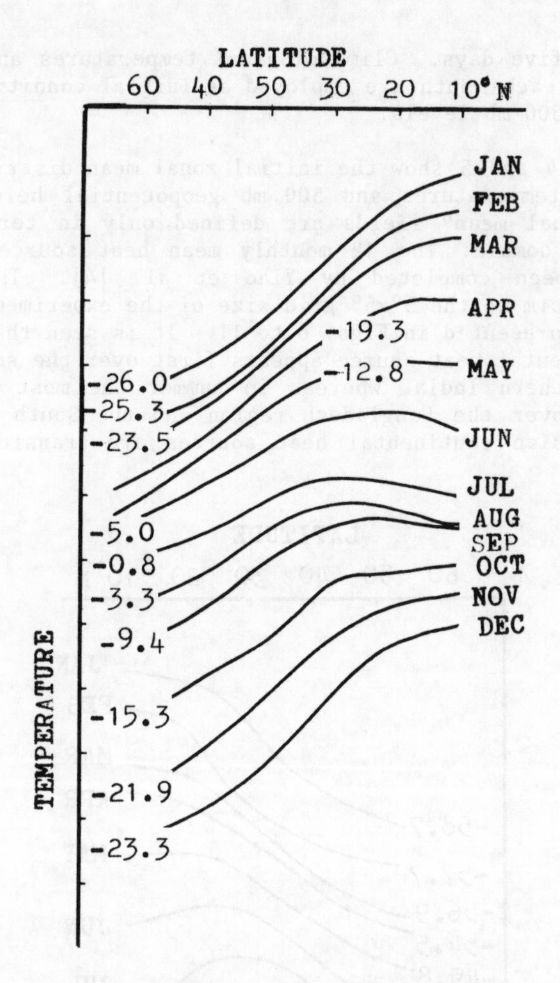

Fig. 4 As in Fig. 3, except for 700 mb temperatures (°C).

sinks except for the southern Tibetan Plateau and northern India; in October no average heat source exists anywhere over the Asian continent.

When we introduce the heat sources into the model we have to distribute the totals vertically in some logical but parameterized manner between the two model layers. The principles for this distribution are as follows: (a) If there is a heat sink, the cooling rates are equal in the upper and lower layers. (b) If there is positive heating during a rainy season, the source is considered as a result of only latent heat release, and the heating is assumed to take place mainly in the upper atmosphere. (c) If a heat source exists in a dry season, we introduce heating primarily in the lower atmosphere under the assumption that it is sensible heat.

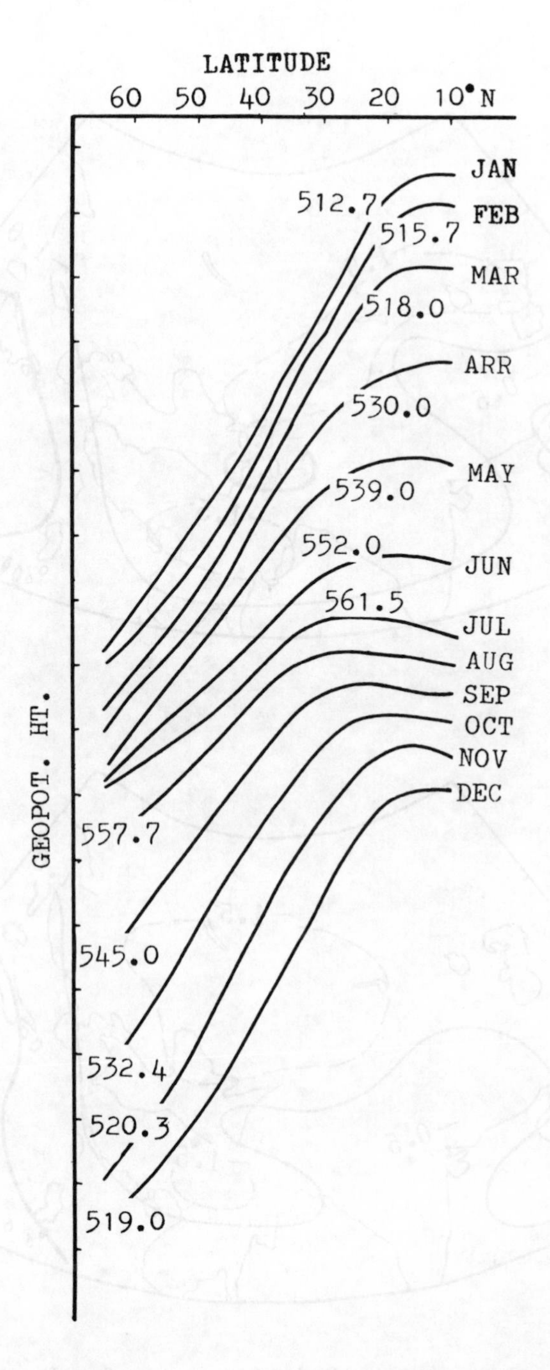

Fig. 5 As in Fig. 3, except for 500 mb geopotential height (decameters; marks on ordinate correspond to 10 decameters).

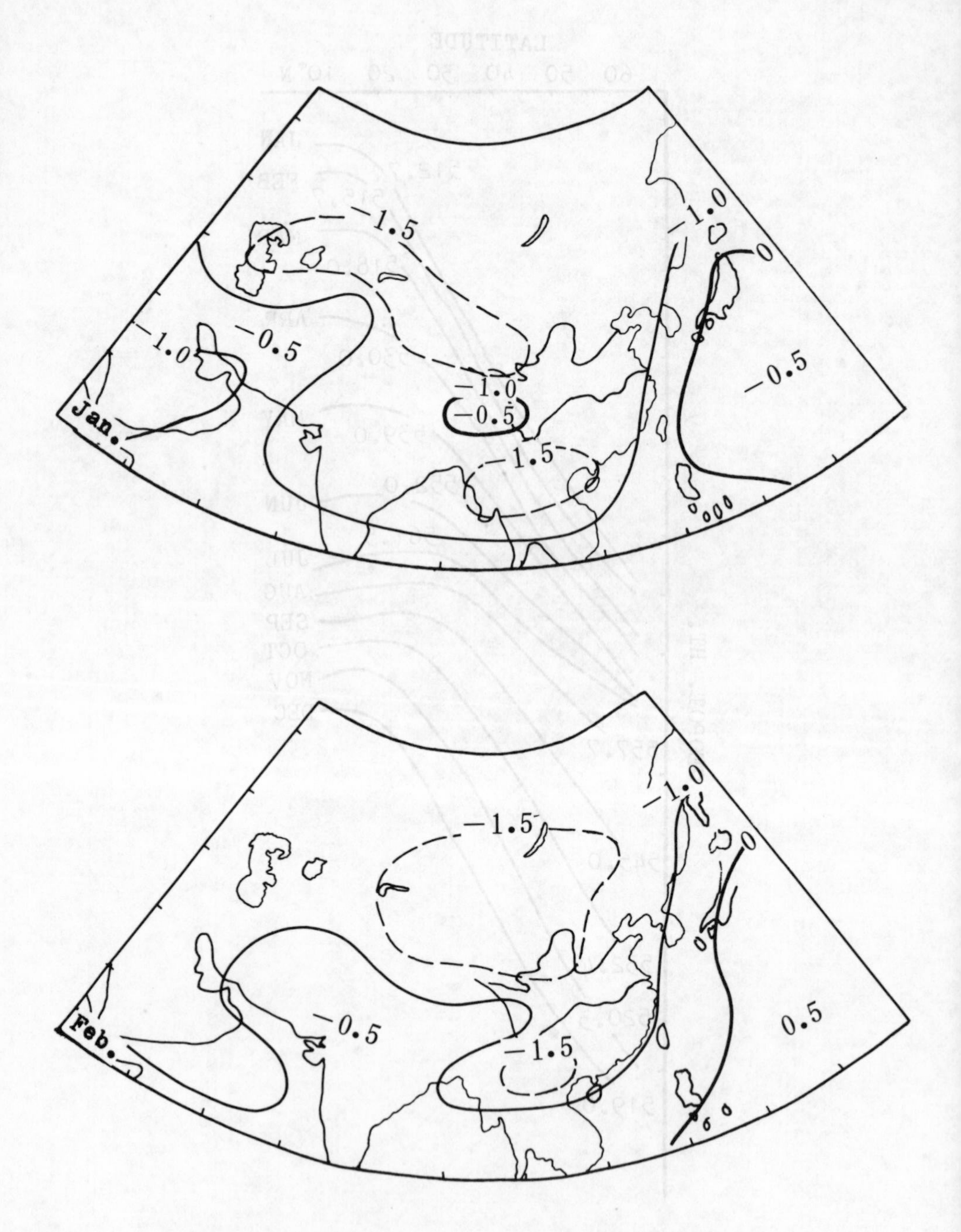

Fig. 6 Mean monthly atmospheric heating rates for January and February. Positive region: heat source region. Negative region: heat sink region. Unit: °C/day.

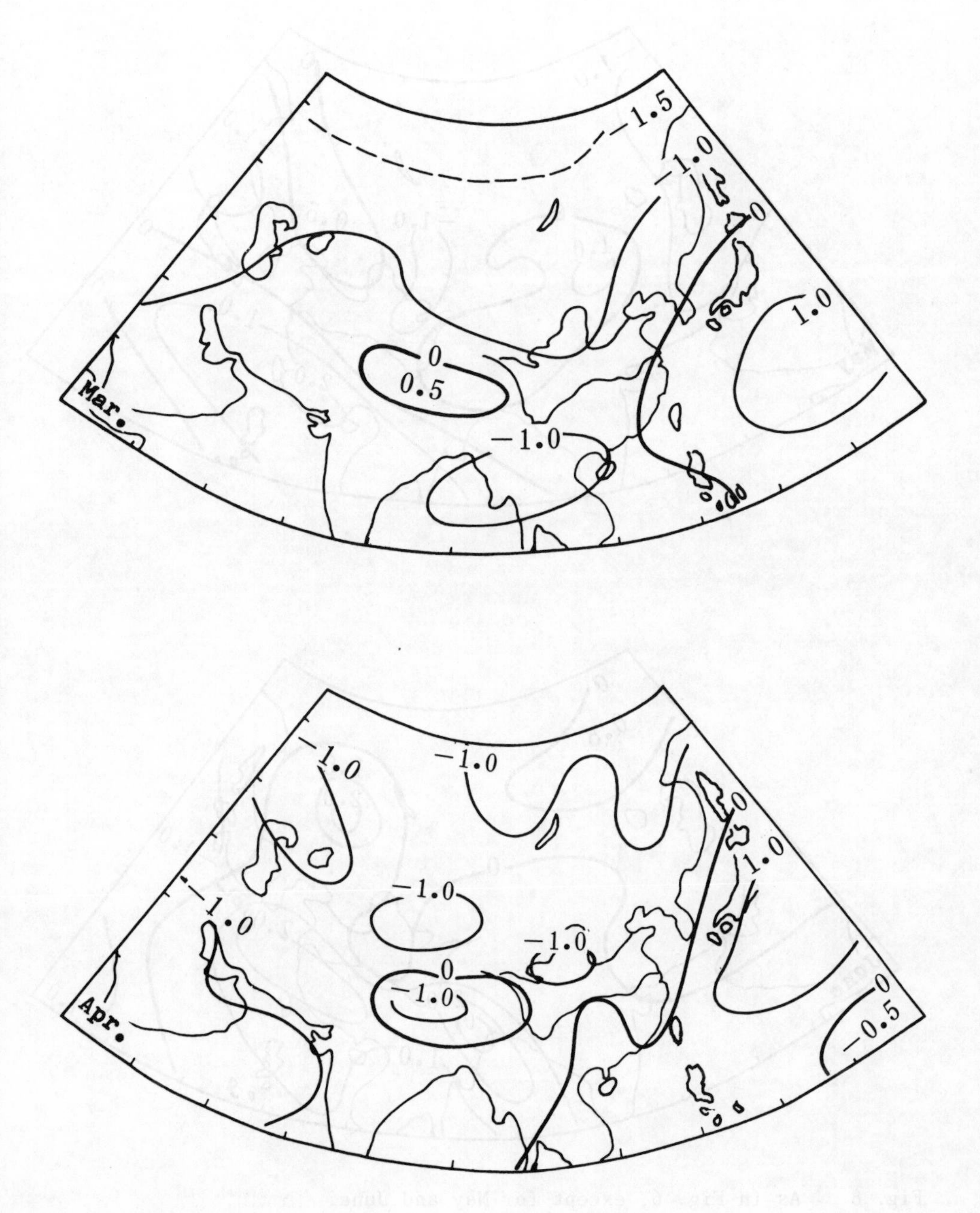

Fig. 7   As in Fig. 6, except for March and April.

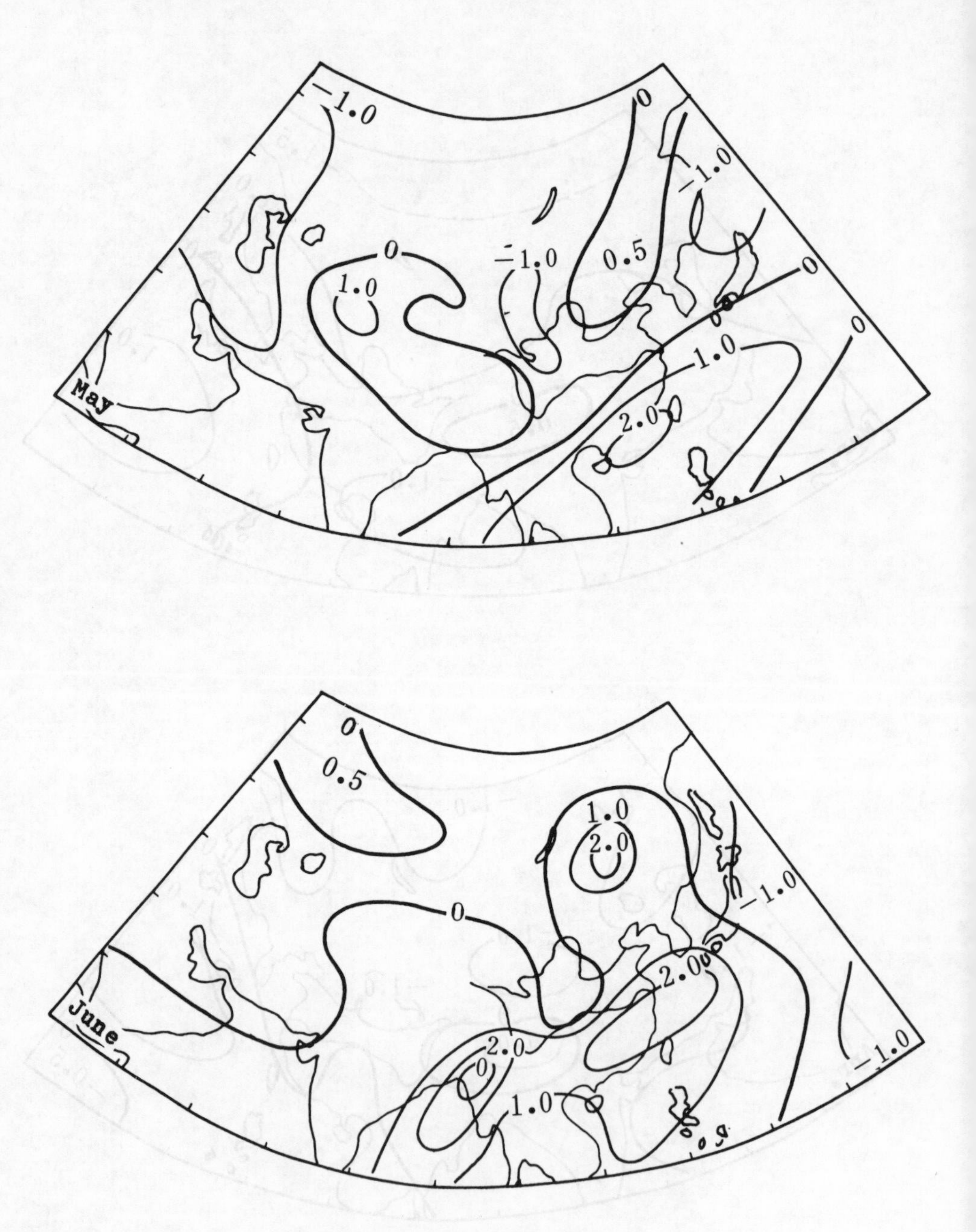

Fig. 8 As in Fig. 6, except for May and June.

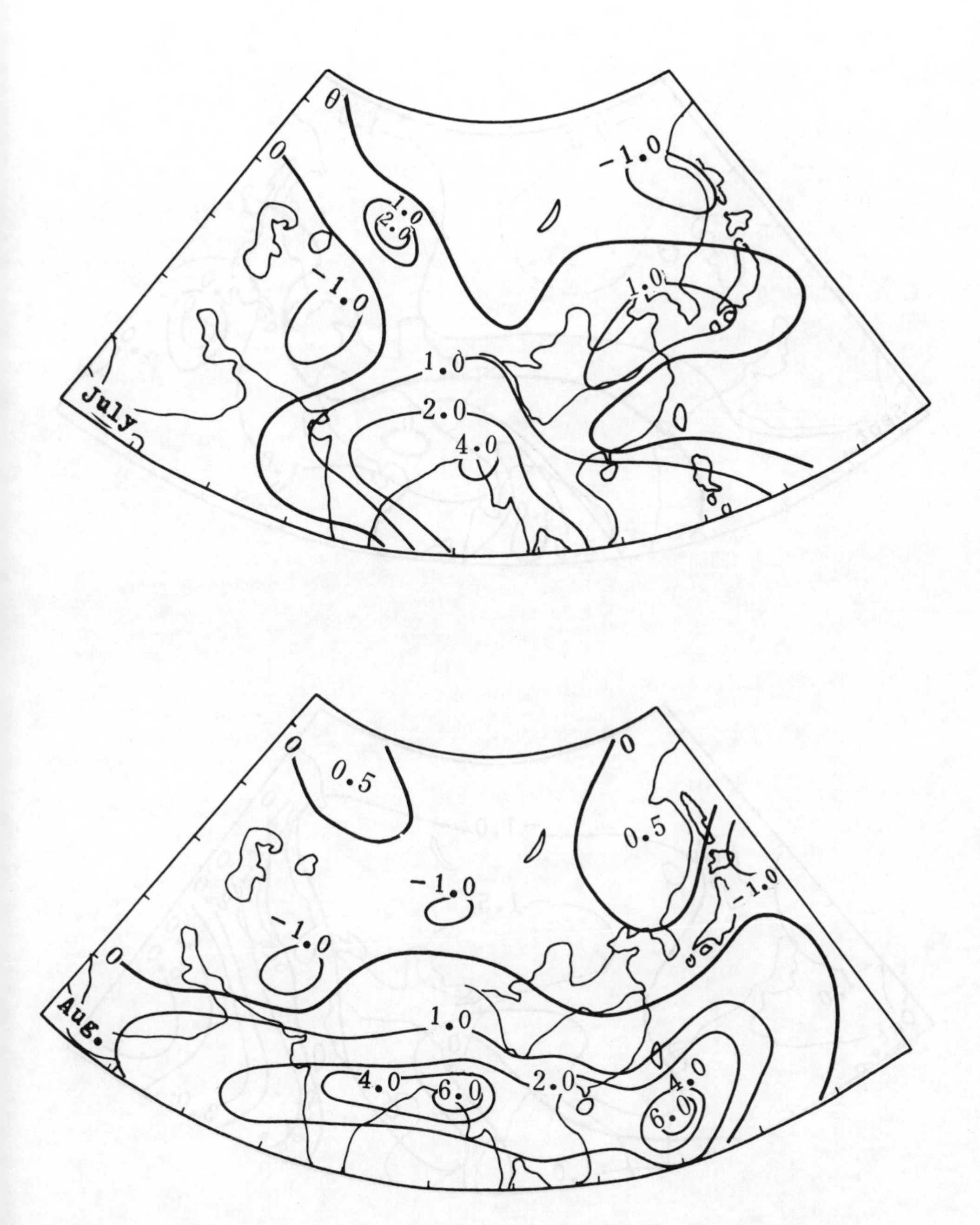

Fig. 9 As in Fig. 6, except for July and August.

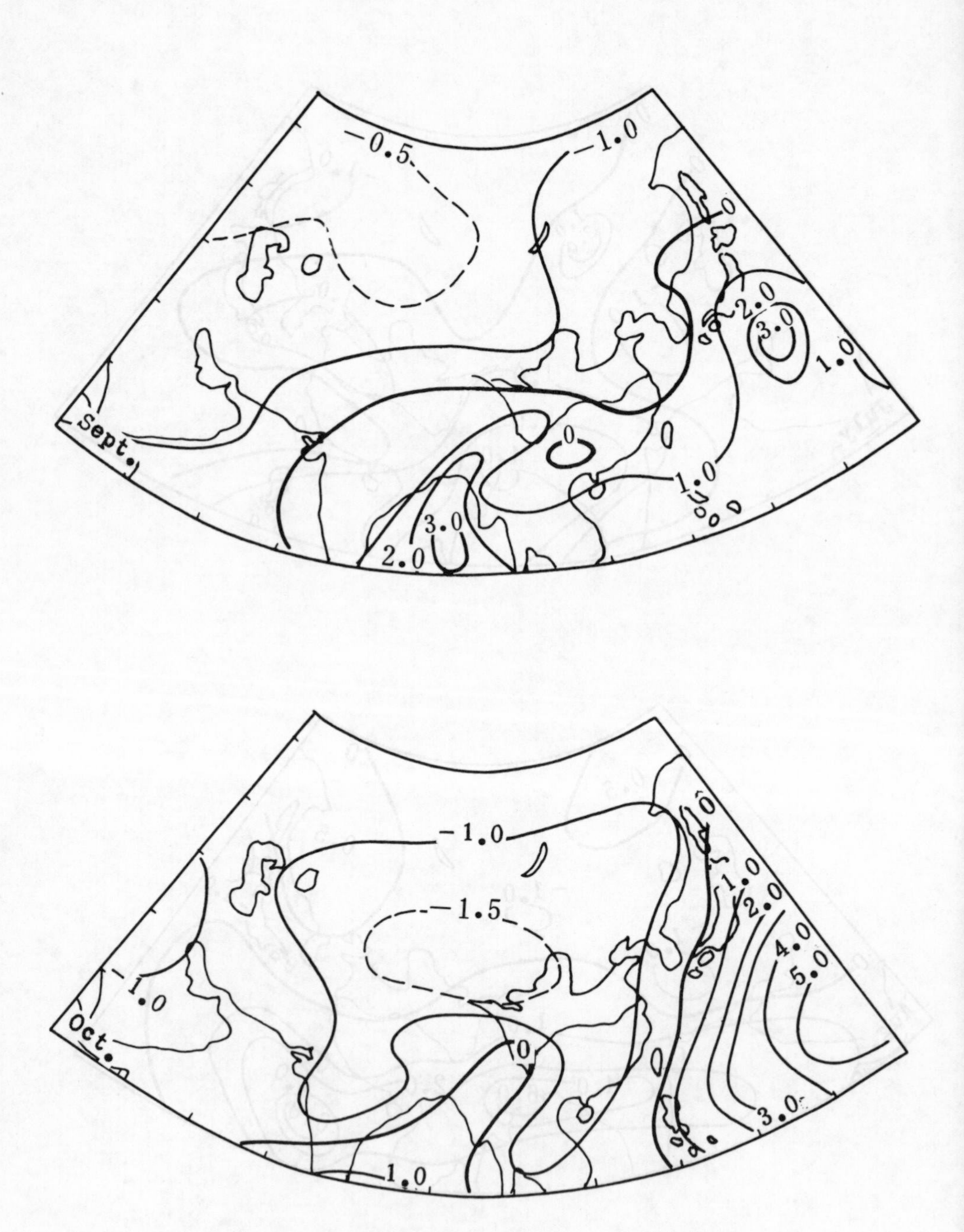

Fig. 10 As in Fig. 6, except for September and October.

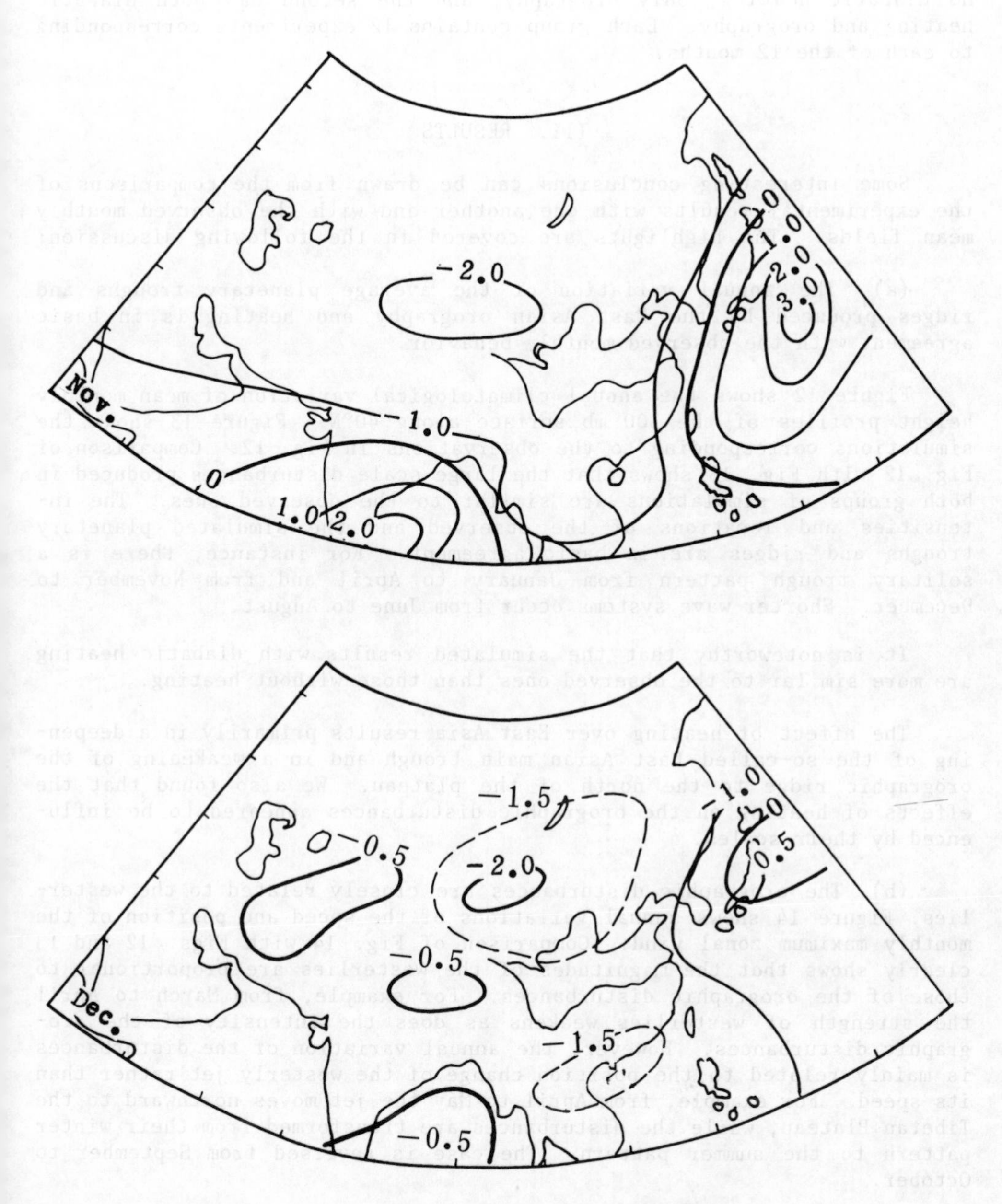

Fig. 11   As in Fig. 6, except for November and December.

Two basic types of experiments were performed: The first group has no diabatic heating, only orography, and the second has both diabatic heating and orography. Each group contains 12 experiments corresponding to each of the 12 months.

III. RESULTS

Some interesting conclusions can be drawn from the comparisons of the experimental results with one another and with the observed monthly mean fields. The highlights are covered in the following discussion:

(a) The annual variation of the average planetary troughs and ridges produced by the East Asian orography and heating is in basic agreement with the observed monthly behavior.

Figure 12 shows the annual climatological variation of mean monthly height profiles of the 300 mb surface along 40°N. Figure 13 shows the simulations corresponding to the observations in Fig. 12. Comparison of Fig. 12 with Fig. 13 shows that the large-scale disturbances produced in both groups of simulations are similar to the observed ones. The intensities and locations of the observed and the simulated planetary troughs and ridges are in basic agreement. For instance, there is a solitary trough pattern from January to April and from November to December. Shorter wave systems occur from June to August.

It is noteworthy that the simulated results with diabatic heating are more similar to the observed ones than those without heating.

The effect of heating over East Asia results primarily in a deepening of the so-called East Asian main trough and in a weakening of the orographic ridge to the north of the plateau. We also found that the effects of heating on the orographic disturbances appeared to be influenced by their scales.

(b) The orographic disturbances are closely related to the westerlies. Figure 14 shows annual variations of the speed and position of the monthly maximum zonal wind. Comparison of Fig. 14 with Figs. 12 and 13 clearly shows that the magnitudes of the westerlies are proportional to those of the orographic disturbances. For example, from March to April the strength of westerlies weakens as does the intensity of the orographic disturbances. However, the annual variation of the disturbances is mainly related to the position change of the westerly jet rather than its speed. For example, from April to May the jet moves northward to the Tibetan Plateau, while the disturbances are transformed from their winter pattern to the summer pattern. The case is reversed from September to October.

(c) The seasonal variation of temperature disturbances in middle and high latitudes produced by heating in East Asia is similar to that of climatological mean monthly temperature fields.

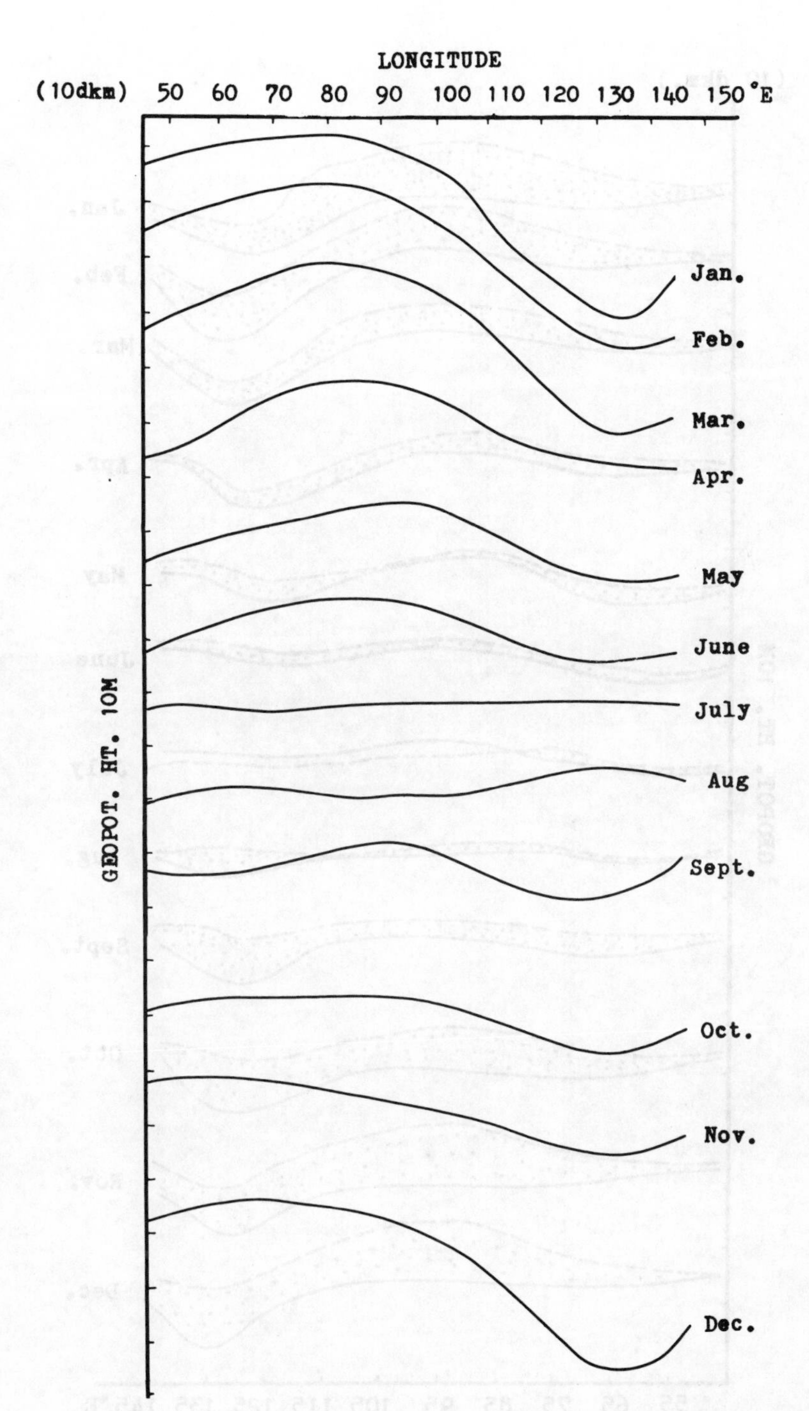

Fig. 12 Observed annual climatological variation of mean monthly height profiles of 300 mb along 40°N. (Marks on ordinate correspond to 10 decameters.)

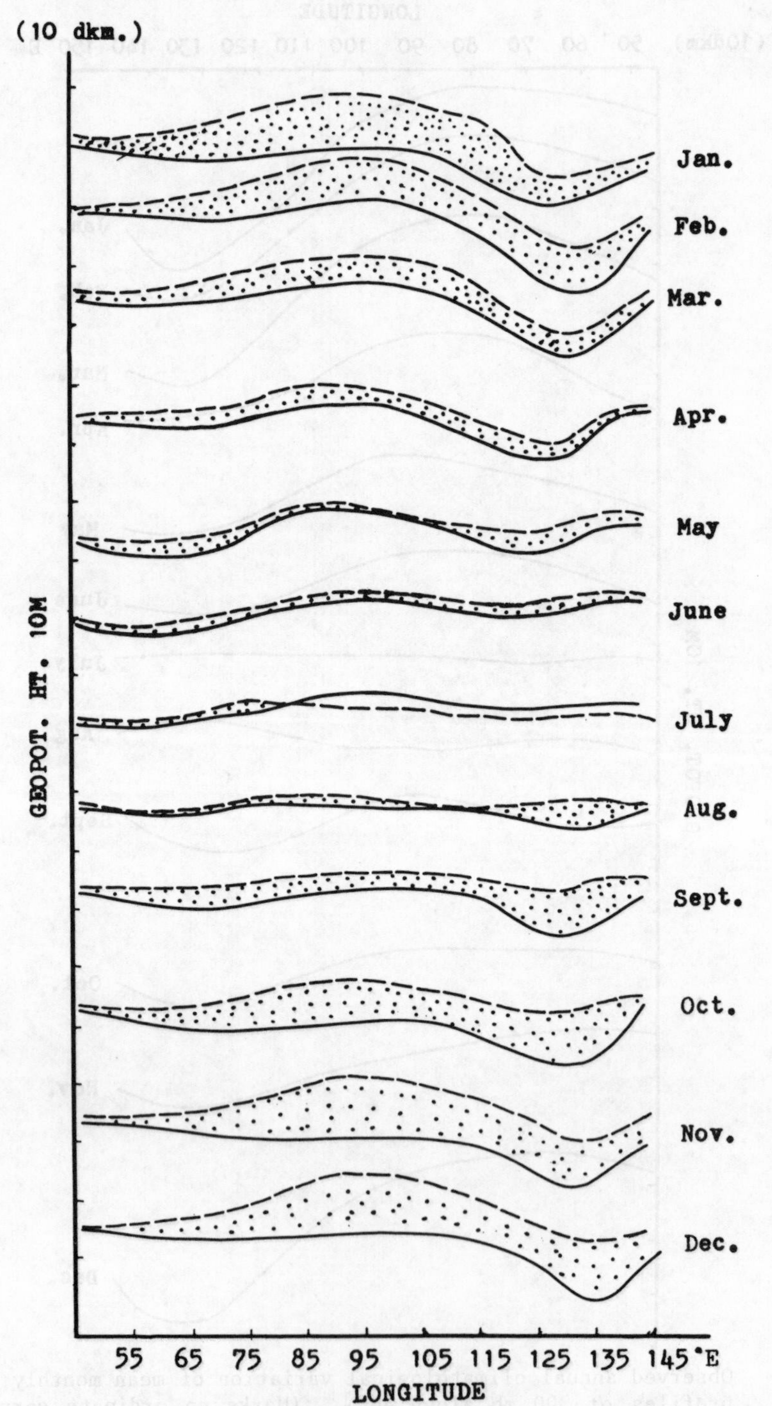

Fig. 13 As in Fig. 12, except for model simulation. Solid lines: Simulated results with diabatic heating and orography. Dashed lines: Simulated results without diabatic heating.

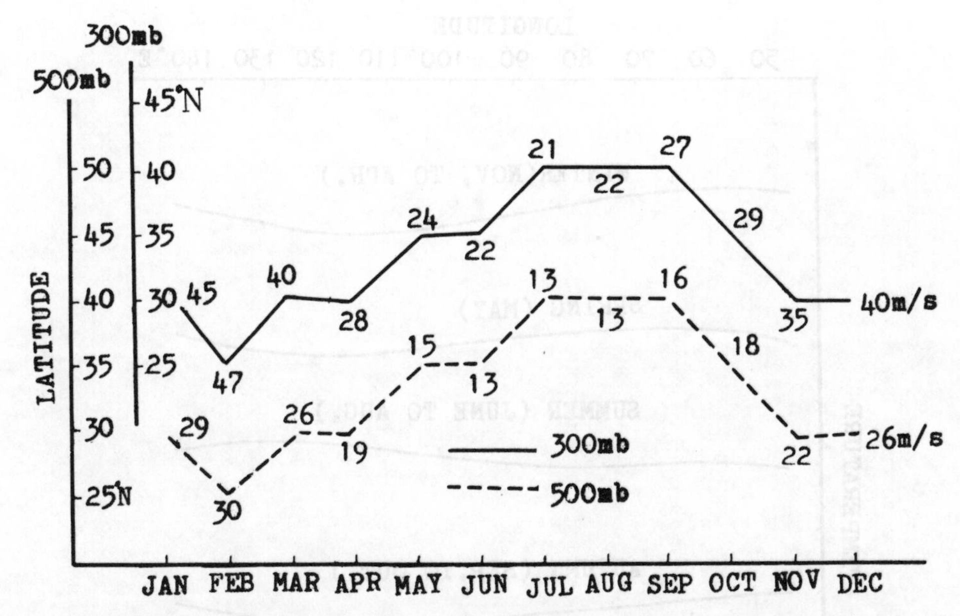

Fig. 14 Annual variation of the initial maximum zonal wind speeds ($m\ s^{-1}$) and their positions.

Figure 15 shows the observed seasonal variation of 300 mb temperature profiles along 50°N. Figure 16 shows comparable data from the simulations. From Fig. 16 it can be seen that the temperature disturbance produced by orography is mainly a weak planetary-scale ridge, and its seasonal variation is poorly defined. Comparison of Figs. 15 and 16 shows that the temperature disturbances created by orography alone are not similar to climatology. The temperature disturbances produced by both heating effect and orography in East Asia are more similar to climatology and exhibit more seasonal variation. For example, there is a trough ridge pattern in winter (November to April) and autumn (September and October) and two weaker ridges and troughs in spring (May) and summer (June to August). One can note in the experiments with diabatic heating (Figs. 13 and 16) that the locations of the temperature disturbances are in accord with the locations of pressure disturbances. In addition, it is noteworthy that the configuration of the temperature and pressure fields can contribute significantly to the development of existing disturbances, for example, by deepening the so-called East Asian main trough in winter.

(d) Heating plays an important role in the southward progression and withdrawal of the westerly jet in the East Asian area. Figure 17 shows the simulated zonal wind speed profiles and their annual variation on the 300-mb surface along 90°E for each month. One notes that in March and April the heating effect results first in an increase of the westerlies over the plateau and thereafter the maximum climbs northward across the plateau. In August and September, the westerlies over the plateau

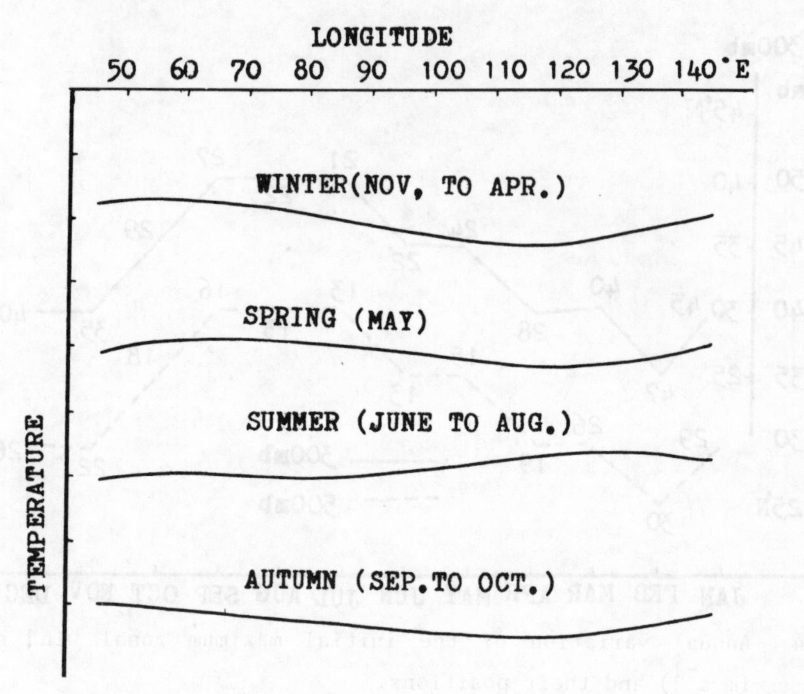

Fig. 15 Climatological seasonal variation of 300 mb temperature profiles along 50°N (°C, marks on ordinate correspond to 5°C).

are strengthened while weakening occurs to the north. Tropical easterlies dominate the region south of the plateau throughout the period. In October westerlies begin to advance to the south and thereafter hold their position over the southern slopes throughout the winter.

In addition it should be noted that in the simulated results with diabatic heating, the maximum speed of westerlies is always larger than that without diabatic heating, i.e. the East Asian heating provides a contribution to the formation of jets in the troposphere.

(e) The heating over East Asia has an important contribution to the formation of the Tibetan summer anticyclone. In East Asia the Tibetan summer anticyclone is a most important atmospheric activity center in the upper troposphere. It has a close relationship to the annual variation of the atmospheric circulation.

Figure 18 shows the mean 300 mb charts in July and August for the Tibetan Plateau and its surrounding area. The upper couplet shows the simulated results without diabatic heating, the middle couplet the simulated results with diabatic heating and the bottom the observed patterns. One notes that without heating the Tibetan anticyclone with a warm core does not exist, whereas with heating the anticyclone appears similar to the climatological one. It must be pointed out that the formation of the

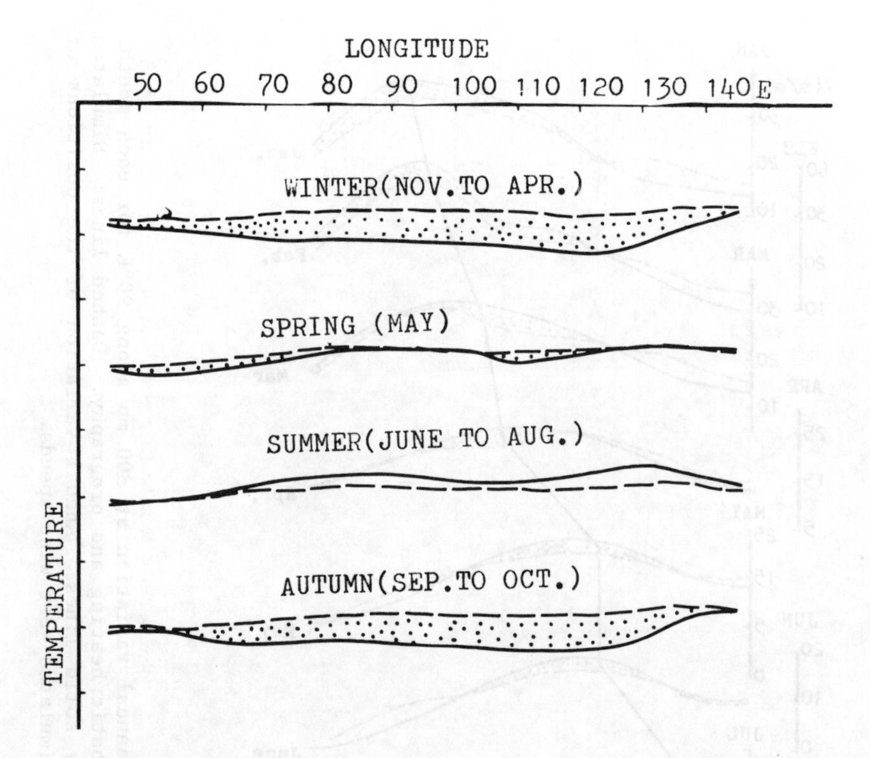

Fig. 16 As in Fig. 15, except model for simulation. Solid lines: Simulated results with diabatic heating and orography. Dashed lines: Simulated results without diabatic heating.

Tibetan anticyclone has a close relation with the background circulation as well.

From Figs. 6 to 11, 13 and 18, it can be seen that in July and August the most intensive heat source in East Asia is located at the boundary between the westerlies and easterlies near the warm core of the Tibetan anticyclone. It appears that a necessary condition for the formation of the Tibetan anticyclone is the positioning of the heat source at the boundary between light tropospheric westerlies and easterlies. This phenomenon is easily explained by the thermal wind principle.

This conclusion is consistent with the results obtained from rotating dishpan experiments by Ye [3] and Fu [1]. Ye pointed out that if the background flow is westerly, it seems impossible to form the anticyclone even if the diabatic heating rates are increased many times. Fu suggested that a critical condition for the formation of the anticyclone is that the heat source should be located in a minimum temperature gradient area.

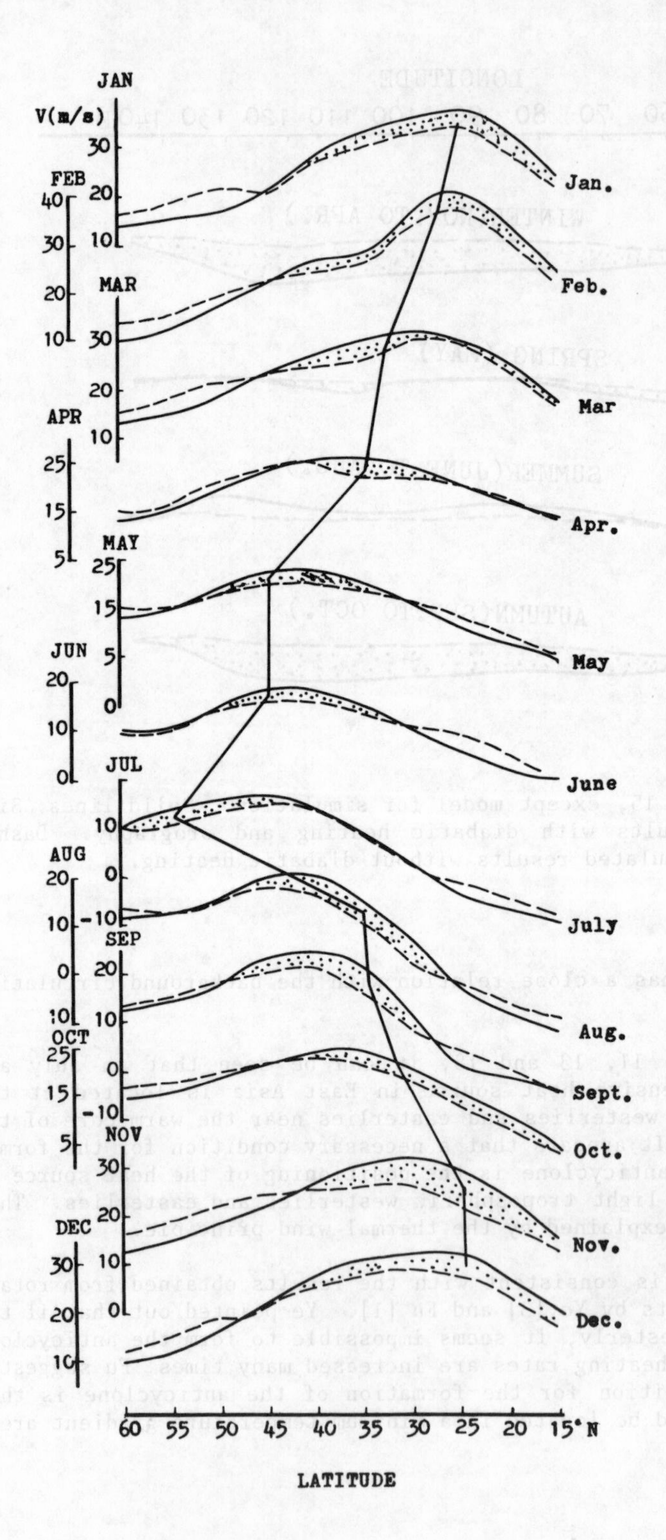

Fig. 17 Simulated wind speed profiles and their annual variation at 300 mb along 90°E for each month. Solid lines: Simulated results with diabatic heating and orography. Dashed lines: Simulated results without diabatic heating. Heavy solid lines: Annual variation of the positions of maximum differences of the two kinds of simulated west wind speeds.

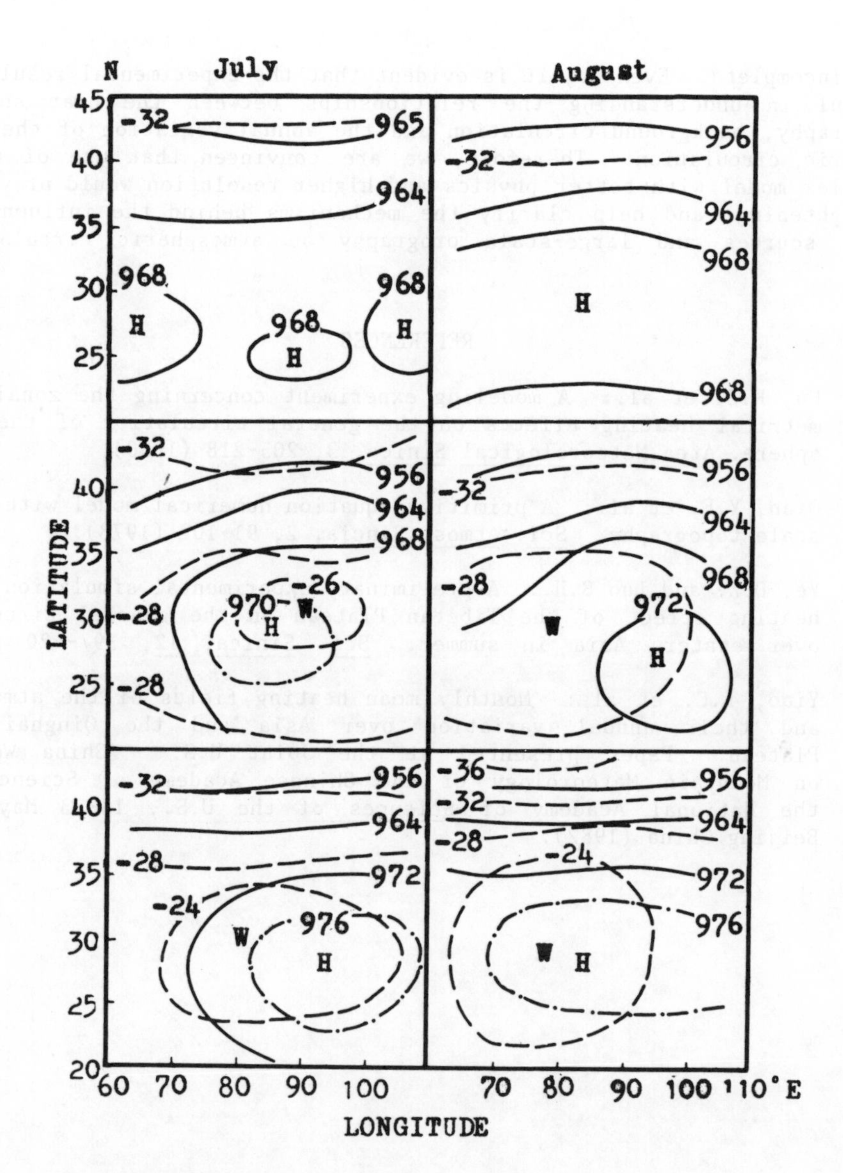

Fig. 18 Mean monthly 300 mb charts for July and August. Upper: Simulated results without diabatic heating. Middle: Simulated results with diabatic heating and orography. Bottom: Climatological monthly mean patterns.

IV. CONCLUDING REMARKS

Our model is not yet sufficiently sophisticated to simulate various physical processes, especially those involving feedback between dynamical and thermal processes. Thus, our results must be considered preliminary and incomplete. Even so, it is evident that the experimental results are useful in understanding the relationships between the heat sources, orography, background circulation and the annual variation of the atmospheric circulation. Therefore, we are convinced that use of a more complex model with better physics and higher resolution would prove more enlightening, and help clarify the mechanisms behind the influences of heat sources and large-scale orography on atmospheric circulations.

REFERENCES

1. Fu, K.Z. et al.: A modeling experiment concerning the zonal asymmetrical heating effects on the general circulation of the atmosphere. Atca Meteorological Sinica, 3, 205-218 (1980).

2. Qian, Y.F. et al.: A primitive equation numerical model with large-scale topography. Sci. Atmos. Sincia, 2, 91-102 (1978).

3. Ye, D.Z. and Luo S.H.: A preliminary experimental simulation of the heating effect of the Tibetan Plateau on the general circulation over eastern Asia in summer. Sci. Sinica, 12, 397-420 (1974).

4. Yiao, L.C. et al.: Monthly mean heating fields of the atmosphere and their annual variation over Asia and the Qinghai-Xizang Plateau. Paper presented at the Joint U.S. - China Workshop on Mountain Meteorology of the Chinese Academy of Sciences and the National Academy of Sciences of the U.S., 18-23 May 1982, Beijing China (1982).

3.4 OROGRAPHICALLY INDUCED VORTEX CENTERS

Isidoro Orlanski
Geophysical Fluid Dynamics Laboratory/NOAA
Princeton University
Princeton, New Jersey 08540 USA

ABSTRACT

A review of some processes by which orography may induce cyclogenesis in the atmosphere has been presented. The discussion focusses on two basic processes of flow interacting with the orography and which are relevant to synoptic and mesoscale; i) the instability of flows along sloping boundaries and ii) the effects of mountain chains on moving fronts.

It was suggested that since the effect of negative and positive mountain slopes on baroclinic instability is to change the scale of the maximum unstable scale, the summer monsoon sees the Tibetan Plateau on its southeastern flank to exhibit a positive slope influence. On the other hand, during the same season the westerlies on the northern plateau will have a negative slope influence from which shorter mesocyclogenesis may be expected. In winter with the westerlies south of the plateau, orographic influence is reversed producing large baroclinic vortices.

A case study on downslope winds over Argentina revealed a mechanism by which geostrophic flow over large mountain barriers can set pressure differences on both sides of the mountain and in the presence of gaps in the barrier may produce strong downstream winds. The mesoscale wind described will operate on a larger scale than the classical downslope winds.

I. INTRODUCTION

It is a well known fact that orography exerts an important influence on all scales of atmospheric circulation. We can see examples of such effects in orographically excited standing ultralong waves, the generation of vortex centers by instabilities of winds blowing over sloping terrain, lee cyclogenesis induced by wind flows over topography, as well as gravity waves and smaller-scale convective systems which clearly illustrate the orographic influence in our daily weather. The fact that major centers exist where orography plays an important role in the general circulation of the atmosphere has been recognized for some time. As we can see in Fig. 1, which shows contours of the earth's orography, areas such as western North and South America, the Antarctic Plateau, the Greeland Plateau, and the Tibetan Plateau exhibit the most predominant orographic features in the world. Other areas, perhaps to a lesser degree, yet still significant, are the mountain chains in southern Europe and eastern Africa. A measure of the activity that these orographic

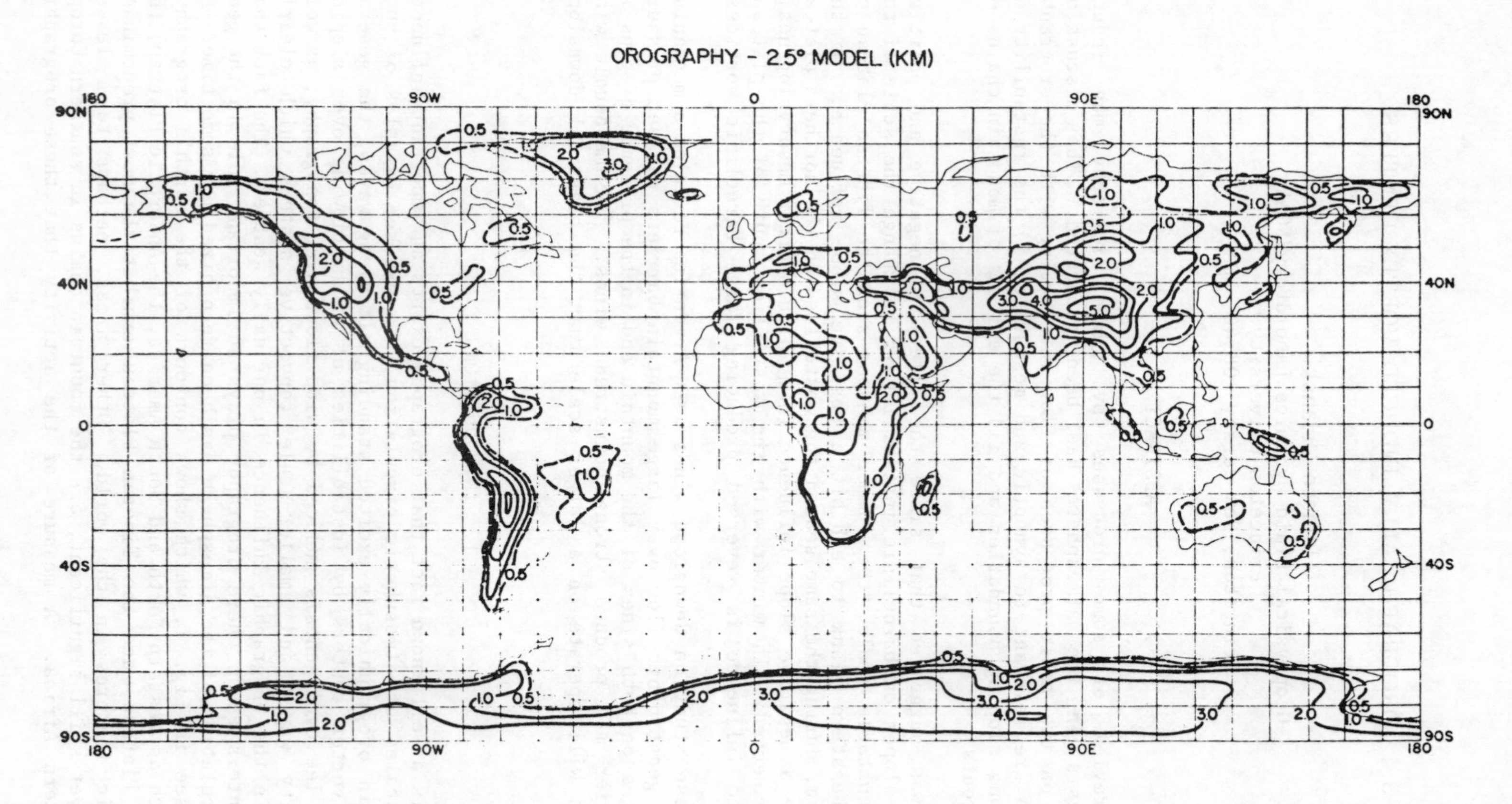

Fig. 1 Earth's orography with contour intervals of 1 km [18].

centers might produce can be estimated by the extent and intensity of the areas of cyclogenesis. An example of this can be seen in Fig. 2 which shows the cyclogenesis regions for the southern hemispheric summer. One should note that Antarctica is an extremely large heat sink in itself, a strong baroclinic zone is produced around it and, as a consequence, cyclogenesis would be expected on the periphery of this continent even in the absence of the orographic effect. Nonetheless, the remarkable feature of the large cyclogenetic region east of Argentina is unquestionably

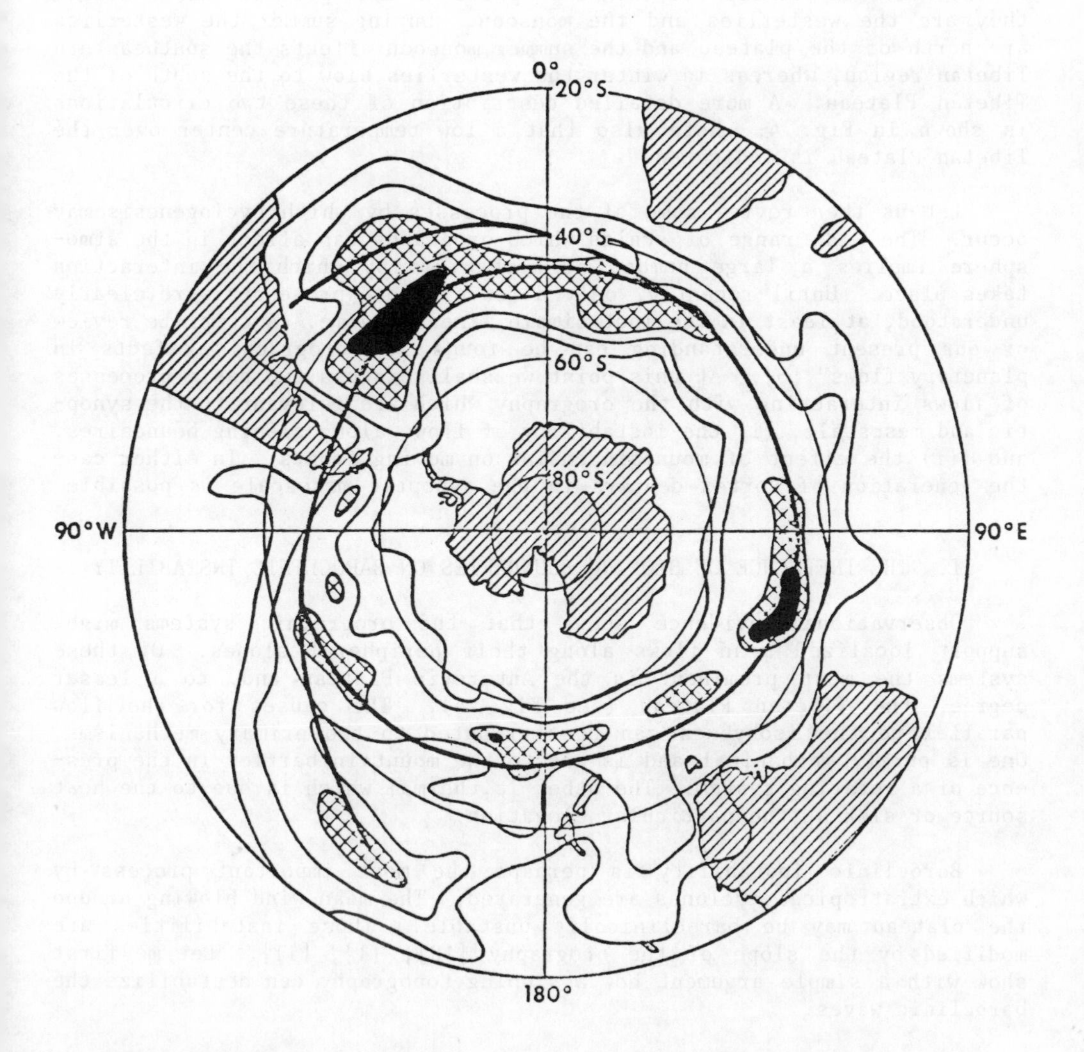

Fig. 2 Regions of cyclogenesis of the southern hemisphere for summer according to Streten and Troup [17]. The solid line indicates the position of the polar front in summer.

related to the disturbance of the westerlies due to the Andes Mountain chain. It is also noteworthy to mention that the baroclinically unstable eddies around Antarctica have already been enhanced by the large plateau of this continent, as will soon be discussed. Another important area of cyclogenesis, one more relevant to this conference, is the region affected by the Tibetan Plateau. In more detail we can see (Fig. 3) that most of the cyclogenesis is located east of the plateau. We can recognize two centers; in summer one is to the northeast of the plateau and the other to the south of the plateau, whereas in winter the northern center moves to the east. There are two main flow patterns that produce this feature; they are the westerlies and the monsoon. During summer the westerlies are north of the plateau and the summer monsoon affects the southeastern Tibetan region, whereas in winter the westerlies blow to the south of the Tibetan Plateau. A more detailed description of these two circulations is shown in Fig. 4. Note also that a low temperature center over the Tibetan Plateau is indicated.

Let us then review some of the processes by which cyclogenesis may occur. The wide range of scales which orography can affect in the atmosphere implies a large number of mechanisms by which the interaction takes place. Until recently, only a few of these processes were clearly understood, at least in the most simple linear regime. A complete review of our present understanding can be found in "orographic effects in planetary flows" [6]. At this point we shall discuss two basic processes of flows interacting with the orography which are relevant to the synoptic and mesoscale, (i) the instability of flows along sloping boundaires, and (ii) the effect of mountain chains on moving fronts. In either case the generation of vortex centers in the synoptic mesoscale is possible.

II. THE INFLUENCE OF SLOPING BOUNDARIES ON BAROCLINIC INSTABILITY

Observational evidence shows that the orographic systems might support localized mean flows along their peripheral slopes. Of these systems the most prominent is the Antarctic Plateau and, to a lesser degree, the Tibetan Plateau (see Fig. 4). The causes for the flow parallel to the isohypses can be attributed to two primary mechanisms. One is purely mechanical and is due to the mountain barrier in the presence of a rotating system. The other is thermal which is due to the heat source or sink at the particular elevation.

Baroclinic instability is perhaps the most important process by which extratropical cyclones are generated. The mean wind blowing around the plateau may be baroclinically unstable. Those instabilities are modified by the slope of the orography [10], [1], [11]. Let me first show with a simple argument how a sloping topography can destabilize the baroclinic waves.

The argument is not new [3], [16], [13]. It has been used to describe the instability mechanism for large-scale quasi-geostrophic atmospheric waves. Here, we shall point out that we do not restrict ourselves to waves that have very low frequencies (quasi-geostrophic waves).

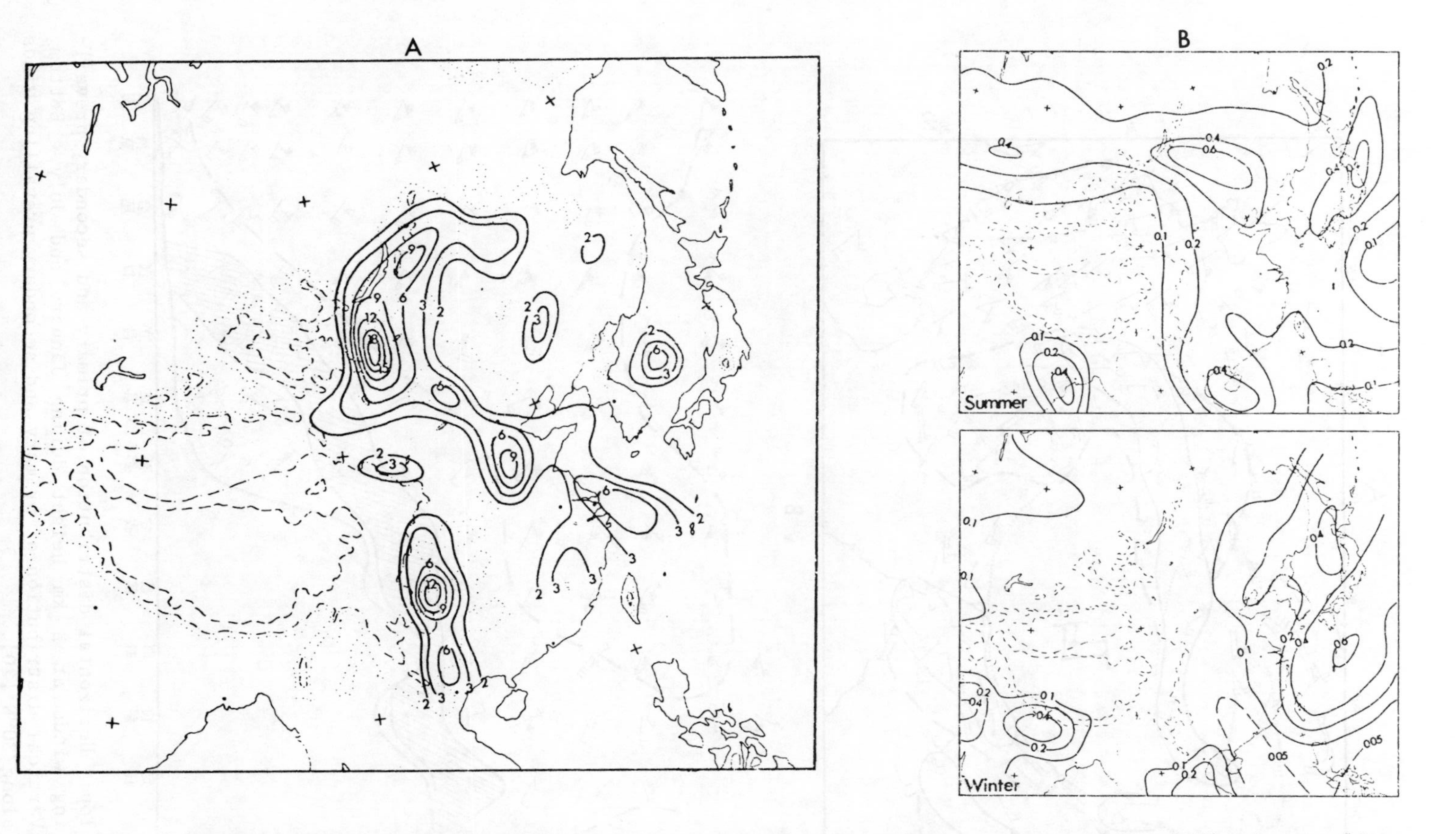

Fig. 3 (a) Frequencies of cyclogenesis in the lee of the East Asian mountains for 1958, in sampling areas of 77,200 km^2 (2.5° latitude x 2.5° longitude). From [5]. (b) Percentage frequency of occurrence of cyclogenesis in squares of 100,000 km^2 in summer and winter. After [14] and [15], p. 267. From [5].

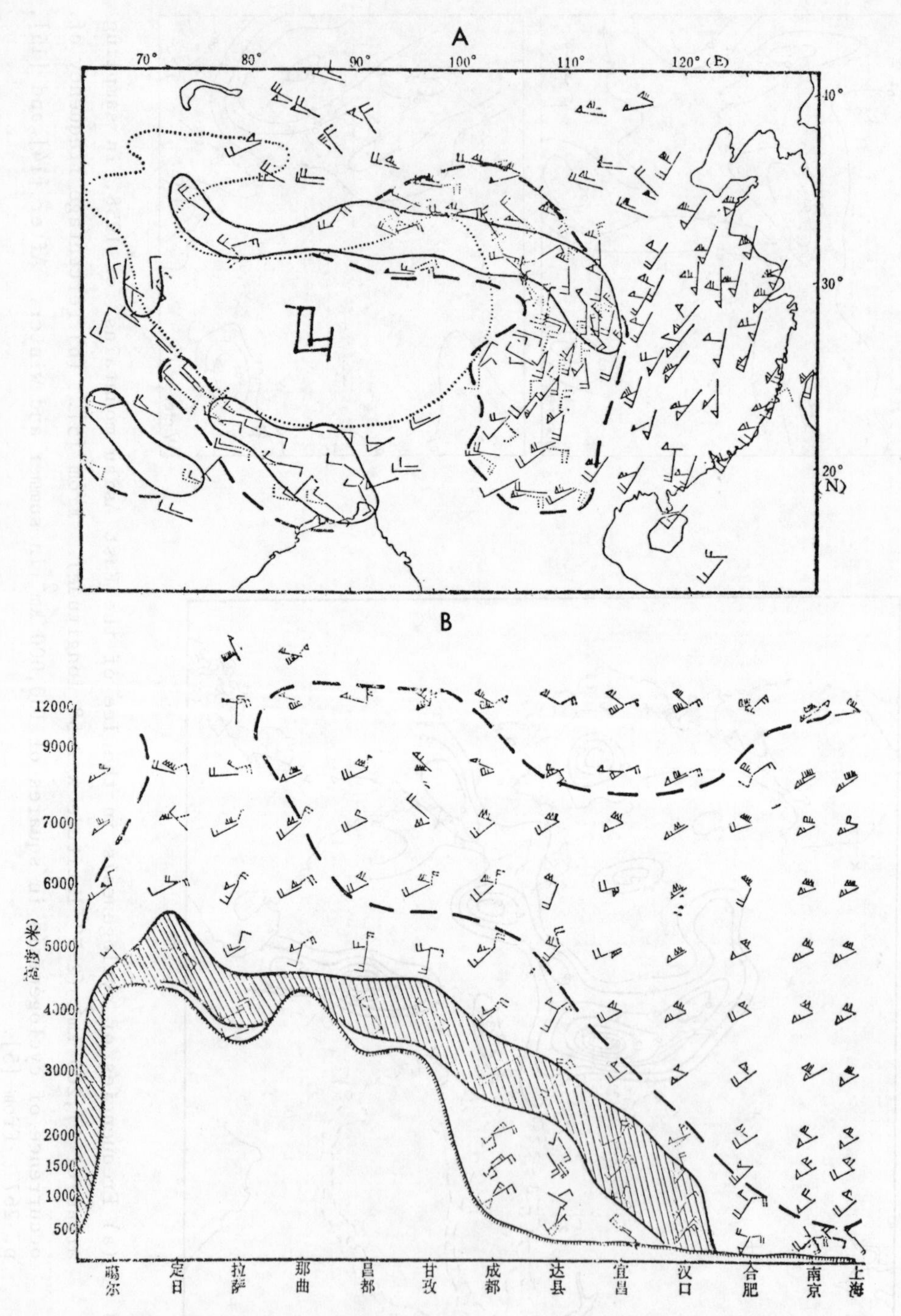

Fig. 4 Top: Horizontal distribution of primary and secondary prevailing winds at 3 km height during January and July. Bottom: Vertical distribution of primary and secondary prevailing winds along 30°N [20].

In fact, they can even have frequencies close to the inertial frequency; but we shall return to this point later.

Assume that we have a cross section of the density field as shown in Fig. 5a where the constant density lines have an angle α with the horizontal. Since any type of motion with any type of slope is possible, we will look at four different trajectories to see when the system can be unstable.

Suppose we force a particle to move in a plane, A to A'. Since the particle is heavier at A than when it is displaced to A' where the environment is lighter, it will feel a restoring force proportional to the difference in density. Consequently, the particle will return to its original position, and if the fluid is inviscid, it will overshoot the original position and oscillate. This type of motion can be called an internal gravity wave.

If the particle is displaced in the plane, B to B' where the particle density at B is heavier than at B', but the gravity force due to this difference acts perpendicular to the plane, the particle will feel no restoring force and will remain in its new position.

Now, if we consider the plane C to C', again the particle is heavier at C than at C'. However, in this case the gravity force will act to accelerate the particle past C' and conversely, a particle in C' that moves to C will feel the gravity force lifting the particle. Therefore, motion in these planes will be amplified. It is possible to see that in this case the velocity as well as the kinetic energy are continuously increasing, but since lighter fluid moves to regions of heavier fluid and vice versa, on the average the horizontal density gradient will diminish. Accordingly, the available potential energy will decrease and a clear conversion from potential energy to kinetic energy can be deduced by these unstable processes (Baroclinic Instability).

Finally, since the particles in the last plane, D to D', have the same density, they will not feel any restoring force and the plane will be neutral.

We can deduce then from this simple argument that for any motion between B and D, or along trajectory planes which have an angle less than α, the system must be unstable regardless of the frequency or the horizontal scale of the motion. The only restrictions for frequencies or scales are that the time variation of the mean density pattern, which we have described, is larger than the wave period and also that the uniformity in the horizontal scale must be larger than the wavelength of the unstable waves. On the basis of these restrictions, the most unstable slope for the waves will be given by:

$$\frac{w}{u} \approx -\frac{1}{2}\alpha$$

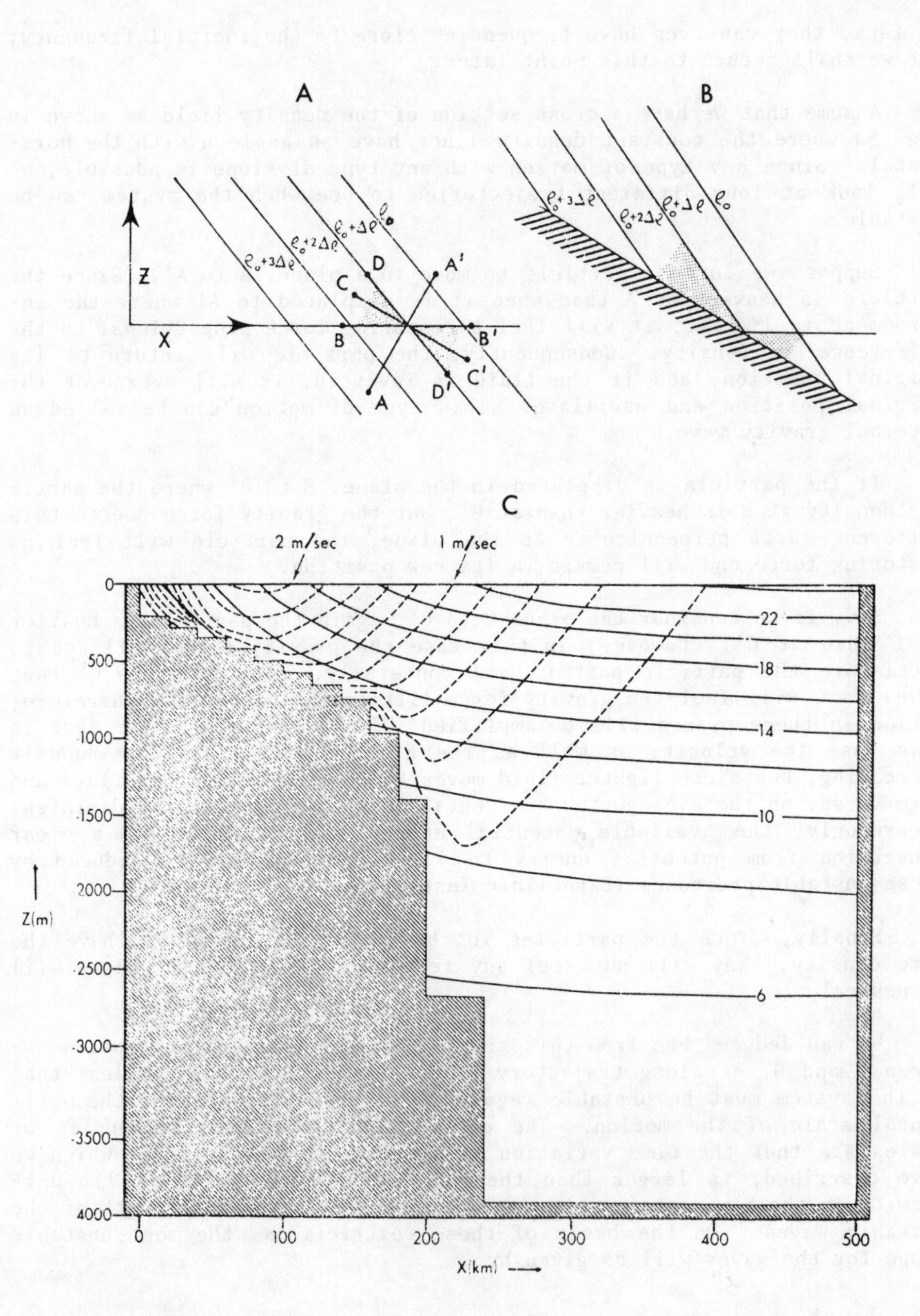

Fig. 5 (a) Different trajectories that could be unstable (shaded area) in a baroclinic fluid. (b) The bottom slope in relation to slope of the maximum instability. (c) A cross section of V and T and the bottom topography in the region considered at the initial time. From [11].

where w and u are the vertical and horizontal velocity components and α is the slope of constant density surfaces and can be written as

$$\alpha = \rho_x/\rho_z \ .$$

Charney [3] and Pedlosky [13] demonstrated that either the variation in the Coriolis parameter (β) or a slope in the topography, i.e. in the lower boundary, will reduce the angle α at which the waves can be unstable. They conclude that if these elements are included, the system will be more stable. Orlanski [10] previously discussed this type of instability for the Gulf Stream showing that, in the opposite sense, the bottom topography can destabilize the system. Furthermore, from the argument we are following, if we assume that the ground has a certain slope, as in Fig. 5b, it will seem natural to say that any type of horizontal motion will induce a vertical motion proportional to the bottom slope by the continuity equation:

$$w = \vec{v}_H \ \nabla h \approx uh_x .$$

Then, if h_x is equal to $(1/2)(\rho_x/\rho_z)$, which is the slope of maximum instability, the slope of the bottom topography will induce the waves to always have a plane that corresponds to the maximum unstable waves. The study by Orlanski and Cox [11], a numerical simulation of baroclinic instabilities in ocean currents, illustrates the effect of sloping boundaries on baroclinic instabilities. Their model was a 15-layer channel flow in which the initial conditions had a baroclinic jet flowing over the topography as shown in Fig. 5c and the solution spanned a time period of 10 days. The time sequence solutions for the east-west velocity and perturbation temperature at two depths are shown in Figs. 6 and 7. Note that after the seventh day the solution exhibits very strong eddies along the baroclinic jet that flows over the topography.

A recent study by Mechoso [9] has shed some light on the process by which the boundary slope can influence baroclinic instabilities. He based his study on two cases; one was a two-layer model and the other was the continuous, stratified Eady model. Let me first review his analysis for the two-layer model. This model contained a topography in which

$$\delta_B = \frac{S_B}{S_I}$$

where S_B is the slope of the bottom boundary and S_I is the slope of the interface. If the potential vorticities of the two layers are

$$\frac{\partial \bar{q}_2}{\partial y} = - \frac{fS_I}{H} (1-\delta_B) \text{ and } \frac{\partial q_1}{\partial y} = \frac{fS_I}{H} ,$$

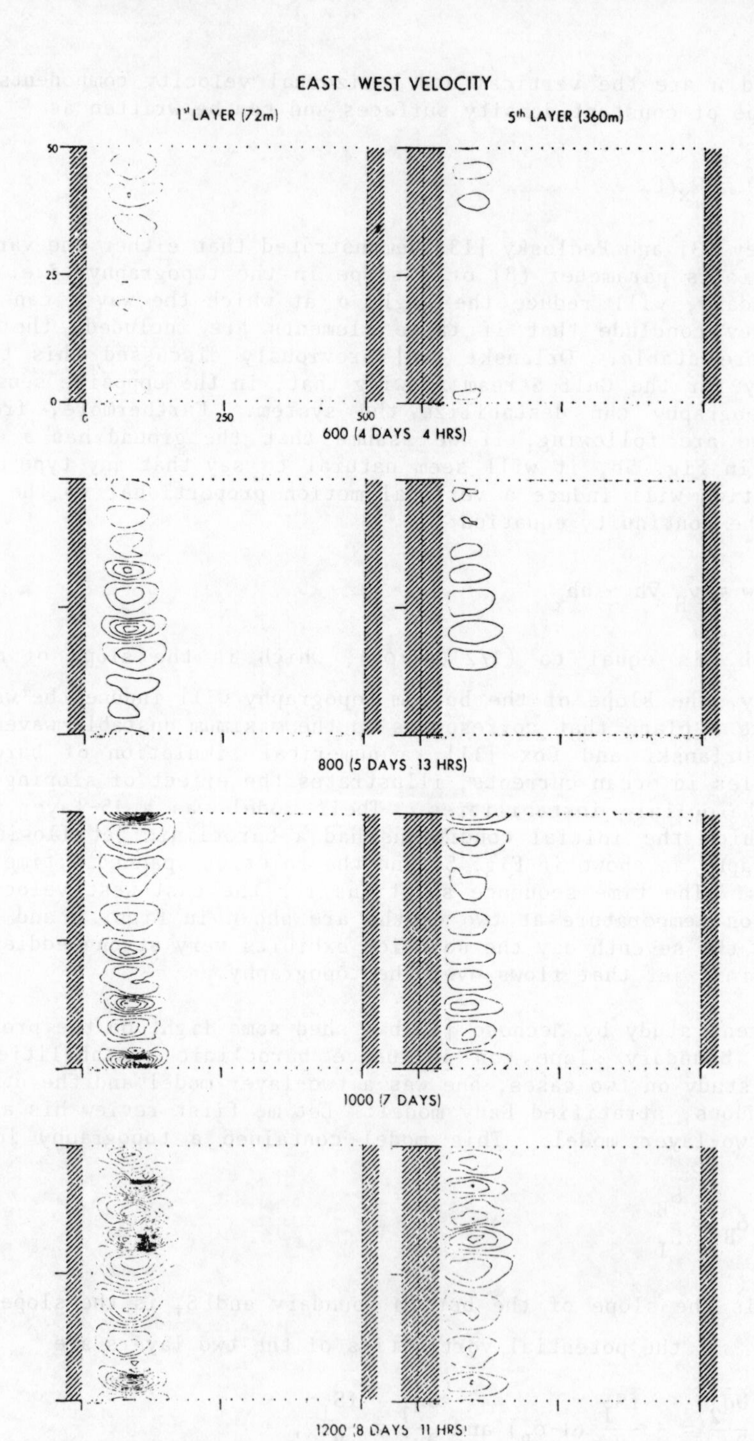

Fig. 6 The east-west velocity patterns as a function of depth (72 m and 360 m) and time integrations (600-1200). From [11].

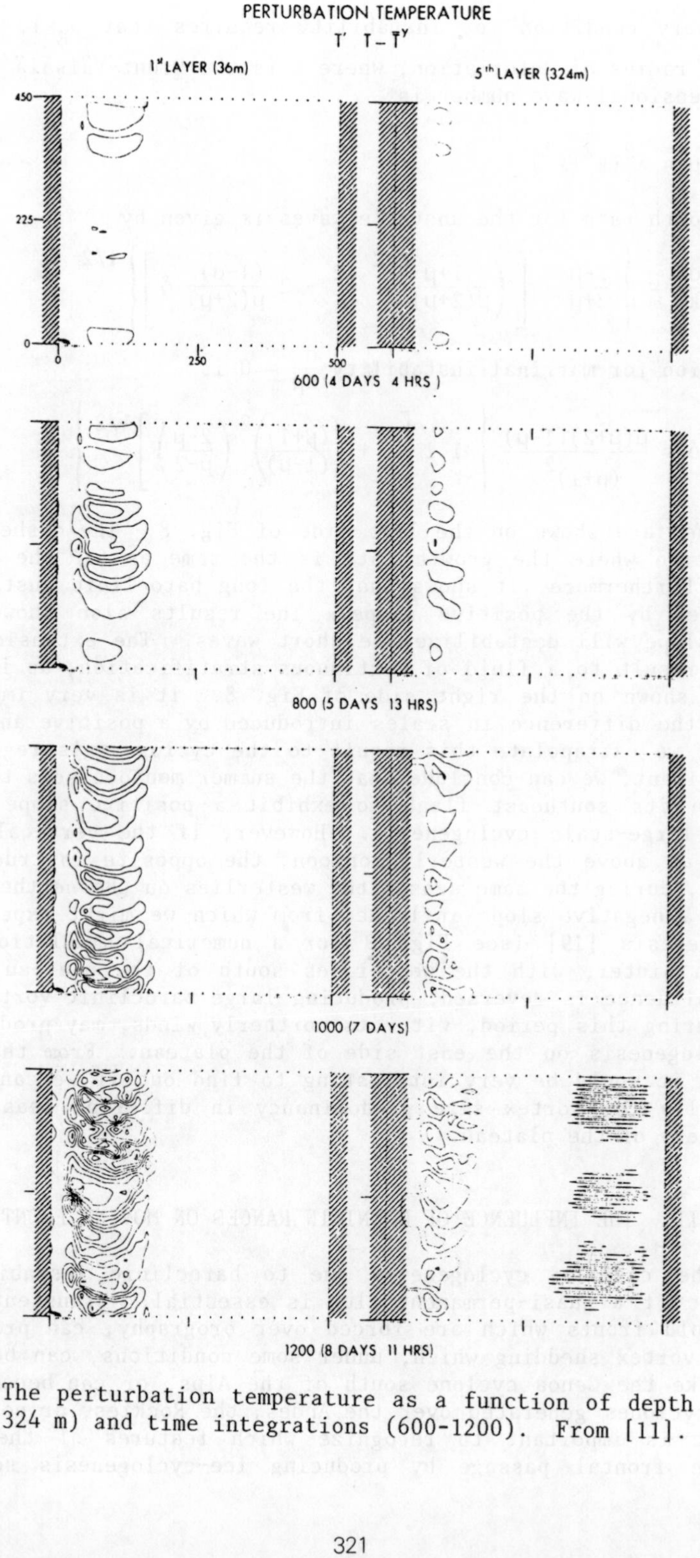

Fig. 7 The perturbation temperature as a function of depth (36 m and 324 m) and time integrations (600-1200). From [11].

the necessary condition for instability requires that $\delta_B<1$. If $\lambda = \frac{NH}{f}$ the Rossby radius of deformation, where N is the Brunt-Väisälä frequency, the nondimensional wave number is

$$\mu = \lambda^2(k^2+\ell^2)$$

and the growth rate for the unstable waves is given by

$$\frac{\sigma}{k\mu} = \left\{ \frac{2-\mu}{2+\mu} - \left[\left(\frac{1+\mu}{\mu(2+\mu)} \right)^2 \delta^2 - 2\, \frac{(1-\mu)}{\mu(2+\mu)}\, \delta \right] \right\}^{1/2}$$

The criterion for marginal instability $\frac{\sigma}{k\mu} = 0$ is

$$\delta = \frac{\mu(\mu+2)(1-\mu)}{(\mu+1)^2} \left\{ 1 \pm \left[1 + \left(\frac{(\mu+1)}{(1-\mu)} \right)^2 \left(\frac{2-\mu}{\mu-2} \right) \right]^{1/2} \right\}$$

Two branches are shown on the left side of Fig. 8. The dashed curve is the condition where the growth rate is the same as for the nonsloping terrain. Furthermore, it shows that the long baroclinic instability is destabilized by the positive slope. The results also showed that a negative slope will destabilize the short waves. The extension of this important result to a fluid of continuous stratification, as in the Eady model, is shown on the right side of Fig. 8. It is very important to recognize the difference in scales introduced by a positive and negative slope. If we extrapolate this result to the cyclogenesis region on the Asian continent, we can conclude that the summer monsoon sees the Tibetan Plateau on its southeast flank to exhibit a positive slope influence producing large-scale cyclogenesis. However, if the vertical shear is negative, as above the westerly monsoon, the opposite is true. On the other hand, during the same season the westerlies on the northern plateau will have a negative slope influence from which we might expect shorter mesocyclogenesis [19] (see Fig. 9 for a numerical simulation of July 1981). In winter, with the westerlies south of the plateau, the orographic influence is reversed, producing large baroclinic vortices. The monsoon during this period, with its northerly winds, may produce small-scale cyclogenesis on the east side of the plateau. From the previous discussion it would be very interesting to find out if we can determine statistically the vortex size predominancy in different seasons around the periphery of the plateau.

III. THE INFLUENCE OF MOUNTAIN RANGES ON MOVING FRONTS

In the case of cyclogenesis due to baroclinic instability, the requirement of a quasi-permanent flow is essential. Transient systems, such as cold fronts which are forced over orography, can produce considerable vortex shedding which, under some conditions, can become stationary like the Genoa cyclone south of the Alps, or can become migrant like the cyclones generated over the Andes, the Rockies, or northeastern China. It is important to recognize which features of the mountain affect the frontal passage by producing lee-cyclogenesis so that an

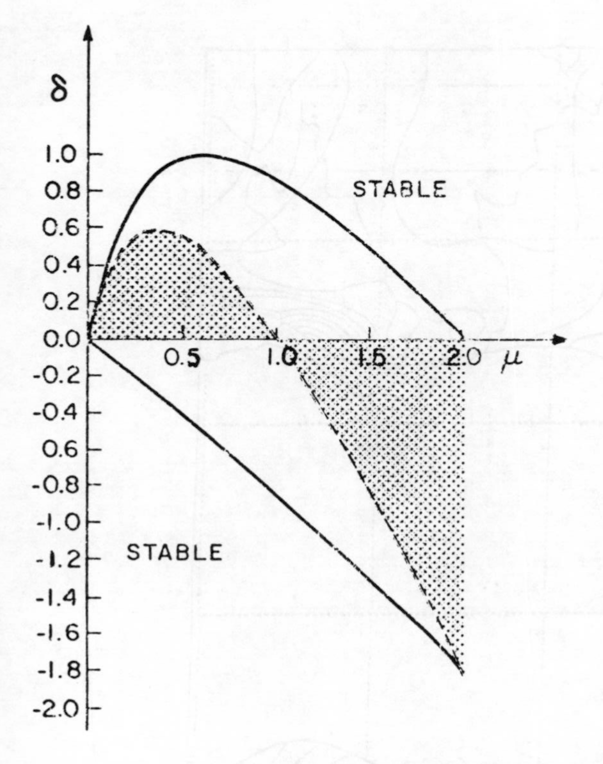

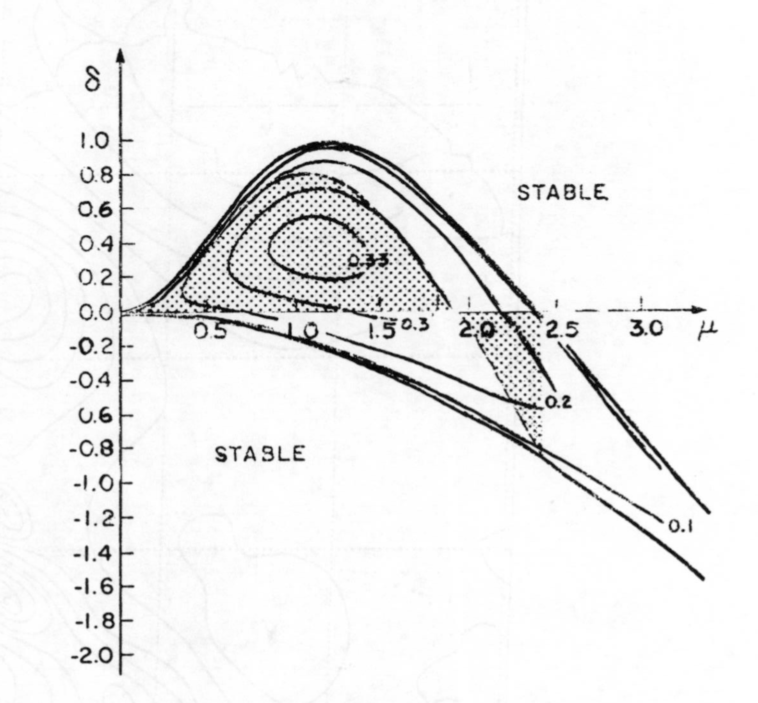

Fig. 8 (a) Criteria of marginal stability for the wave on the slope to have the same growth rate as in the horizontal bottom case. The symbols are explained in the text. (b) As in (a), except for the Eady model with sloping bottom.

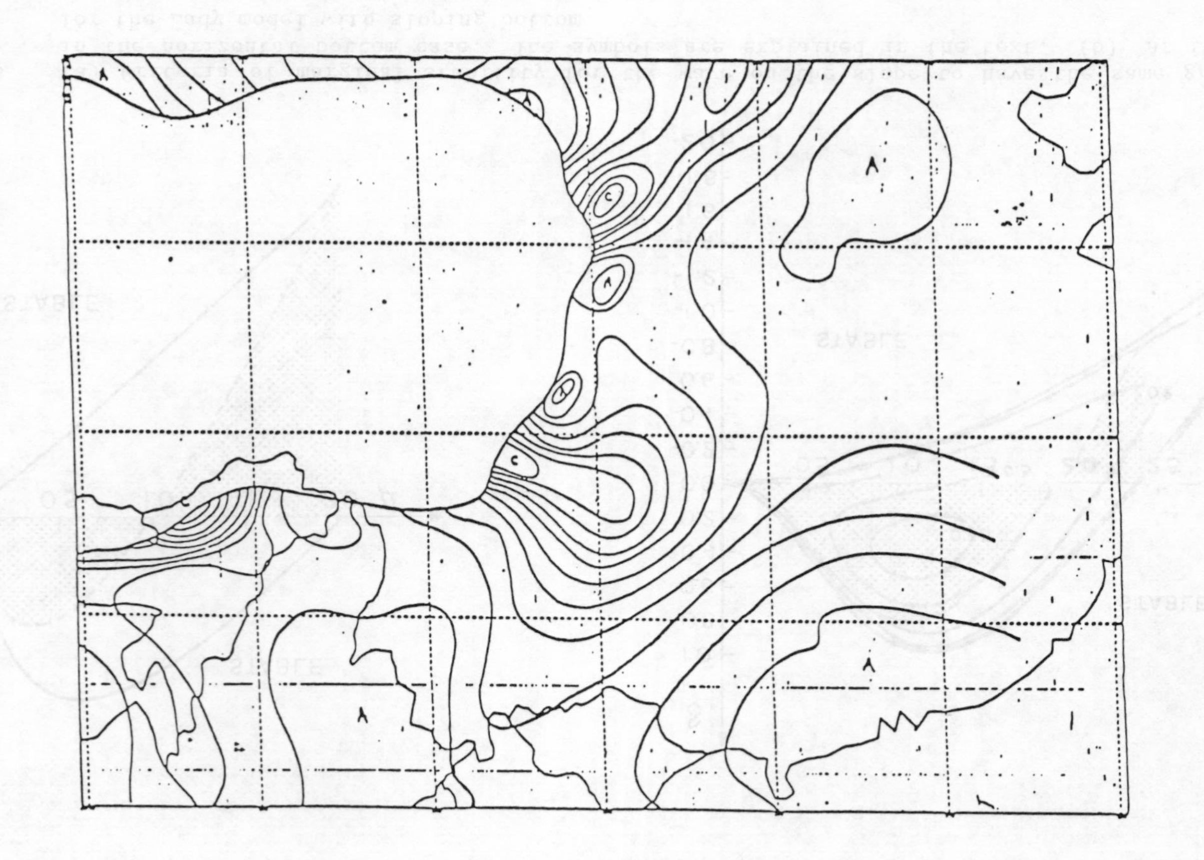

Fig. 9 Vorticity field at 700 mb after 12-hour time integration with no latent heating included; contour interval 10^{-5} sec^{-1}. Numerical experiments with the Sichuan flooding catastrophe (11-15 July 1981). After [7].

accurate forecast of these events can be made. In this context, we shall discuss how small-scale orographic features may produce large disturbances. Two basically different influences of the orography on cold fronts will be discussed:

(1) cold fronts parallel to the mountains
(2) cold fronts perpendicular to the mountains.

(1) Certainly different combinations of these configurations occur all over the globe. However, there are distinct geographic locations in which one or the other occurrence is prevalent. For instance, it is well-recognized that cold fronts moving through northern Europe, upon encountering the Alps, have their surface front parallel to the mountains (Fig. 10). A similar configuration, but perhaps more extreme, can be found on the northern mountain range (Tien Shan) of the Tibetan Plateau due to its height (∿ 6000 m as compared to ∿ 2000 m for the Alps).

A cold front has its primary geostrophic balance between the sharp surface temperature gradients and the vertical shear of the wind along the front and a smaller but not less important ageostrophic cross-stream circulation that maintains the geostrophic balance along the front. When the system moves over the mountains, the surface thermal field and cross-stream circulation is highly altered. Depending on the steepness of the orography, the frontal jet may not suffer any large distortion in crossing the barrier since it is usually located at higher altitudes. In that case a large ageostrophic imbalance between the thermal wind and horizontal temperature gradient is produced, with the possible development of cyclogenesis.

(2) The orthogonal situation is, 2, the case of fronts moving perpendicular to the orography. Typical examples of these weather patterns can be found in many locations around the world such as the Rocky Mountains, the Eastern Asian mountains, and the Andes in South America.

The Andes are a formidable barrier with an average height of 3000 m in Patagonia (south of Argentina) and up to 6000 m in the northern part of Argentina, and lie perpendicular to the westerlies. For that reason they are a strong source of lee-cyclogenesis as was discussed by Y.S. Chung [4] (Fig. 11) and more recently by Mechoso [9]. It should be pointed out, however, that a considerable amount of cyclogenesis is also produced by cold fronts as they move north from the periphery of Antarctica and encounter the mountain ranges.

The present case study for a typical spring season (May 7 - May 8, 1977) developed when a cold front sweeping from Antarctica spilled over the South American continent. The surface maps in Fig. 12 are shown for a 12-hour sequence from May 7, 00Z through May 8, 12Z. We can observe a cold front in the southern part of South America which is clearly seen through 12Z of May 7. Note also the division produced by the Andes mountain barrier. The front moves northward until 00Z of May 8 when it is located at 35°S. Twelve hours later we observe strong cyclogenesis. A complementary pattern can be seen on the 700-mb charts of Fig. 13. Note the low on the east coast of Argentina at 00Z on May 8. On May 7, 00Z the 250-mb chart clearly shows the orographically induced disturbance in the westerlies (Fig. 14). At 12Z the low is east of the Andes and on

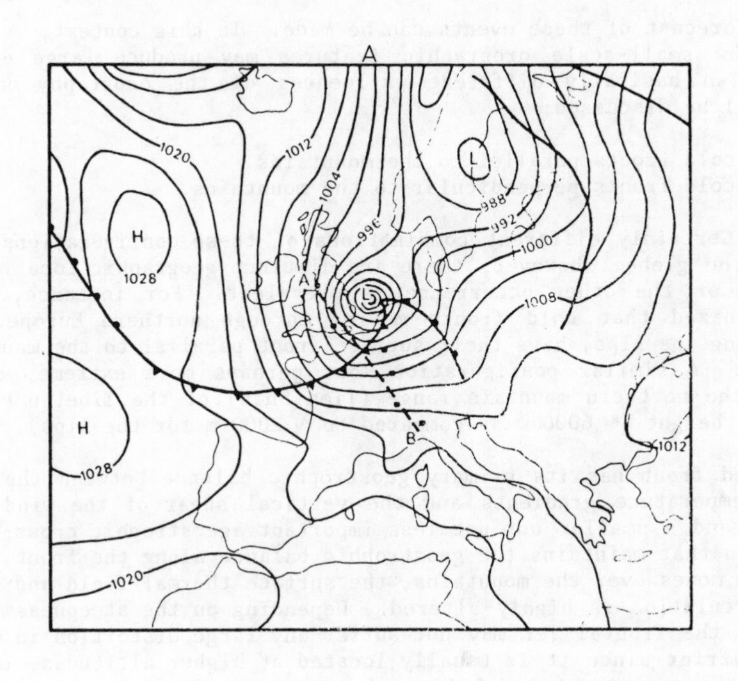

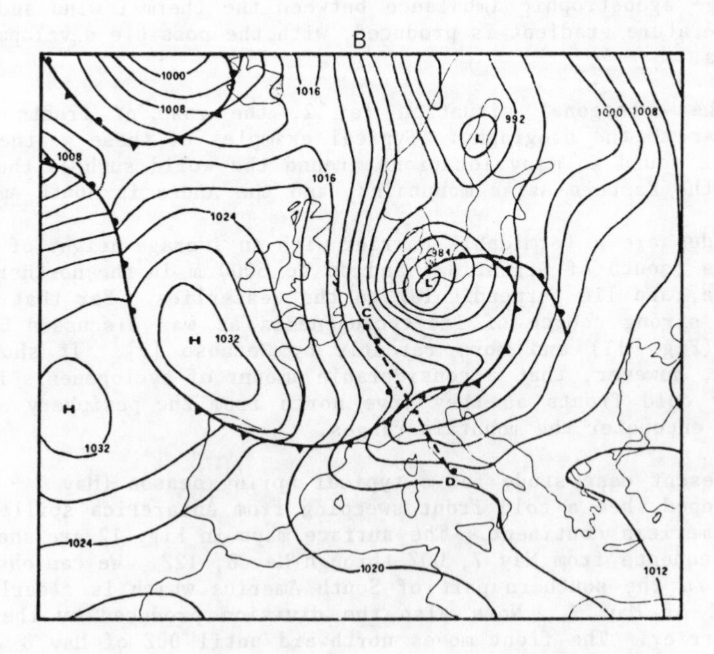

Fig. 10 (a) MSLP, 2 April 1973, 1200 GMT. (b) MSLP, 3 April 1973, 0000 GMT. (c) MSLP, 3 April 1973, 1200 GMT. (d) MSLP, 1 April 1973, 0000 GMT. After [2].

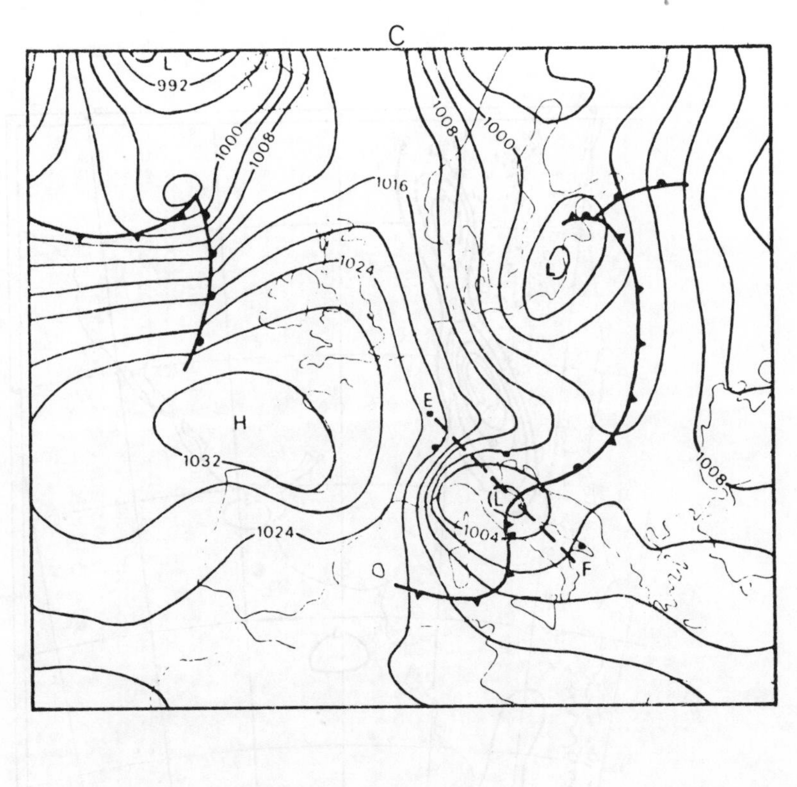

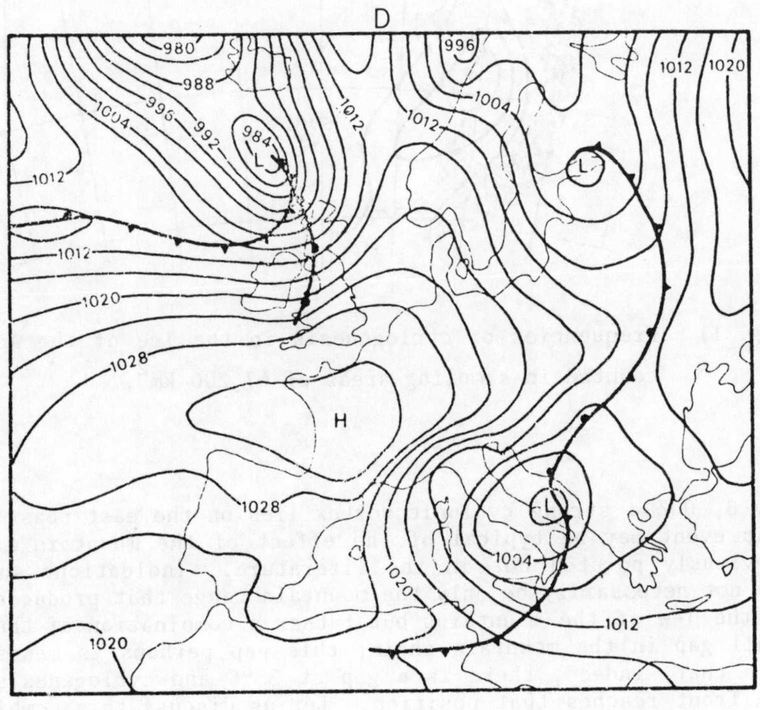

Fig. 10 (Continued)

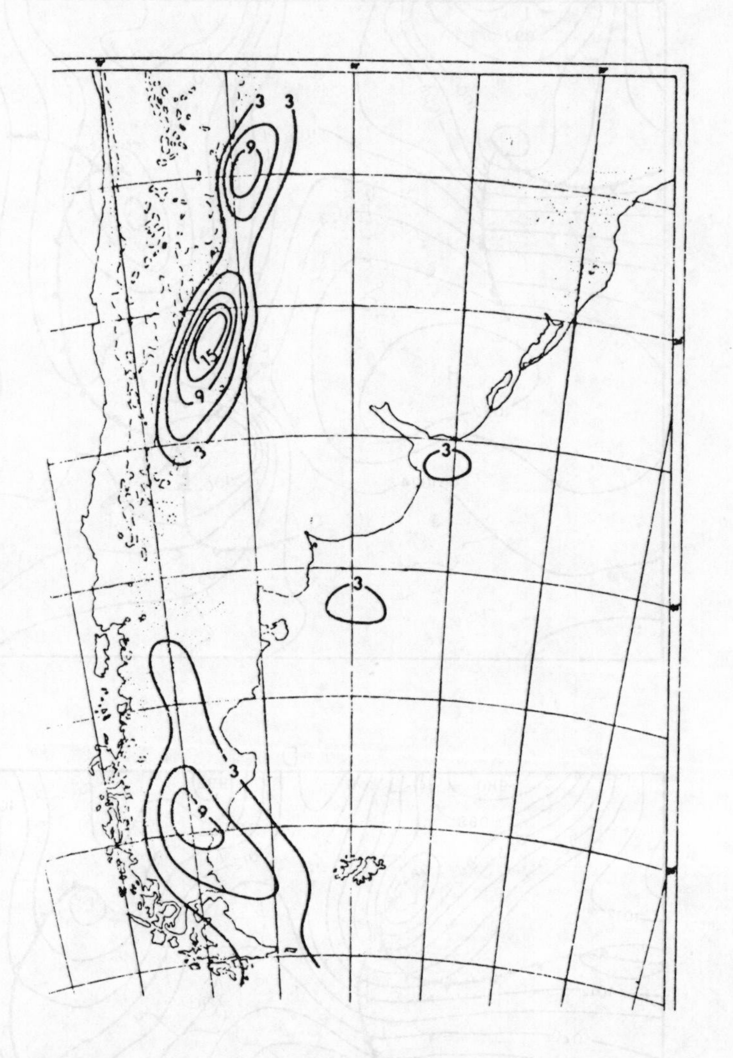

Fig. 11 Frequencies of cyclogenesis in the lee of the Andes for 1973, counted in sampling areas of 67,200 km^2.

May 8, 00Z a strong cyclonic vortex lies on the east coast of Argentina. This event may be typical of the effect of the mountain disturbance not previously pointed out in the literature. Indications suggest that it may not necessarily be only the mountain range that produces cyclogenesis on the lee of the mountains but rather a combination of the latter and a small gap in the mountain chain; this gap perhaps is mesoscale in size. Note that, indeed, there is a gap at 35°S and cyclogenesis occurs after the front reaches that position. Let us discuss this combined effect of the planetary scale disturbance and a mesoscale circulation.

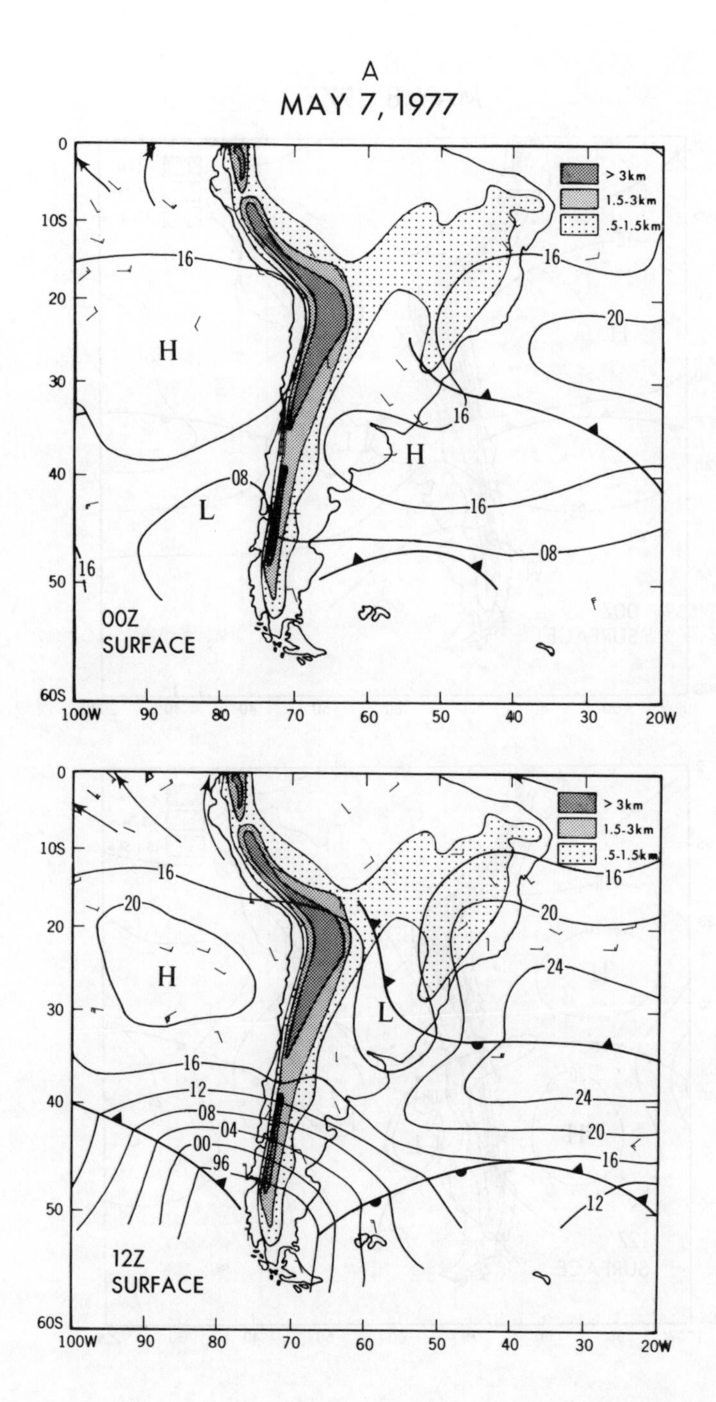

Fig. 12 (a) Surface weather maps at 12-hour intervals for May 7, 1977 showing the advance of a cold front in the southern part of South America and the orographic effects. (b) As in Fig. 12a, but for May 8, 1977.

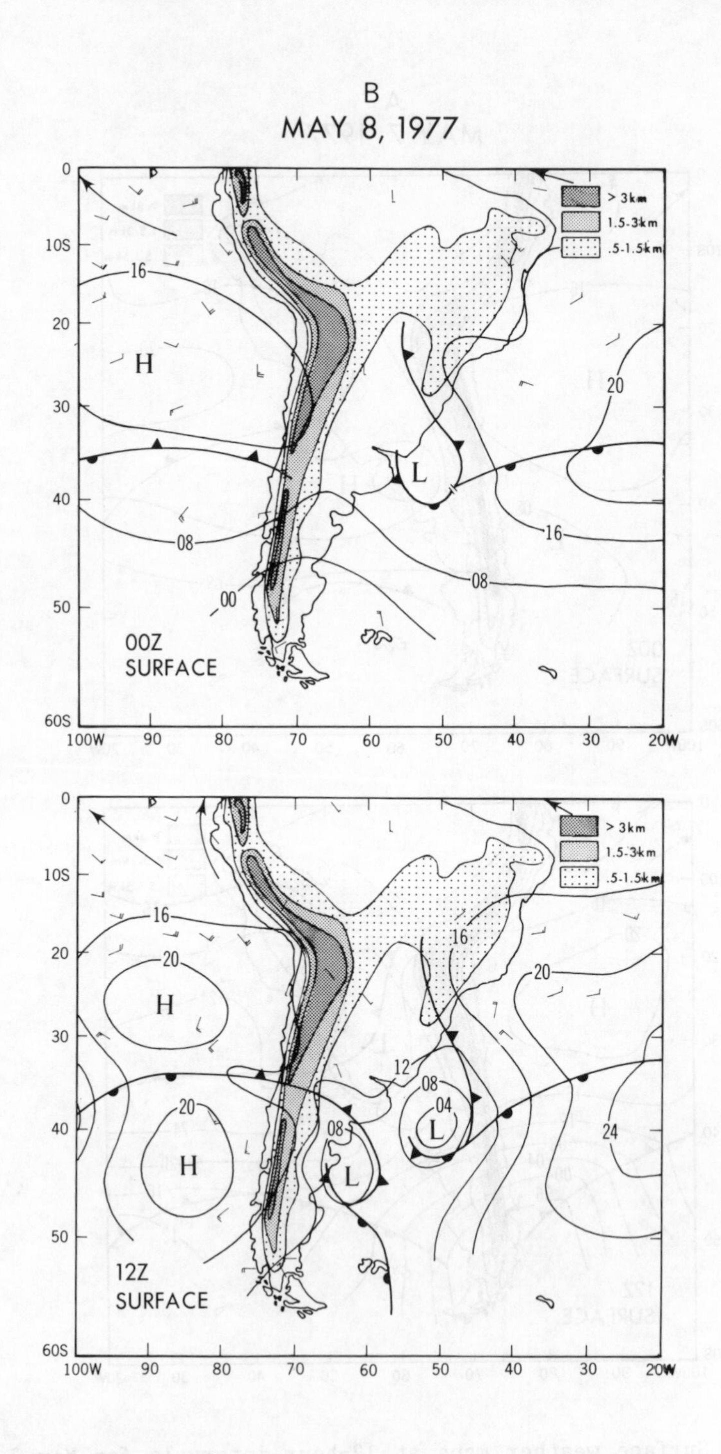

Fig. 12 (Continued)

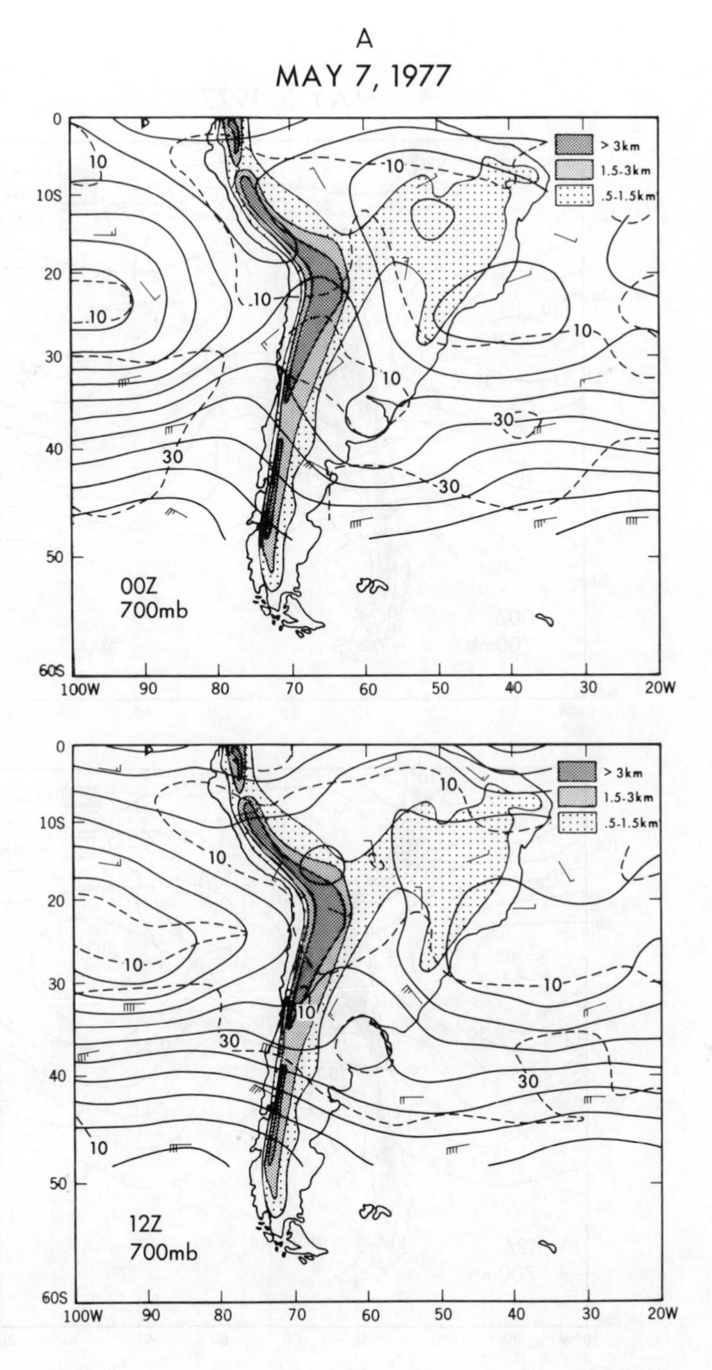

Fig. 13 (a) 700 mb charts for the period described in Fig. 12 showing the upper air cyclogenesis patterns associated with the frontal system and orography. (b) As in Fig. 13a, but for May 8, 1977.

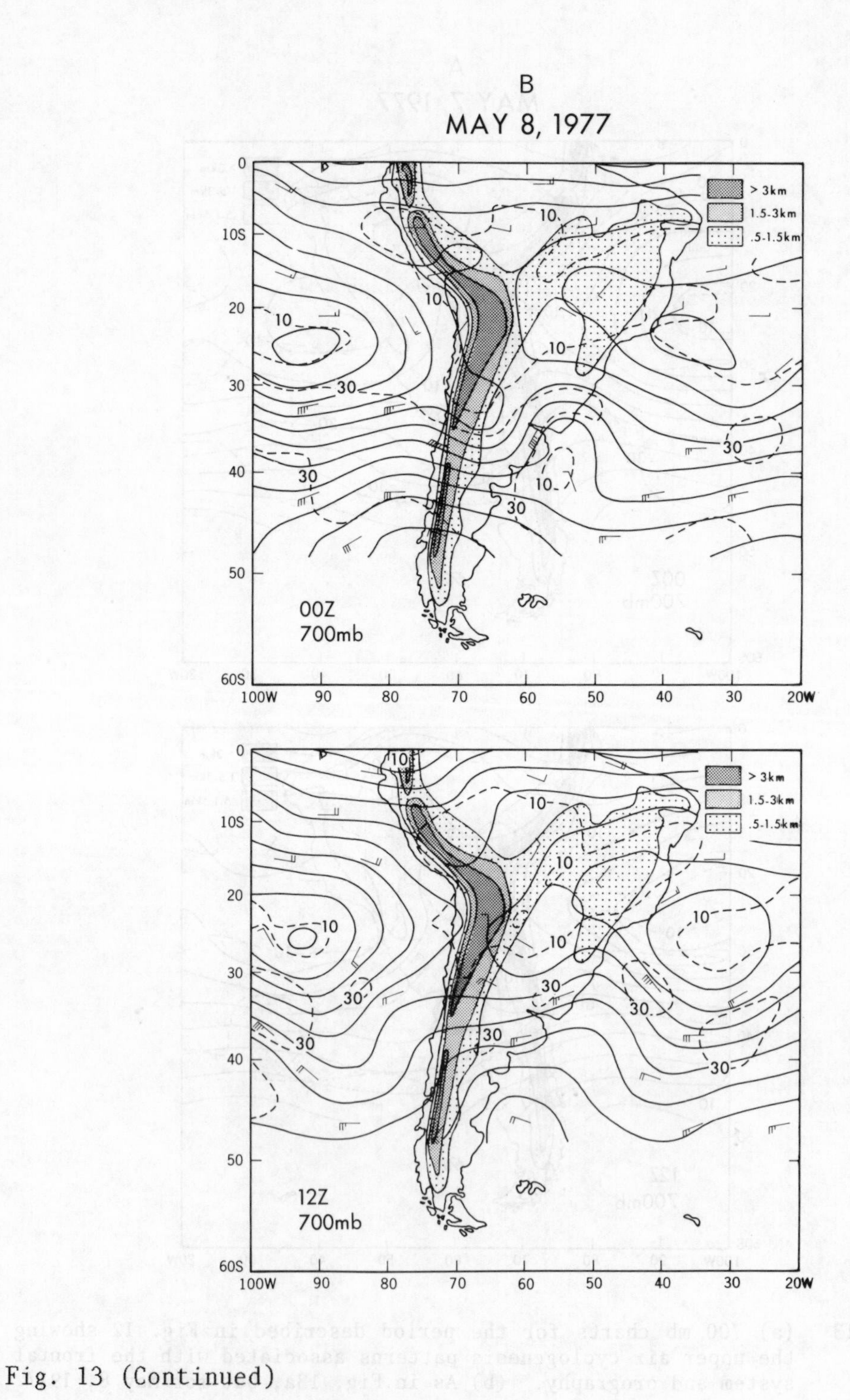

Fig. 13 (Continued)

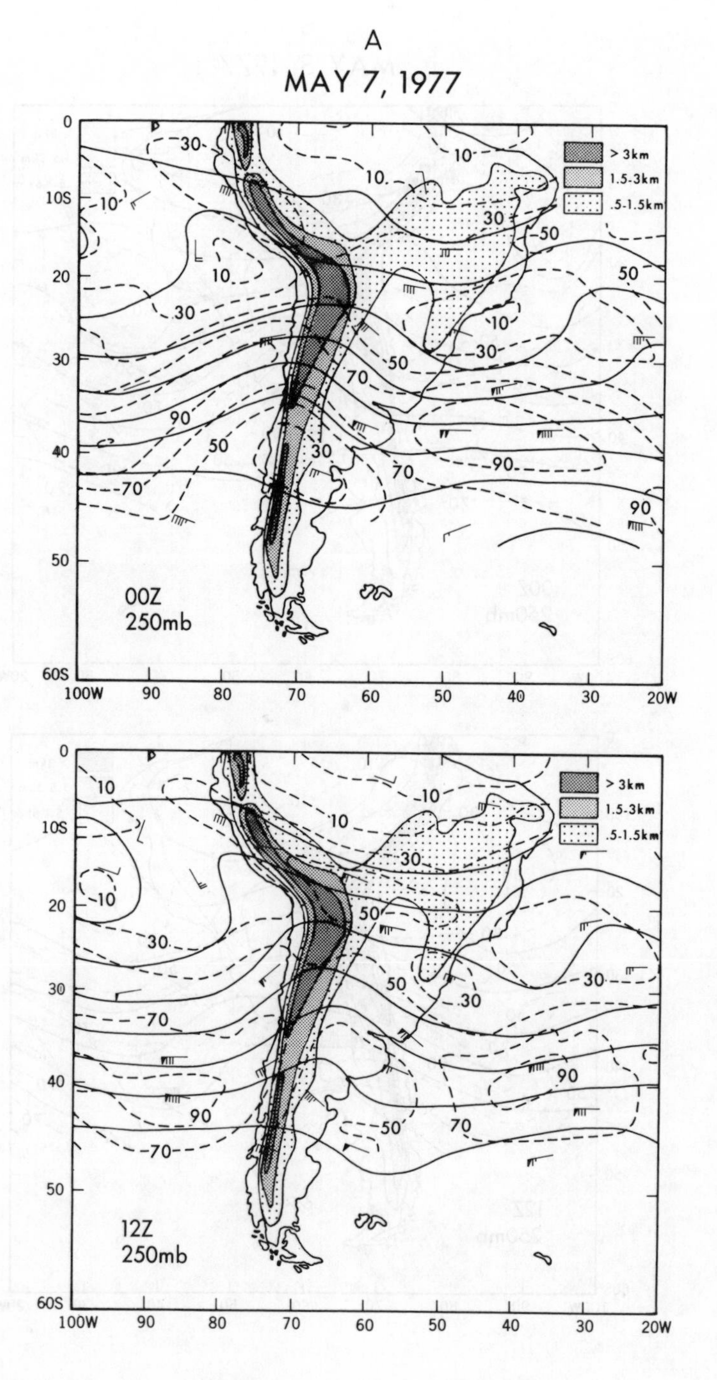

Fig. 14 (a) 250 mb charts shown for the same area and time as in Fig. 12. (b) As in Fig. 14a, but for May 8, 1977.

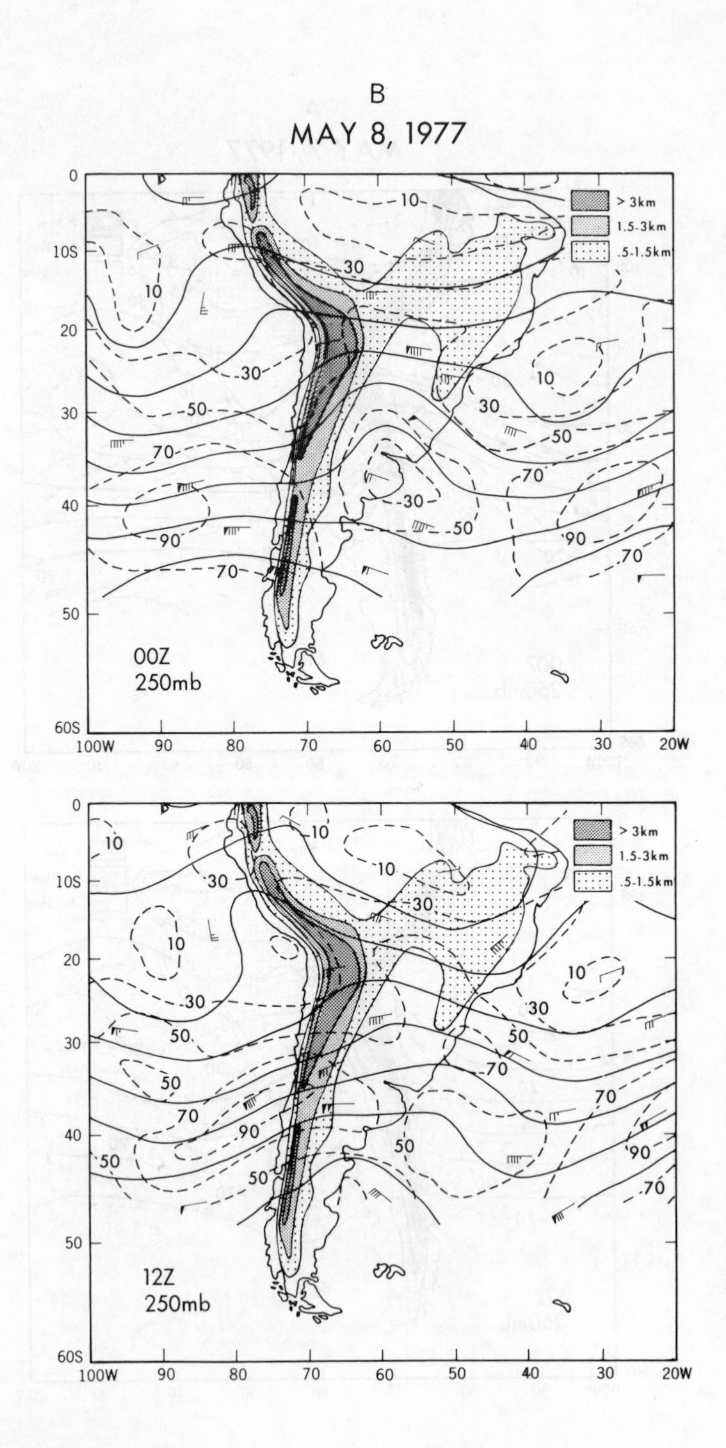

Fig. 14 (Continued)

IV. THE CORIOLIS PUMPING

It is well known that a westerly flow over an obstacle will produce anticyclonic vorticity on the top of the mountain, but due to the differential earth rotation or stratification (β effect) cyclonic vorticity will be generated on the lee side of the mountain. As shown by H. Huppert and K. Bryan [8] temporally varying currents over bottom topography can redistribute vorticity and temperature in such a way that a relatively cold fluid with anticyclonic vorticity exists over the topographic feature while fluid shed from above the topographic feature sinks, thereby inducing a warm anomaly with cyclonic vorticity. They found that for sufficiently strong oncoming flows, the shed fluid continually drifts downstream in the form of a relatively warm eddy. If the oncoming flow is relatively weak, however, the interaction between the anticyclone and cyclonic vorticity distributions traps the warm eddy and it remains in the vicinity of the topographic feature (see Fig. 15).

In fact, the low pressure system in the lee of the mountain barrier can be produced by many different synoptic scale disturbances like the passage of a frontal system as previously discussed. We can see that the flow at 250 mb (Fig. 14) shows cyclonic tendencies on the lee side of the

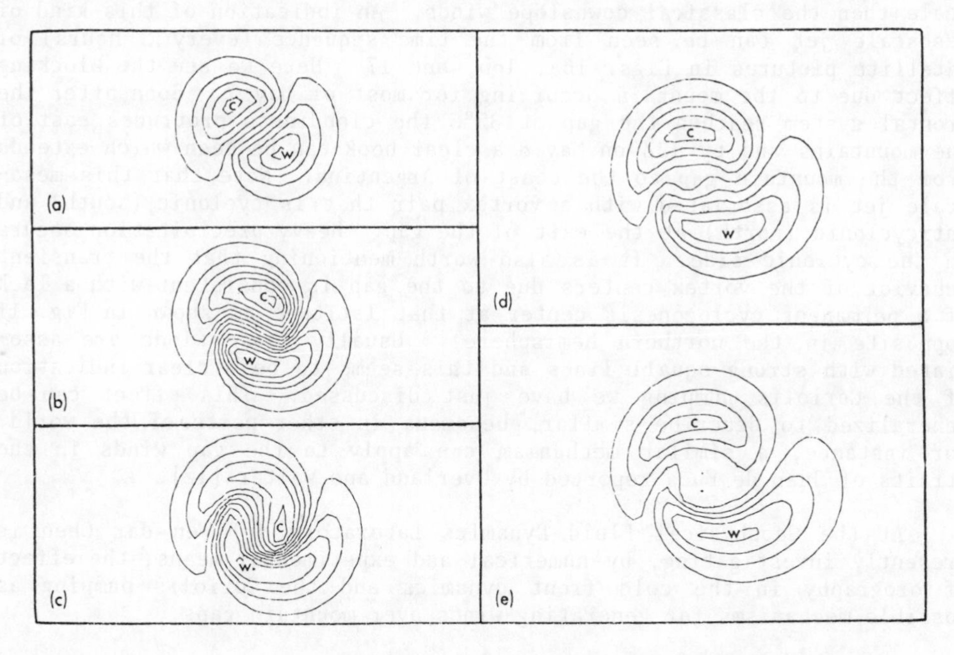

Fig. 15 The isopycnals with a contour interval of $\Delta\rho/\rho_o = 4 \times 10^{-6}$ at a depth of 3720 m for $U_o = 1.0$ cm s^{-1} and $h_m = 20$ m. (a) 4.6 days, (b) 9.3 days, (c) 13.9 days, (d) 23.1 days and (e) 34.7 days. From [8].

mountain. This vorticity field is associated by geostrophic balance with a high-pressure field on the upstream side of the mountain and a low-pressure field on the downstream side of the mountain. Some indication of this effect can be seen on the surface maps as well. Since this pressure is maintained by the large-scale flow, it seems plausible that a mesoscale gap in a mountain barrier would not alter this balance very much. If so, an ageostrophic flow might be set up across the gap due to the differential pressure of the large-scale flow. From a scaling argument one can see that only a 1-mb pressure difference over the length of the gap, say 100 km, might produce a crosswind,

$$u = \left(\frac{\Delta p}{\rho}\right)^{1/2} ,$$

$$u = 10 \text{ m/s} .$$

It seems that a value of 1 mb is highly conservative for the expected difference in pressure due to the mountain range. In fact, an effect of this sort under more severe conditions could produce strong mesoscale wind patterns. I should point out that this is a completely different phenomenon than that described in the classical downslope wind theories which are more in the category of finite standing internal gravity wave patterns. The mesoscale winds described here will operate on a larger scale than the classical downslope winds. An indication of this kind of mesoscale jet can be seen from the time sequence (every 3 hours) of satellite pictures in Figs. 16a, 16b, and 17. Here we see the blocking effect due to the mountain occurring for most of May 7. Soon after the frontal system reaches the gap at 35°S the cloudiness protrudes east of the mountains and by 12Z on May 8 a clear hook can be seen which extends from the mountain gap to the coast of Argentina. Note that this mesoscale jet is associated with a vortex pair that is cyclonic (south) and anticyclonic (north) at the exit of the gap. Heavy precipitation occurs in the cyclonic side. It is also worth mentioning that the transient behavior of the vortex centers due to the gap is consistent with a lack of a permanent cyclogenesis center at that latitude as shown in Fig. 11 (opposite in the northern hemisphere). Usually those winds are associated with strong squall lines and this seems to be a clear indication of the Coriolis pumping we have just discussed. This effect can be generalized to describe similar phenomena in other parts of the world. For instance, a similar mechanism can apply to the gap winds in the Straits of Juan de Fuca reported by Overland and Walter [12].

At the Geophysical Fluid Dynamics Laboratory Mr. Wen-dar Chen is presently investigating, by numerical and experimental means, the effect of orography in the cold front dynamics and the Coriolis pumping as possible mechanisms for generating winds over mountain gaps.

A

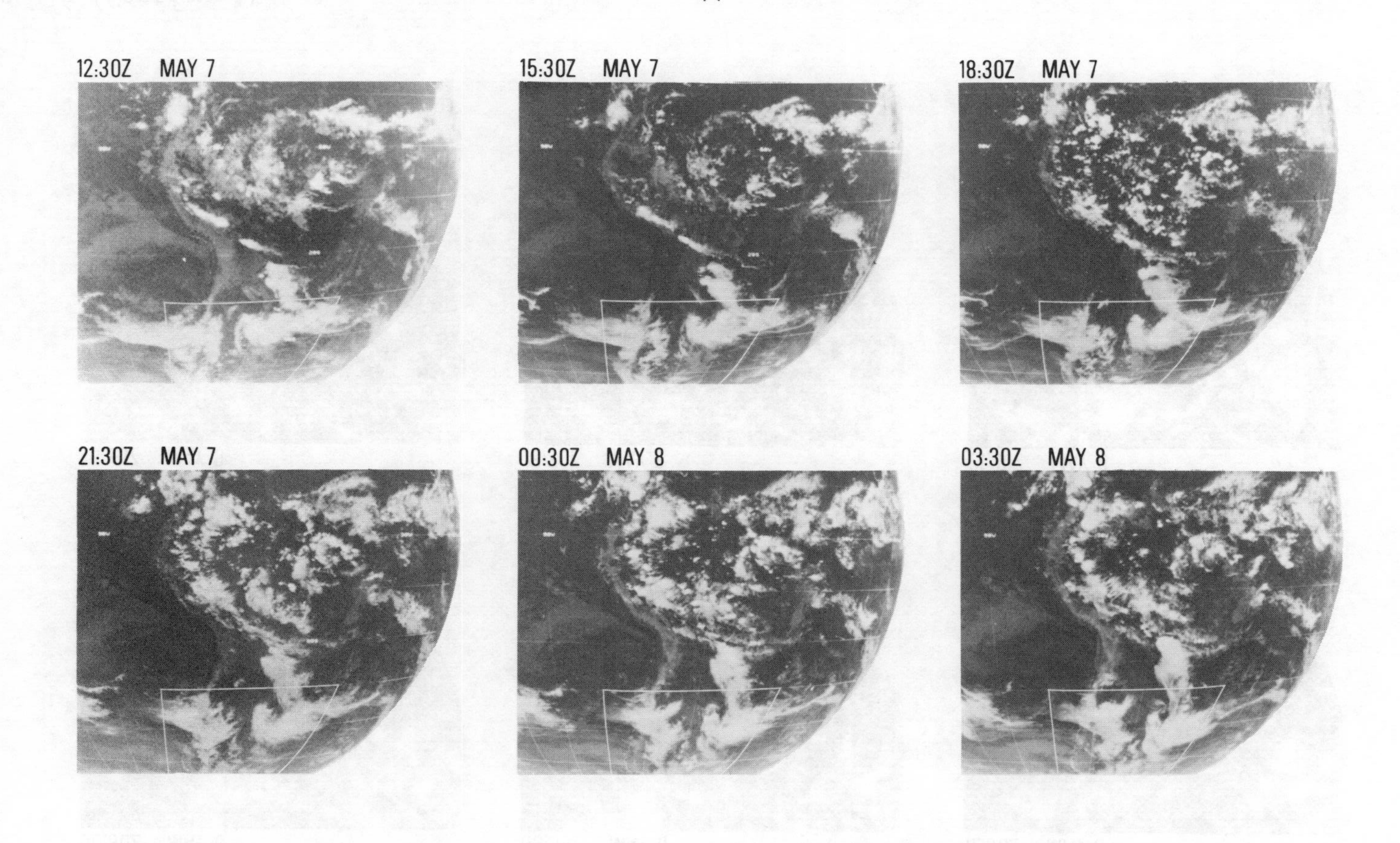

Fig. 16 (a) Satellite pictures every three hours on May 7 and 8 showing the mesoscale jet.
(b) Satellite pictures every three hours on May 8 showing the mesoscale jet.

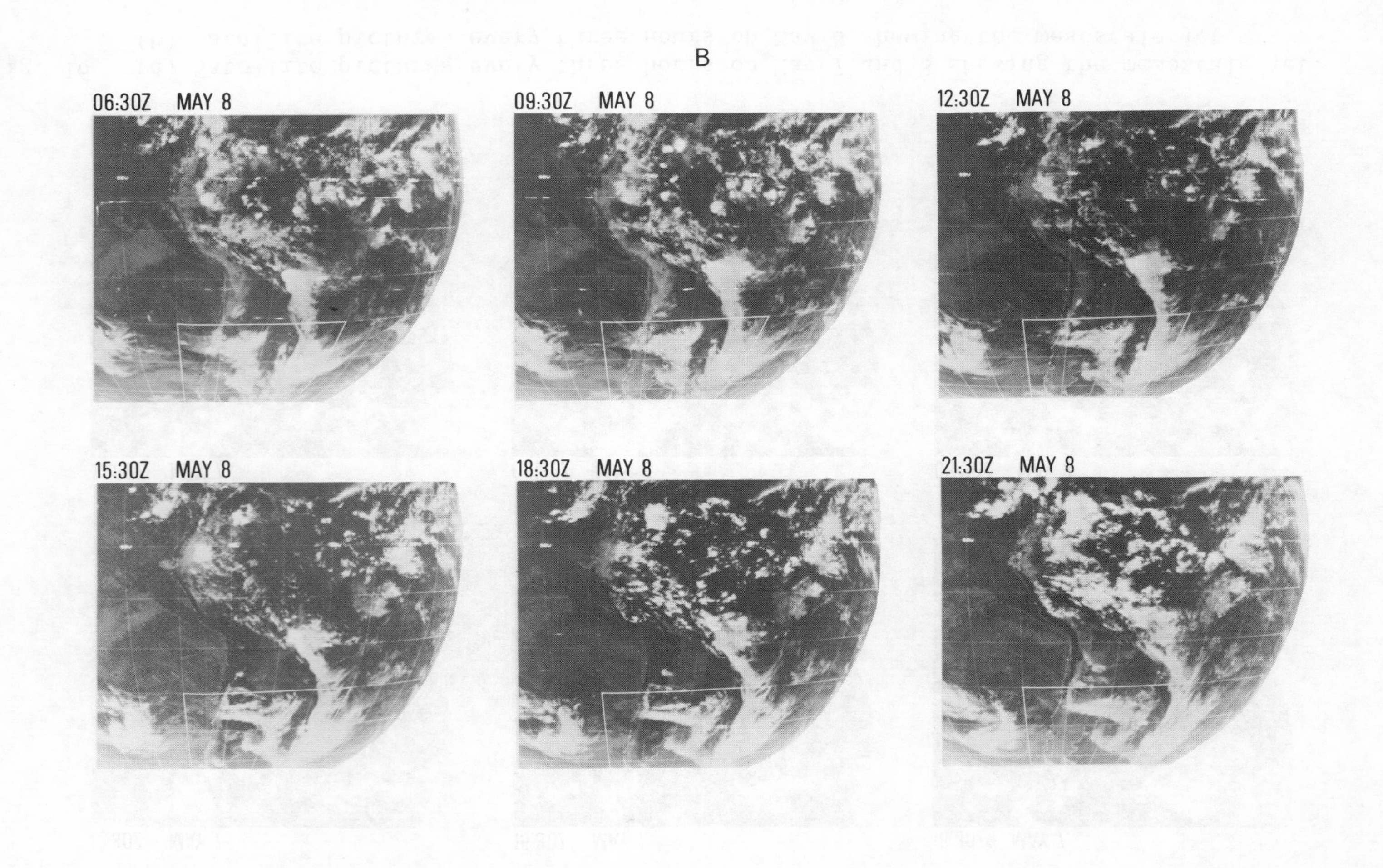

Fig. 16 (Continued)

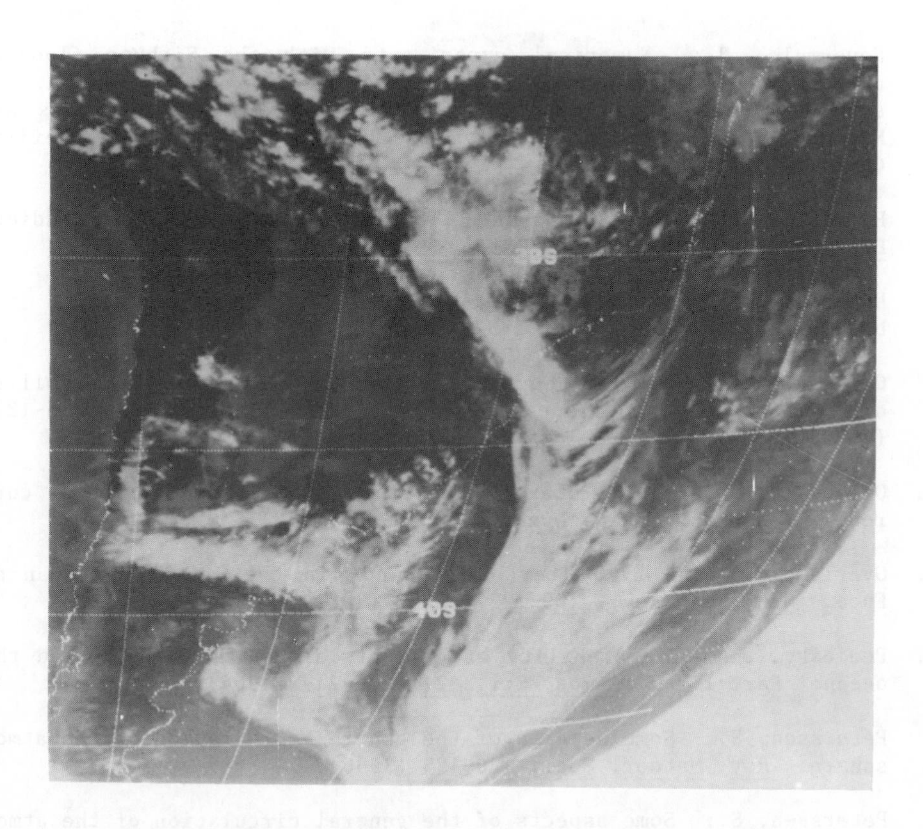

Fig. 17 An enlarged satellite photo showing the mesoscale jet.

REFERENCES

1. Blumsack, S. and P.J. Gierasch: Mars: the effects of topography on baroclinic stability. J. Atmos. Sci., 29, 1081-1089 (1972).

2. Buzzi, A. and S. Tibaldi: Cyclogenesis in the lee of the Alps: A case study. Quart. J. R. Met. Soc., 104, 271-287 (1978).

3. Charney, J.C.: The dynamics of long waves in a baroclinic westerly current. J. Meteor., 4, 135-162 (1947).

4. Chung, Y.S.: On the orographic influence and lee cyclogenesis in the Andes, the Rockies and the East Asian Mountains. Arch. Met. Geoph. Biokl. Ser. A, 26, 1-12 (1977).

5. Chung, Y.-S., K.D. Hage and E.R. Reinelt: On lee cyclogenesis and airflow in the Canadian Rocky Mountains and the East Asian Mountains. Mon. Wea. Rev., 104, 879-891 (1976).

6. G.A.R.P. Report: Orographic effects in planetary flows. Vol. 23 (1980).

7. Hovermale, J.: Numerical experiments with the Sichuan flooding catastrophe (11-15 July 1981). Proceedings, Joint U.S.-China Workshop on Mountain Meteorology of Chinese Academy of Sciences and National Academy of Sciences of the U.S., 18-23 May 1982, Beijing China (1982).

8. Huppert, H.E. and K. Bryan: Topographically generated eddies. Deep-Sea Res., 23, 655-679 (1976).

9. Mechoso, C.R.: Baroclinic instability of flows along sloping boundaries. J. Atmos. Sci., 37, 1393-1399 (1980).

10. Orlanski, I.: The influence of bottom topography on the stability of jets in a baroclinic fluid. J. Atmos. Sci., 26, 1216-1232 (1969).

11. Orlanski, I. and M.D. Cox: Baroclinic instability in ocean currents. Geophys. Fluid. Dyn., 4, 297-332 (1973).

12. Overland, J. and B. Walter: Gap winds in the Straits of Juan de Fuca. Mon. Wea. Rev., 109, 2224-2233 (1981).

13. Pedlosky, J.: The stability of currents in the atmosphere and the ocean: Part I. J. Atmos. Sci., 21, 201-219 (1964).

14. Peterssen, S.: Some aspects of the general circulation of the atmosphere. Roy. Meteor. Soc., 120-155 (1950).

15. Peterssen, S.: Some aspects of the general circulation of the atmosphere. Vol. I, 2nd Ed., McGraw Hill, Chaps. 3, 12, 13 and 16 (1956).

16. Smagorinsky, J.: General circulation experiments with the primitive equations. Mon. Wea. Rev., 91, 99-164 (1963).

17. Streten, N.A. and A.J. Troup: A synoptic climatology of satellite observed cloud vortices over the southern hemisphere. Quart. J. R. Met. Soc., 99, 56-72 (1973).

18. Washington, W.M., B. Otto-Beisner, and G. Williamsen: January and July simulation experiments with 2.5° latitude-longitude version of the NCAR general circulation model. NCAR Technical Note. NCAR/TN-123 + STR, National Center for Atmospheric Research, Boulder, CO, Vol. 1 (text) 39 pp, Vol. 2 (figures) 61 pp (1977).

19. White, G.H.: An observational study of the northern hemsiphere extratropical summertime general circulation. J. Atmos. Sci., 39, 24-40 (1982).

20. Yeh, T.C.: Meteorology of the Tibetan Plateau. Peking, Science Press (1979).

3.5 NUMERICAL SIMULATION OF DEVELOPMENT OF MONSOON CIRCULATION IN JULY

H.L. Kuo
Department of Geophysical Sciences
The University of Chicago
Chicago, IL 60637 U.S.A.

Qian Yongfu
Lanzhou Institute of
Plateau Atmospheric Physics
The People's Republic of China

ABSTRACT

Using numerical experiments we assessed the relative importance of various physical processes, such as radiative warming and cooling, especially diurnal variation of solar radiation, large-scale and deep cumulus condensation, thermal and pure dynamic effects of orography, and the initial states, involved in the development of the monsoon circulation in summer. The model used for the monsoon simulations is a limited-domain, primitive-equation numerical model with five layers in the atmosphere and one layer below the underlying surface. The details of the model can be found in a previous paper [1].

Eight different experiments were carried out with integration times running from 8 to 20 days. The land-sea contrast and the large-scale condensation are included in all of the experiments. We utilized two different initial states which corresponded to the June monthly mean and the June monthly zonal mean fields of the meteorological elements. We compared the results from different numerical simulations with one another and also compared model output with the observed July monthly mean field. These comparisons have yielded some interesting conclusions. These are:

(1) The mean sea-level pressure distribution and low-level flow pattern produced by these simulations are found to be determined mainly by the diabatic heating distribution and are only slightly influenced by orography, but are almost independent of the initial states.

(2) The low-pressure systems develop faster when diurnal variation of solar radiation is taken into consideration.

(3) The precipitation distribution is found to be critically influenced by radiative heating, orography, and initial conditions. For example, without orography, maximum precipitation occurs in the coastal region of East Asia, instead of the region around 100°E, 25°N as with

orography. The precipitation rate is greatly reduced, both over land and over ocean, without radiative heating.

I. INTRODUCTION

It is commonly believed that various monsoon circulations are attributable to the different thermal properties of land and sea, but the presence of large mountains and plateaus over some continents certainly adds another element to the problem, not only from their barrier effect but also from the large differential heating between elevated surfaces of the plateaus and mountains and the surrounding free atmosphere. These effects give rise to large diurnal temperature changes and daytime convection over the plateaus, thereby also influencing the circulation and precipitation in far away regions, as has been demonstrated by our previous work on the influence of the Tibetan Plateau on climate and weather [1]. Here we would like to assess the relative importance of various physical processes involved in the development of the monsoon circulations in summer. These processes include radiative warming and cooling and related sensible heating, diurnal variation of solar radiation, large-scale and deep cumulus condensation, thermal and pure dynamic effects of orography, land and sea distribution, and the initial states. The method of assessment is to compare results from different numerical experiments with observed flow conditions. Since the prominent monsoon areas are located between latitudes 25°S and 35°N and longitudes 15°W and 170°E, our numerical model covers the domain between 0 to 180°E and 25°N to 55°N. The model which we used for the monsoon simulations is a primitive-equation numerical model with five layers in the atmosphere and one layer below the underlying surface. In the vertical it uses a combined pressure-sigma coordinate system. The details of the model can be found in a previous paper [1]. Here we give only the schematic illustration of the vertical structure of the model in Fig. 1. Figure 2 is the smoothed topography used in our model with a unit of 1 km. The time step and the grid size are 15 min and 5 degrees in both latitude and longitude, respectively. The time integration scheme is the so-called combined Euler's backward and leap-frog scheme.

We completed eight different experiments of monsoon simulation. In all of the eight experiments, land-sea contrast and large-scale condensation were included. When the radiative warming and cooling were considered, we also took into account cumulus parameterization, sensible heating, and evaporation from ocean surfaces and elevated plateau surfaces above 3 km. Here we shall designate the two different initial states as I and II, which correspond to the June monthly mean and the June monthly zonal mean fields of the meteorological variables, respectively. Figures 3a and 3b show the initial sea-level pressure distributions of states I and II. We use A_1 and A_2 to designate the radiation conditions with and without diurnal change, and O and N the orographic conditions with and without mountains. The letter B always means no radiative heat sources in the model; hence, the only heating comes from the large-scale condensation. The titles of the eight experiments could, thus, be encoded as A_1OI, A_1OII, A_2OI, A_2OII, A_1NI, A_1NII, BOI, and BOII.

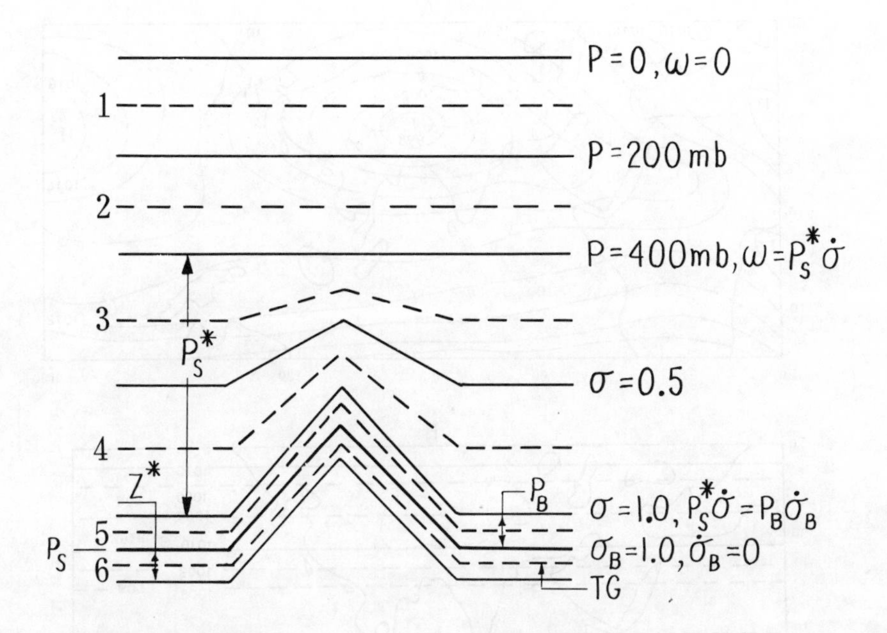

Fig. 1 Schematic illustration of the vertical structure of the model.

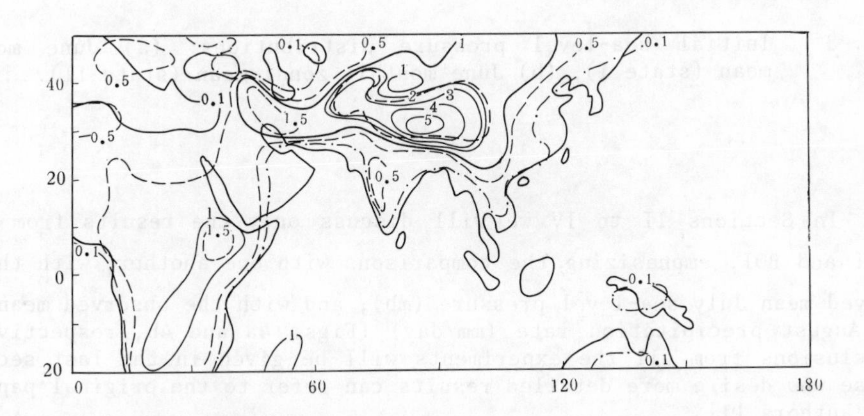

Fig. 2 Smoothed topography used in the model (unit: km).

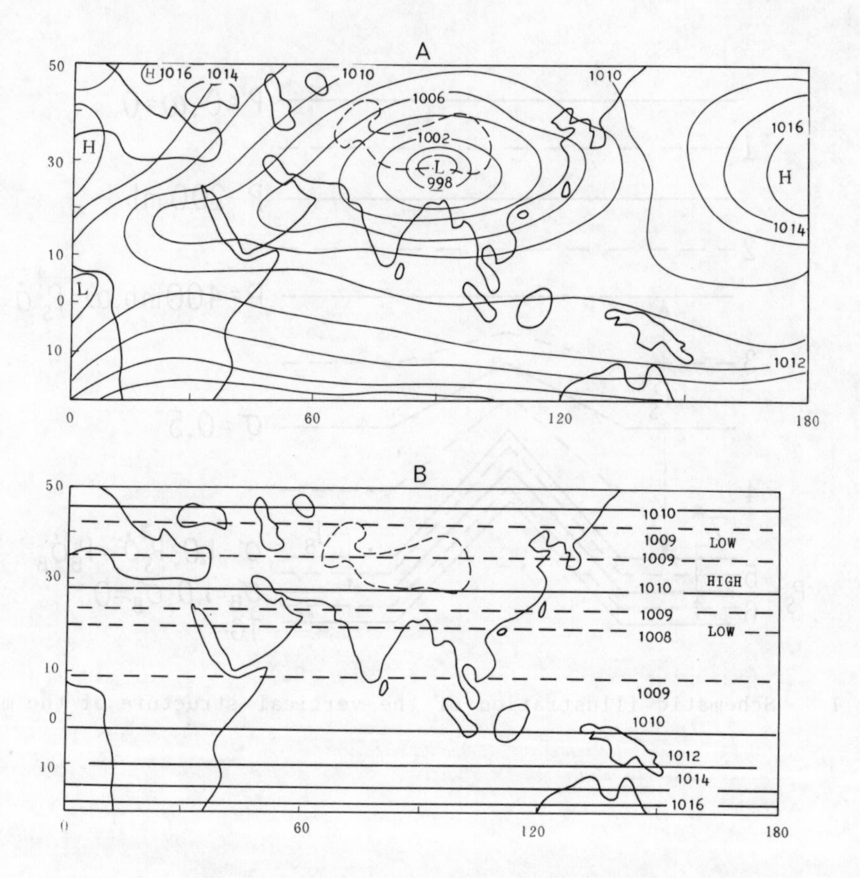

Fig. 3 Initial sea-level pressure distributions: (a) June monthly mean (state I), (b) June monthly zonal mean (state II).

In Sections II to IV we will discuss only the results from A_1OI, A_1NI and BOI, emphasizing the comparisons with one another, with the observed mean July sea-level pressure (mb), and with the observed mean June to August precipitation rate (mm/day) (Figs. 4a and 4b, respectively). Conclusions from all the experiments will be given in the last section. Those who desire more detailed results can refer to the original paper by the authors [1].

II. EXPERIMENT A_1OI

This experiment includes orography and diurnal changes in solar radiation. The experiment was carried out for eight days of integration time and the essential results are given below.

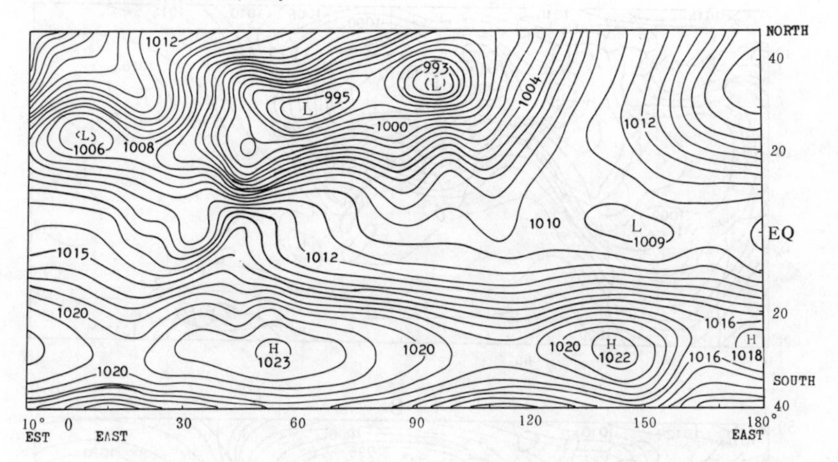

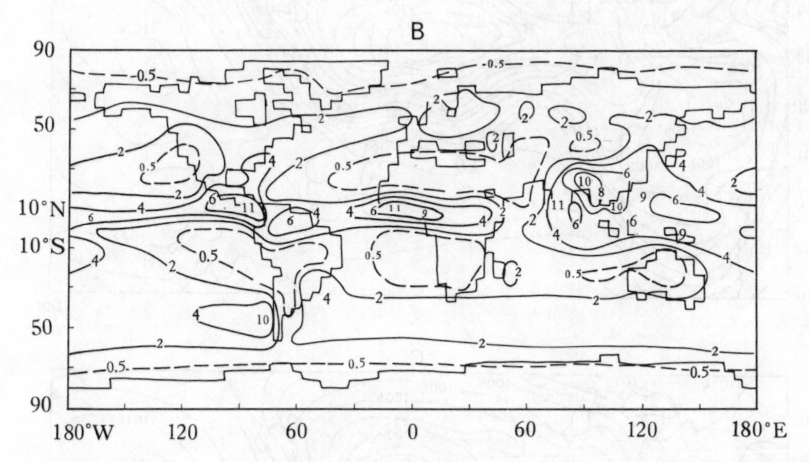

Fig. 4 (a) Observed mean July sea-level pressure (mb). (b) Observed mean June to August precipitation rate (mm/day).

1. Sea-Level Pressure

The sea-level pressure patterns can be obtained by extrapolation. The fifth day, the eighth day and eight-day average sea-level pressure distributions from this simulation are illustrated in Figs. 5a, b, and c, respectively. We can see that they are all quite different from the initial sea-level pressure field in Fig. 3a, but they agree very well with the observed mean July sea-level pressure distribution in Fig. 4a, even including the extent and strength of the western-Pacific high. From Figs. 5a, b, and c we clearly see the southwestward extension of the high system in the 160°E, 30°N region, the two troughs along 85°E and 110°E

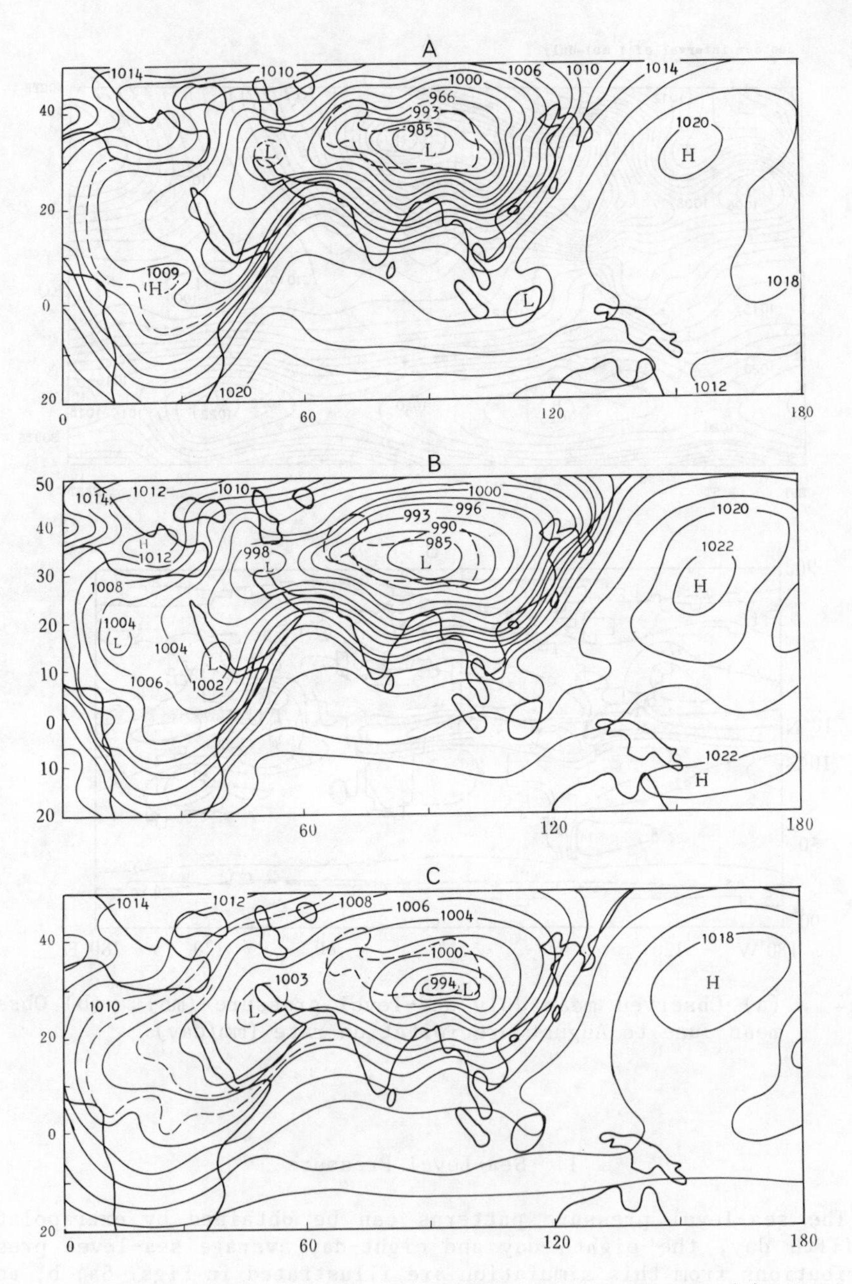

Fig. 5 The sea-level pressure distributions of A_1OI: (a) fifth day, (b) eighth day, (c) eight-day mean.

associated with the extensive and deep heat low centered over the Tibetan Plateau, and the low-pressure belt extending from the Iranian highland to the coastal region of northeast Africa with very steep pressure gradients between that belt and the high-pressure ridge over the Arabian Sea. This gradient is evidently connected with the Somalian jet. All these results are in good agreement with the observed features illustrated in Fig. 4a. Additionally, from Figs. 5a and b we see that the sea-level pressure distribution changed very little from the fifth day to the eighth day, indicating that it has almost reached a quasi-equilibrium state after five days' integration. In addition, the extent and the strength of the simulated mean Pacific high in Fig. 5c agrees very well with that in Fig. 4a, but further south the simulated sea-level pressure is somewhat too high in comparison with that in Fig. 4a. This discrepancy is the result mainly of the cyclic west-east boundary condition.

2. Surface and 300-mb Flow Patterns

The eight-day average flow patterns and isotachs given by the simulation in the surface boundary layer and the 300-mb level are illustrated in Figs. 6a and 6b. It is clearly seen that the Somalian jet and the

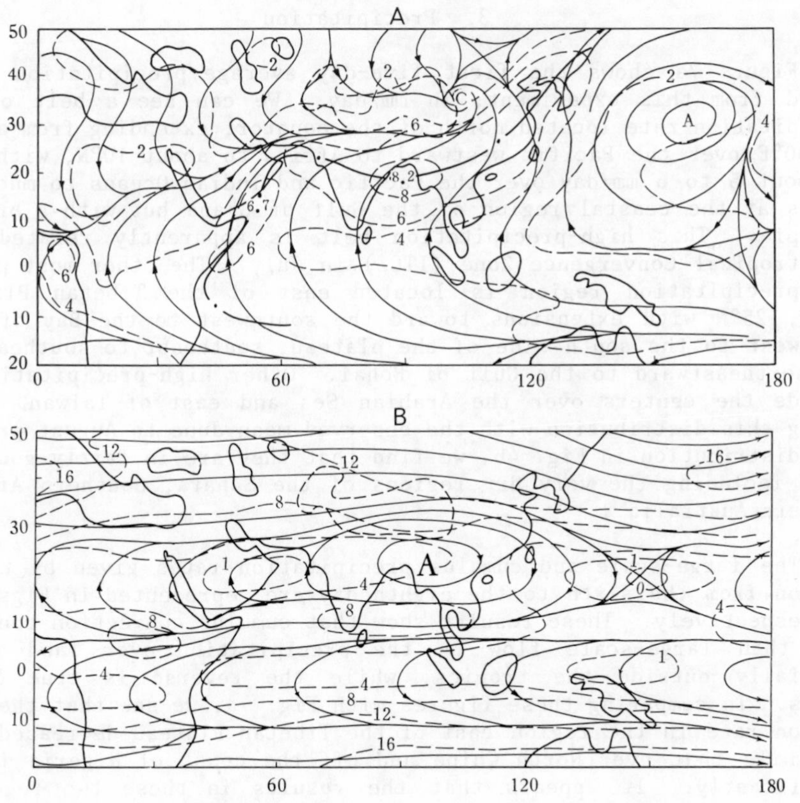

Fig. 6 The simulated A$_1$OI eight-day average flow patterns (solid) and isotaches (dashed) (a) in the surface boundary layer and (b) at the 300 mb level.

Indian southwest jet are both correctly created by the simulation, together with the gradual change from the southeast flow south of 10°S to the southwest flow north of 15°N and the convergence toward the Tibetan Plateau. The strong-wind belt begins with a west-east direction at about 20°S and becomes narrower along the east coast of Africa and then changes to a west-southwest to east-northeast direction and spreads wider as it passes through the Arabian Sea, the Indian Peninsula, the Bay of Bengal, and South China to the coastal area of the East China Sea. A strong convergence zone exists to the north of this jet, while anticyclonic circulation prevails over the western Pacific. In comparing the simulated mean surface flow with the observed mean July surface flow given by Ramage [2], we see they agree very well.

On the 300-mb flow pattern chart we see that a strong elongated anticyclonic cell exists in the belt around 20°N with its center located east of the Tibetan Plateau, and another counterclockwise curved flow belt occupies the area between 15°S to 5°N which corresponds to a high-pressure belt. These flow features are generally similar to the observed mean July 300-mb flow pattern given by Van de Boogaard [3], except that the simulated velocity is somewhat lower than the observed one.

3. Precipitation

Figure 7a shows the first five-day average precipitation rate obtained from this experiment, in mm/day. We can see a belt of higher precipitation rate located north of the equator, extending from about 5°N at 160°E over the Pacific westward to Africa to about 10°N, with centers of about 5 to 6 mm/day over the Pacific and Indian Oceans to much higher values at the coastal region of the Gulf of Ghana but with a break over Ethiopia. This high-precipitation belt is apparently related to the Intertropical Convergence Zone (ITCZ) in July. The other most prominent high-precipitation region is located east of the Tibetan Plateau at 100°E, 25°N with extensions toward the southwest to the Bay of Bengal, northwest to the south side of the plateau, southeast to southeast China and northeastward to the Gulf of Bohai. Other high-precipitation areas include the centers over the Arabian Sea and east of Taiwan. In comparing this distribution with the observed mean June to August precipitation distribution in Fig. 4b, we find that they are in fairly good agreement, including the very dry regions of the Sahara, Southern Africa and Northern Australia.

The large-scale and cumulus precipitation rates given by this simulation from the sixth to the eighth day are represented in Figs. 7b and 7c, respectively. These results show that cumulus convection contributes more than large-scale flow to the precipitation over land in July, especially outside the tropics, while the reverse is true over the oceans. In comparing these figures with Fig. 7a, we see that the precipitation rate in the region east of the Tibetan Plateau decreased slightly, while that over North China and off the coast of Nigeria increased significantly. It appears that the results in these two regions are increasingly influenced by the side-boundary conditions as time goes on and become unreliable at later times. Therefore, we stopped our integration after the eighth day.

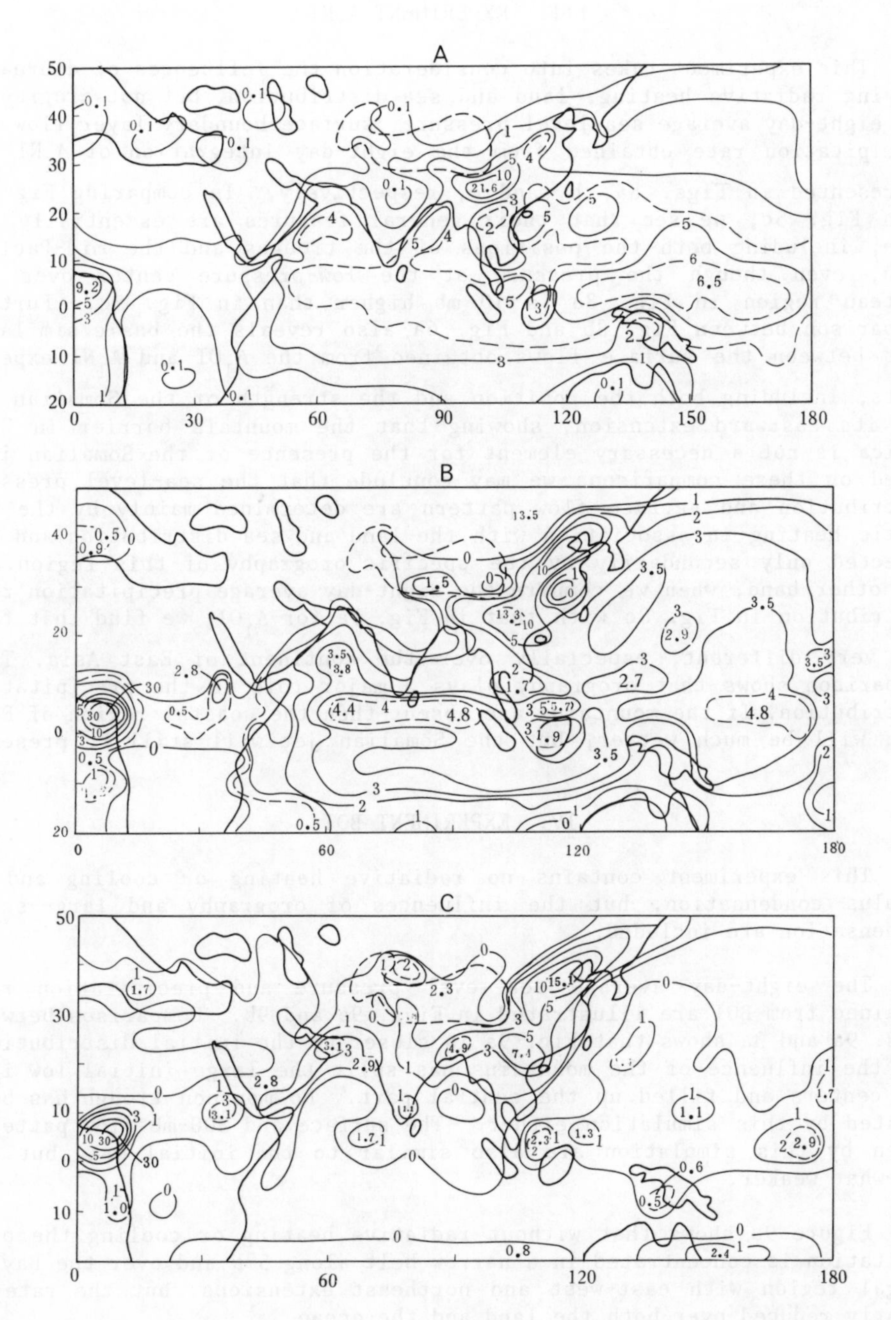

Fig. 7 The simulated A_1OI precipitation distributions (mm/day): (a) the first five-day average, (b) large-scale precipitation rate from the sixth to the eighth day, (c) the same as b, except for cumulus precipitation.

III. EXPERIMENT A_1NI

This experiment takes into consideration the influences of diurnally varying radiative heating, land and sea distribution, but not orography. The eight-day average sea-level pressure, surface boundary layer flow and precipitation rate obtained from the eight-day integration of A_1NI are represented in Figs. 8a, b, and c, respectively. In comparing Fig. 8a with Fig. 5c, we see that their general features are essentially the same, including both the positions of the troughs and the mid-Pacific high, even though the pressure at the low-pressure center over the plateau region in Fig. 8a is 10 mb higher than in Fig. 5c. Further comparison between Fig. 8b and Fig. 6a also reveals the basic similarities between the surface flows obtained from the A_1OI and A_1NI experiments, including both the position and the strength of the Somalian jet and its eastward extension, showing that the mountain barrier in East Africa is not a necessary element for the presence of the Somalian jet. Based on these comparisons we may conclude that the sea-level pressure distribution and surface flow pattern are determined mainly by the diabatic heating in association with the land and sea distribution and are affected only secondarily by the specific orography of this region. On the other hand, when we compare the eight-day average precipitation rate distribution in Fig. 8c with that in Fig. 7a for A_1OI, we find that they are very different, especially over the continent of East Asia. This comparison shows that orography plays a major role in the precipitation distribution. If the mountains are absent then the coastal region of East Asia will be much wetter, but the Somalian jet will still be present.

IV. EXPERIMENT BOI

This experiment contains no radiative heating or cooling and no cumulus condensation, but the influences of orography and large-scale condensation are included.

The eight-day average sea-level pressure and precipitation rate obtained from BOI are illustrated in Figs. 9a and 9b. Comparison between Figs. 9a and 3a shows that Fig. 9a is closer to the initial distribution, but the influence of the mountains has split the large initial low into two centers and filled up the central part. No monsoon trough has been created by this simulation either. The surface and 300-mb flow patterns given by this simulation are also similar to the initial data but are somewhat weaker.

Figure 9b shows that without radiative heating or cooling the precipitation is concentrated in a narrow belt along 5°N and over the Bay of Bengal region with east-west and northeast extensions, but the rate is greatly reduced over both the land and the ocean.

V. SUMMARY AND CONCLUDING REMARKS

Eight numerical experiments have been carried out for periods of 8 to 20 days with a model having five atmospheric layers and a resting

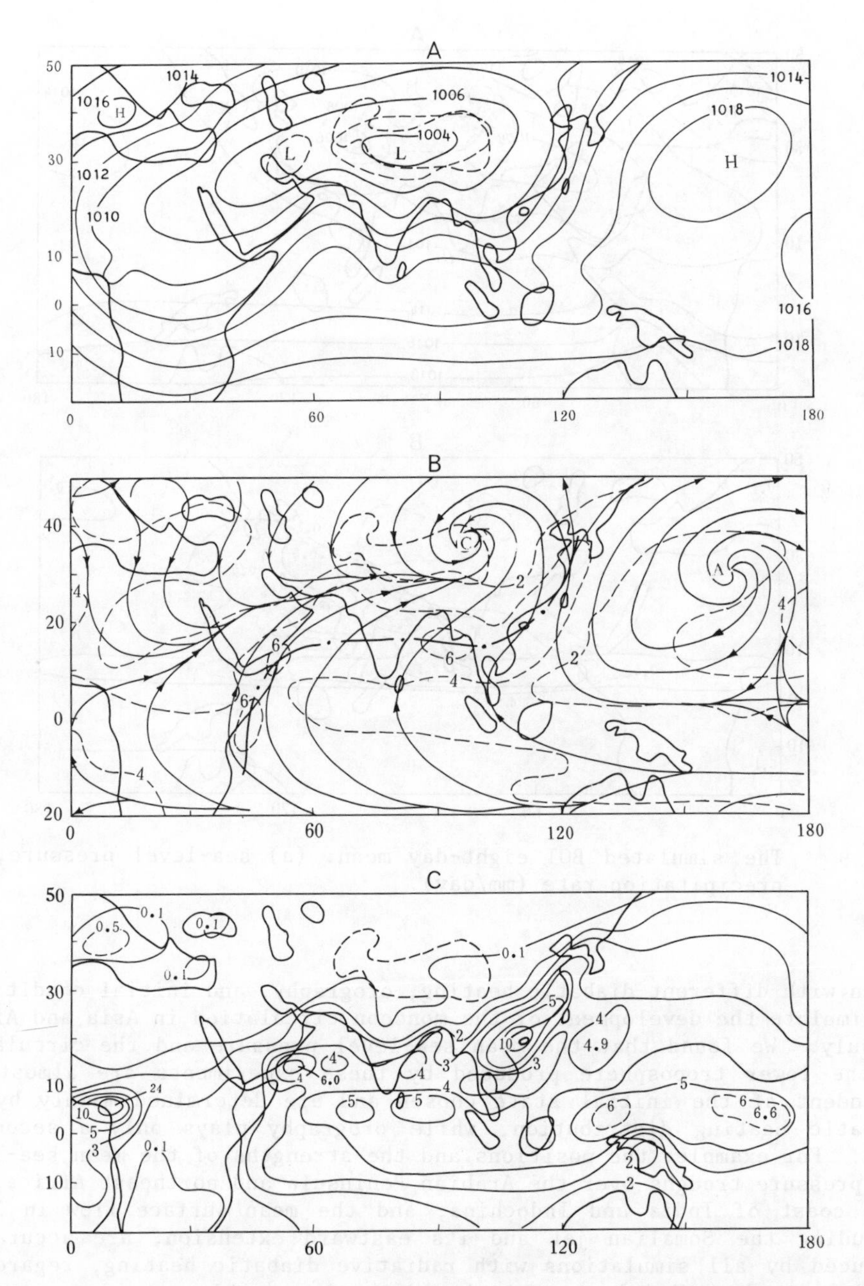

Fig. 8 The simulated A_1NI eight-day mean: (a) sea-level pressure, (b) surface boundary layer flow, (c) precipitation rate (mm/day).

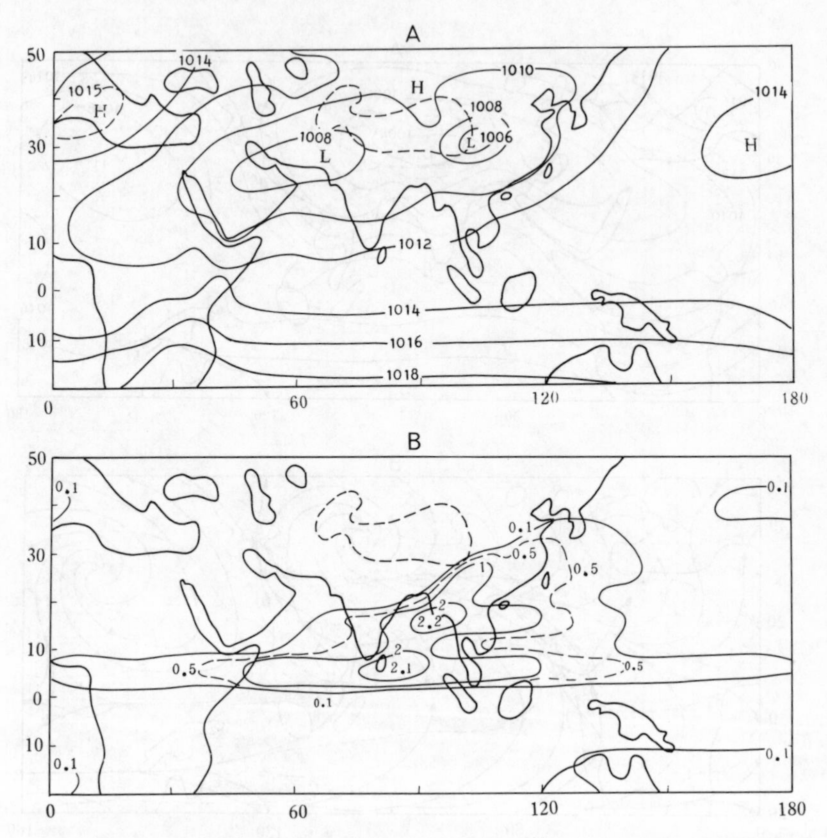

Fig. 9 The simulated BOI eight-day mean: (a) sea-level pressure, (b) precipitation rate (mm/day).

ocean with different diabatic heating, orography, and initial conditions, to simulate the development of the monsoon circulation in Asia and Africa in July. We found that the mean sea-level pressure and the circulation in the lower troposphere produced by these experiments are almost independent of the initial state chosen and are determined mainly by the diabatic heating distribution, while orography plays only a secondary role. For example, the positions and the strengths of the mean sea-level low-pressure troughs over the Arabian Peninsula and northeast Africa, the east coast of India and Indochina, and the mean surface flow in July, including the Somalian jet and its eastward extension, are accurately produced by all simulations with radiative diabatic heating, regardless of whether the initial state is based on the monthly mean June pressure distribution or on the mean June zonal average pressure distribution and also irrespective of the presence of orography. None of these systems are predicted when radiative heating is excluded. On the other hand, precipitation distributions are critically influenced both by diabatic heating, topography, and initial data. Thus, the precipitation distribution over continental Asia will be completely altered when only land and sea distribution and radiative heating are included but no topography, with the maximum precipitation shifted from the 100°E, 25°N region east of the Tibetan Plateau to the coastal region of East Asia. When there is no radiative heating, the precipitation rate is greatly reduced over both land and ocean. The diurnal variation of solar radiation is found to be of importance both for precipitation and for the rate of development of the low-level flow. Its inclusion makes the low-pressure troughs over the east coast of India and over Indochina develop much faster than without it.

REFERENCES

1. Kuo, H.L. and Y.F. Qian: Influence of Tibetan Plateau on cumulative and diurnal changes of weather and climate in summer. Mon. Wea. Rev., 109, 2337-2356 (1981).

2. Ramage, C.S.: Monsoon meteorology. Academic Press, New York and London, 296 pp (1971).

3. Van de Boogaard, H.: The mean circulation of the tropical subtropical atmosphere, July. NCAR Technical Note TN-118 + STR, 48 pp. (1977).

3.6 A NUMERICAL EXPERIMENT ON THE DYNAMIC AND THERMAL EFFECTS OF THE QINGHAI-XIZANG (TIBET) PLATEAU IN SUMMER

Ji Liren, Shen Rujin and Chen Yuxiang
Institute of Atmospheric Physics
Academia Sinica
The People's Republic of China

ABSTRACT

This paper discusses the influence of the Tibetan Plateau on circulation systems of various scales over Asia, Africa, and the Pacific regions in summer.

Starting from the summer mean zonal flow, a three-level primitive-equation model is integrated in σ-coordinates. It includes both the mechanical blocking action and friction caused by the plateau, and thermal forcing. A comparative test, using several kinds of topography and heating, is made to investigate the influences of dynamic and thermal forcing of the plateau on the formation and changes of the monsoonal circulation over Asia-Africa, with emphasis on tropical and subtropical areas. It is found that the major features of the summer monsoonal circulation over Asia-Africa, such as the cyclonic circulation around the plateau, the east-west oriented, narrow high-pressure belt to the north of the plateau, and the southwest airflow to the south in the lower troposphere, the anticyclone over Tibet, and the tropical easterlies in the upper troposphere, can be simulated. The disturbances to the zonal flow due only to the dynamic effect of the plateau are shallower, marked only in the lower part of the troposphere, and confined to a limited extent. The effect of the thermal contrast between continent and ocean plays a dominant role in the formation of broad-scale monsoons, and the influence of the heated plateau should be considered as one component superimposed on the effect caused by continental heating. For the regional flow pattern near the plateau, especially in the lower troposphere, the effects of topography and differential heating between the plateau and its adjacent area appear to be more essential.

I. INTRODUCTION

The Qinghai-Xizang (Tibet) Plateau, which is the highest and largest plateau in the world and acts as a heat source in the northern hemisphere in summer, plays an important role in the formation and evolution of the summer monsoon circulation over Asia-Africa. It also affects the global general circulation. Great attention has been paid to its effects over the years. Several studies on the effects of the plateau have been conducted, and many interesting results have been obtained. Among them are those showing either the dynamic forcing or the thermal forcing to be important. Other investigations took the combined effects of thermal and orographic forcing into account. Now research is in progress on the effects of the plateau, incorporating realistic topography and zonally asymmetric heating.

Many researchers have noticed from the observed mean circulation over Asia and its adjacent regions that there are circulation systems of various scales which involve not only large-scale phenomena, such as the huge anticyclone over the Asian-African region, which extends to the upper troposphere and stratosphere (sometimes called the Tibetan high or South Asian high), and the cross-equatorial current that couples the two hemispheres, but also unique regional systems, such as the tilted trough to the east of the plateau and a narrow, but extended, belt of high pressure to the north and a secondary vertical circulation superimposed on the monsoonal circulation. They have also noticed the fact that the plateau, as an elevated heat source in the midtroposphere in summer, can cause differential heating between the mountain ranges and surrounding areas. On the other hand, the plateau can enhance the land-sea contrast as well. These features represent thermal forcing of different scales. It is the primary purpose of our study to see how the different-scale phenomena are linked and to examine the various effects of the Tibetan Plateau.

In this paper, we use a three-level primitive-equation model in σ coordinates, originally developed by the Group of Medium-Range Forecasting of the Institute of Atmospheric Physics [4], to study the aforementioned problem. For this initial study we have obtained a modified version of the model which is applicable to lower latitudes and suitable for a long-term integration period. Based on this model several cases have been tested, including different topographies and heating. By comparing the results obtained from these cases, the effects of topographic and heat forcing of the plateau are discussed.

II. A BRIEF DESCRIPTION OF THE MODEL

1. Basic Equations, Their Initialization, and Boundary Condition.

The prognostic equations of a primitive-equation model which has three levels in the vertical in σ coordinates are:

$$\frac{\partial U}{\partial t} = - L(u) + fV - \frac{P_*}{m} fv_g + \frac{P_*}{m} F_u + \frac{P_*}{m} D_u, \quad (1)$$

$$\frac{\partial V}{\partial t} = - L(v) - fU + \frac{P_*}{m} fu_g + \frac{P_*}{m} F_v + \frac{P_*}{m} D_v, \quad (2)$$

$$\frac{\partial \dot{\Sigma}}{\partial \sigma} = - \frac{1}{m} [m^2(\frac{\partial U}{\partial x} + \frac{\partial V}{\partial y}) + \frac{\partial P_*}{\partial t}], \quad (3)$$

$$\frac{\partial P_*}{\partial t} = - \int_0^1 m^2 (\frac{\partial U}{\partial x} + \frac{\partial V}{\partial y})\, d\sigma, \qquad (4)$$

$$\frac{\partial}{\partial t}(\frac{P_*}{m} C_p T) = - L(C_p T+\phi) - \frac{1}{m}\frac{\partial \phi\sigma}{\partial \sigma}\frac{\partial P_*}{\partial t} + (UfV_g - VfU_g) + \frac{P_*}{m}\dot{Q} + \frac{P_*}{m} D_T, \qquad (5)$$

$$\frac{\partial \phi}{\partial \ell n \sigma} = - RT, \qquad (6)$$

$$U_g = - \frac{m}{f}(\frac{\partial \phi}{\partial y} + RT\, \frac{\partial \ell n P_*}{\partial y}),$$

and

$$V_g = - \frac{m}{f}(\frac{\partial \sigma}{\partial x} + RT\, \frac{\partial \ell n P_*}{\partial x}),$$

where F represents the momentum dissipation due to friction in the boundary layer, D is horizontal eddy diffusion, $\dot{Q}$ is the rate of diabatic heating, m is the map magnification factor (defined by m = sec ϕ on Mercator projection), and ϕ is latitude. $U = P_* u/m$, $V = P_* v/m$, $\dot{\Sigma} = P_* \dot{\sigma}/m$, where (u,v) is the horizontal velocity component, $\sigma = P/P_*$, P is pressure, and P_* is the pressure at the surface. Other symbols are commonly used and no more explanation is needed.

We define the operator $L(a) \equiv m[\frac{\partial}{\partial x}(Ua) + \frac{\partial}{\partial y}(Va)] + \frac{\partial}{\partial \sigma}(\dot{\Sigma} a)$.

In the above σ-coordinate system, the blocking effect and the surface friction due to topography are included. The thermal effects of the plateau and of the continent of Asia-Africa on the atmosphere is described by $\dot{Q}$. Therefore, three basic kinds of effects involving orographic and thermal forcing are included in this model.

The model atmosphere is divided into three layers in the vertical which have equal thickness, $\Delta\sigma = 1/3$. The domain of integration covers an area from 37.1°S to 62.2°N, and from 0°E eastward to 165°W, consisting of 40 x 25 grid points on a Mercator projection map. Europe, Asia, Africa, and the west-central Pacific Ocean are included. The horizontal grid distance is 555 km at the equator (Fig. 1). The Euler-backward difference scheme is used for time integration with a time step of 15 minutes.

The boundary conditions are as follows: a cyclic condition in the east-west direction generates identical conditions at the eastern and western boundaries, and fixed values at the northern and southern boundaries. $\dot{\sigma} = 0$ is adopted at the surface and at the top of the model atmosphere.

All experiments are started from the mean global zonal flow in summer (Fig. 2). Given the geopotential height at 28.7°N, the geopotential height at various latitudes and levels in the vertical can be obtained by means of the geostrophic relation in the subregion to the north of 28.7°N, and by the balance equation in the subregion to the south of

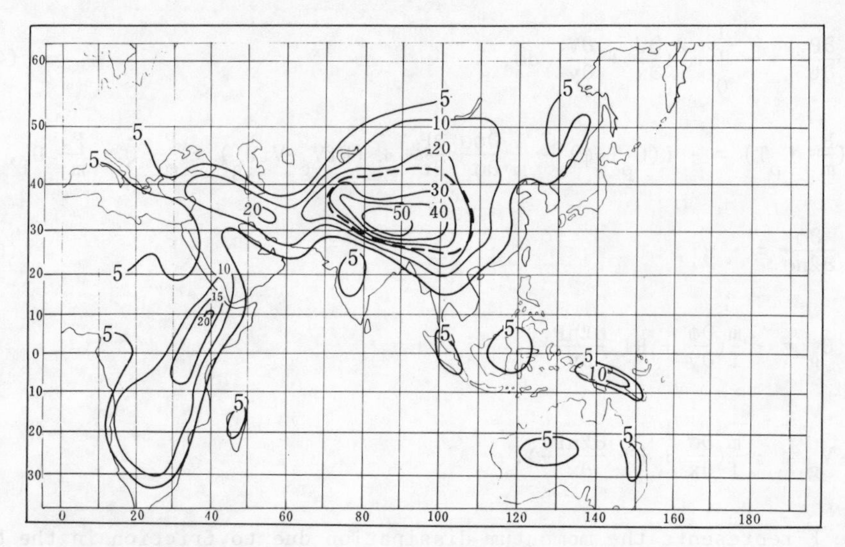

Fig. 1 The integration domain and the contours of the smoothed topography (solid lines) in units of 100 m; the dashed line is the outline of the isolated plateau.

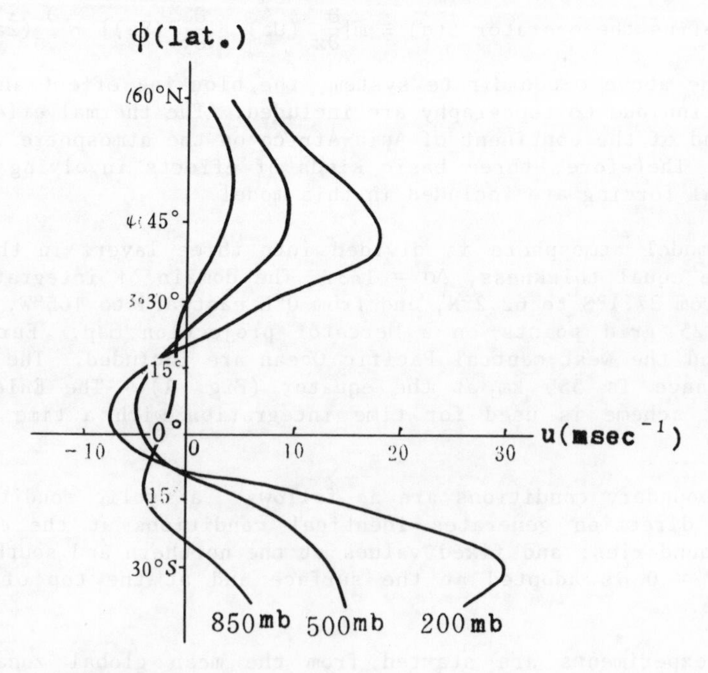

Fig. 2 Mean zonal wind profile in summer (June, July, and August).

28.7°N. The initial temperature field is obtained by differentiating the geopotential with respect to σ and by using the hydrostatic equation [5].

2. Topographic Features Incorporated in the Model.

For comparison, three kinds of topographic features have been used in the experiments (as can be seen in Fig. 1). One is an ideal and isolated plateau with lower-than-realistic elevation, its maximum height being 3150 m. To obtain these heights, the heights of the smoothed orography are first taken from Berkofsky and Bertoni [2], then reduced by 2000 m, and finally multiplied by 1.5. If the original height of the topography is less than 2000 m, the resulting height is set equal to zero. Thus, all regions other than the plateau are treated as flat plains. Another plateau configuration is similar to the first but multiplied by 2. In this case the maximum height is 4200 m. The third configuration is more realistic and includes topographic features in both hemispheres, with the Tibetan Plateau included. In this case the smoothed orographic heights are multiplied by 0.8. The maximum height is 4320 m, but the slope of the plateau is reduced.

The cases of isolated plateaus are studied to enhance and manifest the contrast between the plateau and the remainder of the continent.

Table 1 Parameters characterizing topography.

Case	Method	Max. Height of the Plateau	Mean Slope Along 90°E	
			North Side	South Side
1. An ideal isolated plateau with a relatively low height	$Z_*=(\tilde{Z}_*-2000)\ 1.5$	3150 m	-0.0024	0.032
2. Plateau with a higher height	$Z_*=(\tilde{Z}_*-2000)\ 2$	4200 m	-0.0033	0.0043
3. A more realistic topography including the plateau	$Z_*=0.8\ \tilde{Z}_*$	4320 m	-0.0016	0.0028

Note: Z_* is the orographic height used in this model. $\tilde{Z}_*$ is smoothed orographic height.

3. Heating in the Model.

For the heating function in the thermodynamic equation, only the sensible heat flux from the surface is taken into account. The effect of heating is treated in two stages: the heat flux from the surface is first computed, and then the heat flux is transported to the middle or upper troposphere by such mechanisms as turbulence and free convection. The details of the model's heat-flux treatment are:

(1) Sensible heat flux from the surface is expressed as:

$$F_* = -P_* C_p C_D |\vec{V}_s| (T_s - T_*), \tag{7}$$

where T_s is the air temperature near the surface, T_* is the temperature at the surface of the earth, and $|\vec{V}_s|$ is the magnitude of wind near the surface. Climatic mean values in summer taken from 22 stations that are over 2500 m in the Qinghai-Xizang (Tibet) region, are specified for the above three variables and taken as constant. C_D is the drag coefficient, usually taken as a function of the height of topography [1]:

$$C_D = 5 \times 10^{-3} + 6.45 \times 10^{-3} \left(\frac{Z_*}{1000+Z_*}\right), \tag{8}$$

where Z_* is the height of topography (unit: meters). The heat flux at various orographic heights reflects itself in the parameter C_D.

The heat flux from the sea is much less than that from the land. Here it is simply set equal to zero over the sea and for the area south of 10°S in the case of the more realistic topography, and for the area to the south of 15°N in the cases of isolated topography.

(2) The vertical flux of heat above the boundary layer is expressed as:

$$F = -PC_p K_{TV} \left[\frac{\partial\theta}{\partial Z} - \left(\frac{1000}{P}\right)^{R/C_p} \gamma_c\right] = -PC_p K_{TV}\left[-\left(\frac{1000}{P}\right)^{R/C_p} \frac{g}{RT}\sigma\frac{\partial T}{\partial\sigma} + \tilde{\gamma}_c\right]$$

$$= -PC_p K_{TV} \gamma_{\sigma}, \tag{9}$$

where

$$\tilde{\gamma}_c = \left(\frac{1000}{P}\right) \quad (\gamma_d - \gamma_c).$$

γ_c is an adjustment factor which indicates the characteristic process of heat transport, similar to the counter-gradient parameter, and is taken to be 0.003°K/m [6].

$$K_{TV} = \begin{cases} A_1 & \gamma_\sigma \geq 0 \\ A_1 + A_2[1.0 - \exp(A_3\, \gamma_\sigma)] & \gamma_\sigma < 0. \end{cases} \quad (10)$$

Therefore, the heating function we adopt is a fixed constant with respect to time, but to some degree it may involve the thermal differences between sea and land, and between plateau and plain.

III. RESULTS OF THE EXPERIMENT

All cases of simulation are started from the mean zonal wind profile for summer as mentioned above, representing a planetary zonal wind field in summer.

Based on the topographic and heating effects mentioned above, six tests are performed (see Table 2). The main results from these tests are given as follows:

i. The dynamic effect of the plateau: Tests 1 to 3 are adiabatic cases without heating on the plateau; that is, only the blocking effect and lateral nonuniformity of friction due to the plateau are taken into account. Figure 3 shows streamlines for day 5 of integration for the isolated plateau with higher height (Test 2). No essential differences are noted between the two cases of isolated plateaus. We can see in the lower troposphere a quasi-stationary trough near the coast of the continent, tilted from northeast to southwest. To the north and northeast of the plateau there is a belt of high pressure which is similar to the observed mean field. Climatologically, it corresponds to the arid area in northwest China. We might infer that the mechanism responsible for the high-pressure belt over northwest China is partly dynamic in nature. The southwest airflow to the south of the plateau does not appear.

In the middle and/or upper troposphere, the disturbance tilts to the west with height and damps upward quickly. There is only a very weak

Table 2: Numerical test configurations.

No. of Test	Topography	Heating
1	An isolated plateau with a low height	
2	An isolated plateau with a higher height	No heating
3	A more realistic topography including the plateau	
4	Same as 1	Including
5	Same as 2	sensible
6	Same as 3	heating

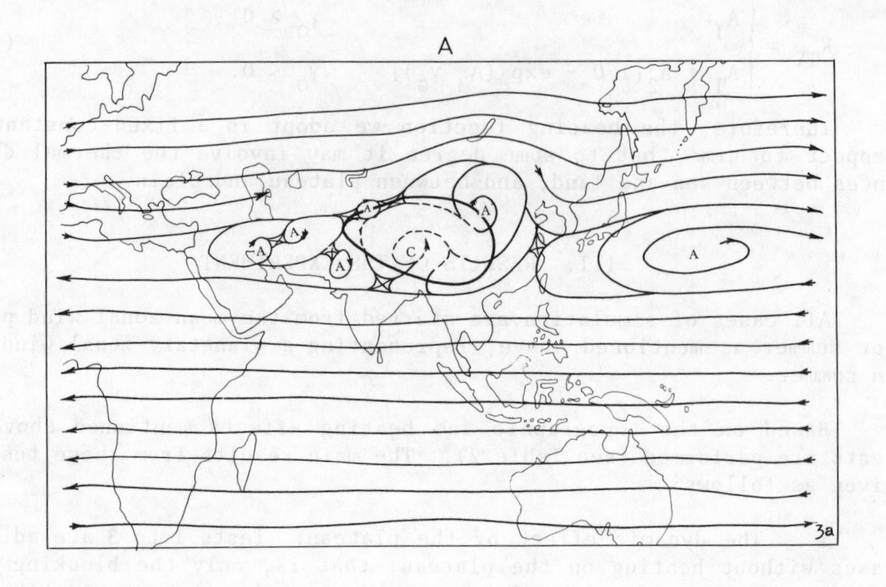

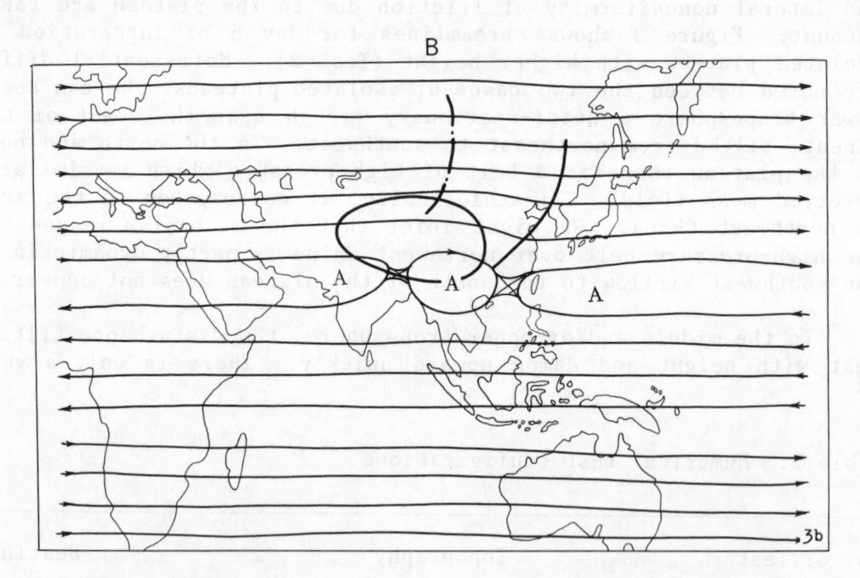

Fig. 3 Streamlines for day 5 of integration for the isolated plateau with higher height (Test 2). (a) 850 mb, (b) 500 mb, (c) 200 mb.

C

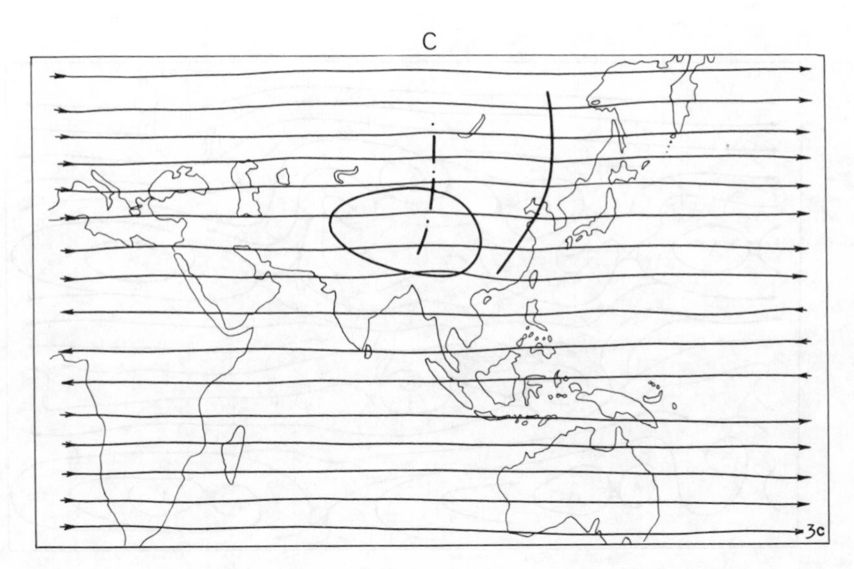

Fig. 3 (Continued)

trough at the 200-mb level. The observed large upper anticyclone over the plateau, one of the main monsoon circulation systems, is not reproduced at all. From this figure we gain the impression that for weak zonal flow and weak shear, as they prevail in summer, the disturbances caused by the dynamic effects of the plateau alone are rather shallow in the vertical and are mainly confined in the horizontal to the vicinity of the plateau.

The results obtained from a more realistic topography which includes all mountains in the integration domain are nearly the same near the plateau as the previous results. These results suggest that the formation of local features over the plateau and its adjacent region is in direct response to the dynamic effect of the plateau. Differences show up mainly at lower latitudes and in the southern hemisphere (Fig. 4).

In the southern hemisphere, there is a splitting of the subtropical high. Both high centers, one over the Mascarene Islands and the other over Australia, are well simulated after the continents of Africa and Australia are introduced in the model. This fact indicates the dynamic nature of the anticyclones. There also appear an Intertropical Convergence Zone (ITCZ), an equatorial buffer zone and a cross-equatorial current from the southern hemisphere along the east coast of Africa, though very weak. However, no large upper anticyclone over the plateau is simulated.

ii. Tests in the cases of an isolated plateau with sensible heating: In Tests 4 and 5 the effect of heating is incorporated for the isolated plateau. The striking differences from the adiabatic cases are the occurrence of cyclonic circulation around the Tibetan Plateau and

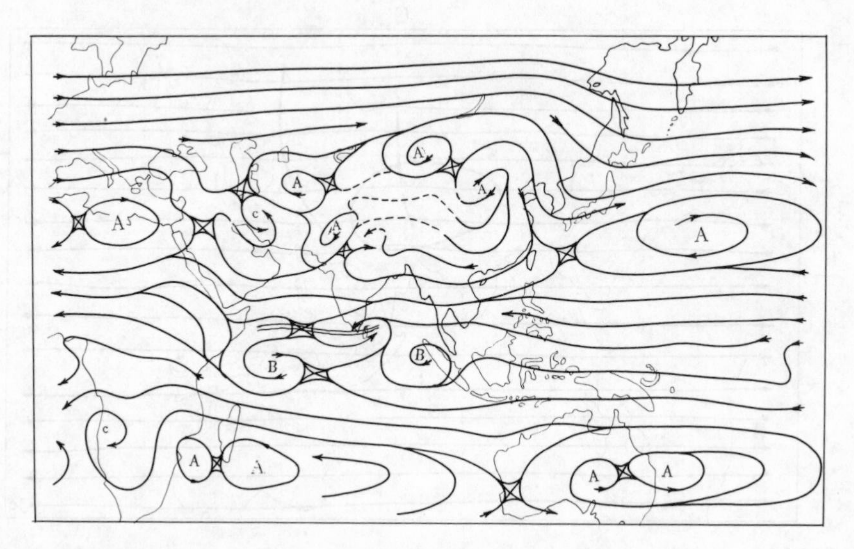

Fig. 4 Streamlines at 850 mb for day 5 of integration for a more realistic topography (Test 3). The double line is the position of the ITCZ.

southwest airflow to the south of the plateau. The latter enables the subtropical high over the West Pacific in the lower troposphere to be strengthened. In the upper troposphere, the upper-level anticyclone over Tibet is also simulated (Fig. 5), but its strength is weaker than that observed. Note especially that the subtropical high is weak and located at 28°N, farther to the south than its actual position. Another apparent problem is that the simulation fails to reproduce the monsoon trough over India.

What happens if the height of the plateau is raised? The results show that there are no essential differences between the case of the lower plateau and the case of the higher plateau. It should be noted that in the early stages of model integration the circulation features in the heating cases are similar to those in Tests 1 to 3. Only after four days of integration does the heating effect become apparent and the main features of the monsoon circulation over Asia-Africa in summer appear. The modelling results show that enhancing the height of the isolated plateau alone cannot improve the simulation of the broad-scale monsoon circulation.

The integration of Test 5 lasts for sixteen days. In the first eight days heating is incorporated, and then it is removed to examine the dynamic effect of the plateau on the simulated circulation in the early stage. During the period of heating, the southwest airflow to the south of the plateau in the lower troposphere and the upper easterlies become increasingly apparent (Fig. 6).

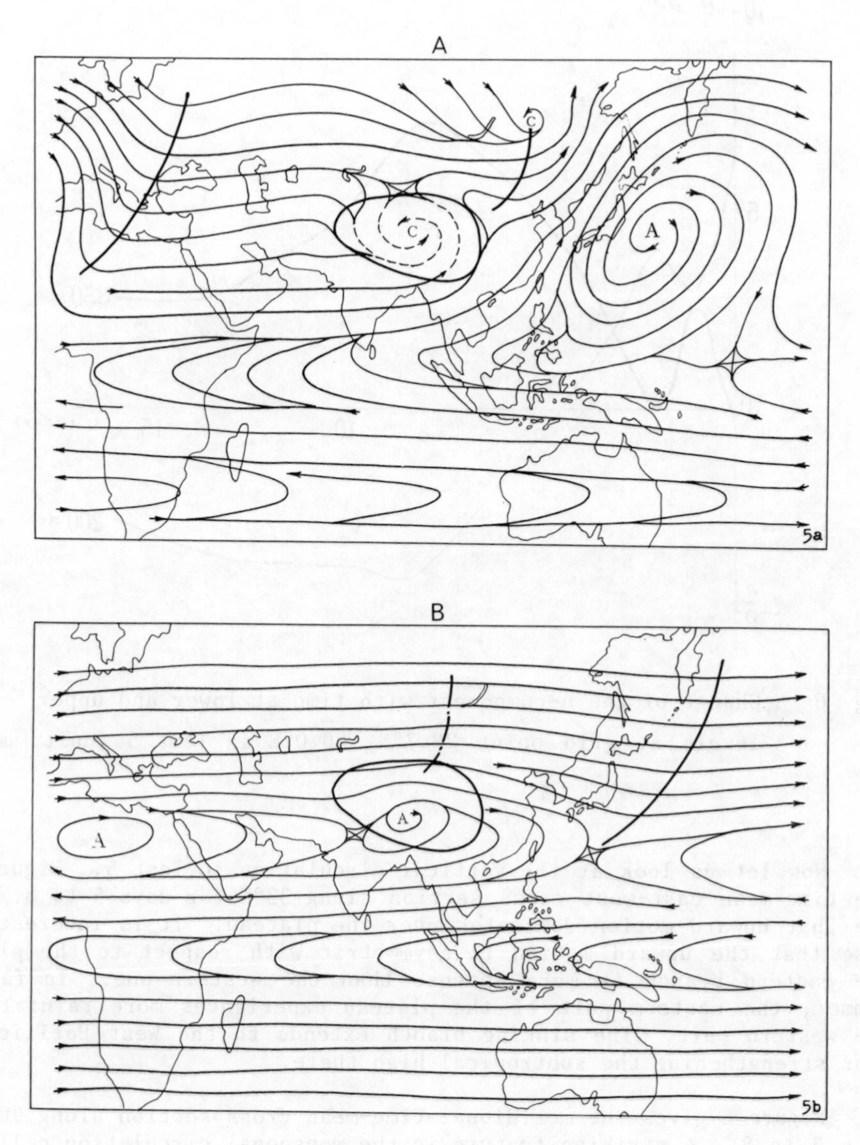

Fig. 5 Streamlines for day 6 of integration for isolated plateau and sensible heating (Test 4). (a) 850 mb, (b) 200 mb.

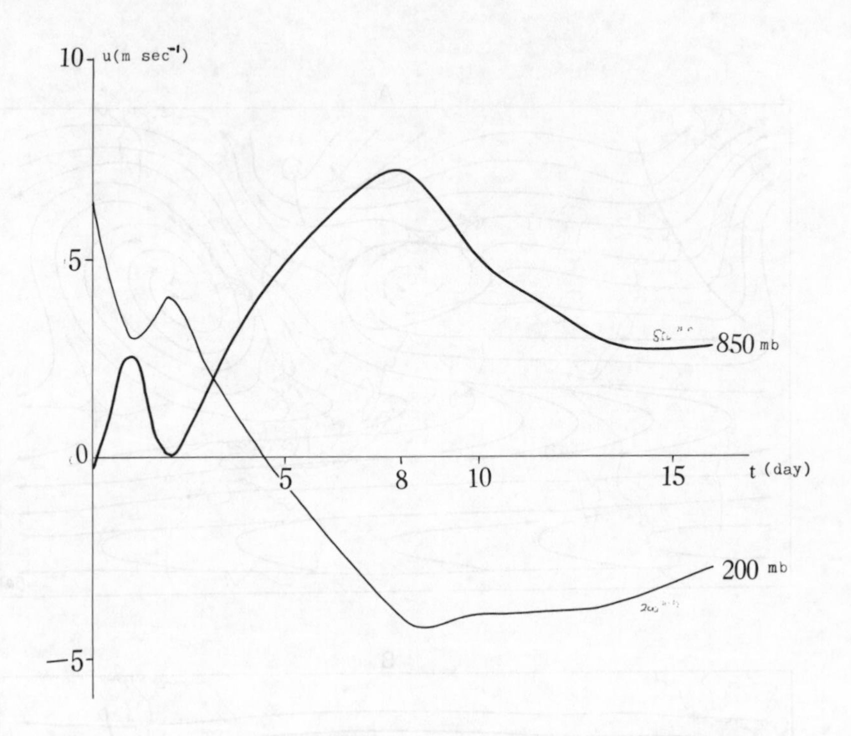

Fig. 6 Change of the u-component with time at lower and upper levels at grid point 28.7°N, 90.0°E in Test 5 (unit: m s^{-1}).

Now let us look at the vertical circulation in Test 5. Figure 7 is the time-mean east-west cross section along 33°N for days 5 to 8. We can see that upward motion dominates over the plateau. It is interesting to note that the upward motion is asymmetric with respect to the plateau. The eastern branch is more intense than the western one. In fact, in summer, the eastern part of the plateau experiences more rainfall than the western part. The sinking branch extends to the West Pacific, perhaps strengthening the subtropical high there.

Figure 8 gives the meridional time-mean cross section along 90°E for days 5 to 8. A striking feature is the monsoonal circulation cell whose sinking branch is located near the equatorial region and even extends into the southern hemisphere. This result suggests the probable extent of the influence of the heated plateau. Another sinking branch to the north of the plateau corresponds to the narrow high-pressure belt at the lower level and the arid climate there.

Figure 9 gives the meridional circulation along 90°E for days 9 to 12. During this period, the heating is removed; the most apparent difference from the case of heating is that the monsoonal circulation is not clear. But on the south slope of the Tibetan Plateau a downdraft appears

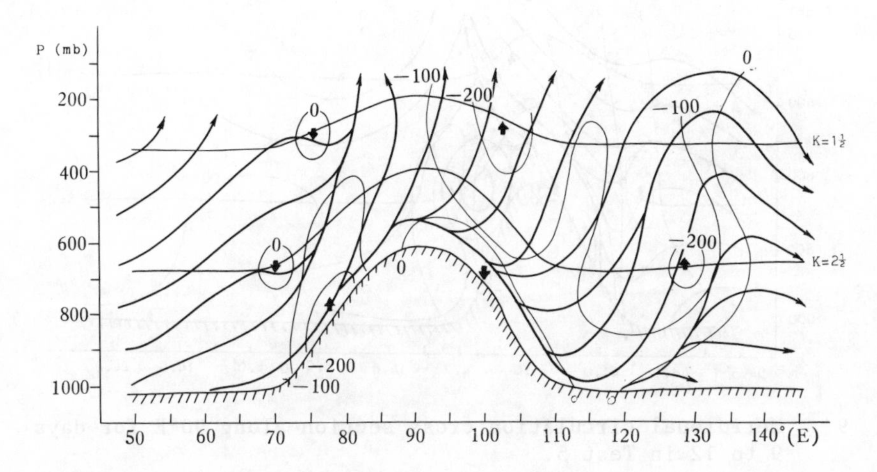

Fig. 7 East-west zonal circulation cross section along 33°N for days 5 to 8 in Test 5. Fine full lines denote the p-vertical velocity in units of 10^{-6} mb s^{-1}.

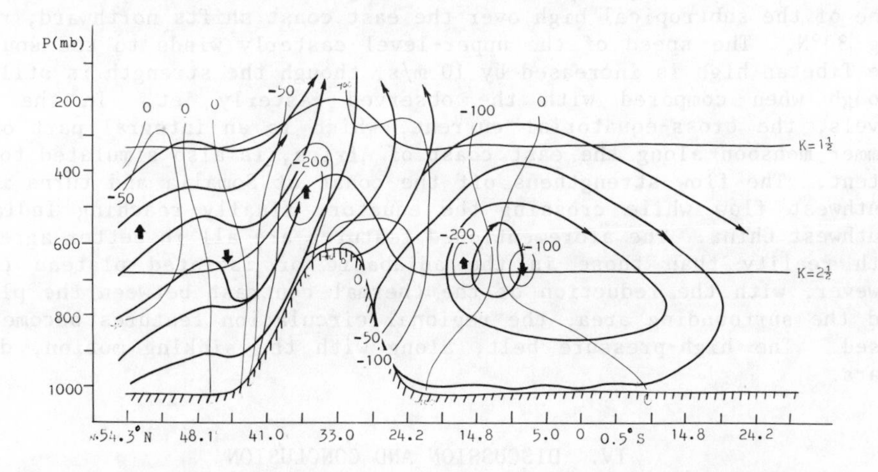

Fig. 8 Meridional circulation cross-section along 90°E for days 5 to 8 in Test 5. Lines are similar to those in Fig. 7.

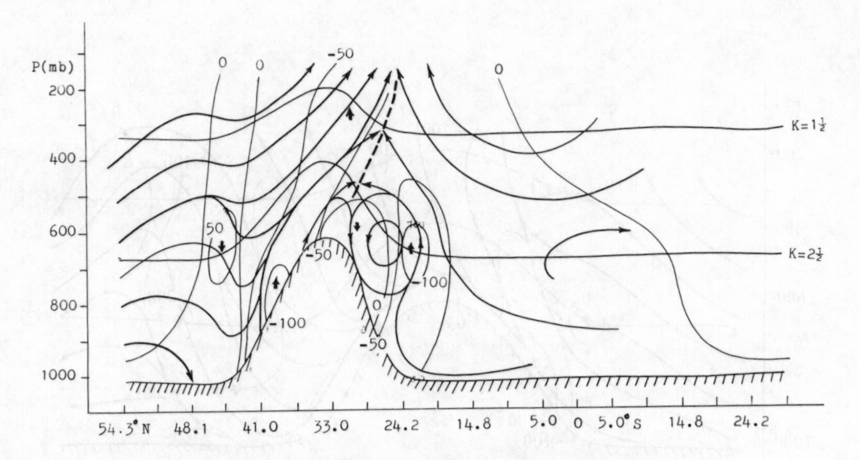

Fig. 9 Meridional circulation cross section along 90°E for days 9 to 12 in Test 5.

below 500 mb, and a secondary vertical circulation is in evidence there, indicating the dynamic effect of the plateau on the regional circulation.

iii. Tests of more realistic topography and heating: Figure 10 gives the results based on an enlarged, realistic topography of planetary scale. The improvement in the upper circulation simulation is obvious. The extent and strength of the Tibetan high is more realistic. The ridge line of the subtropical high over the east coast shifts northward, reaching 30°N. The speed of the upper-level easterly winds to the south of the Tibetan high is increased by 10 m/s, though the strength is still not enough when compared with the observed easterly jet. In the lower levels, the cross-equatorial current, which is an integral part of the summer monsoon along the east coast of Africa, is also simulated to some extent. The flow strengthens off the coast of Somalia and turns into a southwest flow while crossing the equator, finally reaching India and southwest China. The aforementioned features are all in better agreement with reality than those in the adiabatic or isolated plateau cases. However, with the reduction of the thermal contrast between the plateau and the surrounding area, the regional circulation features become confused. The high-pressure belt, along with the sinking motion, disappears.

IV. DISCUSSION AND CONCLUSION

Based on the three-level primitive-equation model in σ-coordinates, incorporating the dynamic and heating effects of the Tibetan Plateau, the major features of monsoons over Asia-Africa have been simulated, such as the cyclonic circulation around the Plateau; the west-east narrow high-pressure belt to the north of the plateau; the southwest airflow to the south; the cross-equatorial current off the coast of East Africa; western Pacific subtropical highs; some features near the equator in the low troposphere; the large anticyclone over Tibet; the easterlies to the

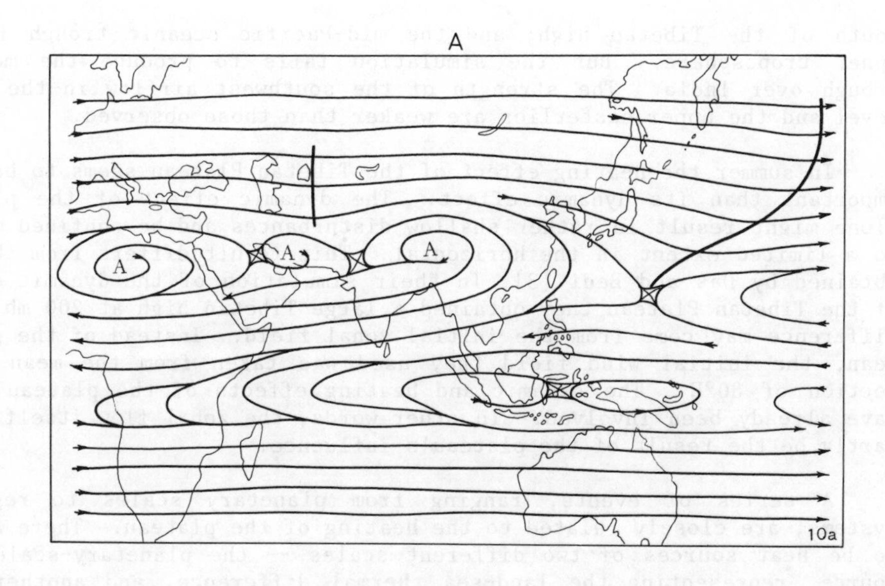

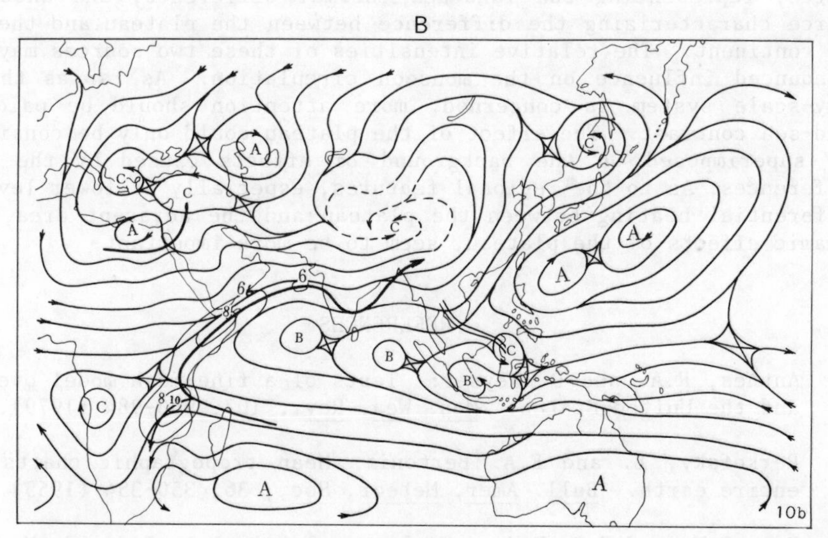

Fig. 10 Streamlines for day 7 of integration for more realistic topography and heating. Double lines denote the position of the ITCZ, fine full lines denote isotachs of wind in units of m s^{-1}, bold thick lines denote the axes of strong airflow. (a) 200 mb, (b) 850 mb.

south of the Tibetan high; and the mid-Pacific oceanic trough in the upper troposphere. But the simulation fails to produce the monsoon trough over India. The strength of the southwest airflow in the lower level and the upper easterlies are weaker than those observed.

In summer the heating effect of the Tibetan Plateau seems to be more important than its dynamic effect. The dynamic effect of the plateau alone might result in rather shallow disturbances and be confined mainly to a limited extent in the horizontal. This result differs from the one obtained by Das and Bedi [3]. In their simulation of the dynamic effect of the Tibetan Plateau they obtained a large Tibetan high at 200 mb. The difference may come from the initial zonal field. Instead of the global mean, the initial wind field they used was taken from the mean cross section of 80°E. The dynamic and heating effects of the plateau might have already been involved. In other words, the zonal flow itself might partly be the result of the plateau's influence.

A series of events, ranging from planetary scales to regional systems, are closely related to the heating of the plateau. There appear to be heat sources of two different scales -- the planetary-scale heat source, representing the land-sea thermal difference, and another heat source characterizing the difference between the plateau and the rest of the continent. The relative intensities of these two sources may exert a pronounced influence on the monsoon circulation. As far as the planetary-scale system is concerned, more attention should be paid to the land-sea contrast. The effect of the plateau could only be considered as one superimposed on the background of effects caused by the land-sea differences. As to the regional features, especially at lower levels, the differential heating between the plateau and the adjacent area, and the dynamic effects of the plateau, seem to be more important.

REFERENCES

1. Anthes, R.A. and D. Keyser: Tests of a fine-mesh model over Europe and the United States. Mon. Wea. Rev., 107, 963-984 (1979).

2. Berkofsky, L. and E.A. Bertoni: Mean tropographic charts for the entire earth. Bull. Amer. Meteor. Soc., 36, 350-354 (1955).

3. Das, P.K. and H.S. Bedi: Title not available. India J. Met. Hydrol. Geophys., 29, 375-383 (1978).

4. The Group of Medium-Range Forecast of the Institute of Atmospheric Physics. A computational scheme for P.E. model and its test. Scientia Atmospherica Sinica, 2, 39-51 (in Chinese) (1976).

5. Shen, Ru-jin: The rationality of the initialization of temperature and the calculation of the pressure gradient term. (To be published in Scientia Atmospherica Sinica) (in Chinese) (1982).

6. Washington, W.M. and D. Williamson: General circulation models of the atmosphere. Methods in Computational Physics, 17, 111-169 (1977).

3.7 THE MEAN WIND STRUCTURE OF THE BAROCLINIC, CONVECTIVE BOUNDARY LAYER

J.C. Wyngaard
National Center for Atmospheric Research[1]
Boulder, Colorado 80307 USA

ABSTRACT

Generalizing and extending the work of Young [19] and Mahrt [10], we develop a simple approach to calculating the wind profile in a baroclinic, convective boundary layer which is subject to very general influences--entrainment, subsidence, spatial and temporal variations in the geostrophic wind, and spatial and temporal variations in the surface conditions.

Along the way, we study the nature of the turbulent shear stress balance in these very general conditions. If the PBL is sufficiently convective ($-z_i/L > 30$, say) our preliminary findings indicate that there is a scalar eddy viscosity which relates mean wind shear and turbulent shearing stress. This viscosity has vertical structure, but scales with $w_* z_i$, and hence is independent of shear; this structure gives linear equations for the wind profile. We do not solve these equations completely, but suggest that their structure is such that mean wind shear is of the order of $m^{-1} = f z_i/w_*$ times the imposed geostrophic wind shear. Thus, m is an important boundary layer parameter.

I. INTRODUCTION

The diurnal heating and cooling cycle which strongly modulates the structure of the planetary boundary layer (PBL) over horizontal terrain has additional effects over mountainous or sloping terrain. There a diurnal upslope/downslope flow cycle, driven by the buoyancy acceleration parallel to the slope, is induced as well. This buoyancy acceleration and the mean pressure gradient combine to produce an effective pressure gradient in the along-slope direction which depends on distance from the surface.

Under these conditions, it follows that when we choose Cartesian axes oriented along, across, and perpendicular to the slope, the momentum equations in the plane parallel to the slope are those for a baroclinic

---

[1] The National Center for Atmospheric Research is sponsored by the National Science Foundation.

boundary layer over a horizontal surface. Thus, the structure of the baroclinic boundary layer is of fundamental interest in mountain meteorology.

As boundary-layer flow evolves in time it can modify the mean temperature distribution and hence change the effective pressure gradient. In this paper we will consider this mean temperature field to be prescribed; however, we will consider prescriptions which do have time dependence.

We will consider only the convective case, where the turbulence is driven primarily by the buoyancy forces generated through the upward flux of temperature (and moisture) at the surface. This case occurs when the stability index ($-z_i/L$) is sufficiently large, say greater than 30. As we will see, this condition makes the internal dynamics of the boundary layer simpler.

We will be discussing the PBL structure which emerges after suitable averaging, for example area averaging (in the horizontally homogeneous case), time averaging (in the stationary case), or ensemble averaging (in the general case). It is generally necessary to average many observations in order to make meaningful comparisons with this type of theory [15].

A useful framework for studying the baroclinic, convective PBL is sketched in Fig. 1. The interfacial layer extends from the PBL top (height h_2) to height h_1; below that is the mixed layer; its lower surface is h_s, the top of the surface layer. Of the three layers, the surface layer is the best understood because of extensive tower-based research over the past three decades.

Individual temperature and moisture profiles indicate that the transition between the mixed layer and the free atmosphere is often quite abrupt. However, there is considerable spatial and temporal variation in the position of this interface, partly because of the large, intense, and random turbulent eddies in the convective PBL. Therefore, after averaging, an interfacial layer of considerable thickness emerges. Deardorff [5] indicates that its depth ranges from 20 percent to 50 percent of the mixed-layer depth.

In this paper we will take the mixed-layer top at h_1 as identical to z_i, the height of the lowest inversion base; z_i is often clearly evident in temperature and moisture profiles. It tends to increase with time due to turbulent entrainment, and to decrease due to mean subsidence. It typically ranges from a few hundred meters to a few kilometers, depending on the conditions.

The surface layer is usually thought to have simpler scaling (Monin-Obukhov similarity) than the mixed layer. It is convenient to think of it as the lower 10 percent, say, of the PBL; thus if the turbulent fluxes have roughly linear profiles in the PBL, falling to zero at the

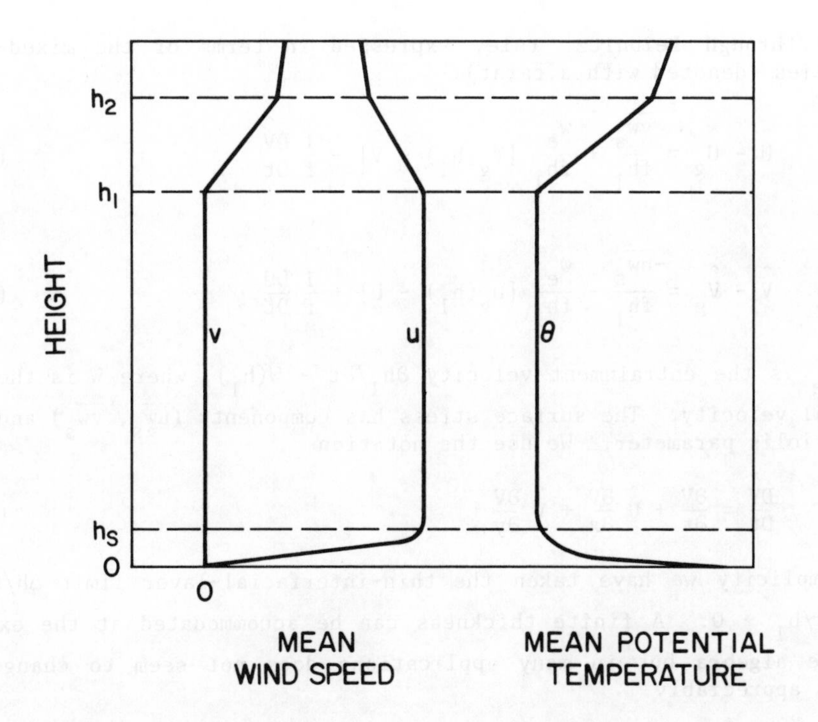

Fig. 1 An idealization of the convective PBL, from [7]. The surface layer extends to h_s; the shear-free mixed layer is between h_s and h_1; the interfacial layer extends from h_1 and h_2.

PBL top, then the fluxes in the surface layer remain within about 10 percent of their surface values. Thus the surface layer is often called the "constant flux" layer. Its structure has been reviewed by Busch [2], Businger [3], and Wyngaard [15].

Within this framework one can construct convective, baroclinic PBL models of varying sophistication. We will discuss two, starting with the simpler.

II. MIXED LAYER WITHOUT SHEAR

The simpler three-layer model assumes that the convectively-induced turbulence creates flat wind and temperature profiles in the mixed layer, Fig. 1. Linear wind profiles are assumed between the mixed-layer values at h_1 and the geostrophic values at h_2. The geostrophic wind components U_g and V_g are assumed to vary linearly between the surface and h_2. For these conditions Garratt et al. [7] have shown that the mean horizontal momentum equation can be integrated over height z and, through Leibnitz' rule, expressed in terms of the mixed-layer quantities (denoted with a carat):

$$\hat{U} - \hat{U}_g = \frac{\overline{vw}_s}{fh_1} + \frac{w_e}{fh_1}[V_g(h_1) - \hat{V}] - \frac{1}{f}\frac{D\hat{V}}{Dt} \tag{1}$$

$$\hat{V} - \hat{V}_g = \frac{-\overline{uw}_s}{fh_1} - \frac{w_e}{fh_1}[U_g(h_1) - \hat{U}] + \frac{1}{f}\frac{D\hat{U}}{Dt} \tag{2}$$

Here w_e is the entrainment velocity $\partial h_1/\partial t - W(h_1)$, where W is the mean vertical velocity. The surface stress has components $(\overline{uw}_s, \overline{vw}_s)$ and f is the Coriolis parameter. We use the notation

$$\frac{D\hat{V}}{Dt} = \frac{\partial\hat{V}}{\partial t} + \hat{U}\frac{\partial\hat{V}}{\partial x} + \hat{V}\frac{\partial\hat{V}}{\partial y} \tag{3}$$

For simplicity we have taken the thin-interfacial-layer limit $\Delta h/h_1 = (h_2-h_1)/h_1 \rightarrow 0$. A finite thickness can be accommodated at the expense of more algebra but in many applications does not seem to change the physics appreciably.

A surface-stress closure is

$$(\overline{uw}_s, \overline{vw}_s) = -C_D S(\hat{U}, \hat{V}) \tag{4}$$

where C_D is a drag coefficient and $S = |(\hat{U}, \hat{V})|$ is the wind speed at the edge of the surface layer. Thus the layer-averaged Equations (1) and (2) become

$$\hat{U} = \hat{U}_g - C\hat{V} + \frac{w_e}{fh_1}[V_g(h_1) - \hat{V}] - \frac{1}{f}\frac{D\hat{V}}{Dt} \tag{5}$$

$$\hat{V} = \hat{V}_g + C\hat{U} + \frac{w_e}{fh_1}[U_g(h_1) - \hat{U}] + \frac{1}{f}\frac{D\hat{U}}{Dt} \tag{6}$$

where $C = C_D S/(fh_1)$ is the new, scaled drag coefficient. If $S = 10\ m\ s^{-1}$, $f = 10^{-4}\ s^{-1}$, and $h = 10^3$ m, then $C = 10^2\ C_D$. Thus we expect C to vary between 0.1 and 1.0 under typical conditions.

Equations (5) and (6) are the two components of what is essentially the mixed-layer mean horizontal momentum balance. The left side is proportional to the Coriolis force; the terms on the right are proportional to the mean pressure gradient, the surface drag, the drag at h_1 due to entrainment, and the sum of local and advective accelerations, respectively. If the last two terms vanish, as is often idealized, then the solution to Eqs. (5) and (6) is simply

$$\hat{U} = \frac{\hat{U}_g - C\hat{V}_g}{1+C^2}; \quad \hat{V} = \frac{C\hat{U}_g + \hat{V}_g}{1+C^2} \tag{7}$$

If we orient the axes along the mean mixed-layer wind, then Eq. (7) reduces to $\hat{U} = \hat{U}_g$, $\hat{V} = 0$, a generalization of the usual mixed-layer solution.

Let us now examine the nature of the mean momentum balances Eqs. (5) and (6) when entrainment and acceleration effects are not negligible. Taking the acceleration term first, we assume that there are velocity and horizontal length scales Q and L, respectively, such that the $f^{-1}D/Dt$ terms in Eqs. (5) and (6) are of order Q/fL times the leading terms. We assume that Q/fL, which we call R since it is a Rossby number, is small, so that the acceleration terms are small. If, for example, $f = 10^{-4}\ s^{-1}$, $Q = 10\ m\ s^{-1}$, and $L = 10^6$ m, then $R = 10^{-1}$.

Turning now to the entrainment term, we note that since by continuity $W(h_1)/h_1 \sim Q/L$, and since w_e is typically of the order of $W(h_1)$, it follows that w_e/fh_1 will typically be of the order of R. Thus we will consider cases where acceleration and entrainment terms are small and of the same order.

Formally, then, we scale the velocities in Eqs. (5) and (6) by Q, scale x and y by L, and scale time by L/Q. We denote dimensionless time by t*, but multiply the equations by Q so that we can retain dimensional velocities. The scaled forms of Eqs. (5) and (6) then are

$$\hat{U} = \hat{U}_g - C\hat{V} + R\left\{\frac{w_e}{fh_1R}\left[V_g(h_1) - \hat{V}\right] - \frac{D\hat{V}}{Dt^*}\right\} \tag{8}$$

$$\hat{V} = \hat{V}_g + C\hat{U} - R\left\{\frac{w_e}{fh_1R}\left[U_g(h_1) - \hat{U}\right] - \frac{D\hat{U}}{Dt^*}\right\} \tag{9}$$

Note that the scaled entrainment parameter $w_e/(fh_1R)$ is of the order of 1.

The solutions to Eqs. (8) and (9) can be expressed as expansions in R; these have been used previously by Young [19] for the isallobaric wind in the Ekman layer and by Mahrt [10] for momentum advections in a mixed layer. We write

$$\hat{U} = \hat{U}_o + R\hat{U}_1 + \ldots \tag{10}$$

$$\hat{V} = \hat{V}_o + R\hat{V}_1 + \ldots \quad (11)$$

If we substitute Eqs. (10) and (11) into Eqs. (8) and (9) and collect terms of O(1), we have

$$\hat{U}_o = \hat{U}_g - C\hat{V}_o \quad (12)$$

$$\hat{V}_o = \hat{V}_g + C\hat{U}_o \quad (13)$$

whose solution is given in Eq. (7). Collecting O(R) terms gives

$$R\hat{U}_1 = -CR\hat{V}_1 + \{\frac{w_e}{fh_1}[V_g(h_1) - \hat{V}_o] - \frac{D\hat{V}_o}{fDt}\} \quad (14)$$

$$R\hat{V}_1 = CR\hat{U}_1 - \{\frac{w_e}{fh_1}[U_g(h_1) - \hat{U}_o] - \frac{D\hat{U}_o}{fDt}\} \quad (15)$$

Clearly $\hat{U}_1$ and $\hat{V}_1$ are nonzero only if there is entrainment or acceleration. These forcing effects are calculated directly from the lowest-order solution; we will illustrate this with simple examples.

1. Effects of Time-Varying Geostrophic Wind

Perhaps the simplest departure from the ideal is a surface geostrophic wind which depends on time. This causes acceleration in the mixed-layer winds, and thereby induces corrections $R\hat{U}_1$ and $R\hat{V}_1$ to the lowest-order solution, Eq. (7).

For simplicity we consider the nonentraining limit $w_e/fh_1 \to 0$. In this case the first-order Equations (14) and (15) become

$$R\hat{U}_1 = -CR\hat{V}_1 - \frac{\partial\hat{V}_o}{f\partial t} \quad (16)$$

$$R\hat{V}_1 = CR\hat{U}_1 + \frac{\partial\hat{U}_o}{f\partial t} \quad (17)$$

A comparison of Eqs. (16) and (17) with Eqs. (12) and (13) shows that the local time accelerations $-\partial\hat{V}_o/\partial t$ and $\partial\hat{U}_o/\partial t$ play the same role in the first-order equations that the geostrophic wind components play in the zero-order equations.

The first-order solution $R\hat{U}_1$ follows from Eqs. (16) and (17) as

$$R\hat{U}_1 = \frac{-\frac{\partial \hat{V}_o}{f\partial t} - C\frac{\partial \hat{U}_o}{f\partial t}}{1 + C^2} \tag{18}$$

with $R\hat{V}_1$ following from Eq. (18) through the Ekman-layer transformation $U \rightarrow V$, $V \rightarrow -U$.

The local time accelerations $\partial\hat{U}_o/\partial t$ and $\partial\hat{V}_o/\partial t$ are found by differentiating the zero-order solutions Eqs. (12) and (13). If we assume that only the surface geostrophic wind (denoted with a subscript s) varies with time, its vertical shear remaining constant, then we obtain

$$\frac{\partial\hat{U}_o}{\partial t} = \frac{\partial U_{gs}}{\partial t}\left(1 - C\frac{\hat{U}_o\hat{V}_o}{S^2}\right) - C\frac{\partial V_{gs}}{\partial t}\left(1 + \frac{\hat{V}_o^2}{S^2}\right) \tag{19}$$

The solution for $\partial\hat{V}_o/\partial t$ can be found from Eq. (19) by the Ekman-layer transformation.

2. Effects of Entrainment

Another simple departure from the ideal is the growth of h_1 with time through entrainment. For simplicity we take the geostrophic wind as constant in space and time. In this case $\hat{U}_1$ and $\hat{V}_1$ are driven by the entrainment and local-acceleration terms in Eqs. (14) and (15), and both can be calculated directly from the lowest-order solutions Eqs. (12) and (13). The entrainment term is straightforward, and for the acceleration term we carry out the differentiation and assume that $C \ll 1$ so that terms in C^2 can be dropped, giving

$$\frac{\partial\hat{U}_o}{\partial t} = \frac{1}{h_1}\frac{\partial h_1}{\partial t}\left[\left(\frac{h_1}{2}\frac{\partial U_g}{\partial z} + C\hat{V}_o\right)\left(1 - \frac{C\hat{U}_o\hat{V}_o}{S^2}\right) - \left(\frac{h_1}{2}\frac{\partial V_g}{\partial z} - C\hat{U}_o\right)\left(1 + \frac{\hat{V}_o^2}{S^2}\right)\right] \tag{20}$$

The solution for $\partial\hat{V}_o/\partial t$ again follows from Eq. (20) by the Ekman-layer transformation. Equation (20) is rather complicated in detail, but one important aspect is clear: the time scale for entrainment-induced accelerations in the mixed layer is $h_1/(\partial h_1/\partial t)$. If $h_1 = 10^3$ m, and $\partial h_1/\partial t = 10^{-2}$ m s^{-1}, this time scale is 10^5 sec, or about one day.

The solution (20) and its counterpart for $\partial\hat{V}_o/\partial t$ can be used in Eqs. (14) and (15) to calculate the entrainment-induced, first-order corrections to the mixed-layer profiles, Eqs. (10) and (11). By inspection we see that these corrections are of the order of w_e/fh_1 times the mixed layer velocity, although again their precise forms are complicated.

In this way one can calculate the mixed-layer response to a number of influences, such as temporal and spatial variations in geostrophic wind; spatially-varying surface roughness, heat transfer, and boundary-layer depth; and subsidence with entrainment. The underlying assumptions are that these forcing effects are weak, and that the mixed-layer wind profile remains flat.

3. Ekman Pumping

Our lowest-order solution for mixed-layer winds is given by Eqs. (12) and (13). If we define a vertical velocity W_o which satisfies the incompressible continuity equation, i.e.

$$\frac{\partial W_o}{\partial z} = -\frac{\partial \hat{U}_o}{\partial x} - \frac{\partial \hat{V}_o}{\partial y} \tag{21}$$

then, since $\hat{U}_o$ and $\hat{V}_o$ are independent of z, we have

$$W_o(h_1) = -h_1\left(\frac{\partial \hat{U}_o}{\partial x} + \frac{\partial \hat{V}_o}{\partial y}\right) \tag{22}$$

which, from Eqs. (12) and (13), is

$$W_o(h_1) = h_1\left(\frac{\partial}{\partial x}(C\hat{V}_o) - \frac{\partial}{\partial y}(C\hat{U}_o)\right) \tag{23}$$

The right side of Eq. (23) is h_1 times the curl of the surface stress, and gives what is known as Ekman pumping. It follows from our expansions, Eqs. (10) and (11), for $\hat{U}$ and $\hat{V}$ that W also has an expansion, i.e.

$$W = W_o + RW_1 + \ldots \tag{24}$$

with W_o being given by Eq. (21) and its value at h_1 by Eq. (23). The correction to Ekman pumping is therefore

$$RW_1(h_1) = - h_1 R(\frac{\partial \hat{U}_1}{\partial x} + \frac{\partial \hat{V}_1}{\partial y}) \tag{25}$$

If we write Eqs. (14) and (15) as

$$R\hat{U}_1 = - CR\hat{V}_1 + F_1 \tag{26}$$

$$R\hat{V}_1 = CR\hat{U}_1 + F_2 \tag{27}$$

then we have, from Eq. (25)

$$RW_1(h_1) = h_1 R(\frac{\partial (C\hat{V}_1)}{\partial x} - \frac{\partial (C\hat{U}_1)}{\partial y}) - h_1(\frac{\partial F_1}{\partial x} + \frac{\partial F_2}{\partial y}) \tag{28}$$

The first pair of terms on the right side of Eq. (28) represents a correction to the curl of the surface stress. The second pair, however, represents a different mechanism of Ekman pumping, one which is a combination of entrainment and acceleration effects.

III. MIXED LAYER WITH SHEAR

Our simple model of Section II has a shear-free wind profile in the mixed layer. However, observations in the baroclinic, convective PBL (e.g. those from AMTEX, Fig. 2) often show significant wind shear. One suspects that departures from the idealized shear-free state can have important effects on mixed-layer structure, perhaps as important as the entrainment and acceleration effects we discussed in Section II.

We achieved closure in Section II by assuming a form for the wind profile. If instead we wish to calculate this profile, we need another approach to closure. We will consider two alternatives in this section.

1. Closure Through the Stress Budget

The stress budget in the atmospheric surface layer has been studied by Wyngaard et al. [18], but there have been only fragmentary studies of its behavior in the mixed layer.

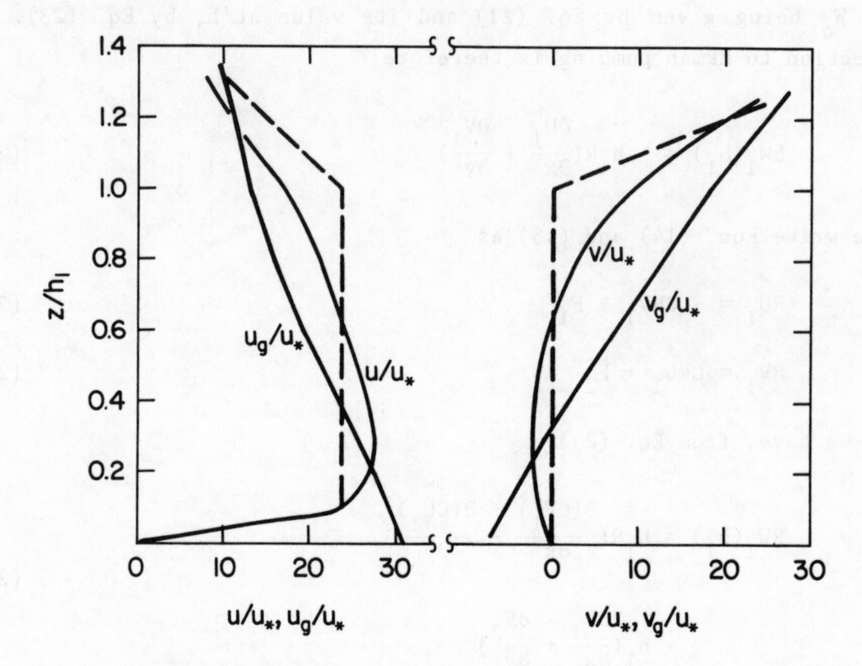

Fig. 2 Observations of the AMTEX winds, from [7]. The dashed lines are the idealized profiles from the model of Fig. 1.

Consider a convective, baroclinic PBL subject to the influences discussed in Section II -- entrainment, time changes, and horizontal inhomogeneity. The appropriate form of the $\overline{uw}$ budget is

$$\frac{\partial}{\partial t}\overline{uw} + U\frac{\partial\overline{uw}}{\partial x} + V\frac{\partial\overline{uw}}{\partial y} + W\frac{\partial\overline{uw}}{\partial z} + \overline{w^2}\frac{\partial U}{\partial z} + \frac{\partial}{\partial z}\overline{uw^2}$$
$$= -\overline{\frac{\partial p}{\partial x}w} - \overline{\frac{\partial p}{\partial z}u} + \frac{g}{T}\overline{\theta u} \tag{29}$$

where (U, V, W) and (u, v, w) are the mean and turbulent wind vectors, and p is the kinematic pressure fluctuation. The $\overline{vw}$ budget follows from Eq. (29) under $u \rightarrow v$, $U, \rightarrow V$.

Equation (29) describes the evolution of $\overline{uw}$. Its terms represent, in order, time change; mean advection; shear production; turbulent transport; pressure loss; and buoyant production. These identifications are discussed in detail by Wyngaard et al. [18].

An important simplification occurs in Eq. (29) and its counterpart for $\overline{vw}$. To illustrate, consider the relative magnitudes of the time change and shear production terms. If $\overline{w^2} = 1\ m^2\ s^{-2}$, $\partial U/\partial z = 10^{-3}\ s^{-1}$

and $\overline{uw} = 10^{-1}\ m^2\ s^{-2}$, all typical values for a convective PBL, then in order for $\partial\overline{uw}/\partial t$ to be equal to $w^2\ \partial U/\partial z$, $\overline{uw}$ would have to change by 100 percent in 100 seconds! Clearly this is all but impossible, and we conclude that the time change terms will be negligible under typical conditions. The same argument shows that mean advection will also typically be quite small.

One can generalize this argument to show that the time-change and mean advection terms scale with $\overline{uw}\ Q/L$, while the remainder of the terms scale with $\overline{uw}\ q/\ell$; here q and ℓ are velocity and length scales of the turbulence. Thus, these terms have magnitudes in the ratio $Q\ell/qL$. Although q is of the order of (but less than) the mean velocity scale Q, the three-dimensionality of the turbulence keeps $\ell \simeq h_1$, while L is much larger; hence $Q\ell/(q/L) << 1$. Thus we have the important result that a simplified stress budget should hold in most baroclinic, convective PBL's, even those with time changes or horizontal inhomogeneity. The simplified component equations are

$$\frac{\partial}{\partial t}\ \overline{uw} = -\ \overline{w^2}\ \frac{\partial U}{\partial z} - \frac{\partial}{\partial z}\ \overline{uw^2} + \frac{g}{T}\ \overline{\theta u} - \overline{\frac{\partial p}{\partial x}\ w} - \overline{\frac{\partial p}{\partial z}\ u} \simeq 0 \qquad (30)$$

$$\frac{\partial}{\partial t}\ \overline{vw} = -\ \overline{w^2}\ \frac{\partial V}{\partial z} - \frac{\partial}{\partial z}\ \overline{vw^2} + \frac{g}{T}\ \overline{\theta v} - \overline{\frac{\partial p}{\partial y}\ w} - \overline{\frac{\partial p}{\partial z}\ v} = 0 \qquad (31)$$

The measurement of all but the pressure terms in these equations is now routine in PBL research programs, so these pressure terms can be inferred from Eqs. (30) and (31) by difference.

It is difficult to study Eqs. (30) and (31) with data from the midlevels of typical convective PBL's with small wind shear, however, because their terms are apt to be lost in the "noise" of the convective turbulence, as discussed by Wyngaard and Arya [17]. This "noise" also causes large scatter in measured stress profiles (e.g. the AMTEX profiles, Fig. 3). Nonetheless, we did find two sets of Minnesota runs with enough lateral mean wind shear to allow the $\overline{vw}$ budget to be resolved, and four AMTEX days which had sufficient streamwise wind shear to enable the $\overline{uw}$ budget to be determined.

The Minnesota $\overline{vw}$ budgets are presented in Fig. 4. In runs 2A1 and 2A2 (top panel) the shear production term could be calculated because the lateral wind shear $\partial V/\partial z$ was fairly large; during the 2.5 hours of observations it averaged $-4\times10^{-3}\ s^{-1}$ between the surface and 1100 m. The stability index $-z_i/L$ averaged 36, and z_i averaged 1430 m. The results clearly show that over the bulk of the PBL shear production is the dominant source of $\overline{vw}$, so that by inference the pressure term is the dominant loss term. In run 5A1 (bottom panel) $\partial V/\partial z$ averaged $3.4\times10^{-3}\ s^{-1}$

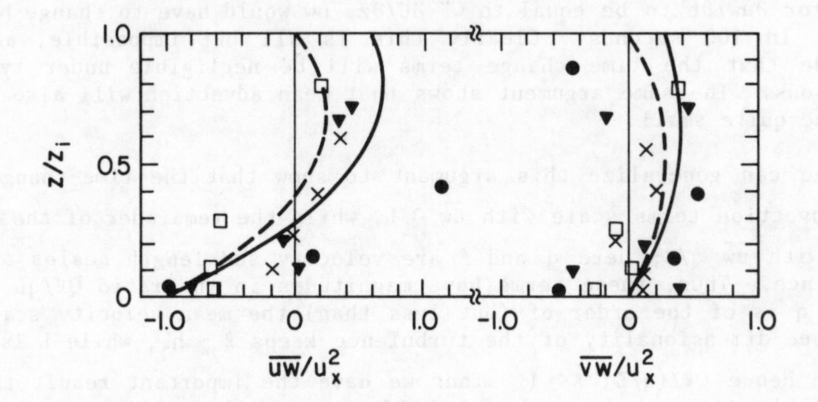

Fig. 3 AMTEX stress profiles [9].

during the 1.25 hour run; z_i was 1100 m and $-z_i/L$ was 150. Since $\overline{vw}$ was negative, gain terms are also negative in this run. Again shear production is the dominant source of $\overline{vw}$, with the pressure term the balancing loss term.

The streamwise wind shear $\partial U/\partial z$ in the AMTEX runs of 15, 16, 18 and 22 February 1975 [9] was smaller, averaging $-1.6\text{x}10^{-3}\ s^{-1}$ between 100 m and 1000 m, and thus it was necessary to average over all four days in order to compute the $\overline{uw}$ budget, Fig. 5. Again, however, the results indicate that shear production is the dominant source term and the pressure term is the dominant sink.

One of the oldest and simplest parameterizations for the pressure terms in Eqs. (30) and (31) is that due to Rotta [12],

$$\overline{\frac{\partial p}{\partial x}w} + \overline{\frac{\partial p}{\partial z}u} = \frac{\overline{uw}}{\tau} \tag{32}$$

$$\overline{\frac{\partial p}{\partial y}w} + \overline{\frac{\partial p}{\partial z}v} = \frac{\overline{vw}}{\tau} \tag{33}$$

where τ is an energy-containing range turbulent time scale. This parameterization implies that the pressure terms are sinks of stress, which is consistent with our results here and with previous surface-layer results. This is the standard parameterization in second-order turbulence models, and has received some theoretical justification from Weinstock [14] in homogeneous turbulence.

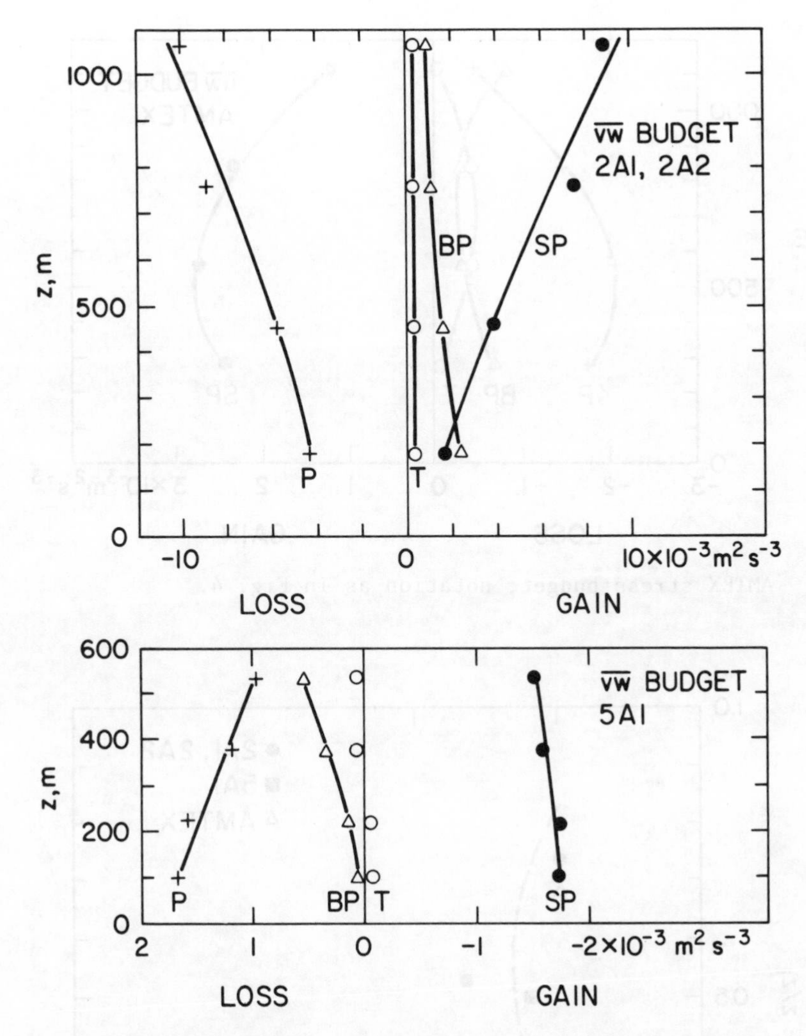

Fig. 4 Stress budgets from the 1973 Minnesota experiments [8]. SP, shear production; BP, buoyant production; T, turbulent transport; P, pressure term, all in Eqs. (30) and (31).

Figure 6 is a test of the Rotta parameterizations Eqs. (32) and (33) with the Minnesota and AMTEX stress budgets. In each case the inferred pressure term, Figs. 4 and 5, was used with Eqs. (32) and (33) and the measured stress to deduce the time scale τ. Figure 6 shows that τ scales with z_i/w_*, where $w_* = (g/T\, \overline{\theta w}_s\, z_i)^{1/3}$, is the convective PBL velocity scale. Much of the scatter in Fig. 6 is due to the scatter in the stress measurements, and possibly some is due to the neglect of "rapid" effects in Eqs. (32) and (33), [20]. Nonetheless Fig. 6 provides encouraging

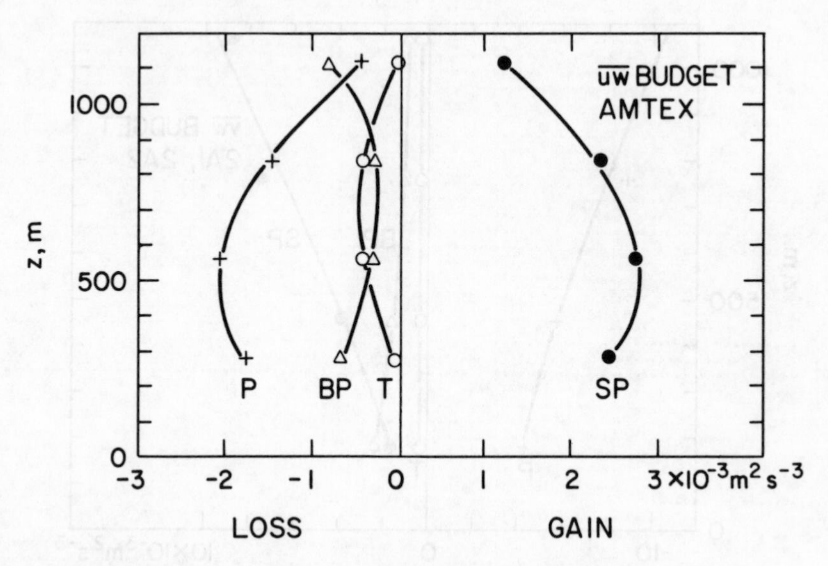

Fig. 5 AMTEX stress budget; notation as in Fig. 4.

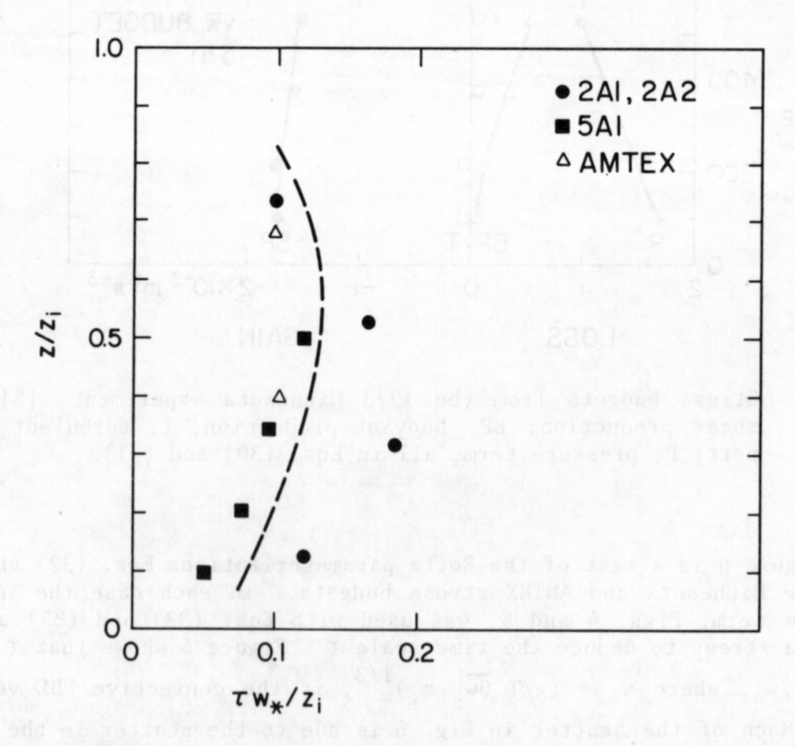

Fig. 6 A test of the Rotta [12] parameterization for the pressure terms in the stress budgets.

support for the Rotta parameterization in a convective PBL with mean wind shear.

These results allow us to write the stress budgets Eqs. (30) and (31) as

$$\overline{uw} = - \overline{w^2}\, \tau \frac{\partial U}{\partial z} \left[1 + \frac{\frac{\partial}{\partial z}\overline{uw^2} - \frac{g}{T}\overline{\theta u}}{\overline{w^2}\frac{\partial U}{\partial z}}\right] \tag{34}$$

$$\overline{vw} = - \overline{w^2}\, \tau \frac{\partial V}{\partial z} \left[1 + \frac{\frac{\partial}{\partial z}\overline{vw^2} - \frac{g}{T}\overline{\theta v}}{\overline{w^2}\frac{\partial V}{\partial z}}\right] \tag{35}$$

which are eddy-diffusivity expressions. In mid-PBL ($z/z_i \simeq 0.5$), $\overline{w^2} \simeq 0.4\, w_*^{\,2}$ [8] [9], and from Fig. 6 we take $\tau \simeq 0.12\, z_i/w_*$. With Eqs. (34) and (35), these give at $z \simeq 0.5\, z_i$

$$K_{\overline{uw}} \simeq 0.05\, w_* z_i \left[1 + \frac{T+B}{SP}\right]_{\overline{uw}} \tag{36}$$

$$K_{\overline{vw}} \simeq 0.05\, w_* z_i \left[1 + \frac{T+B}{SP}\right]_{\overline{vw}} \tag{37}$$

where the bracketed terms represent the full expressions written in Eqs. (34) and (35). We tested Eqs. (36) and (37) with measurements of K at $z \simeq 0.5\, z_i$ in runs 2A1, 2A2 and 5A1 of the Minnesota experiment, from the four AMTEX days, and from hour 15 in Deardorff's [4] three-dimensional simulation of day 33 of the Wangara experiment with the results shown in Fig. 7. In the latter case we could not compute the bracketed terms in Eqs. (36) and (37); we took them as 1.0. Allowing for the scatter, Eqs. (36) and (37) fit the results well.

This K scaling for the convective PBL is remarkably similar to that found previously for neutral cases. Brost et al. [1] cite measurements in laboratory boundary layers and in the inversion-capped neutral PBL which indicate that midlayer K values are of the order of $0.03\, q\, z_i$, where in the laboratory cases z_i is taken as Townsend's [13] δ_o. Here $q^2 = \langle \overline{u^2} + \overline{v^2} + \overline{w^2} \rangle$, where the brackets denote averaging over the layer. For the convective PBL, $q \simeq w_*$ [9] so that our results imply midlayer K values of the order of $0.05\, q\, z_i$.

The bracketed terms in Eqs. (36) and (37) vary from about 0.6 to 1.25 in the data we have analyzed here; we will take them as 1.0. While it might appear that they could be calculated more accurately via second-order closure, recent evidence [16] shows that many of the typical closure parameterizations in these models are unrealistic in some portion of the convective PBL. Thus, second-order model calculations of the eddy viscosity carry their own uncertainties. In view of these uncertainties

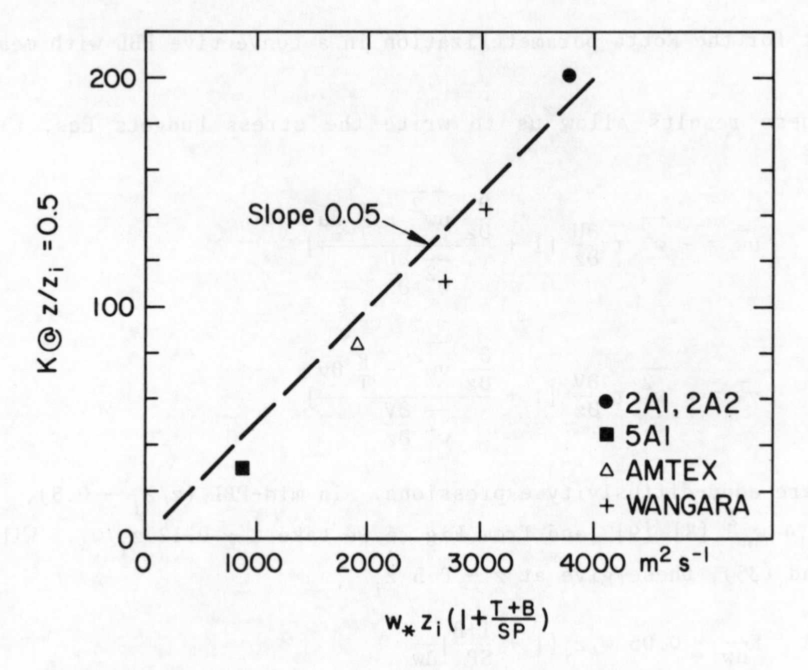

Fig. 7 A comparison of midlayer K values inferred from stress budgets with those directly measured.

and the complication of second-order models we prefer to use here our simple finding that the mid-PBL eddy viscosity has a magnitude of the order of 0.05 $w_* z_i$, which implies that to first approximation it is independent of the mean wind shear. Thus this "convective viscosity" approximation leads to linear equations, as we will show.

2. Closure Through Large-Eddy Simulation

Three-dimensional, time-dependent numerical modelling of turbulence, also called large-eddy simulation, is another way to obtain information on stress maintenance in the convective PBL. The method is generally considered reliable to the extent that it resolves the dynamics of the energy-containing eddies; thus in PBL applications it is necessary to use a grid scale of the order of 100 m. The smallest eddies are, of course, unresolvable and are parameterized.

Space does not allow more than a brief description of some large-eddy simulations that my colleague, R.A. Brost, and I are doing for the baroclinic, convective PBL. The domain size is 5 km x 5 km x 2 km deep, and the numerical grid 40 x 40 x 40; the grid interval is therefore 125 m in the horizontal and 50 m in the vertical. The model is basically a dry version of that discussed by Deardorff [6].

The flow is made very baroclinic by imposing a geostrophic wind which varies linearly from zero at the surface to 18 m s^{-1} at 2 km. For the preliminary results presented here the inversion base z_i was about 1300 m, and the stability index $-z_i/L$ was about 30. The simulations are still underway, and will be reported in detail in the future. In the meantime, we can discuss some preliminary results.

Figures 8 and 9 are the horizontally-averaged wind and shear-stress profiles at a given time, and in coordinates aligned with the surface-layer wind. The mean wind shear is generally smaller than the geostrophic wind shear. In order to study the stress-mean shear relationship we rotated the axes, aligning them with the direction of mean shear averaged over the mixed layer. Figure 10 shows that in these axes the along-shear stress component is much the larger; the shear stress and mean shear were within about 30° of parallelism throughout the mixed layer. Figure 11 shows the resulting eddy diffusivity profile for the along-shear stress component. Like the other plots, it shows some scatter due to insufficient averaging, but it is gratifying that the midlayer

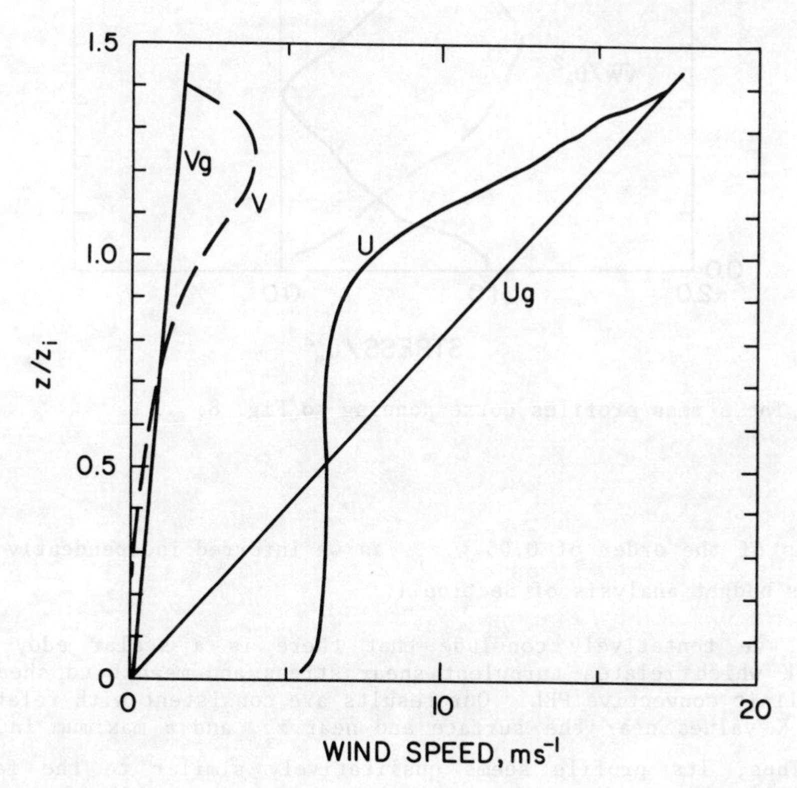

Fig. 8 Mean wind profiles in a very baroclinic, convective PBL from large-eddy simulations.

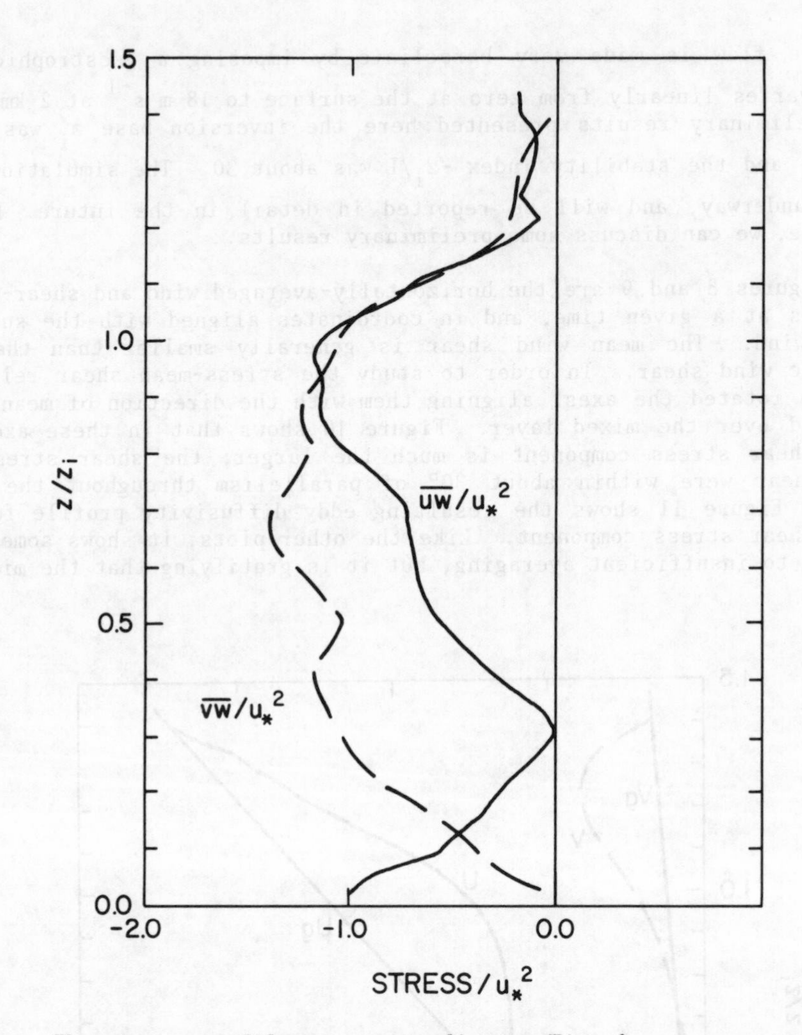

Fig. 9 The stress profiles corresponding to Fig. 8.

K value is of the order of 0.05 $w_* z_i$, as we inferred independently from the stress budget analysis of Section 1.

Thus, we tentatively conclude that there is a scalar eddy diffusivity K which relates turbulent shear stress and mean wind shear in the baroclinic convective PBL. Our results are consistent with relatively small K values near the surface and near z_i, and a maximum in midlayer. Thus, its profile seems qualitatively similar to the famous O'Brien [11] third-order polynomial. One important difference seems clear, however; its midlayer values scale with $w_* z_i$, while the O'Brien profile does not. This difference could be quite important.

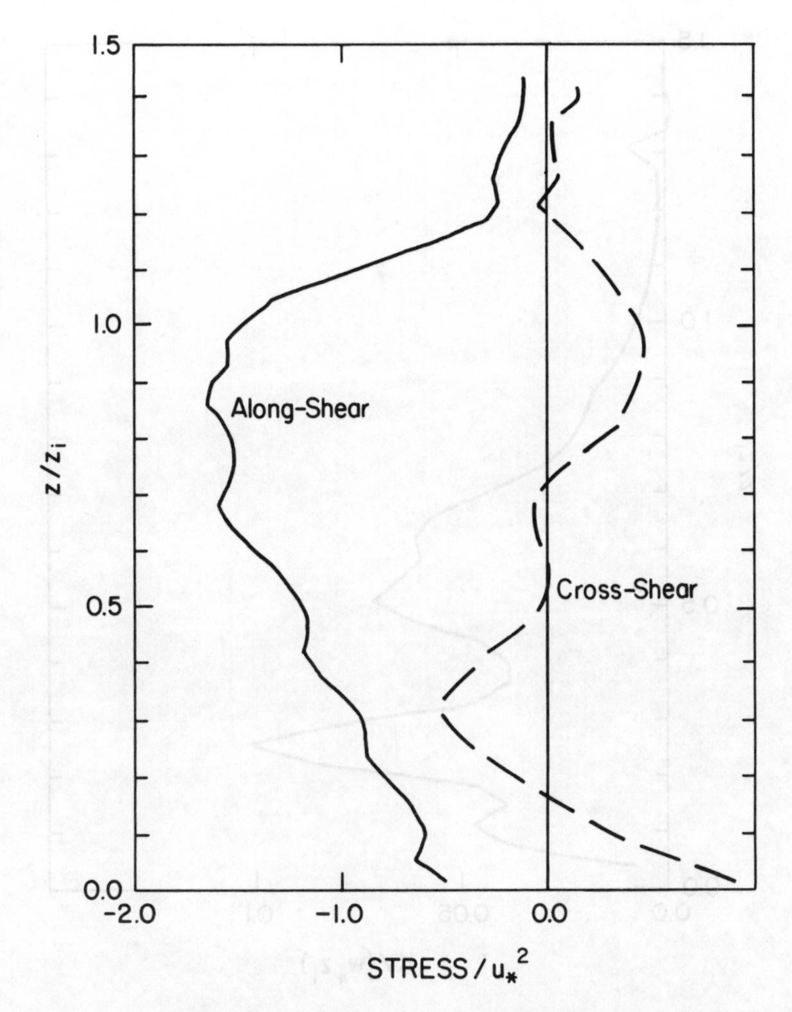

Fig. 10 Stress components of Fig. 9 in coordinates aligned with the layer-averaged wind shear.

3. Outline of a Solution for the Mixed-Layer Wind Profile

From our results to this point we can write the equations of mean horizontal momentum in the general case as:

$$\frac{\partial U}{\partial t} + U\frac{\partial U}{\partial x} + V\frac{\partial U}{\partial y} + W\frac{\partial U}{\partial z} - \frac{\partial}{\partial z}K\frac{\partial U}{\partial z} = f(V-V_g) \tag{38}$$

$$\frac{\partial V}{\partial t} + U\frac{\partial V}{\partial x} + V\frac{\partial V}{\partial y} + W\frac{\partial V}{\partial z} - \frac{\partial}{\partial z}K\frac{\partial V}{\partial z} = f(U_g-U) \tag{39}$$

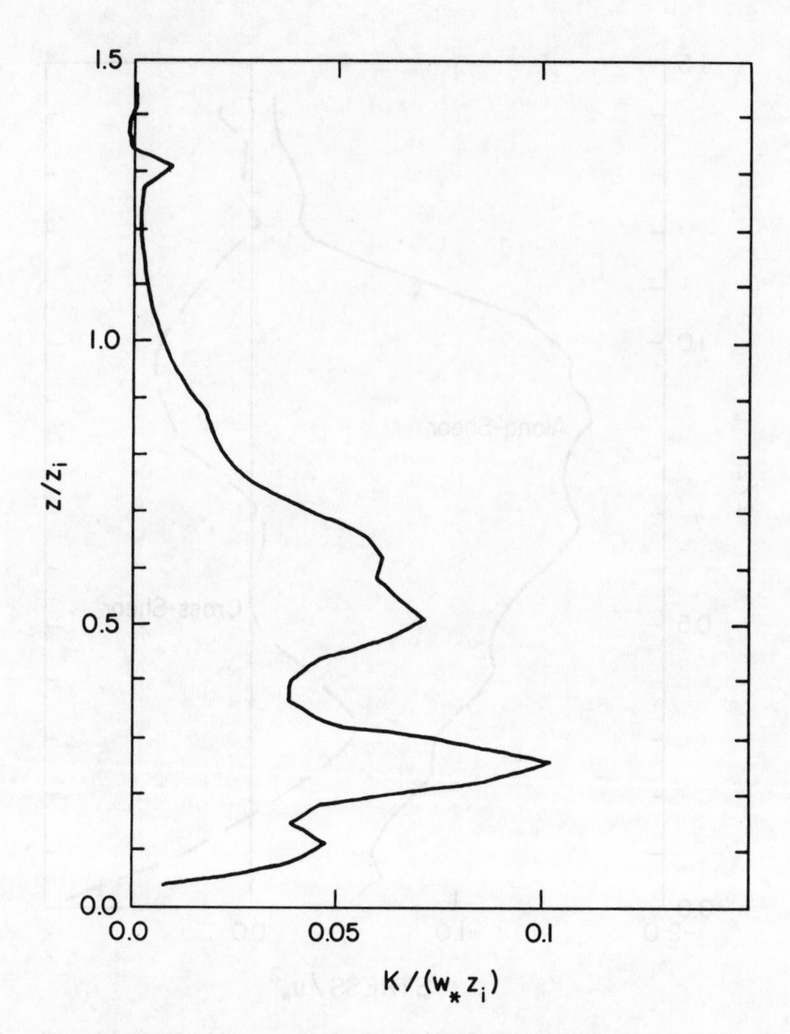

Fig. 11 The eddy diffusivity for the along-shear stress component of Fig. 10.

We assume the PBL is sufficiently convective ($-h_1/L > 30$, say) that in the mixed layer ($h_s < z < h_1$) our "convective viscosity" model holds:

$$K = w_* h_1 \; D(z/h_1) \tag{40}$$

We make no distinction here between the height h_1 in the three-layer model (Fig. 1) and the inversion-base height z_i.

If, as in Section II, we scale Eqs. (38) and (39) with velocity scale Q and horizontal length scale L, and require that $R = Q/fL << 1$, we find that the time change and mean advection terms in Eqs. (38) and (39) are small. Thus if we again write the fields as expansions in R, i.e.

$$U = U_o + RU_1 + \dots \tag{41}$$

$$V = V_o + RV_1 + \dots \tag{42}$$

then the lowest-order equations become

$$\frac{\partial}{\partial z} K \frac{\partial U_o}{\partial z} + fV_o = fV_g \tag{43}$$

$$\frac{\partial}{\partial z} K \frac{\partial V_o}{\partial z} - fU_o = -fU_g \tag{44}$$

which represent a balance among friction, Coriolis, and pressure gradient forces. Note that, because of the closure Eq. (40), these are linear differential equations.

Boundary conditions can be found from the three-layer idealization, Fig. 1. There is little point in applying lower boundary conditions at the surface, since the structure of the surface layer is now so well understood. Instead, we take our lower surface to lie within the surface layer and use "free-slip" conditions.

Upper boundary conditions could be applied at the interfacial layer top, h_2, or at its bottom, h_1. Since we know so little about K in the interfacial layer, and since its value there probably depends on wind shear (in contrast to our hypothesis for the mixed layer) it would seem preferable to apply them at h_1. One could then assume profile forms (linear, say) between h_1 and the geostrophic values at h_2.

At this point it is convenient to combine our governing equations into one by introducing the complex velocity $\tilde{W} = U + iV$. We also define a scaled height $\eta = z/h_1$. Equations (43) and (44) then become

$$\left(\frac{w_*}{fh_1}\right) \frac{d}{d\eta} D(\eta) \frac{d\tilde{W}_o}{d\eta} - i\tilde{W}_o = - i\tilde{W}_g \tag{45}$$

The lower boundary condition is, following Eq. (4),

$$\left.\frac{d\tilde{W}_o}{d\eta}\right|_0 = \left\{\frac{h_1 C_D S \tilde{W}_o}{K}\right\}\Bigg|_0$$

$$= \frac{fh_1}{w_*} \left\{\left(\frac{C_D S}{h_1 fD}\right) \tilde{W}_o\right\}\Bigg|_0 \qquad (46)$$

$$= \frac{fh_1}{w_*} B \tilde{W}_o (0)$$

This is a nonlinear boundary condition, since B (for bottom) depends on $S = |\tilde{W}|$; however, one can linearize by taking B to be a constant of O(1).

The upper boundary condition is obtained by integrating between h_1 and h_2. If we assume that both $\tilde{W}$ and $\tilde{W}_g$ have linear profiles in the interfacial layer, then we find from Eq. (45)

$$\left.\frac{d\tilde{W}_o}{d\eta}\right|_1 = \left\{\frac{ifh_1^2 \Delta h}{2Kh_1} (\tilde{W}_g - \tilde{W}_o)\right\}\Bigg|_1$$

$$= \left(\frac{fh_1}{w_*}\right)\left\{\left(\frac{i\Delta h}{2Dh_1}\right)(\tilde{W}_g - \tilde{W}_o)\right\}\Bigg|_1 \qquad (47)$$

$$= \frac{fh_1}{w_*} T (\tilde{W}_g - \tilde{W}_o)\Bigg|_1$$

which is linear.

An important feature of our lowest-order equation (45) and its boundary conditions Eqs. (46) and (47) is that the convective parameter $m = w_*/fh_1$ appears in each. Let us summarize the set:

$$m \frac{d}{d\eta} D \frac{d\tilde{W}_o}{d\eta} - i\tilde{W}_o = - i\tilde{W}_g \qquad (48)$$

$$m \left.\frac{d\tilde{W}_o}{d\eta}\right|_0 = B\tilde{W}_o\Big|_0 \qquad (49)$$

$$m \left.\frac{d\tilde{W}_o}{d\eta}\right|_1 = T(\tilde{W}_g - \tilde{W}_o)\Big|_1 \qquad (50)$$

where the parameters are defined by

$$m = w_*/fh_1 \tag{51}$$

$$B = \frac{C_D S(0)}{h_1 f D(0)} \tag{52}$$

$$T = \frac{i(h_2-h_1)}{2D(1)h_1} \tag{53}$$

and the dimensionless convective viscosity is

$$D(\eta) = \frac{K}{w_* h_1} \tag{54}$$

The general solution to the set of Eqs. (48) to (50) can be written as

$$\tilde{W}_o = C_1 H_1 + C_2 H_2 + \tilde{W}_p \tag{55}$$

where H_1 and H_2 are the two independent solutions of the homogeneous form of Eq. (48), $\tilde{W}_p$ is a particular solution of Eq. (48), and C_1 and C_2 are constants. Because of the structure in D, analytical solutions H_1 and H_2 might prove elusive, but they can easily be found numerically, and then $\tilde{W}_p$ can be generated from them. In the meantime, we can get some insight into the behavior of the set Eqs. (48) to (50) by examining a particular solution for a simple D.

We take D to be the second-order polynomial

$$D = 0.2\ \eta(1-\eta) + C \tag{56}$$

where the 0.2 is chosen to give midlayer K values of the order of 0.5 $w_* h_1$, and C is a small constant which gives nonzero K values as $\eta \rightarrow 0$ and $\eta \rightarrow 1$. Then it is easy to verify that a particular solution to Eq. (48) has a derivative such that

$$\frac{d\tilde{W}_p}{d\eta} = \frac{i\,\dfrac{d\tilde{W}_g}{d\eta}\,(0.4\ m-i)}{(0.4\ m)^2 + 1} \tag{57}$$

so that the component wind shears are

$$\frac{dU_p}{d\eta} = \frac{\frac{dU_g}{d\eta} - 0.4\ m \frac{dV_g}{d\eta}}{(0.4\ m)^2 + 1} \tag{58}$$

$$\frac{dV_p}{d\eta} = \frac{0.4\ m \frac{dU_g}{d\eta} + \frac{dV_g}{d\eta}}{(0.4\ m)^2 + 1} \tag{58}$$

Note that here the stability parameter is $0.4\ m = 0.4\ w_*/fh_1$. Under very convective conditions (say $w_* = 2.5\ m\ s^{-1}$, $f = 10^{-4}\ s^{-1}$, and $h_1 = 10^3$ m) then $0.4\ m = 10$, and Eqs. (58) and (59) show that this particular solution has mixed-layer wind shear values much smaller than the imposed geostrophic wind shear.

Because the boundary conditions Eqs. (49) and (50) also indicate that the wind shear at $\eta = 0$ and $\eta = 1$ is of order 1/m, it seems likely that the homogeneous solution is of the order of the particular solution, so that both are important in the full solution Eq. (55). Thus while Eq. (57) suggests how the mixed-layer wind shear relates to the geostrophic shear, we cannot take it as representative of the full mixed-layer solution.

Nonetheless, we note that the wind shear occurs in the set of Eqs. (48) to (50) only when multiplied by m. Since the wind itself will not continue to vary significantly as m becomes large, and since the geostrophic shear occurs explicitly in Eq. (48), we conclude that the mixed-layer wind shear is of the order of m^{-1} times the geostrophic shear.

This allows us to scale the shear production rate of turbulent energy in the baroclinic mixed layer, and thereby to test our convective viscosity approximation. This shear production rate is the dot product of stress and mean wind shear, and is thus of the order of K times the square of wind shear. K scales with $w_* z_i$, and mean shear with $m^{-1}\Delta G/z_i$, where ΔG is the change in geostrophic wind speed across the mixed layer. Thus we have

$$\text{Shear production of energy} \sim \frac{w_* (\Delta G)^2}{m^2 z_i} \tag{60}$$

From the definition of w_*, the buoyant production rate of turbulent energy scales with w_*^3/z_i. Thus we have

$$\frac{\text{Shear production}}{\text{Buoyant production}} \sim \left(\frac{\Delta G}{m w_*}\right)^2 \tag{61}$$

Our convective viscosity model Eq. (40) assumes that this ratio Eq. (61) is sufficiently small so that K is unaffected by wind shear. We do not yet know the numerical factor in Eq. (61), but the ratio does become small for m sufficiently large; thus physical grounds do exist for our convective viscosity model.

IV. ACKNOWLEDGMENTS

This work has evolved over a period of time, and during that period several colleagues have helped in various ways. I am particularly grateful to S.P.S. Ayra and J.R. Garratt for productive discussions; to B. Stankov for assisting in the AMTEX data analysis; to R. Brost for carrying out the large-eddy simulations; to M. Hampy for assistance in the computations; and to D. Howard for typing the manuscript.

REFERENCES

1. Brost, R.A., J.C. Wyngaard, and D.H. Lenschow: Marine stratocumulus layers. Part II: Turbulence budgets. J. Atmos. Sci., 39, 818-836 (1982).

2. Busch, N.E.: On the mechanics of atmospheric turbulence. Workshop on Micrometeorology, D.A. Haugen, Ed., Boston, Amer. Meteor. Soc., 1-65 (1973).

3. Businger, J.A.: Turbulent transfer in the atmospheric surface layer. Workshop on Micrometeorology, D.A. Haugen, Ed., Boston, Amer. Meteor. Soc., 67-98 (1973).

4. Deardorff, J.W.: Three-dimensional numerical study of the height and mean structure of a heated planetary boundary layer. Boundary-Layer Meteor., 7, 81-106 (1974).

5. Deardorff, J.W.: Prediction of convective mixed-layer entrainment for realistic capping inversion structure. J. Atmos. Sci., 36, 424-436 (1979).

6. Deardorff, J.W.: Stratocumulus-capped mixed layers derived from a three-dimensional model. Boundary-Layer Meteor., 18, 495-527 (1980).

7. Garratt, J.R., J.C. Wyngaard, and R.J. Francey: Winds in the atmospheric boundary layer -- prediction and observation. To appear J. Atmos. Sci. (1982).

8. Kaimal, J.C., J.C. Wyngaard, D.A. Haugen, O.R. Coté, Y. Izumi, S.J. Caughey, and C.J. Readings: Turbulence structure in the convective boundary layer. J. Atmos. Sci., 33, 2152-2169 (1976).

9. Lenschow, D.H., J.C. Wyngaard, and W.T. Pennell: Mean-field and second-moment budgets in a baroclinic, convective boundary layer. J. Atmos. Sci., 37, 1313-1326 (1980).

10. Mahrt, L.: The influence of momentum advections on a well-mixed layer. Quart. J. Roy. Meteor. Soc., 101, 1-11 (1975).

11. O'Brien, J.J.: A note of the vertical structure of the eddy exchange coefficient in the planetary boundary layer. J. Atmos. Sci., 27, 1213-1215 (1970).

12. Rotta, J.C.: Statistiche theorie nichthomogener turbulenz. Arch. Phys., 129, 547-572 (1951).

13. Townsend, A.A.: The structure of turbulent shear flow. Cambridge Univ. Press, Cambridge, 429 pp. (1956).

14. Weinstock, J.: Theory of the pressure-strain rate. Part 1. Off-diagonal elements. J. Fluid. Mech., 105, 369-395 (1981).

15. Wyngaard, J.C.: On surface layer turbulence. Workshop on Micrometeorology, D.A. Haugen, Ed., Boston, Amer. Meteor. Soc., 101-149 (1973).

16. Wyngaard, J.C.: The atmospheric boundary layer -- modelling and measurements. Turb. Shear Flows II, Springer-Verlag, Berlin, 352-365 (1980).

17. Wyngaard, J.C. and S.P.S. Arya: Some aspects of the structure of convective planetary boundary layers. J. Atmos. Sci., 31, 747-754 (1974).

18. Wyngaard, J.C., O.R. Coté, and Y. I. Izumi: Local free convection, similarity, and the budgets of shear stress and heat flux. J. Atmos. Sci., 28, 1171-1182 (1971).

19. Young, J.A.: A theory of isallobaric airflow in the planetary boundary layer. J. Atmos. Sci., 30, 1584-1592 (1973).

20. Zeman, O.: Progress in the modelling of planetary boundary layers. Ann. Rev. Fl. Mech., 13, 253-272 (1981).

3.8 THE ROLE OF STATIONARY AND TRANSIENT WAVES IN SEASONAL FORCING OF THE BUILDUP OF THE TIBETAN HIGH

Zhu Baozhen, Sung Chengshan and Luo Meixia
Institute of Atmospheric Physics
Academia Sinica
The People's Republic of China

ABSTRACT

Preliminary results from observational studies as well as theoretical approaches are presented for the seasonal buildup of the Tibetan high. We find that the interaction between the midlatitude quasi-stationary westerly ridge over the highlands and the moving subtropical high from Iran and India-Pakistan, plus transient troughs in the subtropical westerlies, might contribute to the establishment of the Tibetan high. Nonlinear wave interactions appear to play an important role in the sudden changes of the zonal flow and of the high-pressure belt.

I. INTRODUCTION

The seasonal variation of the planetary-scale circulation over East Asia in early summer is one of the most important characteristic features of the global circulation. There is a sudden change in the general circulation over East Asia which has often been identified with some concomitant features:

(i) During this period there is often a major adjustment of long waves over Eurasia. The mean trough originally located over the Asiatic coast moves toward the interior of the continent.

(ii) The subtropical westerly currents and their associated jet stream suddenly jump northward, causing a rapid disappearance of the westerly jet stream and the buildup of an upper easterly jet stream in low latitudes over South Asia.

(iii) Meanwhile, in the troposphere, the so-called Tibetan high at 100 or 200 mb also suddenly becomes established over the plateau and then stabilizes there.

(iv) At the same time, the Indian southwest monsoon precipitation over Tibet and the Mei-yü in the Yangzi River starts.

These points have been confirmed by many authors since the 1950's [4]. The mechanism of the seasonal change is often studied by considering it as a response to the thermal and dynamical effects of the plateau. However, we have pointed out that if we examine the upper air data for different years, a considerable interannual variability is found. Thus, it is reasonable to infer that the thermal effect perhaps only provides a physical background, and the sudden change in seasonal circulation may be caused by the complicated nonlinear dynamics of the topographic forcing and the atmospheric circulation [5].

In this presentation, we will summarize two of our recent papers on this subject.

II. OBSERVATIONAL ASPECTS

1. Climatic Analysis

The Tibetan high has been known to move continuously on monthly climatic mean charts (Fig. 1) from the eastern Philippine Sea in April, through the Peninsula of Indochina in May, and finally northwestward to the eastern part of the Tibetan Plateau in June. By looking at a sequence of daily synoptic maps, however, we can see some details of the buildup of the Tibetan high.

Figure 2 shows a composite chart of the establishment of the Tibetan high in the western plateau. From 20 to 25 May, 1969 the western plateau

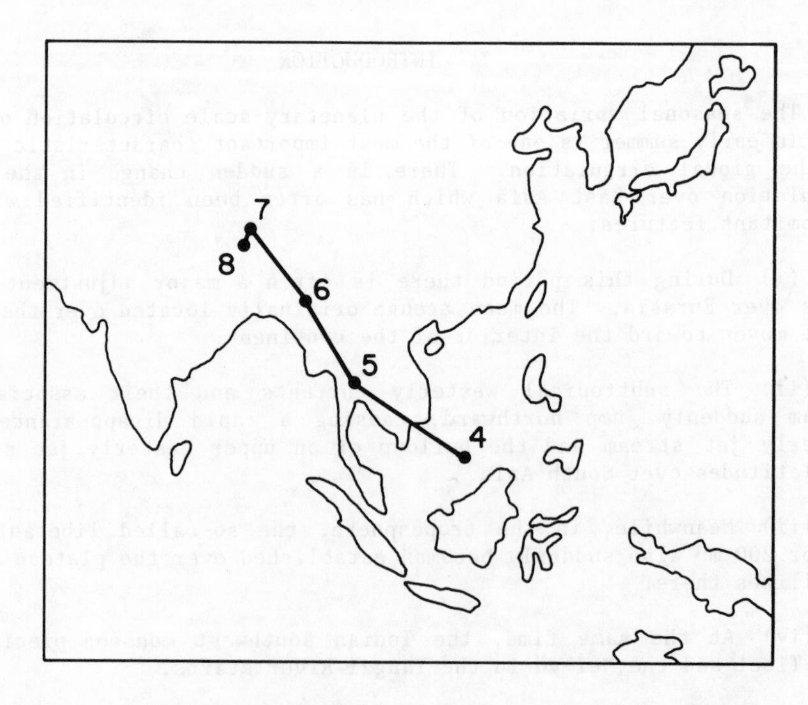

Fig. 1 Track of the anticyclonic center on monthly mean 200 mb streamline charts, April to August.

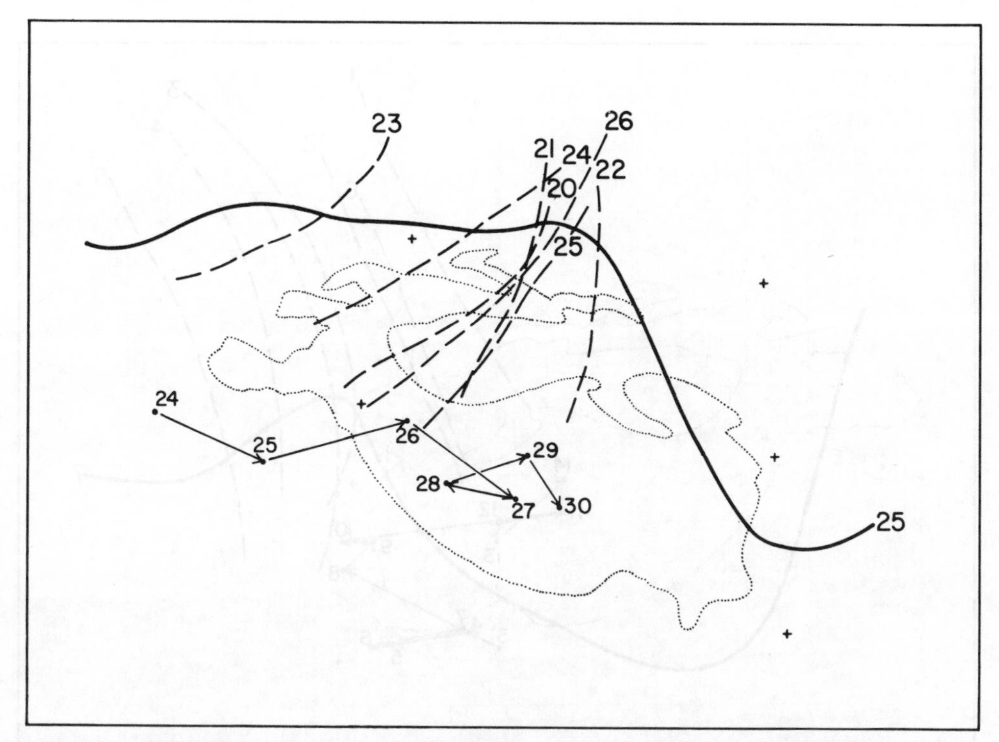

Fig. 2 Composite chart for May 1969. Westerly ridges (dashed lines), a special contour at 500 mb (solid line), and track of the anticyclonic center at 100 mb are shown. Numbers indicate the dates.

was dominated by a semistationary westerly ridge. The contour shown represents broad-scale distributions of trough and ridge at 500 mb just prior to the buildup of the Tibetan high. At that time, the subtropical high moved east from Iran and reached the western plateau, leading to the establishment of the Tibetan high on 26 May.

Figure 3 indicates another case of the establishment of the Tibetan high over the eastern plateau. From 7 June 1974 a semistationary ridge prevailed over the eastern plateau, and a subtropical high stagnated over North Indochina and Yunnan Provinces, thus leading to the buildup of the Tibetan high over the eastern plateau by 8 June.

From an examination of daily weather maps for a ten-year period, it is found that 70 percent of the Tibetan highs originated over Iran or India-Pakistan and then became established over the western part of the plateau, and only 30 percent of them came from Indochina and built up over the eastern part. Why is it that the buildup of the Tibetan high occurs in most cases over the western plateau? This prevalence may be attributed to the maintenance of a semistationary westerly ridge with the superposition of the moving subtropical high.

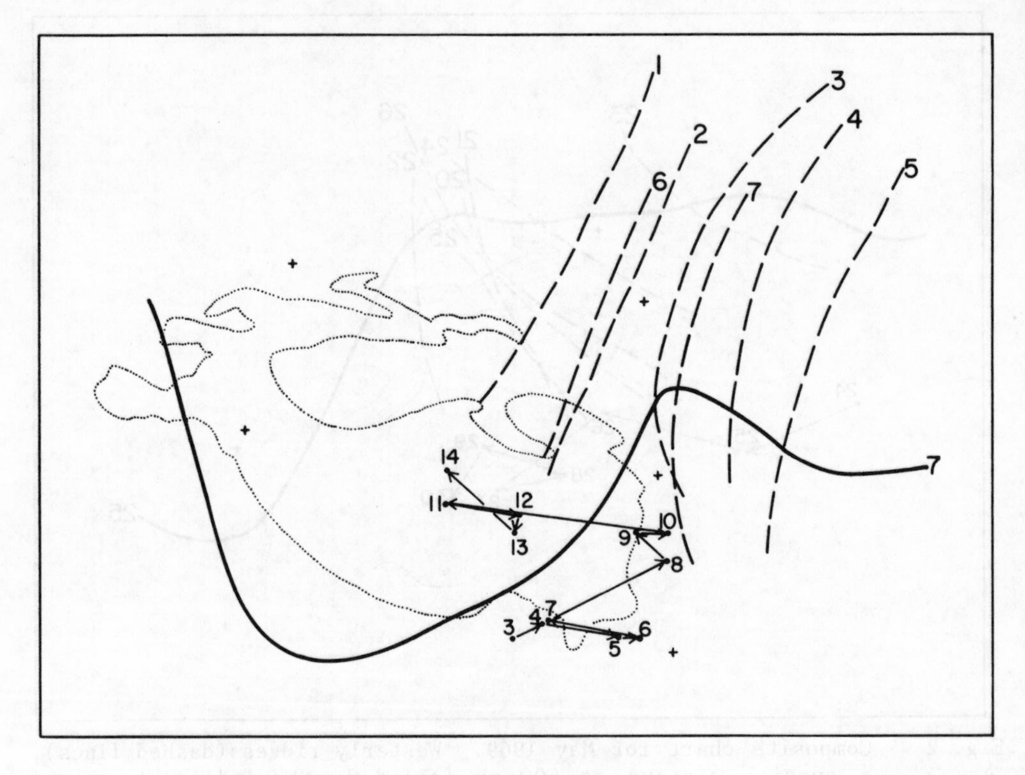

Fig. 3 As in Fig. 2, except for June 1974.

Every year, from spring to summer, the nonzonal distribution of heating along the subtropical belt is very obvious, according to the heat balance chart by Budyko [1]. Over Arabia there is a maximum of sensible heat on a global basis. In the Asiatic summer monsoon regions, there are thermal lows over the Arabian Peninsula and the Tibetan Plateau, but the former precedes the latter by about one month [2]. The most northerly position of the pronounced subtropical high pressure feature is usually reached over Arabia in May.

Figure 4 shows the climatic five-day mean latitudinal positions of the subtropical high-pressure over Arabia and the West Pacific from May to June. We can see that the Arabian high is 5 degrees latitude farther north than the West Pacific high. The Arabian continental subtropical high is well developed and, in general, continuously shifts toward the east. At this time, the dynamic effect of the Tibetan Plateau is more favorable for the formation of a quasi-stationary ridge over the western region of the plateau. On the other side of the continent, retrogression of a semistationary trough off the Asiatic eastern coast, which is at least partly due to the continental thermal effect as well as the mechanical effect of topography, may make the eastern highlands a region of

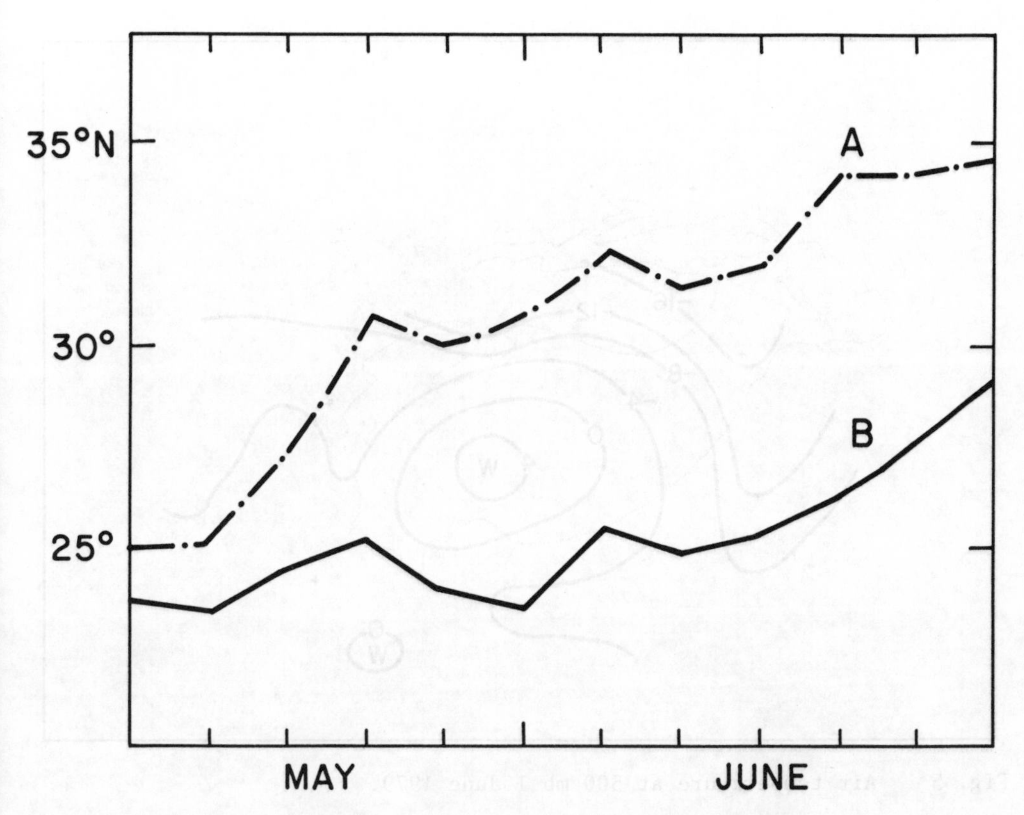

Fig. 4 Climatic five-day mean latitudinal positions of the 500 mb subtropical high over Arabia (a) and the West Pacific (b) from May to June.

cyclonic vorticity genesis, which is unfavorable for stimulating Tibetan highs from Indochina.

When the subtropical high moves to a preferred position of the planetary ridge in the westerlies, the combined effects of the transient subtropical high and the quasi-stationary ridge at midlatitude may intensify the interaction of the Tibetan thermal field and the planetary waves so as to favor a sudden establishment of the Tibetan high over the western region of the plateau.

2. Case Study: June 1979

The above-mentioned points are strongly affected by the westerly current. In the first half of June 1979, the period of summer MONEX, the temperature over the plateau increased in early June, and a warm center formed there. On 1 June, a warm center (4°C at 500 mb, Fig. 5) was located over the eastern plateau. Although the warm region persisted in

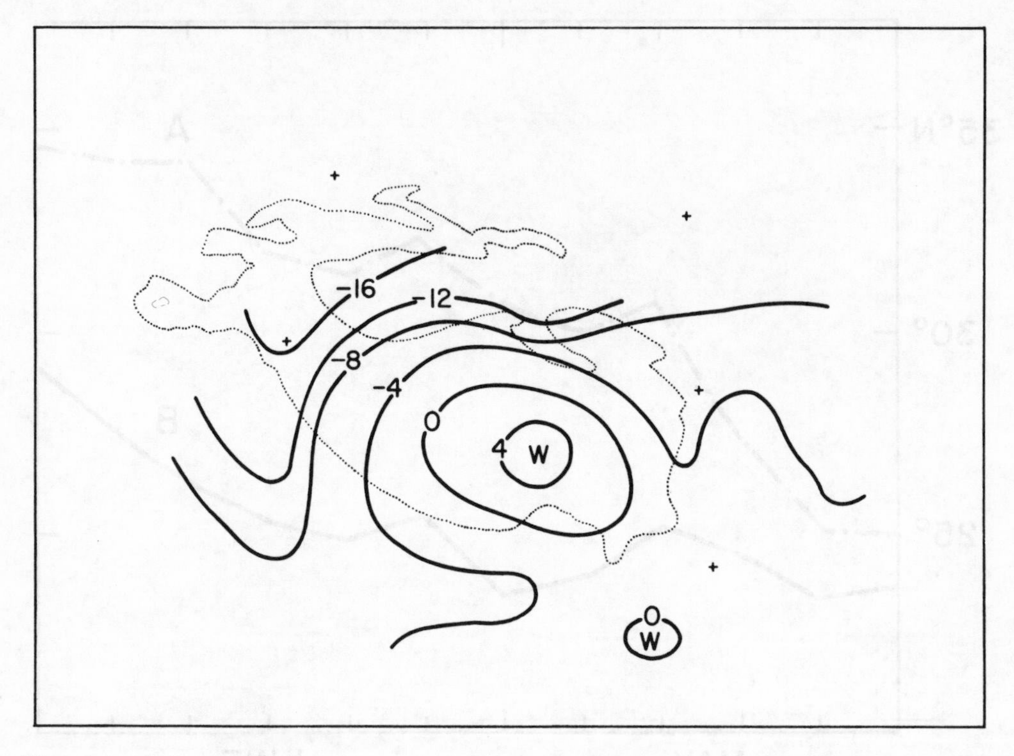

Fig. 5 Air temperature at 500 mb 1 June 1979.

the first half of June, a remarkable feature of the high-level anticyclone is that it moved slowly along the foothills of the Himalayas. Figure 6 shows a composite chart of the track of the Tibetan high centers at 100 mb and the distribution of westerly trough lines at 500 mb. Anticyclonic centers can be divided into four regions of location while they were still over North India and the southern periphery of the plateau during 1 to 17 June. From 18 June, a high center started moving northward to 30°N, leading to the buildup of the Tibetan high over the plateau, but its position was farther south than normal.

This behavior is strongly dependent on the nature of the midlatitude circulation systems. For the whole month of June there were blocking processes prevailing over Europe, including stable flow regimes characterized by a quasi-stationary planetary vortex over western Siberia and by a large amplitude trough that migrated southward from the vortex and continuously intruded into the main body of the plateau. Figure 7 gives the daily contour lines of 12,240 gpm at 200 mb (the vertical center line in the diagram passes through the middle of the plateau). Before 18 June, the intensified trough continuously affected the plateau region. When the trough weakened and the western plateau was located in the rear part of the trough, the Tibetan high finally moved northward to 30°N, thus leading to the abnormal establishment of the Tibetan high and a

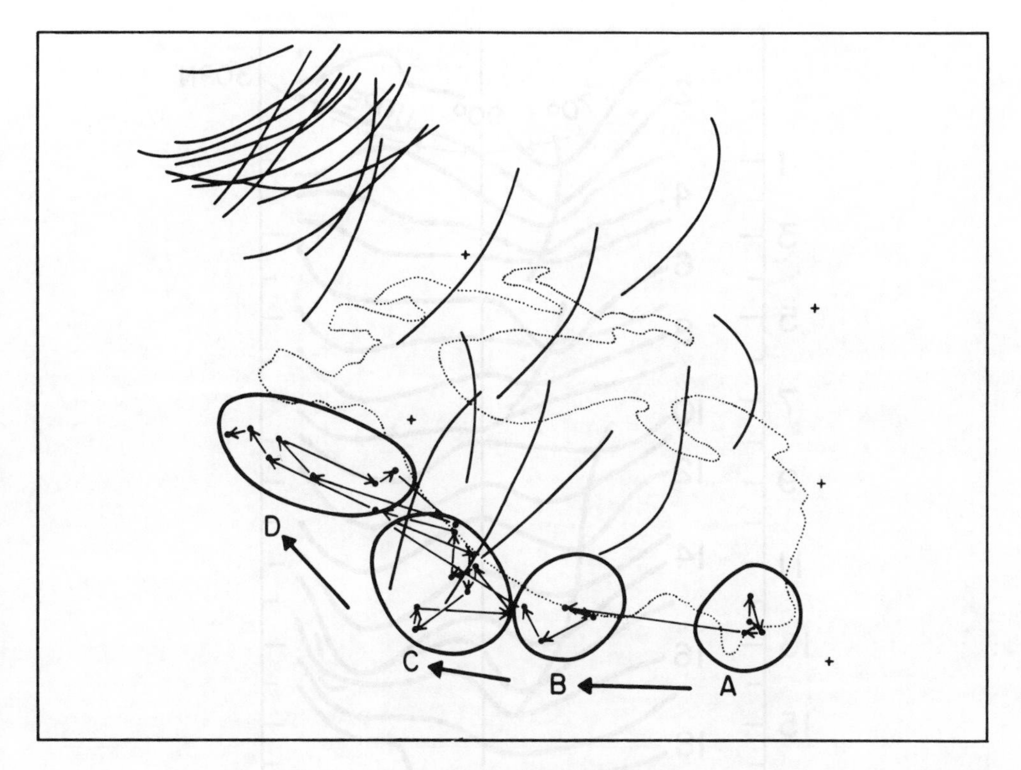

Fig. 6 Composite chart for June 1979. Westerly troughs at 500 mb (1-17 June) and anticyclonic centers at 100 mb. (a) 1-4, (b) 5-8, (c) 9-18, 23-24, (d) 19-22, 25-28 June.

delay of the onset of the summer monsoon with below-normal rainfall in North India. This phenomenon has already been pointed out by Fein and Kuettner [2]. By this time, it was the intensification of this Tibetan high and the development of westerlies on its northern side that finally brought the southern branch of the jet to the north of the plateau. This process has earlier been interpreted as a disappearance of the jet stream in the southern branch of westerlies whose formation was considered to be due to the splitting effect of the topography.

III. THEORETICAL APPROACHES

From the above observational studies, it is possible that a non-linear interaction between a quasi-stationary westerly wave and a moving subtropical high might contribute to characteristics of seasonal forcing of the buildup of the Tibetan high. A theoretical approach toward the understanding of this problem follows.

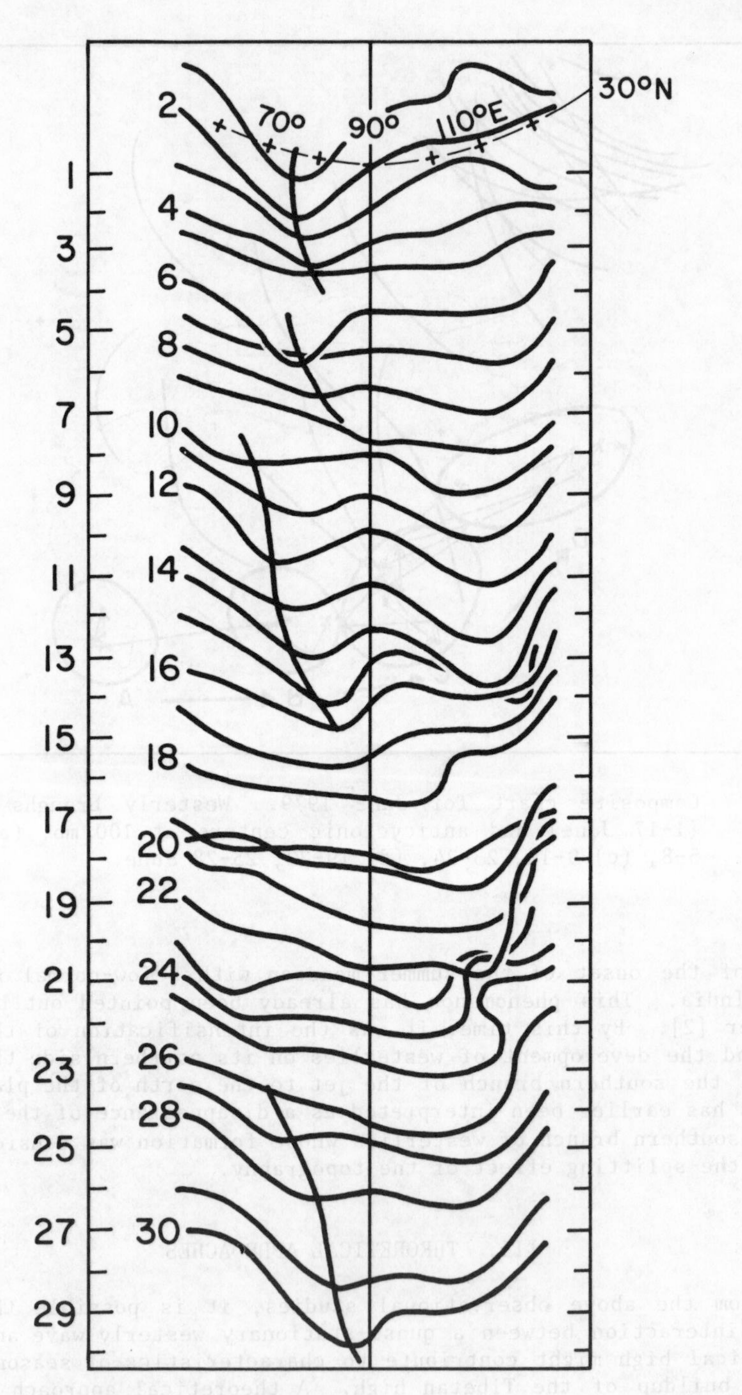

Fig. 7 Daily contour of 12, 240 gpm at 200 mb. The middle line indicates 90°E.

1. The Model

For simplicity, we take the two-layer model of geostrophic equations employed by Phillips [3]:

$$\left(\frac{\partial}{\partial t} + u_1 \frac{\partial}{\partial x} + v_1 \frac{\partial}{\partial y}\right)(\Delta\Phi_1 + f\beta y) = \frac{f^2}{P}\omega_2,$$

$$\left(\frac{\partial}{\partial t} + u_3 \frac{\partial}{\partial x} + v_3 \frac{\partial}{\partial y}\right)(\Delta\Phi_3 + f\beta y) = -\frac{f^2}{P}(\omega_2 - \omega_4), \quad (1)$$

$$\left(\frac{\partial}{\partial t} + u \frac{\partial}{\partial x} + v \frac{\partial}{\partial y}\right)(\Phi_1 - \Phi_3) = \frac{f^2}{\lambda^2 P}\omega_2 + \frac{R}{\rho c_p}\varepsilon \ .$$

where ω_4 indicates the orographic effect; ε, the heating rate per unit mass of air; λ^2, the static stability.

The variables are split up into zonal mean, quasi-stationary forcing and the transient perturbations, i.e.

$$\Phi(x,y,t) = \Phi(y,t) + \bar{\Phi}(x,y) + \Phi'(x,y,t),$$

$$\vec{v}(x,y,t) = \vec{V}(y,t) + \bar{\vec{v}}(x,y) + \vec{v}'(x,y,t), \quad (2)$$

$$\omega(x,y,t) = \Omega(y,t) + \bar{\omega}(x,y) + \omega'(x,y,t),$$

$$\varepsilon(x,y,t) = \tilde{\varepsilon}(y,t) + \bar{\varepsilon}(x,y).$$

Assuming steady waves are due to orography and nonadiabatic heating and transient waves are small perturbations superimposed on a basic zonal flow, then we have three systems of equations for different motions.

The equations forced by topography and heating are written in the steady form

$$U_1 \frac{\partial}{\partial x}\Delta\bar{\Phi}_1 + \left(\beta - \frac{\partial^2 U_1}{\partial y^2}\right)\frac{\partial\bar{\Phi}_1}{\partial x} = \frac{f^2}{P}\bar{\omega}_2,$$

$$U_3 \frac{\partial}{\partial x}\Delta\bar{\Phi}_3 + \left(\beta - \frac{\partial^2 U_3}{\partial y^2}\right)\frac{\partial\bar{\Phi}_3}{\partial x} = -\frac{f^2}{P}a\rho_o g\left(u_3\frac{\partial h}{\partial x} + \frac{H_o F}{f^2}\Delta\bar{\Phi}_3\right) - \frac{f^2}{P}\bar{\omega}_2 \ , \quad (3)$$

$$U_1 \frac{\partial}{\partial x}(\bar{\Phi}_1 - \bar{\Phi}_3) - (U_1 - U_3)\frac{\partial\bar{\Phi}}{\partial x} = \frac{f^2}{\lambda^2 P}\bar{\omega}_2 + \frac{R}{\rho c_p}\bar{\varepsilon} \ .$$

Let the stationary wave solutions and the orographic and nonadiabatic forcing have the form

$$\begin{bmatrix} \bar{\Phi}_1 \\ \bar{\Phi}_3 \\ h \\ \bar{\varepsilon} \end{bmatrix} = \begin{bmatrix} A_1 \cos(kx+\psi_1) \\ A_3 \cos(kx+\psi_3) \\ H \cos(kx+\psi_H) \\ \varepsilon_o \cos(kx+\psi_\varepsilon) \end{bmatrix} \cos \mu y \tag{4}$$

If H and $\bar{\varepsilon}$ are known functions, we can determine A_1, A_3, ψ_1, and ψ_3.

If the velocities of basic current, U_1 and U_3, are only initially independent of y, we can derive perturbation equations for transient waves:

$$\begin{aligned} &\left(\frac{\partial}{\partial t} + U_1 \frac{\partial}{\partial x}\right)[\Delta\phi_1 - \lambda^2(\phi_1-\phi_3)] + [\beta+\lambda^2(U_1-U_3)]\frac{\partial\phi_1}{\partial x} = 0 \\ &\left(\frac{\partial}{\partial t} + U_3 \frac{\partial}{\partial x}\right)[\Delta\phi_3 + \lambda^2(\phi_1-\phi_3)] + [\beta-\lambda^2(U_1-U_3)]\frac{\partial\phi_3}{\partial x} = 0 \end{aligned} \tag{5}$$

Linear solutions of baroclinic, amplifying waves have been obtained as follows:

$$\begin{aligned} \phi_1^{(j)} &= \sigma^{(j)}\, D\, e^{i[k(x-c^{(j)}t)+\psi_1^{(j)}]} \cos j\mu y, \\ \phi_3^{(j)} &= D\, e^{i[k(x-c^{(j)}t)]} \cos j\mu y, \end{aligned} \tag{6}$$

where j=1,2 for different meridional wavelengths and σ, ψ and c are found as suggested by Phillips [3].

Finally, we have the system of equations for the change of the basic current:

$$\begin{aligned} \frac{\partial}{\partial t}\Delta\Phi_1 - \lambda^2 \frac{\partial}{\partial t}(\Phi_1-\Phi_3) &= -\overline{(\bar{\vec{v}}_1+\vec{v}_1')\cdot\nabla[\Delta(\bar{\Phi}_1+\Phi_1')]} \\ &\quad + \lambda^2\overline{[(\bar{\vec{v}}_1+\vec{v}_1')\cdot\nabla(\bar{\Phi}_1-\bar{\Phi}_3+\Phi_1'-\Phi_3')]} - \lambda^2 \frac{R}{\rho c_p}\tilde{\varepsilon}, \\ \frac{\partial}{\partial t}\Delta\Phi_3 + \lambda^2 \frac{\partial}{\partial t}(\Phi_1-\Phi_3) &= -\overline{(\bar{\vec{v}}_3+\vec{v}_3')\cdot\nabla[\Delta(\bar{\Phi}_3+\Phi_3')]} \\ &\quad + \lambda^2\overline{[(\bar{\vec{v}}_3+\vec{v}_3')\cdot\nabla(\bar{\Phi}_1-\bar{\Phi}_3+\Phi_1'-\Phi_3')]} + \lambda^2 \frac{R}{\rho c_p}\tilde{\varepsilon}, \end{aligned} \tag{7}$$

which are controlled by second-order changes in the basic current due to the interaction of steady, forcing waves and transient wave components with different meridional wavelength.

Assuming meridional radiational heating as

$$\tilde{\varepsilon} = -\varepsilon_1(t)\sin\mu y$$

and using Eqs. (4) and (6) to evaluate the right side of Eq. (7), we find that Eq. (7) may be written as

$$\frac{\partial^2}{\partial\zeta^2}\frac{\partial\Phi_1}{\partial t} - \nu\left(\frac{\partial\Phi_1}{\partial t} - \frac{\partial\Phi_3}{\partial t}\right) = -\frac{1}{2}E_1\sin 2\zeta - E_2\sin 4\zeta$$

$$+ (E_3+E_4)(3\sin 3\zeta + \sin\zeta) + E_6\sin\zeta$$

$$\frac{\partial^2}{\partial\zeta^2}\frac{\partial\Phi_3}{\partial t} + \nu\left(\frac{\partial\Phi_1}{\partial t} - \frac{\partial\Phi_3}{\partial t}\right) = \frac{1}{2}E_1\sin 2\zeta + E_2\sin 4\zeta \tag{8}$$

$$- (E_4+E_5)(3\sin 3\zeta + \sin\zeta) - E_6\sin\zeta$$

where

$$E_1 = \frac{\lambda^2 k}{f\mu}[\sigma^{(1)}D^2 e^{2kc_i^{(1)}t}\sin\psi_1^{(1)} + A_1A_3\sin(\psi_1-\psi_3)$$

$$- \sigma^{(1)}DA_3 e^{kc_i^{(1)}t}\sin(\psi_3-\psi_1+kc_\gamma^{(1)}t) + DA_1 e^{kc_i^{(1)}t}\sin(\psi_1+kc_\gamma^{(1)}t)],$$

$$E_2 = \frac{\lambda^2 k}{f\mu}\sigma^{(2)}D^2 e^{2kc_i^{(2)}t}\sin\psi_1^{(2)},$$

$$E_3 = \frac{3k\mu}{4f}\sigma^{(2)}D[A_1 e^{kc_i^{(2)}t}\sin(\psi_1-\psi_1^{(2)} + kc_\gamma^{(2)}t)$$

$$+ \sigma^{(1)}De^{k(c_i^{(1)}+c_i^{(2)})t}\sin\{\psi_1^{(1)}-\psi_1^{(2)}-k\Delta c_\gamma t\}],$$

$$E_4 = \frac{\lambda^2 k}{4f\mu}[D^2 e^{k(c_i^{(1)}+c_i^{(2)})t}\{\sigma^{(2)}\sin(-\psi_1^{(2)}-k\Delta c_\gamma t) - \sigma^{(1)}\sin(\psi_1^{(1)}-k\Delta c_\gamma t)\}$$

$$+ De^{kc_i^{(2)}t}\{\sigma^{(2)}A_3\sin(\psi_3-\psi_1^{(2)}+kc_\gamma^{(2)}t) - A_1\sin(\psi_1+kc_\gamma^{(2)}t)\}],$$

$$E_5 = -\frac{3k\mu}{4f}\sigma^{(2)}D[A_3 e^{kc_i^{(2)}t}\sin(\psi_3+kc_\gamma^{(2)}t) - \sigma^{(1)}De^{k(c_i^{(1)}+c_i^{(2)})t}\sin k\Delta c_\gamma t],$$

$$E_6 = \frac{R\varepsilon_1}{\rho c_p}\nu = \frac{R\varepsilon_0}{\rho c_p}\nu(1 + \frac{1}{4}\sin\frac{2\pi}{T}t),$$

$$\Delta c_\gamma = c_\gamma^{(1)} - c_\gamma^{(2)},$$

$$\nu = \frac{\lambda^2}{\mu^2}, \quad \zeta = \mu y.$$

The solutions of Eq. (8) satisfying the lateral rigid-wall boundary conditions are

$$\frac{\partial\Phi_1}{\partial t} = \frac{E_1}{2(2+\nu)}\left(\frac{\text{Sh}\sqrt{2\nu}\,\zeta}{\sqrt{2\nu}\,\text{Ch}\sqrt{2\nu}\,\frac{\pi}{2}} + \frac{1}{2}\sin 2\zeta\right) - \frac{2E_2}{8+\nu}\left(\frac{\text{Sh}\sqrt{2\nu}\,\zeta}{\sqrt{2\nu}\,\text{Ch}\sqrt{2\nu}\,\frac{\pi}{2}} - \frac{1}{4}\sin 4\zeta\right)$$

$$-\nu(E_3-E_5)\left(\frac{\sin 3\zeta}{3(9+2\nu)} + \frac{\sin\zeta}{1+2\nu}\right) - (E_3+E_4)\left(\frac{3\sin 3\zeta}{9+2\nu} + \frac{\sin\zeta}{1+2\nu}\right) - E_6\frac{\sin\zeta}{1+2\nu}$$

$$\frac{\partial\Phi_3}{\partial t} = -\frac{E_1}{2(2+\nu)}\left(\frac{\text{Sh}\sqrt{2\nu}\,\zeta}{\sqrt{2\nu}\,\text{Ch}\sqrt{2\nu}\,\frac{\pi}{2}} + \frac{1}{2}\sin 2\zeta\right) + \frac{2E_2}{8+\nu}\left(\frac{\text{Sh}\sqrt{2\nu}\,\zeta}{\sqrt{2\nu}\,\text{Ch}\sqrt{2\nu}\,\frac{\pi}{2}} - \frac{1}{4}\sin 4\zeta\right) \tag{9}$$

$$-\nu(E_3-E_5)\left(\frac{\sin 3\zeta}{3(9+2\nu)} + \frac{\sin\zeta}{1+2\nu}\right) + (E_5+E_4)\left(\frac{3\sin 3\zeta}{9+2\nu} + \frac{\sin\zeta}{1+2\nu}\right) + E_6\frac{\sin\zeta}{1+2\nu}$$

and changes in the zonal basic current are given by

$$\frac{\partial U_1}{\partial t} = \frac{\mu}{f}\frac{\partial}{\partial\zeta}\frac{\partial\Phi_1}{\partial t},$$

$$\frac{\partial U_3}{\partial t} = \frac{\mu}{f}\frac{\partial}{\partial\zeta}\frac{\partial\Phi_3}{\partial t}. \tag{10}$$

Integrating Eqs. (9) and (10) with respect to t gives the basic zonal current and geopotential height as functions of time.

2. Numerical Results

The model includes a domain bounded laterally by 12.5° and 57.5°N. We introduce a fixed perturbation to simulate the stationary wave forced by topography and heating, and slowly amplifying transient perturbations with a different meridional wavelength.

Taking $\lambda^2 = 1.17 \times 10^{-12}\ m^{-2}$, $a = \frac{1}{3}$, $D = 500\ m^2 s^{-2}$, $H = 8$ km, $k = 0.393 \times 10^{-6}\ m^{-1}$, $\mu = 0.628 \times 10^{-6}\ m^{-1}$, $U_1 = 40\ ms^{-1}$, $U_3 = 5\ ms^{-1}$, we can obtain features of the zonal potential height and momentum day by day.

At day 4, a tilted trough running northwest to southeast (northeast to southwest) in the northern (southern) part is formed by these waves. At this time, the convergence of momentum transport tends to intensify the westerly current in the middle part of the β plane and the high-pressure belt is located at a low latitude (Fig. 8).

At day 10, when a moving ridge approaches the stationary ridge, the resulting trough becomes tilted northeast to southwest (northwest to southeast) in the northern (southern) part. The middle part of the domain is a region of divergence of momentum transport, leading to weakening westerlies.

When the moving ridge of the transient wave coincides with the stationary wave ridge, the westerly jet, zonal wind and high-pressure belt suddenly shift northward (Fig. 9).

North-south shifts and abrupt changes of the zonal wind and high-pressure belt are primarily due to the momentum transport by the wave nonlinearities. From numerical calculations we find that the role played by the momentum transport is more significant than the effect of heat transport and meridional radiation heating.

IV. SUMMARY REMARKS

In this paper we have examined the dynamics of the buildup of the Tibetan high during the early summer. Daily maps have given some interesting insight into the interannual variability of the large-scale fields. There are some indications that the establishment of the Tibetan high appears to be related mainly to the eastward or northward drift of the subtropical high that moves from Iran or India-Pakistan and merges with the semistationary westerly ridge over the plateau.

We believe that major influences on the dynamics are the mechanical effect of mountains, the differential heating over Arabia for the formation of the westerly ridge over the Tibetan highlands and the subtropical high pressure over Arabia. There are also many kinds of transient waves causing sudden changes in the general circulation. Interactions between stationary and transient waves lead to a more complicated behavior of planetary-scale circulations.

These properties were investigated using a weakly nonlinear analysis. As a result, we suggest that the dynamic mechanism of the establishment of the Tibetan high in relation to seasonal changes in the general circulation may be a kind of instability of the basic current due to the interaction of stationary waves forced by topography and heating and large-scale transient waves with different meridional wavelength. Both of them vary from year to year, thus leading to considerable interannual variability in seasonal changes in the general circulation.

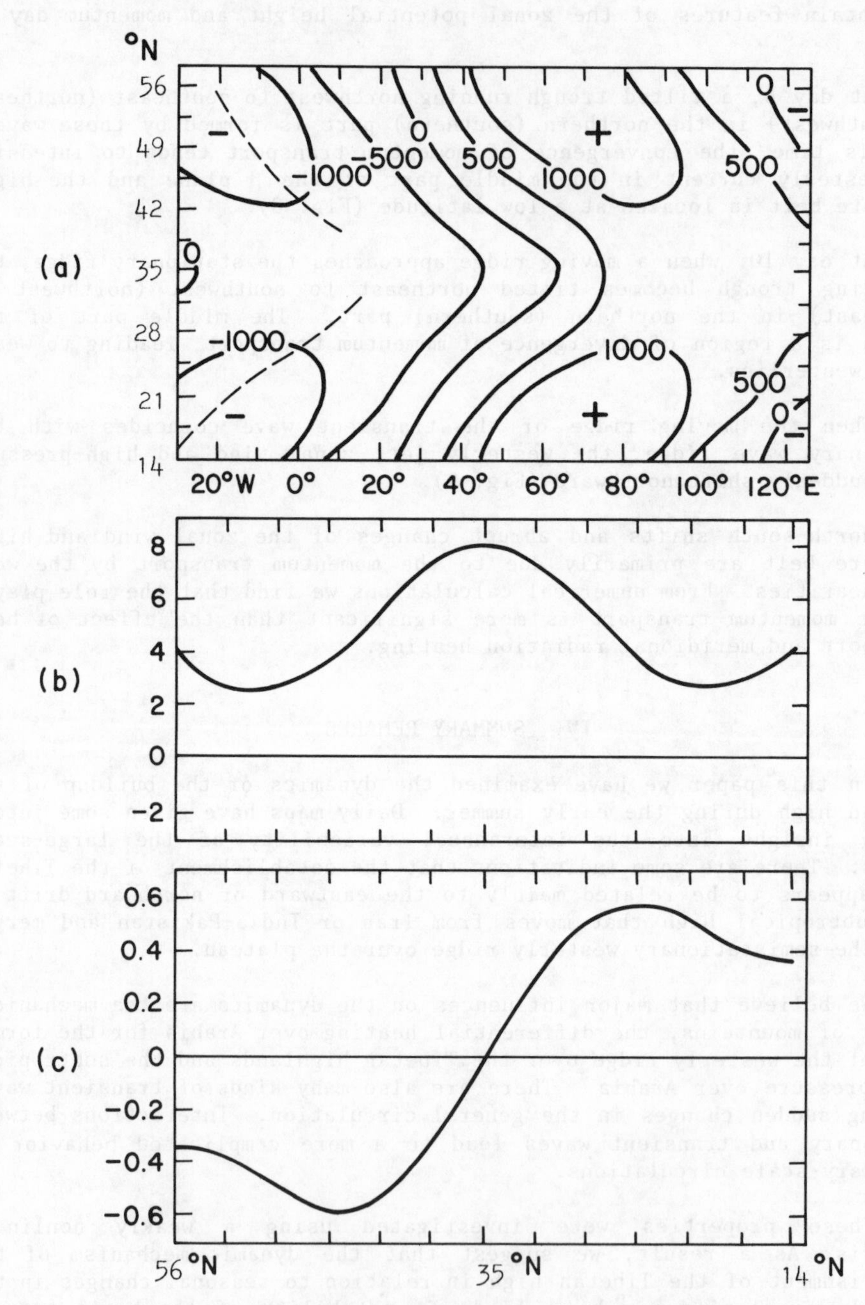

Fig. 8 (a) Horizontal distribution of stationary and transient perturbation heights in units: $m^2 s^{-2}$; (b) meridional distribution of basic zonal current in units: 10 ms^{-1}; (c) zonal geopotential height changes in units: $m^2 s^{-2}$, at an upper level for t = day 4.

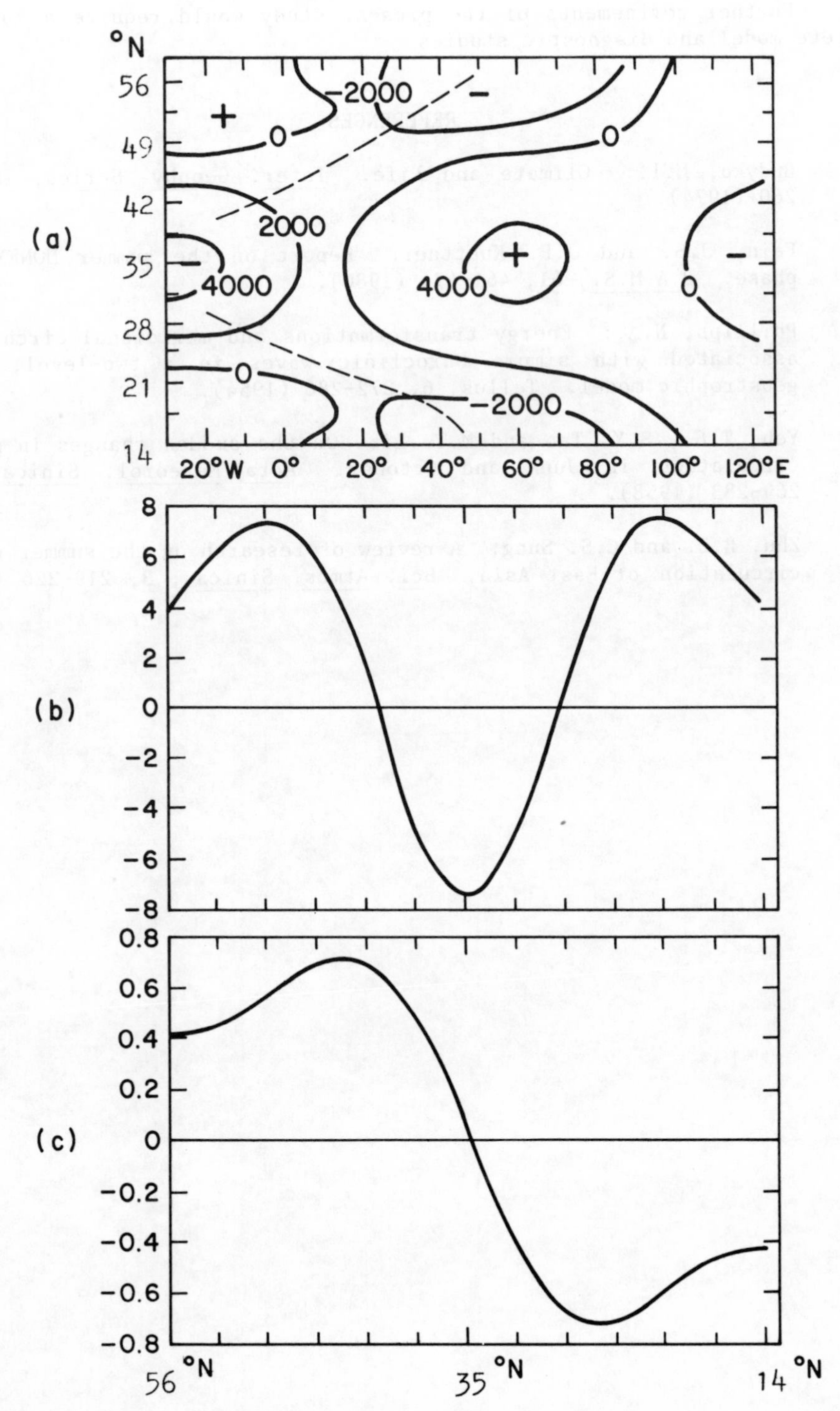

Fig. 9 As in Fig. 8, except for t = day 10.

Further refinements of the present study would require a more complete model and diagnostic studies.

REFERENCES

1. Budyko, M.I.: Climate and Life. Inter. Geophy. Series, 18, 140-260 (1974).

2. Fein, J.S. and J.P. Kuettner: Report on the summer MONEX field phase. B.A.M.S., 61, 461-474 (1980).

3. Phillips, N.A.: Energy transformations and meridional circulations associated with simple baroclinic waves in a two-level, quasi-geostrophic model. Tellus, 6, 272-286 (1954).

4. Yeh, T.C., S.Y. Tao and M.T. Li: On the sudden changes in general circulation in June and October. Acta Meteorol. Sinica., 29, 269-283 (1958).

5. Zhu, B.Z. and C.S. Sung: A review of research on the summer general circulation of East Asia. Sci. Atmos. Sinica., 3, 219-226 (1979).

3.9 MONTHLY MEAN HEATING FIELDS OF THE ATMOSPHERE AND THEIR ANNUAL VARIATION OVER ASIA AND THE QINGHAI-XIZANG PLATEAU

Yao Lanchang and Luo Siwei
Lanzhou Institute of Plateau Atmospheric Physics
Academia Sinica
The People·s Republic of China

ABSTRACT

New results are summarized which give details of the monthly-mean heating fields over Asia and the Qinghai-Xizang Plateau. We find a heat source over the plateau from April through September and a heat sink during the other months. We suggest that this heat source may play an important role in the mechanism triggering the southeast monsoon.

I. INTRODUCTION

It is important, for understanding the circulation and its annual variation over the plateau and its surroundings, to calculate the monthly average heating fields and the three-dimensional airflow over these regions. In the past, calculated heating fields were not sufficiently accurate. For example, Zhu et al. [6] calculated only the heating field below 500 mb, and the data over the plateau were inadequate. Chen et al. [1] calculated the heating for only January and July, and the results obtained by Ye et al. [5] were the average over the plateau. By using recent field data and satellite data, the authors [3] have calculated monthly heating fields over Asia and have corrected some of the above results. One of us has also calculated the average three-dimensional airflow and the average vertical distribution of the heating field over the plateau [2]. This paper summarizes these new results.

II. DATA AND METHODS

In this paper the monthly mean (January through December) radiation and heat budgets of the atmosphere over Asia (45°E to 150°E, 10°N to 60°N) are calculated with the mean data of 1961 to 1970 for more than 180 meteorological stations. Mean vertical velocity, divergence and vertical distribution of the heating field over the plateau are also calculated. Characteristics of the heating fields and their annual variations are analyzed and some relations between the circulation and the heating field are also discussed briefly.

We calculated radiation and heating quantities, including the short-wave radiation absorbed by the atmosphere (Q_{sr}), long-wave radiation from the earth's surface ($Q_{1r}^{\uparrow}$), atmospheric long-wave back radiation ($Q_{1r}^{\downarrow}$), surface effective radiation ($Q_{elr}=Q^{\uparrow}_{lr}-Q^{\downarrow}_{lr}$), outgoing long-wave radiation at the top of the atmosphere ($Q^{\uparrow}_{olr}$), net long-wave radiation ($Q_{lr}=Q_{elr}-Q_{olr}$), the atmospheric radiation balance ($Q_r=Q_{lr}+Q_{sr}$), latent heat (Q_l) and sensible heat (Q_s), and total air column heat ($Q=Q_s+Q_l+Q_r$). If $Q > 0$, we consider it a heat source, and if $Q < 0$, a heat sink. We used Qian's method but made some improvements [4] in the calculation of heat flux, especially for sensible heat over the plateau. Results with these improvements agree better with the observed values than the results obtained by other authors, especially over the plateau.

III. HEATING FIELD

1. Winter

Figure 1 shows that in winter there is a heat source over the ocean, but in the area from the east coast of Asia to Japan there is a heat sink which is strongest in February. Over the continent there is a heat sink whose center is located to the north of the plateau. The heat sink situated in the latitude zone 30° to 37°N over the plateau is weaker than those south and north of this area.

2. Summer

From Fig. 2 we can see that there is a heat source over the ocean, its axis extending from the east coast of China to Japan; there is a weak heat sink situated to its south. Contrary to the winter situation, there are generally heat sources, but heat sinks are still maintained over the Iranian Plateau, the Asian part of the USSR and the western and central part of the People's Republic of Mongolia. The heat source whose center is situated in the area from Bengal to Assam is strongest in July, when it averages about 8.0°C/day. The intensity of heat sources over the plateau is about 1.0° to 1.5°C/day, stronger than over other plains at that same latitude.

3. Transitional Season

We will not give the monthly average figures (as in Fig. 1) for the transitional season but will give instead the total monthly mean heating rate of some representative regions (see Table 1).

(1) The plateau. Since radiative cooling rates in the atmosphere are usually about -1.0°C to -1.5°C/day, regional variations of heating fields depend mainly on latent or sensible heat.

From Table 1 we can see that there is a heat source over the Qinghai-Xizang Plateau during the period from April to September. This source is stronger than that situated over the flat lands of the same latitude, such as eastern China and Iran. A cold source exists over the plateau in the other months which is weaker than that situated in its surroundings at the same latitude. Temperature changes over the plateau resulting from a heat source and sink in the atmosphere are about ± 1° to

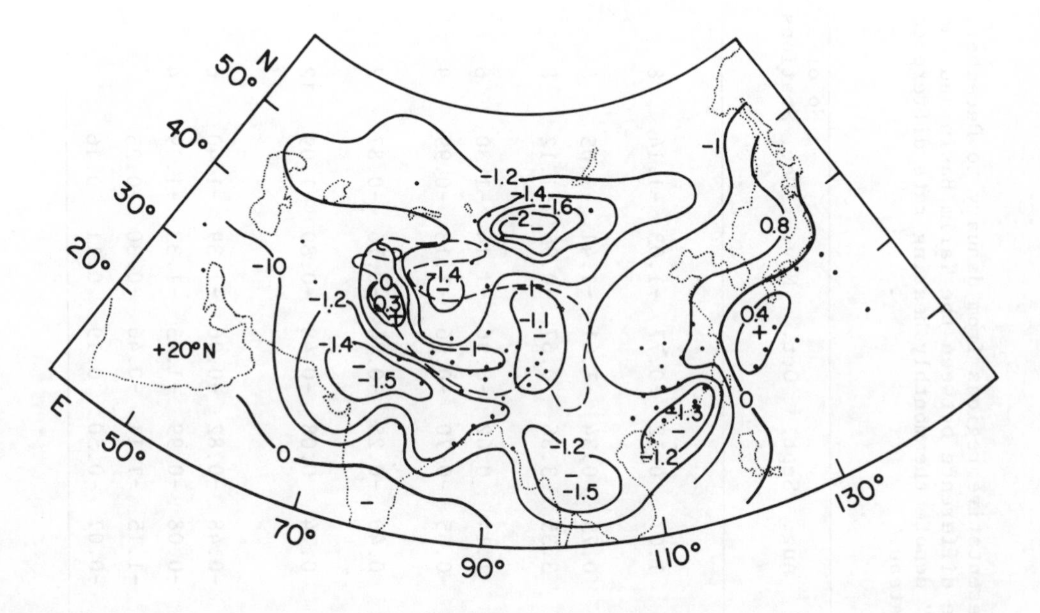

Fig. 1 Heat source distribution (°C/day) for winter.

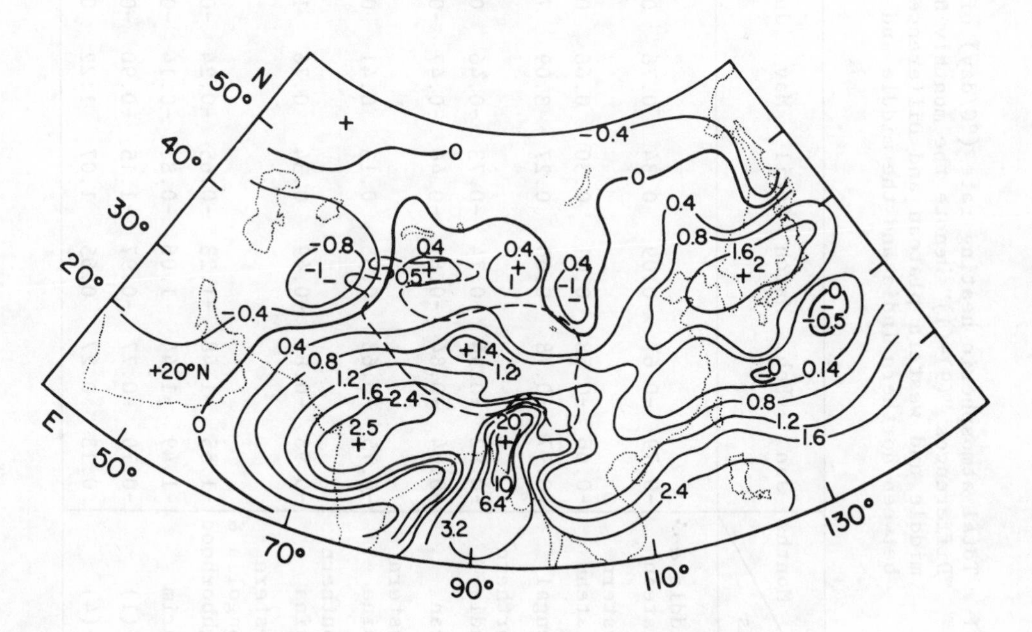

Fig. 2 Heat source distribution (°C/day) for summer.

Table 1 Total atmospheric heating rate (°C/day) of some representative regions from January to December. Differences "(9)-(1)" denote the monthly heating rate difference between the Tarim Basin and the middle and western plateau and Differences "(1)-(4)" denote the monthly heating rate difference between northern India and the middle and western plateau.

Regions \ Months	Jan.	Feb.	March	April	May	June	July	Aug.	Sept.	Oct.	Nov.	Dec.	No of Stations
(1) Middle & Plateau	-0.79	-0.67	-0.09	0.34	0.76	0.69	1.17	1.07	0.12	-0.77	-1.03	-1.14	8
(2) Eastern Plateau	-0.66	-0.58	0.01	0.30	0.66	0.83	0.91	0.75	0.34	-1.61	-1.41	-1.05	8
(3) Bengal	-0.97	-0.78	-0.60	0.27	3.09	7.24	7.82	3.35	3.36	0.67	-0.95	-1.12	8
(4) Northern India	-0.92	-0.94	-0.74	-0.73	-0.46	0.36	1.58	1.14	0.62	-0.96	-1.14	-1.30	6
(5) Iran	-0.64	-0.84	-0.61	-0.46	-0.47	-0.37	-0.25	-0.65	-0.70	-0.96	-1.13	-0.99	9
(6) Eastern China	-0.75	-0.68	-0.38	0.13	0.41	0.66	0.49	0.40	-0.25	-0.74	-0.76	-0.87	7
(7) Southern China	-0.98	-0.81	-0.49	0.34	0.76	1.91	0.67	0.74	0.08	-0.48	-0.89	-1.09	12
(8) Western Mongolia & Neighborhood	-1.55	-1.48	-1.23	-0.65	-0.24	-0.05	-0.17	-0.48	-0.82	-0.96	-1.39	-1.40	6
(9) Tarim	-1.49	-1.44	-1.06	-0.81	-0.14	-0.15	0.14	-0.08	-0.99	-1.23	-1.33	-1.39	4
(9)-(1)	-0.70	-0.77	-0.97	-1.15	-0.90	-0.84	-1.03	-1.15	-1.11	-0.46	-0.30	-0.25	
(1)-(4)	0.13	0.27	0.65	1.07	1.22	0.33	-0.41	-0.07	-0.50	0.19	0.11	0.16	

2°C/day, with the highest value in the layer 600 to 500 mb, the median value in the upper troposphere (300 to 100 mb) and the lowest value at intermediate levels. When the sensible heat strengthens abruptly over the plateau in March, the sink becomes a heat source in April, and when the sensible heat increases again in May the rainy season begins in June. The heat source becomes a heat sink in October when the sensible heat weakens abruptly. The annual variation of the heating field over the plateau is mainly dependent upon the sensible heat change, except that the heating field over the east plateau depends upon both latent heating and sensible heating in June through September.

Our results correspond approximately to those obtained by Ye et al. [5], but our sensible heat values over the plateau are weaker than his. Our results should be more reliable because we calculated them from recent field data.

(2) The Regions around the plateau. Since the sum of sensible and latent heat is always smaller than the radiation cooling, there is a heat sink for the whole year over the areas situated on the northern and western sides of the plateau, including the Tarim Basin, the People's Republic of Mongolia, and the Iranian Plateau.

As sensible heat and its annual variation over the area of Bengal, the northern part of India and southern China, are very small and the heating rate is about 0.1° to 0.3°C/day. The heating fields and their annual variation there depend mainly on the latent heat. When rainfall increases abruptly, the heat sink becomes a heat source and vice versa. The changes of the heating fields depend on the seasonal changes of rainfall.

There is a heat source over the northern part of India in June through September, and a heat sink in the other months. Its intensity is stronger than that over the plateau. A heat source appears over Assam or Bengal in April, but its intensity is weaker and its area is small. Although the heat source center becomes stronger in May, its area is still much smaller than that over the plateau.

The intensity of this heat source is about 8.0°C/day in June and July and, therefore, is the largest center in Asia. It weakens in September and becomes a heat sink in November. There is a heat source in April through September in southern China, and it is stronger than that over the plateau in May and June. There is a heat source in April through August in the middle and lower reaches of the Yangzi River (East China), but it is weaker than that of the plateau. There is a heat source over the sea for the whole year except for February.

IV. THE EFFECTS OF THE HEATING FIELD

1. Average Flow Field Over the Plateau

In winter the airflow converges toward the plateau in the upper troposphere (200 to 100 mb), then descends in heat sinks and diverges outward in the 600 to 500 mb layer. In summer, on the contrary, the airflow converges toward the plateau at the 600 to 500 mb layer, then ascends in the heat source and diverges at the 100 mb surface. There is a typical Hadley cell in January and February whose descending branch is situated over the plateau. The Hadley cell moves southward in May and the plateau is occupied by the monsoon circulation in June. In October the monsoon circulation is replaced by the Hadley cell.

2. 300 to 500 mb Thicknesses

From Table 2 we can see that the monthly mean thicknesses increase from March to May (or from March to June) and are largest over the middle and western plateau, the eastern plateau and the Tarim Basin, being 1500, 1800 and 1500 (or 2300, 2100 and 2000) geopotential meters, respectively. Decreases are largest from September to November, being -1900, -1800 and -2100 geopotential meters, respectively. These changes may be related to the heating field changes but occur one month later. For example, over the middle and western plateau, the heating field increases abruptly in March, but the thickness increases one month later (comparing Table 1 to Table 2). As further examples, monthly changes of the heating field become very weak in June, but the thickness increases are still evident. The heat sink change is very weak in November but there still exists a large value of thickness decrease. Thus, the 300 to 500 mb thickness change seems to be the result of the heating field change.

It is worthwhile to note that monthly changes of the heating field in Bengal are the largest in Asia, but the corresponding thickness changes are almost the smallest. Over the plateau (especially over the western plateau) monthly changes of the heating field are not very large, but corresponding thickness changes are the largest in Asia. It is not clear why this is so, and it may be worthwhile to study this problem further.

3. The Jet Stream

Table 1 shows that in winter the heat sink over the middle and western parts of the plateau is weaker than that over the northern part of India. This means that the subtropical jet stream situated to the south of the plateau in the upper troposphere is not maintained mainly by the thermodynamic effect but by the dynamic effect of the plateau. In spring and summer there is a heat source in the middle and western parts of the plateau and a sink in the Tarim Basin to the north of the plateau. When the sensible heat of the plateau abruptly strengthens in spring, the sinks over the Tarim Basin and northern India remain and change slowly. This distribution increases the slope of the isobaric surface and, hence, the velocity of the westerlies north of the plateau and weakens the westerlies south of the plateau. Consequently, these changes result in a northward jump of the subtropical jet from the southern side to the northern side of the plateau.

4. A Trigger Mechanism

When there are sinks surrounding the plateau (except for Bengal), the heat source over the plateau strengthens abruptly in May. Although the heat source of Bengal is stronger than that over the plateau at this

Table 2 Monthly changes of 500 to 300 mb thicknesses of some representative regions (10^{-1} meter/month).

Regions \ Months	Jan.-Dec.	Feb.-Jan.	March-Feb.	April-March	May-April	June-May	July-June	Aug.-July	Sept.-Aug.	Oct.-Sept.	Nov.-Oct.	Dec.-Nov.
West Mongolia & Surroundings	-30	30	10	45	55	75	40	-35	-80	-75	-80	-30
Tarim Basin	-10	30	10	100	50	50	70	-50	-20	-110	-100	-20
Middle & Western Plateau	15	0	5	75	75	80	25	0	-45	-95	-95	-40
Eastern Plateau	10	-10	-10	80	100	30	40	-10	-30	-90	-90	-20
Iranian Plateau	-20	-5	25	25	45	100	15	25	-95	-65	-50	-30
Northern Plateau	-20	-5	20	20	65	70	5	5	-35	-50	-40	-35
Southern & Eastern China	-45	-45	40	80	60	10	30	10	-5	-115	-10	-100
Bengal	-25	-15	-5	55	55	40	10	10	-25	-40	-35	-5

time, the area of the former is much smaller than that of the latter and the total thermal effect of Bengal is smaller than that of the plateau. The strong heat source in June over Bengal may be a result of the onset of the rainy season but may not be its trigger mechanism, because the onset of the summer monsoon is at the beginning of June.

V. CONCLUDING REMARKS

(i) There is a heat source over the Qinghai-Xizang Plateau in April through September and a heat sink for the other months of the year. The formation of a heat source over the eastern plateau is determined by an abrupt increase of surface sensible heat from March through May, but it is determined by both latent heat and sensible heat from June through September. The formation of heat sources and sinks and their annual variation over the middle and western plateau is determined by a change of only surface sensible heat. The heat source over the plateau appears early and abruptly. There are two abruptly, increasing heating fields in March and May, respectively, which contribute to the northward jump of the subtropical jet stream and the onset of the rainy season over the plateau in June. In winter the heat sink over the middle and western plateau does not contribute to the maintenance of the subtropcial jet stream situated to the south of the plateau; it may be maintained by the dynamic effect of the plateau, because the heat sink over the plateau is weaker than that over North India.

(ii) There is a heat sink for the entire year over the region situated to the north and west of the plateau, because the sensible and latent heat there are very weak each month and their total is smaller than the radiation cooling. In July a high over Iran at 500 mb is dynamical in origin.

(iii) There is very weak sensible heating to the south and east of the plateau, and latent heat release and its annual variation there are strong. The heating fields and their annual variations over these regions are determined solely by precipitation and its annual variations. There is a heat source over North India in June through September and a heat sink for the other months. The beginning of a heat source over Bengal appears in April, and is weaker than that over the plateau. It strengthens in May and is largest in June and July when it is 7 to 8 times larger than that over the plateau. The heat source over the plateau may play an important role in a triggering mechanism of the onset of the southeast monsoon, and a large heat source over Bengal in June and July may be its result, not cause, because the onset of the summer monsoon appears at the beginning of June.

ACKNOWLEDGMENTS

The authors are indebted to comrades Zhang Mingjuan, Pan Ruinian and Chu Zhenshan for their assistance in data processing and final editing of the original manuscript. We would also like to thank comrade Lu Shihua for his programming.

REFERENCES

1. Chen Longxun et al,. Radiation budget of the atmosphere in South-east Asia, (1) (II). Q. J. Meteor. Soc. of China, 34 (164), 146-160, 320-343.

2. Luo Siwei et al,. Average heating field and three-dimensional flow pattern over the Qinghai-Xizang Plateau and its surroundings. Q. S. Plateau Meteor. China, 4 (1982).

3. Yao Lanchang et al,. Monthly mean heating fields of the atmosphere and their annual variation over Asia. Q. J. Plateau Meteor. China, 3 (1982).

4. Yao Lanchang et al,. The study of monthly mean heat sources and sinks over the Qinghai-Xizang Plateau and its surroundings in summer. Symposium of Plateau Meteor. of China, 1 (1982).

5. Ye Duzhen, Gao Youxi et al,. Meteorology of the Qinghai-Xizang Plateau. Academia Sinica, Science Press (1980).

6. Zhu Baozhen et al,. Annual variations of heat sources and sinks and general circulation of the atmospere in the northern hemisphere. Symposium of Dynamical Meteorology, Institute of Geophysics and Meteorology, Academia Sinica, Science Press (1961).

SESSION IV

MESOSCALE EFFECTS OF MOUNTAINS ON HEAVY RAINFALL AND SEVERE LOCAL STORMS: OBSERVATIONAL ASPECTS

4.1 AN EPISODE OF MESOSCALE CONVECTIVE COMPLEXES THAT FORMED TO THE LEE OF THE COLORADO ROCKY MOUNTAINS

William R. Cotton
Colorado State University
Fort Collins, Colorado 80523 USA

ABSTRACT

An episode of 14 mesoscale convective complexes (MCCs) is described. The synoptic, meso-α and meso-β structure of the system is analyzed. It is shown that western MCCs have meso-β components which can be traced to having an origin in mountain convection. The eastern MCCs also exhibit meso-β building blocks. However, few of the meso-β components of the eastern systems have a distinct origin in the mountains.

It is shown that the MCCs exhibit features which are similar to tropical cloud clusters including:

1) convective organization as revealed by radar and satellites,
2) an anticyclonic outflow layer near the tropopause,
3) a "warm core" storm structure,
4) a subsiding clear region of suppressed convection surrounding the systems.

I. INTRODUCTION

During the summer of 1977 an episode of mesoscale convective complexes formed to the east of the Colorado Rocky Mountains. Many of these large convective complexes satisfied Maddox's [5] definition of a mature mesoscale convective complex (MCC) based on enhanced infrared (IR) satellite imagery. Table 1 summarizes Maddox's definition of an MCC. The episode occurred during the period 3-10 August 1977 during which the eastern two-thirds of the United States was dominated by an approximately steady-state pattern of moist southwesterly low-level flow south of a weak, roughly stationary front, which extended from Colorado across northern Illinois and into New England. North of the front the air mass was slightly cooler and drier than the air mass to the south, with generally weak and variable low-level flow. Mesoscale convective complexes were observed to develop daily near the Colorado Rocky Mountains and move generally eastward along this stationary front, sometimes maintaining identity for several days.

The tracks of 14 MCCs in this episode are shown in Fig. 1, based on GOES infrared (IR) satellite-defined centroids of convection associated with each MCC. (The darkened tracks indicate periods when the convection met Maddox's [5] areal and thermal criteria for a mature MCC.) There are two general genesis regions evident: one along the eastern slopes of the

Table 1

1. A continuous apparent black-body temperature less than -32°C covering an area greater than 100,000 km^2.

2. An apparent black-body temperature less than -53°C with an area greater than 50,000 km^2.

3. Area requirements 1 and 2 must be continuously met for at least six hours.

4. At the time of maturity, the width/length ratio of the -32°C isotherm must be greater than 0.7. This is to distinguish MCCs from squall lines. The time of maturity is defined as the time at which the coldest apparent black-body temperature is reached during the period described in 3.

Rocky Mountains and High Plains, and the other further east in the Missouri and Mississippi River basins. Other notable features in Fig. 1 are that the MCCs tended to occur in the vicinity of, and track along, a quasi-stationary surface front, and that the remnants of decayed complexes persisted as identifiable regions of loosely organized convection for long periods (up to three days) that occasionally reintensified into mature complexes.

The eight-day total precipitation distribution for the United States east of the Continental Divide (Fig. 2), based on the average accumulation within each one-degree latitude/longitude block, illustrates the significance of the MCC activity during the episode. A coherent band of relatively heavy precipitation (exceeding 20-40 mm) resembles the MCC track density in Fig. 1: It originates over an extensive region along the eastern slopes of the Rocky Mountains, converges into a more intense and concentrated swath across the central plains and midwest, and then broadens and weakens to the south and east of the Great Lakes. Precipitation in this band from the High Plains to the eastward extent of the 40 mm isohyet range from about 130 to 500 percent of the mean amount for an eight-day early-August period, with the rest of the country having below or near normal amounts (means based on Visher [13]). The narrow band of excessive precipitation (exceeding 60-100 mm) from western Kansas to New York was aligned just to the south of the quasi-stationary surface front and marked the axis to numerous severe weather reports, including tornadoes, hail, and damaging winds. Minor flooding occurred in a number of locations along this band; for example, on 6 August, Parke County in westcentral Indiana received 175 to 225 mm of precipitation in six hours and a total of 250 to 325 mm in 24 hours.

Table 2 gives the percent of severe weather reports occurring in various stages of the MCC life cycle, stratified by western and eastern system, and also for all systems. The pre-MCC period is defined as up to the time of MCC initiation; the early, middle, and late periods divide the mature MCC into approximate thirds; and the post-MCC period is after MCC termination.

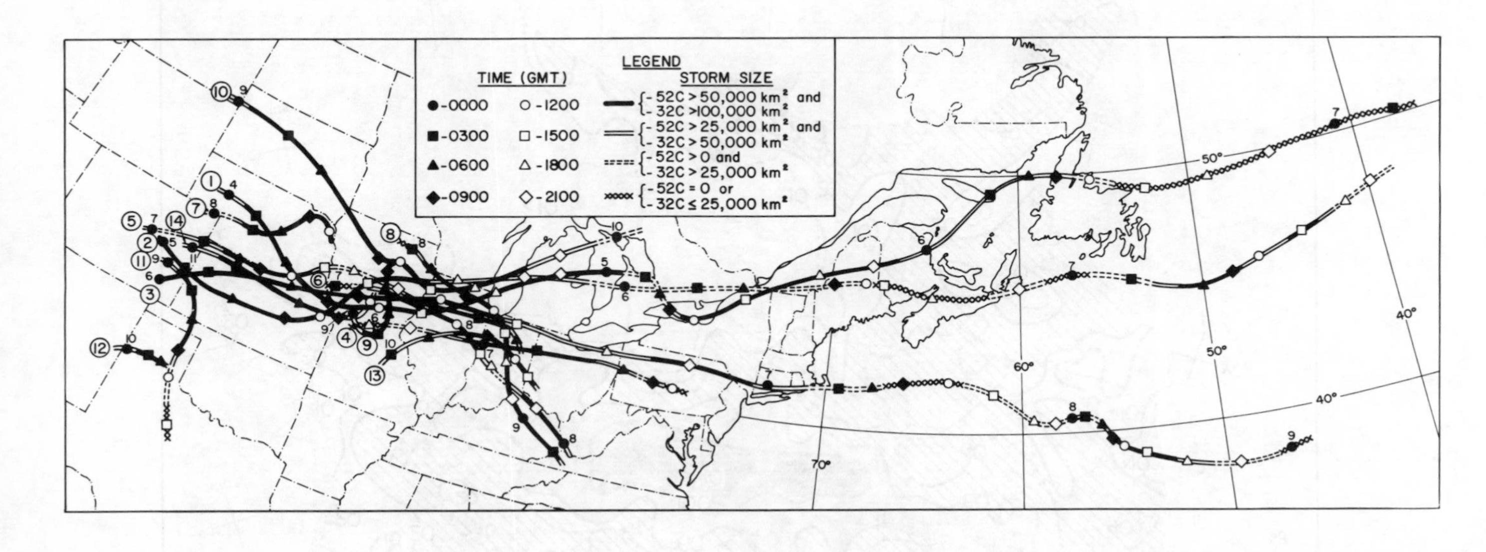

Fig. 1 Tracks of the centroid of the fourteen MCCs which developed during the episode of 3-10 August 1977. The circled sequential numbers chronologically identify the systems, with the date given near the 0000 GMT symbol.

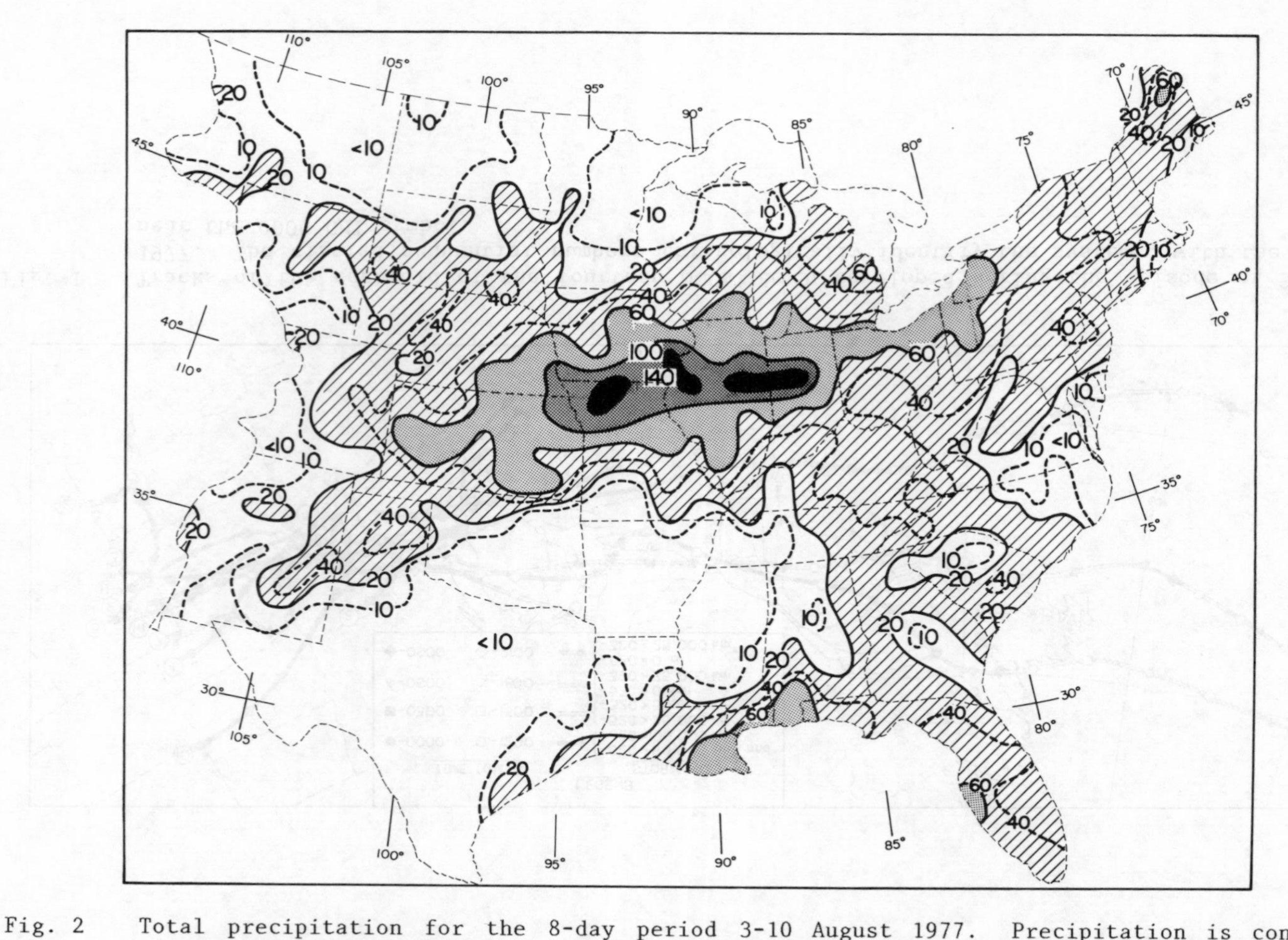

Fig. 2 Total precipitation for the 8-day period 3-10 August 1977. Precipitation is contoured and shaded in intervals of 40 mm, beginning at 20 mm. Intermediate contours of 10 and 40 mm are dashed.

Table 2 Percent of MCC-related severe weather reports occurring by period.

	Pre	Early	Mid	Late	Post
West	28	9	35	6	22
East	5	56	7	2	30
Total	19	28	24	5	25

The approximate one-fourth of MCC-related severe weather events occurring in the post-MCC period is due to the extended persistence (up to two days) of the rather unorganized mesoscale remnants of the decayed mature systems, with a larger proportion of these reports (over one-half) being high winds.

Of the pretermination events, most occur prior to the late-MCC period, which is seen to be relatively benign for both eastern and western systems. The intense convection occurring in the pre-MCC and early-MCC periods produce many of the severe weather events, particularly for the eastern systems. The occurrence of severe weather proceeds well into the mid-MCC period for the western systems, while the eastern systems become rather quiet in terms of severe weather. This may reflect the more complex meso-β substructure associated with the western systems, as opposed to the more uniform meso-α structure of the eastern systems as discussed in Section IV.

During the eight-day period, fourteen cloud systems that reached MCC dimensions at some time during their existence, developed and propagated along similar paths across the United States. An IR image of a large MCC at its mature state over eastern Kansas is shown in Fig. 3. That system, which developed on the western High Plains on the afternoon of 3 August, was the first in the series of fourteen.

Nine of the 14 systems shown in Fig. 1 developed to the immediate lee of the Rocky Mountains with at least one developing each afternoon of the episode. The remaining five systems developed west of the Mississippi River in Missouri or Iowa. The first three systems of the episode, including one to be discussed in later sections, survived intact for at least three days, and can be tracked well into the Atlantic Ocean. Based on the analysis of satellite IR isotherm areas shown, these systems appeared to reintensify over the Appalachian Mountains on the afternoon of the second day of their life, and/or over the Atlantic Ocean, possibly in the vicinity of the Gulf Stream.

The third MCC of the episode, which developed on 5 August (system 3 in Fig. 1) followed a somewhat more southerly path than the first two, and seemed to follow the scenario for ideal evolution of a long-lived convective complex. It survived for at least three days, reintensifying

Fig. 3 Enhanced infrared image of United States at 0900 GMT 4 August 1977, from GOES satellite at 75°W longitude. The stepped gray shades of medium gray, light gray, dark gray, and black are thresholds for areas with apparent black-body temperatures colder than -32, -42, -53, and -59°C, respectively. Temperatures progressively colder than -63°C appear as a gradual black to white range. The cloudiness over Utah indicates the midlevel moisture source for convective development on 4 August over Colorado. The meso-α scale convective complex (MCC) in eastern Kansas originated in the eastern Rockies and western plains the previous evening.

three times in response to favorable convective conditions, first over the High Plains, second over the Appalachian Mountains, and third apparently near the Gulf Stream. Bosart and Sanders [4] documented a similar mesoscale disturbance over a five-day period, during which it underwent three intensification cycles over land and another over the Atlantic. Such long-lived systems are probably not common events since they require a rather specific combination of environmental factors to sustain them. Most MCCs presumably survive only one, or occasionally two days. Others may become modified by baroclinic features and survive for longer periods but lose their primarily convective nature.

Nearly all the MCCs reached their mature stage between 0600 and 1200 GMT, early morning local time (Central Standard Time = GMT - 6). A similar finding for a much larger sample of MCCs was reported by Maddox [5]. Bosart and Sanders [4] noted a similar tendency in the intensification cycle of the disturbance they studied. One explanation for the strong nocturnal preference of these systems is the enhanced convergence produced by a low-level jet. Evidence of these jets was apparent in radiosonde data south of many of the storms. Past studies have noted the link between the increased low-level inflow afforded convective storms by the low-level jet, and the nocturnal thunderstorm phenomenon [8], [10], [2], [11]. The correlation, both in space and time, between the climatological peak intensity of the low-level jet [3] and the mature stage of many of the MCCs suggests a link between the two phenomena, although the physical relationship of cause and effect requires further study.

In this paper I shall describe the initiation phase of several convective elements of the episode. I will then summarize the structure of the mature MCCs. Then I will describe the meso-β (or 25-250 km) scale structure of the MCCs.

II. THE MOUNTAIN-GENERATED COMPONENT

Here I shall describe the evolution of convection that was directly associated with the formation of an MCC during the episode. To illustrate the mountain component, we will look at one day in the episode, namely 4 August 1977.

On the morning of 4 August 1977 the synoptic pattern over North America was dominated by a broad closed cyclonic circulation centered over Hudson Bay at 50 kPa (500 mb) (Fig. 4). Imbedded within this flow were a series of weak short waves, as indicated by the heavy dashed lines. One of these, located over the upper Mississippi valley, was associated with a large MCC which developed in eastern Colorado and Wyoming on the evening of 3 August (see Fig. 3). Farther to the northwest, at this time, two short waves, one of which was very weak, were entering the northern Rocky Mountain states. Both of these waves were linked to southward surges of cool polar air at the surface as shown in the mesoscale surface analyses, Figs. 5a-g. The first of the cold fronts, initially the stronger of the two, pushed into Colorado about midmorning. This surge of air was too shallow to climb the Front Range and directly affect cumulus formation over South Park, but its northeasterly flow enhanced low-level moisture convergence over northeastern Colorado in the afternoon and helped intensify the cumulonimbus systems emanating from the Front Range onto the High Plains. By 1800 GMT, as the first front was weakening and stalling in central Colorado, the second front approached Wyoming. This second surge of cool air, supported by the stronger short-wave trough at 50 kPa, became the dominant synoptic feature by the afternoon of 4 August, and continued to push southeastward, affecting the later evolution of the convective system.

Also indicated on the surface analyses (Fig. 5) are two features in Kansas which affected the later development of the 4 August convection. The first is the remnant of the previous day's MCC, which left a slightly cooler and more moist air mass over most of Kansas. The boundary between this air and the hot, drier air to the south formed a discontinuity along which convection developed later in the day. Secondly, a weak surface low pressure trough, which moved through Kansas on 4 August, might have served to enhance the local convergence in the area of the discontinuity in southern Kansas. In any case, convection also developed late on 4 August in eastern Kansas in the vicinity of this trough.

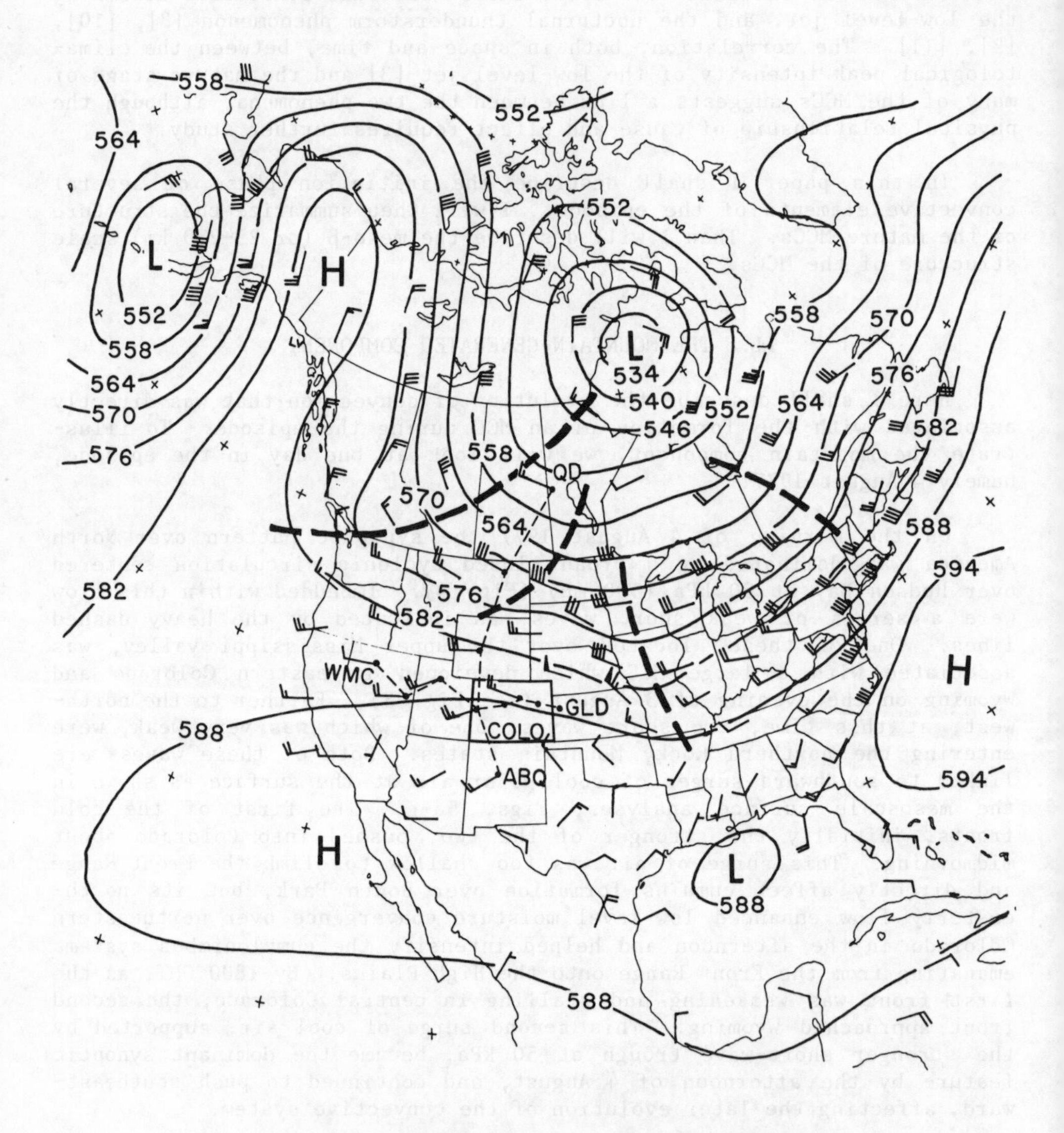

Fig. 4 50 kPa (500 mb) analysis for 1200 GMT 4 August 1977. Heavy dashed lines denote short-wave troughs. Height contours are labelled in decameters. Wind pennants, full barbs, and half barbs denote 25, 5, and 2.5 m/s, respectively. (The dotted lines WMC-GLD and QD-ABQ refer to cross sections that do not appear in this paper.)

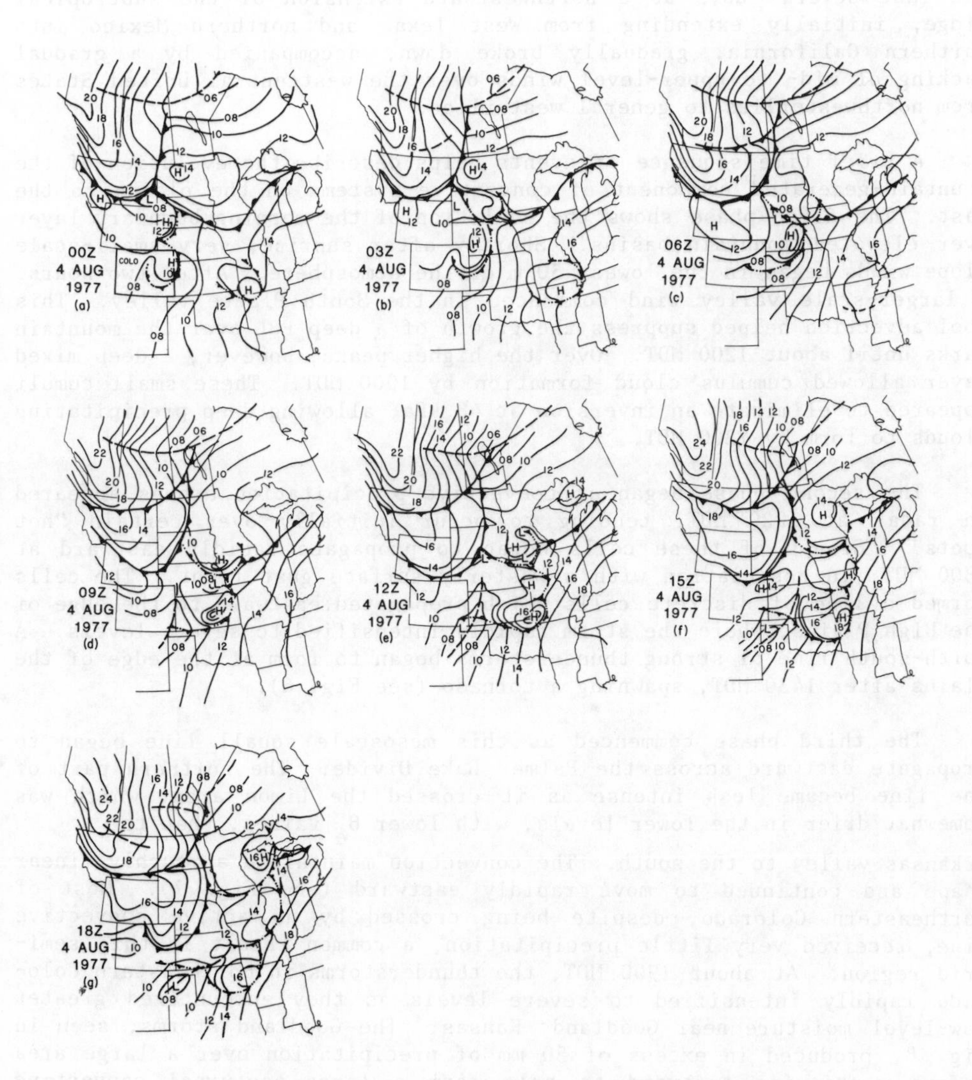

Fig. 5a-g Mesoscale analyses of surface pressure reduced to sea level (contours labelled in 10^{-1} kPa, with prefix of "10" omitted) and surface discontinuities (fronts, troughs, and mesohigh boundaries). (a) to (g) cover the period 0000 GMT to 1800 GMT, 4 August 1977, at three-hour intervals.

Over western Colorado and much of Utah early on 4 August an area of midlevel convective cloudiness seen on the satellite image (Fig. 3) suggests a moist, rather unstable air mass in a region of weak anti-cyclonic flow at 50 kPa. Dew points in this region had increased over the last several days as a northwestward extension of the subtropical ridge, initially extending from West Texas and northern Mexico into northern California, gradually broke down, accompanied by a gradual backing of mid- to upper-level winds over the westcentral United States from northwesterlies to general westerlies.

A brief time sequence of events helps describe the evolution of the mountain-generated component of convective systems on the plains to the east. The first phase shows the evolution of the morning boundary layer over elevated mountain basins. Shortly after sunrise, very small-scale slope winds began in the lowest 50 m of the atmosphere. After two hours, a larger-scale valley wind commenced in the South Platte valley. This cool advection helped suppress the growth of a deep PBL over the mountain parks until about 1200 MDT. Over the higher peaks, however, a deep mixed layer allowed cumulus cloud formation by 1000 MDT. These small cumuli appeared to eliminate an inversion at 48 kPa, allowing deep precipitating clouds to form by 1200 MDT.

The second phase began as convective precipitation echoes appeared on radar at 1220 MDT, tending to occur initially over certain "hot spots". Groups of these cells began to propagate rapidly eastward at 1300 MDT, in association with a westerly surface gust front. The cells formed a line of discrete cells which propagated eastward to the edge of the High Plains, where the storm rapidly intensified to severe levels. A north-south line of strong thunderstorms began to form at the edge of the plains after 1430 MDT, spawning a tornado (see Fig. 6).

The third phase commenced as this mesoscale squall line began to propagate eastward across the Palmer Lake Divide. The northern part of the line became less intense as it crossed the Limon area, which was somewhat drier in the lower levels, with lower θ_e values, than the Arkansas valley to the south. The convection maintained a roughly linear shape and continued to move rapidly eastward (see Fig. 7). Most of northeastern Colorado, despite being crossed by an active convective line, received very little precipitation, a common result in this semi-arid region. At about 1900 MDT, the thunderstorms in northeastern Colorado rapidly intensified to severe levels as they encountered greater low-level moisture near Goodland, Kansas. The Goodland storms, seen in Fig. 8, produced in excess of 50 mm of precipitation over a large area before continuing eastward to help form a large nocturnal convective complex over the central plains.

Figure 9 illustrates the eastward progression of the convective systems as viewed by infrared satellite images. Quite apparent is the up-scale intensification of the systems as they encounter greater low-level moisture. Also evident in Figs. 9(f) and (g) is the development of an east-west oriented line of convective cells along the frontal boundary left by the remnant of the previous day's MCC shown in Figs. 5(d) to (g).

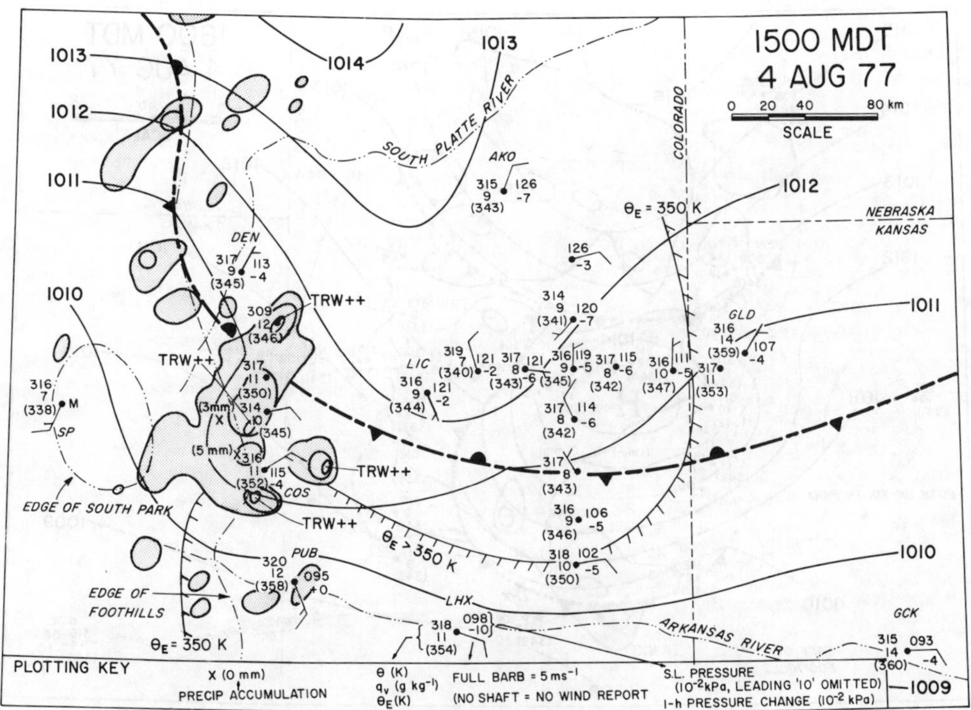

Fig. 6 SPACE/HIPLEX surface map for 1500 MDT 4 August 1977, superimposed with Limon radar summary for 1530 MDT. Dashed lines indicating South Park and foothills boundaries are based on more detailed topography (Palmer Divide, which extends from LIC westward to the foothills, is omitted for clarity). Experimental mesonetwork data are plotted at the unlabelled locations. Supplemental data from standard hourly observations at Denver (DEN), Colorado Springs (COS), Limon (LIC), Akron (AKO), Fort Collins (FCL), Pueblo (PUB), and La Junta (LHX), Colorado and from Goodland (GLD) and Graden City (GCK), Kansas, and from beyond the map boundaries were also used in the analysis. Winds, potential temperature, mixing ratio, equivalent potential mixing ratio, pressure reduced to sea level, and one-hour pressure changes are plotted according to the key in the figure. Altimeter settings from the standard observations were reduced in a manner consistent with the experimental pressures (diurnal effects are not removed). Isobars are analyzed at intervals of 10^{-1} kPa. Successive radar echo contours (beginning with VIP level 1) and maximum thunderstorm intensity designations conform to National Weather Service standards. Hatched line separates region to the south and east which has θ_e exceeding 350 K. Precipitation values (plotted in mm) are one-hour accumulations ending at map time, based on a sparse hourly recording network.

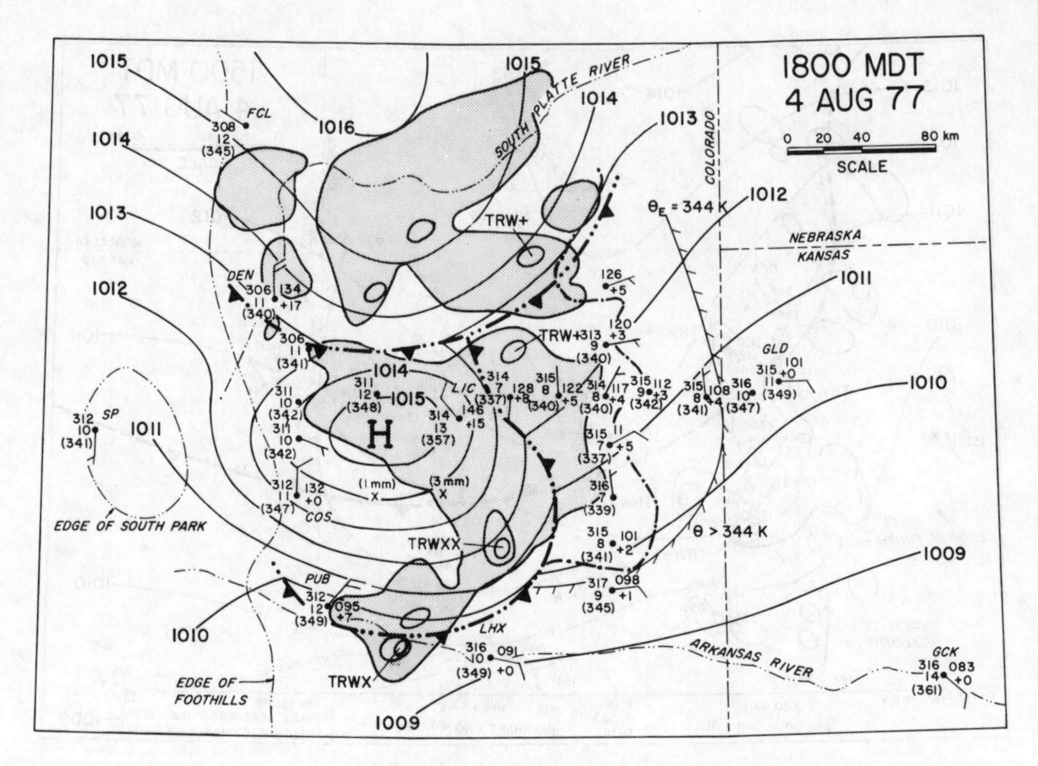

Fig. 7 Similar to Fig. 6, for 1800 MDT 4 August 1977. Hatched line delineates high θ_e air (> 350 K) to the east of the storm. Precipitation values are three-hour accumulations ending at map time. The weak stationary front in Fig. 6 has been dropped with the more dominant mesohigh boundaries now included.

III. STRUCTURE OF THE MATURE MCC

This section examines some aspects of the interaction of the MCC which developed late on 4 August, with its environment, as deduced from rawinsonde observations at 1200 GMT 5 August 1977, about three hours after the most intense phase of the storm. Inferences about the nature of the circulation generated by the system will be made, based on the evidence presented.

The isentropic analysis on the north/south cross section in Fig. 10 reveals that the surface cold front in the vicinity of the MCC was associated with the polar front jet core located near the United States/ Canada border at about 25 kPa. A moisture analysis (not shown) suggests that the inversion feature at 40 kPa in the vicinity of the convection was a result of subsidence. Although the Topeka (TOP) sounding was contaminated by nearby convection, the soundings from Monett (UMN) and

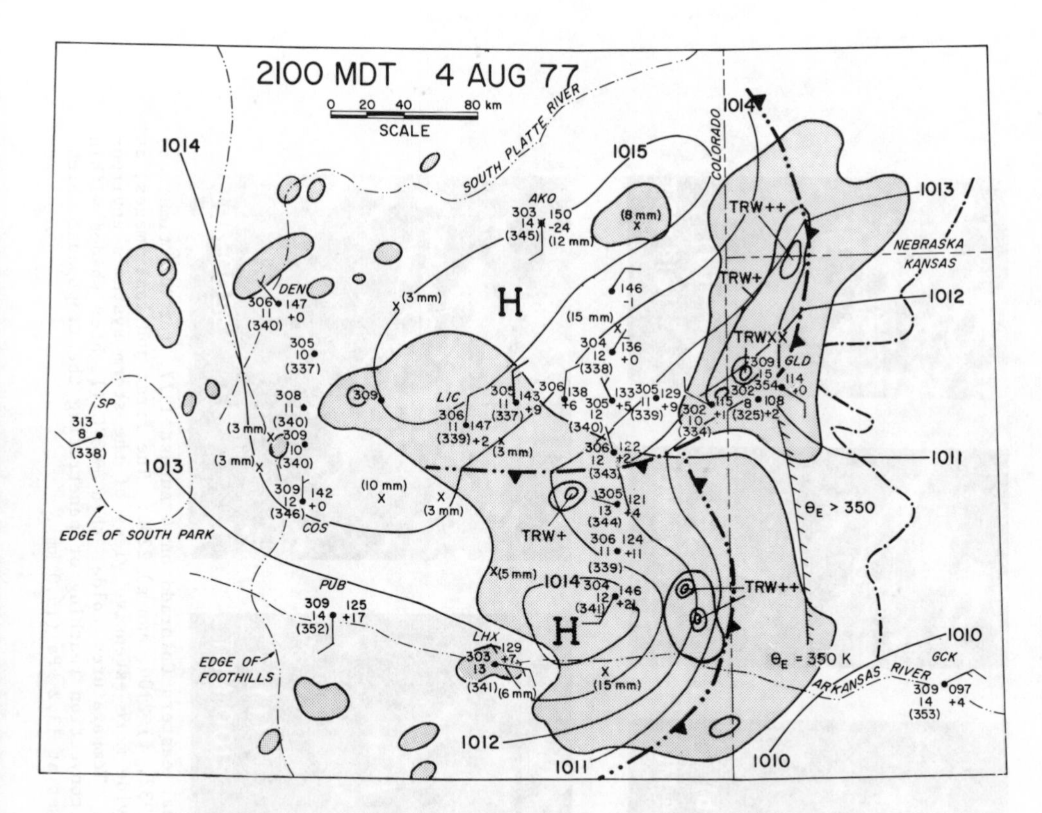

Fig. 8 Similar to Fig. 6, for 2100 MDT 4 August 1977. Hatched line delineates high θ_e air (> 350 K) to the east of the storm. Precipitation values are three-hour accumulations ending at map time.

Omaha (OMA) reveal a significantly drier air mass above the inversion than below it. Further evidence of this subsidence can be seen in Fig. 11, an east/west cross section through Topeka for the same time. Between 20 and 40 kPa west of Topeka the isentropes slope sharply downward toward the east. Since the wind was from the west in that vicinity, subsidence is the most likely result. East of the storm, in the upper troposphere over Peoria (PIA), one finds a rather shallow layer of high speed westerly flow. Satellite film loops show that cirrus clouds spreading outward from the MCC anvil covered a broad area including most of the state of Illinois by 1200 GMT 5 August, leading to the conclusion that the MCC was the source of the high speed air in the wind maximum. Further evidence to support this point is presented later.

Focusing further on the upper-level structure in the vicinity of the MCC, we look at the 20 kPa analysis as it evolved between 0000 GMT and

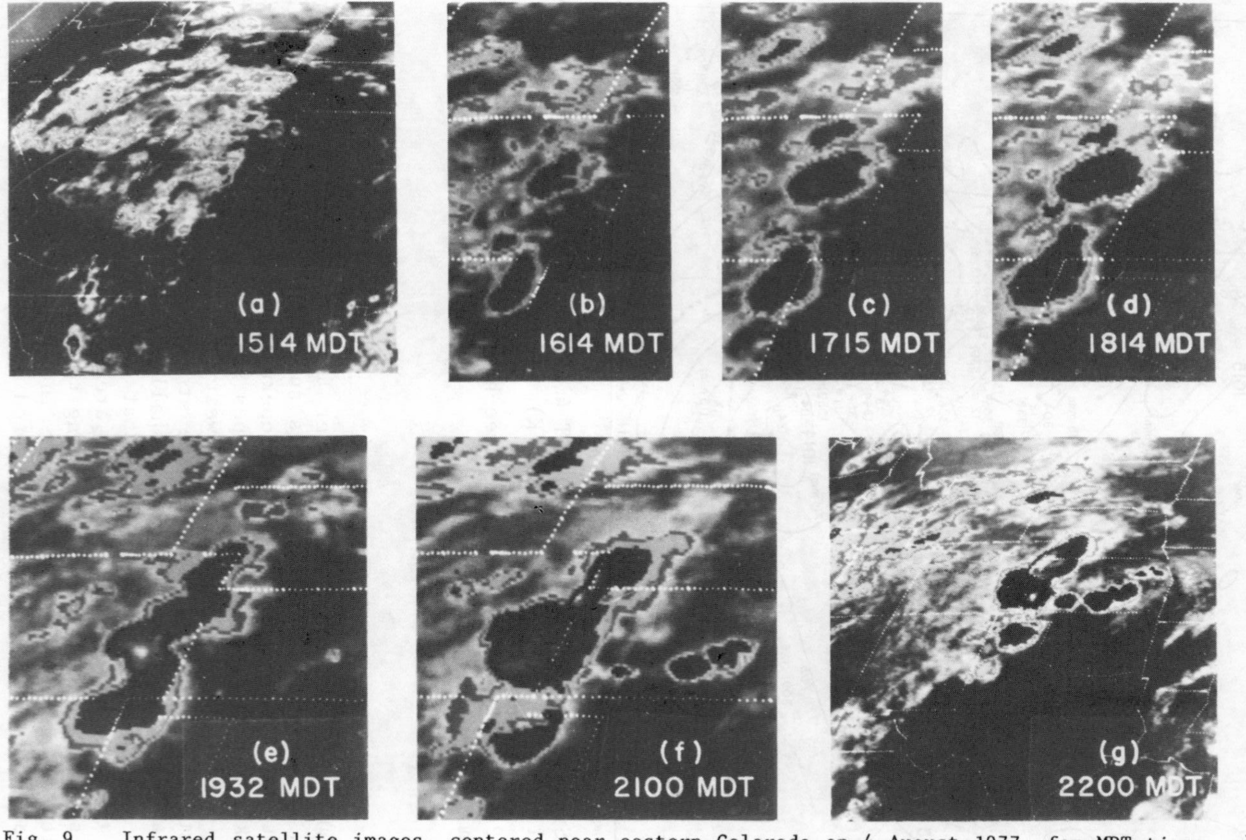

Fig. 9 Infrared satellite images, centered near eastern Colorado on 4 August 1977, for MDT times a) 1514, b) 1614, c) 1715, d) 1814, e) 1932, f) 2100, and g) 2200. The first and last images, at one-half the scale of the others, provide more extensive views of the storm system's environment. Gray shades are as in Fig. 3. Temperatures colder than -63°C (lighter shades within the interior black contours) indicate convection reaching or penetrating the tropopause which, based on the 1700 MDT Limon sounding, was at 13.2 kPa (15.0 km).

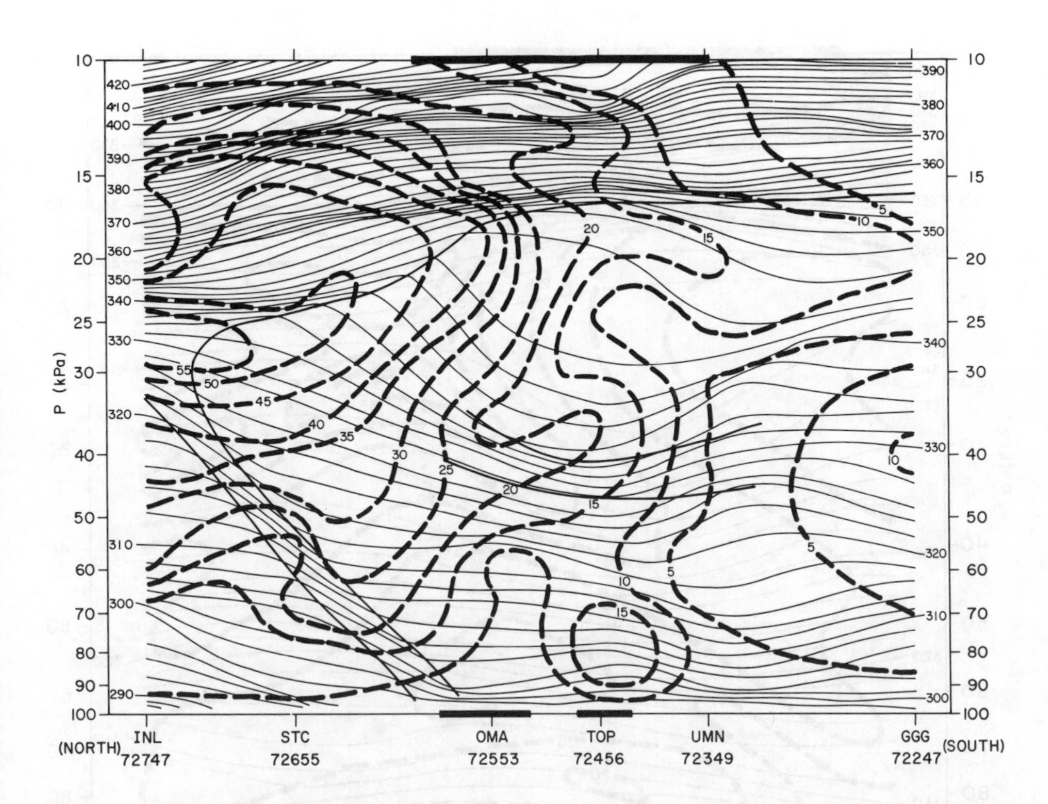

Fig. 10 North/south cross section at approximately 95°W for 1200 GMT 5 August 1977. Analysis of wind (dashed lines) as in m s^{-1}, and of potential temperature (thin solid lines) in degrees K. Significant inversion features are located with heavy solid lines. The extent of the cross section covered by the MCC cloud shield (IR temperatures colder than -32°C) is indicated by the bar along the top, whereas the bars along the bottom show the extent of the meso-β radar features. Stations included are: (INL) International Falls, Minnesota; (STC) St. Cloud, Minnesota; (OMA) Omaha, Nebraska; (TOP) Topeka, Kansas; (UMN) Monett, Missouri; (GGG) Longview, Texas. Their WMO numbers are indicated.

1200 GMT on 5 August. Figure 12 shows that the pre-MCC convection in eastern Colorado was developing in a region of large-scale, weak anticyclonic flow, with weak vertical wind shear. Twelve hours later (Fig. 13) a closed anticyclone had developed over the MCC, with a length scale similar to that of the storm. Also apparent from comparison of Figs. 12 and 13 is an acceleration of the flow north of the storm. This manifests itself both in increased wind speeds and in the slight cross-contour component of the flow toward lower heights at stations north of the convection in Fig. 13. Such acceleration has also been observed in association with other MCCs (e.g. [6]). The anticyclone developed over

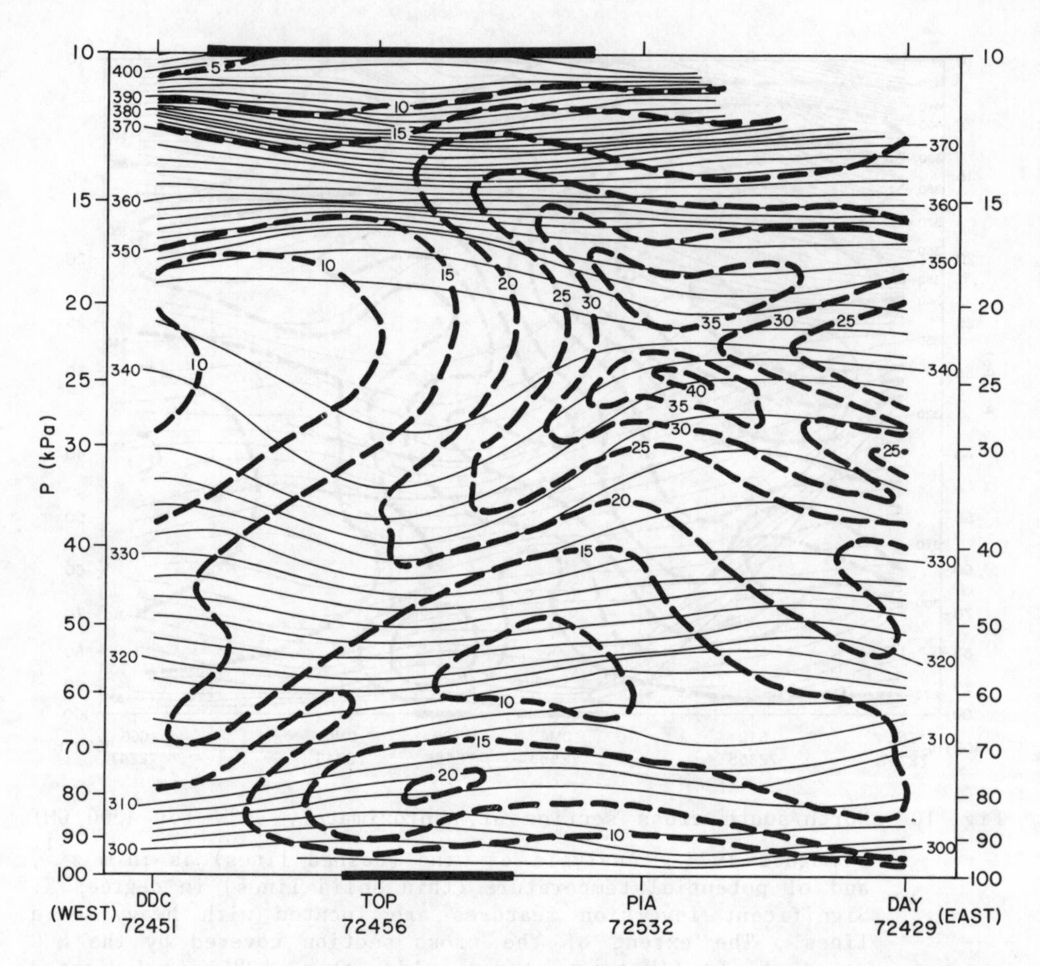

Fig. 11 East/west cross sections at approximately 40°N for 1200 GMT 5 August 1977, with potential temperature (solid lines) in degrees K, wind speed (dashed) in m s^{-1}. Bars along top and bottom are as in Fig. 10. Stations included are: (DDC) Dodge City, Kansas; (TOP) Topeka, Kansas; (PIA) Peoria, Illinois; and (DAY) Dayton, Ohio. Their WMO numbers are indicated.

the MCC as a result of a significant warming of the upper troposphere. This development is evident from the 12-hour change in the 50 to 25 kPa thickness (not shown).

Figure 14 shows the divergence profiles in the vicinity of two of the MCCs of the episode. These profiles were calculated in 5 kPa layers using full resolution tracking winds obtained from stations arranged in a polygon surrounding each MCC, using the technique discussed by Bellamy

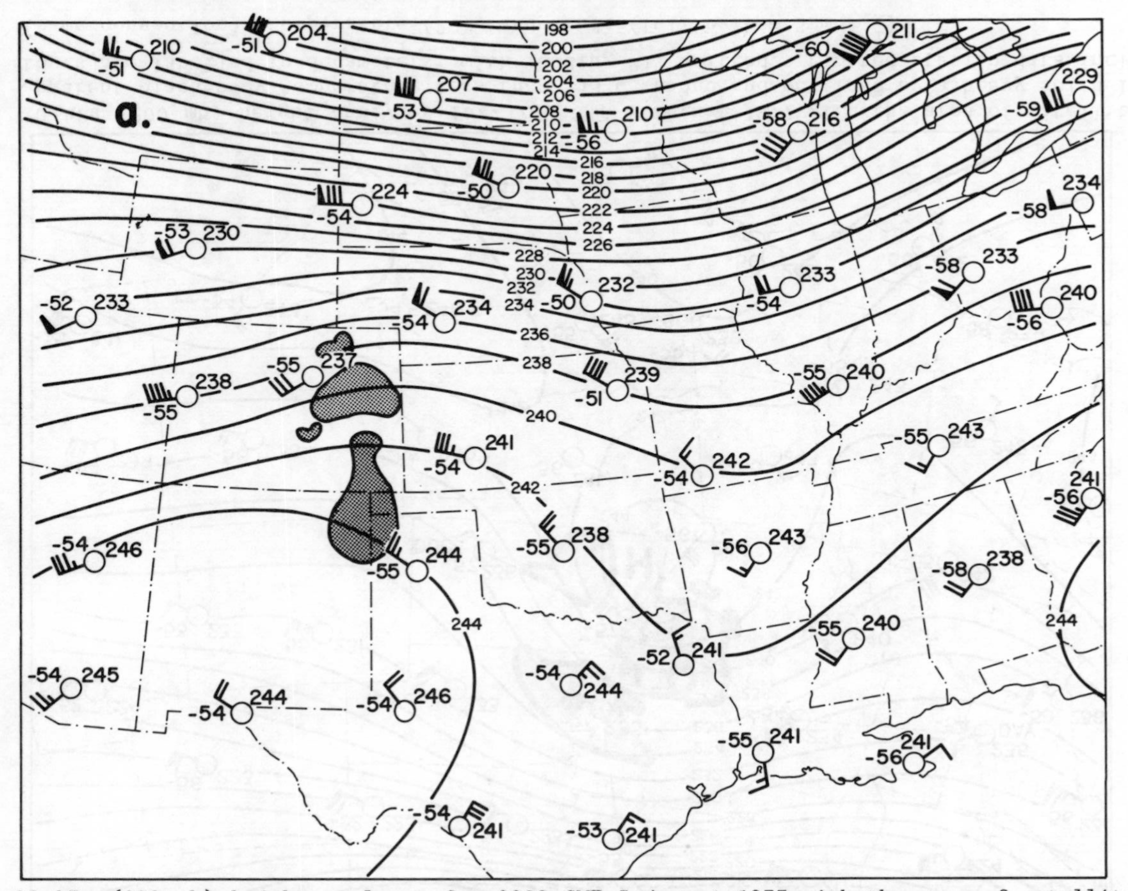

Fig. 12 20 kPa (200 mb) height analyses for 0000 GMT 5 August 1977 with the area of satellite observed apparent black-body temperature less than -53°C shaded and outlined by a heavy solid line. Contours are labelled in decameters, with leading "1" omitted. Plotted station data include height (same convention), temperature (°C), and winds (full barb = 5 m s^{-1}).

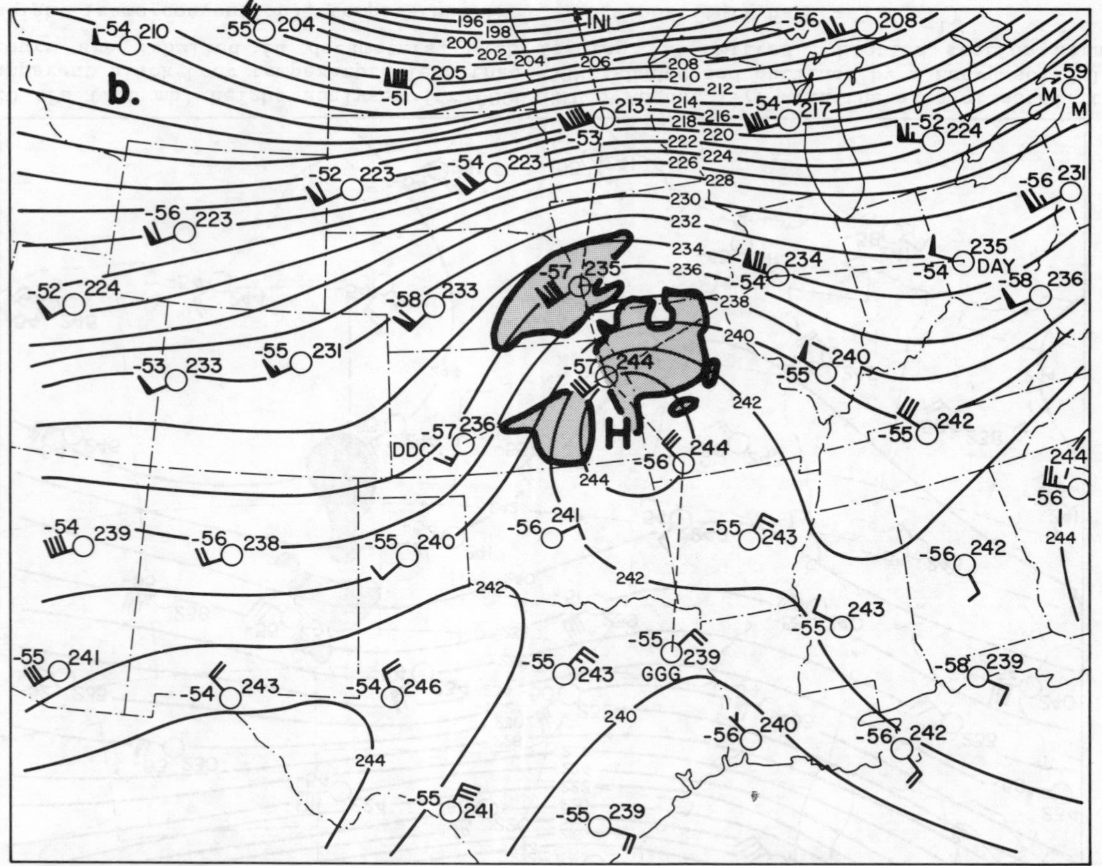

Fig. 13 20 kPa (200 mb) height analyses for 1200 GMT 5 August 1977 with the area of satellite observed apparent black-body temperature less than -53°C shaded and outlined by a heavy solid line. Contours are labelled in decameters, with leading "1" omitted. Plotted station data include height (same convention), temperature (°C), and winds (full barb = 5 m s^{-1}).

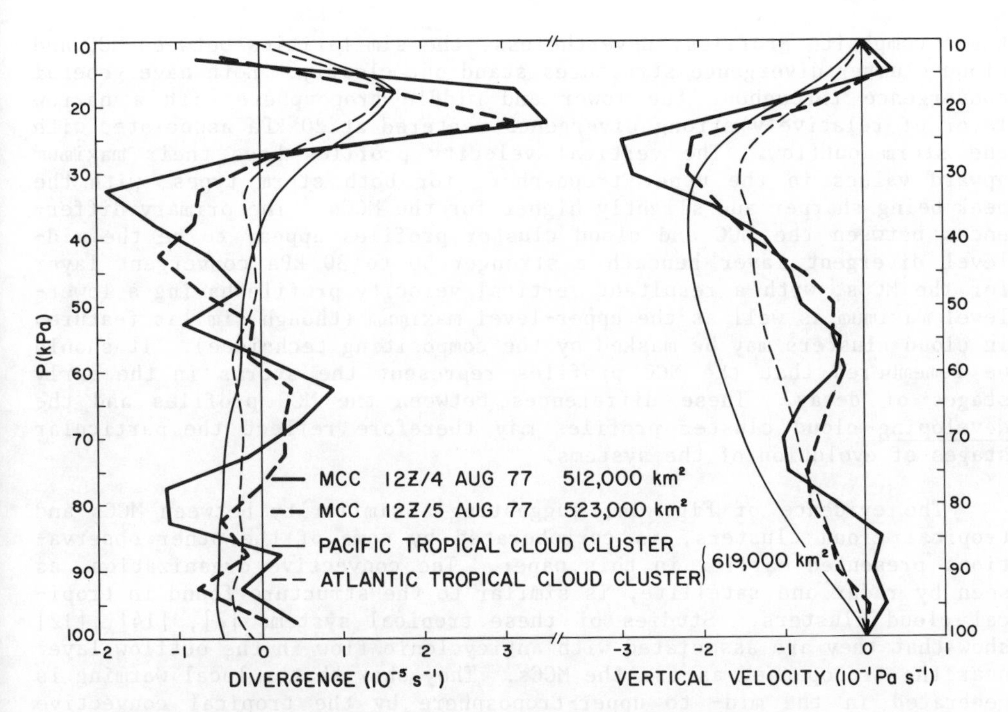

Fig. 14 Vertical profiles of divergence and p-velocity as reported by McBride and Zehr [7] for developing Pacific and Atlantic tropical cloud clusters, and within polygons surrounding the two MCCs studied here. For 1200 GMT 4 August 1977 the polygon used has an area of 512,000 km^2 and has at its boundary the following stations: Oklahoma City (WMO number 72353); Monett, Missouri (72349); Salem, Illinois (72433); Peoria, Illinois (72532); Omaha, Nebraska (72553); North Platte, Nebraska (72562); and Amarillo, Texas (72363). For 1200 GMT 5 August 1977 the polygon used has an area of 523,000 km^2 and has at its boundary the following stations: Oklahoma City; Monett; Peoria; Huron, South Dakota (72655); North Platte, Nebraska; and Dodge City, Kansas (72541).

[1]. Also shown on Fig. 14 for comparison are composite calculations of divergence and vertical velocity in developing tropical cloud clusters as presented by McBride and Zehr [7] for two separate data sets, one from the western Atlantic and one from the western Pacific Oceans. The area within which McBride and Zehr calculated divergence is about 20 percent larger than that used for the MCCs, similar enough so that qualitative comparison of these curves is quite reasonable.

The smoother, less detailed nature of the cloud cluster can be attributed to the large number of soundings and storms combined into these composite profiles; nevertheless, the similarities between MCC and cloud cluster divergence structures stand out clearly. Both have general convergence throughout the lower and middle troposphere with a narrow layer of relatively strong divergence centered at 20 kPa associated with the storm outflow. The vertical velocity profiles have their maximum upward values in the upper troposphere for both storm types, with the peak being sharper and slightly higher for the MCCs. The primary differences between the MCC and cloud cluster profiles appear to be the mid-level divergent layer beneath a stronger 50 to 30 kPa convergent layer for the MCCs, with a resultant vertical velocity profile having a lower-level maximum as well as the upper-level maximum (though similar features in cloud clusters may be masked by the compositing technique). It should be remembered that the MCC profiles represent the storms in the early stages of decay. These differences between the MCC profiles and the <u>developing</u> cloud cluster profiles may therefore reflect the particular stages of evolution of the systems.

The evidence of Fig. 14, suggesting a similarity between MCCs and tropical cloud clusters, is corroborated by some of the other observations presented earlier in this paper. The convective organization, as seen by radar and satellite, is similar to the structure found in tropical cloud clusters. Studies of these tropical systems [7], [14], [12] show that they are associated with anticyclonic flow in the outflow layer near the tropopause, as are the MCCs. They show that a local warming is generated in the mid- to upper-troposphere by the tropical convective system, resulting in a "warm core" storm structure. Similar warming associated with the MCCs is evidenced by the upper tropospheric thickness increases that we have found as the storms intensified. They also show tropical clusters to be surrounded by a subsiding clear region of suppressed convection. Evidence for a similar subsidence phenomenon associated with two MCCs has been found in our analysis.

Of course, tropical cloud clusters exist in an essentially barotropic environment and rely primarily on buoyant accelerations to drive them. We have shown that the MCCs of this episode favored a proximity to east/west surface frontal discontinuities. The north/south temperature gradient associated with these weak summertime fronts is not large, and by definition it is confined to the cool side of the boundary. While the MCC convection in the episode was concentrated on the warm side of the discontinuities, the fronts must nevertheless be considered as a possible source of energy for the MCCs. They may provide a mechanism to enhance upward vertical motion, which could initiate, strengthen, or maintain convection. The vertical motion values at 30 kPa for the MCCs in Fig. 14 are equivalent to a uniform 5 to 6 cm sec^{-1} upward motion over a square 700 km on a side. This is a rather large synoptic-scale value, such as one might find in the warm sector of a developing midlatitude wintertime cyclone (e.g. [9]). However in this case, if the surface front and the baroclinic processes associated with cyclogenesis along a front are important to the production of these large MCC-generated vertical velocities, one would expect to see some of the manifestations of cyclogenesis, such as falling surface pressures and the north/south exchange of heat that develops as the front deforms and the storm "wraps up". Our analyses have shown, however, that there is no consistent drop in surface pressure associated with the MCCs. The fronts do not appear to deform in the vicinity of the storms except where they intersect the outflow boundaries of meso-β scale convective clusters.

To address more quantitatively the question of a north/south heat exchange, a process which is necessarily associated with the production of vertical velocity in baroclinic storms (when the temperature gradient is concentrated in the north/south direction), we have calculated the meridional heat transport for two MCCs of the episode. Figure 15 presents the vertical profiles of average meridional sensible heat transport by the two storms along with the profile from a mature winter cyclone centered at the same location. Each profile was constructed by calculating the nine-station mean temperature for each 50 kPa layer from the surface to 10 kPa. Each individual station's deviation from that mean, T', was then multiplied by the meridional wind component V (equivalent to V' since the zone average, $\bar{V}$, is zero), and the nine values of VT' were averaged arithmetically. The results show that the winter cyclone generated a strong northward transport of heat through the troposphere while virtually no northward heat transport was produced by the MCCs. In fact, throughout most of the troposphere, except near the surface, there was a slight southward transport of heat by these systems. The low-level transport can probably be better attributed to the effects of the low-level jet than to the surface front. One is thus led to the conclusion that a weather system which displaces as much mass vertically as a baroclinic system without transporting significant sensible heat poleward must be operating in a predominantly barotropic environment and must be basically driven by buoyant accelerations (e.g. the tropical cloud cluster).

IV. MESO-β STRUCTURE

We have thus far concentrated on the early origin of MCCs which have a well-defined meso-β structure, and mature MCCs which have well-defined meso-α (250 to 2500 km) structure. We will now investigate the meso-β structure of MCCs throughout their life cycles.

Figure 16 illustrates several important evolutionary features of the second system shown in Fig. 1, which we described in earlier sections. The system reached its maximum extent in the Kansas/Nebraska region. Included in this figure are GOES-east IR apparent black-body isotherms of -40 and -62°C, meso-β clusters and lines of relatively strong convection, and approximate displacements of those features since the previous map time or over the previous three hours (if in existence for that long). The identification of meso-β features was subjectively based primarily on National Weather Service (NWS) radar film which was generally of a poor quality as far as quantitative intensity resolution was concerned. However, the temporal resolution of the PPI data was generally excellent, so an amalgamation of intensity inferences, using all data sources, allowed the reliable identification of relatively strong meso-β scale echo entities that persisted for well over an hour. Each meso-β feature depicted may represent near-solid echo coverage, or alternately, simply a cluster or line of discrete echoes that maintains a meso-β cohesiveness.

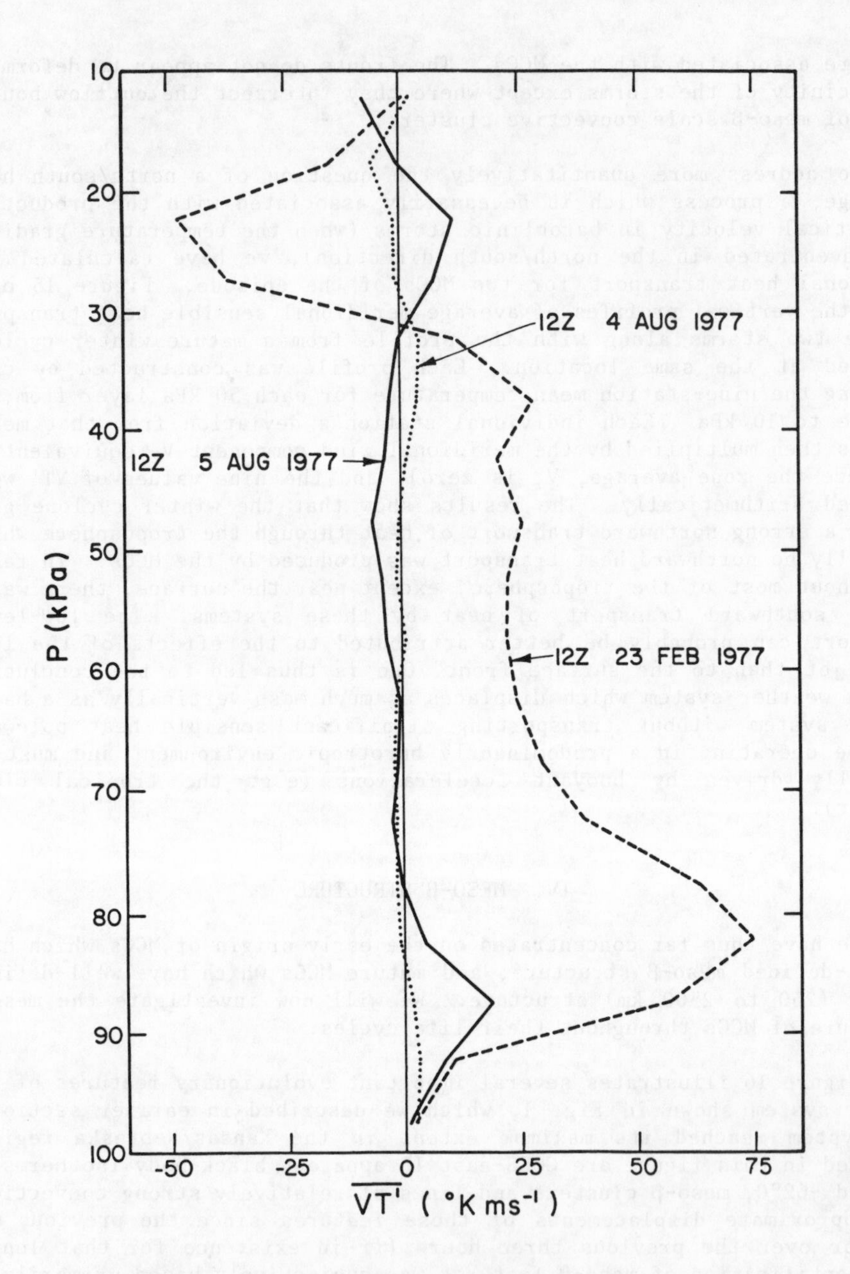

Fig. 15 Vertical profiles of nine-station average meridional heat transport for the two MCCs and for a mature winter cyclone centered at the same point. The nine stations used for each profile are the following: North Platte, Nebraska; Omaha, Nebraska; Peoria, Illinois; Oklahoma City; Monett, Missouri; and Little Rock, Arkansas. Their WMO numbers are indicated in previous figures.

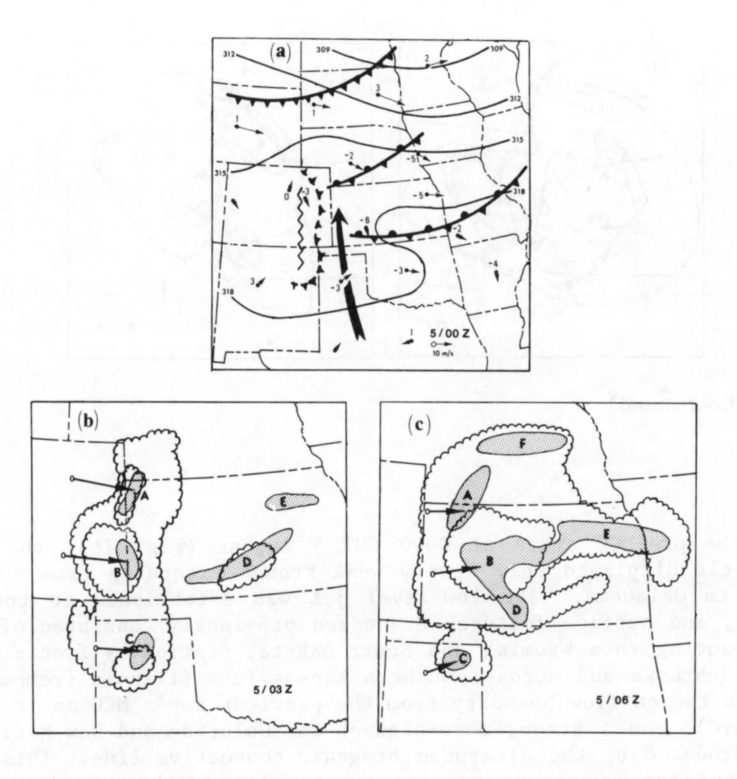

Fig. 16 (a) 0000 GMT, 5 August 1977, 70 kPa (700 mb) height contours (solid and thin dashed lines, labelled in tens of meters) and wind vectors (magnitude shown as three-hour displacement), 85 kPa maximum wind axis (broad arrow), lifted index (plotted) with surface features including east/west fronts, and north/south mesoscale ridge and bubble-high boundary. (b) Infrared satellite and radar composite chart for 0300 GMT 4 August. Shaded regions (identified by letters) denote significant meso-β scale convective lines or clusters, with the solid vectors giving the displacement of their most intense centers over the previous three hours or since the previous depicted chart time in the sequence. The heavy dashed segments extending from the shaded regions are axes of weaker convection along which the more intense meso-β features are imbedded. The outer and inner scalloped lines are the IR apparent black-body isotherms of -40 and -62°C, respectively. (c) Similar to (b), except for 0600 GMT. (d) Similar to (b), except for 0900 GMT. (e) Similar to (b), except for 1200 GMT. The dotted vectors are moist-layer to 50 kPa wind shear (defined in text) at the selected stations (denoted by large dots) at 1200 (magnitude shown gives a three-hour displacement at that speed).

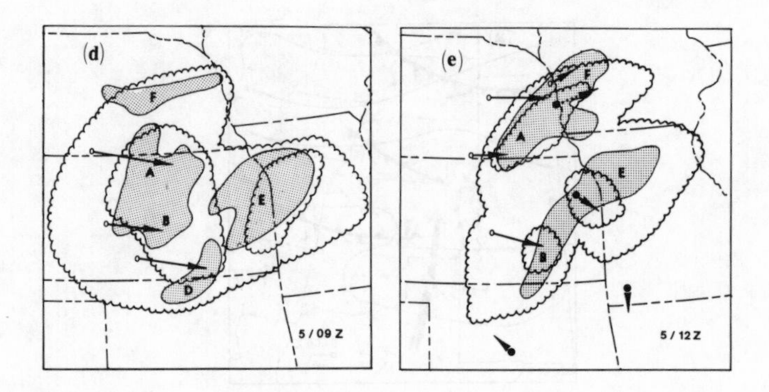

Fig. 16 (Continued)

At the pre-MCC stage at 0000 GMT 5 August (Fig. 16), the 70 kPa height field displayed only a very weak trough extending from northwest Colorado to Oklahoma. The low-level jet was established in the Texas panhandle, and surface features discussed previously consisted of a cold front advancing into Wyoming and South Dakota, stationary fronts through southern Nebraska and across southern Kansas into Illinois (remember the latter was the outflow boundary from the previous day's MCC as it tracked on eastward), and a strong meso-high on the Colorado and New Mexico High Plains, produced by the afternoon orogenic convective line. This meso-α north/south line of storms was similar to, but farther south than, the previous day's orogenic line.

By 0300 GMT (Fig. 16), the orogenic line, consisting of three meso-β clusters (A, B, C), had moved eastward, with B producing severe surface winds and hail in southeast Colorado. New meso-β lines (D, E), oriented east-west, developed in Kansas, the southern one (D) occurring along the stationary front separating the cooler, moister wake of the previous day's MCC from the unmodified maritime tropical air to the south. This development was one hour prior to MCC initiation, as determined by IR cloud shield area.

As the eastward moving meso-α orogenic line (A, B, C) intersected the western end of the newer meso-α east-west line (D, E), the meso-β entities at the region of intersection (B, D) intensified rapidly, as was the case the day before. By 0600 GMT (Fig. 16), the MCC was well established and centered on cluster B, which merged with the western end of D, developed rapidly towards line E to the east, and produced local hourly rainfalls of 60 mm. A new meso-β line, F, developed in central Nebraska along the east-west stationary front shown in Fig. 16. Throughout the genesis stage and early mature stage, the meso-β features maintained a general confluent tendency.

The MCC had achieved maximum intensity and was beginning to weaken by 0900 GMT (Fig. 16). A and B had merged and expanded into a large precipitation region, but not too intense. Line D split off of B and displayed a southeastward propagation. Bands E and F, both originating as east-west lines, remained essentially stationary in the westerly steering flow as individual cells propagated eastward along the complexes. E developed in areal extent as the A-B cluster approached it. In the dissipation stage (1200 GMT, Fig. 16), the A-B cluster split, with B merging into E and A merging into F to give two weak meso-β bands with a squall-like configuration. The A-F band propagated east-northeastward while the E-B band moved southeasterly. This pattern of diffluence motion of the meso-β components of a dissipating MCC has been found to be a characteristic of this stage of the MCC life cycle.

All the High Plains MCCs that we have analyzed were characterized by their upscale intensification being centered on a meso-β component in a north/south meso-α orogenic line that intersected another meso-α feature to the east. While this quite regularly observed orogenic component is instrumental in many of the western plains MCCs, the eastern genesis region does not have the strong topographic forcing to help generate its MCCs. I will now briefly summarize the evolution of system 6 in order to point out similarities and differences to the western plains systems.

Figure 17 shows some synoptic features at and around 0000 GMT 7 August that help explain all three of the MCCs formed on this day (systems 4, 5, and 6). The 70 kPa height field has a short-wave trough through the eastern Dakotas, with a weaker, shorter wave through Colorado into the Texas panhandle. At 0300 GMT, when the initial convection of system 6 appeared, surface features consisted of a warm front extending from Colorado through central Iowa and a trough from South Dakota to northeast Colorado. The western MCC (system 5) developed at the triple intersection of an orogenic meso-α line and the surface front and trough in northeastern Colorado. Nine hours earlier, at 1800 GMT 6 August, the position of the surface warm front (broken warm front symbol) was across northern Missouri, where it was intersected by a north/south band of residual moisture from the dissipated system 3 (line of circles). MCC number 4 initiated at this intersection at that time. At 0300 GMT 7 August, system 4 was a mature MCC centered over Indiana and had left a wake boundary trailing to the northwest where it intersected the warm front in western Iowa. This feature was barely evident in high resolution visible satellite imagery, but it was also supported by surface observations. Warm moist air to the west of the outflow boundary and to the south of the warm front was being advected northward toward the intersection.

The initial convection leading to the MCC appeared after 0300 GMT in extreme northwest Missouri within the warm moist surface air. By 0600 GMT (Fig. 17), this cluster (labeled A) had gradually produced a large meso-β cloud shield centered over southeastern Iowa. Another meso-β echo feature, line B, had first appeared about an hour earlier. Line B developed within a broad weak band of convection that was apparently triggered by the weak outflow boundary in Fig. 17.

The mature MCC apparently developed as a consequence of meso-β scale interaction involving clusters A and B and the intersecting of the outflow boundary and the warm front.

Maximum intensity of the MCC occurred about 1100 GMT. By 1200 GMT (Fig. 18), an east-west meso-β convective line (C) had developed and

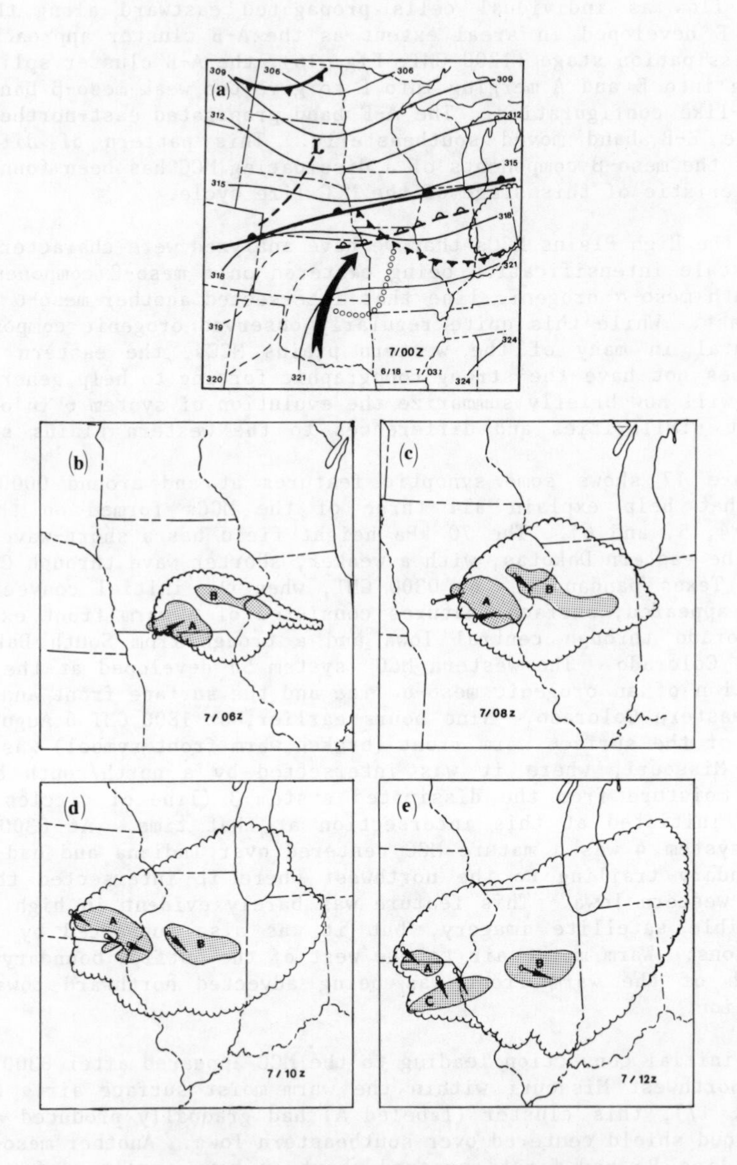

Fig. 17 (a) Similar to Fig. 16a, except for 0000 GMT 7 August 1977. Also included is the six-hour earlier position of the warm front at 1800 GMT 6 August (broken warm front symbol) and line of residual midlevel moisture, and the 0300 GMT 7 August positions of an MCC over Indiana (scalloped -32°C IR contour) and the boundary of its wake to the west. (b) to (e) Similar to Fig. 2b, except for 0600, 0800, 1000, and 1200 7 August 1977, respectively.

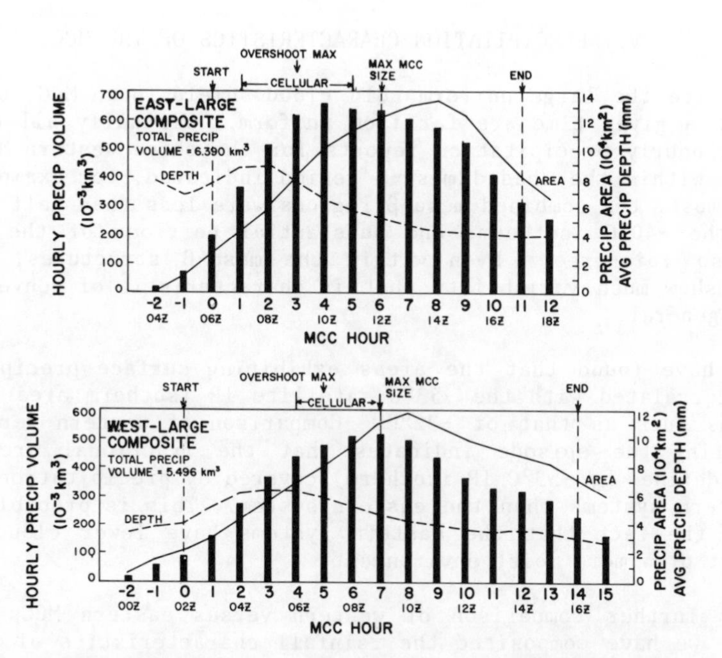

Fig. 18 Hourly precipitation characteristics for composite eastern-large MCC (top) and western-large MCC (bottom). The abscissa's time scale units are in MCC-relative hours, where the individual MCCs were composited relative to their beginning time (hour 0), time of maximum IR extent, and ending time, all based on the area colder than -53°C as depicted in IR satellite data. The times of these IR features are shown along the top, along with the period when the MCC appears most "cellular", and the time within that period when the overshooting tops appear coldest and most extensive. An idealized GMT time scale is also shown, which closely corresponds to the IR features of the individual MCCs in each composite. The vertical bars represent the hourly storm precipitation volume (left ordinate scale) for the hour ending at the time shown. The solid curve gives the total area receiving MCC precipitation each hour, whereas the dashed curve shows the average depth of the hourly precipitation within that area (right ordinate scale for both). The depth multiplied by the area yields the precipitation volume.

separated from the parent cluster A. The propagation of line C away from its parent cluster, as the MCC weakened and assumed more of an east to west line or structure, was a difluent motion between meso-β entities, also observed in the western MCCs as they began to decay.

V. PRECIPITATION CHARACTERISTICS OF THE MCCs

Despite the large uniform cold cloud shield in an MCC, the rainfall rates at a given time are far from uniform. Virtually all of the measureable hourly precipitation reports for the three western MCCs studied occurred within the shaded meso-β region indicated, for example, in Fig. 16. At most, the combined meso-β regions were less than half of the area within the -40°C isotherm, and substantial portions of the -62°C area were also rain-free. Even within the meso-β structures, the hourly reports show much variability that is characteristic of convective rainfall in general.

We have found that the areas exhibiting surface precipitation are better correlated with the -53°C satellite IR isotherm area than warmer isotherms such as that of -32°C. Comparison of western versus eastern MCCs during the episode indicates that the fractional area of cloud shield (defined by -53°C IR isotherm) covered by precipitation is less in the western systems than the eastern system. This is probably a reflection of the fact that the eastern systems have lower cloud bases and build out of a more moist environment.

For further comparison of western versus eastern MCCs during the episode, we have composited the rainfall characteristics of the largest of the eastern and western MCCs during the period. These are systems having a maximum -53°C IR isotherm area greater than 200×10^3 km^2 and which produce an estimated precipitation volume greater than 3 km^3. Of the systems illustrated in Fig. 1, the first three western systems meet these criteria, and eastern systems 6, 8, and the merger of 9 and 10 meet these criteria.

Figure 18 illustrates the composited precipitation depth, precipitation area, and precipitation volume for the large eastern and western MCCs. Both the eastern and western, large MCCs produce their maximum precipitation rate early during the growth period. This is also the period during which these storms are most likely to produce severe weather. It is also interesting to note that for both eastern and western storms, 50 percent of the storm precipitation falls prior to the time that the maximum IR isotherm area occurs.

Once initiated, the eastern, large MCCs exhibit faster growth rates and produce a greater volume of precipitation than the western systems, but with about three hours less of maturity. Further examination of surface rainfall rates suggests that a greater fraction of the surface rainfall in the eastern systems comes from anvil-type precipitation as observed in tropical clusters than it does in the western systems where much of the anvil precipitation is lost by evaporation.

An interesting feature for comparison is that the western systems become mature (0-hour) nearly four hours earlier (02Z) than the eastern systems (06Z).

VI. CONCLUDING REMARKS

I have summarized some features of a family or episode of MCCs which have either distinct mountain-related contributions to their initiation and structure or less distinct mountain-related contributions.

The western MCCs of the episode have distinct meso-β substructures which can be identified throughout their life cycles. Many of these meso-β components can be traced to having an origin in mountain convection. These meso-β convective elements frequently organize into a nearly circular mass of convection on the meso-α scale once they encounter the greater moisture in the central High Plains, thereby becoming MCCs.

The eastern MCCs also have well-defined meso-β building blocks which can be identified throughout their life cycles. However, few of these meso-β components have a distinct origin in the mountains. Some of the meso-β components can be traced to the remnants of orogenic MCCs (i.e., surface meso-highs, wet ground versus dry ground discontinuities, etc.) but many of the meso-β components do not have any apparent relationship to the mountains.

One must also consider the role of the Rocky Mountains and the Great Basin in setting up conditions that favor the formation of individual MCCs and such an episode. Episodes such as this frequently occur, but rarely are they of the duration and extent of this particular episode. The presence of a nearly stationary front extending from the Colorado Rocky Mountains northeastward across the United States clearly favors the formation of MCCs. To the south of the front is a generally southerly moist flow which provides the fuel for MCCs. That such midsummer fronts frequently stall out with their tail extending through the low-lying terrain of the Great Divide north of Colorado suggests a role of the mountain barrier in determining the position of the frontal surface. Also orogenic convection is favored when southwesterly flow over the mountain barrier brings moist air into the high country. We locally refer to this period as the Colorado monsoon. The extent to which a monsoonal-type circulation is involved needs to be investigated.

Finally, there seems to be a meso-α scale influence of the Rocky Mountains which favors the formation of both the orogenic MCCs and the eastern MCCs. We have noted, for example, that the meso-β orogenic cumulonimbi frequently propagate or translate off the mountains as a coherent line of convection of linear dimensions on the meso-α scale. Thus, there is the suggestion that there exists a mountain-related wave disturbance which provides an organizing influence on convection on the meso-α scale. We have also noted that the eastern MCCs reach maturity some four hours later than the western MCCs during the episode. This delay is in spite of the fact that the diurnal heating cycle in the eastern region is about one hour ahead of the western region. This finding is consistent with an eastward propagating wave disturbance playing an organizing role on convection in the eastern region as well. It is thus hypothesized that solar heating in the Rocky Mountain barrier and/or the Great Basin generates a convectively-reinforced wave disturbance which organizes convection on the MCC scale in both the western High Plains and eastern High Plains. Wherever this eastward propagating wave disturbance interrogates low-level conditions capable of supporting moist deep convection on the mesoscale, locally generated meso-β clusters converge and organize as a meso-α MCC, thereby reinforcing the deep-tropospheric wave disturbance.

Clearly there is the need for theoretical and observational studies which can further elucidate the link between large convective systems on the high plains and the Rocky Mountain region.

REFERENCES

1. Bellamy, J.C.: Objective calculations of divergence, vertical velocity and vorticity. Bull. Amer. Meteor. Soc., 30, 45-49 (1949).

2. Bonner, W.D.: Thunderstorms and the low-level jet. Meso-meteorology Research Paper No. 22, University of Chicago, 5734 S. Ellis Ave., Chicago, IL 60637, 23 pp., [NTIS No. AD-602 540] (1963).

3. Bonner, W.D.: Climatology of the low-level jet. Mon. Wea. Rev., 96, 833-850 (1968).

4. Bosart, L.F. and F. Sanders: The Johnstown flood of July 1977: A long-lived convective system. J. Atmos. Sci., 38, 1616-1642 (1981).

5. Maddox, R.: Mesoscale convective complexes. Bull. Amer. Meteor. Soc., 61, 1374-1387 (1980).

6. Maddox, R.A., D.J. Perkey and J.M. Fritsch: Evolution of upper tropospheric features during the development of a mesoscale convective complex. J. Atmos. Sci., 38, 1664-1674 (1981).

7. McBride, J.L. and R. Zehr: Observational analysis of tropical cyclone formation. Part II: Comparison of nondeveloping versus developing systems. J. Atmos. Sci., 38, 1132-1151 (1981).

8. Means, L.L.: On thunderstorm forecasting in the central United States. Mon. Wea. Rev., 80, 165-189 (1952).

9. Palmén, E. and C.W. Newton: Atmospheric circulation systems. Academic Press, New York, 390-425, 523-560 (1969).

10. Pitchford, K. and J. London: The low-level jet as related to nocturnal thunderstorms over midwest United States. J. Appl. Meteor., 1, 43-47 (1962).

11. Raymond, D.J.: Instability of the low-level jet and severe storm formation. J. Atmos. Sci., 35, 2274-2280 (1978).

12. Ruprecht, E. and W.M. Gray: Analysis of satellite-observed tropical cloud clusters. I. Wind and dynamics fields. Tellus, 28, 391-413 (1976).

13. Visher, S.S.: Climatic atlas of the United States. Howard University Press, Cambridge, Massachessetts, Plate 554 (1954).

14. Williams, K.T. and W.M. Gray: Statistical analysis of satellite observed trade wind cloud clusters in the western North Pacific. Tellus, 25, 313-336 (1973).

4.2 SOME OBSERVATIONAL EVIDENCE OF THE INFLUENCE OF TOPOGRAPHY ON SEVERE RAINSTORMS IN CHINA

Cai Zeyi and Li Jishun
Institute of Atmospheric Physics
Academia Sinica
The People's Republic of China

ABSTRACT

Analyses and descriptions of the influence of topography on the four largest among all of the heavy rain events in China are presented. A narrow belt of heavy rainfall is often elongated along the upwind side of a mountain against the prevailing lower-level wind, and the maximum centers are located inside the bell-shaped valleys on these isohyetal maps. The mesoscale rainfall cells always developed into a larger cell following its slowdown and merger with its predecessor along the foothills. Orographical conditions associated with these heavy rainfall events were found to be very similar.

I. INTRODUCTION

Although most heavy rains in China occur over the plains of northern China, eastern China, and of the coastal areas in southern China on a climatological basis compared to hailstorms, the local effects of some mountain ranges and elevated geographical features on the amplification of precipitation rate is often stressed.

Topography in China is very complex, with many mountains, valleys and plateaus. In general, western China is higher than eastern China. Figure 1 shows the annual average numbers of heavy rain days and 1000 m terrain contour above sea level. To the west of the contour, the annual number of days with heavy rain is less than one, but over almost all the eastern region it is more than two days. One day of heavy rains can occur everywhere in eastern China. Such rains are particularly frequent over the Nanling Mountain Range and its surroundings. The maximum is centered along the coastal areas of southern China where heavy rains occur more than ten days per year. The main characteristics and distributions of the heavy rain days are determined by both the southwesterly and southeasterly monsoons. The monsoon circulation of southern Asia provides much of the moisture for heavy precipitation that falls along the windward slope of some mountain ranges.

Based on the most famous severe rainstorms in China [9], [10], [5], we have analyzed the influence of topography, and selected seven cases which are presented in Table 1. The most severe rainfall in China occurred at Xinliao in Taiwan Province on 17 October 1967. It produced the largest precipitation amount in 24 hours in China and next to the

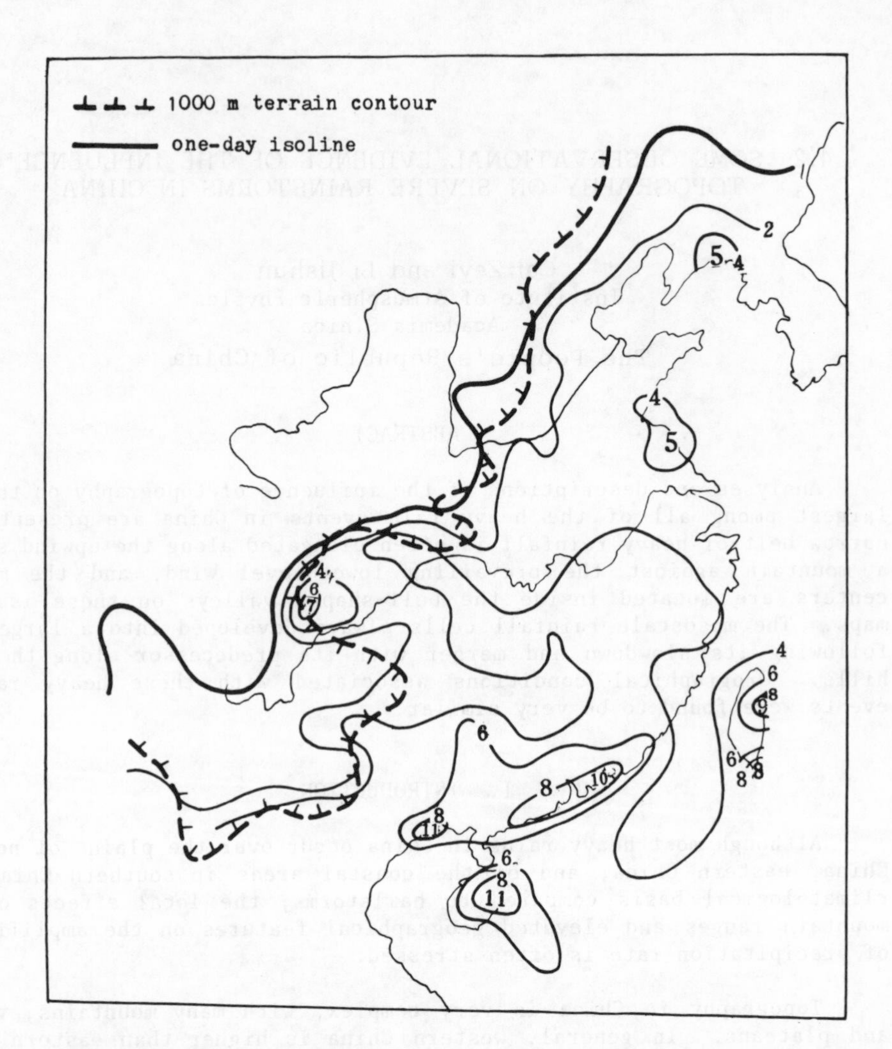

Fig. 1 Distribution of average annual numbers of heavy rain days which are defined as rainfall amounts greater than 50 mm/day (1960-1970). Isolines of days are drawn at intervals of two days.

world record for a three-day rainfall. The 24-hour rainfall amount was 1672 mm, and the three-day amount was 2749 mm. The second most severe rainfall event in China occurred at Paishin in Taiwan Province on 11 September 1963. The 24-hour recorded rainfall was the largest amount of rainfall directly produced by a typhoon in Asia and the West Pacific. The 24-hour and three-day precipitation amounts were 1248 mm and 1684 mm, respectively. The third most severe rainstorm in China was the unprecedented heavy rainfall at Linzhung in Henan Province on 7 August 1975. The 24-hour and three-day amounts were 1060 mm and 1605 mm, respectively.

Table 1 The maximum recorded rainfall at some stations in China.

No.	Province	Station	Longitude	Latitude	Date	Rainfall (mm) 24 Hrs	Rainfall (mm) 3 Days	Surface Wind	Mountain
1	Taiwan	Xinliao	24 35	121 45	17 Oct. 1976	1672	2749	ENE 12	Taiwan
2	Taiwan	Paishin	24 33	121 13	11 Sept. 1963	1248	1684	N 24	Taiwan
3	Henan	Linzhung	33 03	113 28	7 Aug. 1975	1060	1605	NE 4	Funiu
4	Hebei	Zhangmo	37 22	114 13	4 Aug. 1963	950	1458	ENE 8	Taihang
5	Beijing	Zaoshulin	40 22	116 16	27 July 1972	479	518	SE 4	Jundu
6	Shanghai	Tangqia	31 20	121 24	21 Aug. 1977	591	592	E 8	Delta of Yangzi
7	Inner Mongolia	Shilanaohai	39 03	109 05	1 Aug. 1977	1050	1050	S 4	Maowusu Desert

The fourth most severe rainstorm was a persistent heavy rainfall which occurred at Zhangmo in Hebei Province on 4 August 1963. It produced the maximum 24-hour and three-day rainfall ever recorded in North China. The ten-day precipitation amount was 2052 mm. These above-mentioned heavy rains are the four largest among all of the heavy rains in China. The fifth, sixth and seventh heavy rainstorms occurred in the provinces (or districts) indicated in Table 1. Interestingly, most of these rainfalls were closely related to the local features of mountain ranges. In the following, a brief survey of the effects of topography is given.

II. THE MESOSCALE EFFECTS OF MOUNTAINS ON SEVERE RAINSTORMS

1. Forced Lifting Effects on the Upwind Side

On a topographic chart (chart of isohyet) 24-hour total rainfall in a severe rainstorm is superimposed. A narrow belt of heavy rainfall is often found to be elongated along the upwind side of a mountain against the prevailing lower-level wind. Main terrain features and total rainfall isohyets are depicted in Fig. 2a for the period of 0000 GMT (GMT is converted to Local Standard Time by adding 8 hours), 1 to 10 August 1963. The wind field (right bottom of Fig. 2a) at low levels is easterly [11]. The rainstorm produced very heavy rains in a long, north-south band along the east side of the Taihanshan Mountain Range from the northeastern corner of Henan Province to Beijing. About 8 years later, a similar severe rainstorm occurred at Linzhang in Henan Province. Figure 2b is a frequency distribution of intense rainfall for this rainstorm from 0400 to 2000 GMT 7 August 1975. The frequency distribution is defined as the number of hours with rainfall of more than 50 mm/hr. During this time, the strongest rainfall of rainstorm No. 3 occurred and the maximum axis of the number of hours with intense rainfall appeared on the windward slope of the Funiushan Mountain Range, oriented from the southeast to the northwest. The maximum number of hours reached 7 hours. This orographic feature that produced the heavy rain events in China is very similar to that associated with the Big Thompson and Black Hills flash floods in the United States [6], [4], [7].

Figure 3 shows the frequency of wind direction at a height of 600 m at many rainfall stations in the Beijing area during heavy rainfall days. (Heavy rainfall days are defined as rainfall amounts greater than 25 mm in 12 hours.) For example, Changping, Mentougou, and Xiayunling are located on the southeast side of westward facing hills. In this hilly area heavy rainfall is often related to the southeasterly wind which is perpendicular to the mountain slope. However, Pingu and Gubeikou are located at the southwest side of the Yanshan Mountains. There, southwesterly winds prevail on heavy rainfall days. Huairou and Miyun are located to the south slope of the Jundu Mountains. There, southerly winds prevail during periods of heavy rainfall. It is well known that these mountains influence the distribution and intensity of precipitation which is greater where the prevailing wind is nearly perpendicular to the hillside, and moist air is forced to ascend along the windward slope.

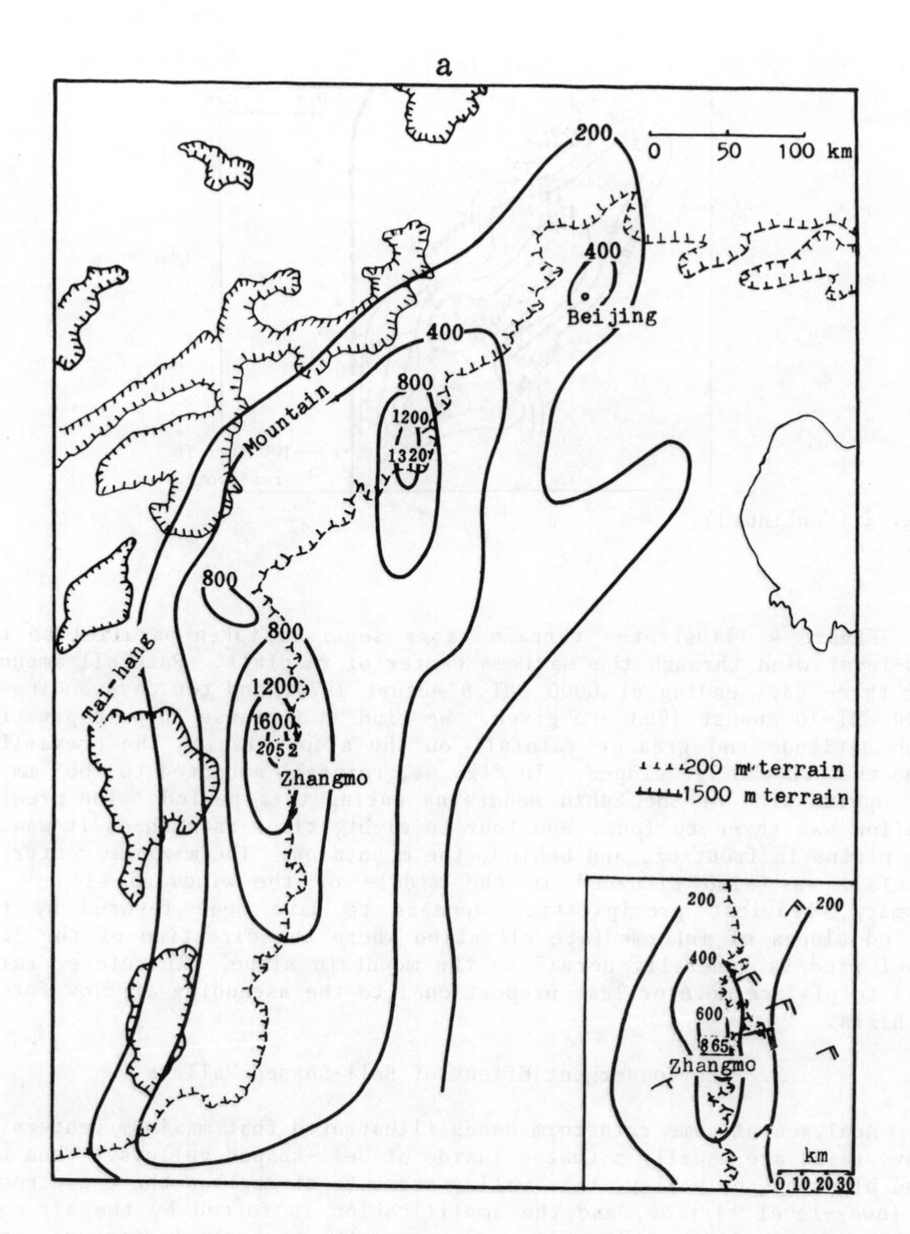

Fig. 2 (a) Total rainfall amount in mm, 0000 GMT 1-10 August 1963. The wind field at 0000 GMT, 5 August is inserted in the right bottom of this figure. The isohyet is drawn at intervals of 200 mm. (b) Distribution of frequency of occurrence of intense heavy rains which are defined as rainfall amounts greater than 50 mm/hr.

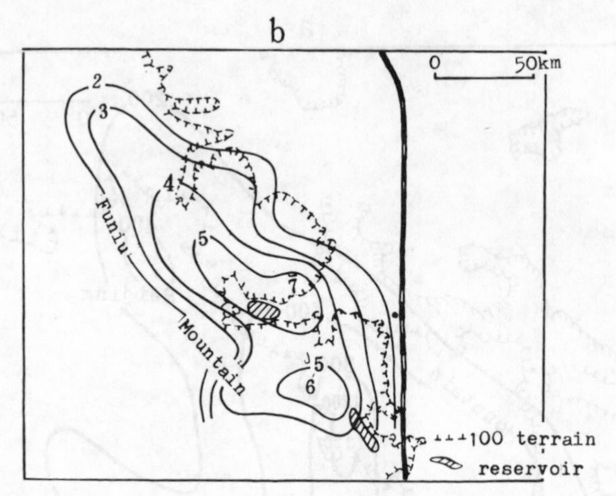

Fig. 2 (Continued)

Figure 4 illustrates terrain cross sections taken parallel to the low-level wind through the maximum center of rainfall. Rainfall amounts for three days ending at 0000 GMT 8 August 1975, and ten days ending at 0000 GMT 10 August 1963 are given. We find an increase in precipitation with altitude and greater rainfall on the slopes facing the prevailing wind than on the lee slopes. In Fig. 4a, rainfall amounted to 1605 mm on the upwind side of the Funiu Mountains during this period. The precipitation was three to four, and four to eight, times as much as it was on the plains in front of, and behind, the mountains. The maximum center of rainfall was also situated in the middle of the windward slope. In summary, heaviest precipitation appears to have been favored by the upwind slopes of intermediate elevation where the direction of the low-level wind is generally normal to the mountain slope. Therefore, rainfall totals are more or less proportional to the ascending airflow forced by hills.

2. The Convergent Effect of Bell-Shaped Valleys

Analyses of some rainstorm cases illustrated that maximum centers of heavy rains are usually situated inside of bell-shaped valleys. When the wind blows up the valley, this valley seems to strengthen the convergence of lower-level airflow, and the amplification is forced by the air mass convergence in this area. Figure 5a shows the topographic features near the maximum center of such rainstorms. Xinliao Valley can be seen to be narrowing from east to west. When Typhoon Carla crossed the northern Philippines, a large-scale easterly airflow formed to the north of that typhoon, extending from the West Pacific to Taiwan Province. With easterly winds blowing up the valley, upward motion is produced and intensified by terrain convergence. Precipitation then concentrates over a very limited area during the heavy rains. Linzhung Valley and Zhangmo Valley are also facing eastward (Fig. 5b and Fig. 2b), but Zaoshulin is located

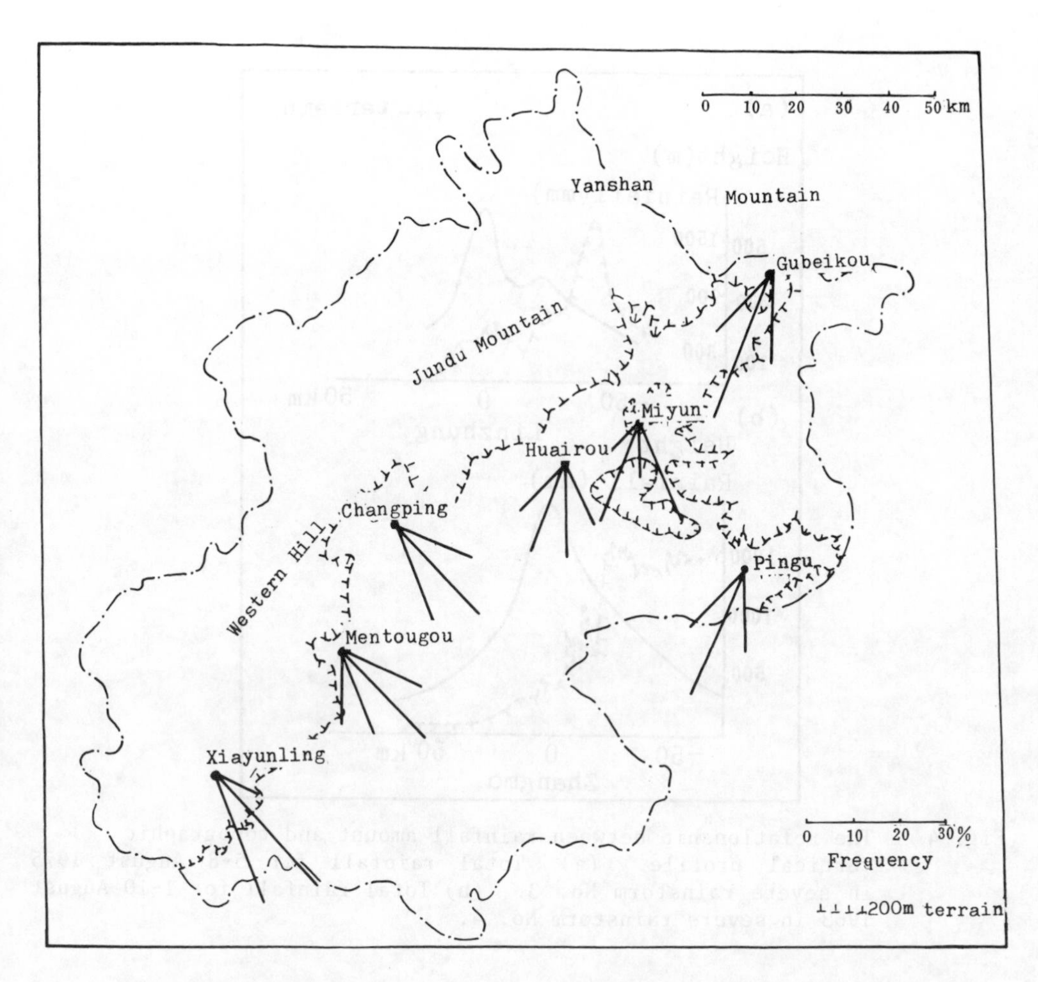

Fig. 3 Frequency of wind direction at a height of 600 m in the Beijing region during heavy rains (1960-1974 from June to August).

in a broad, bell-shaped valley facing southeastward (Fig. 5c). When Typhoon Rita moved inland into Hebei Province and decayed, the southeasterly wind related to the northeastern sectors of this typhoon blew up the valley. Orographic convergence appeared to intensify the rainfall amount. Therefore, location of the maximum precipitation center depends on the direction of the low-level wind. Another case occurred in the northern Beijing area on 23 July 1976 when southwesterly winds prevailed in North China. The heaviest rainfall occurred at Gubeikou within the deep part of this bell-shaped valley facing towards the southwest (Fig. 5c). The 24-hour cumulative rainfall was 250 mm in this location and only 49.5 mm at Zaoshulin at the same time.

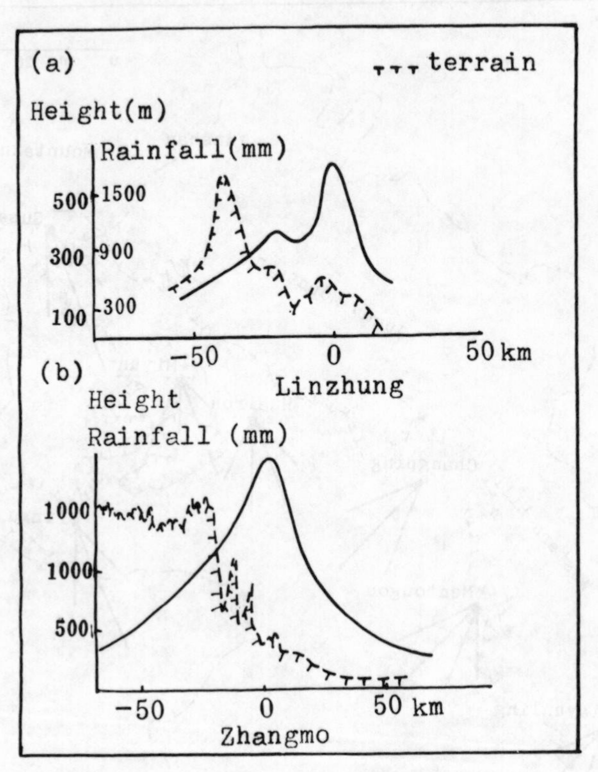

Fig. 4 The relationship between rainfall amount and topographic vertical profile. (a) Total rainfall for 5-8 August 1975 in severe rainstorm No. 3. (b) Total rainfall for 1-10 August 1963 in severe rainstorm No. 4.

Rainstorm No. 2, Paishin, is located in a bell-shaped valley facing toward the north. Precipitation accumulation in six hours at Paishin and the northerly component of the surface wind at Taibei, which is 65 km to the north, is shown in Fig. 6. As the speed of the northerly wind increased so did the rainfall, except the phase lagged a little behind. The wind speed reached its greatest strength of 24 m/s at 0000 GMT and precipitation was largest, being 429 mm in six hours [8] from 1800 GMT on the 10th to 0000 GMT on the 11th. When the wind speed dropped rapidly, the rainfall amount decreased markedly, too. Finally, when the wind direction turned southwesterly, the rainfall ended. Therefore, the topographic effect became most pronounced when the speed of the low-level wind was strong and the wind direction was the same as the orientation of the bell-shaped valley.

3. Blocking Effect of the Mountains as a Barrier

From hourly isohyetal maps we found that, as the mesoscale rainfall cells arrive in mountain regions, they often slow down and even become

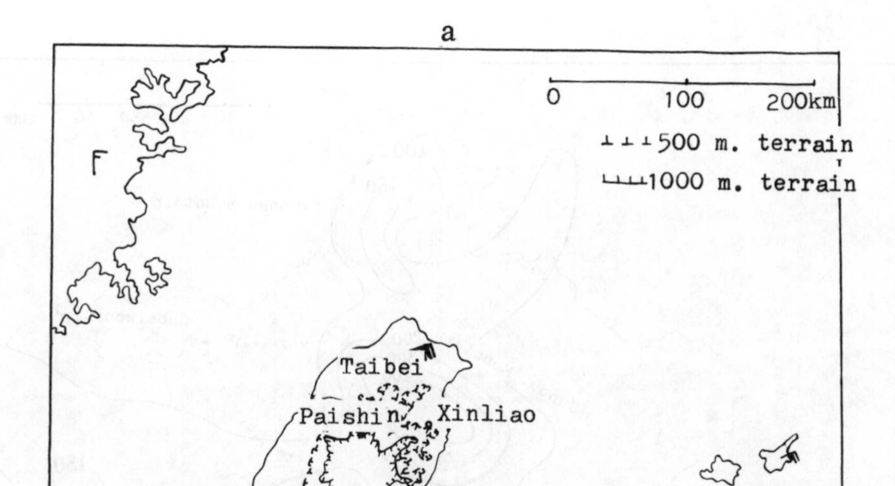

Fig. 5a Topographic features around Xinliao valley, Tiawan Province.

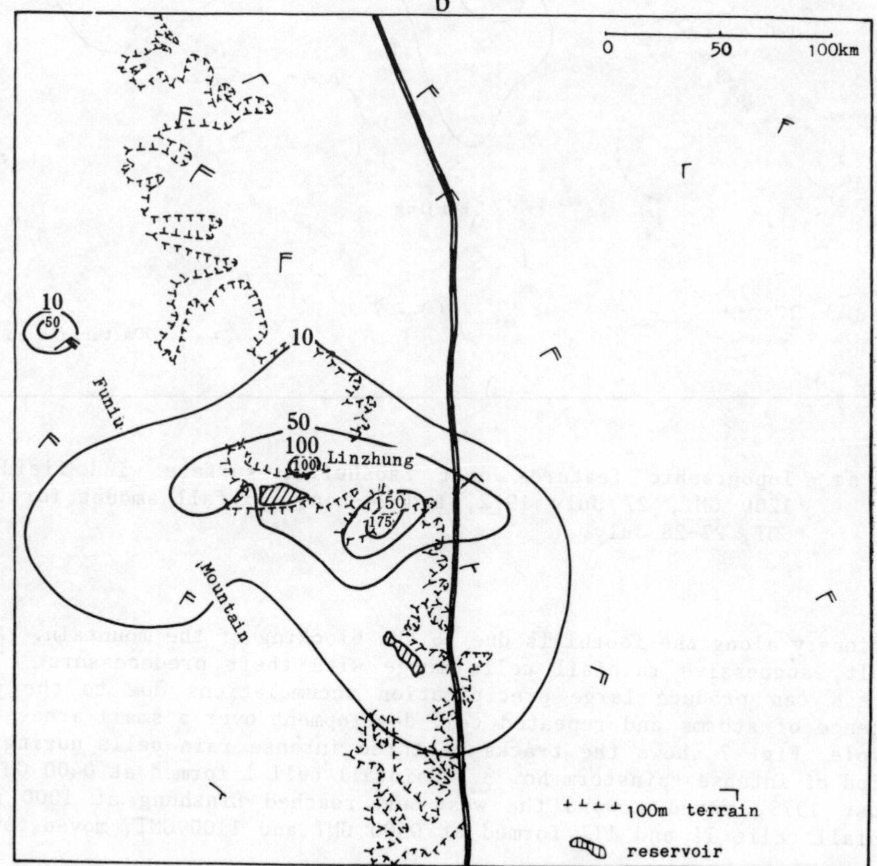

Fig. 5b Cumulative rainfall isoplethes for one hour on 5 August 1975 (black lines), 100 m terrain contour is dashed.

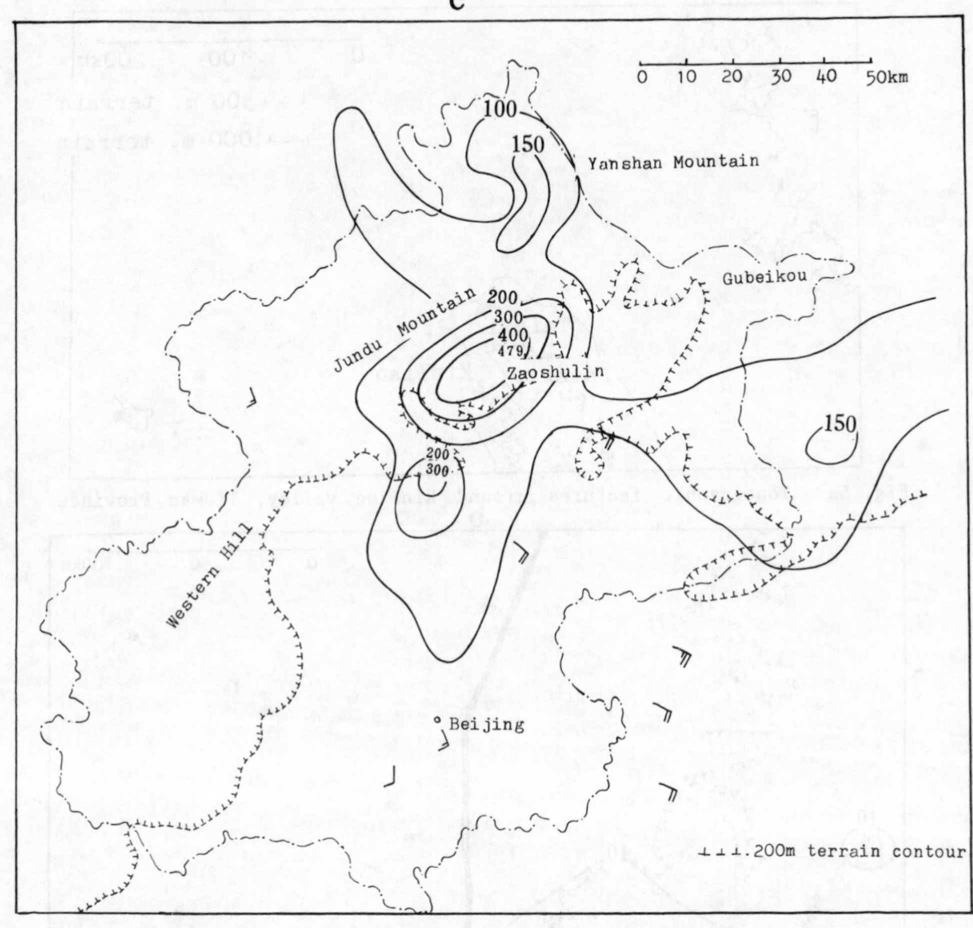

Fig. 5c Topographic features near Zaoshulin, surface wind field at 1200 GMT, 27 July 1972, and 24-hour rainfall amount for 0000 GMT, 27-28 July.

stationary along the foothills due to the blocking of the mountain. As a result, successive rainfall cells merge with their predecessors. This process can produce large precipitation accumulations due to the persistence of storms and repeated cell development over a small area. For example, Fig. 7 shows the tracks of three intense rain cells during the period of intense rainstorm No. 3. Rainfall cell I formed at 0400 GMT, 7 August 1975, moved toward the west and reached Linzhung at 1000 GMT. Rainfall cells II and III formed at 0500 GMT and 1100 GMT, moved toward

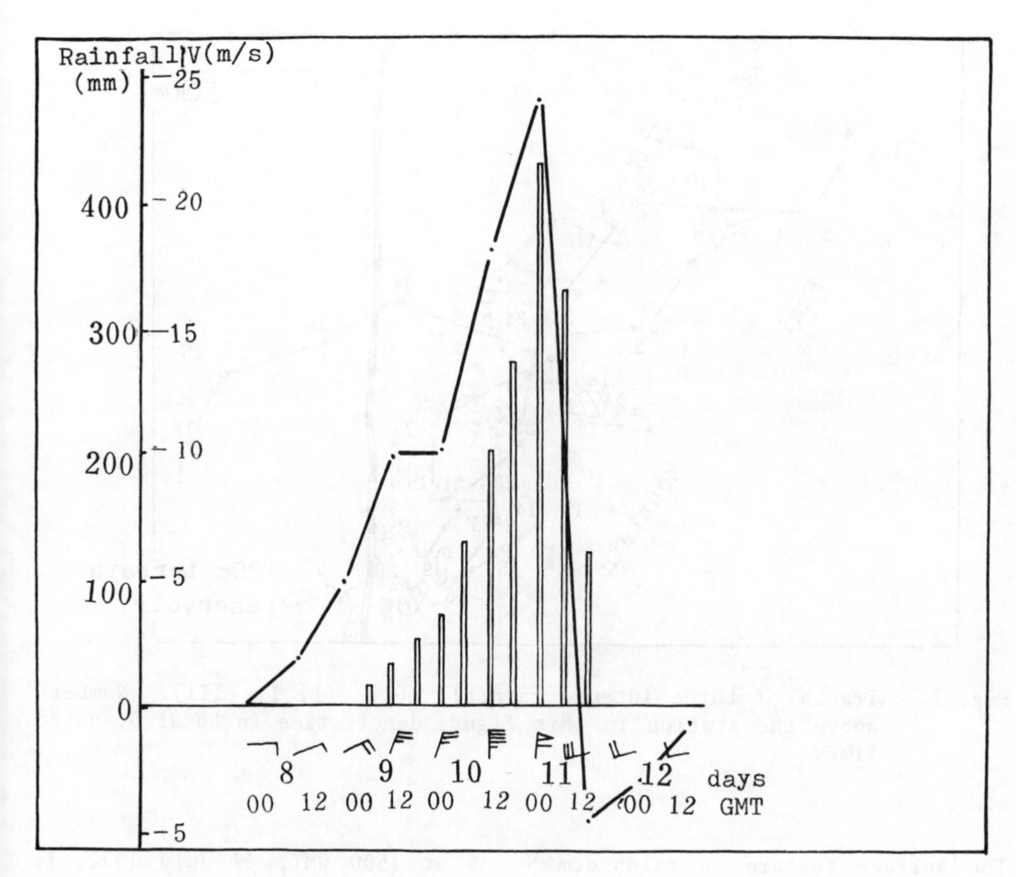

Fig. 6 Rainfall amount in six hours at Paishin and time series of the northern component of surface wind in Taibei.

the north and also reached Linzhung one after the other. They then stagnated and merged with each other. Large, strong rainfall cells then developed. Maximum rainfall exceeded 100 mm in one hour (Fig. 5b), and remained in Linzhung Valley for seven hours. However, sometimes the rain cells moved eastward and slowed down at the foothills. It is concluded that on such occasions the rain cells decayed and ended quickly due to the nonmerger of successive cells.

4. Formation and Enhancement of Mesoscale Synoptic Systems Near Mountainous Regions

In some special mountain regions, for example mountain ranges having shapes like an arc, cyclonic shear in the lower-level airflow may favor the increase of low-level convergence in some mesoscale systems. These cases were always found in North China, when the airflows strikes the Taihang and Yanshan mountain regions. In this case the low-level easterly wind was forced to curve cyclonically and strong rainfall was formed.

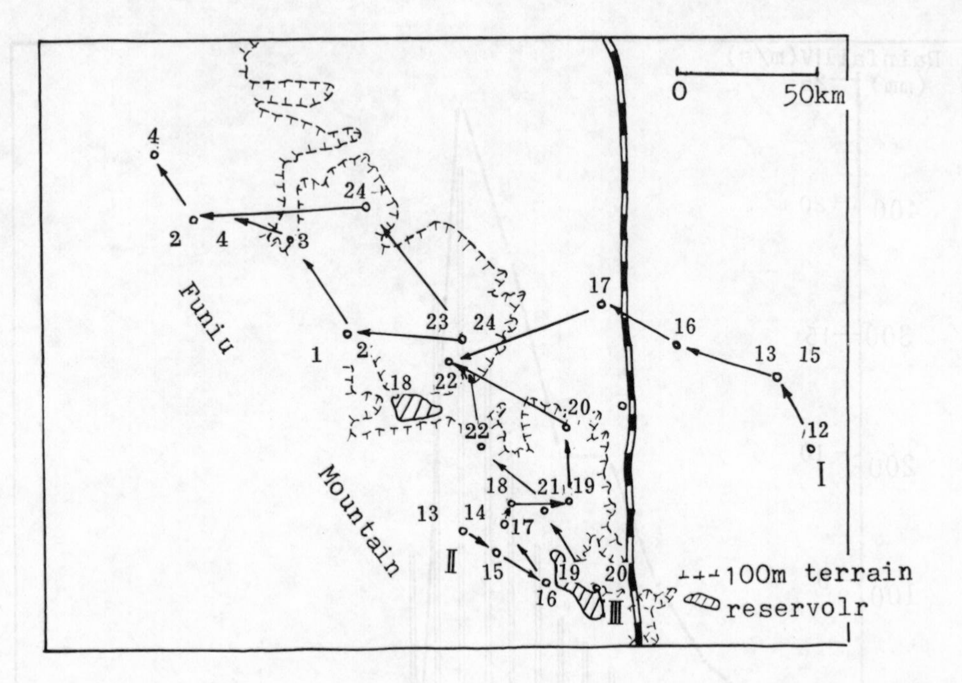

Fig. 7 Tracks of three intense rainfall cells (I, II, III). Numbers above the station in this figure denote time in Local Standard Time.

The surface feature in rainstorm No. 5 at 1500 GMT, 27 July 1972, is shown in Fig. 8. When easterly flow prevails over the plains in front of the Yanshan Mountains, the airflow is forced to curve cyclonically as it encounters the Taihang Mountains next to the Yanshan Mountains. This flow produces two quasi-stationary lines of low-level convergence between the easterly or northerly airflow and a weak southerly airflow over the plains of North China. Both of the convergence lines intersect each other at the north ends, and a mesoscale vortex forms in this small area. The strongest rainfall produced there was 87 mm in an hour at Zaoshulia.

5. The Interaction of Cool and Warm Air Masses with Canyon and Mountain Gaps

In China many mountain ranges have a west to east orientation. When a cold front moves southward and weakens, the cool air remains in front of the large mountains. Usually a portion of the air mass pours down through the canyons or gaps of the mountains. On occasion, the cool air lifts warm air ahead of the cold front and triggers a severe rainstorm. In addition, warm and moist air masses can move northward through the gaps to transport heat and moisture to southwest China where the generate heavy rain. Such a circumstance is illustrated in Fig. 9 in South China. A quasi-stationary front with a marked temperature gradient was backed up against the foothills of the Nanling Mountain Range at 0600 GMT, 20 June

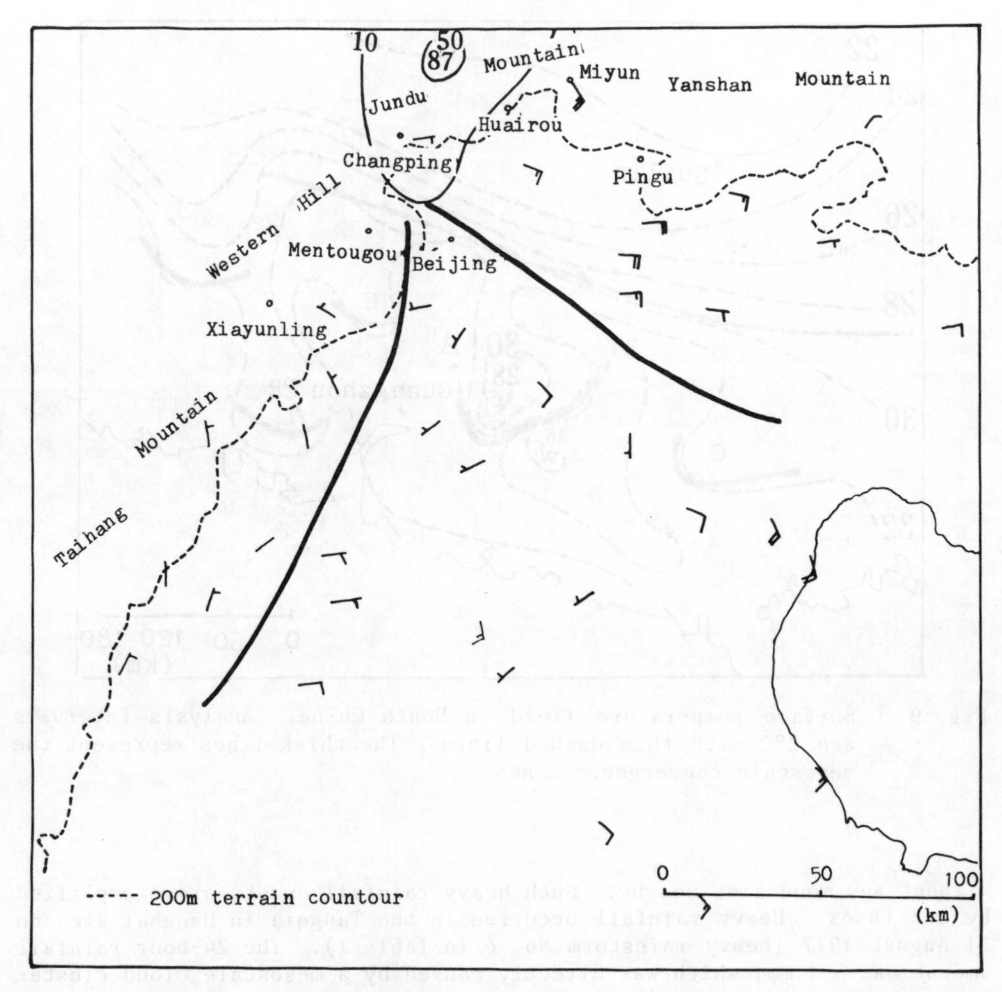

Fig. 8 Surface wind field at 1500 GMT, 27 July 1972, in the Beijing area. Thin lines represent rainfall amount in the same hour and thick lines represent the convergence line.

1977. A portion of the cool air mass moved through the gaps and formed a mesoscale convergence line at the southern edge of that air mass. It lifted the potentially unstable air and triggered severe heavy rainstorms in a local area near convergence line A (Fig. 9).

III. DISCUSSION

In the preceding sections we have demonstrated the effects of topography on heavy rains. It must be remembered, however, that there are a considerable number of heavy rains on the plains, even in desert areas,

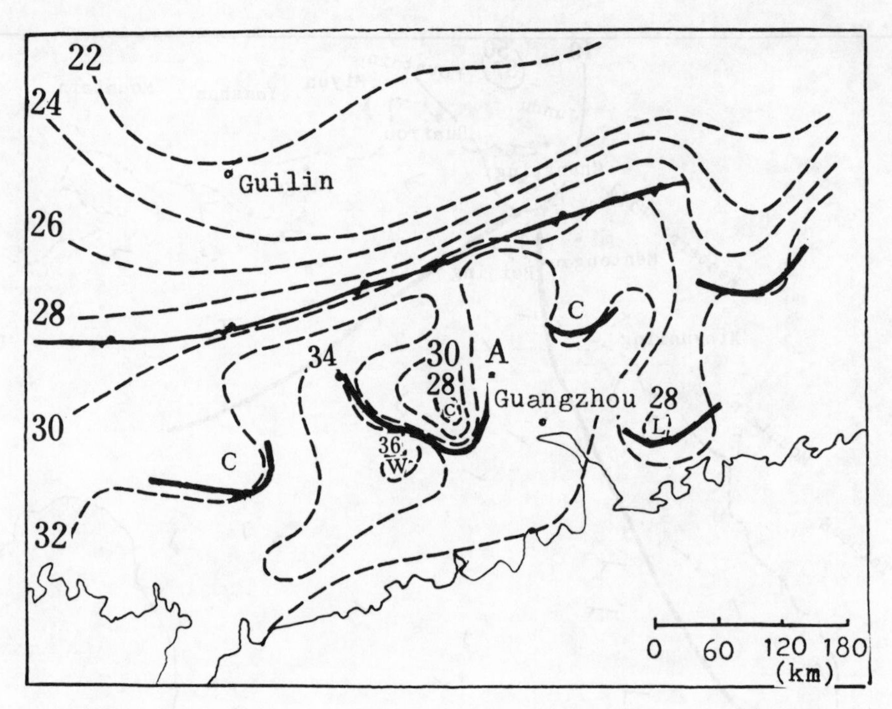

Fig. 9 Surface temperature field in South China. Analysis intervals are 2°C with thin dashed lines. The thick lines represent the mesoscale convergence lines.

without any mountains nearby. Such heavy rainfall events are exemplified by two cases. Heavy rainfall occurred in the Tangqia in Hanghai area on 21 August 1977 (heavy rainstorm No. 6 in Table 1). The 24-hour rainfall amount was 591 mm, which was directly caused by a mesoscale cloud cluster embedded in an easterly flow. When the cloud cluster moved westward from the East China Sea and landed in the delta of the Yangzi River, intense precipitation amounted to 540 mm in eight hours in a local area (Fig. 10). Another case which is noted for its rarity is the intense convective rainfall which occurred in the desert area of the border region between Inner Mongolia and Shansi Province in early August 1977. A total rainfall amount of 1050 mm fell in eight hours. It was oriented in a rain band which stretched from west to east on the Maowushu Desert. Total rainfall at some points exceeded 1000 mm with cellular features (black points in Fig. 11). The persistence of this heavy rainfall was not long, however, otherwise damage caused by the total rainfall amount would have been increased by orders of magnitude.

Summarizing these facts, we believe that on the one hand topography has an important effect on heavy rain. On the other hand, the topographical action works only on the basis of interaction with a synoptic weather system. In other words, some vigorous synoptic or mesoscale

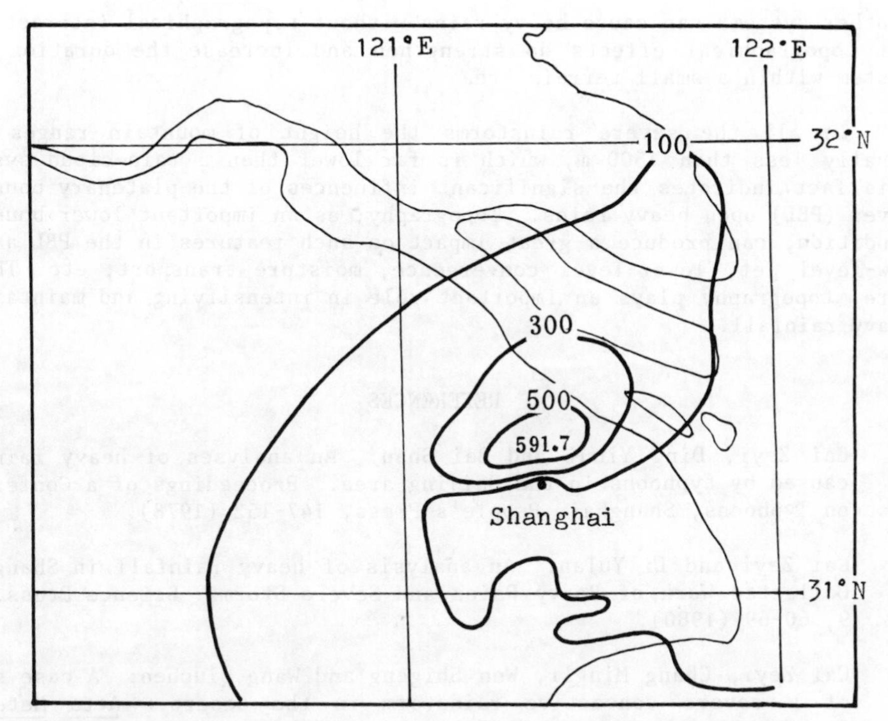

Fig. 10 Total precipitation in mm, 0000 GMT, 21-23 August 1977.

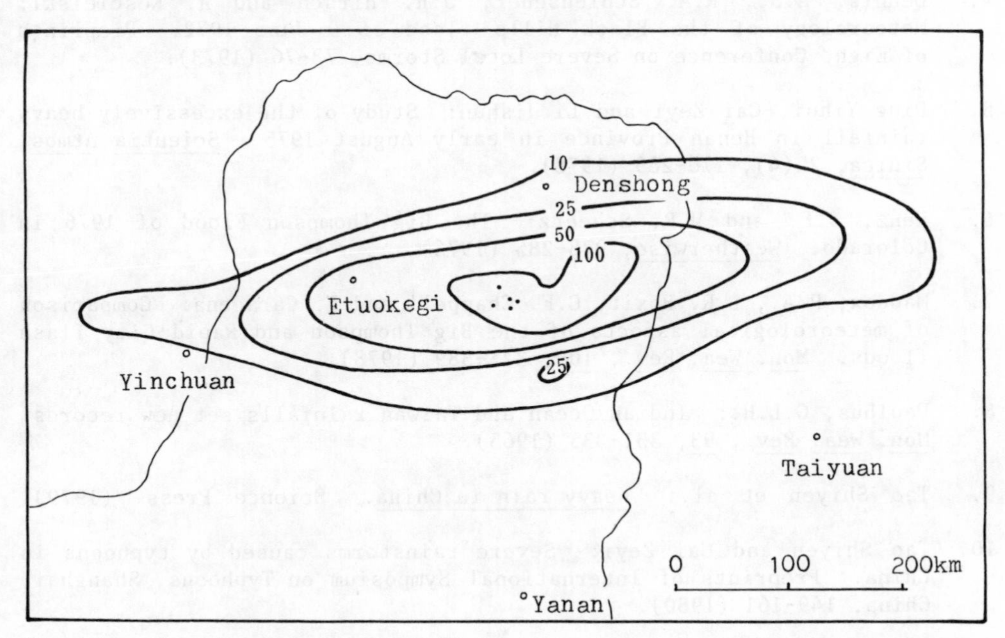

Fig. 11 Total precipitation in mm, 0000 GMT, 1-2 August 1977 based on meteorological and hydrologic raingauge network, supplemented by unofficial reports.

weather systems can cause heavy rains without topographical interactions. But topographical effects do strengthen and increase the duration of a system within a small terrain area.

In all the severe rainstorms the height of mountain ranges were usually less than 1500 m, which is far lower than a rain-cloud system. This fact indicates the significant influences of the platenary boundary layer (PBL) upon heavy rains. Topography, as an important lower boundary condition, can produce a great impact on such features in the PBL as the low-level jet, lower-level convergence, moisture transport, etc. Therefore, topography plays an important role in intensifying and maintaining heavy rainfall.

REFERENCES

1. Cai Zeyi, Ding Yihui and Bai Shan: An analyses of heavy rainfall caused by typhoons in the Beijing area. Proceedings of a Conference on Typhoons, Shanghai, People's Press, 147-152 (1978).

2. Cai Zeyi and Li Yulan: An analysis of heavy rainfall in Shanghai. Collected Work of Heavy Rains and Severe Storms. Science Press, No. 9, 60-69 (1980).

3. Cai Zeyi, Chang Mingli, Wen Shigeng and Wang Qiuchen: A case study of a severe convective rainstorm in the desert. Acta Meteoro. Sinica, 39 (1), 110-117 (1980).

4. Dennis, A.S., R.A. Schleusener, J.H. Hirsch and A. Koscielski: Meteorology of the Black Hills flood of 9 June 1972. Preprints of Eight Conference on Severe Local Storms, 73-76 (1973).

5. Ding Yihui, Cai Zeyi and Li Jishun: Study of the excessively heavy rainfall in Henan Province in early August 1975. Scientia Atmos. Sinica, 2 (4), 276-289 (1978).

6. Henz, J.F. and V.R. Scheetz: The Big Thompson Flood of 1976 in Colorado. Weatherwise, 278-285 (1976).

7. Maddox, R.A., L.R. Hoxit, C.F. Chappell and F. Caracena: Comparison of meteorological aspects of the Big Thompson and Rapid City flash floods. Mon. Wea. Rev., 106, 375-389 (1978).

8. Paulhus, G.L.H.: Indian Ocean and Taiwan rainfalls set new records. Mon. Wea. Rev., 93, 331-335 (1965).

9. Tao Shiyen et al.: Heavy rain in China. Science Press. (1979).

10. Tao Shiyen and Cai Zeyi: Severe rainstorms caused by typhoons in China. Preprints of International Symposium on Typhoons, Shanghai, China, 149-161 (1980).

11. You Jingyan: Some mesoscale systems within a heavy rainfall belt. Acta Meteoro. Sinica, 35 (3), 293-304 (1965).

4.3 THE INFLUENCE OF MT. LIUPAN ON HAIL PROCESSES OVER THE PINGLIANG DISTRICT IN CHINA

Qu Zhang[1]
Lanzhou Institute of Plateau Atmospheric Physics
Academia Sinica
The People's Republic of China

ABSTRACT

Pingliang District is located in the eastern part of Gansu Province, on the northwest plateau of China. Famous Mt. Liupan with a length of 300 km in the north-south direction and a width of 50 to 60 km in the west-east direction passes through the northwest plateau. The southern part of this mountain range passes through the western area of Pingliang District which has an altitude of 1350 to 1950 m above sea level and exhibits one of the highest frequency of hailstorms in China (Fig. 1).

In 1972 a 3 cm wavelength weather radar was installed in Pingliang County to the east, about 40 km away from the southern part of Mt. Liupan. Since that year, we have been conducting observations and analyses of radar echoes and synoptic weather systems for every hail season in summer.

In this paper we discuss (1) the climatic effect of Mt. Liupan on hail processes and (2) a case study of a squall line with synoptic scale in its horizontal length on a special day.

I. THE CLIMATIC EFFECT OF MT. LIUPAN ON HAIL PROCESSES

The area with a radius of 10 to 40 km around the radar station was divided into a number of small squares. The number (frequency) of intense thunderstorm echoes with central intensity of Ze $\geq$ 46 dbZ was counted in each square for the years 1973 to 1978 in summer. The frequency distribution of intense radar echoes (Fig. 2) was obtained by drawing isolines of the numbers in all squares. The distribution diagram is characterized by one high frequency belt and two secondary high frequency belts, one of which is near the mountain edge.

During the hailstorm seasons of 1973 to 1978 continuous observations of intense convective clouds around our radar station were made. The cloud being observed should be within a region with a horizontal radius about 50 m from the station and its central intensity must be 36 dbZ or more. The paths of hailstorms were determined from the central intensity positions and from damage reports. Altogether 82 paths of hailstorm

[1] With collaborations of Mr. N.H. Gong, Mr. Q.M. Cai, Mrs. Yuan and some students from Lanzhou University.

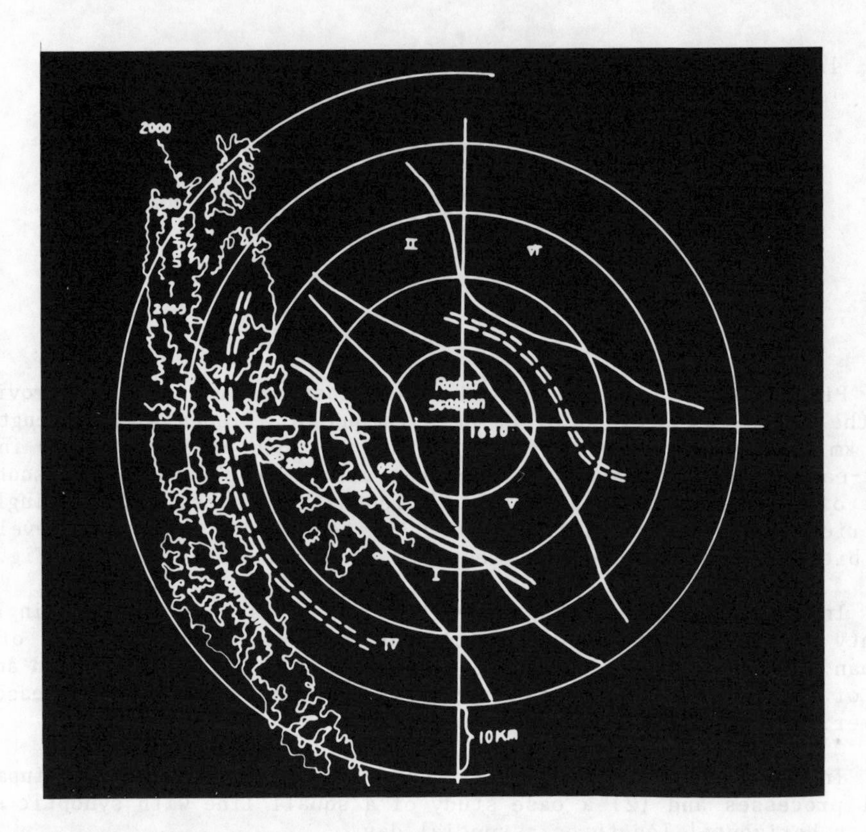

Fig. 1 Topography and range circles covered by the 3 cm radar located in Pingliang County.

echoes (Fig. 3) were collected, among which 40 paths had damaging hailstorms. The 82 paths belong to 5 hailstorm zones (Table 1). Zone I has the largest number (40) of hailstone falls which is 48.8 percent of the total hailstorms. However, damaging hailstorms occurred only 9 times, or 22.5 percent of the total damaging hailstorms. Zone II has 28.0 percent of the total hailstorms (23) but has 19 damaging hailstorms which correspond to 47.5 percent of the total damaging hailstone falls. Zone I coincides with the maximum frequency belt of intense thunderstorm echoes and Zone II with one of the two secondary high frequency belts. Zone I appears about 30 km to the east of Mt. Liupan crest and Zone II is located about 30 km eastward of Zone I. It seems that Zone I and Zone II to the lee of Mt. Liupan have some properties of lee waves and are representative of the climatic effect of Mt. Liupan. We plan to do some further dynamic studies on the effect of the mountain in the future.

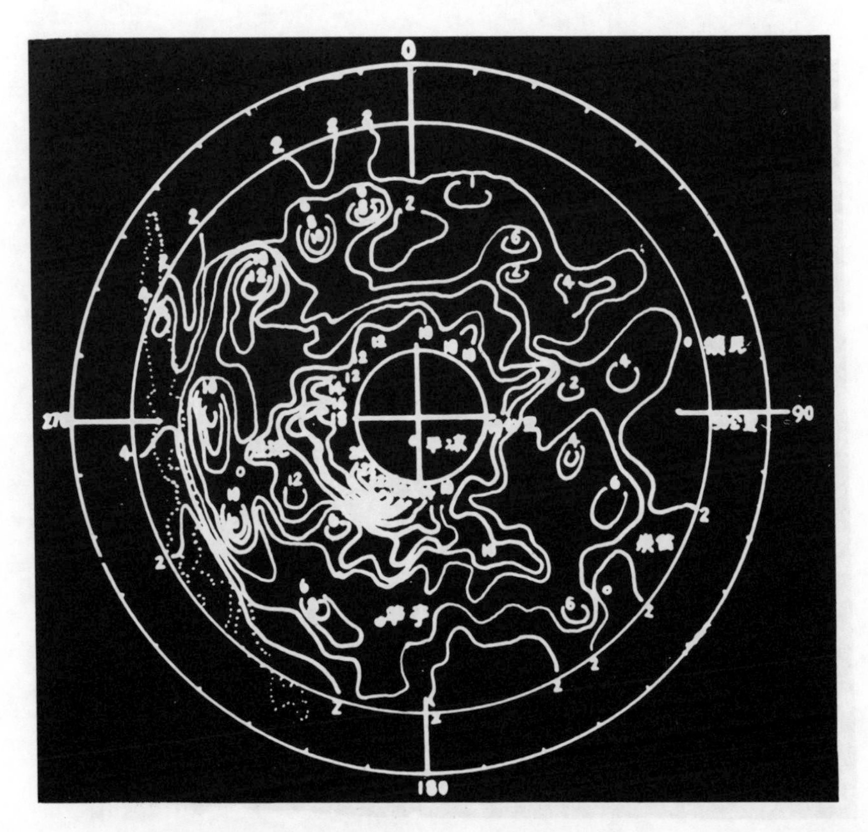

Fig. 2 Frequency distribution of intense radar echoes 10 to 40 km from the radar station from 1973 to 1978.

II. THE INFLUENCE OF THE LIUPAN MOUNTAIN RANGE ON A SQUALL LINE FROM A CASE STUDY

In this part, we shall concentrate our analysis on a severe squall line having synoptic scale dimensions. In order to study the orographic effect of Mt. Liupan, the strip-chart records of temperature, relative humidity and wind collected from 104 observational stations in this region are used.

1. The Effects of Mt. Liupan on the Moving Speed of the Squall Line

Figure 4 illustrates the topography of Mt. Liupan and its eastern and western flanks (interval of the isopleths is 200 m). It can be seen from Fig. 4 that the Mt. Liupan ridge is oriented in a north-south direction in the area of 105°E to 108° 30'E and 34°N to 36°N. Hereafter we will refer to this area as "the mountain ridge region". To the east of this ridge region there is a low valley extending southeastward.

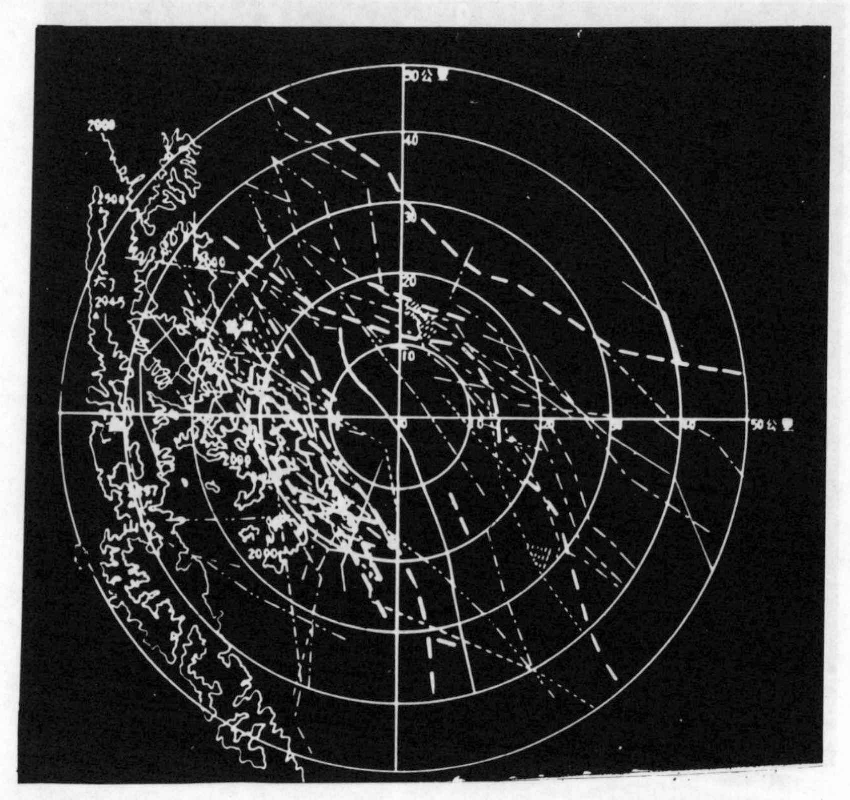

Fig. 3 Paths of 82 intense radar echoes observed in Pingliang County from 1973 to 1978.

Table 1 Hailstorm occurrence in five different zones.

Hailstorm Zone	I	II	III	IV	V	Total
Hailstorm No.	40	23	8	8	3	82
Percent	48.8	28.0	9.8	9.8	3.6	100
Damaging Hailstorm No.	9	19	5	5	2	40
Percent	22.5	47.5	12.5	12.5	5.0	100

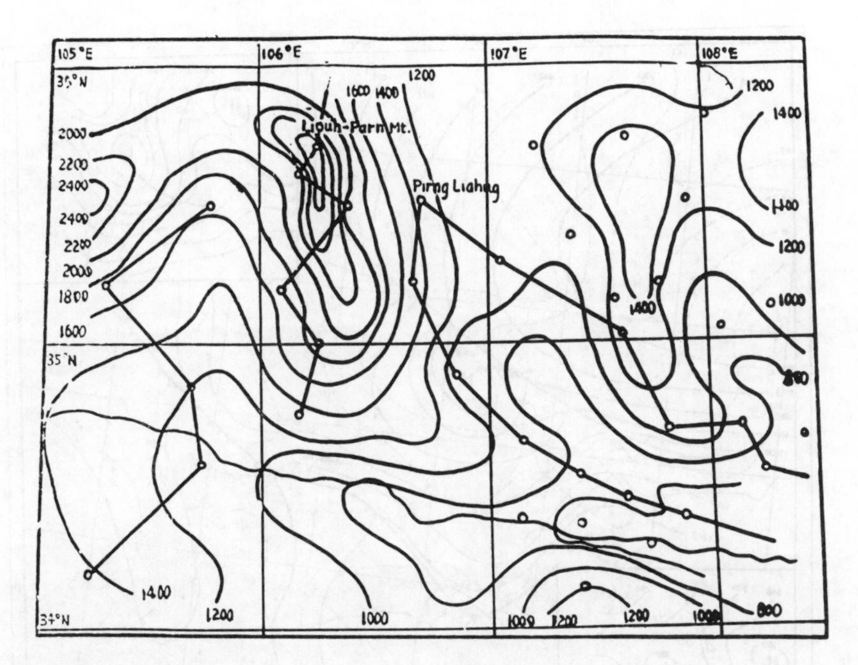

Fig. 4 Topography of Mt. Liupan and its eastern and western flanks (interval of the isopleth is 200 m).

On May 27, 1973 a cold front moved over the Liupan Mountain Range from northwest to southeast with a squall line ahead of it (Fig. 5). When the squall line passed over the Liupan Mountain region, hailstorms were observed at several weather stations and some hailstones were of egg size.

Figure 6 is the isochrone chart of the thunderstorm high moving over each station. It is clear that the high pressure at both sides of the mountain moved faster than over the mountain ridge region.

Figure 7 is the isochrone chart for hailstone fallout time at each station on that day. We can see that the initial time of hailfall in the mountain ridge region is later than that in the valley, with a maximum time lag of two hours or more.

2. The Effect of Mt. Liupan on the Intensity of Weather Development

When the squall line passed over Mt. Liupan, variations of meteorological elements were very different between the mountain ridge region and both the western and eastern flanks. Figure 8 illustrates the strip-chart recordings of relative humidity at some stations which are denoted

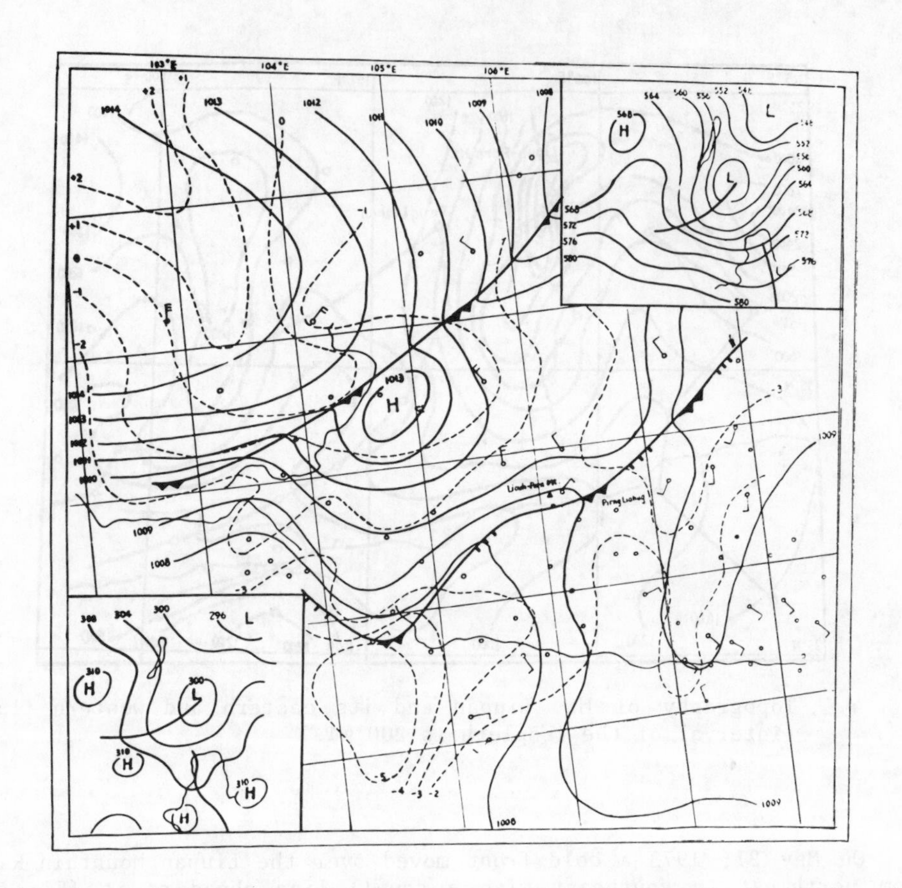

Fig. 5 Surface weather chart for 14 h (Beijing Standard Time). The map in the upper right is a 500 mb map and the one in the lower left is a 700 mb map for 8 h (Beijing Standard Time) on 27 May 1973.

by heavy full lines in Fig. 4. In Fig. 8, the first row to the left represents one of the stations on the western flank. The second row is from one of the stations in the mountain ridge region, and the record of stations on the eastern flank are presented by two rows on the right side. During the passage of the squall line from north to south, the relative humidities observed at both flanks evidently increased. Since the disturbance amplitudes were very large, the appearance time of the peak value may be used to identify the motion of the squall line. In comparison with both flanks, the disturbance amplitudes of the relative humidity over the mountain ridge region decreased with no evidence of a peak value. The time lag is not obvious from north to south.

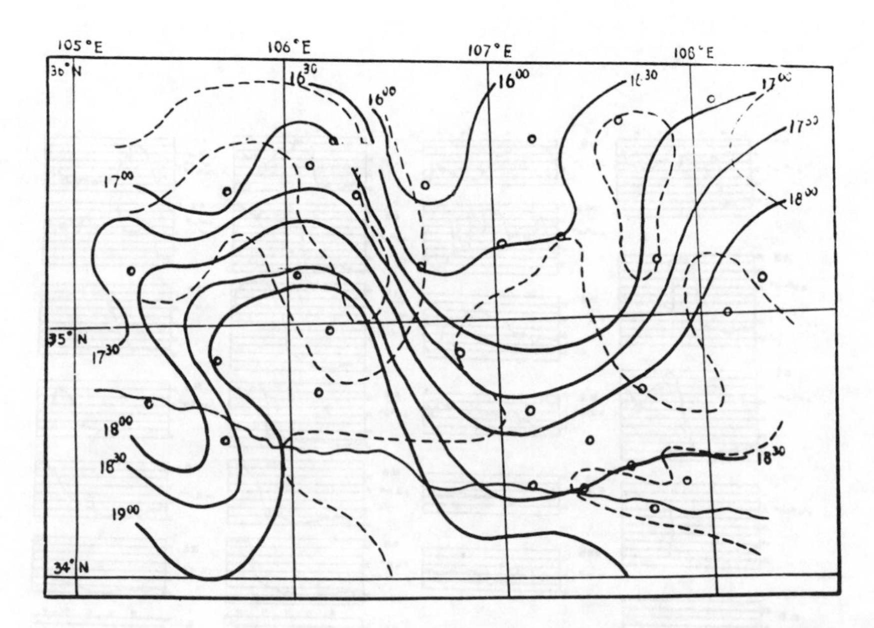

Fig. 6 Isochrone chart for the thunderstorm high on 27 May 1973.

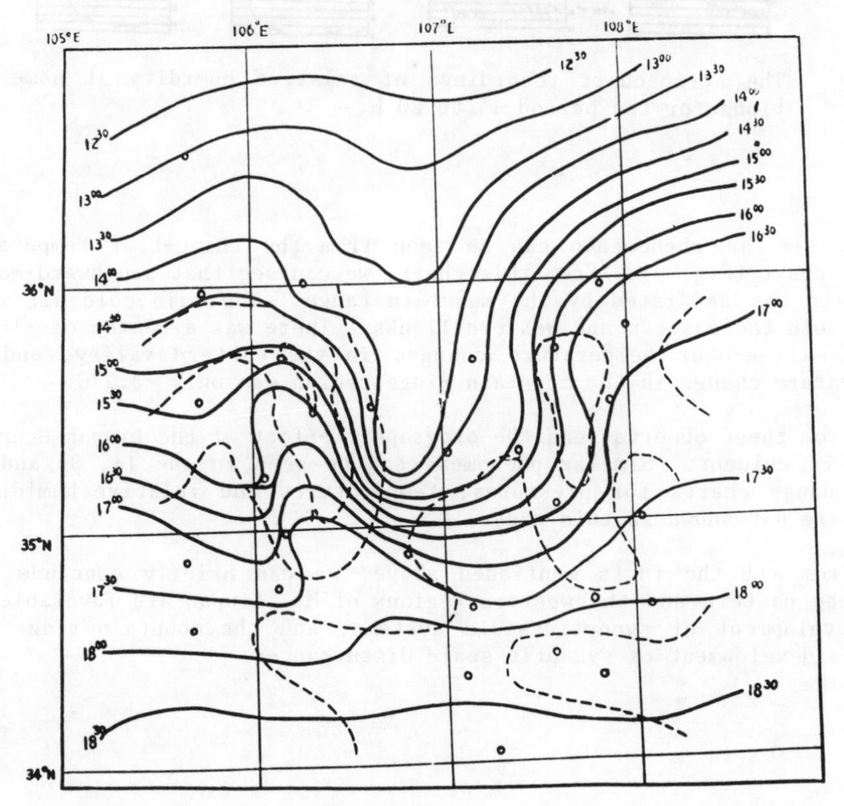

Fig. 7 Isochrone chart for hailstone fallout on 27 May 1973.

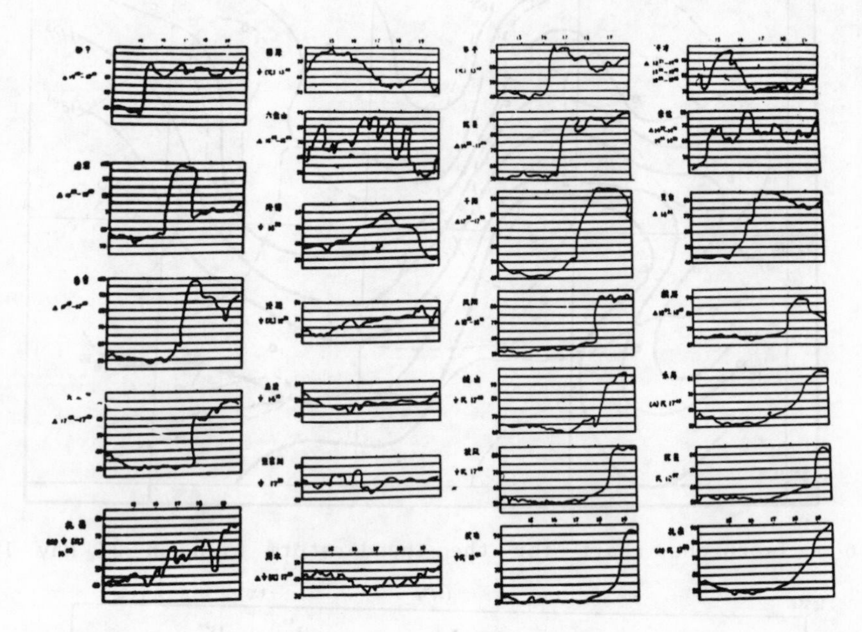

Fig. 8 The strip-chart recordings of relative humidity at some stations for the period 14 to 20 h.

A similar phenomenon can be seen from the three-hour temperature change chart (Fig. 9). From this chart, we can see that southward-moving cold air was separated by the mountain range. The main cold air moved along both the eastern and western flanks. There was a center of -14.6°C of the three-hour temperature changes in the eastern valley, and the temperature change in the mountain ridge region was only -3.2°C.

From these observations the orographic effect of the Liupan Mountain Range is evident. Similar phenomena can be seen in the 1-, 3- and 24-hour change charts for pressures, temperatures and relative humidities which are not shown in this paper.

From all the facts mentioned above, we can briefly conclude that both the eastern and the western regions of Mt. Liupan are favorable for the development of synoptic-scale systems, and the mountain ridge suppresses development of synoptic-scale disturbances.

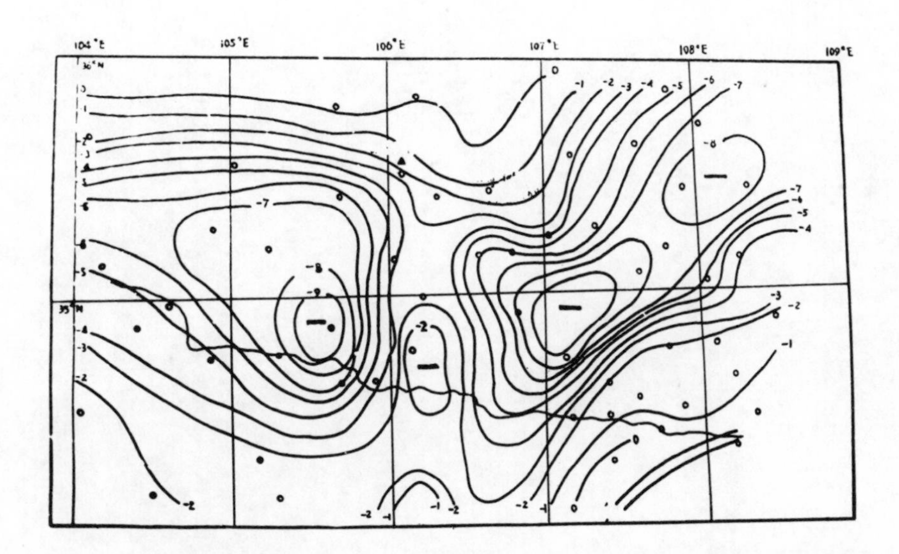

Fig. 9 The three-hour temperature change chart, 18 h (Beijing Standard Time) on 27 May 1973.

4.4 CONCEPTUAL AND NUMERICAL MODELS OF THE EVOLUTION OF THE ENVIRONMENT OF SEVERE LOCAL STORMS

Richard A. Anthes
National Center for Atmospheric Research[1]
Boulder, Colorado 80307 USA

Toby N. Carlson
The Pennsylvania State University
University Park, Pennsylvania 16802 USA

ABSTRACT

A conceptual model of the evolution of the severe local storm environment is discussed in conjunction with two case studies drawn from data obtained during the 1979 SESAME field program. In the conceptual model the formation of severe local storms is brought about by a particular distribution of topography which favors the formation of a lower-tropospheric inversion called a lid. The lid originates from the differential advection of a hot, dry mixing layer from an elevated plateau over a cooler, moister layer which flows northward ahead of a developing cyclone. This type of stable temperature stratification suppresses release of convective instability while, nevertheless, allowing the sensible and latent energy of the boundary layer to increase with time. Intense convection occurs along the lateral boundary of the lid partly because the mechanism for suppressing convection is suddenly removed as the latently unstable air in the boundary layer flows out from beneath the lid (a process called underrunning), but also because the vertical circulation is established in the baroclinic zone along the lid boundary which promotes mesoscale ascent in the area immediately outside the lid.

The development of mesoscale features in numerical model forecasts of the environment of severe local storms is examined for the two SESAME-1979 cases. The results show that a 10-layer model with a horizontal resolution of about 100 km, simple physics and initialized with essentially synoptic-scale data is capable of generating and maintaining mesoscale phenomena in the 0- to 24-hour time period. These results indicate that some mesoscale phenomena are predictable for periods of time longer than the lifetime of the mesoscale feature itself. Mesoscale features produced in the forecasts of the 10-11 April and 25-26 April cases include low-level jets, mesoscale convective complexes, upper-level jet streaks, cold and warm frontogenesis, drylines, mountain waves and capping inversions (lids). The development and structure of these phenomena in the model forecast are examined in detail and the interactions among the phenomena are emphasized.

[1] The National Center for Atmospheric Research is sponsored by the National Science Foundation.

I. INTRODUCTION

Severe local thunderstorms occur in various parts of the world, but the fact that individual convective outbreaks tend to cluster in small areas and favor particular geographic locations at certain times of the year, such as the southern Great Plains of the United States during spring, suggests that the topography exerts a profound influence on the distribution and intensity of convective disturbances. Carlson and Ludlam [7] and Carlson et al. [8] have proposed a conceptual model illustrating how a unique combination of terrain, boundary-layer processes and synoptic-scale circulation results in the enhancement and release of latent instability.

In the conceptual model of Carlson and Ludlam [7], the wet-bulb potential temperature (θ_w) in the boundary layer increases with time in the southerly flow ahead of a large-scale trough. The low level θ_w is increased by surface heating in the presence of intense sunshine and the advection of warm, moist air from the south. When the small-scale convection in the boundary layer is confined beneath a restraining inversion, called a <u>lid</u>, which is formed by the advection of hot, dry air aloft from a neighboring arid plateau, deep convection is prevented from occurring until the low-level air reaches a point where the lid is no longer present or has been lifted sufficiently to allow deep convection to take place. Removal of the lid can occur through a combination of surface heating and moistening, large-scale vertical motion or by the moist air flowing into a region where the lid is absent.

In this paper, two case studies from the 1979 SESAME field program will be used to illustrate aspects of the conceptual model. In addition, preliminary numerical simulations by a mesoscale model illustrate that many of the mesoscale features present in the SESAME cases can be predicted by a numerical model initialized with synoptic-scale data.

II. A REVIEW OF THE CONCEPTUAL MODEL

A general view of the airflow pattern associated with lid formation is presented in Fig. 1. The evolution of the synoptic-scale pattern is as follows. With the approach of a cyclone from the west, air in the low levels (represented by the streamline labeled M in the figure) moves northward to the east of a high, arid plateau. At upper levels, relatively dry air with a higher potential temperature than the moist air also moves toward the north and east, but in so doing the airstream is strongly heated over the plateau and the potential temperature near the base of this airstream is raised significantly throughout a deep mixed layer. When this heated air flows away from the plateau, it forms a dry, warm lid over the moist airstream M along the streamline labeled CD in Fig. 1. Both the dry and moist airstreams experience large-scale ascent as they flow poleward on the east side of the upper-level trough; consequently, the base of the lid rises toward the north, with the greatest vertical motion and vertical displacement occurring along the western side of the airstreams. These airstreams also become confluent with air

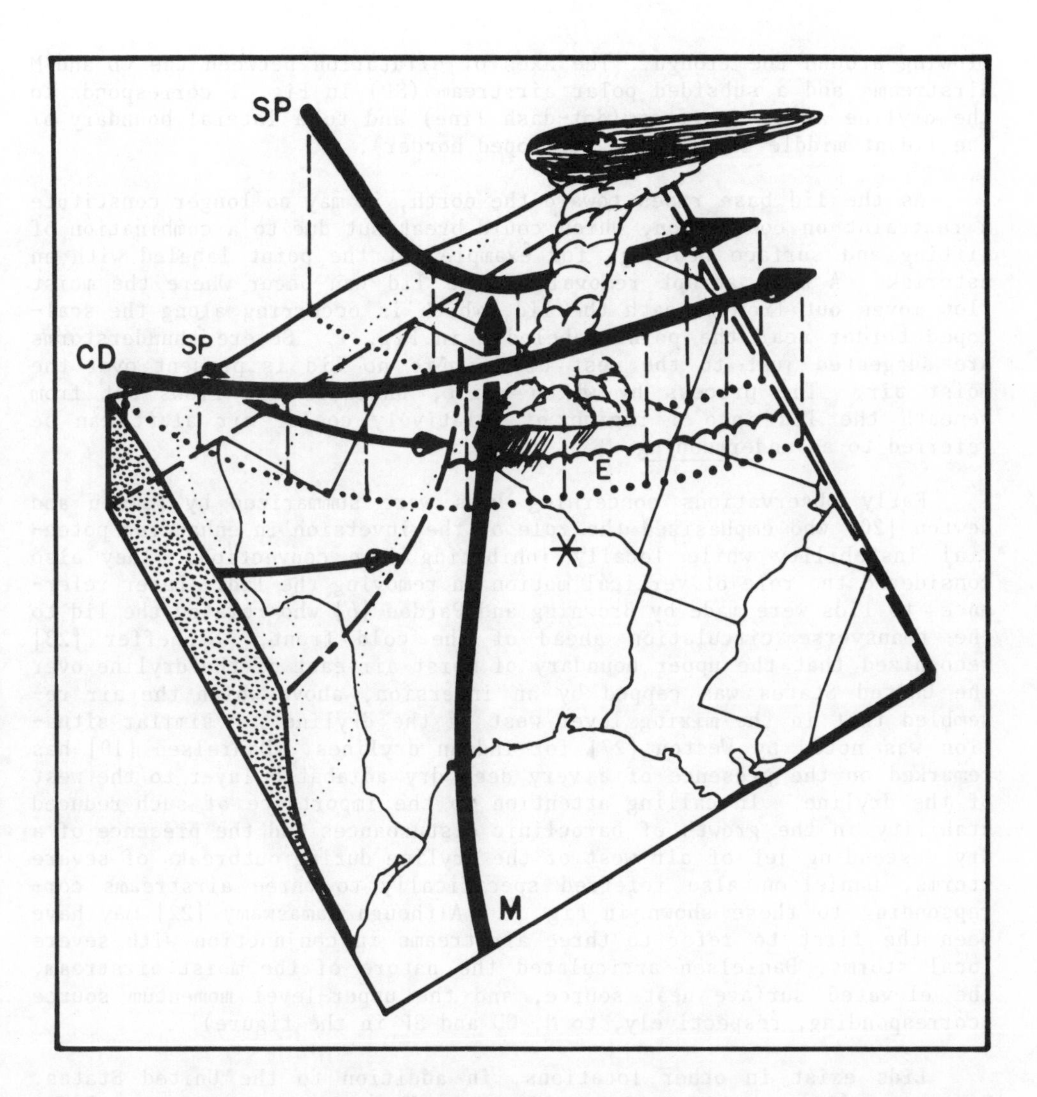

Fig. 1 Schematic flow diagram, representing in three dimensions airstreams M, CD and SP, is shown in perspective against topography of southern Great Plains of the United States and Mexico. Thin dotted lines denote surface terrain elevation contours and illustrate the gradient of surface elevation north and east of the high Plateau of Mexico (cut-away section). Left edge of the moist airstream (M) is bounded by the dryline (dot-dash line); left edge of the Mexican airstream labeled CD, forming a lid over moist air, is denoted by scalloped border. Underrunning (see text) occurs at the location of the thunderstorm near E, most favored area for violent release of latent instability in underrunning process. Thunderstorms can also occur beneath the lid at the location of the asterisk where large-scale ascent coupled with surface heating may be removing the lid. Third airstream SP is subsiding polar air which originates west of the trough.

flowing around the trough. The axes of dilatation between the CD and M airstreams and a subsided polar airstream (SP) in Fig. 1 corresponds to the dryline at the surface (dot-dash line) and to a lateral boundary of the lid at middle levels (the scalloped border).

As the lid base rises toward the north, it may no longer constitute a restraint on convection, which could break out due to a combination of lifting and surface heating, for example, at the point labeled with an asterisk. A more abrupt removal of the lid can occur where the moist flow moves out from beneath the lid, which is occurring along the scalloped border near the point labeled E in Fig. 1. Severe thunderstorms are suggested just to the west of E where no lid is present over the moist air. The process by which moist, unstable air flows out from beneath the lid into a region of relatively cooler air aloft can be referred to as underrunning.

Early observations concerning lids were summarized by Palmén and Newton [20] who emphasized the role of the inversion in enhancing potential instability while locally inhibiting deep convection. They also considered the role of vertical motion in removing the lid. Later references to lids were made by Browning and Pardoe [6] who related the lid to the transverse circulation ahead of the cold front. Schaeffer [23] recognized that the upper boundary of moist air east of the dryline over the United States was capped by an inversion, above which the air resembled that in the mixing layer west of the dryline. A similar situation was noted by Weston [27] for Indian drylines. Danielsen [10] has remarked on the presence of a very deep dry adiabatic layer to the west of the dryline. In calling attention to the importance of such reduced stability in the growth of baroclinic disturbances and the presence of a dry descending jet of air west of the dryline during outbreaks of severe storms, Danielsen also referred specifically to three airstreams correpsonding to those shown in Fig. 1. Although Ramaswamy [22] may have been the first to refer to three airstreams in conjunction with severe local storms, Danielsen articulated the nature of the moist airstream, the elevated surface heat source, and the upper-level momentum source (corresponding, respectively, to M, CD and SP in the figure).

Lids exist in other locations, in addition to the United States. Ramaswamy [22], in identifying three confluent airstreams over India during periods when intense cumulonimbus convection was present, recognized that a dry southwesterly flow aloft formed a capping inversion over the shallow, potentially unstable moist air over northeastern India and Bangladesh. He noted that a similar configuration could be found in conjunction with severe local storms over New South Wales and southern Queensland, Australia, over South Africa and southeastern Brazil, Uruguay and northeastern Argentina. Colon [9] discussed the same type of stratification which he observed over the Arabian Sea and western India. He attributed the origin of the air above the inversion as either Africa or Arabia because of its nearly isentropic lapse rate and high dust content. Weston [27] studied the Indian dryline and noted a similar type of isentropic stratification above the lid. Carlson and Ludlam [7] showed that air from Spain and North Africa formed a capping inversion in the lee of the Pyrenees which permitted an increase in Θ_w to occur at low levels over France, leading to the release of the convection in severe storms over southern England. Peterson et al. [21] have described a tornado outbreak over Australia where there was a strong capping inversion above the moist maritime air near the coast. The lapse rate above the inversion was nearly dry adiabatic throughout a deep layer, suggesting an origin for the lid over the interior desert.

III. SYNOPTIC ANALYSIS

Case studies are presented in this section for the Red River Valley severe storm outbreak of 10-11 April 1979 (the Wichita Falls tornado) and for the convective rainfall situation of 25-26 April 1979, respectively referred to as SESAME I and III.

1. SESAME I, 10-11 April 1979

This outbreak of severe local storms proved to be the most spectacular and devastating of the 1979 SESAME field program [1]. Severe weather developed over the northern Texas panhandle after 1900 GMT and moved rapidly eastward where a second and more intense outbreak began after 2100 GMT just south of the Oklahoma border. The second series of storms, the Red River outbreak, was manifested by a series of tornadoes and large hail which moved in a narrow swath across northern Texas and Oklahoma. Heavy rainfall during the 24-hour period occurred in a narrow swath over Oklahoma in association with the Red River outbreak of storms.

In many respects, the large-scale weather pattern on 10-11 April was typical of that found during many other severe storm episodes over the southern Great Plains of the United States. A well-developed surface low pressure center located over southern Colorado (Fig. 2) was associated with a strong cold front which moved rapidly eastward and reached extreme western Texas by 0000 GMT on 11 April. East of the low was a low-level southerly jet with wind speeds in excess of 20 m s^{-1} at 1 km over Texas. A dryline at the surface moved rapidly eastward and northeastward across Texas during the day. The first thunderstorm cells formed south of Amarillo, Texas (AMA) in the moist air just north of the rapidly advancing portion of the dryline.

Aloft, a trough in the westerly flow located over the western U.S. at 1200 GMT moved eastward in association with a strong jet streak whose exit region extended across the trough axis over northern Texas and Oklahoma. A well-developed lid, present across eastern Texas to Louisiana (Fig. 3) is identifiable at 800 mb on the 0000 GMT sounding at Stephenville (Fig. 4), which was located a short distance east of the dryline. On this sounding, a nearly isentropic layer ($\theta \sim$ 40-42°C) above the lid extended to 500 mb.

Although the earliest severe storm activity was initiated south of Amarillo (AMA), the primary tornadic outbreak was confined to a relatively narrow zone, with most of the severe incidents occurring within a

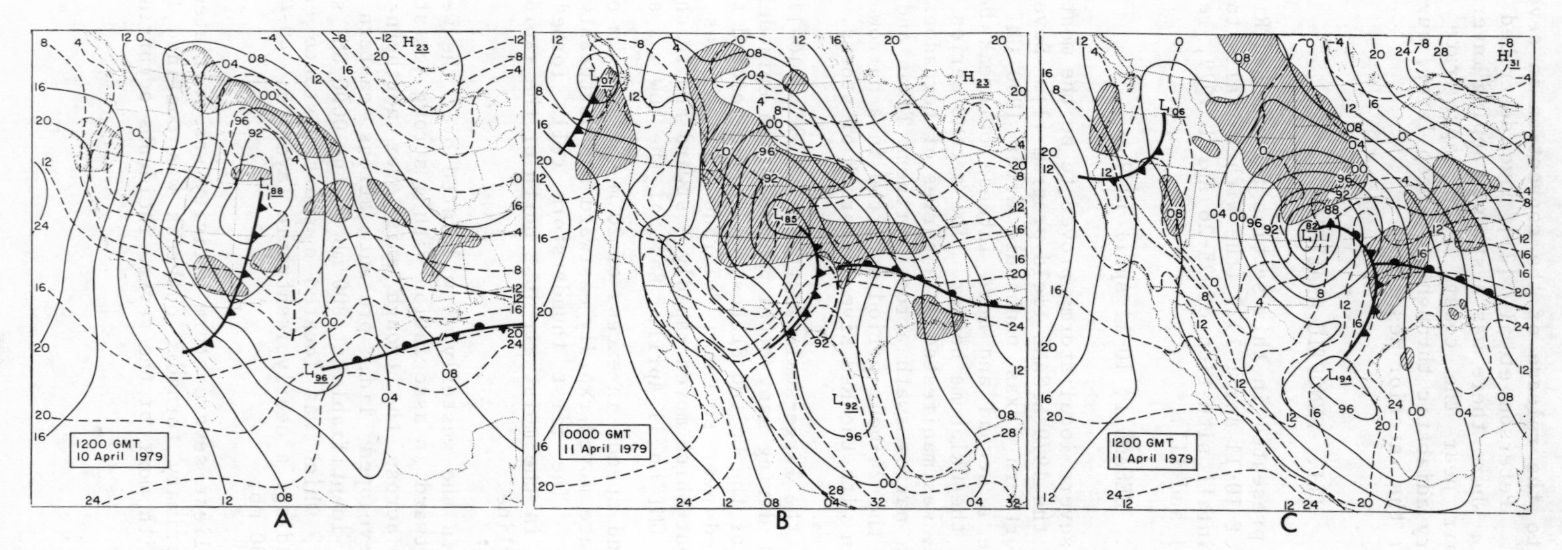

Fig. 2 Surface synoptic charts for (a) 1200 GMT 10 April 1979, (b) 0000 GMT 11 April, and (c) 1200 GMT 11 April. Solid lines are sea-level isobars (in mb above 1000), dash lines are isotherms (°C). Shading denotes precipitation. Dash-dot line denotes dryline.

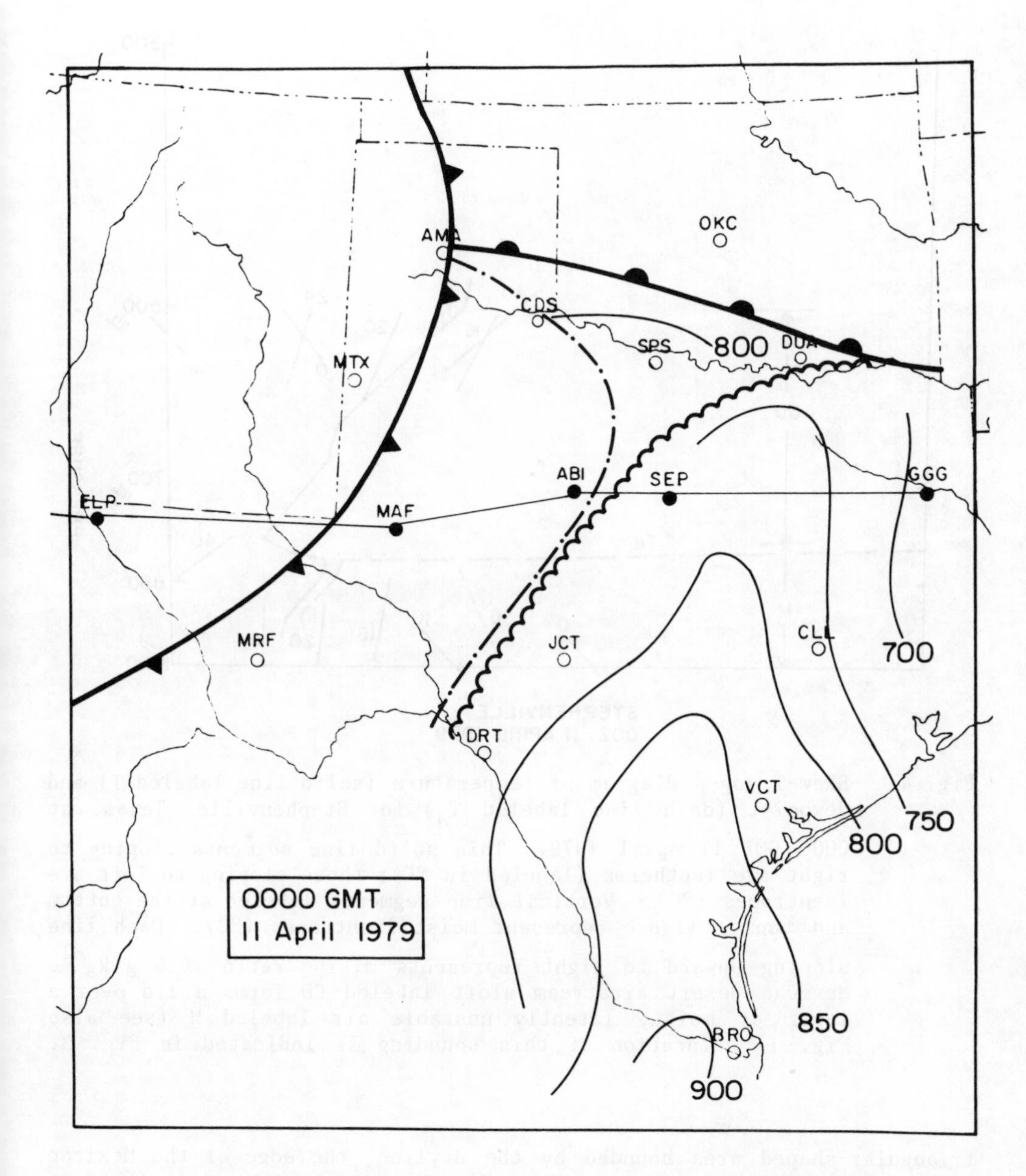

Fig. 3 Lid base analysis for 0000 GMT 11 April 1979. Solid lines represent isopleths of lid base (mb). Scalloped border and dot-dash lines represent, respectively, the western edge of the Mexican airstream (lid) and the surface position of the dryline. Surface fronts are represented by conventional symbols.

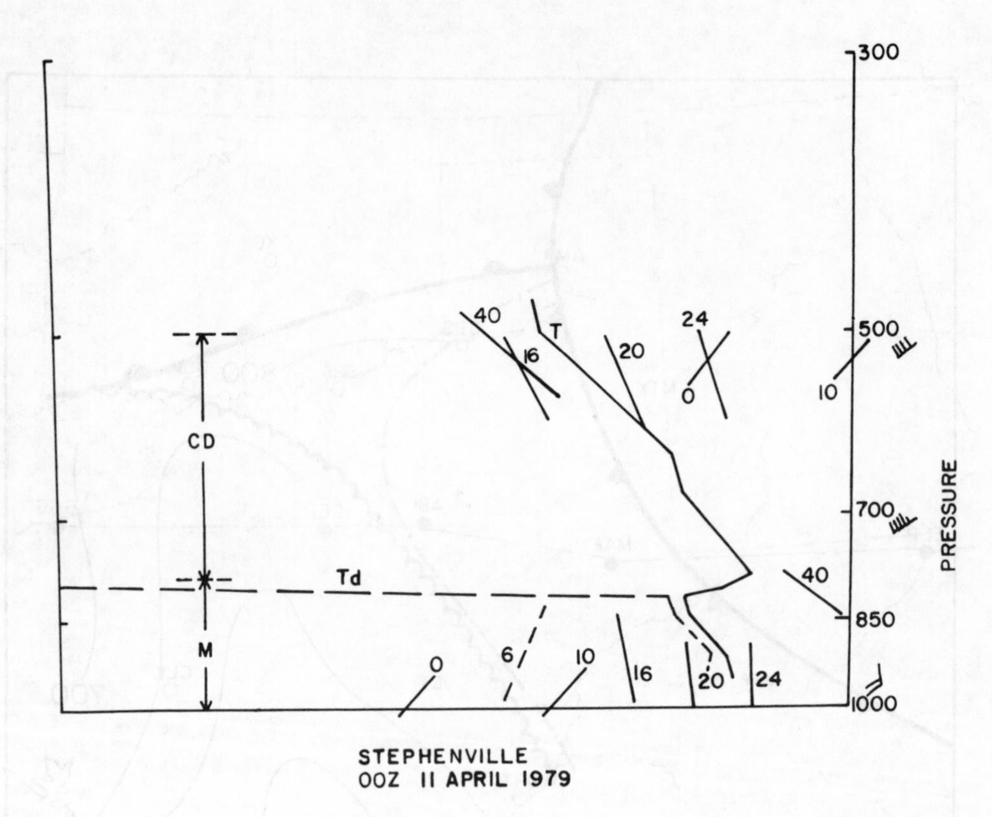

Fig. 4 Skew-T log-p diagram of temperature (solid line labeled T) and dewpoint (dash line labeled T_d) for Stephenville, Texas, at 0000 GMT 11 April 1979. Thin solid line segments sloping to right are isotherms (labeled in °C); those sloping to left are isentropes (°C). Vertical line segments labeled at the bottom and top of figure represent moist isentropes (°C). Dash line sloping upward to right represents mixing ratio of 6 g kg^{-1}. Mexican desert airstream aloft labeled CD forms a lid over a layer of moist, latently unstable air labeled M (see also Fig. 1). Location of this sounding is indicated in Fig. 3.

triangular-shaped area bounded by the dryline, the edge of the Mexican lid and the surface warm front (Fig. 3). However, considerable thunderstorm activity resulted from overrunning of moist, unstable air north of the warm front over Oklahoma. During the 10th, this triangular area moved northeastward and diminished rapidly in size due to the encroachment of dryline and lid edge. Most of the precipitation occurred west and north of the Mexican lid or to the east where the inversion had been lifted and was ineffective in suppressing convection.

A west-east vertical cross section (Fig. 5) illustrates the stratification between the moist air (M) and the elevated mixing layer (CD) which formed the Mexican lid (see Fig. 3 for the orientation of the cross section and the location of the stations). The horizontal temperature and moisture gradients at the dryline folded back toward the east to form the lid above the moist air.

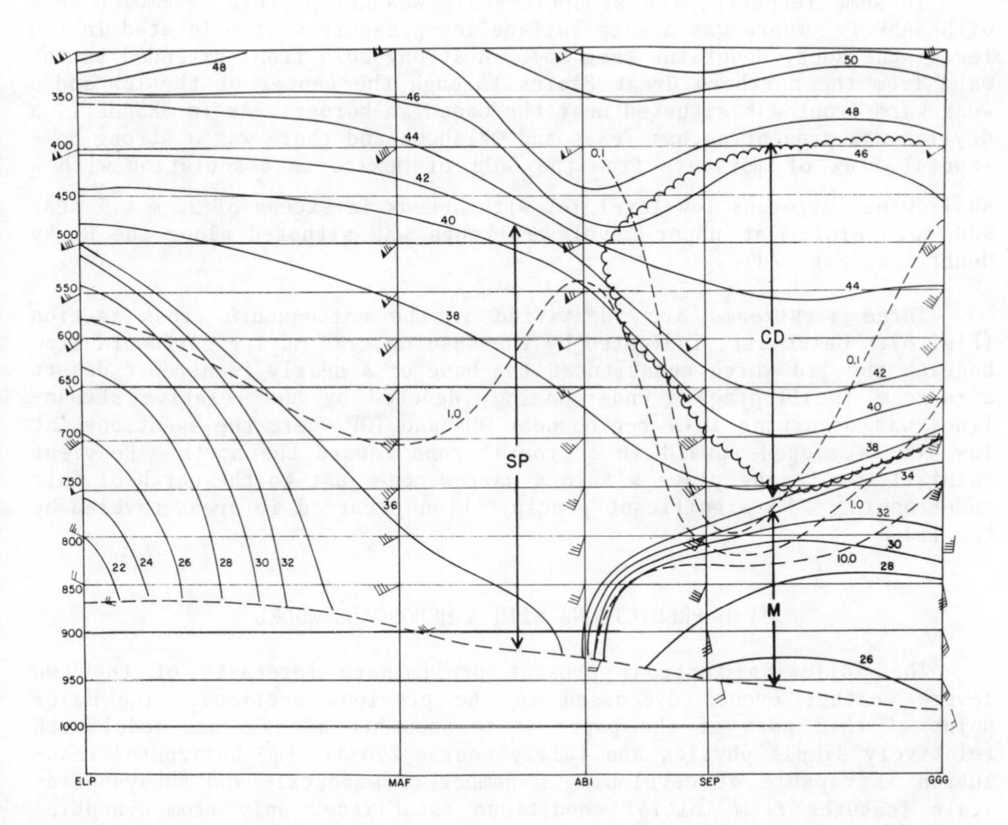

Fig. 5 Vertical cross section from El Paso, Texas (ELP) to Longview, Texas (GGG) along the line of stations in Fig. 3 for 0000 GMT 11 April 1979. Solid lines correspond to isentropes (°C), dashed line to isopleths of mixing ratio (g kg^{-1}) and stippling to mixing ratios in excess of 10 g kg^{-1}. Scalloping is intended to offset the Mexican desert air (CD), which is forming the lid, from the moist (M) air below and from the modified subsided polar air (SP) to the west. Conventional wind barbs are included at the station locations (1 flag = 5 m s^{-1}). The surface dryline was near Abilene (ABI).

2. SESAME III, 25-26 April 1979

Heavy convective rainfall and a few hail episodes characterized the thunderstorm outbreak in the SESAME III case. Between 0000 GMT and 1800 GMT on the 25th, a cluster of thunderstorms formed over southwestern Nebraska and moved eastward to western Illinois. During this period, the precipitation pattern exhibited a sharp southern edge, with the heavy accumulations occurring within a remarkably narrow tongue orientated west to east across Nebraska (Fig. 7a).

In some respects, the synoptic-scale weather pattern resembled that of SESAME I. There was a deep surface low-pressure system located in the lee of the Rocky Mountains (Fig. 6). A strong cold front extended southward from the northern Great Plains through the center of the low and a weak warm front was situated near the Canadian border. As in SESAME I, a dryline was present across Texas and Oklahoma and there was a strong horizontal flux of moisture from the Gulf of Mexico in association with a shallow but vigorous low-level jet with speeds in excess of 20 m s^{-1} near 900 mb. Aloft, at upper levels a trough was situated along the Rocky Mountains (Fig. 7d).

Three airstreams are identified in the north-south cross section (Fig. 8). Moist air, indicated by the shading, was moving northward from beneath the lid which constituted the base of a nearly isentropic desert airstream (scalloping). Underrunning, denoted by the relative streamline, was occurring in a region near OMA and TOP where the isentropes at low levels sloped upward in a frontal zone toward the north. Heaviest rainfall was taking place within a narrow zone just to the north of this underrunning. No significant precipitation occurred in areas covered by the lid.

IV. PREDICTIONS WITH A MESOSCALE MODEL

The following sections present preliminary forecasts of the two severe weather events discussed in the previous sections. The major point of this part of the paper is to show how a mesoscale model with relatively simple physics and fairly coarse (90-111 km) horizontal resolution is capable of developing a number of mesoscale and subsynoptic-scale features from initial conditions established only from synoptic-scale data. Features to be investigated include: (1) the development and decay of a low-level jet, (2) the formation of a mesoscale convective complex (MCC) and its subsequent modification of the lower- and upper-level circulation, (3) the formation of mesoscale regions of heavy precipitation, (4) the intensification of surface warm and cold fronts, (5) the formation of drylines, (6) the dynamic coupling of an upper-level jet (ULJ) and a low-level jet (LLJ), (7) the formation of a mountain wave and (8) the formation and maintenance of capping inversions (lids).

Although the tendency in some previous studies has been to discuss these features as distinct entities, they are part of a strongly interacting system of dynamical and thermodynamical processes, as emphasized

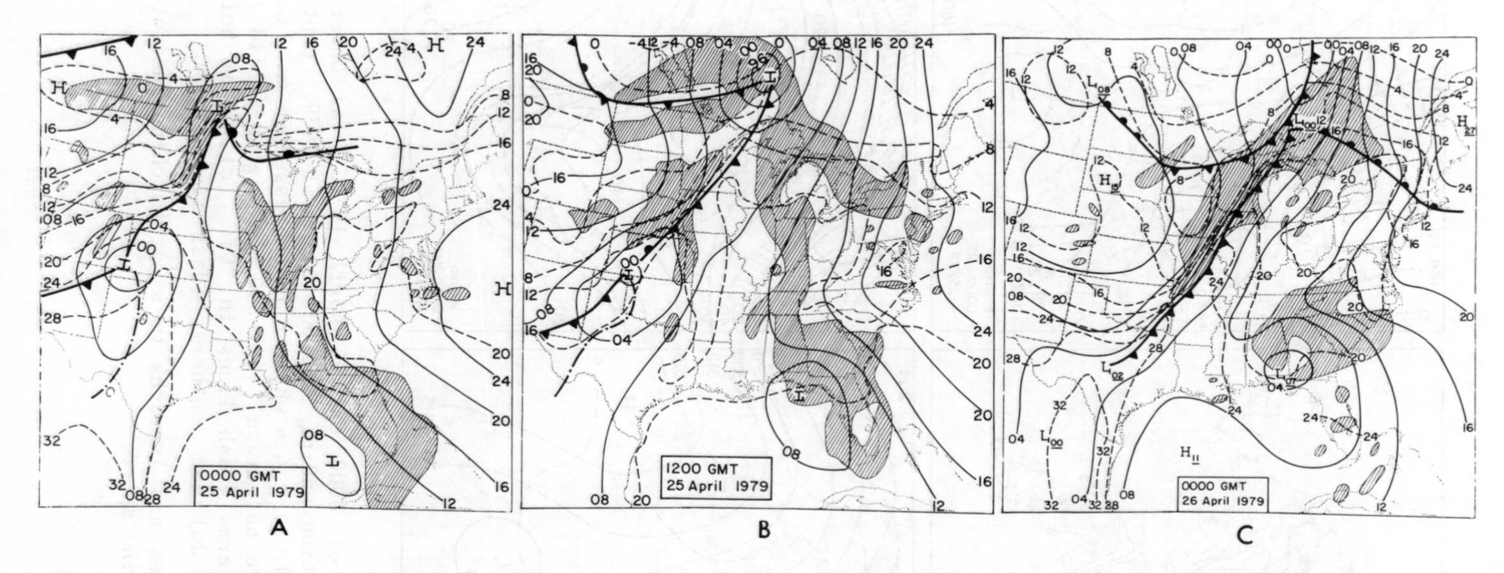

Fig. 6 Surface synoptic chart for (a) 0000 GMT 25 April 1979, (b) 1200 GMT 25 April and (c) 0000 GMT 26 April. Solid lines are sea-level isobars (in mb above 1000), dashed lines are isotherms. Shading denotes locations of precipitation. Dot-dash line denotes dryline.

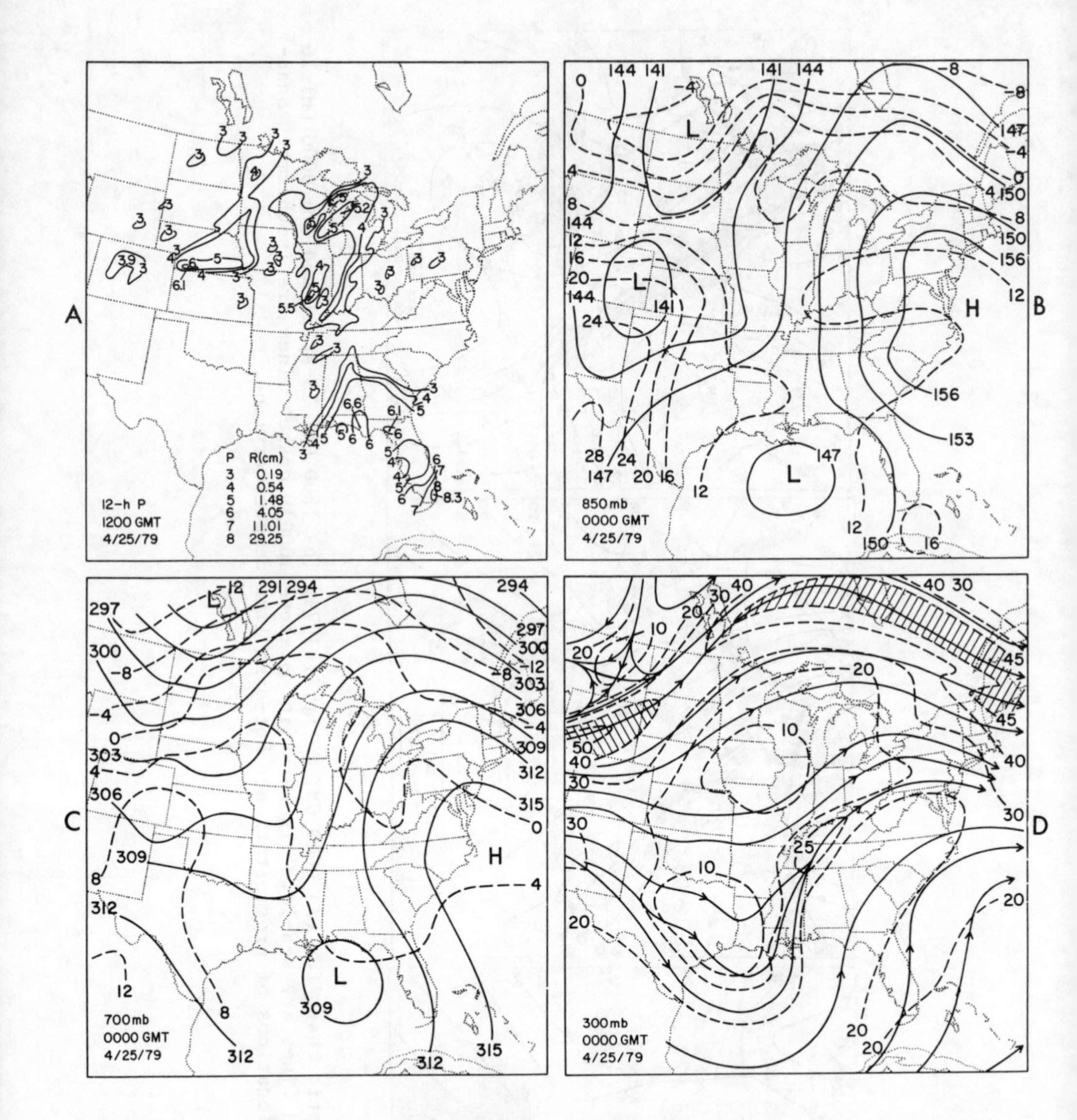

Fig. 7 Analyses of 0000 GMT 25 April 1979. (a) Observed 12-hour precipitation ending 1200 GMT 25 April 1979. Contours of P = ℓn (R + 0.01) + 4.6 where R is expressed in cm. Contours less than 3 are not drawn. (b) 850 mb heights (solid lines in dm) and isotherms (dash lines in °C). (c) 700 mb heights (solid lines in dm) and isotherms (dash lines in °C). (d) 300 mb streamlines and isotachs (m s^{-1}); hatching indicates bands of maximum wind speed.

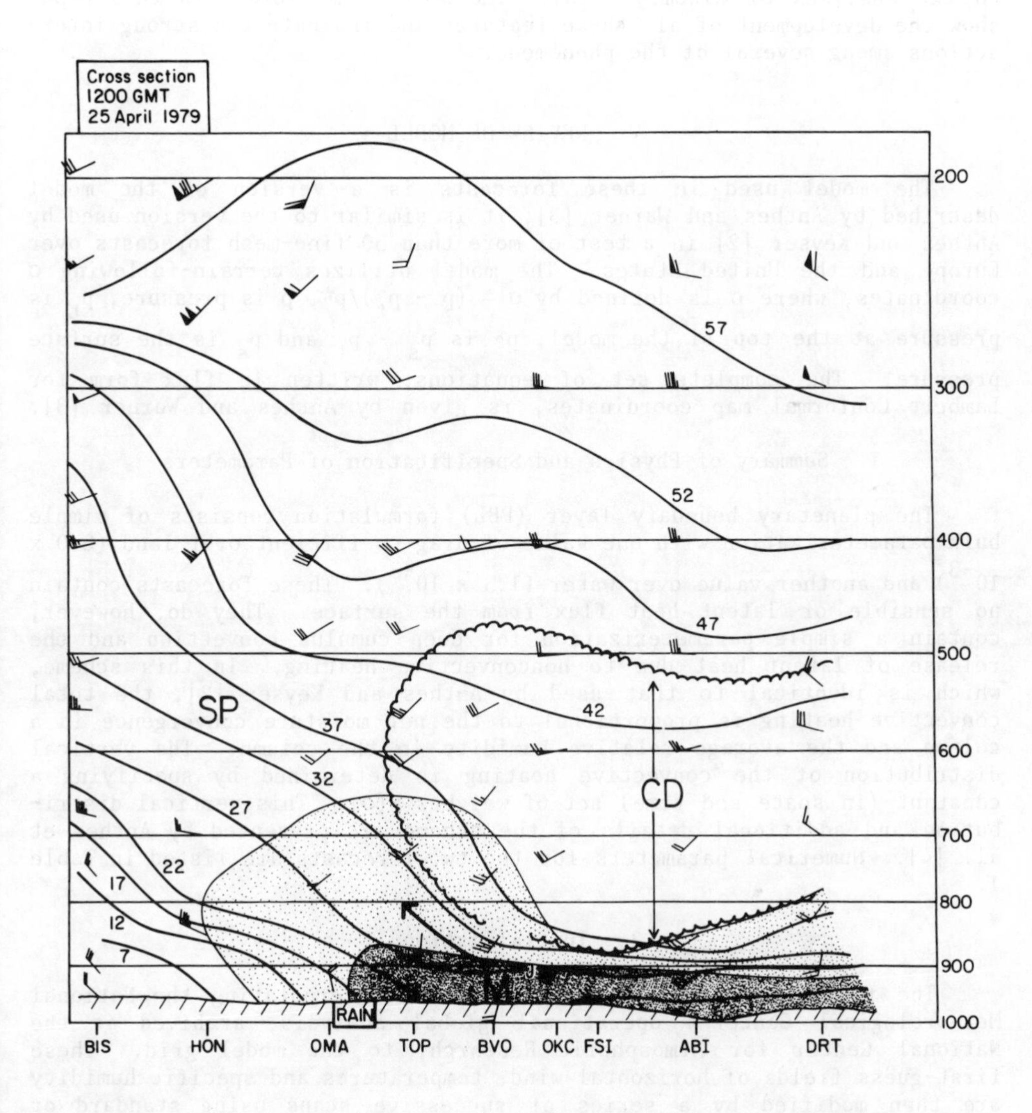

Fig. 8 Vertical cross section from Bismarck, North Dakota (BIS) to Del Rio, Texas (DRT) for 1200 GMT 25 April 1979. Solid lines correspond to isentropes (°C), light stippling to mixing ratio in excess of 5 g kg^{-1}, heavy shading to mixing ratios in excess of 10 g kg^{-1}. Conventional wind barbs are indicated (full flag = 5 m s^{-1}). Letter J near OKC marks the core of the low-level jet in the cross section. Arrow near the surface between BVO and TOP suggests relative wind motion in cross section.

in the analyses of Ninomiya [18]. The model simulations in this paper show the development of all these features and indicate the strong interactions among several of the phenomena.

V. REVIEW OF MODEL

The model used in these forecasts is a version of the model described by Anthes and Warner [3]; it is similar to the version used by Anthes and Keyser [2] in a test of more than 30 fine-mesh forecasts over Europe and the United States. The model utilizes terrain-following σ coordinates, where σ is defined by $\sigma = (p - p_t)/p^*$, p is pressure, p_t is pressure at the top of the model, p^* is $p_s - p_t$ and p_s is the surface pressure. The complete set of equations, written in flux form for Lambert Conformal map coordinates, is given by Anthes and Warner [3].

1. Summary of Physics and Specification of Parameters

The planetary boundary layer (PBL) formulation consists of simple bulk parameterization with one value of drag coefficient over land (2.0×10^{-3}) and another value over water (1.5×10^{-3}). These forecasts contain no sensible or latent heat flux from the surface. They do, however, contain a simple parameterization for deep cumulus convection and the release of latent heat due to nonconvective heating. In this scheme, which is identical to that used by Anthes and Keyser [2], the total convective heating is proportional to the net moisture convergence in a column and the average relative humidity in the column. The vertical distribution of the convective heating is determined by specifying a constant (in space and time) set of weights $N(\sigma)$. This vertical distribution and additional details of the scheme are presented by Anthes et al. [4]. Numerical parameters for the two forecasts are listed in Table 1.

2. Initialization

The forecasts are initialized by first interpolating the National Meteorological Center's operational global analysis, archived at the National Center for Atmospheric Research, to the model grid. These first-guess fields of horizontal wind, temperatures and specific humidity are then modified by a series of successive scans using standard or special rawinsonde data or artificial soundings created from careful subjective analyses, satellite pictures and surface data. After the objective analyses are complete, the vertical mean divergence is removed as suggested by Washington and Baumhefner [26].

VI. 25-26 APRIL SIMULATION

Figure 9 summarizes the results of the first 12 hours of the forecast, verifying at 1200 GMT 25 April. The surface map (Fig. 9a) shows a reasonably accurate prediction of the observed features (Fig. 6b),

Table 1. Parameters for 24-hour simulations.

	10-11 April 1979 Case	25-26 April 1979 Case
p_t (mb)	100	150
Horizontal grid size Δs (km)	111.1	90
Time step Δt (s)	236	191
Horizontal array size	37 x 37	41 x 41
Initial time and date	1200 GMT 10 April 1979	0000 GMT 25 April 1979
Center of model domain	34.5 N 105.0 W	38.0 N 88.0 W
Maximum K_H ($m^2\ s^{-1}$)	35×10^4	35×10^4
CRAY-1 time for 24-hour forecast (min and sec)	4:22	6:36

including the cyclone north of Lake Superior, a cold front trailing southwestward into a second low over Kansas and Oklahoma and a dryline extending southward from this low across Texas. The isotherm pattern shows the intensity of the cold front and also indicates a broad baroclinic zone across the northeastern portion of the domain. The weak low in the eastern Gulf of Mexico is well-forecast.

The low-level ($\sigma = 0.96$) wind field (Fig. 9b) shows a strong southerly flow ahead of the frontal zone with a sharp wind shift across the front and a strong northerly flow behind the front. An LLJ, with maximum speed 17 m s^{-1}, has formed over Texas and Oklahoma, as observed. The strong southerly flow has advected moisture northward into the upper Midwest. A sharp horizontal gradient in specific humidity exists across the dryline and also across the cold front (Fig. 9b).

At 300 mb, a jet streak has developed over Minnesota, where the winds have strengthened locally from 20 m s^{-1} at the initial time to 46 m s^{-1} (Fig. 9c). This intensification is consistent with the frontogenesis in this region. The relative humidity at 300 mb shows three main areas of deep convection. The greatest area is over the southeastern United States in the southerly flow east of the cutoff low over the Gulf of Mexico. Smaller areas exist over the Great Lakes and Nebraska. Figure 9d depicts and predicted 12-hour precipitation ending at 1200 GMT. Maximum amounts of 1.91 cm, 2.38 cm and 4.21 cm were predicted over Nebraska, the Great Lakes and Florida, respectively. Each of these regions received heavy rainfall (Fig. 7a). The sharp southern edge to the precipitation maximum over Nebraska in the model prediction as well

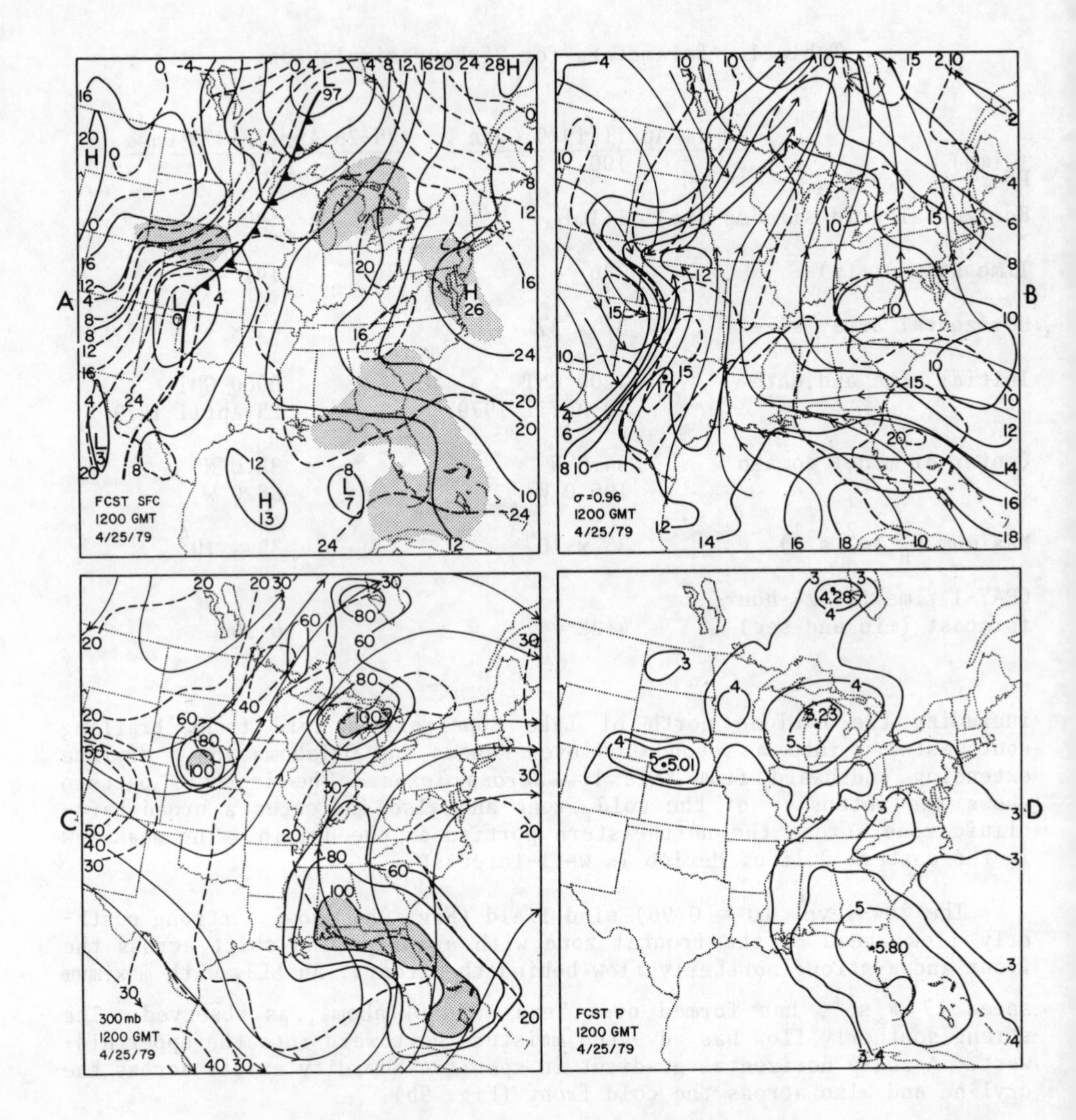

Fig. 9 12-hour forecast ending at 1200 GMT 25 April 1979. (a) Sea-level isobars (solid lines in mb) and isotherms (dashed lines in °C). Areas of precipitation exceeding 0.19 cm in past 3 hours indicated by stippled area. Front and dryline are located subjectively. (b) Streamlines and isotachs (dash lines in m s^{-1}) and specific humidities (solid lines in g kg^{-1}) at the $\sigma = 0.96$ level (boundary layer). (c) 300 mb streamlines and isotachs (dash lines in m s^{-1}) and relative humidity (solid lines). Regions over 100 percent are shaded to indicate strong convection. (d) 12-hour rainfall depicted as in Fig. 7a.

as the observations reflects the sudden ascent of the northward-flowing air in the frontal zone.

During the last 12 hours of the forecast, the northern low moved rapidly eastward and the cold front moved across Wisconsin, Iowa and Nebraska (Fig. 10a). The Oklahoma low weakened considerably, as observed (Fig. 6c). Heavy rain continued over portions of Iowa, Missouri and Kansas in association with the strong front. Heavy precipitation was also predicted over the southeastern United States and over the Bahamas as the Gulf of Mexico low moved northeastward.

There were two areas of frontogenesis during the 24-hour forecast. The low-level horizontal temperature gradient associated with the cold front in the Midwest increased from a typical value of approximately 3 K $(100\ km)^{-1}$ to greater than 9 K $(100\ km)^{-1}$ over the period (compare Figs. 6a and 10a). In addition, a warm front developed over southeastern Canada, with the temperature gradient increasing from 2.5 K $(100\ km)^{-1}$ to 4.5 K $(100\ km)^{-1}$.

The horizontal gradient of specific humidity in the PBL remained strong over Texas and strengthened over the upper Midwest (Fig. 10b). As shown by Anthes et al. [4], both confluence and shearing deformation contributed to this increase in gradient of moisture.

The wind field at 300 mb (Fig. 10c) shows an increase in the intensity of the jet streak over the Lake Superior region, from 46 m s^{-1} at 1200 GMT to nearly 57 m s^{-1}. This increase was associated primarily with the MCC over Nebraska (an experiment without latent heating showed a maximum wind speed at 24 hours of 49.1 m s^{-1}).

In summary, the large-scale features are predicted well by the model at all times of the 24-hour forecast. The next sections investigate in greater detail the mesoscale aspects of the forecast, including the formation of the low-level jet, the frontogenesis, the formation of the dryline and the development of the precipitation area over Nebraska.

1. Development and Structure of the Low-Level Jet

As discussed earlier, a low-level jet (LLJ) was present over Oklahoma at 1200 GMT 25 April. Figure 9b shows the jet maximum over north-central Texas at the $\sigma = 0.96$ level at 1200 GMT. From an initial velocity over Oklahoma of less than 5 m s^{-1} from the south, the winds increase to about 11 m s^{-1} at 3 hours, 14 m s^{-1} at 6 hours and 17 m s^{-1}. After 12 hours, the jet moves northeastward and by 24 hours is located over the Arkansas-Missouri border and has weakened to about 12 m s^{-1}.

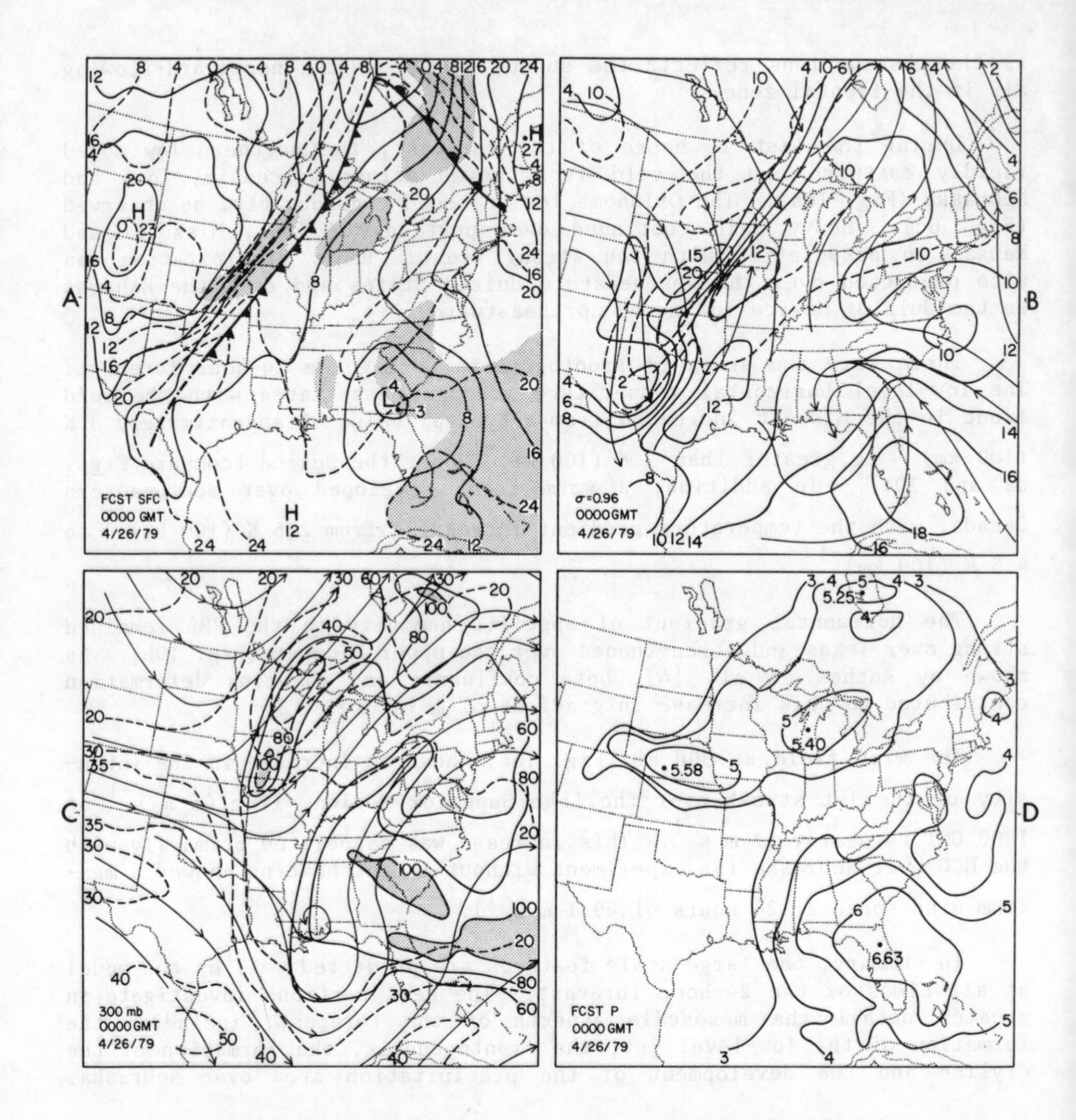

Fig. 10 24-hour forecast ending at 0000 GMT 26 April 1979. (a) As in Fig. 9a. (b) As in Fig. 9b. (c) As in Fig. 9c. (d) 24-hour rainfall depicted as in Fig. 7a.

The development of the jet in this model forecast is associated with a maximum in the geostrophic wind at the lowest model level ($\sigma = 0.96$) rather than with diurnal variations in the PBL [5], because this forecast contains only a simple bulk-PBL formulation with no heating. This maximum is associated with the strong baroclinic zone at the eastern edge of the lid, as shown by the 850 mb potential temperature map at 6 hours (Fig. 11) and as discussed by Means [16].

Figure 12a shows a west-east cross section along the line A-B in Fig. 11 and across the LLJ at 1200 GMT. This cross section also shows the lid as resolved by the model (the layer of stable air between 800 mb and the surface) and the northern portion of the dryline. From Fig. 12a, it is obvious that the LLJ has a major role in transporting moisture northward, since the maximum v component occurs near the surface where the specific humidity is highest. Above 800 mb, the v component is weak and the air is very dry.

The observed west-east cross section of θ and v at 1200 GMT along the line A-B in Fig. 11 is presented in Fig. 12b for comparison with the model-predicted cross section in Fig. 12a. The location of the jet is predicted well; however, the model jet is more diffuse in the horizontal direction and is weaker than observed. Likely reasons for this defect include coarse horizontal resolution, absence of a high-resolution PBL model and neglect of a diurnal heating cycle.

The role of the LLJ, the lid and the front to the north in producing the precipitation area over Nebraska can be seen in Fig. 12c, which is a north-south cross section along the line C-D in Fig. 11 and parallel to the LLJ. The low-level air flows northward under the restraining inversion and ascends abruptly when it encounters the steeply sloping frontal zone. Vertical velocities in this mesoscale zone reach a maximum of 25 $\mu b\ s^{-1}$ at about 600 mb.

Figure 12d shows the relative humidity and the u component of the wind along the north-south cross section C-D shown in Fig. 11. Typical relative humidities beneath the lid are 50 percent, but immediately above the lid inversion, the relative humidities drop to 20-30 percent. North of the cold front the low-level relative humidities are high (85-100 percent). In the frontal zone, the humidity is also high throughout the entire troposphere, ranging from 75 percent to over 100 percent. The humidity in the middle troposphere has not yet reached 100 percent in spite of the fact that moderate convective precipitation is occurring at this time. The convective parameterization does not require saturation at the grid scale (90 km) to produce rainfall.

2. Frontogenesis and Formation of the Dryline

Anthes et al. [4] studied the kinematics of the intensification of the cold front and the dryline using the equation for the temporal change of the horizontal gradient of potential temperature and specific humidity. This equation, applied to the model data in the boundary

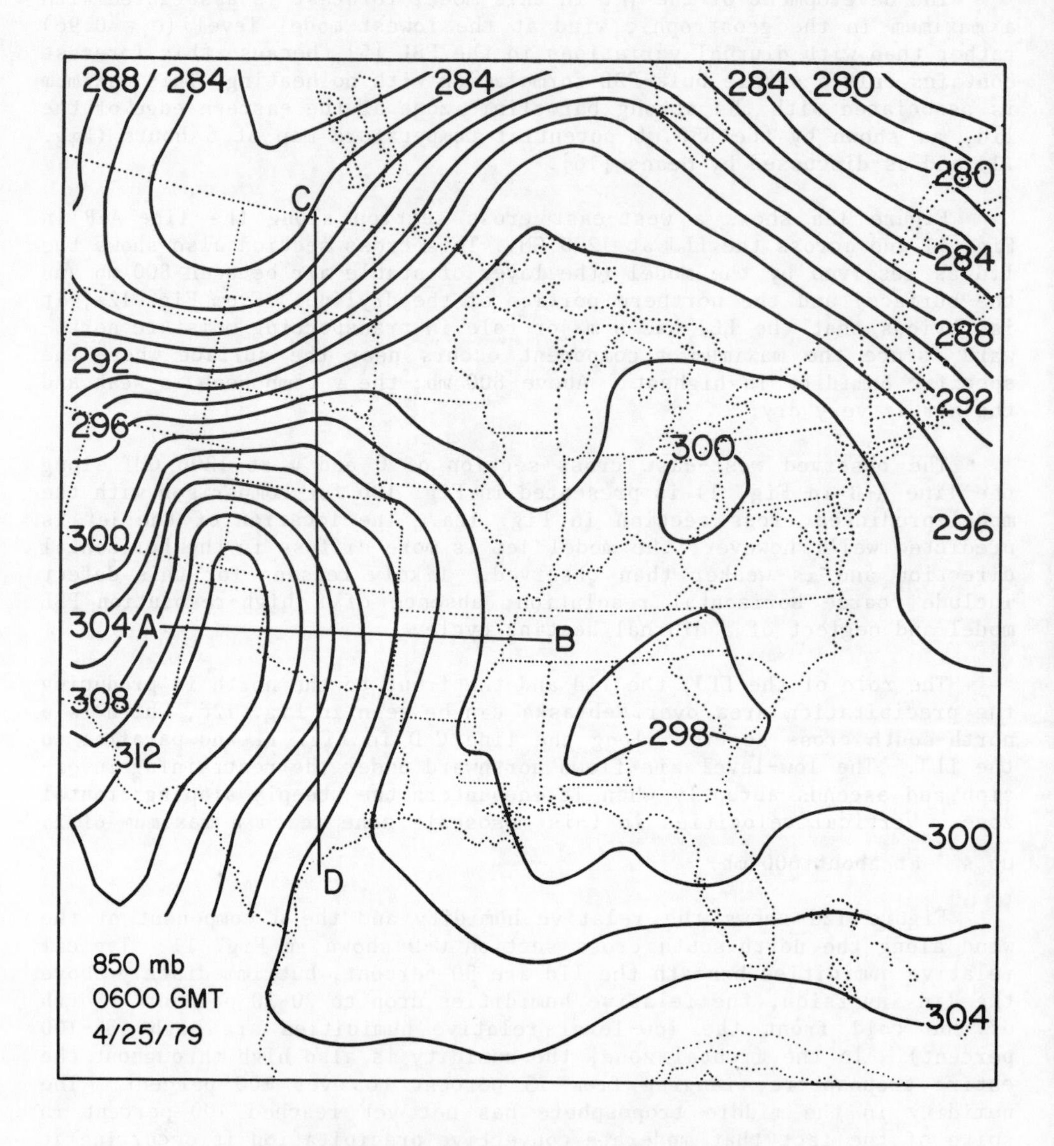

Fig. 11 850 mb potential temperature (K) at 0600 GMT 25 April 1979 in model forecast. Lines A-B and C-D indicate positions of cross sections shown in Fig. 12.

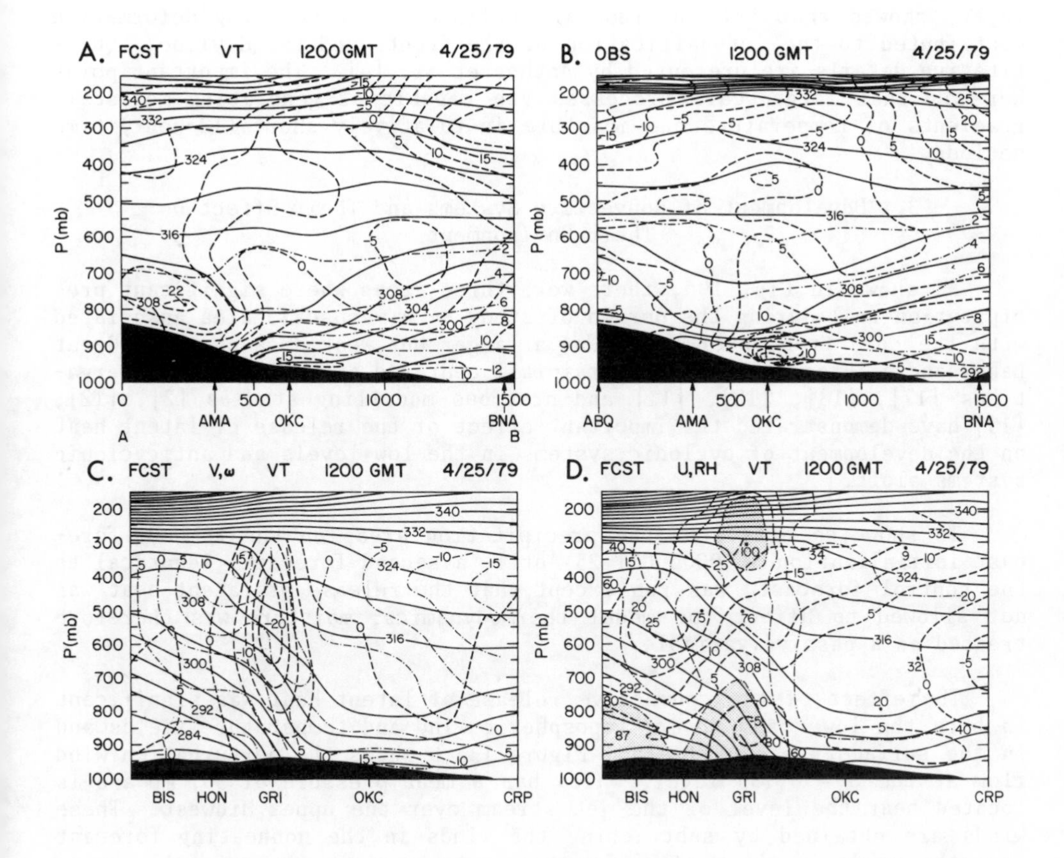

Fig. 12 Vertical cross sections at 1200 GMT 25 April 1979 along lines shown in Fig. 11. (a) 12-hour model forecast, solid lines are isentropes (K), dashed lines are v component of wind (perpendicular to plane of cross section) in m s^{-1}, dash-dot line are specific humidity (g kg^{-1}). (b) As in 12a, but observed. (c) Model forecast, solid lines are isentropes (K), dashed lines are vertical velocity (ω) in μb s^{-1}, dash-dot lines and u component of wind (parallel to plane of cross section) in m s^{-1} (dashed lines). (d) As in Fig. 12c, but showing relative humidity (dash-dot lines) and u component of wind (perpendicular to plane of cross section) in m s^{-1} (dash lines).

layer, showed that both horizontal confluence and shearing deformation contributed to the intensification of the fronts and the dryline. Quantitative details are presented by Anthes et al. [4]. The important point here is that large-scale processes are capable of generating mesoscale gradients of temperature and moisture in relatively short (12-hour) time periods.

3. Development of Convective Systems and Their Effect on Their Environment

As shown in Fig. 10d, there were three areas where significant precipitation fell during the period of study -- a mesoscale area associated with the cold front over Nebraska, a larger but weak area over the Great Lakes and a large area of heavy rainfall centered over Florida. Observations [17], [18], [15], [11] and previous modelling studies [2], [12], [15] have demonstrated the important effect of the release of latent heat on the development of cyclonic systems in the low levels and anticyclonic systems aloft.

To show the effect of the precipitation areas on the 24-hour forecast initialization at 0000 GMT 25 April, a second forecast, identical to the control forecast, was run except that the release of latent heat was not allowed to affect the model thermodynamics; moisture was therefore treated as a passive variable.

The effect of the convective release of latent heat was significant in both the lower and upper tropospheric wind and temperature fields and in the surface pressure field. Figure 13a,b shows the perturbation wind flow at the $\sigma = 0.185$ level, which has a mean pressure of 307 mb and is located near the level of the jet stream over the upper Midwest. These winds are obtained by subtracting the winds in the nonheating forecast from the winds in the control forecast. Areas where the rainfall during the previous 3 hours exceeded 0.19 cm are also depicted. Early in the forecast, the effect of latent heating is to generate a primarily divergent outflow aloft over the main precipitation areas (Fig. 13a). A maximum perturbation velocity of over 10 m s^{-1} from the east is associated with the maximum 3-hour rainfall of 1.64 cm over Florida. Weaker divergent circulations are centered over the rainfall maxima over Lake Michigan, Minnesota and N. Dakota.

Twelve hours later, the longer period of heavier precipitation and the earth's rotation have generated divergent anticyclonic circulations over the main precipitation areas over Nebraska and Florida (Fig. 13b). The maximum perturbation winds over the Nebraska MCC and over the southeastern United States approach 20 m s^{-1}. The perturbation wind circulations generated by the latent heating are associated with temperature perturbations which reach their maximum values in the upper troposphere. For example, maximum temperature perturbations of 4°C at 300 mb are produced over Wisconsin and Minnesota, while 9°C anomalies occur in association with the southeastern precipitation system.

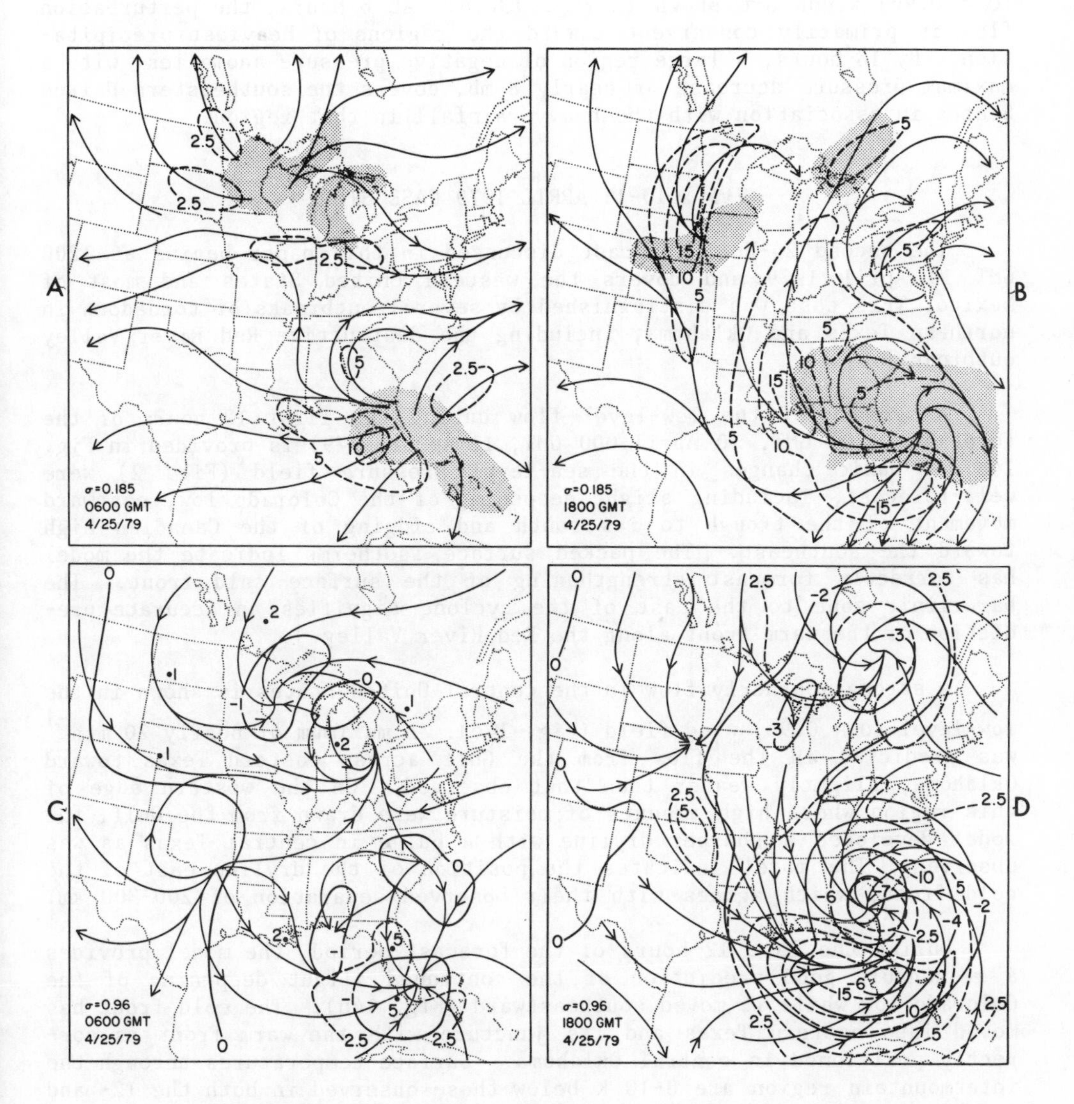

Fig. 13 Difference fields (experiment with latent heating minus experiment without latent heating). Shaded areas depict regions which received more than 0.19 cm of rain in the previous 3 hours (a) Winds (m s^{-1}) at σ = 0.185 level (p $\cong$ 307 mb), 6 hours into forecast. (b) As in a, but 18 hours into forecast. (c) Boundary-layer (σ = 0.96) winds (m s^{-1}) and surface pressure difference (mb) at 0600 GMT. Data points are local maxima in wind speed. (d) As in c, except for 1800 GMT.

The effects of latent heating on the surface pressure and low-level (σ = 0.96) winds are shown in Fig. 13c,d. At 6 hours, the perturbation flow is primarily convergent toward the regions of heaviest precipitation. By 18 hours, a large region of negative pressure anomalies, with a maximum pressure decrease of nearly 8 mb, covers the southeastern United States in association with the heavy rainfall in that region.

VII. 10-11 APRIL 1979 CASE STUDY

The second 24-hour forecast discussed in this paper begins at 1200 GMT 10 April 1979 and covers the western United States and most of Mexico. This case was distinguished by several outbreaks of tornadoes in northern Texas and Oklahoma, including the destructive Red River Valley outbreak [19].

A summary of the low-level flow during the first 12 hours of the forecast (1200 GMT, 10 April-000 GMT, 11 April 1979) is provided in Fig. 14a,c. Basic changes in the sea-level pressure field (Fig. 2) were well-forecast, including slight deepening of the Colorado low, eastward movement of the trough to its south and ridging of the Canadian high toward the southeast. The packed surface isotherms indicate the model has correctly forecast strengthening of the surface cold front. The baroclinic zone to the east of the cyclone signifies an accurate prediction of the warm front along the Red River Valley.

A strong southerly flow in the central United States is shown in the low-level (σ = 0.96) wind field (Fig. 14c). A maximum of nearly 20 m s^{-1} was predicted in the flow from the Gulf across eastern Texas toward Oklahoma, slightly weaker than that observed. On the western edge of this region where high amounts of moisture were drawn from the Gulf, the model developed a sharper dryline with a bulge in central Texas as was observed. The model indicates the position of the dryline east of the cold front, which agrees with their observed separation of 200-300 km.

During the last 12 hours of the forecast period, the model provides a reasonably good prediction of the continued slight deepening of the Colorado low while it moved southeastward (Fig. 14b). The cold front has moved into eastern Texas and its juncture with the warm front is correctly positioned in central Oklahoma. Surface temperatures through the intermountain region are 8-10 K below those observed in both the 12- and 24-hour forecasts due to the lack of surface heating in the model PBL parameterization. The model forecast a continued eastward movement of the dryline across northern Texas, while moister air remained in southern Texas along the Rio Grande Valley, owing to the low center in northeastern Mexico (Fig. 14d). The maximum southerly low-level wind (σ = 0.96) was forecast to move eastward to the Mississippi Valley by 24 hours (Fig. 14 d) as was observed.

Warm frontogenesis continued during the second half of the forecast period, and the maximum horizontal temperature gradient increased to 4 K

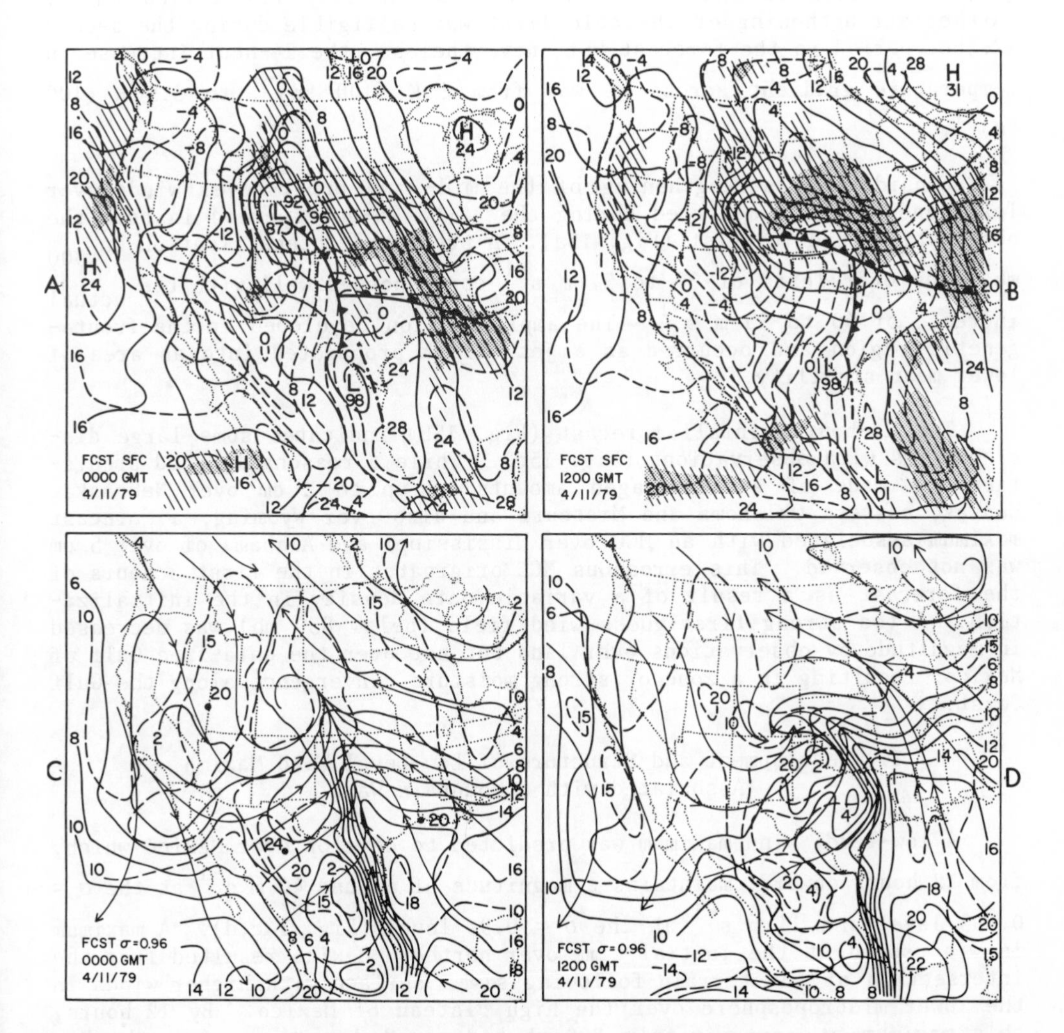

Fig. 14 (a) 12-hour forecast ending at 0000 GMT 11 April 1979 of sea-level pressures (solid lines in mb), temperature (dash lines, °C). Hatched area indicates previous 3-hour rainfall in excess of 0.07 cm; stippled area indicates previous 3-hour rainfall in excess of 0.19 cm. Fronts and dryline are located subjectively. (b) As in a, except for 24-hour forecast ending at 1200 GMT 11 April 1979. (c) 12-hour forecast ending 0000 GMT 11 April of $\sigma = 0.96$ streamlines and isotachs (dash lines in m s^{-1}) and specific humidity (solid lines in g kg^{-1}). Data points are local maxima in wind speed. (d) As in c, except for 24-hour forecast ending at 1200 GMT 11 April 1979.

(100 km)$^{-1}$ from an initial condition value of only 1.5 K (100 km)$^{-1}$. Further strengthening of the cold front was negligible during the second 12-hour period in the forecast but, nevertheless, the 24-hour increase in temperature gradient across it was from 2 K (100 km)$^{-1}$ to 3.5 K (100 km)$^{-1}$.

Considerable strengthening of the mid- and upper-level winds over Texas and Oklahoma occurred during the 12-to 24-hour period in both the observations and forecast (Fig. 15a). A predicted intensification in 500 mb winds over Texas of 39 to 57 m s^{-1} corresponded well with the actual increase of 40 to 55 m s^{-1}. The amplification developed in the frontogenetic region and occurred as a jet streak propagated into an area of lower mean stability.

The 24-hour rainfall forecast (Fig. 15b) indicates some large discrepancies with observations (Fig. 15c). While a reasonably good prediction was made of area-averaged amounts of up to 2 cm over Nebraska, Kansas, eastern Oklahoma and Missouri and also over Wyoming, a forecast maximum associated with an MCC over Mississippi and Alabama of over 5 cm was not observed. This erroneous MCC originates in the first 3 hours of the forecast as a result of a variable data density in the initialization. A too strong first-guess wind field (below 850 mb) was decreased in magnitude by observations over land but not over the data-free Gulf of Mexico, resulting in a zone of strong moisture convergence along the Gulf Coast.

1. Development and Structure of Low-Level Wind Maxima Associated with a Mountain Wave

A low-level wind maximum was predicted to develop over Mexico which, from 12 hours onward, maintains a magnitude of around 40 m s^{-1} at the σ = 0.895 level and 25 m s^{-1} at the σ = 0.96 level (Fig. 14c,d). A maximum in wind speed at the initial time over northern Mexico resulted from the intersection of the terrain-following sigma surface with higher winds in the lower midtroposphere over the high Plateau of Mexico. By 12 hours, this maximum at approximately 800 mb had moved downstream over the Rio Grande Valley (Fig. 16a), where it remained nearly stationary for the rest of the forecast period. This stationary characteristic and the proximity to the strong terrain gradient at the eastern edge of the Mexican Plateau suggested that this wind maximum was associated with a standing mountain wave.

Accordingly, a cross section of the 12-hour model forecast results (Fig. 16a) was produced from south of Baja, California, to southeastern Texas (along line A-B in Fig. 15b). The basic large-scale conditions for mountain waves were present -- strong winds throughout much of the troposphere, perpendicular to a large mountain barrier. (The magnitude of the low-level wind maximum was decreased by 40 percent in another simulation of this case with much weaker winds crossing Mexico in the initial conditions.)

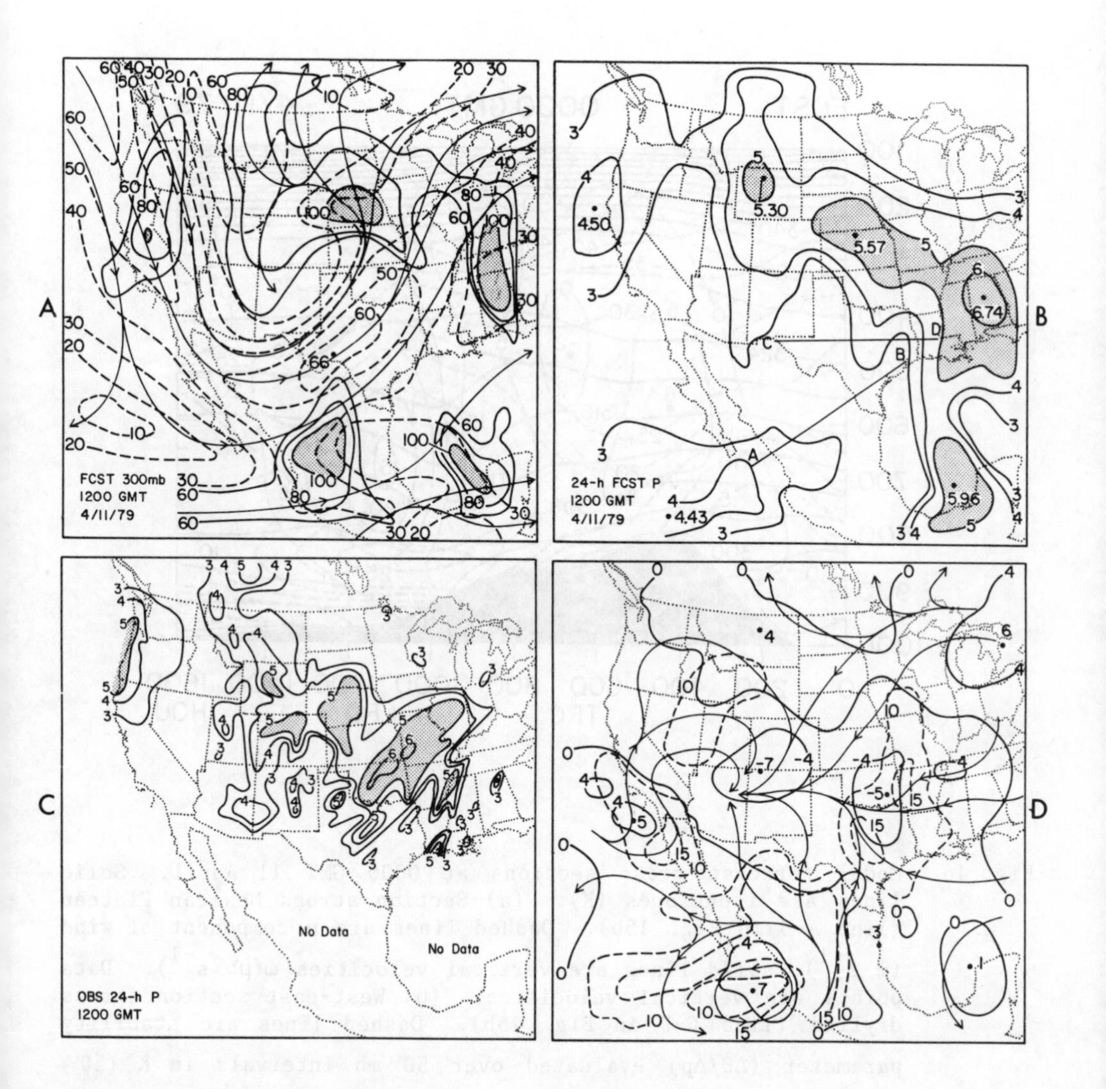

Fig. 15 (a) 24-hour forecast ending at 1200 GMT 11 April of 300 mb streamlines and isotachs (dashed lines in m s^{-1}) and relative humidity (solid lines). Regions over 100 percent are shaded to indicate strong convection. (b) 24-hour forecast rainfall ending at 1200 GMT 11 April, as in Fig. 7a. (c) As in b, except observed. (d) 12-hour surface pressure change (solid lines in mb) ending at 0000 GMT 11 April and ageostrophic flow ($\underset{\sim}{V} - \underset{\sim}{V}_g$) in m s^{-1} at $\sigma = 0.96$ in 12-hour model forecast. Data points are 12-hour pressure changes.

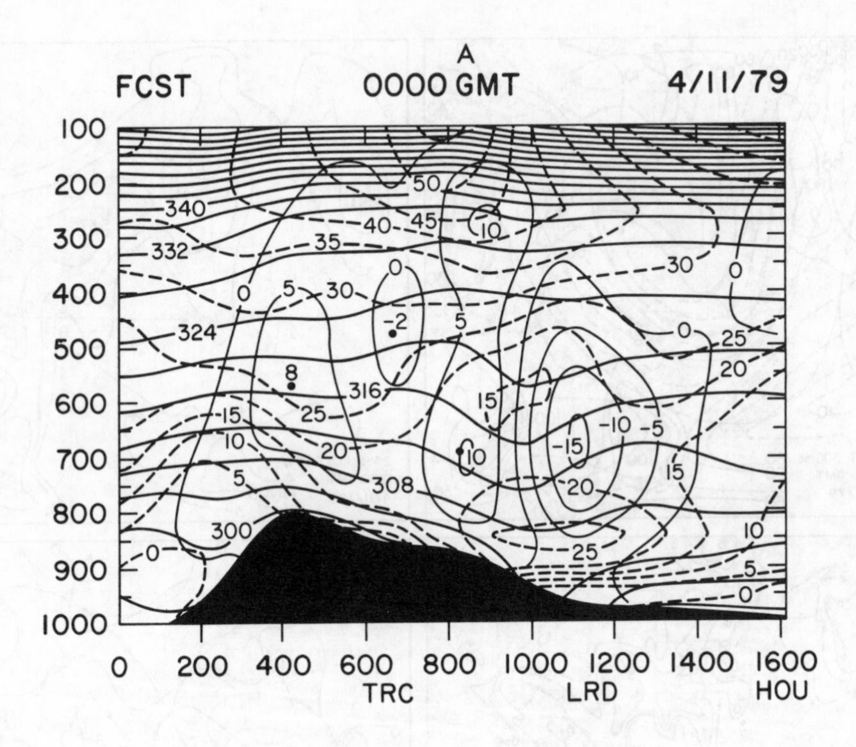

Fig. 16 Model forecast cross sections at 0000 GMT 11 April. Solid lines are isentropes (K). (a) Section across Mexican Plateau (line A-B in Fig. 15b). Dashed lines are u component of wind (m s^{-1}), solid lines are vertical velocities ω(μb s^{-1}). Data points are vertical velocities. (b) West-east section across dryline (line C-D in Fig. 15b). Dashed lines are stability parameter ($\Delta\theta/\Delta p$) evaluated over 50 mb intervals in K (50 $mb)^{-1}$. Solid lines are vertical motion ω in μb s^{-1}.

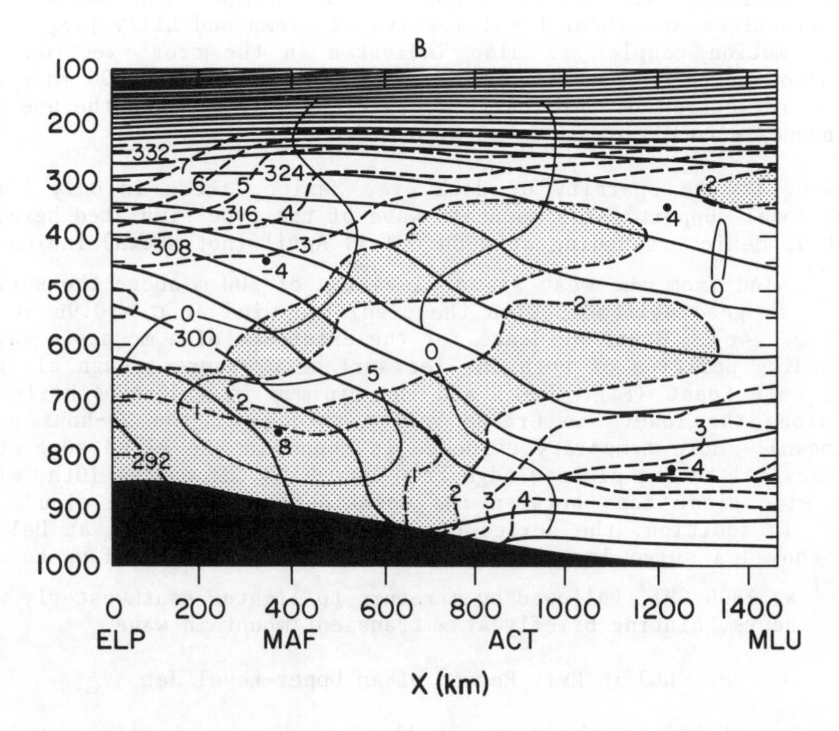

Fig. 16 (Continued)

The model mountain wave wind maximum covers a large horizontal area (400-500 km). This is much larger than that of observed Colorado downslope winds (25-100 km). However, the difference in mountain half-width (much larger for the model terrain on the downslope side of Mexican Plateau than for Colorado front range [20-30 km]) is a likely explanation for this discrepancy. Numerical simulations of mountain waves performed by Klemp and Lilly [14] for a ramp-shaped mountain (resembling a plateau edge) showed strong surface velocities over an area of two times the mountain half-width. Thus, the horizontal extent of the model mountain wave LLJ is in good agreement with their theoretical results.

The perturbation in potential temperature (Fig. 16a) associated with the mountain wave generally resembles those of earlier studies [13, 14]. A lee trough associated with downward motion in the wave is apparent in the model cross section. The absence of a pronounced upward tilt of the wave with height in the model simulation may be related to the low horizontal and vertical resolutions of the model and the use of a reflecting upper boundary condition.

The mountain wave produced by the model most closely resembles the structure of earlier studies in the cross section of westerly (cross-barrier) wind component. A minimum in the middle troposphere is apparent slightly upwind of the low-level maximum, a configuration noted in both the observations and theoretical results of Klemp and Lilly [13, 14]. A vertical motion couplet is also indicated in the cross section. The model also indicates a maximum wind speed between 100 and 200 mb resulting from reflection of vertically propagating waves due to the use of an upper boundary condition of $\dot{\sigma} = 0$ at $\sigma = 0.0$.

Owing to the sparsity of data over Mexico, there is only limited observational support for a mountain wave of the type simulated here. The Chihuahua, Mexico, sounding at 0000 GMT 11 April (not shown) indicates a 25 m s^{-1} wind from the west at an elevation of 500 m above the surface, which is in good agreement with the predicted wind at $\sigma = 0.96$ at this time (Fig. 14c). Other evidence for the existence of a mountain wave is the standing position of both the lee surface pressure trough along the Mexican Gulf Coast (Fig. 2a-c) and the minimum in midtropospheric wind speed along the lower Rio Grande Valley throughout the 24-hour period (not shown). More temporary evidence is found in the cloud edge strikingly parallel to the plateau edge from 1600-2100 GMT on the 10th, with a 300 km wide clear zone between the downslope area and the clouds (not shown). In addition, the series of 3-hour SESAME soundings at Del Rio, Texas, showed a surge in westerly momentum at low levels of up to near 15 m s^{-1} at 1800 GMT, followed by a return to lighter southwesterly winds within 3 hours, hinting briefly at a transient mountain wave.

2. LLJ in Exit Region of an Upper-Level Jet

The third LLJ in the forecast that we discuss is the maximum in southerly flow at $\sigma = 0.96$ which translates from central and eastern Texas at 12 hours and to the Mississippi River Valley by 24 hours. While it develops near the low-level jet in this region at $\sigma = 0.895$ previously discussed, its nonamplifying nature from 12 to 24 hours and different location and direction by 24 hours lead us to treat it separately.

Synoptic analyses indicate that this LLJ is coupled by a transverse circulation to the exit region of the mid- to upper-level jet streak in a manner similar to that described by Uccellini and Johnson [25] and Uccellini [24]. A cross section across the jet streak exit region from the 12-hour model forecast (not shown) revealed evidence of a transverse circulation, with strong rising motion in the midtroposphere over Oklahoma and weak sinking over southeastern Texas. Figure 15d shows the ageostrophic wind field at 12 hours superimposed on the 12-hour surface pressure change. A maximum in the ageostrophic wind (southeasterly in direction) in north-central Texas closely resembles the observed ageostrophic surface flow shown in Fig. 17. The 12-hour change in model surface pressure depicts an area of maximum pressure falls in the vicinity of the observed maximum falls. Both results are consistent with the explanation that the low-level jet underrunning the lid edge in central Texas is an ageostrophic response to the isallobaric field in the exit region of an upper-level jet streak.

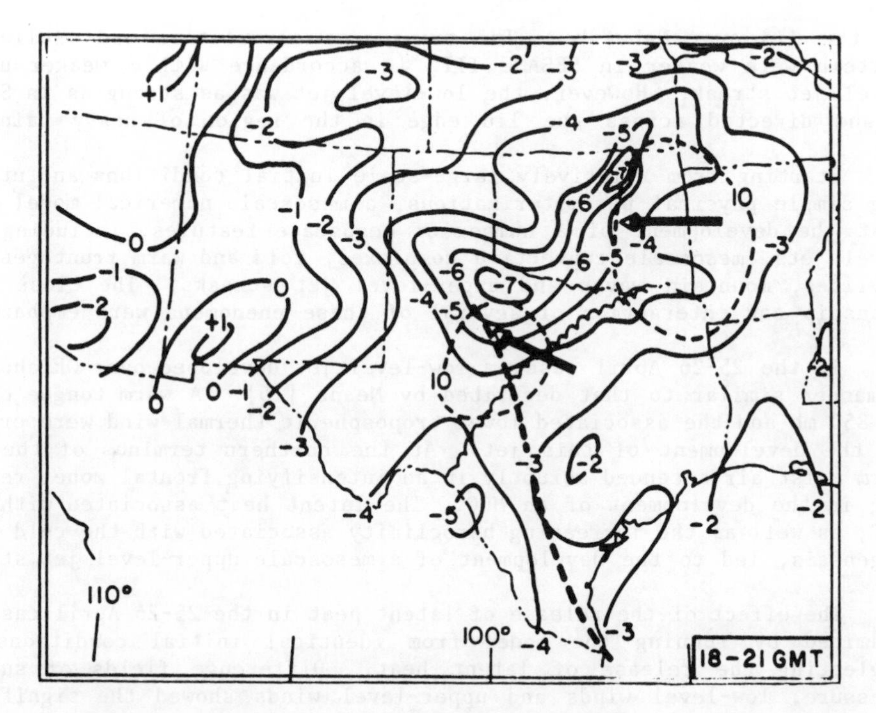

Fig. 17 Observed 3-hour pressure tendencies [solid lines in mb $(3h)^{-1}$] between 1800 and 2100 GMT 10 April 1979. The instantaneous 10 m s^{-1} ageostrophic wind velocity ($\underset{\sim}{V} - \underset{\sim}{V}_g$) isotach for regions east of the dryline at 2100 GMT is dashed, and the direction of the ageostrophic flow is indicated by arrows.

VIII. SUMMARY

Two SESAME cases were analyzed to illustrate aspects of a conceptual model in which orography and differential heating play a dominant role in shaping the severe storm environment. In the conceptual model, the presence of a lid allows the potential instability in the surface layers to increase until the unstable air reaches a point where the lid has been removed. Violent release of that instability occurs not only because the low-level air finds itself in a region of greatly enhanced convective instability where the lid is no longer present, but also because the underrunning process of lid removal is created by transverse circulations across the lid edge and the accompanying vertical motions there serve as a trigger.

In the case of SESAME I, the lid was formed by air previously heated over the Mexican Plateau, whereas in SESAME III the convective activity occurred farther north and the lid was formed by air heated over the southwestern United States desert regions of western Texas, New Mexico and Arizona. Underrunning in SESAME III occurred along the northern edge of the lid over Nebraska. The transverse circulation and isollabaric pattern were weaker in SESAME III, in accordance with a weaker upper-level jet streak. However, the low-level jet was as strong as in SESAME I and directed across the lid edge in the region of heavy rainfall.

Starting from relatively large-scale initial conditions and utilizing simple physical parameterizations, a mesoscale numerical model forecast the development of a number of mesoscale features, including low-level jets, mesoscale convective complexes, cold and warm frontogenesis, drylines, mountain waves and upper-level jet streaks. The close relationship and interaction of several of these phenomena were emphasized.

In the 25-26 April case, a low-level jet developed over Oklahoma in a manner similar to that described by Means [16]. A warm tongue of air at 850 mb and the associated lower tropospheric thermal wind were crucial to the development of this jet. At the northern terminus of the jet, warm moist air ascended abruptly in an intensifying frontal zone, resulting in the development of an MCC. The latent heat associated with this MCC, as well as the increasing baroclinity associated with the cold frontogenesis, led to the development of a mesoscale upper-level jet streak.

The effect of the release of latent heat in the 25-26 April case was examined by running the model from identical initial conditions but neglecting the release of latent heat. Difference fields of surface pressure, low-level winds and upper-level winds showed the significant impact of latent heating for this case. Latent heat resulted in a low-level cyclonic perturbation and an upper-level anticyclonic perturbation in regions of heavy precipitation. The minimum pressure anomaly associated with a perturbation circulation over Florida was 8 mb. Perturbation wind speeds exceeded 15 m s^{-1} in both the lower and upper troposphere.

Strong westerly winds around the base of an intense trough in the 10-11 April case produced a mesoscale mountain wave over the Mexican Plateau. The structure of this wave agreed reasonably well with theory and observations, in spite of the relatively coarse vertical and horizontal resolutions. The maximum low-level wind speed associated with this wave was about 40 m s^{-1}. Although detailed observations were not available over Mexico to verify the magnitude of the wave effects produced by the model, the limited observations that did exist were consistent with the presence of a mountain wave.

In general, the results from these experiments indicate the meso-α scale models with relatively simple physics are capable of generating and then maintaining several observed mesoscale phenomena, even when initialized with synoptic-scale data. Deficiencies in the forecasts, which included large errors in precipitation forecasts over some areas and errors in the low-level temperature structure, were attributed to analysis errors over the oceans and neglect of surface physics, including heating and evaporation. With improved physics and incorporation of satellite data over the data-void regions, it is likely that the major dificiencies in these forecasts can be eliminated.

ACKNOWLEDGMENTS

We acknowledge with thanks the contributions to the interpretation and analyses of these cases by Gregory Forbes, Ying-Hua Kuo, Stanley Benjamin, Yu-Fang Li and Nelson Seaman. Ann Modahl capably typed the manuscript. The major portion of the research was supported by the U.S. Air Force under Grant No. AFOSR-79-0125, by the National Aeronautics and Space Administration (NSG 5205) and the National Science Foundation (ATM78-02699). The three-dimensional computations, the drafting of the figures and the preparation of the manuscript were done by the National Center for Atmospheric Research, which is supported by the National Science Foundation.

REFERENCES

1. Alberty, R.L., D.W. Burgess and T.T. Fujita: Severe weather events of 10 April, 1979. Bull. Amer. Meteor. Soc., 61, 1033-1034 (1980).

2. Anthes, R.A. and D. Keyser: Tests of a fine-mesh model over Europe and the United States. Mon. Wea. Rev., 107, 963-984 (1979).

3. Anthes, R.A. and T.T. Warner: Development of hydrodynamic models suitable for air pollution and other mesometeorological studies. Mon. Wea. Rev., 106, 1045-1078 (1978).

4. Anthes, R.A., Y-H. Kuo, S.G. Benajamin and Y-F. Li: The evolution of the mesoscale environment of severe local storms: Preliminary modeling results. Mon. Wea. Rev., 110 (September issue) (1982).

5. Blackadar, A.K.: Boundary layer maxima and their significance for the growth of nocturnal inversions. Bull. Amer. Meteor. Soc., 38, 283-290 (1957).

6. Browning, K.A. and C.W. Pardoe: Structure of low-level jet streams ahead of midlatitude cold fronts. Quart. J. Roy. Meteor. Soc., 99, 619-638 (1973).

7. Carlson, T.N. and F.H. Ludlam: Conditions for the formation of severe local storms. Tellus, 20, 203-226 (1968).

8. Carlson, T.N., R.A. Anthes, M. Schwartz, S.G. Benjamin and D.G. Baldwin: Analysis of and prediction of severe storms environment. Bull. Amer. Meteor. Soc., 61, 1018-1032 (1980).

9. Colon, J.A.: On the interactions between the southwest monsoon current and the sea surface. Indian J. Meteor. Geophys., 15, 183-198 (1964).

10. Danielsen, E.F.: The relationship between severe weather, major dust storms and rapid large-scale cyclogenesis. Subsynoptic Extra-tropical Weather Systems, National Center for Atmospheric Research, Pb-247 285, ISSN 215-241 (1974).

11. Fritsch, J.M. and R.A. Maddox: Convectively driven mesoscale weather systems aloft. Part I: Observations. J. Appl. Meteor., 20, 9-19 (1981a).

12. Fritsch, J.M. and R.A. Maddox: Convective driven mesoscale weather systems aloft. Part II: Numerical simulations. J. Appl. Meteor., 20, 20-26 (1981b).

13. Klemp, J.B. and D.K. Lilly: The dynamics of wave-induced downslope winds. J. Atmos. Sci., 32, 320-339 (1975).

14. Klemp, J.B. and D.K. Lilly: Numerical simulation of hydrostatic mountain waves. J. Atmos. Sci., 35, 78-107 (1978).

15. Maddox, R.A., D.J. Perkey and J.M. Fritsch: Evolution of upper-tropospheric features during the development of a mesoscale convective complex. J. Atmos. Sci., 38, 1664-1674 (1981).

16. Means, L.L.: A study of the mean southerly wind-maximum in low levels associated with a period of summer precipitation in the middle west. Bull. Amer. Meteor. Soc., 35, 166-170 (1954).

17. Ninomiya, K.: Dynamical analysis of outflow from tornado-producing thunderstorms as revelaed by ATS III pictures. J. Appl. Meteor., 10, 275-294 (1971a).

18. Ninomiya, K.: Mesoscale modification of synoptical situations from thunderstorm development as revealed by ATS III and aerological data. J. Appl. Meteor., 10, 1103-1121 (1971b).

19. Ostby, F.P. and L.F. Wilson: Tornado! Weatherwise, 33, 31-35 (1980).

20. Palmén, E. and C.W. Newton: Atmospheric circulation systems. Academic Press, New York, 603 pp. (1969).

21. Peterson, R.E., J.E. Minor, J.H. Golden and A. Scott: An Australian tornado. Weatherwise, 32, 188-193 (1979).

22. Ramaswamy, C.: On the subtropical jetstream and its role in the development of large-scale convection. Tellus, 8, 26-60 (1956).

23. Schaeffer, J.T.: A simulative model of dryline motion. J. Atmos. Sci., 31, 956-964 (1974).

24. Uccellini, L.W.: On the role of upper tropospheric jet streaks and leeside cyclogenesis in the development of low-level jets in the Great Plains. Mon. Wea. Rev., 108, 1689-1696 (1980).

25. Uccellini, L.W. and D.R. Johnson: The coupling of upper and lower tropospheric jet streaks and implications for the development of severe convective storms. Mon. Wea. Rev., 107, 682-703 (1979).

26. Washington, W.M. and D.P. Baumhefner: A method of removing Lamb waves from initial data for primitive equation models. J. Appl. Meteor., 14, 114-119 (1975).

27. Weston, K.J.: The dryline of northern India and its role in cumulonimbus convection. Quart. J. Roy. Meteor. Soc., 98, 519-531 (1972).

4.5 A COMPARATIVE NUMERICAL SIMULATION OF THE SICHUAN FLOODING CATASTROPHE (11-15 JULY 1981)

Richard A. Anthes and Philip L Haagenson

National Center for Atmospheric Research[1]
Boulder, Colorado 80307 USA

ABSTRACT

In a preliminary test of model dependence in forecasts of the Sichuan flooding event described by Hovermale [4], a 24-hour simulation beginning at 1200 GMT 12 July 1981 was made with a version of the mesoscale model described by Anthes and Warner [3]. The version, referred to here as the NCAR model, is very similar to that used to simulate mesoscale features in two case studies over the United States [2]. The modifications to the model in these simulations are made to conform to the model (designated MFM) used by Hovermale. The domain size (shown in Fig. 1) and vertical and horizontal resolutions are identical to the MFM, and the initial data are obtained in the same way. Major differences from the MFM are:

I. MODEL RESULTS

1. Use of high-resolution terrain rather than smooth terrain (compare Fig. 1 with Fig. 2 in Hovermale [4]).

2. Use of a ∇^4 rather than ∇^2 diffusive term on variables.

3. Method of specifying lateral boundary conditions.

4. Use of staggered horizontal grid.

5. Use of flux form rather than advective finite-difference scheme.

6. Differences in the parameterization of cumulus convection.

7. Possible differences in initialization of the moisture field.

Without further study, it is not possible to say which of the above differences contribute most to the differences in forecasts, although it is quite certain that the different terrains have a large impact. The differences in the cumulus parameterizations, though not fundamental, may be quite important, because previous results have shown the distribution and amount of rainfall to be sensitive to minor variations in the vertical distribution of heating [1].

---

[1] The National Center for Atmospheric Research is sponsored by the National Science Foundation.

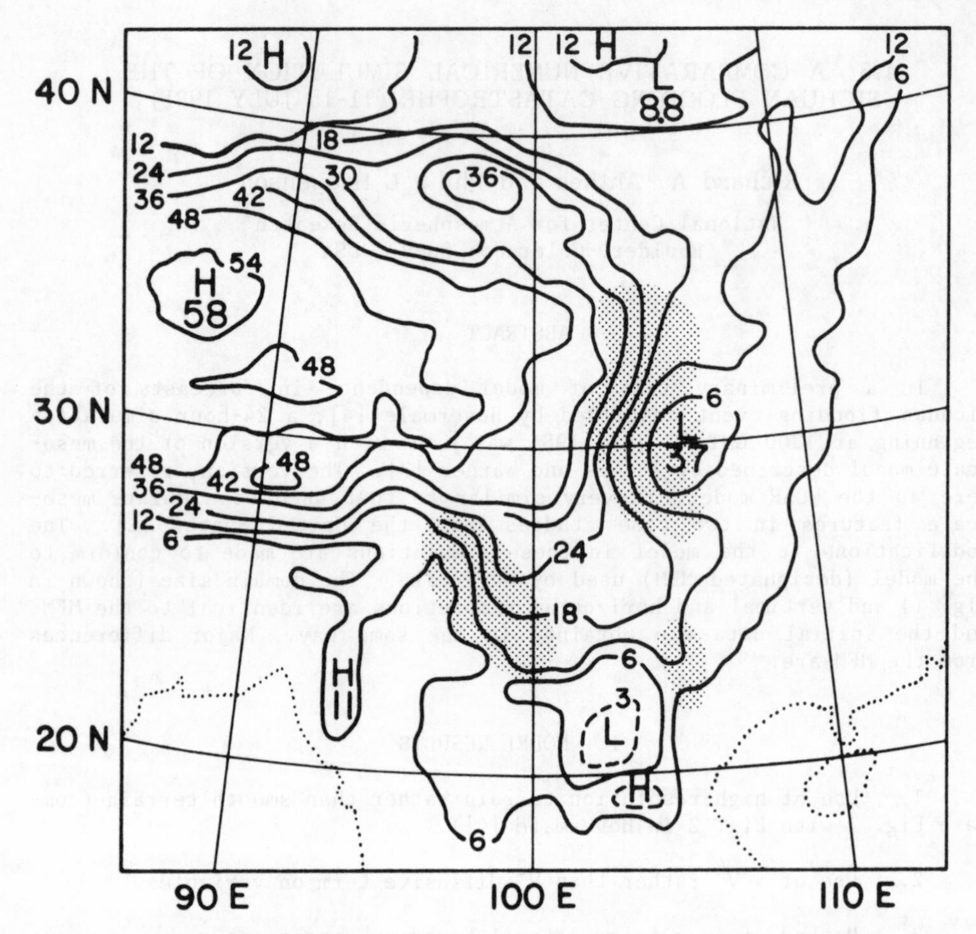

Fig. 1 Terrain elevation (contour interval 600 m) in control simulation. Shaded areas denote regions receiving more than 3 cm of rain in 24 hours in model simulation.

The flow fields in this model develop in much the same way as described by Hovermale [4]. At low levels, hot, very moist (mixing ratios greater than 30 g kg^{-1}) air flows northeastward from the Bay of Bengal into a broad cyclonic circulation over central China (Fig. 2). Heavy rain (greater than 3 cm) is produced in three regions where this air moves upward over sloping terrain. These areas include the slopes along the border between Burma and China, a small region centered at 23°N, 105°E near the border between China and Vietnam, and a large third region over Sichuan Province (Fig. 3). Two of these areas appear to correspond to regions of maximum rainfall predicted by the MFM (shown by dashed lines in Fig. 3). However, the MFM does not predict the small area of heavy rain along the China-Vietnam border. The most likely reason for this difference is the higher resolution in the terrain in the NCAR model; the small ridge in its terrain (Fig. 1) is not present in the MFM model.

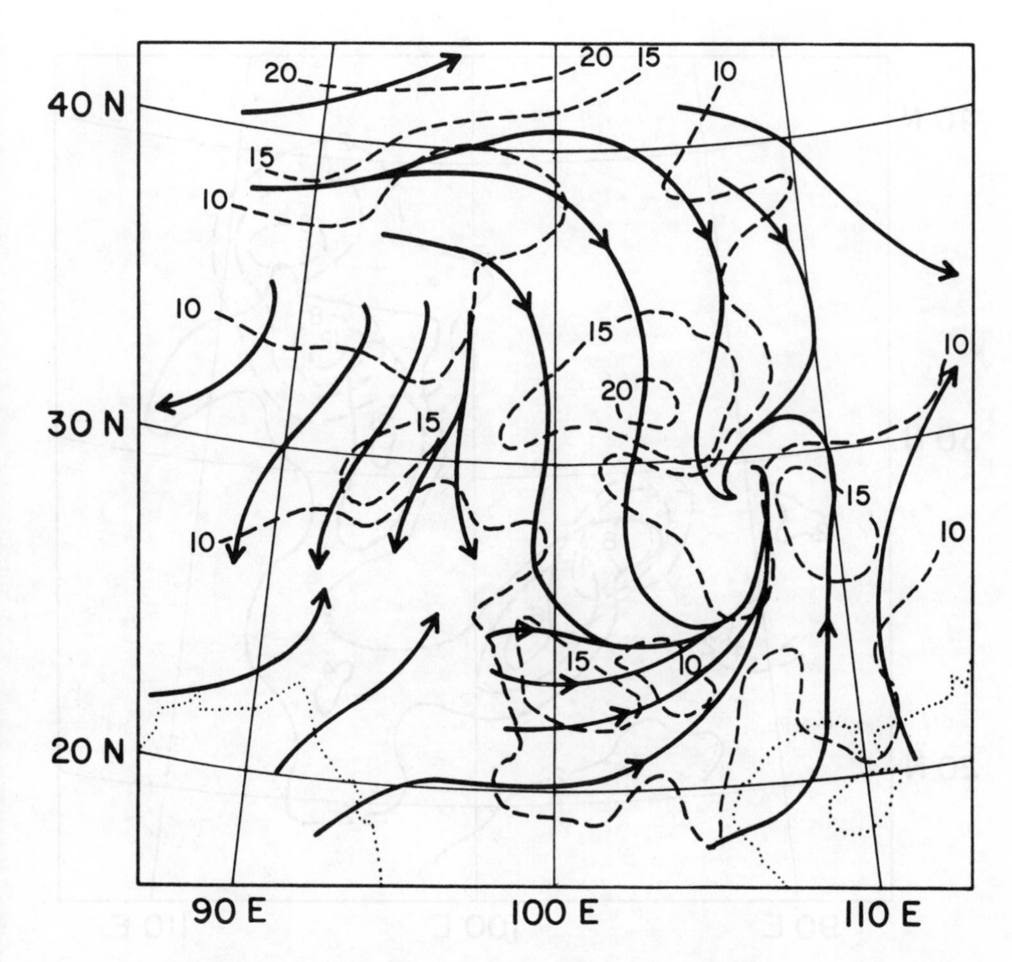

Fig. 2 Streamlines and isotachs (m s^{-1}) in boundary layer (σ = 0.96) at 24 hours of forecast (1200 GMT 13 July 1981).

The differences in location of the other two rainfall maxima are also probably related to the large differences in terrain gradient in the two models. The axes of the elongated rainfall maxima predicted by the NCAR model correspond well with the orientation of the terrain contours in the regions of greatest slope (Fig. 1). Such concentrated zones of steepness are not present in the MFM.

Besides the differences in locations of the rainfall maxima, the two models predict different area-averaged amounts of precipitation, with the NCAR model predicting perhaps two times the amount predicted by the MFM. This is not explained easily by the differences in terrain. In another experiment in which the terrain elevations were reduced to 10 percent of the actual values, the NCAR model predicted similar area-averaged amounts, but with much less mesoscale detail. The differences in total

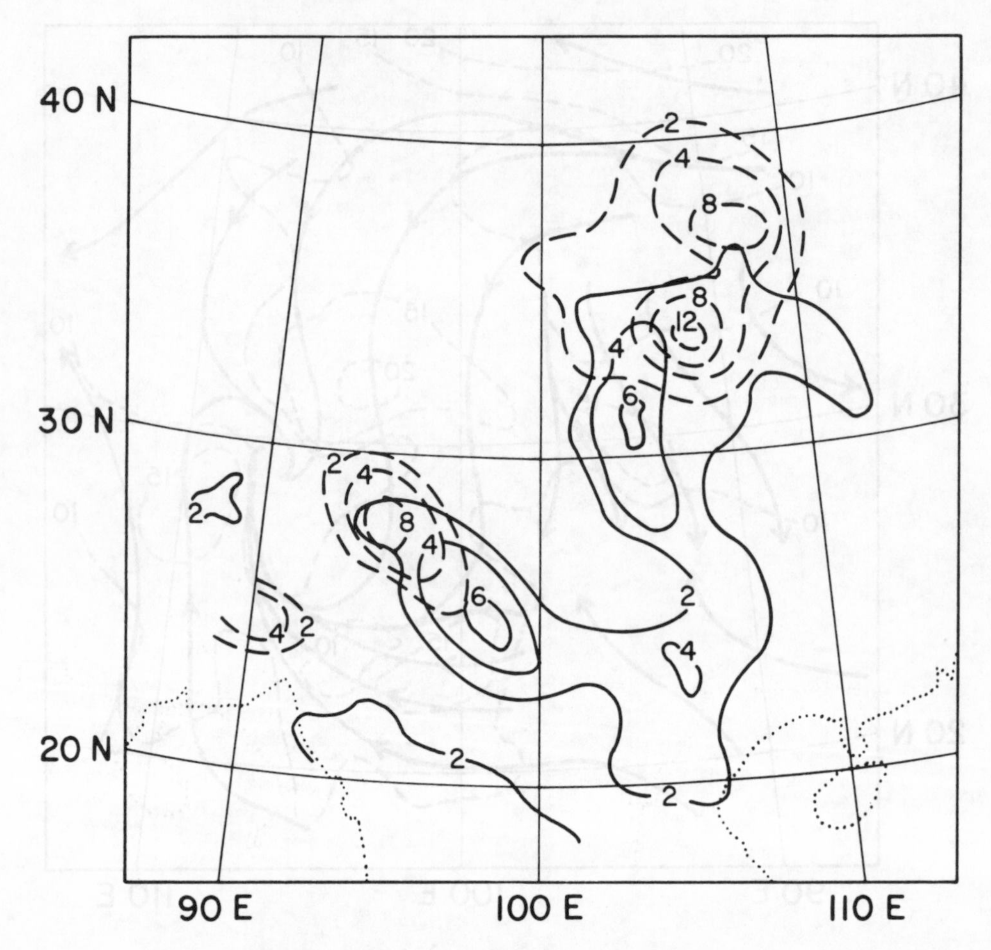

Fig. 3 24-hour precipitation (cm) ending at 1200 GMT 13 July 1981. NCAR model amounts indicated by solid lines, MFM model amounts given by dash lines.

rainfall are more likely related to differences in initial moisture fields and the parameterizations of cumulus convection and friction (both surface drag and horizontal mixing). Further studies should investigate these effects in models which use the same terrain.

An interesting aspect of the simulation in which the terrain elevations were reduced to 10 percent of the values shown in Fig. 1 was the generation of a much more intense cyclone (Fig. 4). The maximum boundary-layer wind was increased from 23.4 to 34.7 m s^{-1}, while the minimum sea-level pressure was reduced from 1001 to 982 mb. Apparently, the high, irregular terrain in the control simulation prevents the organization and intensification of the cyclone.

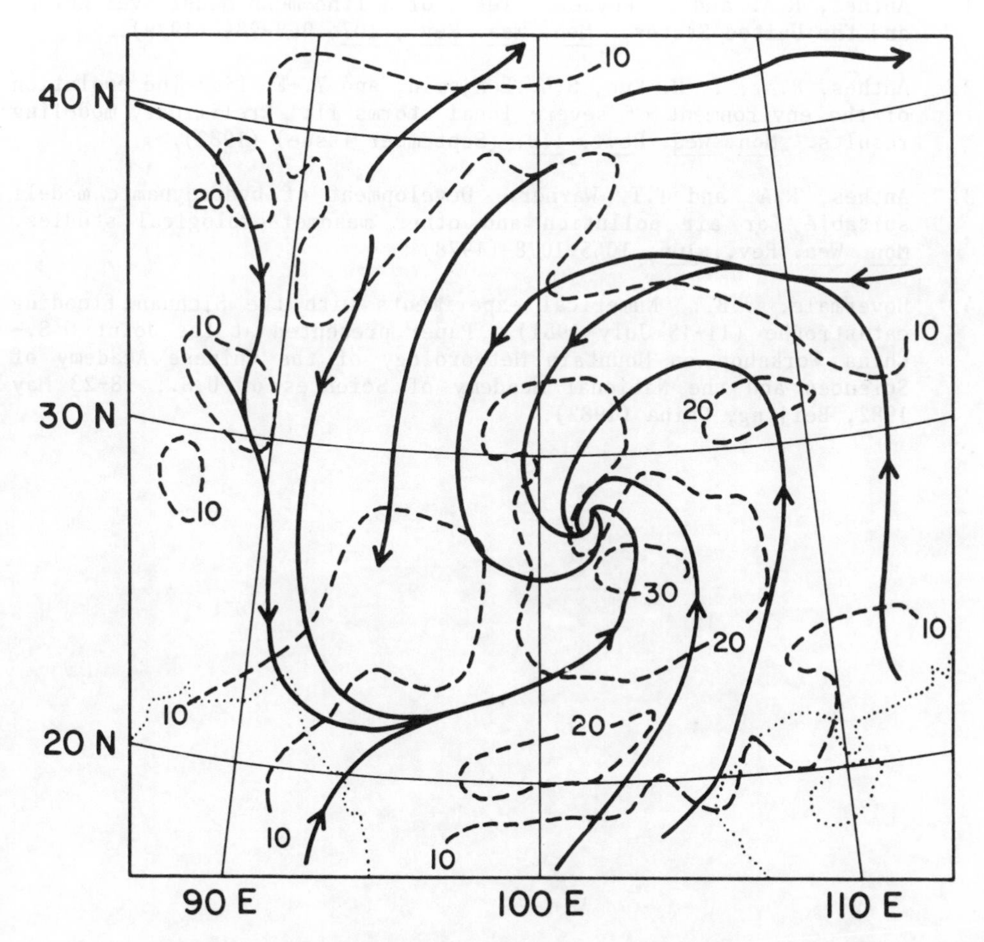

Fig. 4 Streamlines and isotachs (m s^{-1}) in boundary layer (σ = 0.96) at 24 hour of forecast with terrain reduced to 10 percent of elevations in control forecast.

II. SUMMARY

In summary, major differences in terrain and minor differences in numerical and physical aspects between two primitive equation models produce significant differences in location and amounts of heavy precipitation over Asia. At this stage, it is not possible to state which factors besides the terrain are responsible for the differences; it is apparent, however, that terrain plays a major role in this region in determining the rainfall distribution, as well as the intensity and mesoscale structure of the circulations.

REFERENCES

1. Anthes, R.A. and D. Keyser: Tests of a fine-mesh model over Europe and the United States. Mon. Wea. Rev., 107, 963-984 (1979).

2. Anthes, R.A., Y.-H. Kuo, S.G. Benjamin, and Y.-F. Li: The evolution of the environment of severe local storms II: Preliminary modeling results. Mon. Wea. Rev., 110, (September issue) (1982).

3. Anthes, R.A. and T.T. Warner: Development of hydrodynamic models suitable for air pollution and other mesometeorological studies. Mon. Wea. Rev., 106, 1045-1078 (1978).

4. Hovermale, J.B.: Numerical experiments with the Sichuan flooding catastrophe (11-15 July 1981). Paper presented at the Joint U.S.-China Workshop on Mountain Meteorology of the Chinese Academy of Sciences and the National Academy of Sciences of U.S., 18-23 May 1982, Beijing, China (1982).

4.6 TOPOGRAPHIC EFFECTS ON HEAVY RAINFALL IN THE DONGTING LAKE DRAINAGE AREA

Zhang Yan
Institute of Weather and Climatology
Academy of Meteorological Science
National Meteorological Bureau
Beijing, The People's Republic of China

ABSTRACT

The topographic effects on heavy and intense rainfall have been studied by means of the distribution of average annual frequency of heavy rainfall events (≥ 50 mm/day) during April to September from 1951 to 1974, based on the data of 757 rain-guage stations in the Dongting Lake drainage area (24.5°-30°N, 109°-114°E). The following results have been obtained:

1. The low-frequency area of heavy rainfall is located mainly in the lake region.

2. The high-frequency areas are in the mountainous regions. The orientation of high-frequency areas tends to be parallel to the contours of terrain. The centers of high frequency are mostly located on the hillsides of valleys and/or on the highlands. It appears also that the maxima of high frequency increase with elevation from the plains to the high mountain regions about 2500 m above sea level. The area of the highest frequency occurs on the relatively large highlands of 1500-2500 m in the mountainous region northwest of this province.

3. High-frequency centers usually appear near (within 10-30 km) of high mountain peaks over 1000 m elevation.

4. There are some differences between the patterns of heavy and light precipitation in mountain regions. Three types of frequency patterns correspond to the mountain, hill, and lake regions.

5. Besides the airstreams that go around or over the mountains, airstreams through the valleys in mountainous regions of this drainage area can affect the formation of intense rainfall. Therefore, the role of a "through-valley stream" should be considered in the modelling and the computation of airflows in mountainous regions.

I. INTRODUCTION

Dongting Lake, one of the largest lakes in central South China, is located in the middle of the Yangzi Valley. The topographic features are diversified -- mountains, hills, basins, and four large rivers. All four rivers flow into Dongting Lake, which is an important adjustment of the amount of discharge from the Yangzi River.

The Dongting Lake drainage area is situated in the well-known southern Chinese monsoon belt under the influence of flow from the East and South China Seas. These airstreams often contain abundant moisture and, therefore, heavy rainfall frequently occurs, especially in the summer.

From the distribution of precipitation, we find that the heavy rainfall is influenced by topographic conditions, i.e. intense rainfall usually occurs in certain mountainous regions. In contrast, other places, such as basins or rainshadow areas, generally receive lighter rainfall. These mesoscale rainfall patterns are interrelated with the synoptic weather conditions and the topographic features. Some events are related to the distribution of mountain ranges, the slope and steepness of the mountains, the direction of the valley and basins, and the blocking effect of the mountains. Some are associated with orographic wind shear, and coastal fronts and/or mesoscale weather systems, which may also be related to the geomorphological conditions.

These factors are always interlinked with synoptic-scale airflows and weather systems, and it is difficult to differentiate the effect of each component. Many previous studies have considered the atmospheric factors. But a more precise understanding of the genesis of heavy rainfall by simple weather analysis is difficult because of the above interactions. I suspect that topography plays a major role in producing heavy rainfall. In this paper, the distribution and characteristics of the annual frequency of heavy rainfall are studied. The influence of various moving weather systems can be deduced by taking an annual average of frequencies over a long period. In this way, we can isolate the topographic effects on heavy rainfall.

II. DATA

The precipitation records of 757 hydrological and meteorological stations in the Dongting Lake drainage area (24.5°-30°N, 109°-114°E) for a period of 24 years (1951-1974) have been selected for discussion. As the heavy rainfall in this area occurs mainly in the summer half-year (April to September), the total number of heavy rainfall events (⩾ 50 mm/day) in these six months is assumed equal to the annual number of heavy rainfall events. In this paper, "heavy rain" is defined as precipitation of 50-99 mm/day, and "intense rain" is precipitation ⩾ 100 mm/day.

Hydrological stations are spaced every 10 km along the river and 20-30 km apart in other places. Meteorological stations are located every 30-40 km. The time series of available records of each station are not the same, so we use the annual frequency (total number of events/ total number of years) for discussion. The records which are too short in time series, such as two to three years, are used for reference.

III. DISTRIBUTION AND CHARACTERISTICS OF HEAVY RAINFALL EVENTS

According to the areal distribution of the annual frequency of heavy rainfall, the areas of low frequency occur mainly in the lower elevation, such as in the lake area and large valleys and in the Hengyang and Shaoyang Basins. The high-frequency areas are mainly in the higher mountain ranges. The annual frequency of heavy rainfall in mountainous regions is two to three times greater than that of the lower plains. The ratio is still higher for the intense rain events (Figs. 1 and 2).

The high-frequency areas coincide closely with the mountain ranges (Fig. 3a). Over relatively isolated mountains, the isofrequency lines are nearly closed around the mountain. In the mountain ranges, the isofrequency lines are oriented parallel to the topographic contours, as shown in Fig. 3b.

From the lake area to high mountain regions, the annual frequency of heavy rainfall increases with elevation. The highest frequency centers are often near the highest mountain peaks, such as upstream of the Li River in the northwest part and at the southeast slope of Zhenbaoding in the southwest part of the drainage area (Figs. 1 and 2). These areas of high frequency are associated with the altitude and steepness of the mountains.

Near the centers of high frequency, there are always high (above 1000 m) mountain peaks within 10 to 30 km. The exact position of the centers of high frequency may be influenced by local topographical effects or atmospheric conditions and, therefore, do not always coincide with the highest mountain peaks. In addition, the analyzed position may be influenced by the spacing of rain-gauge stations, because at most of the highest mountain peaks there are no such stations and hence no rainfall records.

IV. DISTRIBUTION OF HEAVY RAINFALL IN DIFFERENT TOPOGRAPHIC AREAS

Three different topographic regions exist in the Dongting Lake drainage area -- mountains, hills, and lake and plains. The annual frequency of heavy rainfall in each of these regions is quite different (Fig. 4).

1. Mountain regions are found mainly on the borderlands of Hunan Province, i.e. in the west, south, and east of this drainage area, with an average elevation of 600-1000 m. The highest mountains are located in the northwest, southwest, and east of this drainage area, and the altitude of the peaks is about 1500-2500 m. The annual frequency of heavy rainfall is highest around these peaks. The frequency of heavy rain is about 4 to 6 times per year, and the frequency of intense rain is 1 to 1.8 times per year (Fig. 4). The gradient of isofrequency lines is larger by a factor of two to three than that in the hill regions.

2. Hill regions are located between the mountains and the lake and plains. The average elevation of these regions is 200-300 m, and the altitude of the hill peaks is about 500-1500 m. The frequency of heavy rain is 2.5 to 4 times per year, and the frequency of intense rain is 0.4

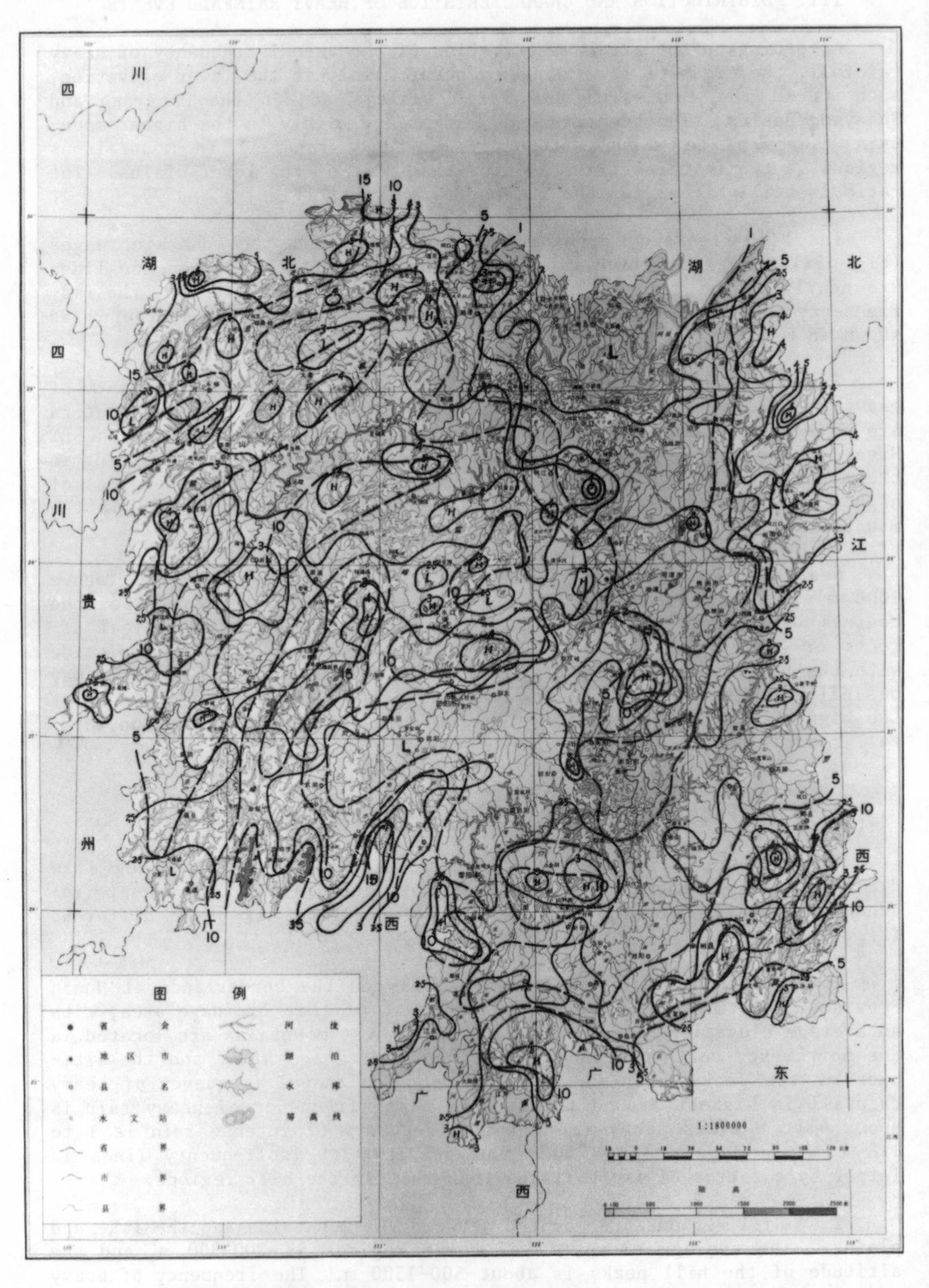

Fig. 1 Average annual frequency of heavy rain (50-99 mm/day) of Dongting Lake drainage area (1951-1974) shown by the solid contours (2.5, 3, 4, 5, and 6). The dashed contours are smoothed terrain elevations in hundreds of meters.

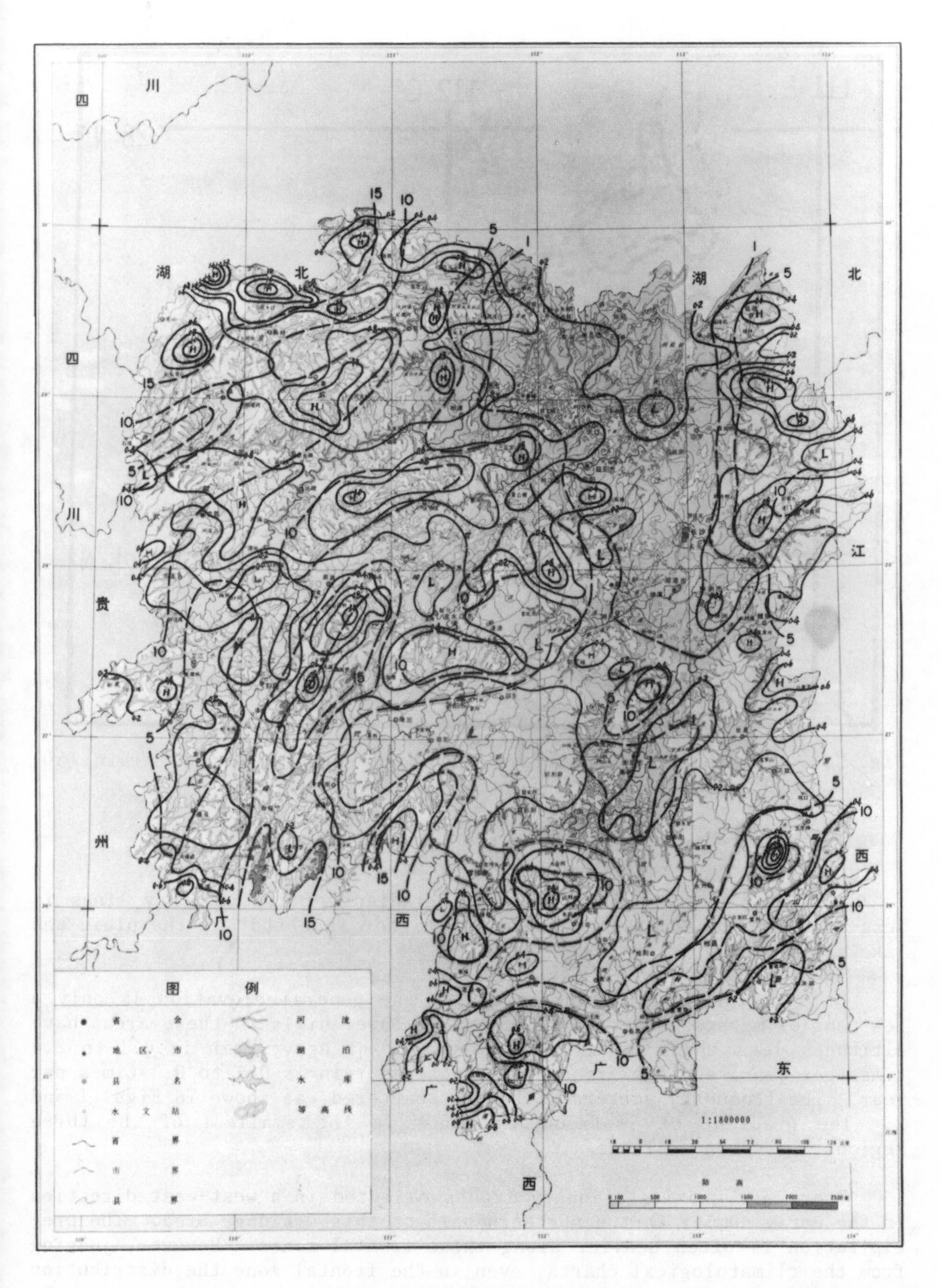

Fig. 2 Average annual frequency of intense rain (≥ 100 mm/day) of Dongting Lake drainage area (1951-1974) shown by the solid contours (0.2 to 1.6 and the contour interval is 0.2). The dashed contours are smoothed terrain elevations in hundreds of meters.

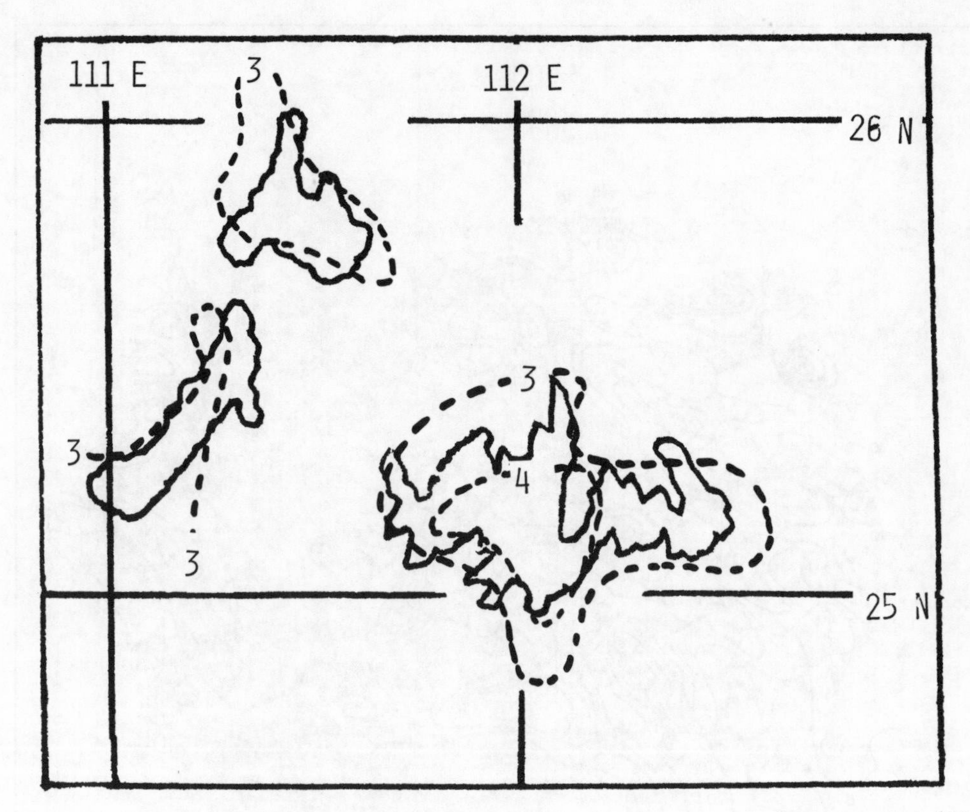

Fig. 3a Distribution of average annual frequency of heavy rain (50-99 mm/day).

to 0.9 time per year (Fig. 4). The gradient of isofrequency lines is less than in the mountain regions but higher than that in the plain and lake areas.

3. In the lake and plain areas, the general elevation is only a few tens of meters above sea level. Some lower hills in these areas have altitudes less than 500 m. The frequency of heavy rain is 0.1 to 2.4 times per year, and the frequency of intense rain is 0.1 to 0.3 times per year. The frequency pattern is rather scattered, as shown in Figs. 1 and 2. The gradient of isofrequency lines is the smallest of the three regions, as shown in Table 1.

There are quasi-stationary fronts oriented in a west-east direction in the early summer in the northern part of this drainage area. The precipitation is often heavier along these frontal zones. However, judging from the climatological charts, even in the frontal zone the distribution of the annual frequency of heavy events is influenced by the topography of this region, with the maxima aligned with the mountains and the lake

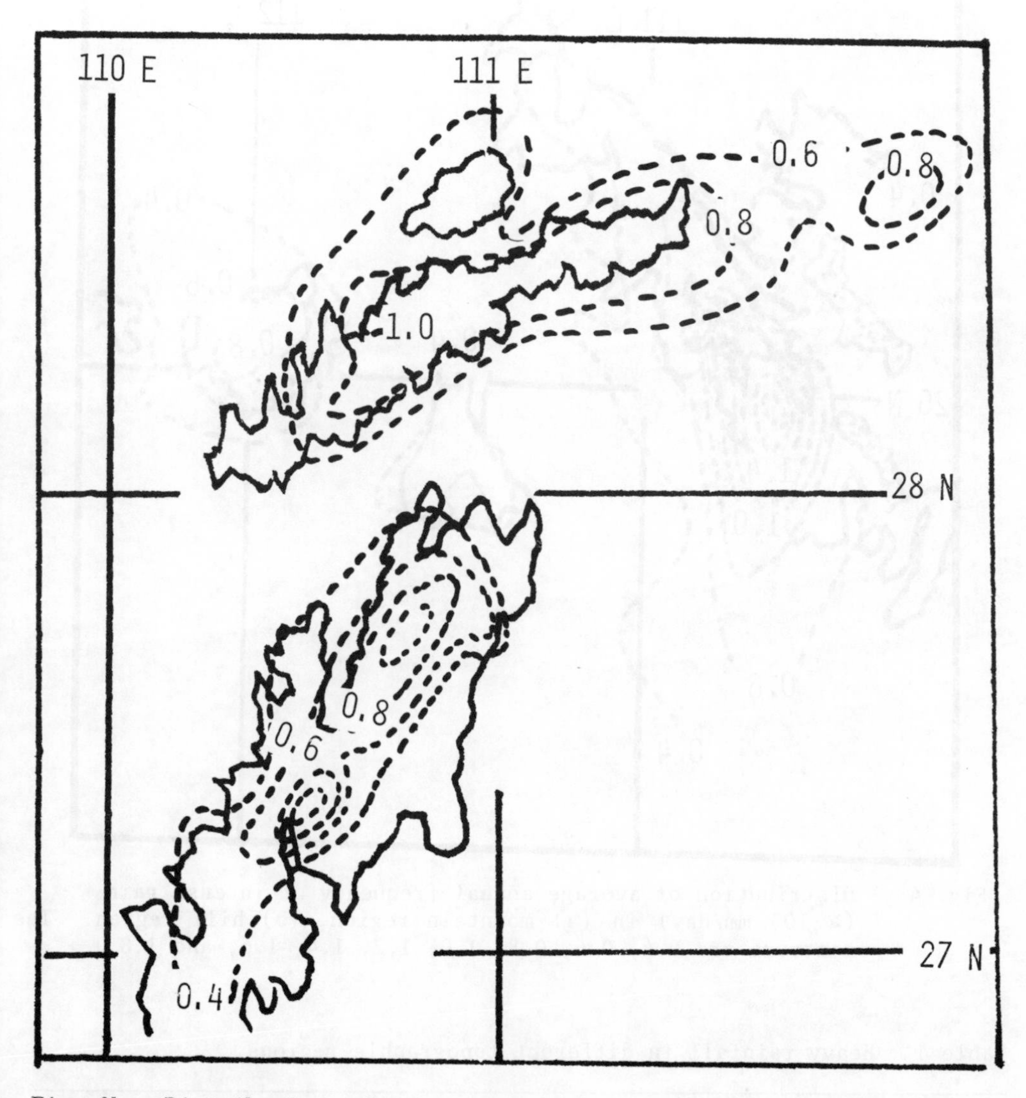

Fig. 3b Distribution of average annual frequency of intense rain (≥ 100 mm/day). The contours are 0.4, 0.6, 0.8, and 1.0.

in the northern part of this drainage area. The maximum frequency occurs in the northwest high mountainous region, and the minimum frequency appears in the northeast part of the great lake. Thus, the maximum does not coincide with the climatological frontal zone, which shows that topography affects the location of heavy rainfall more significantly than the climatic position of the front.

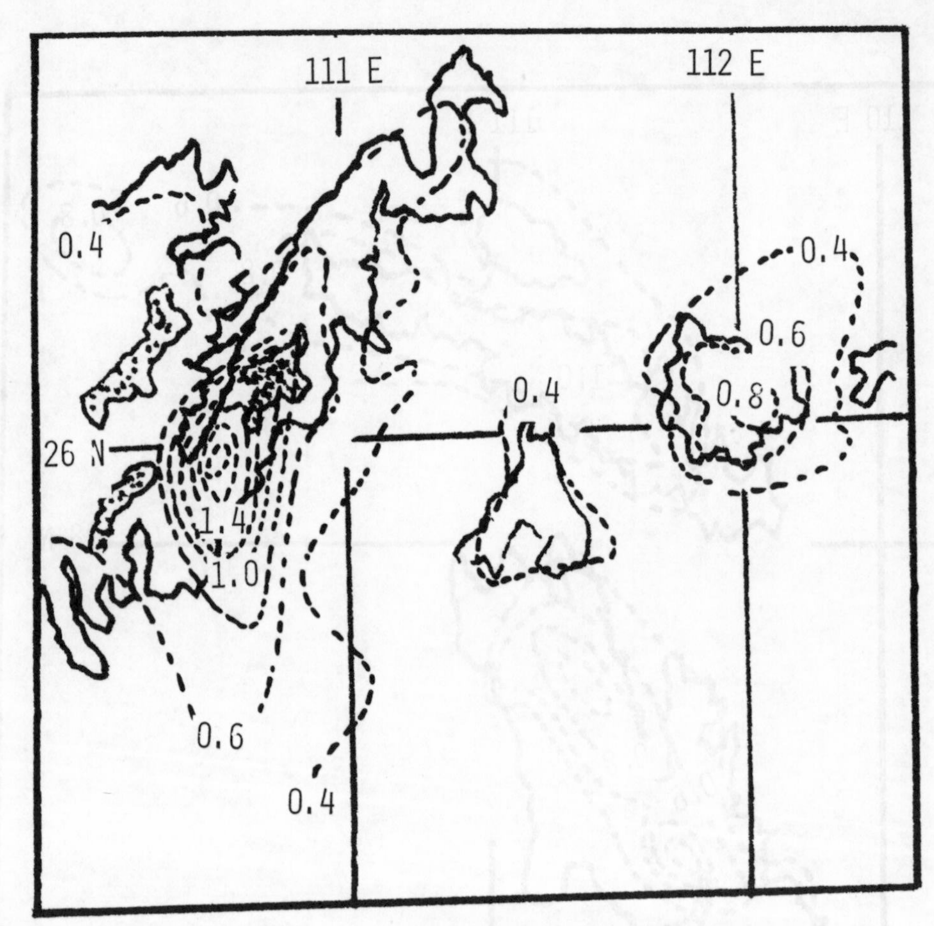

Fig. 4 Distribution of average annual frequency of intense rain (⩾ 100 mm/day) in (a) mountain region, (b) hill region. The contours are 0.4, 0.6, 0.8, 1.0, 1.2, 1.4, 1.6, and 1.8.

Table 1 Heavy rainfall in different topographic regions.

TYPE	ELEVATION OF PEAKS (M)	ANNUAL FREQ. HEAVY RAIN	(TIMES/YR) INTENSE RAIN	GRADIENT OF ISOFREQUENCY LINE
Mountain region	1500-2500	4-6	1.8	Largest
Hill region	500-1500	2.5-4	0.4-0.9	Medium
Lake and plain	500	0.1-2.4	0.1-0.3	Small and haphazard

The topographic effects on the annual frequency of heavy rainfall of different intensity in these three regions are summarized in Table 2. This table indicates that the higher the elevation, the more frequent are intense rainfall events, at least up to 2000 m in this middle latitude.

In some studies on severe rainfall events using the "omega" equation, the results show that the topographic effects are comparatively small. The results of this study strongly contradict those theoretical results, which likely apply only to much larger scales of motion.

Table 2 Proportion of annual frequency of heavy rainfall.

	LAKE AND PLAIN	HILL REGION	MOUNTAIN REGION
Heavy rain	1	2	3
Intense rain	1	3	6

V. "THROUGH-VALLEY STREAMS"

In the mountainous regions, whether on the upwind side or in the lee relative to the large-scale wind field, there are considerable high-frequency areas on the slopes of both sides of the valleys. These features are more conspicuous on the higher and steeper mountain slopes along a relatively long and narrow valley, but still can be observed in low mountain regions. In mountain ranges, high-frequency centers usually occur on the upwind sides, i.e. on the southern slopes. However, higher frequencies also occur on facing slopes of parallel mountain chains, such as by the Li, Yan, and Zhi Rivers. Some of these maxima are more evident than those on the upwind sides (Fig. 5).

By the narrow tube (Bernoulli) effect, the wind velocity might be intensified with airflows passing through a valley. The wind may also be deflected by the mountains, and thus the local wind direction in a valley may differ significantly from the large-scale wind field.

Because the cross-section diameters of the valleys vary, the divergence, convergence, and vorticities are also quite variable. At some places, they may be intensified and hence increase the upward velocity. Therefore, I considered the possibility that the high frequency of heavy rainfall on both sides of mountain slopes along rivers may be caused by the through-valley streams.

The prevailing wind direction at valley stations is mostly along the general trend of the valley (Table 3 and Fig. 6) and is usually oblique to the general flow field during the period of heavy rain.

Table 3 Prevailing wind during heavy rainfall.

STATION	YEAR	NO. OF HEAVY RAINFALL EVENTS	WIND FREQUENCY DURING HEAVY RAINFALL										PREVAILING WIND NO.	%
Anhua	1965-1970	29	C	N	NNE	W	NE	NEE	E	NNW	SEE	NWW	N through NEE	
			71	14	5	4	3	3	3	2	1	1	25	69
Sangzhi	1965-1970	26	C	NE	NEE	NNE	N	SE	SSE	NW	S		N through NEE	
			60	10	7	4	3	3	2	2	2		24	73
Shimen	1965-1969	22	E	C	NEE	NWW	W	NE	SW	SWW	NNE	SEE	NNE through E	
			15	13	10	10	8	6	6	4	3	2	34	53
Yuanling	1964-1970	32	C	NE	NEE	NEE	SW						NEE through NEE	
			60	29	28	2	2						59	97
Wugang	1965-1970	13	C	NNE	N	NNW	SWW	NE	SW				N through NE	
			12	12	10	3	3	2	2				24	75
Xuefeng Shan	1977-1980	15	SEE	SE	S	NNW	SW	NWW					S through SE	
			24	8	7	2	2	2					29	87

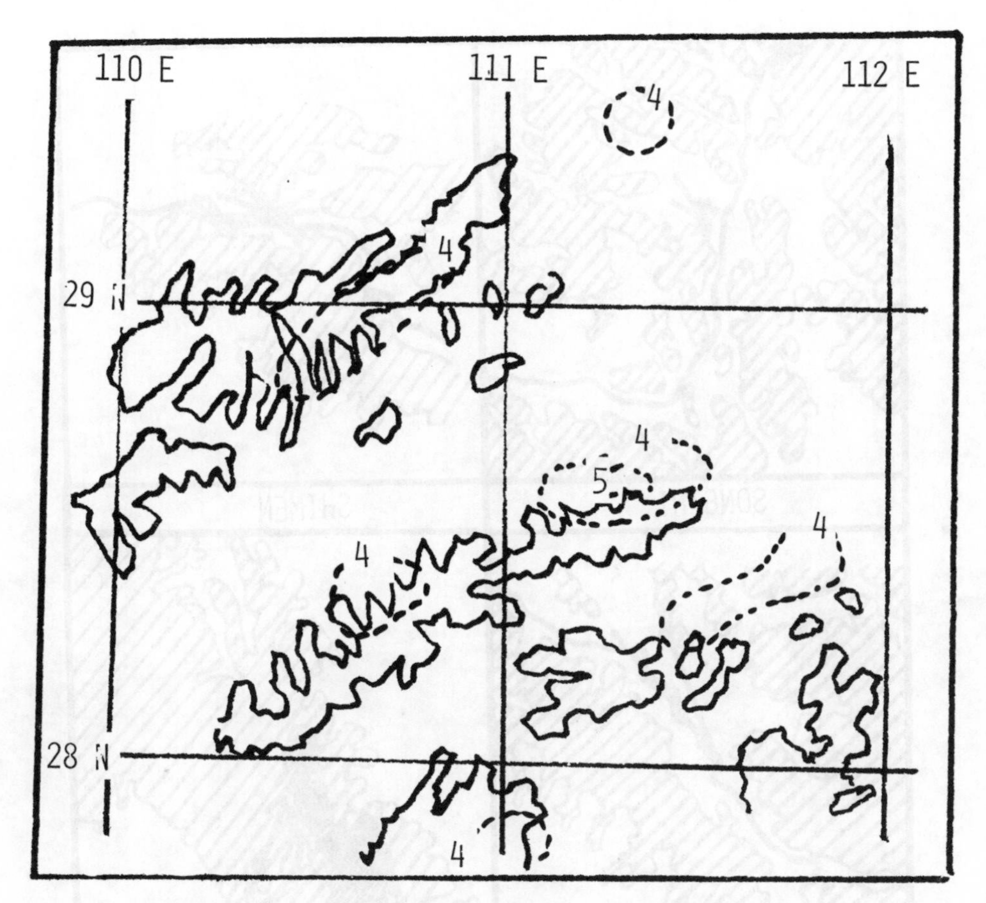

Fig. 5 Distribution of average annual frequency of heavy rain (50-99 mm/day) along rivers. The contours are 4 and 5.

Moreover, from the distribution of the annual frequency along rivers, we found that the frequency of heavy rainfall events increases in the upstream direction where the altitude is higher. Examples include the Li, Xiang, Liuyang, Miluo, Mi, and Lu Rivers (Fig. 7). These upstream increases may be caused by several factors, including through-valley streams, general upslope motion, and the blocking effect of mountains, all preventing the motion of convective clouds and leading to heavy rainfall repeatedly at almost the same place.

These data show that, along the valleys in the inner parts of mountainous regions, there is intense rainfall on both sides of the slopes. This result is meaningful to the forecasting of flooding and the estimation of maximum heavy rainfall in a water reservoir design. Therefore, in addition to the around-mountain and upslope flows as usually discussed, the through-valley flows should be considered in modelling or calculating heavy rainfall in mountainous regions.

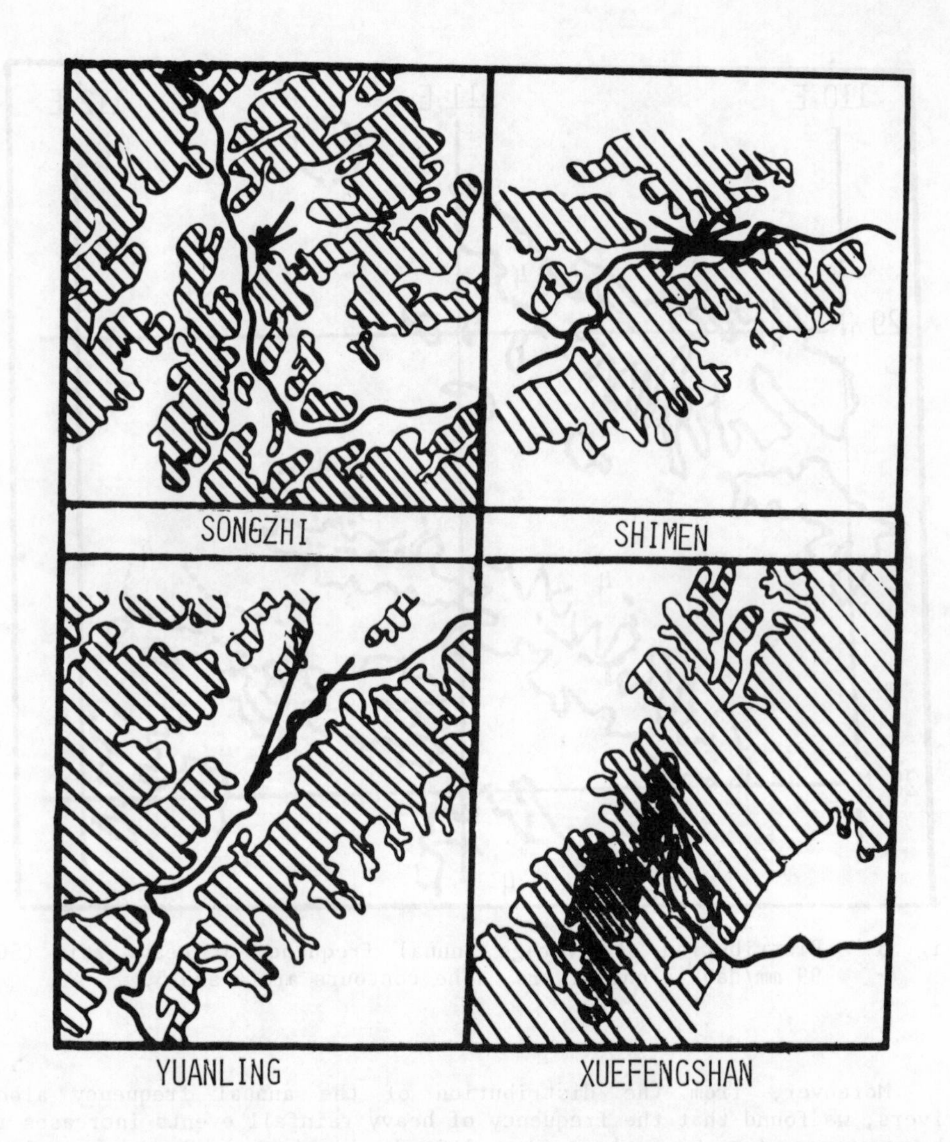

Fig. 6 Prevailing wind at valley stations during periods of heavy rain.

VI. SUMMARY

The main features of the distribution and characteristics of heavy rainfall in the Dongting Lake drainage area are:

1. The distribution of annual frequency of heavy rainfall events is variable in different topographic regions: (a) low in plain and lake areas, (b) medium in hill (low mountain) regions, and (c) high in mountain regions. The proportion of annual frequency of heavy rain events in

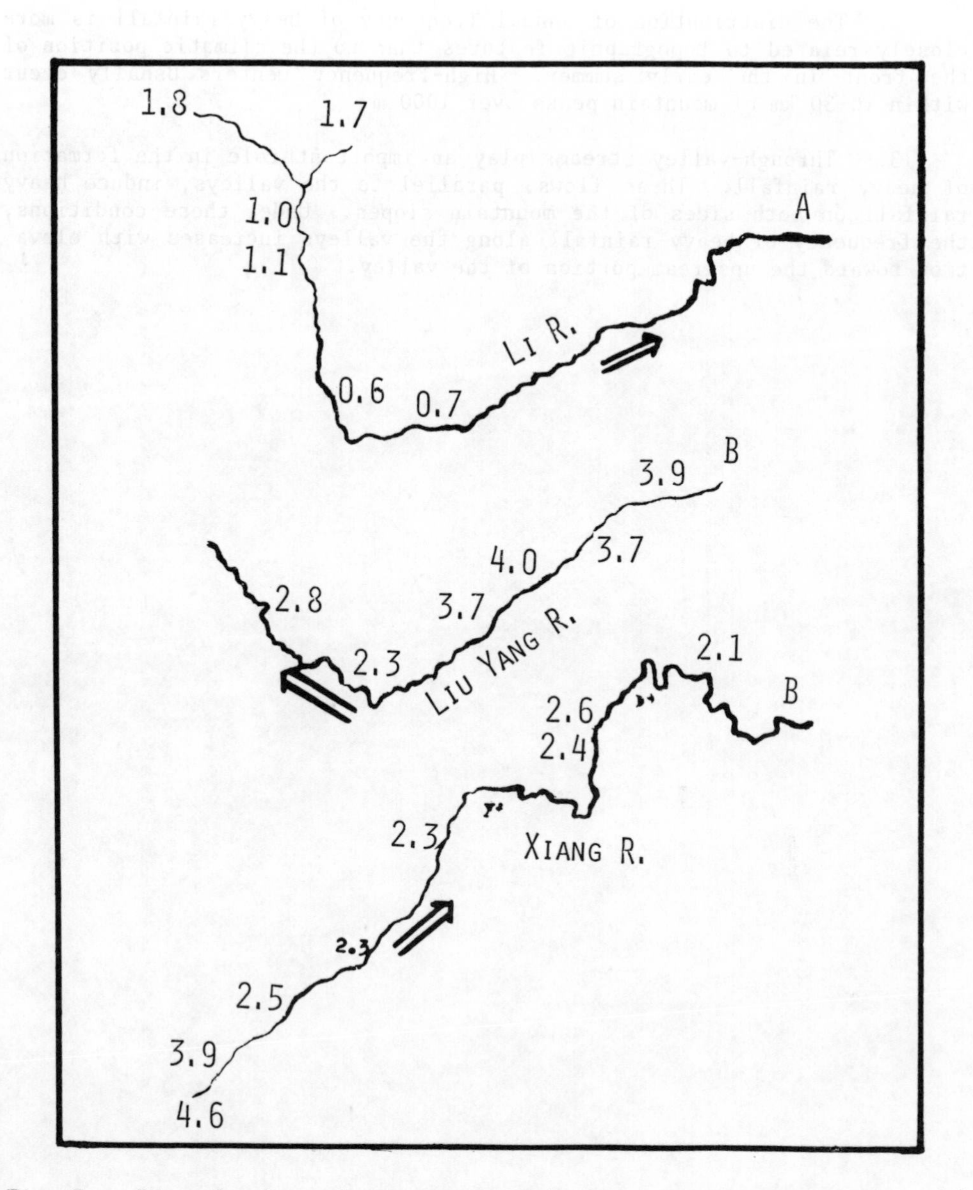

Fig. 7 Distribution of average annual frequency of heavy rain (a) ≥ 100 mm/day, (b) 50-99 mm/day, along rivers.

these three regions is roughly 1:2:3, and the proportion of intense rain events is about 1:3:6. The increase of frequency of intense rain with elevation is thus greater than for the heavy rain events. The gradient of the isofrequency lines is smallest in the plain and lake areas and largest in the high mountain regions.

2. The distribution of annual frequency of heavy rainfall is more closely related to topographic features than to the climatic position of the front in the early summer. High-frequency centers usually occur within 10-30 km of mountain peaks over 1000 m.

3. Through-valley streams play an important role in the formation of heavy rainfall. These flows, parallel to the valleys, induce heavy rainfall on both sides of the mountain slopes. Under these conditions, the frequency of heavy rainfall along the valleys increases with elevation toward the upstream portion of the valley.

4.7 ON THE GRAVITY WAVE OF JET TYPE IN THE PLANETARY BOUNDARY LAYER AND THE FORMATION OF HEAVY RAIN

Yu Zhihao
Department of Meteorology
Nanjing University
The People's Republic of China

and

Chen Liangdong
Institute of Meteorology
Chinese People's Liberation Army Air Force
The People's Republic of China

ABSTRACT

The gravity wave is an important trigger mechanism in the formation of heavy rain. Especially when accompanying a boundary-layer jet, gravity waves can produce convergence and ascent which may trigger the formation of strong convection and heavy rain. In this paper, by using the linear equations governing motion in the boundary layer of a Boussinesq fluid, we decompose air motion into the Ekman frictional flow and the disturbance flow of the gravity wave. Therefore, two sets of equations governing the motion are introduced. Under the condition of the low-level jet, the phase velocity of the gravity wave, including the effects of the boundary layer, is obtained as

$$C_g = -\bar{v} \pm \sqrt{\sigma g/(k^2 + f/2A)}$$

The heavy rain in the Meng-Shan region in Guangxi during 26-27 May 1978 is analyzed. There is good agreement between the theoretical results and the observational analyses.

I. INTRODUCTION

In the formation of heavy rain, the mesoscale gravity wave has been increasingly considered an important trigger mechanism [4]. Recently, Li [2] solved the equations governing the motion of the free atmosphere with the Boussinesq approximation and noted that the phase difference between the pressure wave and vertical velocity wave is $\pi/2$. This result is in agreement with the observation that heavy rain occurs usually between the mesoscale low and high. But this theoretical solution for gravity waves was obtained under the assumption that the stratification of the whole troposphere (depth H_o = 10 km) is conditionally unstable. Also, the order of magnitude of the updraft in the theoretical study is about 10 m s^{-1} which is that of an updraft of violent convection and is not merely a weak value appropriate for a trigger mechanism. Further study is therefore needed.

The occurrence of heavy rain in southern China is always related to the activities of mesoscale systems [5] and to the formation of the boundary layer (300-900 m) jet [1]. It may be expected that the occurrence of a boundary-layer jet will provide at least two favorable conditions for the formation of heavy rain. First, the boundary-layer jet may be an efficient conveyor belt of water vapor to form the heavy rain. Second, the convergence and ascending flow of gravity waves in the boundary layer may be important mechanisms to trigger the strong convection. In this paper, the latter will be discussed. In discussing the wave-CISK in the tropics, Lindzen [3] pointed out that the initial triggering action of convergence and ascent in CISK is not necessarily the frictional convergence proposed by Charney; it may also be associated with a gravity wave. However, the gravity wave proposed by Lindzen is caused by the forcing of latent heat at certain levels in the free atmosphere. In this paper, the gravity wave of jet type in the boundary layer, which is present under adiabatic conditions, is different from Lindzen's [3] and Li's [2]. The gravity wave may also trigger strong convection and heavy rain. Because boundary-layer jets are common, the gravity wave of jet type may occur often if there is a suitable stratification.

The boundary-layer equations include terms of turbulent viscosity such as $A(\partial^2 u/\partial z^2)$. These are second-order partial differential terms. So far, few papers dealing with the problem of waves in the boundary layer have appeared in the literature. Here we use the linear equations governing the motion in the boundary layer for a Boussinesq fluid and decompose the motion into a steady Ekman frictional flow, which includes the frictional and rotational effects and the residual disturbance flow of the gravity wave. We obtain not only the frequency equation for a gravity wave, but also the formula of phase velocity including the factors A and f. In the last part of this paper, we analyze the heavy rain event in the Meng-shan region of the Guangxi autonomous region on 26-27 May 1978 by using the theoretical results obtained in the first part.

II. THE GOVERNING EQUATIONS OF THE BOUNDARY LAYER AND THE SOLUTION FOR THE GRAVITY WAVE OF JET TYPE

Generally, the gravity wave of jet type in the boundary layer is a mesoscale system associated with strong convection and heavy rain, and thus it is not proper to make the hydrostatic approximation. We use the linear boundary-layer equations of the adiabatic flow of a Boussinesq fluid as follows:

$$\frac{\partial u_1'}{\partial t} + \bar{v}\frac{\partial u_1'}{\partial y} - fv_1' = -\frac{\partial p_1'}{\partial x} + A\frac{\partial^2 u_1'}{\partial z^2}$$

$$\frac{\partial v_1'}{\partial t} + \bar{v}\frac{\partial v_1'}{\partial y} + fu_1' = -\frac{\partial p_1'}{\partial y} + A\frac{\partial^2 v_1'}{\partial z^2}$$

$$\frac{\partial w_1'}{\partial t} + \bar{v}\frac{\partial w_1'}{\partial y} - g\theta_1' = -\frac{\partial p_1'}{\partial z} \tag{1}$$

$$\frac{\partial u_1'}{\partial x} + \frac{\partial v_1'}{\partial y} + \frac{\partial w_1'}{\partial z} = 0$$

$$\frac{\partial \theta_1'}{\partial t} + \bar{v}\frac{\partial \theta_1'}{\partial y} + w_1'\sigma = 0$$

where $\bar{v}$ = constant is the basic flow of the southeast jet stream in the boundary layer; all variables with primes are disturbed quantities, and disturbed flow variables (u_1', v_1', w_1') actually are disturbed momentum, i.e.

$$(u_1', v_1', w_1') = \bar{\rho}(u', v', w')$$

where $\bar{\rho}$ is basic density and is only dependent on z. The density profile satisfies the static equation $\partial\bar{p}/\partial z = -\bar{\rho}g$. Similarly, the disturbed potential temperature is

$$\theta_1' = \bar{\rho}\,\frac{\theta'}{\bar{\theta}}$$

where $\bar{\theta}$ is the basic potential temperature corresponding to static stability $\sigma \equiv \partial \ell n\bar{\theta}/\partial z$, $f = 2\Omega\sin\phi$ is the Coriolis parameter and $A/\bar{\rho}$ is the eddy viscosity. The remaining are conventional symbols. For convenience, the superscript "′" of the disturbed variabile in Eqs. (1) is omitted later.

Since Eqs. (1) are linear, the disturbed flow in the boundary layer may consist of the steady Ekman frictional flow and the flow of the gravity wave, i.e.

$$\begin{bmatrix} u' \\ v' \\ w' \\ p' \\ \theta' \end{bmatrix} = \begin{bmatrix} u'_W(t,x,y,z) \\ v'_W(t,x,y,z) \\ w'_W(t,x,y,z) \\ p'_W(t,x,y,z) \\ \theta'_W(t,x,y,z) \end{bmatrix} + \begin{bmatrix} u'_E(x,y,z) \\ v'_E(x,y,z) \\ w'_E(x,y,z) \\ p'_E(x,y,z) \\ \theta'_E(x,y,z) \end{bmatrix} \tag{2}$$

where the variables with subscript "W" and "E" are the flow of the gravity wave and Ekman flow, respectively. The Ekman flow is related to the large-scale flow of the free atmosphere; both have the same time scale. The Ekman flow changes with time slower than the gravity waves, so we can make use of the assumption that the Ekman flow is steady.

$$\frac{\partial}{\partial t}(u', v', w') = \frac{\partial}{\partial t}(u'_W, v'_W, w'_W) \tag{3}$$

and all the frictional effects are considered in the Ekman flow:

$$A\frac{\partial^2}{\partial z^2}(u', v') = A\frac{\partial^2}{\partial z^2}(u'_E, v'_E) \tag{4}$$

In Ekman flow, the Coriolis, pressure gradient and frictional forces are balanced and the advection terms are neglected; then

$$\bar{v}\frac{\partial}{\partial y}(u', v') = \bar{v}\frac{\partial}{\partial y}(u'_W, v'_W) \tag{5}$$

Substituting Eqs. (2) into Eqs. (1) and with Eqs. (3) to (5), we obtain the equations of Ekman flow and equations governing the motion of gravity waves, respectively. The equations of Ekman flow are

$$-fv'_E = -\frac{\partial p'_E}{\partial x} + A\frac{\partial^2 u'_E}{\partial z^2} \tag{6a}$$

$$fu'_E = -\frac{\partial p'_E}{\partial y} + A\frac{\partial^2 v'_E}{\partial z^2} \tag{6b}$$

$$\bar{v}\frac{\partial w'_E}{\partial y} - g\theta'_E = -\frac{\partial p'_E}{\partial z} \tag{6c}$$

$$\frac{\partial u'_E}{\partial x} + \frac{\partial v'_E}{\partial y} + \frac{\partial w'_E}{\partial z} = 0 \tag{6d}$$

$$\bar{v}\frac{\partial \theta'_E}{\partial y} - w'_E\sigma = 0 \tag{6e}$$

As usual, we assume that the horizontal pressure gradient is independent of z in the Ekman layer and is equal to the pressure gradient at the top of the Ekman layer, i.e.

$$- \frac{\partial p_E'}{\partial x} = - \frac{\partial p_E'}{\partial x}\bigg|_{h_E} = - f\bar{\rho} v_g = 0$$

$$- \frac{\partial p_E'}{\partial y} = - \frac{\partial p_E'}{\partial y}\bigg|_{h_E} = f\bar{\rho} u_g = 0$$

where $h_E = \pi(2A/f)^{\frac{1}{2}}$ is the height of the top of the Ekman layer. From the first two equations of Eqs. (6), we obtain the classic solution of the frictional flow in the Ekman layer

$$u_E' = u_{g1}[1 - \exp(- \frac{\pi}{h_E} z) \cos(\frac{\pi}{h_E} z)]$$

$$v_E' = u_{g1} \exp(- \frac{\pi}{h_E} z) \sin(\frac{\pi}{h_E} z) \tag{7}$$

where u_{g1} is the momentum of geostrophic wind at the top of the Ekman layer. Substituting Eqs. (7) into Eqs. (6d), we may easily obtain the vertical motion due to frictional convergence. It can be seen from Eq. (6e) that there must be horizontal temperature advection to balance the adiabatic cooling. We only assume that the horizontal pressure gradient is independent of z. However, p_E' is a function of z and can be calculated from Eq. (6c).

Moreover, if we consider a one-dimensional gravity wave along the direction of the basic flow $\bar{v}$, i.e. $u_E' \equiv 0$, the equations will be

$$\frac{\partial v_W'}{\partial t} + \bar{v} \frac{\partial v_W'}{\partial y} = - \frac{\partial p_W'}{\partial y} \tag{8a}$$

$$\frac{\partial w_W'}{\partial t} + \bar{v} \frac{\partial w_W'}{\partial y} - g\theta_W' = - \frac{\partial p_W'}{\partial z} \tag{8b}$$

$$\frac{\partial v_W'}{\partial y} + \frac{\partial w_W'}{\partial z} = 0 \tag{8c}$$

$$\frac{\partial \theta_W'}{\partial t} + \bar{v} \frac{\partial \theta_W'}{\partial y} + w_W'\sigma = 0 \tag{8d}$$

When we derive these equations from Eqs. (1), the effects of f on the mesoscale gravity wave is neglected. Equations (8) have the following wave-type solution

$$\begin{bmatrix} v'_W \\ w'_W \\ p'_W \\ \theta'_W \end{bmatrix} = \begin{bmatrix} v^*(z) \\ w^*(z) \\ p^*(z) \\ \theta^*(z) \end{bmatrix} \exp[i(\nu t + ky)] \exp(-\frac{x^2}{\alpha^2}) \tag{9}$$

where $\exp(-x^2/\alpha^2)$ represents the profile of disturbance flow, the axis of the jet stream coincides with y, α is the half-width of jet, and ν and k are the frequency and wave number of the gravity wave, respectively.

Substituting Eq. (9) into Eqs. (8), we obtain

$$i(\nu + \bar{v}k)\, v^* = - ikp^* \tag{10a}$$

$$i(\nu + \bar{v}k)\, w^* - g\theta^* = - \frac{dp^*}{dz} \tag{10b}$$

$$ikv^* + \frac{dw^*}{dz} = 0 \tag{10c}$$

$$i(\nu + \bar{v}k)\, \theta^* + \sigma w^* = 0 \tag{10d}$$

From Eqs. (10a-d), we have

$$\frac{d^2v^*}{dz^2} + (\frac{\pi}{H})^2\, v^* = 0 \tag{11}$$

where

$$\frac{\pi^2}{H^2} \equiv k^2\, [\frac{\sigma g}{(\nu+\bar{v}k)^2} - 1] \tag{12}$$

From the no-slip lower boundary condition and Eqs. (1), we get the boundary condition for v'_W as follows:

$$v' = v'_W + v'_E = 0 \quad \text{at } z = 0$$

We have $v'_E = 0$ at $z = 0$ in Eqs. (7), so that

$$v'_W = 0 \quad \text{at } z = 0$$

or (13a)

$$v^* = 0 \quad \text{at } z = 0 .$$

The gravity wave flow should also satisfy the viscous condition corresponding to Eq. (13a) at the lower boundary, although we have previously assigned the viscous effects to the Ekman flow in Eq. (4). Now we assume again that the viscous stress in the gravity wave flow near the lower boundary is proportional to the stress of the Ekman flow

$$\frac{\partial v'_W}{\partial z} \propto \frac{\partial v'_E}{\partial z} \quad \text{at } z = 0$$

or from Eqs. (7), we have

$$\frac{dv^*}{dz} = A\pi \langle u_g \rangle h_E^{-1} \tag{13b}$$

where A is the proportionality constant and $\langle u_g \rangle$ is a mean value of u_g over time and space.

The solutions of Eqs. (11), (10a), (10c) and (10d) under the boundary conditions (13a) and (13b) are

$$v^*(z) = v^*_{max} \cos \frac{\pi}{H}(z - \frac{H}{2}) \tag{14a}$$

$$w^*(z) = - \frac{ikH}{\pi} v^*_{max} \sin \frac{\pi}{H}(z - \frac{H}{2}) \tag{14b}$$

$$p^*(z) = - \frac{(\nu+\bar{v}k)}{k} v^*_{max} \cos \frac{\pi}{H}(z - \frac{H}{2}) \tag{14c}$$

$$\theta^*(z) = \frac{k\sigma H}{\pi(\nu+\bar{v}k)} v^*_{max} \sin \frac{\pi}{H}(z - \frac{H}{2}) \tag{14d}$$

where $v^*_{max} \equiv AH\langle u_g \rangle/h_E$. Equations (14) or (9) represent solutions of the jet-type wave. To represent the solution of the jet type with subscript "J", we have

$$v^*_J(z) = v^*_{max} \cos \frac{\pi}{H}(z - \frac{H}{2}) \tag{15}$$

When we set the axis of the jet stream at the half-height of the boundary layer, i.e.

$$v^*_J = v^*_{max} \quad \text{at } z = \frac{h_E}{2} \tag{16}$$

from Eq. (15)

$$H = h_e .$$

Substituting this expression into Eq. (12), the frequency equation is

$$k^2\left[\frac{\sigma g}{(\nu+\bar{v}k)^2} - 1\right] = \frac{\pi^2}{h_E^2} \tag{18}$$

Because $h_E = \pi(2A/f)^{\frac{1}{2}}$, the phase speed from Eq. (18) may be written

$$C_g = -\bar{v} \pm \sqrt{\sigma g/(k^2+f/2A)} \tag{19}$$

This is a formula for the phase speed of a gravity wave in the boundary layer, because it was obtained from the condition of jet type in the boundary layer, and it includes the depth of the Ekman layer. When the basic flow is absent ($\bar{v} = 0$) and the frictional effect A is neglected, Eq. (19) becomes almost coincident with Li's [2] formula for the phase velocity of a one-dimensional gravity wave without the effect of f. In comparing Eqs. (14a) and (14b), we find that the phase of the horizontal flow v_W' lags behind the vertical motion wave by $\pi/2$. Figure 1 schematically represents the flow of the jet-type gravity wave in the boundary layer. By using Eqs. (14a), (14b), and (15), we can deduce the order of magnitude of the ratio of vertical and horizontal velocity in the gravity wave,

$$|w_J^*(z)|/|v_J^*(z)| \sim \frac{2h_E}{L_y} \tag{20}$$

where $L_y = 2\pi/k$ is the horizontal wave length of the gravity wave. If we treat the gravity wave as a mesoscale system, then $L_y \sim 200$ km and $h_E \sim 1$ km. In this case the order of magnitude of the ratio vertical and horizontal velocity is 10^{-2}. According to Eq. (20), if $|v_J^*| \sim 10\ \mathrm{m\ s^{-1}}$, the $|w_J^*| \sim 10\ \mathrm{cm\ s^{-1}}$, the vertical disturbance flow of jet type in the boundary layer can reach or exceed the order of magnitude of ascent in large-scale systems.

Since

$$k^2 << \pi^2 h_E^{-2} \tag{21}$$

then Eq. (19) is approximately

$$C_g \simeq -\bar{v} \pm \sqrt{\sigma g/\pi^2 h_E^{-2}} \tag{22}$$

Under typical stability stratification in the boundary layer, we have

$$\sigma = \frac{\partial\ \ell n\ \bar{\theta}}{\partial z} \sim 10^{-5}\mathrm{m}^{-1};\ g \sim 10\ \mathrm{m\ s^{-2}};\ \pi^2 h_E^{-2} \sim \pi^2\ 10^{-6}\ \mathrm{m}^{-2}$$

and therefore

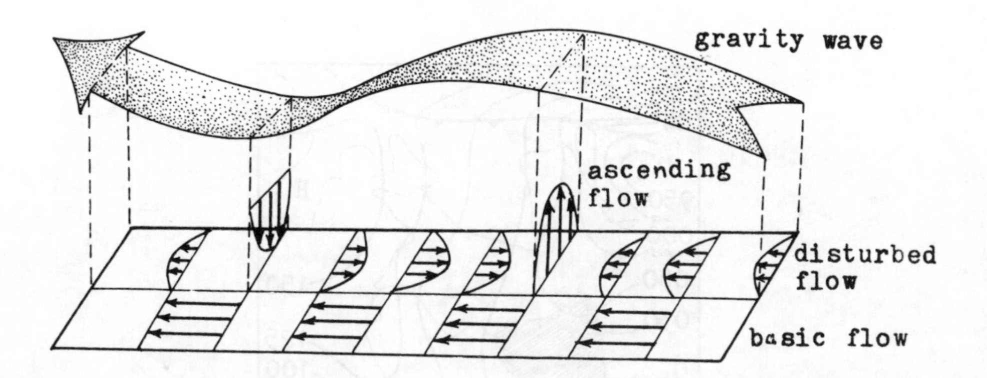

Fig. 1 Schematic representation of the horizontal and vertical flow for the gravity wave of the jet type in the boundary layer.

$$\sqrt{\sigma g/\pi^2 h_E^{-2}} \lesssim 10 \text{ m s}^{-1}, \qquad (23)$$

which indicates that the phase velocity of the gravity wave is of the same order of magnitude as that of the basic flow $\bar{v}$.

III. THE HEAVY RAIN IN THE MENGSHAN REGION IN GUANGXI AUTONOMOUS REGION DURING 26-27 MAY 1978

Chen et al. [1] analyzed the heavy rain event of 26-27 May 1978 in the northern part of Guangxi Autonomous Region. Figure 2 shows the sea-level pressure and precipitation at 0000 GMT, 26 May. A large region of precipitation is occurring around the southeastern part of a cyclone at the 700 and 850 mb level (not shown here) in southwest China and on the eastern part of the surface trough. The heaviest rain region, located in northern Guangxi, is shown in detail in Fig. 3. There are three precipitation centers in excess of 300 mm for the entire rain event. Two of the centers -- Xingan (332.1 mm) and Rongshui (300.9 mm) -- are on the southern slopes of the large-scale topography of northern Guangxi. The third (306.2 mm) is in the Mengshan region to the east of Liuzhou city. In this paper, we emphasize the third region; an analysis of the entire heavy rain event can be found elsewhere [1].

Figure 4 is a time cross section of the wind at the Guilin, Wuzhou and Nanning stations on 26-27 May 1978. It can be seen from this figure that there is quasi-steady southwesterly flow above 1500 m and southeasterly flow below 900 m. This southeast flow in the boundary layer exists for 30-48 hours over Guangxi. Its horizontal and vertical distribution has the characteristics of a jet (Fig. 5). The height of the jet axis is about 600 m, and its axis lies between Liuzhou and Mengshan from south to north close to Mengshan. Although the wind direction of the southeasterly jet in the boundary layer was steady state, its strength and horizontal extent decreased gradually during the

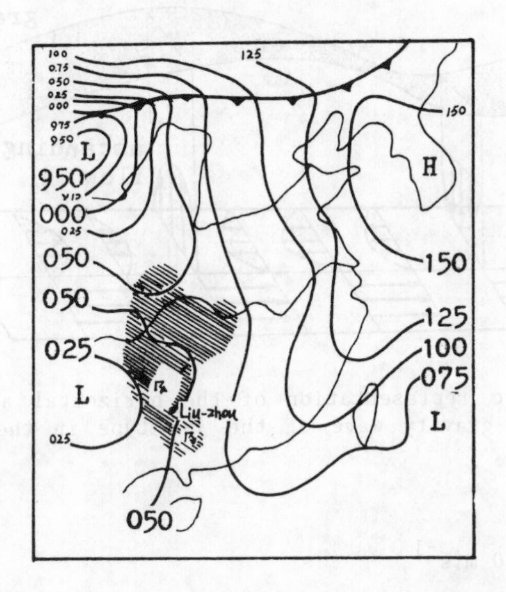

Fig. 2 Sea-level pressure and precipitation region (shaded) at 0000 GMT, 26 May 1978.

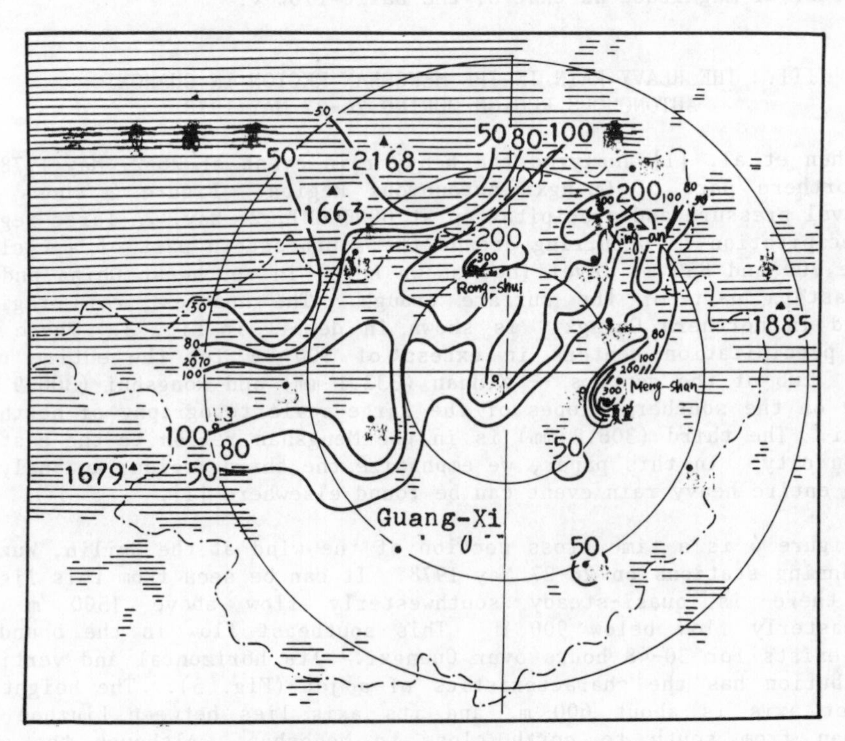

Fig. 3 Total precipitation (mm) on 26-27 May 1978 in Guangxi autonomous region. Hatching indicates terrain elevations above 750 m.

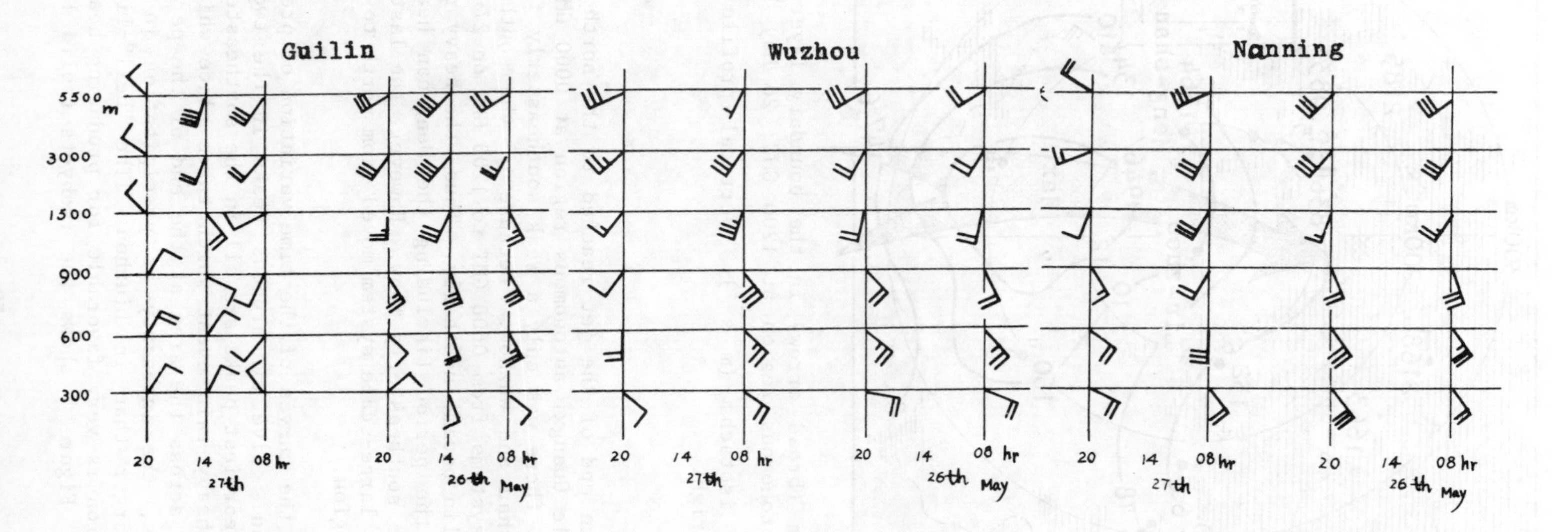

Fig. 4 Time cross section of the wind for Guilin, Wuzhou and Nanning on 26-27 May 1978.

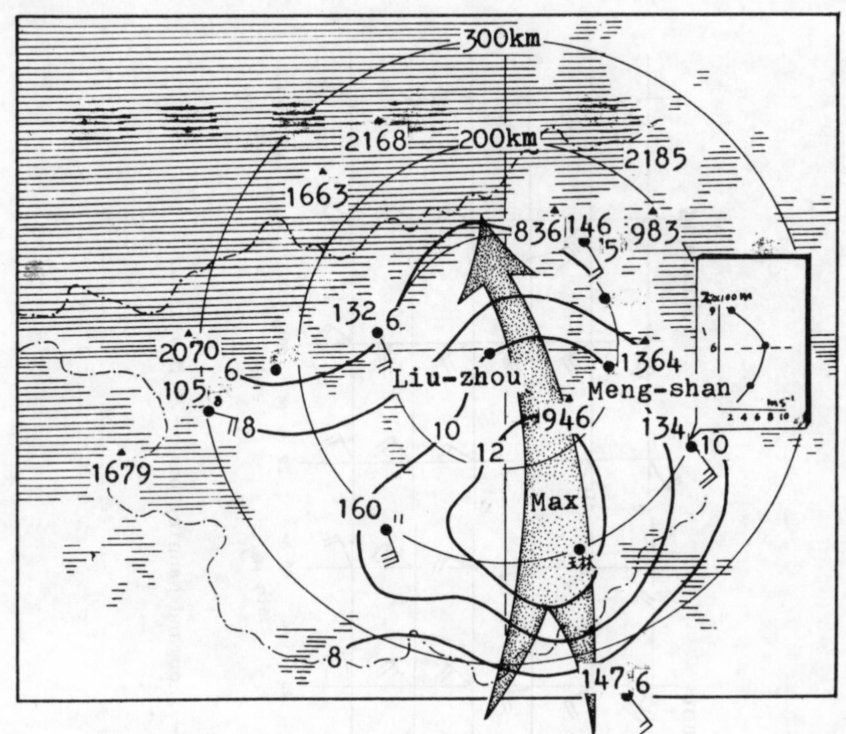

Fig. 5 Jet stream (broad arrow) in the boundary layer at 600 m in Guangxi autonomous region at 1200 GMT, 26 May 1978. Solid lines are isotachs (m s^{-1}). Vertical profile is shown in insert at right.

period. The northern end of the jet reached to the north of Guilin and spread over the whole Guangxi autonomous region at 0000 GMT, 26 May. By 1200 GMT on 27 May, there was only a weak southeasterly flow left over the region of Mengshan and Wuzhow. Because of the southward movement of the large-scale systems from 0600 GMT to 1200 GMT on 27 May, the low-level flow over Guilin became northerly. Thus, the heavy rain of northern Guangxi during the period (including the Mengshan heavy rain) occurred mainly in the southeasterly flow. However, the last of the heavy rain occurred as the large-scale system moved from north to south through the whole Guangxi region.

Figure 6 shows the curves of the time variation of precipitation of Liuzhou and Mengshan. Five synchronous peaks in the two places were observed. The four earliest peaks are all in the southeasterly flow, and the last one is a heavy rain event which takes place while the large-scale system sweeps across the area at the end of the period. For the four earliest peaks, the intensity of precipitation in the Mengshan region is almost larger than in Liuzhou. The "bell"-like topography in the Mengshan region is very favorable for producing heavy rain under southeasterly flow. Figure 7 shows the isohyets (solid lines) and the

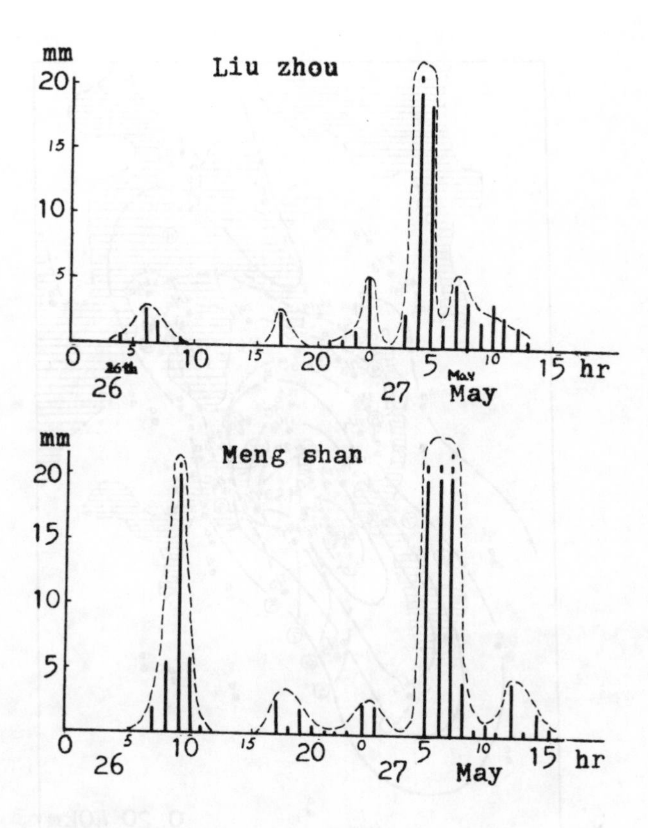

Fig. 6 Hourly variation of precipitation at Liuzhou and Mengshan.

isochrones of the beginning of precipitation (dashed lines). There is a center of heavy rain near Mengshan which is close to the mouth of the bell. Since the zero isohyet contour is almost closed, this heavy rain event in Mengshan is formed by local effects. The precipitation began at 1900 GMT, 25 May and the isochrones of precipitation moved toward the NNE with a mean speed of about 5-10 m s^{-1}. Why are there four synchronous precipitation peaks in the quasi-steady southeasterly flow in the boundary layer of Mengshan region from 0000 GMT, 26 to 1200 GMT, 27 May, and why do the isochrones of the beginning of precipitation move toward the NNE with a speed of 5-10 m s^{-1}? These questions cannot be answered completely because we do not have enough observational data in both time and space. We may reasonably infer that there are some mesoscale disturbances in the southeasterly jet flow around the Liuzhou and Mengshen regions. As these disturbances propagate, they will generate the four successive peaks. The movement of the isochrones reflects the propagation of mesoscale disturbances. Therefore, we hypothesize that the mesoscale disturbance in the boundary-layer jet, as the above analyses reveals, corresponds to the theoretical gravity wave solution discussed in the previous section.

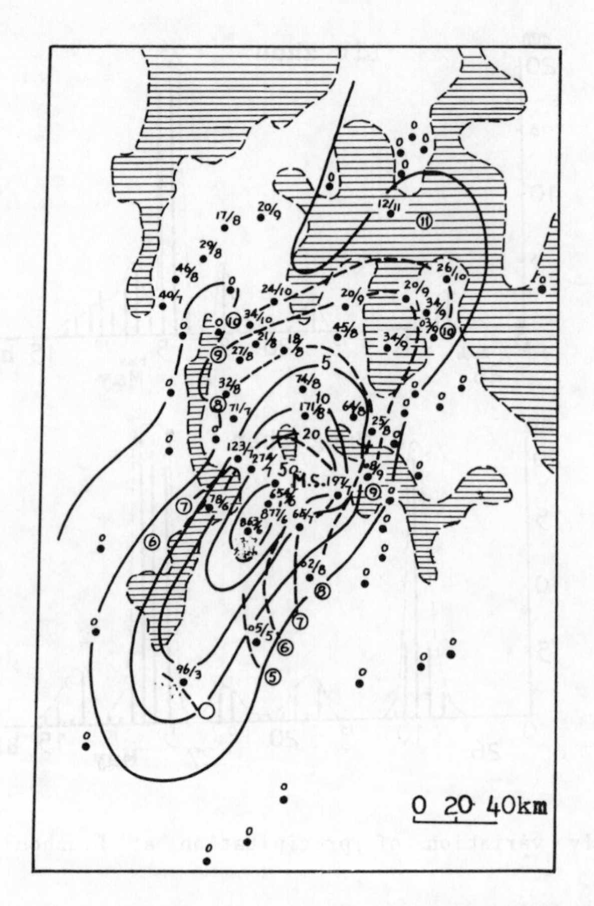

Fig. 7 Isohyets (solid lines in mm) and the isochrones of the beginning of precipitation (dashed lines) from 1600, GMT 25 to 0600 GMT, 26 May 1978; the numerator is the precipitation amount and the denominator is the time of precipitation beginning.

Based on the above evidence, the local heavy rain in Mengshan is associated with the gravity wave of the jet type in the boundary layer. The isochrones and the precipitation radar echoes reflect the movement of the gravity wave. Radar data, in general, reveal that precipitation echoes move along the direction of flow at the mean level of the convection. The height of the echoes is generally about 10 km in the case of the Mengshan heavy rain. Thus, the precipitation echoes move along the direction of the southwesterly flow at 500 mb. The isohypses (Fig. 7) move toward the NNE, which is the mean direction of the 500 mb flow and boundary-layer flow. The velocity at 500 mb is about 7 m s^{-1}, and the theoretical propagation velocity of the gravity wave of the jet type in the boundary layer is about 8 m s^{-1}. These values agree with the mean velocity of the isochrones (5-10 m s^{-1}). This agreement supports the hypothesis that mesoscale disturbances in the boundary-layer jet stream produce the heavy rain episodes over Mengshan.

IV. SUMMARY

In this paper, we have obtained a formula for the phase velocity of the gravity wave in the boundary layer containing a jet stream. Although this approximate solution neglects several important effects, it may be called the gravity wave of the jet type in the boundary layer. The four peaks of heavy rain under the influence of the southeasterly flow in the boundary layer in the Mengshan region during 26-27 May 1978 indirectly verify the theory.

The order of magnitude of the upward velocity is only 10 cm s^{-1} for the gravity wave of the jet type in the boundary layer. Therefore, the gravity wave is only a trigger mechanism for the heavy rain and the strong convection. The movement of this gravity wave can be identified with the isochrone pattern. In this view, the movement of the heavy rain area is due to the propagation of the triggering action.

REFERENCES

1. Chen Liangdong, Xing Maoyin, Yu Zhihao: On the influence of topography upon the formation of heavy rain in north Guangxi. Scientia Meteorologica Sinica, No. 1-2, 53-64 (1980).

2. Li Maicun: The trigger action of the gravity wave for larger heavy rains. Scientia Atmospherica Sinica, No. 3, 201-209 (1978).

3. Lindzen, R.S.: Wave-CISK in the tropics. J. Atmos. Sci., No. 1, 156-179 (1974).

4. Tao Shiyan: Some problems for the analysis and forecasting of the heavy rain. Scientia Atmospherica Sinica, No. 1, 63-72 (1977).

5. The Technical Group of the Heavy Rain Experiment in South China: The technical reports in Longyan meeting, Jan. (1979).

4.8 A BRIEF ANALYSIS AND NUMERICAL SIMULATION OF THE SICHUAN EXTRAORDINARILY HEAVY RAINFALL EVENT IN 1981

Zhou Xiaoping
Institute of Atmospheric Physics
Academia Sinica
The People's Republic of China

and

Hu Xingfang
Sichuan Meteorological Bureau
The People's Republic of China

ABSTRACT

The main period of heavy rain between 12 and 13 July 1981 had its main source of water vapor in the South China Sea. Numerical model results in a σ-coordinate system, using different interpolation schemes, are discussed.

I. INTRODUCTION

From 12 to 15 July 1981, there was an extraordinary, 100-400 mm/day heavy rainfall in the Sichuan Basin covering an area about 38,000 km^2. This event caused the greatest flash flooding since 1949 in the upper reaches of the Changjiang (Yangzi) River. About 1.5 million people were left homeless, 753 dead and 558 missing. The property loss was near 1.2 billion U.S. dollars.

The Qinghai-Xizang (Tibet) Plateau has a significant influence on worldwide and Chinese climate. In particular, the plateau is an important factor in the formation of heavy rainfall in southeastern China. The topography forces cold air from the north to be concentrated in eastern China. When warm and moist currents of the southeast and southwest monsoon flow into South China, they meet the cold air in a narrow region east of the plateau and west of the coastline of China. This convergence is the reason that most of the heavy rainfall in China occurs in the southeast. Correct forecasts of heavy rainfall in China, to a large extent, depend on the correct evaluation of the influences of the plateau. Sichuan Province is located near Tibet. The Sichuan heavy rainfall in 1981 was largely related to a medium-scale vortex [4] which formed over the plateau, moved eastward and developed strongly over the Sichuan Basin by the interaction of three warm and cold currents. In this paper, a simple analysis and some results of a numerical simulation of heavy rain will be presented. The main period of the extraordinarily heavy rainfall occurred between 0000 GMT, 12 and 0000 GMT, 13 July. In Fig. 1, the total precipitation during that 24-hour period is shown by solid lines. We focussed our study on the same period. The model used here is a fine-mesh model developed at the Institute of Atmospheric Physics since 1975 [9]. We also disucss the numerical treatment of the pressure gradient force over the steep slopes of the Tibetan Plateau.

II. THE SYVOPTIC BACKGROUND OF THE SICHUAN HEAVY RAINFALL IN 1981

The general summer rainfall in China is related to the position of the Pacific subtropical anticyclone [7]. The heavy rainfall events often have a strong relationship with low-level (850 mb or lower) jets at the southwest edge of the subtropical high [6], [8]. The position of the subtropical high, when the Sichuan flash flooding occurred, was stretched farther west than normal into the China mainland. At 500 mb (Fig. 2), the 5880 m geopotential height contour (dashed line) was nearly 2000 km west of its normal position (solid line). This position allowed marine warm and moist air to be transported into the continent near the eastern side of the plateau. In this flow, a low-level jet transported very moist air into the Sichuan Province. Meanwhile, the low pressure of the Indian monsoon during this period was located 3° latitude farther north of its normal position [1]. The satellite picture shows that the cloud clusters of the Sichuan rainfall and the clouds of the monsoon were organized in a complete cloud system during this period. It was originally thought that the Indian monsoon brought most of the water vapor to supply the Sichuan heavy rainfall. However, diagnostic analyses showed that the fraction of

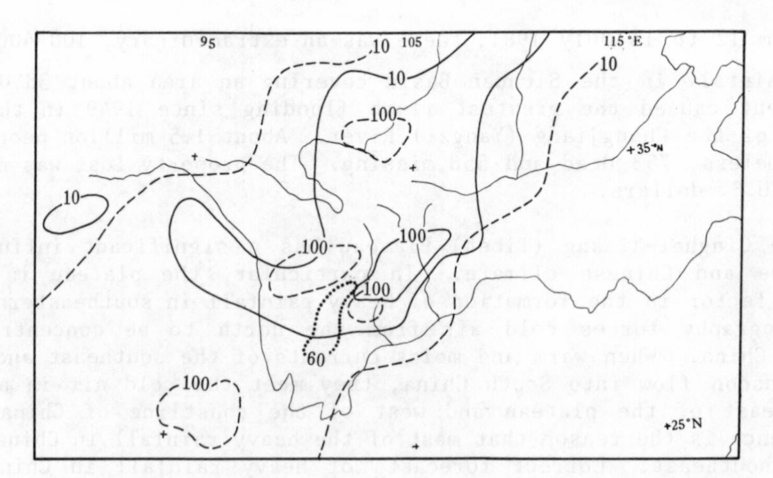

Fig. 1 Total precipitation in mm from 0000 GMT, 12 to 0000 GMT, 13 July. Solid line: observed; dotted line: simulated without topography, but with the effects of the plateau on the initial data; dashed line: simulated with topography, but without the pressure gradient force modification.

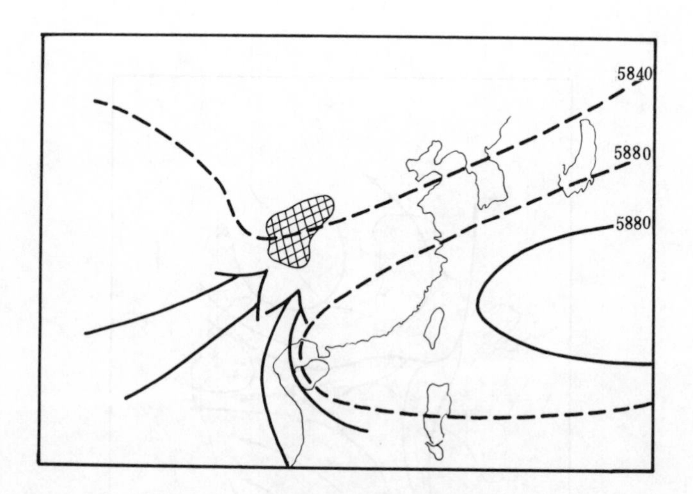

Fig. 2 Position of Pacific subtropical high at 500 mb. Solid line: normal position of the 5880 m geopotential height contour; dashed line: average 5880 geopotantial height contour during the period 11-15 July, 1981; arrows depict the warm and moist currents from India and the South China Sea (courtesy of Y.L. Li).

water vapor transported across the west boundary of Sichuan above 2000 m in two days between 12 and 13 July was only 35 percent [2]. Thus, the main transport of water vapor during this period was from the South China Sea at heights lower than 2000 m.

Many experienced forecasters have pointed out the effect of cold air from the north before the occurrence of the Sichuan heavy rainfall. It is easy to identify cold air to the northwest which moves with an eastward-moving trough. There is also cold air to the northeast which is related to high pressure over the Japan Sea. Many cases of heavy rainfall in China are associated with cold air in this location [7]. These events are more difficult to forecast than when cold air is located to the northwest. The Sichuan heavy rainfall in 1981 was associated with a low-level mass of cold air to the northeast of Sichuan [3]. Figure 3 shows the low-level cold (but not very dry) air coming from North China towards the Sichuan Basin. The success of the Sichuan rainfall forecast partly depends on the extent to which we can simulate this northeastern cold air current.

III. THE MODEL AND THE INITIAL DATA

We briefly introduce the model used in the simulation [9]. The system of governing equations with sigma

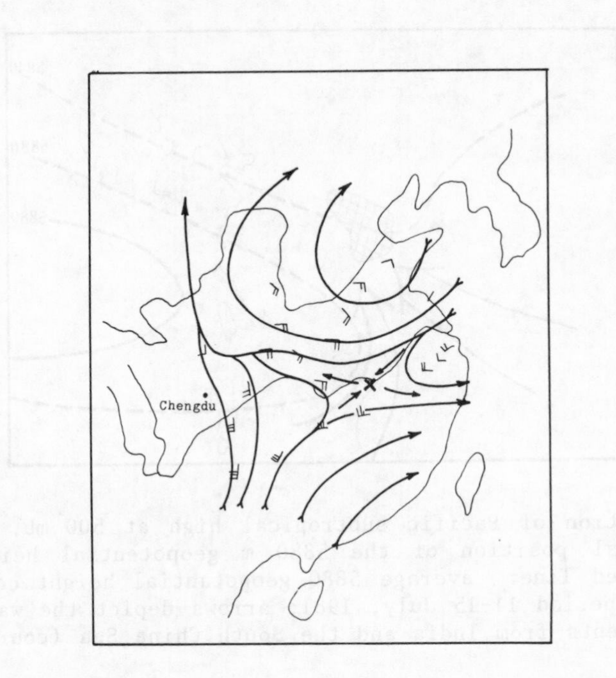

Fig. 3 Streamlines at 850 mb at 0000 GMT 12 July 1981 (after [3]).

$$\sigma = \left(\frac{p}{p_s(x,y,t)}\right)$$

as vertical coordinate is as follows:

$$\frac{\partial p_s}{\partial t} = -\int_0^1 \left(\frac{\partial p_s u}{\partial x} + \frac{\partial p_s v}{\partial y}\right) d\sigma \tag{1}$$

$$\frac{\partial p_s \dot{\sigma}}{\partial \sigma} = -\left(\frac{\partial p_s}{\partial t} + \frac{\partial p_s u}{\partial x} + \frac{\partial p_s v}{\partial y}\right) \tag{2}$$

$$\frac{\partial p_s u}{\partial t} = -\nabla \cdot p_s u \underset{\sim}{V} + p_s f(v - v_g) + p_s F_u \tag{3}$$

$$\frac{\partial p_s v}{\partial t} = -\nabla \cdot p_s v \underset{\sim}{V} + p_s f(u_g - u) + p_s F_v \tag{4}$$

$$\frac{\partial C_p p_s T}{\partial t} = -\nabla \cdot p_s (C_p T + \phi) \underset{\sim}{V} - \frac{\partial \sigma \phi}{\partial \sigma} \frac{\partial p_s}{\partial t} + p_s f(u v_g - v u_g) \tag{5}$$

$$+ L p_s C + C_p p_s F_T$$

$$\frac{\partial p_s q}{\partial t} = -\nabla \cdot p_s q\underset{\sim}{V} + p_s C + p_s F_q \tag{6}$$

$$\frac{\partial}{\partial t}\frac{\partial \phi}{\partial \sigma} = -\frac{R}{\sigma}\frac{\partial T}{\partial t} \tag{7}$$

where u, v, $\dot{\sigma}$ are velocity components of x, y, σ, respectively; u_g and v_g are the geostrophic wind components; φ is geopotential height; T is temperature; q, specific humidity; p_s, pressure at ground surface; F_u, F_v, F_T and F_q represent eddy diffusion; f, R and C_p are the Coriolis parameter, gas constant of air and specific heat at constant pressure; C is the condensation rate; and

$$\nabla \cdot A\underset{\sim}{V} = \frac{\partial Au}{\partial x} + \frac{\partial Av}{\partial y} + \frac{\partial A\dot{\sigma}}{\partial \sigma} \quad \text{(A is an arbitrary function)}$$

$$fu_g = -\left(\frac{\partial \phi}{\partial y} + RT\frac{\partial \ell n p_s}{\partial y}\right)$$

$$fv_g = \left(\frac{\partial \phi}{\partial x} + RT\frac{\partial \ell n p_s}{\partial x}\right)$$

The vertical boundary conditions are $\dot{\sigma} = 0$ at $\sigma = 0$ and $\sigma = 1$. The lateral boundary conditions are steady with time. This system is closed if C and the diffusion terms are parameterized. Written in this form, the total energy conservation is easily satisfied in designing the difference scheme because the kinetic-potential energy term in Eqs. (3), (4), and (5) can be compensated entirely. In addition, we used the derivative with time, of the hydrostatic equation instead of the hydrostatic equation itself. In doing so, we can use both the temperature and geopotential height as initial data. Both height and temperature are important in dynamics and in dealing with the change of phase of water. This method also significantly decreases the truncation error when the model has few layers. For finite differences, we use Lilly's scheme for the $\nabla \cdot A\underset{\sim}{V}$ terms, expressed in Shuman's notation

$$\nabla \cdot A\underset{\sim}{V} = (\bar{A}^x \bar{u}^x)_x + (\bar{A}^y \bar{v}^y)_y + (A\bar{\dot{\sigma}}^\sigma)_\sigma .$$

Simple centered differences are used for the geostrophic wind and for terms like $\nabla \cdot p_s\underset{\sim}{V}$. The leapfrog scheme is used for the time integration, with an Euler-backward step once every hour.

We found that the most effective method to smooth the checkerboard modes was to use two smoothing operators alternately on the tendencies of all variables:

$$\text{(a)} \quad \tilde{A} = (1 - 4K) A + K\overline{\overline{A}^{2x}}^{2y}$$

(This operator damps the modes parallel to lateral boundary lines.)

$$\text{(b)} \quad \tilde{A} = (1 - 4K) A + K(\overline{A}^{2x} + \overline{A}^{2y})$$

(This smoother suppresses the zig-zag modes along a tilted line at a 45° angle with the lateral boundary lines.) If K equals 0.125, the two-grid increment waves are entirely removed. The smoothing operators on the tendencies could not take the place of the eddy diffusion and could be used many times without oversmoothing the longer waves. This smoothing treatment is something like that used by Shuman [5], but the present one is aimed specifically at the checkerboard modes.

The grid size, Δs, used in the model is 100 km. The time step Δt = 180 s. Figure 4 shows the vertical indexing for the five-level model.

The initial data for 0000 GMT, 12 July 1981, are ϕ, T, $T - T_d$ (T_d is the dew point temperature) u, v at 200, 500, 700 and 850-mb constant pressure surfaces and sea level pressure, p_o. In mountain areas, the geopotential height, ϕ, was obtained at the 700- and 850-mb constant pressure surfaces by horizontal interpolation. Figure 5 shows the smoothed topography in the model. The highest point in the model is 4600 m.

We compared results with the observed ground surface pressure, p_{ob}, and with the pressure, p_s, interpolated to the points of the topographic heights from the constant pressure surfaces. The latter procedure caused less error and was used in all our experiments.

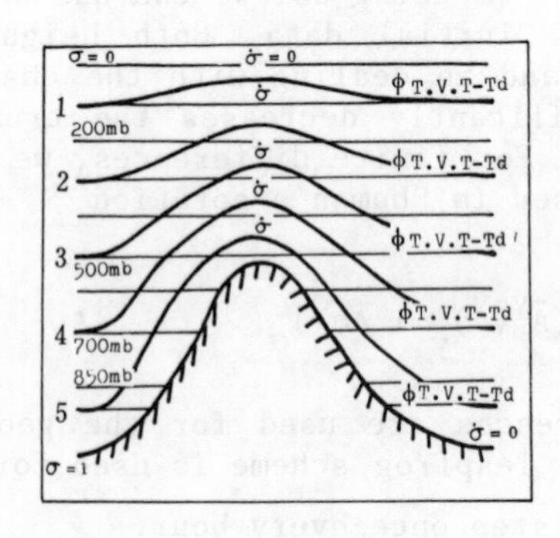

Fig. 4 Vertical indexing of the model and initial data used in the model.

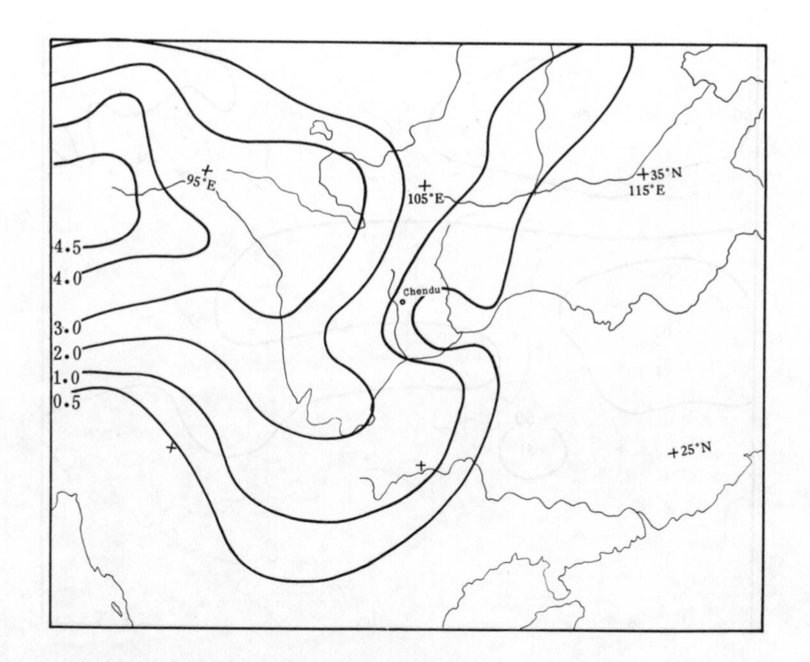

Fig. 5   Smoothed model orography (km).

IV. RESULTS

Usually, the error of the pressure gradient force in σ-coordinates is very large over steep slopes. In our model, the order of the error is the same as that of the actual pressure gradient force, but the error is systematic rather than random. In Fig. 6, the distribution of $p_s fu_g = -p_s((\partial\phi/\partial y) + RT(\partial \ell n p_s/\partial y))$ at $\sigma = 0.1$, at the first time step, shows two areas with absolute values larger than 50 on both sides of the plateau. The calculated westerly component of the geostrophic wind over these areas, between 50 and 100 mb, is 50-100 m s^{-1}. For comparison, we also show the geopotential heights at 200 mb (Fig. 7). The pressure gradients at 200 mb over the southern and northern slopes of the plateau are larger than in other places. However, the calculated maximum geostrophic winds are only 50 m s^{-1}. Thus, although the calculated geostrophic winds are larger than in reality, they do reflect the actual geostrophic winds, i.e. the error does not affect the sign of the pressure gradient force. The distribution of the error could suggest a method to decrease the error caused by the steep slopes.

In Fig. 1, the results of rainfall predicted by two schemes are shown for comparison with the observed precipitation. The area of the observed rainfall amounts larger than 100 mm is located in the Sichuan Basin. The 10-mm contour surrounds a very large area. The dashed lines

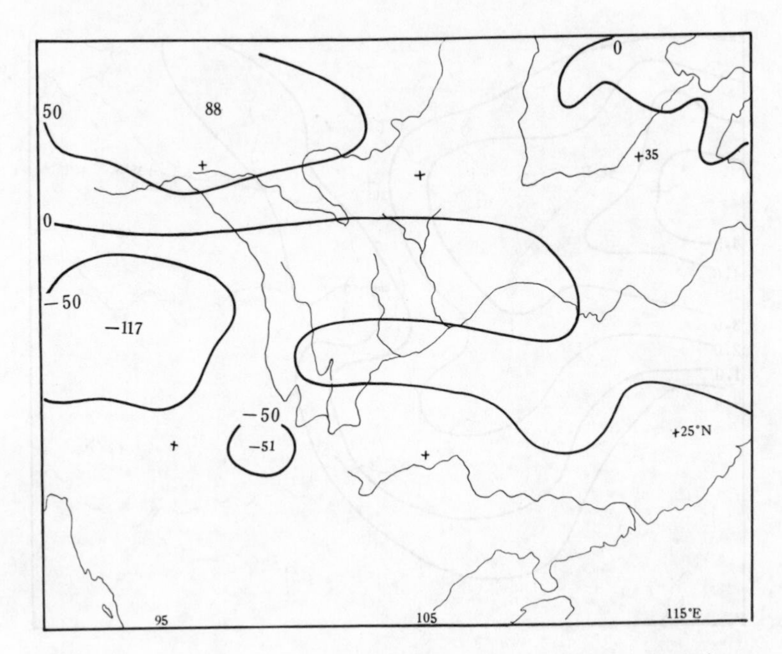

Fig. 6 Distribution of $p_s f u_g = -p_s((\partial\phi/\partial y) + RT(\partial \ln p_s/\partial y)$ at $\sigma = 0.1$ at 0000 GMT, 12 July 1981.

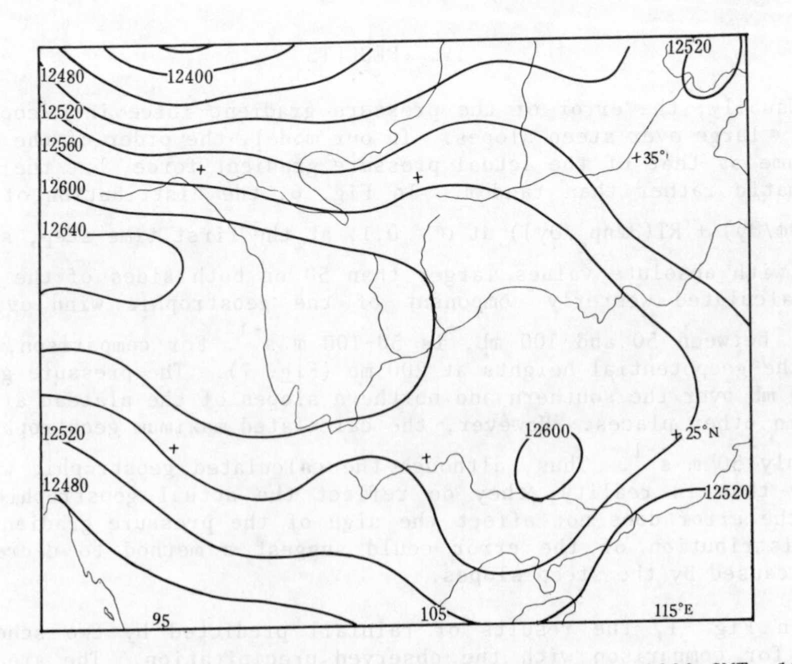

Fig. 7 Geopotential height (m) of 200-mb surface at 0000 GMT, 12 July 1981.

show the predicted rainfall when the plateau topography is present in the model. The area of precipitation larger than 10 mm is a little greater than observed. The predicted 100 mm rainfall area is located west of the real Sichuan heavy rainfall by about 200 km. Moreover, the model predicts two spurious areas with more than 100 mm rainfall over the northeastern and southeastern slopes of the plateau. These errors are most likely caused by the incorrect evaluation of the pressure gradient force which causes an incorrect prediction of the flow of the cold air from the north over the eastern slopes of the plateau. In this case, the warm and moist air coming from the south could move farther west, where fictitious convergence and rainfall took place. The area surrounded by dotted lines shows the 60-mm precipitation predicted without topography in the model. In this case, the initial data still contained some effects of the plateau. The two false, heavy rainfall events predicted with plateau topography do not appear in the case without topography.

Because the wind data were only smoothed by hand and not analyzed objectively at the initial time, a large area of precipitation occurred with the simulated gravity waves. However, in a short time steady rainfall became established in areas where there was a constant convergence maintained by thermodynamic and dynamic processes.

Various methods of interpolation from pressure to σ-coordinates were tested in the initialization phase. Compared to spline and Lagrangian methods, a linear interpolation gave better results. Figure 8 shows the

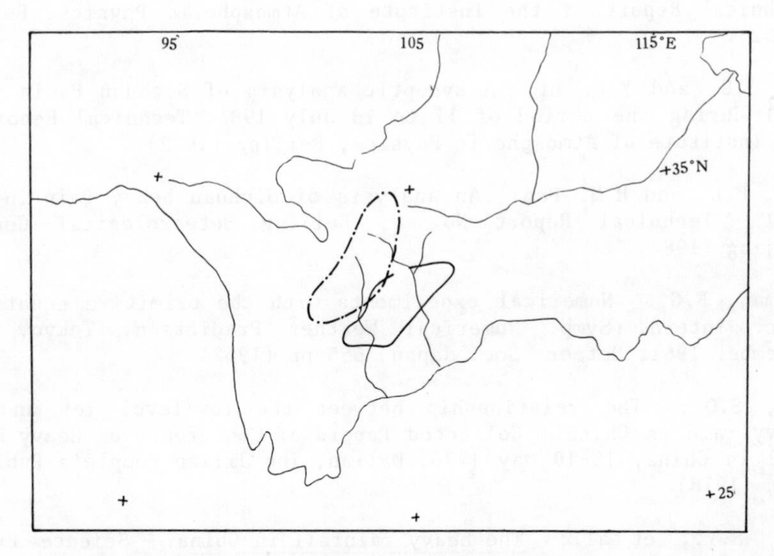

Fig. 8 Predicted precipitation amounts with linear interpolation from $\ln$ p-coordinate to σ-coordinate during the initialization phase. Dashed line shows the predicted 24-hour rainfall larger than 100 mm from 0000 GMT, 12 to 0000 GMT, 13 July 1981. Solid line is observed isohyet.

comparison between the observed rainfall (solid line) and that predicted (dashed) with the linear interpolation method. The center of the predicted heavy rainfall is located about 100 km northwest of the observed. This forecast is our best by numerical methods so far. This is not to say that we have found the way to understand and to simulate numerically heavy rainfall, but we have identified problems, such as the initialization and the treatment of the geostrophic wind over steep slopes similar to those near the Tibetan Plateau.

ACKNOWLEDGMENTS

We are deeply grateful to Profs. D.Z. Ye and S.Y. Tao who initiated this work, and we express our appreciation to Mr. Z.Y. Wang, senior engineer of the Sichuan Meteorological Bureau for his constant support. Thanks are also due to Miss D.W. Chen for her preparation of data and to Ms. D.Y. Zhang and Ms. S.M. Zhao for their help in programming. Without this help we could not have completed these experiments and analysis.

REFERENCES

1. Chengdu Central Observatory: A study of Sichuan extraordinary heavy rainfall in July 1981. Collected Papers of Sichuan Basin Heavy Rainfall in 1981, Sichuan Meteorlogical Bureau, Chengdu, (1982).

2. Fang, Z.Y. and S.C. Tian: The water vapor analysis during the period of Sichuan extraordinary heavy rainfall in July 1981. Technical Report of the Institute of Atmospheric Physics, Beijing (1982).

3. Li, J.S. and Y.L. Li: A synoptic analysis of Sichuan Basin rainfall during the period of 11 to 13 July 1981. Technical Report of the Institute of Atmospheric Physics, Beijing (1982).

4. Shi, F.L. and H.M. Pan: An analysis of Sichuan heavy rain in July 1981. Technical Report No. 9, Beijing Meteorological Center, Beijing (1982).

5. Shuman, F.G.: Numerical experiments with the primitive equations. Proc. Intern. Symp. Numerical Weather Prediction, Tokyo, 7-13 November 1961, Meteor. Soc. Japan, 655 pp (1962).

6. Sun, S.Q.: The relationship between the low-level jet and the heavy rain in China. Collected Papers of Conference on Heavy Rainfall in China, 10-19 May 1978, Dalian, The Jilian People's Publication (1978).

7. Tao, S.Y., et al.: The heavy rainfall in China. Science Press, Beijing (1980).

8. Zhou, X.P.: A tentative conceptual model of heavy rain in China. Technical Report of the Institute of Atmospheric Physics, Beijing (1982).

9. Zhou, X.P., S.X. Zhao, K.S. Zhang, and S.H. Liu: Some results of the fine-mesh model for numerical forecasting of heavy rainfall and severe convective storms. To be published in the Annual Report of the Institute of Atmopspheric Physics, Science Press, 1981 (1982).

SESSION V

MOUNTAIN-VALLEY CIRCULATIONS AND THEIR EFFECTS ON METEOROLOGICAL FIELDS

5.1 STABLY STRATIFIED MOIST AIRFLOW OVER MOUNTAINOUS TERRAIN

Douglas K. Lilly
and
Dale R. Durran
National Center for Atmospheric Research[1]
Boulder, Colorado 80307, USA

ABSTRACT

This paper reviews the results of two separate studies. The first, subtitled "Wet Mountain Waves," is a theoretical numerical analysis of the effects of condensation and evaporation on the stationary waves that form in stably stratified fluid crossing a topographic barrier. Many additional details of the analysis not found here are contained in Durran [6] and in Durran and Klemp [7, 8]. The second study, subtitled "Colorado Upslope Snowstorms", is an observational analysis of a meteorological event which yields the most important winter precipitation produced on the eastern slopes of the central United States Rocky Mountains. An earlier version of this paper appeared as a conference preprint [15].

I. WET MOUNTAIN WAVES

1. Purpose

Although dry mountain waves have been studied extensively for the last 40 years, the influence of moisture on the dynamics of these waves has received little attention. There is reason to believe, however, that moisture can significantly modify the wave flows, since the static stability in the lower and middle troposphere can be greatly reduced or reversed by release of the latent heat of condensation. We have often found that a dry wave model applied to real atmospheric conditions may greatly overpredict actual wave response when the upstream humidity is high. Two previous studies by Barcilon et al. [1, 2] examined the effects of nonprecipitating clouds on linear hydrostatic waves. In both of these the microphysics are very idealized, however, and the flow regimes are limited to conditions in which the dry and moist static stabilities and the windspeed are constant with height.

In order to investigate a wider variety of realistic situations a numerical model is used here, which allows application of more realistic microphysics and also the simulation of nonlinear and nonhydrostatic, compressible, and time-dependent waves produced by mountains of arbitrary shape. The time-dependent simulation approach guarantees that any steady

[1] The National Center for Atmospheric Research is sponsored by the National Science Foundation.

state solutions obtained are stable, and also allows the simulation of convective regimes in which the response is inherently time-dependent. At the same time it is possible to compare the model conditions and results with those of classical dry analytic solutions.

2. Numerical Model Structure

The model is designed to calculate the two-dimensional airflow over an infinitely long, uniform mountain barrier, for which Coriolis terms are neglected. It is closely patterned after the convective cloud model of Klemp and Wilhelmson [13], with three major modifications. It is reduced from three to two dimensions, a terrain-following coordinate system is introduced, and a damping layer is added to the top to simulate a wave-absorbing, or radiation, boundary condition.

The basic momentum, continuity, and thermodynamic equations for a dry atmosphere in the Klemp-Wilhelmson model may be written in two dimensions as follows:

$$\frac{du}{dt} + c_p \theta \frac{\partial \pi}{\partial x} = D_u \tag{1}$$

$$\frac{dw}{dt} + c_p \theta \frac{\partial \pi}{\partial z} = g \frac{\theta - \bar{\theta}}{\bar{\theta}} + D_w \tag{2}$$

$$\frac{d\pi}{dt} + w \frac{\partial \bar{\pi}}{\partial z} + \frac{R}{c_v} (\bar{\pi} + \pi)\left(\frac{\partial u}{\partial x} + \frac{\partial w}{\partial z}\right) = \frac{R}{c_v} \frac{\bar{\pi} + \pi}{\theta} \frac{d\theta}{dt} \tag{3}$$

$$\frac{d\theta}{dt} = D_\theta, \tag{4}$$

where $d/dt = \partial/\partial t + u\partial/\partial x + w\partial/\partial z$, $\bar{\pi} + \pi = (p/p_o)^{R/c_v}$, $\partial\bar{\pi}/\partial z = -g/c_p\bar{\theta}$, with the overbars referring to horizontal averages and other symbols defined conventionally. The D terms on the right signify Reynolds stress and flux derivatives associated with subgrid-scale dynamics. A conventional first order eddy viscosity closure is used incorporating Richardson number effects [14]. Thus, the terms D_u, D_w, and D_θ are evaluated as follows:

$$D_u = (K_M A)_x + (K_M B)_z, \quad D_v = (K_M B)_x - (K_M A)_z,$$

$$D_\theta = (K_H \theta_x)_x + (K_H \theta_z)_z, \tag{5}$$

where $A = u_x - w_z$, $B = u_z + w_x$, $K_M = k^2 \Delta x \Delta z \,|\mathrm{Def}|\, [\max(1 - Ri(K_H/K_M), 0)]$, $R_i = N^2/\mathrm{Def}^2$, $N^2 = g\partial(\ln\theta)/\partial z$, and $\mathrm{Def}^2 = A^2 + B^2$. For our purposes we set the constant $k = 0.21$ and the inverse eddy Prandtl number $K_H/K_M = 3$, both as suggested by Deardorff [4, 5]. Thus turbulent mixing begins when $Ri < 1/3$.

The effects of moisture are incorporated into the system by replacing θ in Eqs. (1) and (2) by a virtual temperature including the contributions of water vapor and liquid, by adding new conservation equations for water vapor, cloud water, and rain water, allowing for latent heat effects in Eq. (4) and the exchange of water between vapor, cloud and rain through condensation, evaporation, and coalescence terms, with the latter given by the Kessler [10] parameterization. Raindrop fall speed is allowed for in the rainwater conservation equation. Ice physics are not considered.

Terrain is incorporated into the model by a transformation of the vertical coordinate [9], i.e.

$$\zeta = \frac{z_T(z - z_S)}{z_T - z_S}, \tag{6}$$

where $z_S(x)$ is the terrain elevation and z_T is the height of the top of the modelling region. This transformation alters the convective and spatial derivative to the forms:

$$\frac{d}{dt} = \frac{\partial}{\partial t} + \frac{u\partial}{\partial x} + (u\partial\zeta/\partial x + w\partial\zeta/\partial z)\,\frac{\partial}{\partial\zeta} \quad ;$$

$$\left.\frac{\partial}{\partial x}\right|_z = \left.\frac{\partial}{\partial x}\right|_\zeta + u\,\frac{\partial\zeta}{\partial x}\frac{\partial}{\partial\zeta} \;; \quad \frac{\partial}{\partial z} = \frac{\partial\zeta}{\partial z}\frac{\partial}{\partial\zeta} \;. \tag{7}$$

The model is integrated using finite difference techniques in time and space, with the variables staggered according to Fig. 1. Since the equations are fully compressible, the usual Courant-Friedrichs-Levy condition requires time steps shorter than the time required for a sound wave to cross a grid point. To minimize this burden, a two-timestep integration scheme devised by Klemp and Wilhelmson [13] is used. The sound wave terms are computed from a linearized form of Eqs. (1) to (3), with the remaining terms and equations updated at time steps appropriate for the slower gravity wave and particle speeds. The resulting integrations are carried out with about the same efficiency as those of a more conventional anelastic system. Since the horizontal grid scale is usually considerably larger than that in the vertical, additional computational efficiency is obtained by making the small time step implicit in the vertical.

The upper boundary condition used is $w = 0$, but this is set at the top of a layer in which Rayleigh damping is applied, similarly to that of Klemp and Lilly [11]. This layer is designed to simulate a radiation condition and absorb upward propagating wave energy. Lateral boundary conditions are chosen to be wave-permeable and are modified versions of those used by Klemp and Lilly [11]. The lower boundary condition is, of course, that the motion is parallel to the terrain, i.e., $w = u\partial z_S/\partial x$ at $\zeta = 0$.

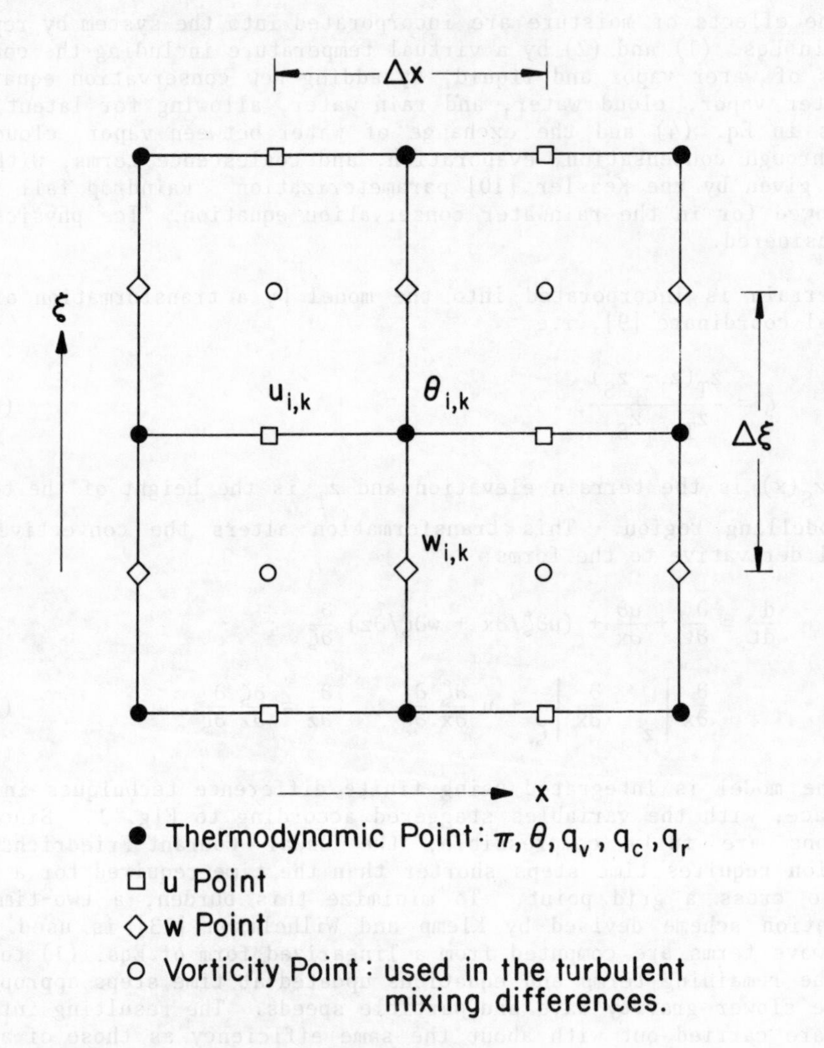

Fig. 1 The numerical grid structure.

3. Model Tests

In order to gain confidence in the use of the rather complex model described above, a number of tests were made in which wave responses were simulated from simple flow conditions and compared with analytic solutions or other simpler simulation model results. For the first test, the mean flow and temperature were set to be constant with height, i.e. $u = 20\ \mathrm{m\ s^{-1}}$, $T = 250$ K. A mountain profile was introduced with the "Witch of Agnesi" form, i.e.

$$z_S(x) = h/(1+x^2/a^2). \quad (8)$$

The quarter-width scale, a, was set to 20 km. The Scorer parameter is the maximum horizontal wave number with which steady linear gravity waves can propagate in the vertical. Neglecting the small effects of compressibility it is given by

$$k_s^2 = N^2/U^2 - U_{zz}/U . \quad (9)$$

Since the product of k_s^2 and the squared width scale is about 100 for this case, the waves should be essentially hydrostatic [12]. The mountain height was set to be 1 m, which should then yield almost exactly linear solutions. Figures 2, 3, and 4 show comparisons of streamlines, horizontal, and vertical velocities for a linear analytic steady state solution, (a), and for the model simulation, (b), at a time t = 60a/U. At this time the simulated flow has become almost steady. Figure 5 shows the momentum flux profile at various simulated times as a fraction of its analytic steady state value. The flow simulations appear to be almost identical to their analytic counterparts, but the momentum flux is reduced by about 6 percent just below the level at which the Rayleigh damping layer begins to diminish it sharply. The difference is not serious, but it may be interpreted as a practical limit on the accuracy of the flux calculations.

A test was also made of the model for a nonlinear wave, using a similar mountain but with an amplitude of 1000 m and with a less stable atmosphere than that of the previous isothermal case. The results were compared with a similar integration by Klemp and Lilly [11], using their hydrostatic θ-coordinate model. The differences between the two models were a little larger than in the linear case. These do not necessarily imply errors in the present model, however, since the steeper parts of a strong wave may not be closely approximated by the hydrostatic assumption.

4. Moisture Effects on Trapped Lee Waves

If the Scorer parameter decreases with height sufficiently abruptly, one or more trapped resonant wave trains (lee waves) can develop. We simulated such waves with and without the effects of moisture condensation. Figure 6a shows a profile of potential temperature and windspeed. The corresponding Scorer parameter profile (solid line on 6b) is suitable for developing strong lee waves in a dry atmosphere. Figure 7 shows the result of a linear (1 m mountain) simulation without moisture. Figure 8 shows comparable results with a saturated atmosphere and 0.2 g/kg of cloud water present in the upstream environment below the 3 km level, while Fig. 9 shows a case when the upstream environment is saturated but with no cloud water present. With cloud water present upstream, the air remains saturated for small upward or downward displacements, while without cloud water, the effects of latent heat exchange are only present when the net vertical displacement is upward.

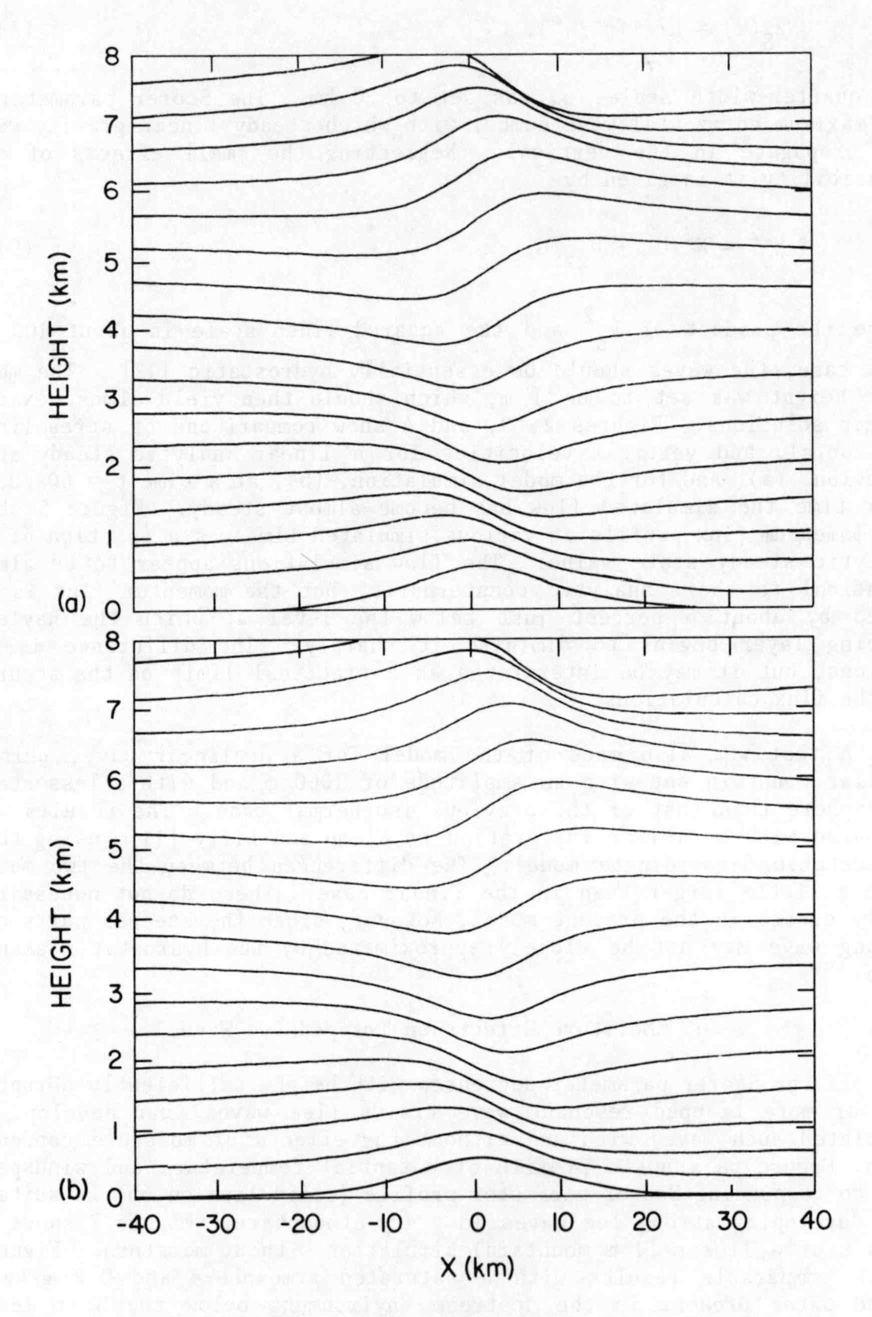

Fig. 2 (a) Steady state streamlines from the linear hydrostatic analytic solution for a 600 m high mountain. (b) Streamlines obtained by numerical simulation for a 1 m high mountain at ut/a=60. Perturbations are multiplied by 600 to normalize the wave amplitude.

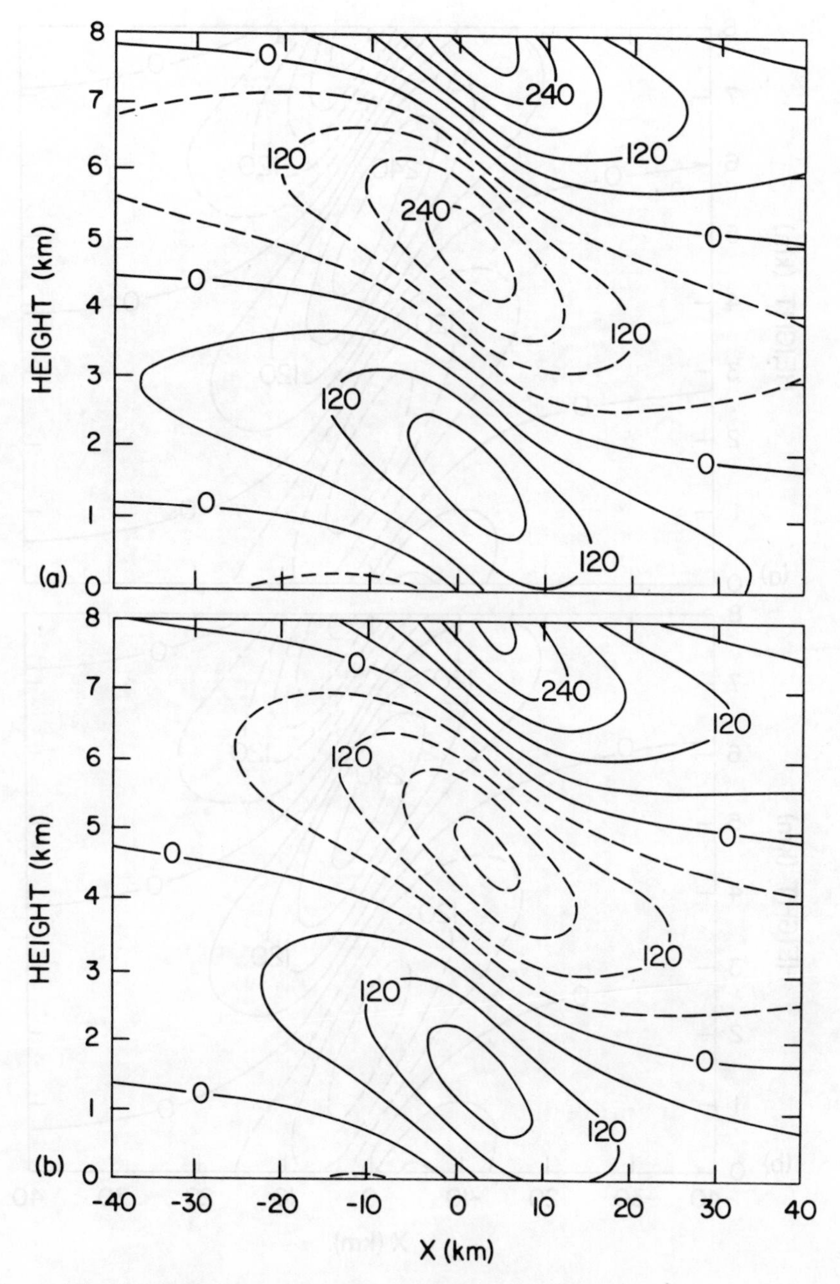

Fig. 3. (a) Steady state perturbation horizontal velocity from the linear hydrostatic solution for a 1 m high mountain. (b) Perturbation horizontal velocity ($\times 10^{-4}$ ms^{-1}) obtained by numerical simulation for a 1 m high mountain at ut/a=60.

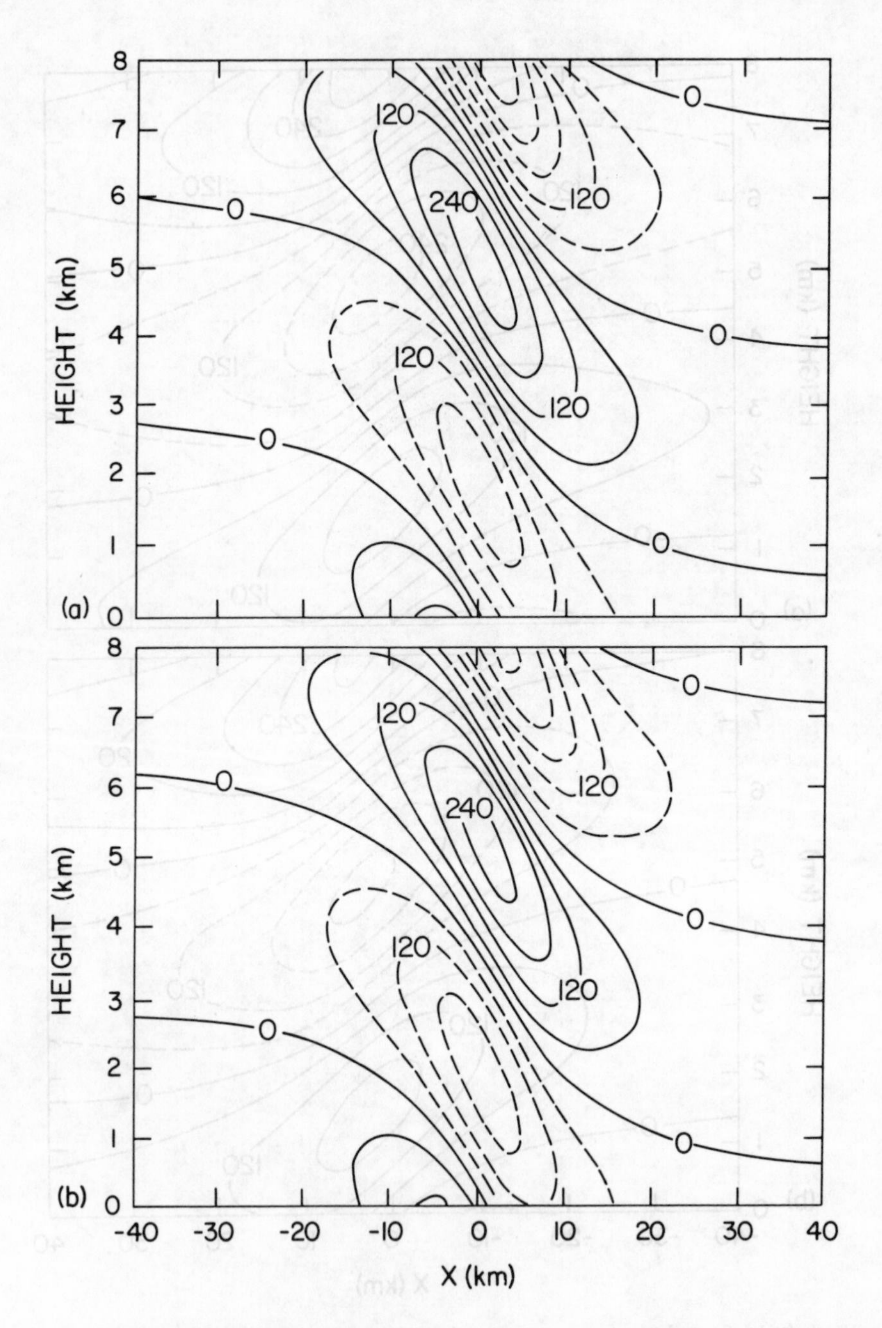

Fig. 4 (a) Steady state vertical velocity from the linear hydrostatic analytic solution for a 1 m high mountain. (b) Vertical velocity ($x10^{-5}$ ms^{-1}) obtained by numerical simulation for a 1 m high mountain at ut/a=60.

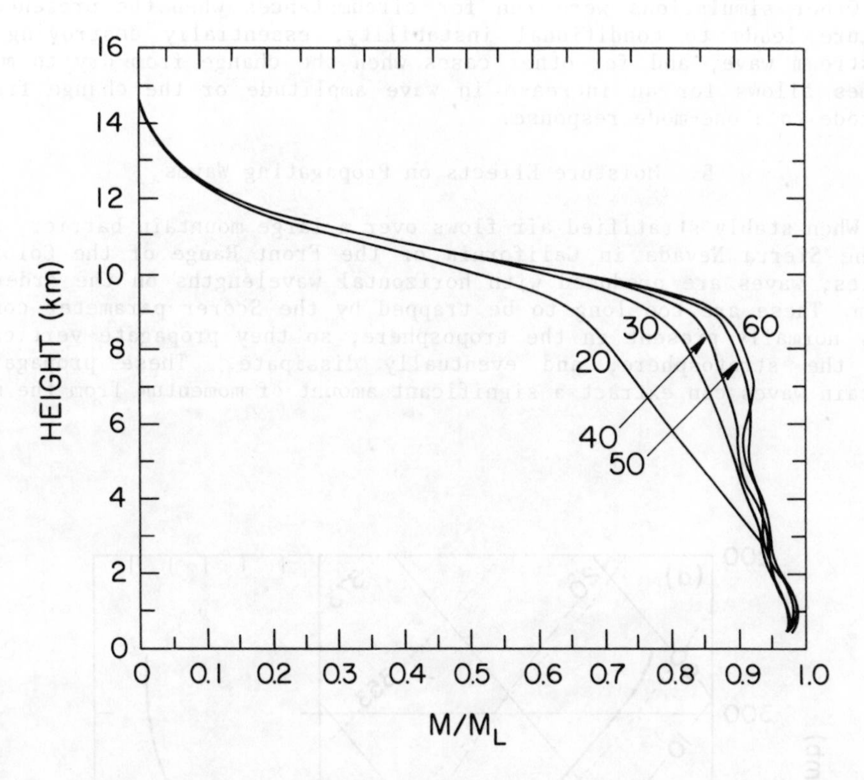

Fig. 5 Momentum flux normalized by its linear hydrostatic value at several nondimensional times ut/a.

The difference in response between Figs. 7 and 8 is obvious and easily explained by the difference between the dry (solid) and moist (dashed) Scorer parameter curves on Fig. 6b. For the dry case a resonant wave develops in agreement with dry analytic theory. For the moist case a trapped wave would not be expected from the profile of moist Scorer parameter. A weak and rapidly decaying downstream wave of greater wavelength is apparent. For the just-saturated upstream environment the results are somewhat intermediate but more similar to the dry case, in that a fully or mostly trapped wave appears of moderately large amplitude. It varies its shape in such a way that its local wavelength in the dry and moist regions varies in a way similar to the wavelengths of the dry and moist solutions individually.

Simulations were also performed using a 300 m mountain, for which some nonlinear effects are expected. Figure 10 shows the resulting streamlines, which are generally similar to those for the linear case. When the upstream humidity was set at 90 percent a fairly realistic lee wave cloud pattern developed, resembling the dry case in amplitude but with longer wavelengths in the cloudy regions and the appearance of a cap cloud over the mountain.

Other simulations were run for circumstances when the presence of moisture leads to conditional instability, essentially destroying the downstream wave, and for other cases when the change from dry to moist regimes allows for an increase in wave amplitude or the change from a two-mode to a one-mode response.

5. Moisture Effects on Propagating Waves

When stably stratified air flows over a large mountain barrier, such as the Sierra Nevada in California or the Front Range of the Colorado Rockies, waves are produced with horizontal wavelengths on the order of 50 km. These are too long to be trapped by the Scorer parameter conditions normally present in the troposphere, so they propagate vertically into the stratosphere, and eventually dissipate. These propagating mountain waves can extract a significant amount of momentum from the mean

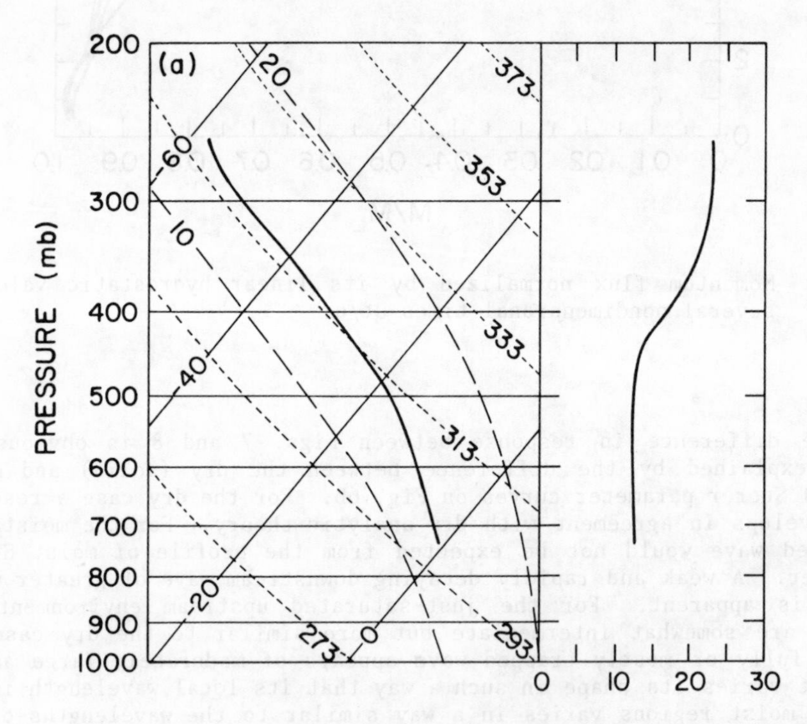

Fig. 6 Absolutely stable atmosphere favorable for the development of dry lee waves. (a) Temperature and windspeed profiles, dry adiabats are marked with a short dashed line, moist psuedo-adiabats with a long dashed line. (b) Scorer parameter (k_s^2 profiles), the dry k_s^2 is marked with a solid line, the equivalent saturated k_s^2 is dashed.

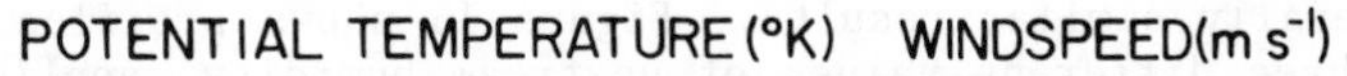

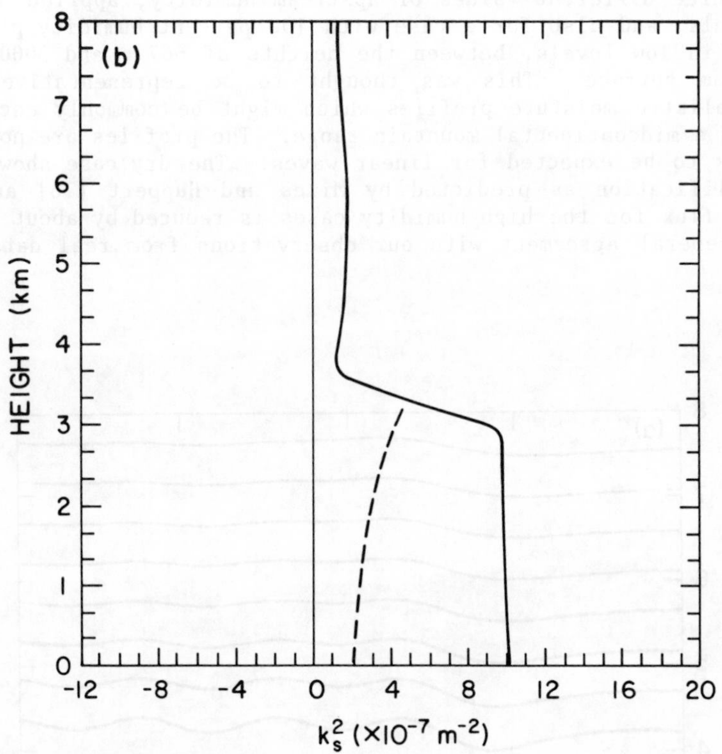

Fig. 6 (Continued)

flow and are often responsible for severe downslope windstorms. Here we describe the effects of moisture on these long propagating waves.

Figure 11 shows the streamlines and horizontal velocities produced by a linear (1 m) mountain in a dry flow with constant Scorer parameter. The same features are shown for upstream flows which are just saturated (Fig. 12) or saturated plus cloud water (Fig. 13). The momentum flux profiles for the three cases are shown in Fig. 14. The maximum horizontal velocity amplitude for a hydrostatic wave with constant Scorer parameter is Nh. Since N is reduced by more than a factor of 2 in the low levels by latent heating effects, the amplitude reduction for the moist cases is easily understood. It is not so clear why the just-saturated case is reduced by about the same amount (and a little more in momentum flux) than that with excess liquid water.

Simulations have been conducted also for flows with constant mean shear in the troposphere. The linear results are generally similar to those shown above. The just-saturated profiles produced results intermediate between the dry and cloud water-containing environments, but closer to the latter. Flows over larger amplitude mountains (1000 m) showed generally similar results. Figure 15 shows profiles of momentum flux for three different values of upstream humidity, applied uniformly at all levels, and also for a case with 100 percent humidity plus cloud water only in low levels, between the heights of 667 m and 3000 m above the upstream surface. This was thought to be representative of the wettest realistic moisture profiles which might be commonly encountered upstream of a midcontinental mountain range. The profiles are normalized by the flux to be expected for linear waves. The dry case shows a non-linear amplification as predicted by Miles and Huppert [16] and Smith [18]. The flux for the high humidity cases is reduced by about a factor of 3, in general agreement with our observations from real data cases.

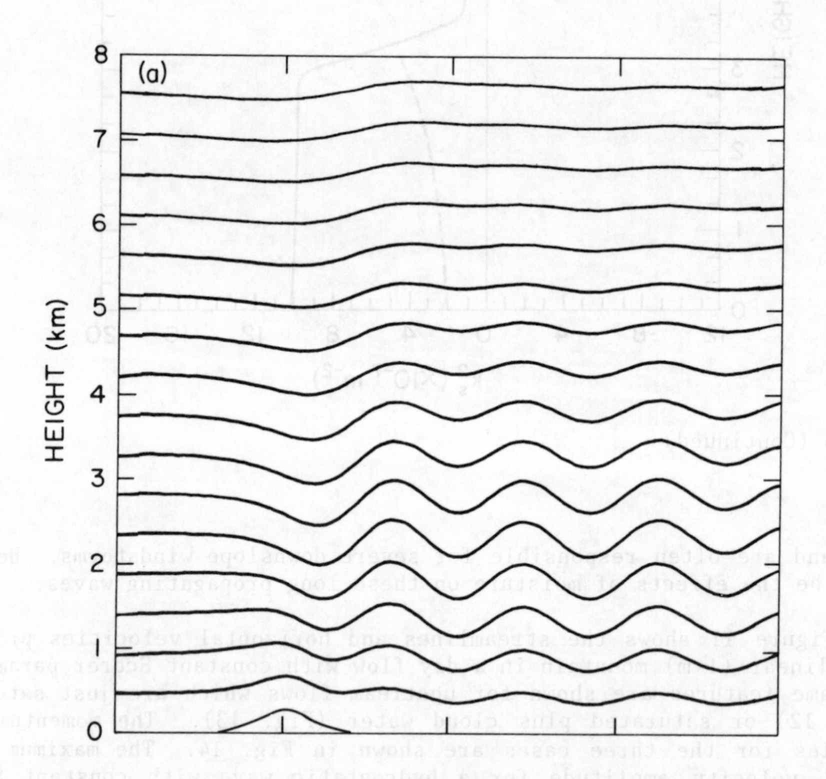

Fig. 7 (a) Streamlines, and (b) vertical velocities ($x10^{-3}$ ms^{-1}) produced by a 1 m high mountain when RH=0 percent upstream. The streamline displacements are multiplied by 300 for display.

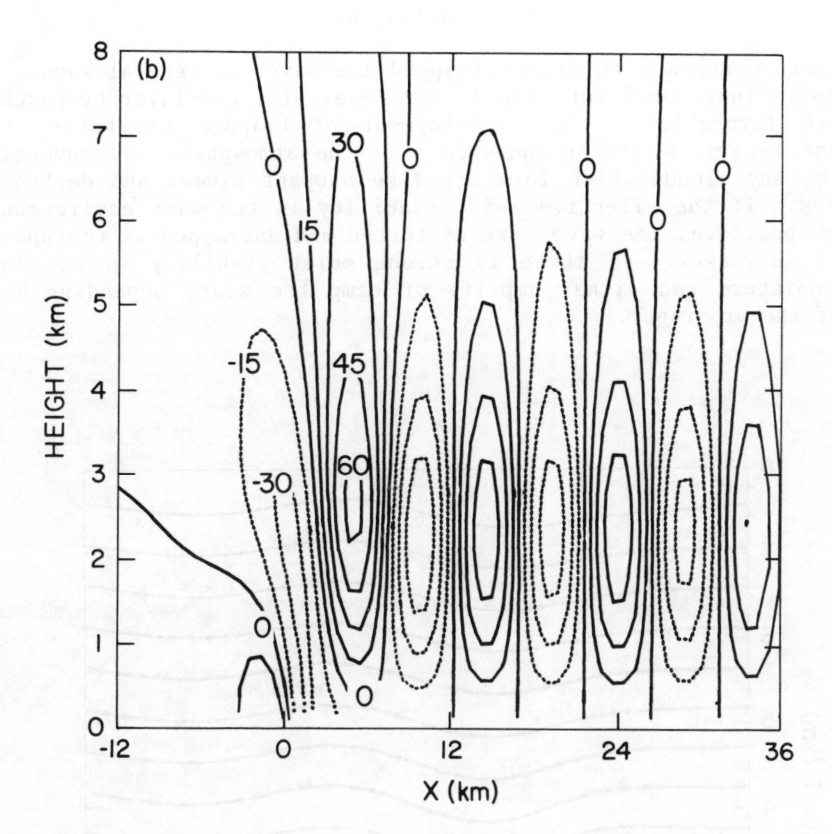

Fig. 7 (Continued)

All the previous results were obtained with the precipitation mechanism suppressed in the model. Precipitation is not generally important for the short trapped waves, but with the longer waves more time is available for microphysical processes to occur. Our study of these effects is not complete. Figure 16 shows examples, however, of the effects of rain on momentum flux profiles obtained for the finite amplitude shear flow case. In general the presence of rain increases the wave and flux amplitudes, though not as much as the presence of clouds decreases them from the dry results. The apparent reason is that rainout reduces the moisture content enough so that the descending lee flow becomes fully dry sooner and more completely. The "instant rainout" case in Fig. 16b is obtained by immediately precipitating all condensed water, instead of allowing some of it to remain as cloud water. This allows a much stronger wave response. The gradients of momentum flux shown at the bottom of the profiles in rainy conditions are associated with the net heat release. In this situation the Eliassen-Palm theorem of constant momentum flux no longer holds.

6. Conclusions

Moisture appears to affect trapped lee waves in several ways. When moisture is introduced into the lowest layer of a two-layer tropospheric structure favorable for the development of trapped lee waves, three different behaviors are encountered. If the atmosphere is convectively unstable, any clouds which form act like buoyant plumes and destroy the lee waves. If the effective moist stability in the wave environment is weak but positive, the waves are distorted and untrapped as the upstream humidity increases. If there is strong moist stability in the lowest layer, moisture can either amplify or damp the wave, depending on the scale of the orography.

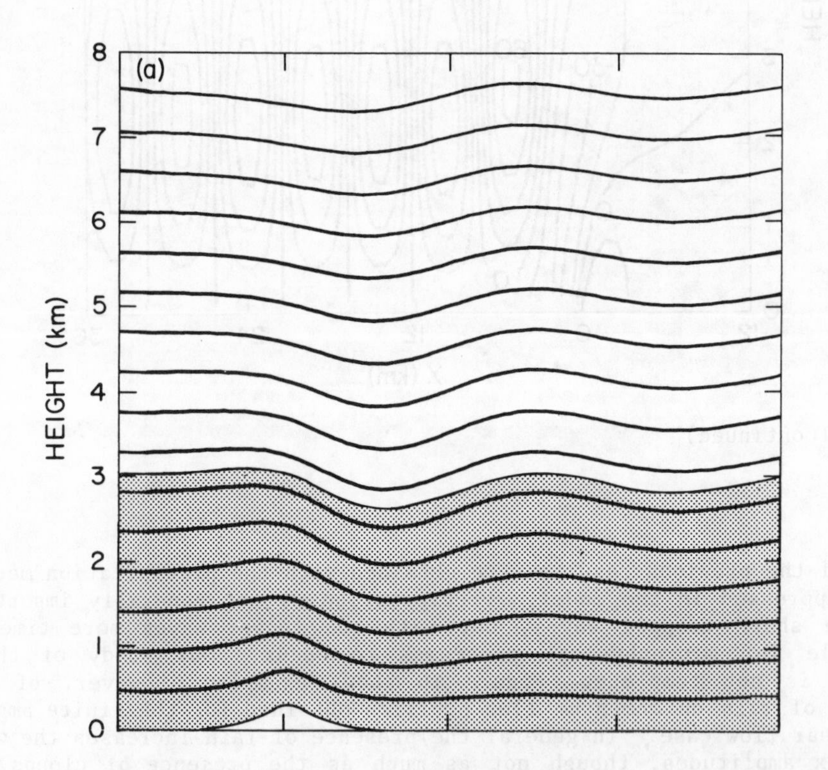

Fig. 8 As in Fig. 7, except that RH=100 percent and there is 0.2 gm/kg of cloud in the lowest layer upstream. Cloudy regions are stippled.

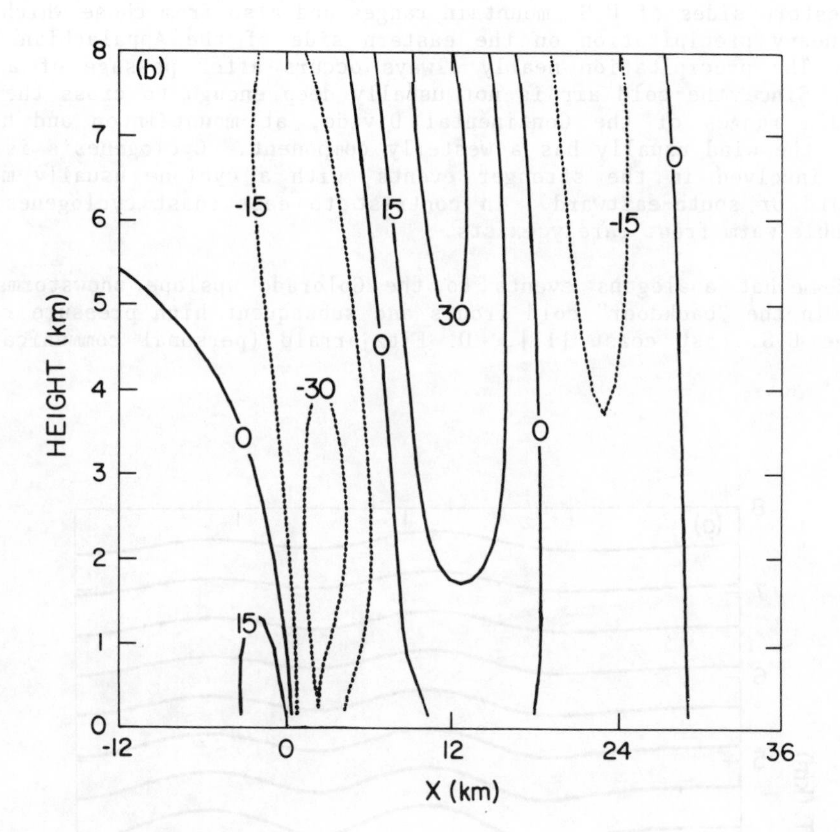

Fig. 8 (Continued)

In the case of propagating waves the only atmospheres examined were those in which the dry Scorer parameter did not change abruptly with height, so there was little downward partial reflection of wave energy. When precipitation is not present, clouds act to damp the waves and reduce the momentum flux. When precipitation occurs, the wave amplitudes and momentum fluxes are stronger than those produced by nonprecipitating flows, but still weaker than in the dry case. The momentum flux in propagating waves with precipitation is not constant with height.

II. OBSERVATIONS OF UPSLOPE SNOWSTORMS

1. Introduction

On the eastern slopes of the Colorado Rocky Mountains most cold season precipitation occurs in association with an easterly flow component at low levels. The synoptic weather conditions favoring these events differ in many ways from those producing upslope precipitation on the western sides of U.S. mountain ranges and also from those which produce heavy precipitation on the eastern side of the Appalachian Mountains. The precipitation nearly always occurs after passage of a cold front. Since the cold air is not usually deep enough to cross the high mountain ranges of the Continental Divide, at mountaintop and higher levels the wind usually has a westerly component. Cyclogenesis is normally involved in the stronger events, with a cyclone usually moving eastward or south-eastward. In contrast to east coast cyclogenesis, a definable warm front rarely exists.

Somewhat analogous events to the Colorado upslope snowstorms are found in the "backdoor" cold fronts and subsequent high pressure ridges of the U.S. east coast [17]. D. Fitzjarrald (personal communication)

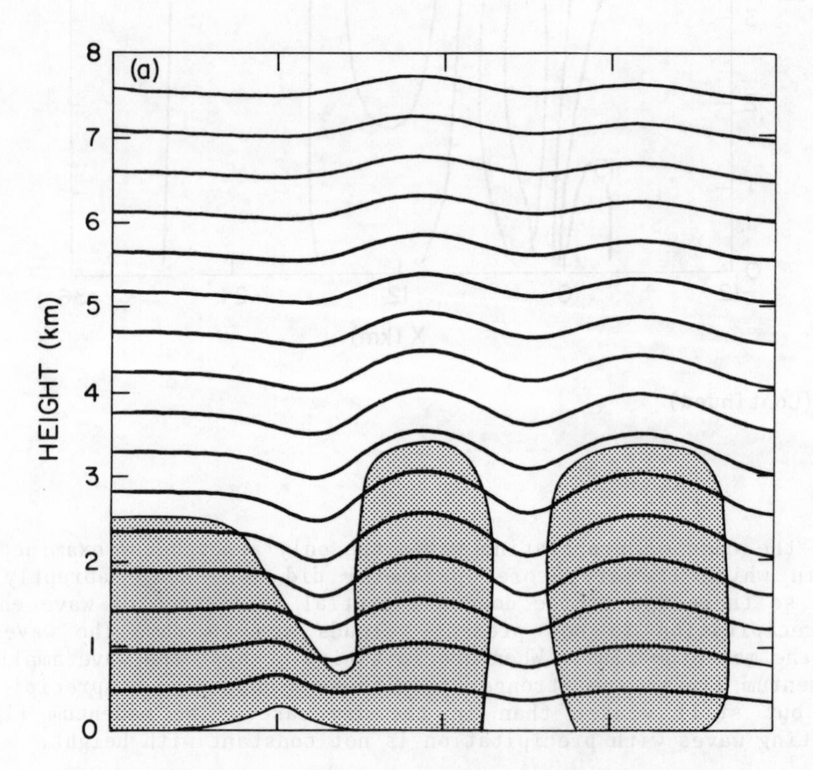

Fig. 9 As in Figs. 7 and 8, except that RH=100 percent in the lowest layer upstream.

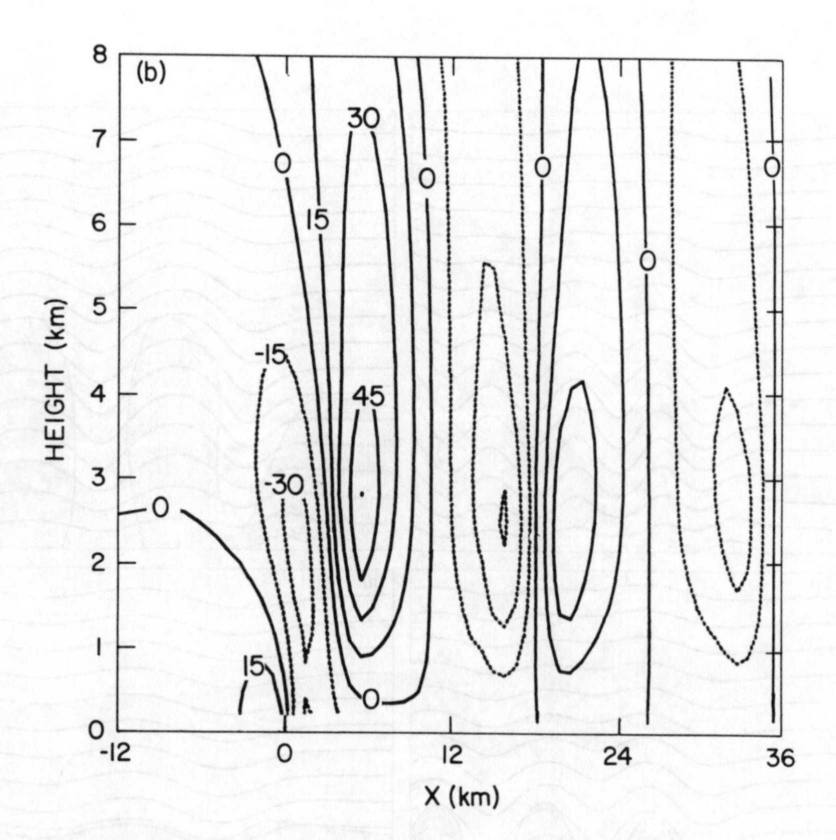

Fig. 9 (Continued)

reports that cloud-topped low-level upslope flow is a climatological feature in the winter on the east coast of Southern Mexico. Some of the upslope precipitation events on the north side of the Alps also resemble those in Colorado, but with most features rotated 90° counterclockwise. There the low-level flow is from the northwest, with southerly component winds passing over the mountains. When cyclogenesis occurs, however, it is usually found on the warm (southern) side of the mountains.

2. Observing Methods

We have conducted observations on several upslope precipitation events, using Doppler radar as the primary observing system. This tool has for many years been used to investigate the structure of precipitation events. Its effectiveness for such purposes has increased greatly with improved display systems and data processing techniques. Wilson et al. [19] have demonstrated the value of a single Doppler radar, so equipped, for defining the character of significant weather events. The NCAR CP-3 radar system, which transmits, receives and records a 5-cm

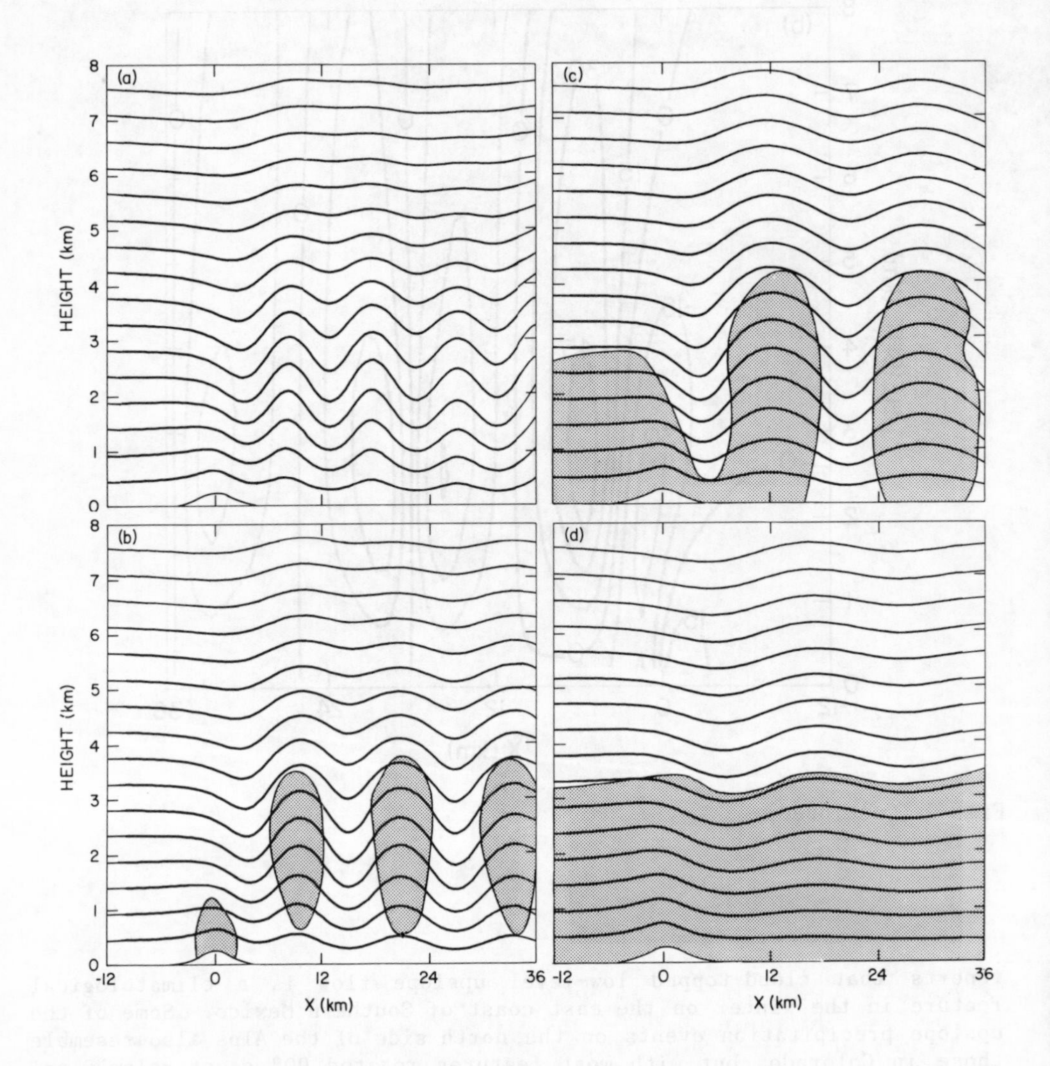

Fig. 10 Streamlines produced by a 300 m high mountain in the flow shown in Fig. 6, for (a) RH=0 percent, (b) RH=90 percent, (c) RH=100 percent, (d) RH=100 percent with 0.2 gm/kg of cloud, in the lowest layer upstream. Cloudy regions are stippled.

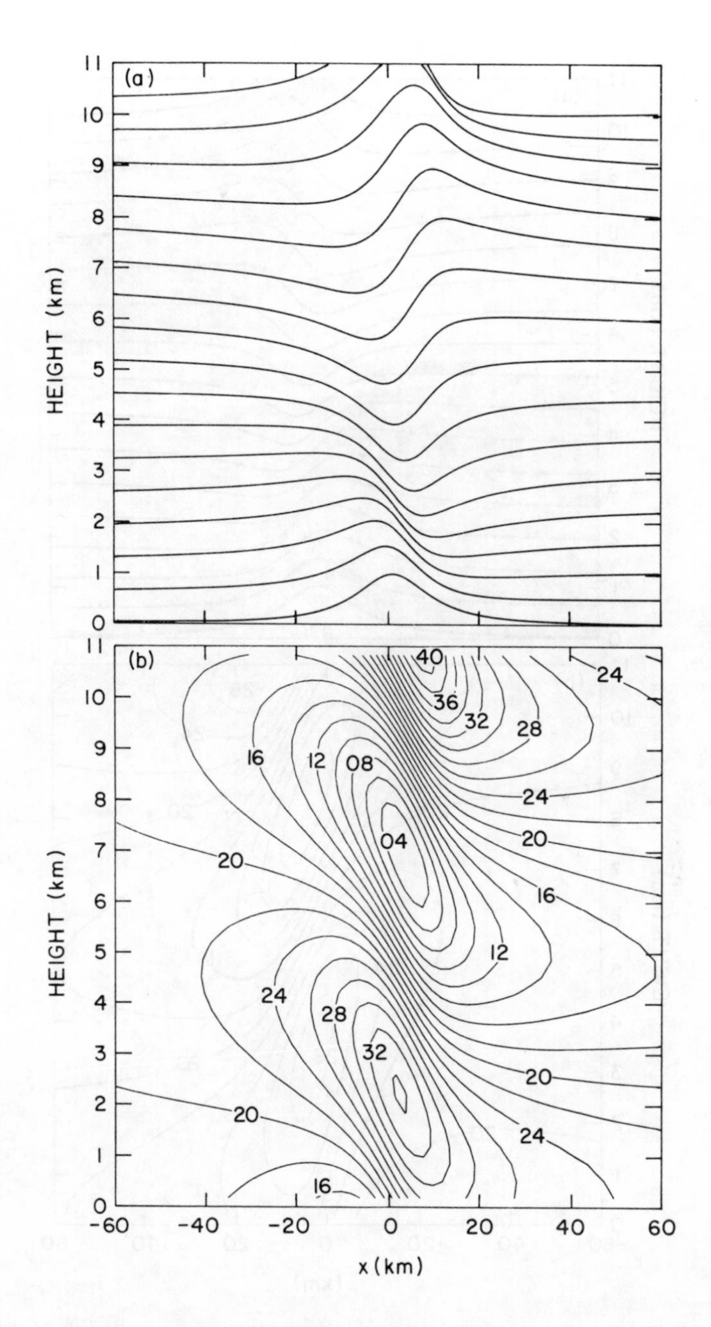

Fig. 11 (a) Streamlines, and (b) horizontal velocities produced by a 1 km high mountain in a flow with constant windspeed and stability when RH=0 percent upstream. The perturbations are multiplied by 1000 for display. As such they constitute a linear numerical solution for a 1 km high mountain.

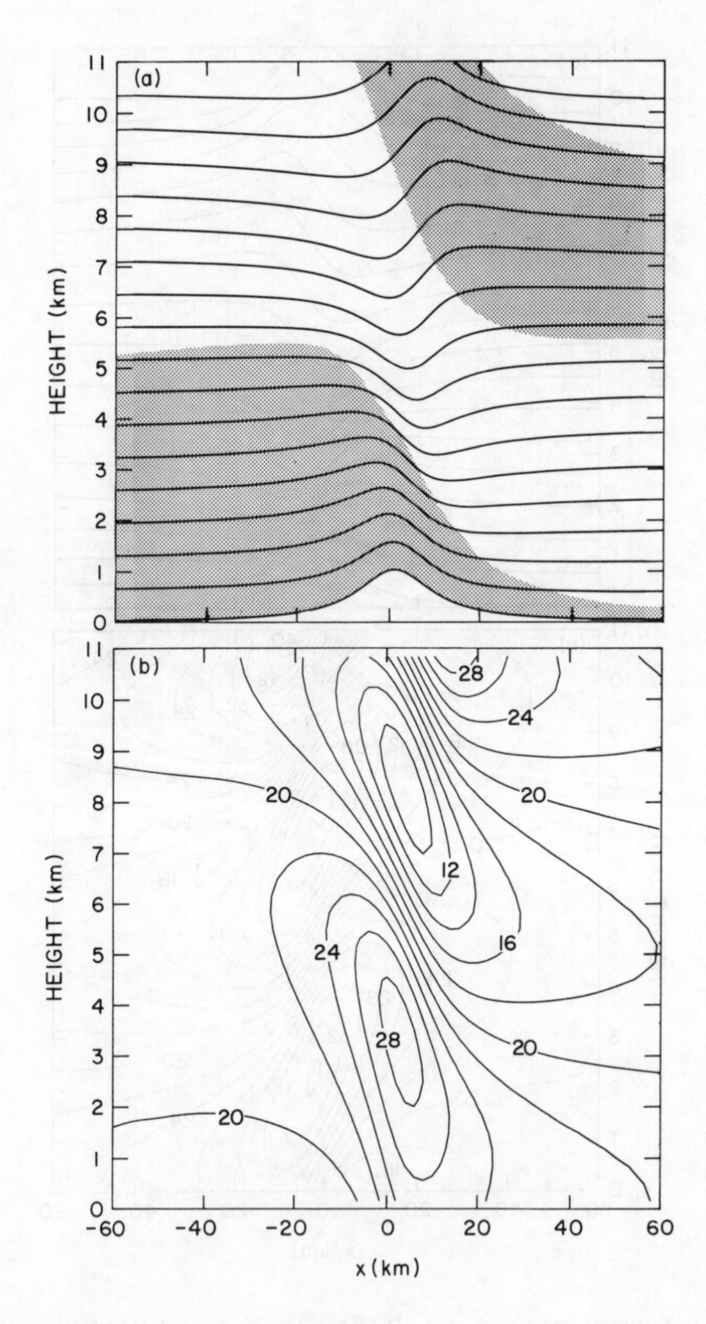

Fig. 12 As in Fig. 11, except that RH=100 percent in the upstream flow. Cloudy regions are stippled.

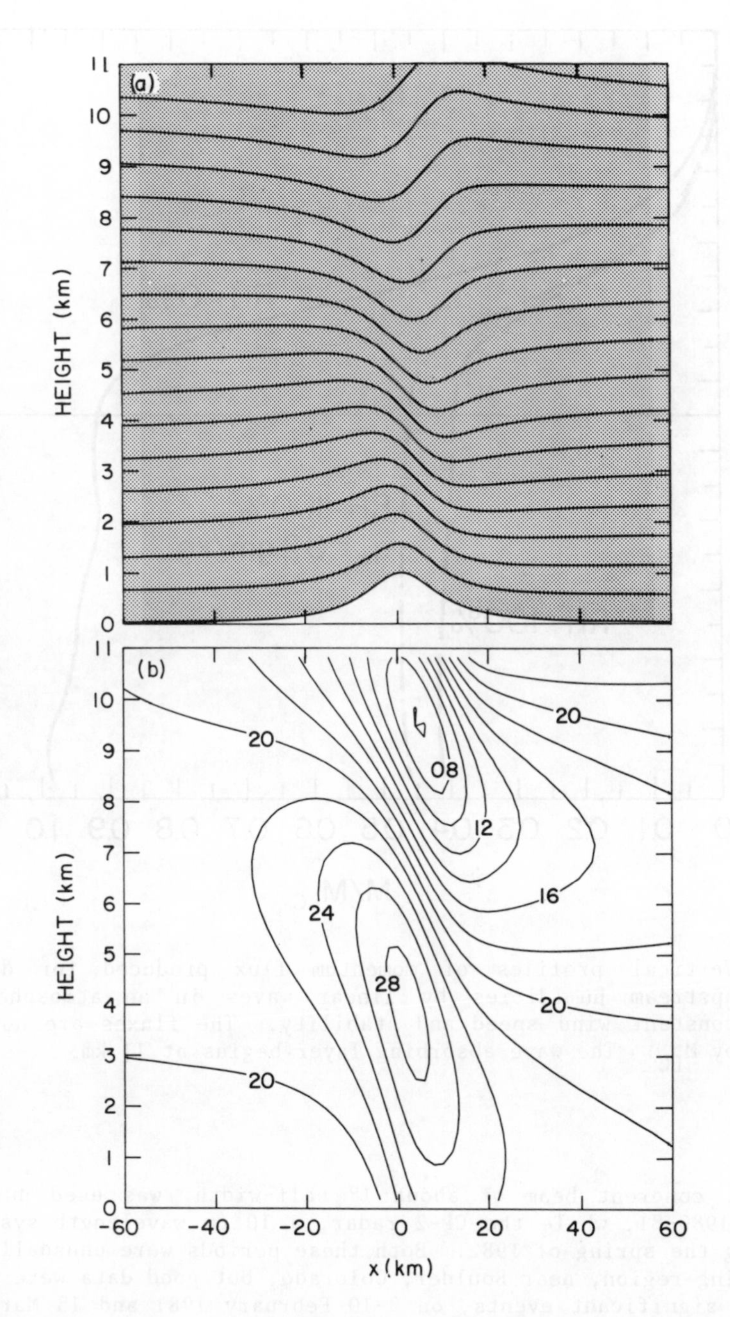

Fig. 13 As in Fig. 12, except that RH=100 percent with 0.2 gm/kg of cloud in the upstream flow.

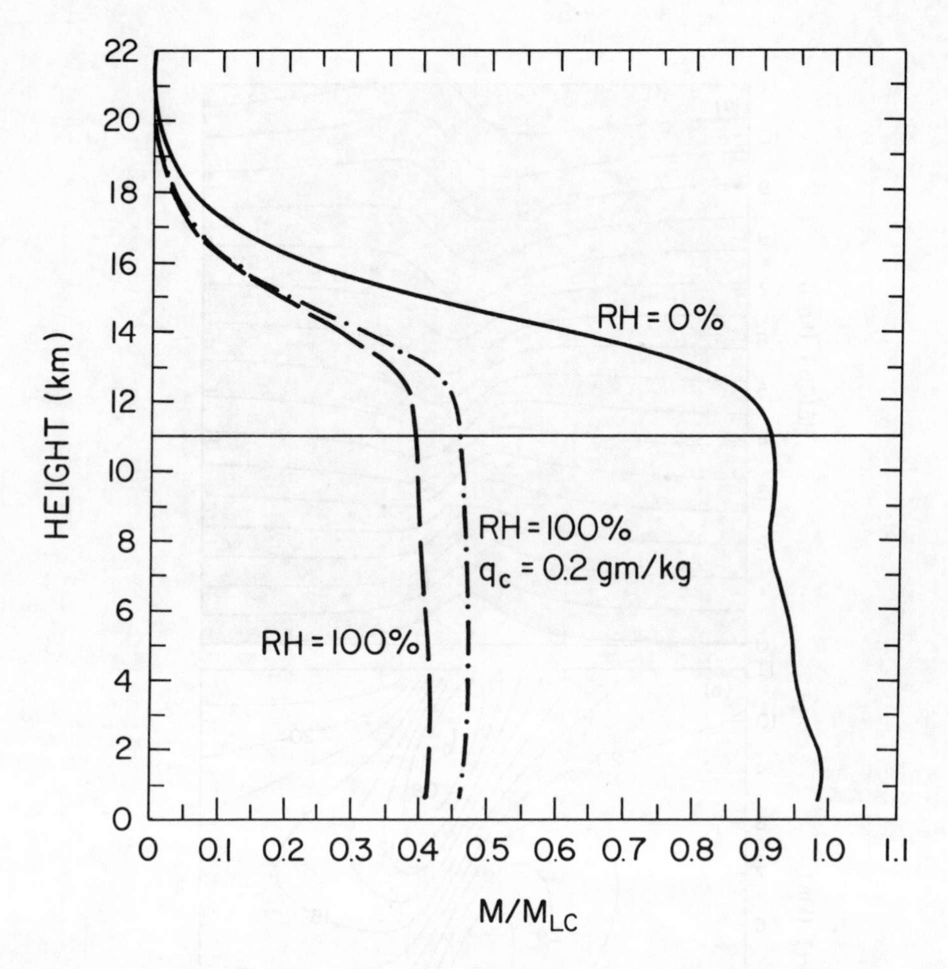

Fig. 14 Vertical profiles of momentum flux produced for different upstream humidities by linear waves in an atmosphere with constant wind speed and stability. The fluxes are normalized by M_{LC}. The wave absorbing layer begins at 11 km.

wavelength, coherent beam of about 1° half-width, was used during the winter of 1980-81, while the CP-2 radar, a 10 cm wavelength system, was used during the spring of 1982. Both these periods were unusually dry in the observing region, near Boulder, Colorado, but good data were obtained during two significant events, on 9-10 February 1981 and 15 March 1982. The data to be shown here were taken during the first event. Many aspects appear to be similar for the second, however, as well as for several weaker or less well-defined events.

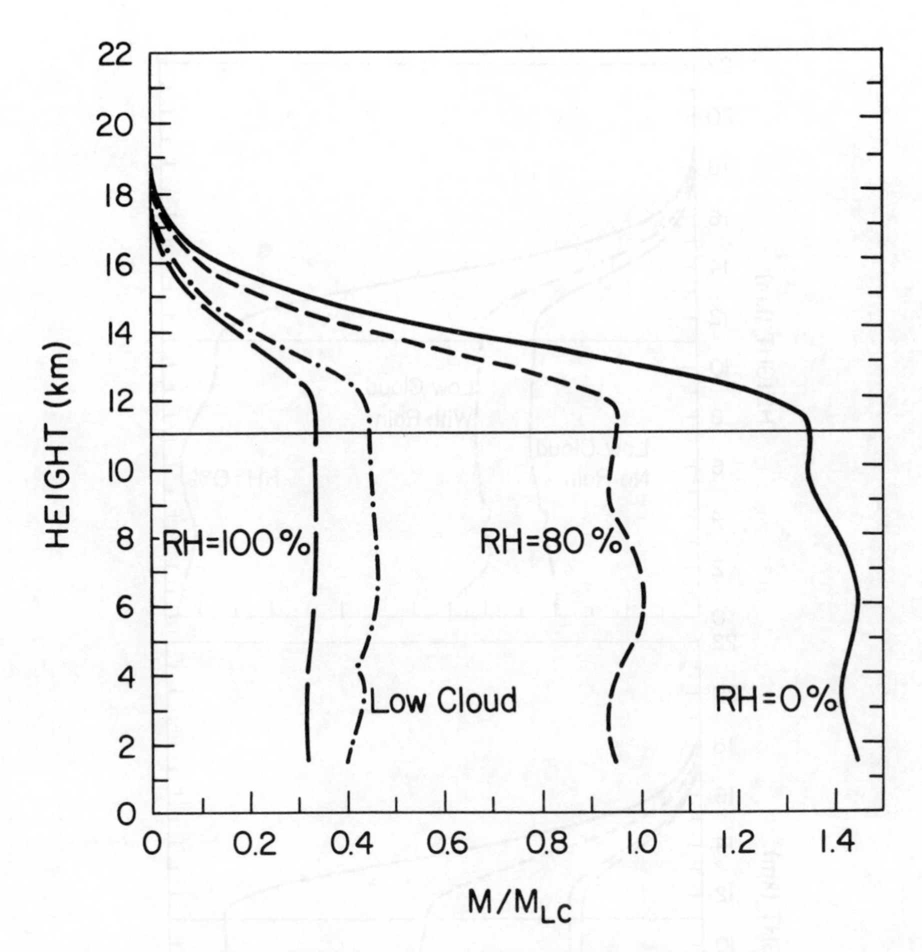

Fig. 15 Vertical profiles of momentum flux produced for different upstream humidities by finite amplitude waves in a flow with constant stability and tropospheric wind shear. The fluxes are normalized by M_{LC}. The wave absorbing layer begins at 11 km.

A single Doppler radar, as used in this program, measures only two properties of the atmosphere -- its reflectivity to microwaves and the radial velocity component of the reflector in the direction of the beam. Because of its high resolution, 1° in azimuth and elevation angles and 150 m in range, and high inherent accuracy in the velocity determination, about 0.2 m s^{-1} in ideal conditions, it is a very powerful tool for mesoscale observation. A variety of limitations, some inherent, some probably removable by further technical improvements, must always be considered, however. The lack of measurement of the tangential velocity

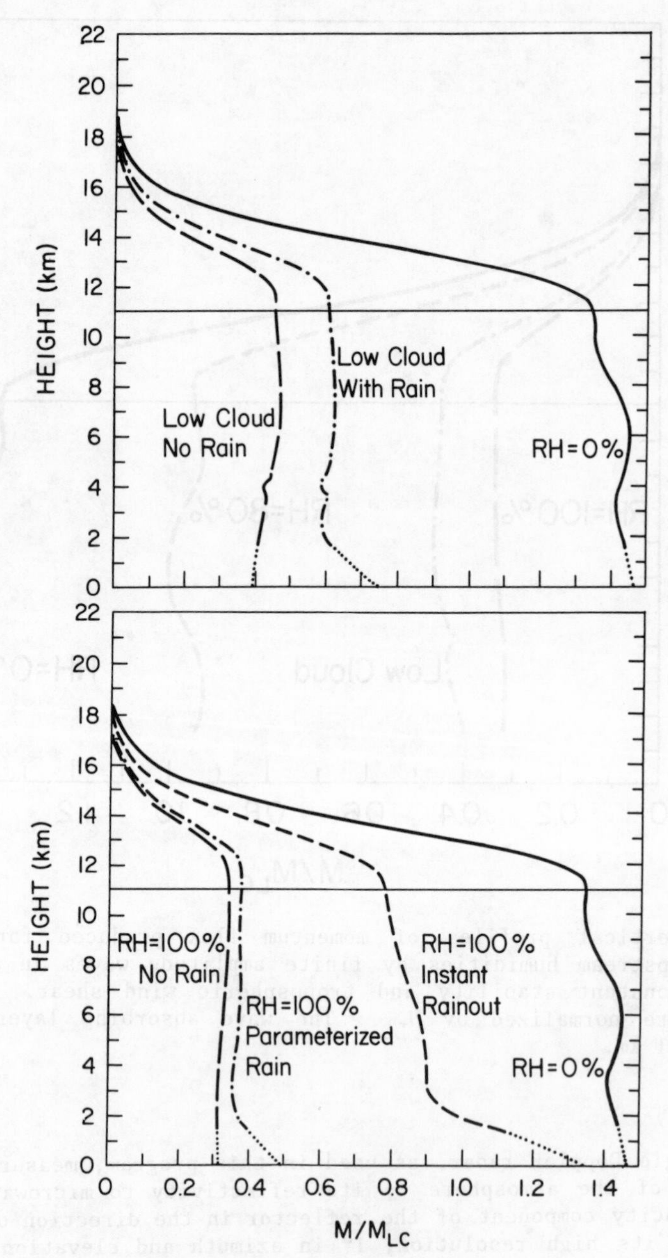

Fig. 16 The effects of rain on the vertical profiles of momentum flux produced by upstream moisture profiles in which (a) there are low clouds between the heights of 667 m and 3000 m, and (b) RH=100 percent everywhere. The fluxes are normalized by M_{LC}. The pressure drag associated with each wave is plotted at z=0 and connected to the appropriate momentum flux profile by a dotted line. The wave absorbing layer begins at 11 km.

components is often a severe deficiency. In the circumstances found in the upslope snow events it is not crucial, since the flow patterns are fairly simple and spatial variability is not great. A more serious problem is the masking and confusion of the weather-produced data by reflections from the lower surface, especially when these "ground clutter" returns are enlarged by antenna side lobes. The clutter problems were more severe in this environment than in some others because of the presence of strong topographic features in various directions and because most of the phenomena of interest were confined to the lowest 2-3 km above the surface. Ground clutter can usually be recognized and removed or ignored in analysis because of its high reflectivity and zero or near zero velocity, but this does not allow restoration of the true weather-produced signals. More effective clutter suppression techniques are needed.

For investigation of upslope storms the Doppler radar data were principally used to determine mean wind profiles, general indications of spatial variability, details of some characteristic secondary flow features, and measurement of turbulent fluctuations by means of spatial structure functions. No "second moment", or velocity variance, data were recorded. The basic mode of radar operation was a series of complete circular azimuthal scans at increasing elevation angles. During some periods r-z profiles were obtained, for which the radar scans in elevation angle, with the azimuth angle changed between scans. For all the data used here the pulse repetition rate was 1428 Hz, which produces a maximum unambiguous range of about 74 km and velocity ambiguities of about 19 m s^{-1}. These ambiguities are an inherent Doppler limitation, but do not severely limit our purposes. Since the real velocity spatial gradients are small, the ambiguity can fairly easily be removed by automated gate-to-gate and scan-to-scan comparison techniques. The range ambiguity is insignificant because distant echo returns are weak and mostly over the horizon.

Besides the radar some other kinds of data were available and valuable for interpretation. A telemetered surface data network developed for the NOAA PROFS (Prototype Regional Observing and Forecasting Services) was in partial operation in north-central Colorado and produced apparently valid measurements of surface winds and temperature fields at 30 min intervals and about 50 km resolution. A second 5-station network, located along a north-south line through Boulder at approximately 5 km spacing, provided one-minute records of wind and temperature. The Boulder Atmospheric Observatory Meteorological Tower, located about 15 km northeast of the radar site, measured mean state profiles to its height of 300 m. No deeper soundings of the thermodynamic structure were available, however, except those taken by the Weather Service in Denver at 00Z and 12Z, mostly before and after the snowstorm event. The vertical thermal structure has been partially determined by means of analysis of the high altitude PROFS stations.

3. Meteorological Evolution

The 9-10 February snowstorm was the only period of extreme cold during the winter in eastern Colorado, but came close to setting records.

A severe Arctic front plunged southward down the eastern slope during the night, accompanied by 2 to 15 cm of snow and temperatures which dropped to -25°C or colder. Figures 17 and 18 show the surface and 500-mb maps as the storm was diminishing in the Denver area but moving eastward as an intensifying lee cyclone.

Figures 19a-c show a sequence of wind and potential temperature data observed by the PROFS surface net. The station altitudes are listed in Table 1. The potential temperatures have, for convenience, been computed using the altitude of the station at Boulder, 1629 m, as a base. The potential temperature based on the 1000-mb reference surface would be about 16° higher. At 2100 local time most of the plains stations measured these "Boulder potential temperatures" of about -5°C to -8°C while the mountain stations to the west were 10°C to 12°C warmer. The plains stations were 3°C to 6°C cooler than the surface temperature of the Denver 00Z radiosonde (1700 local time), shown in Fig. 20. This can probably be attributed to normal diurnal cooling. Surface winds at 2100 were very light and mostly northerly or easterly, except at the mountain stations, where a westerly direction predominated and at the north-eastern four stations, where increased northeasterly winds were observed. Much stronger cooling had also occurred at the northernmost station, Nunn. The Arctic front, which had been dropping through Wyoming during the day, had evidently passed across the Cheyenne Ridge, a row of hills along the Colorado-Wyoming border about 500 m above the plains to the north and south.

After 2100 the front rapidly moved southward, with temperatures dropping about 15°C in its wake, and the colder air also rose through the lower two of the mountain stations. Squaw Mountain, a station about 1900 m above Boulder, only cooled about 5°C during the night and remained much warmer than the Arctic air mass. Westerly or northwesterly flow persisted at this station. Since the lowest point on the Continental Divide is at about the same level as Squaw Mountain, it is apparent that all of the Arctic air was confined to the eastern slopes. Snow began falling immediately after the well-marked frontal wind surge at Boulder and continued during the night, gradually decreasing during the early morning. Similar sequences are believed to have occurred throughout the area, although snow was probably falling before the frontal passage at the mountain stations.

The 12Z Denver sounding (Fig. 20) shows the cold air about 1.3 km deep, surmounted by a strong inversion. Winds in the cold air are from slightly east of north, backing rapidly to west-northwest through the frontal surface. The air above the front, though having moved in from the west, has a similar temperature profile to that of the sounding from the previous afternoon lifted about 1200 m, except for the much lower tropopause.

4. VAD Analysis Results

Most of the conical scan data obtained during the snowstorm were subjected to VAD (velocity-azimuth display) analysis. In this technique,

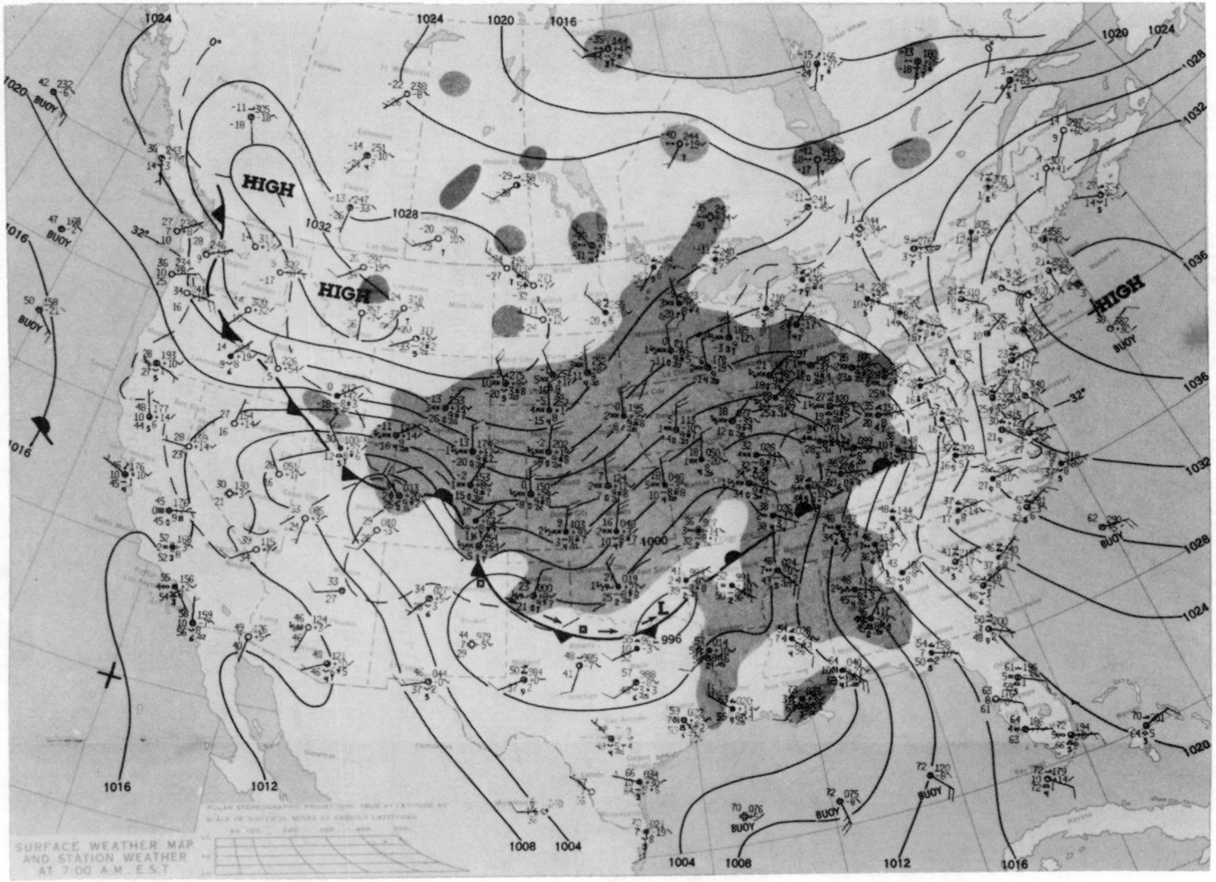

Fig. 17 Surface analysis from National Weather Service at 12Z, 10 February 1981. Station temperatures are in degrees F.

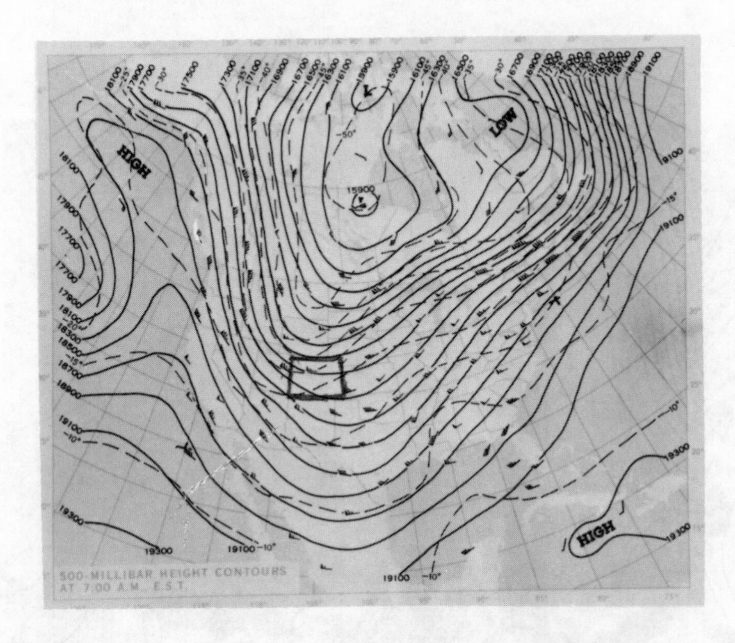

Fig. 18 500-mb analysis at 12Z, 10 February 1981. State of Colorado outlined.

Table 1. Altitudes of PROFS Stations (meters)

Station	Altitude	Station	Altitude
Virginia Dale	2143	Brighton	1518
Nunn	1580	Fritz Peak	2750
Fort Collins	1609	Arvada	1642
Loveland	1526	Aurora	1625
Greeley	1415	Byers	1555
Estes Park	2369	Lakewood	1832
Fort Morgan	1387	Squaw Mountain	3506
Longmont	1533	Littleton	1750
Keenesburg	1482	Louviers	1707
Boulder	1629	Elbert	2132

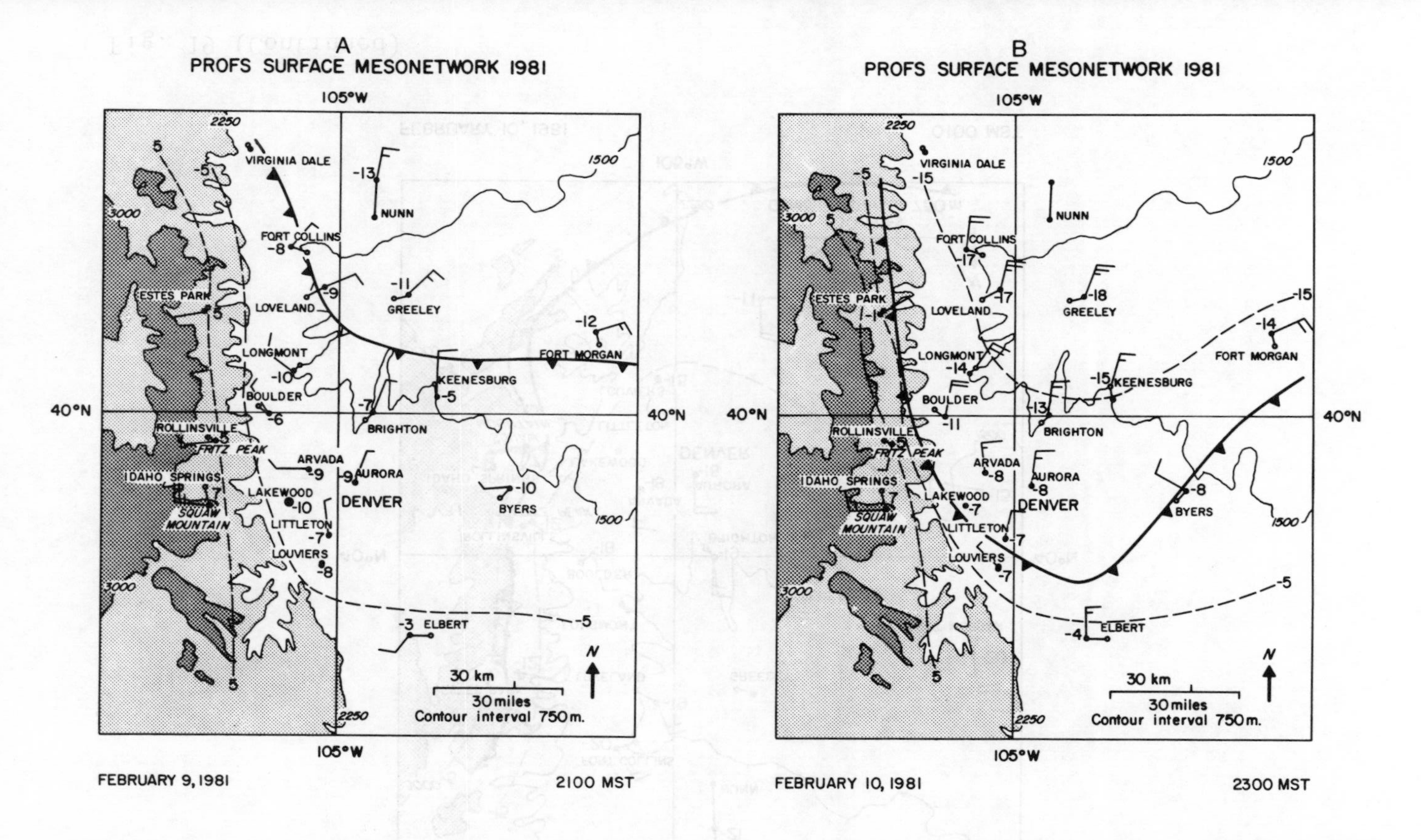

Fig. 19a-c Surface analysis at 2100, 2300, 0100 local time, 9-10 February (04-08Z), from PROFS surface net in north-central Colorado. Altitude contours are in meters, temperatures in degrees C, reduced dry-adiabatically to the altitude of Boulder. A full wind barb is 10 knots. The frontal position is based mainly on wind speed. The station locations are shown as solid dots, the nearest towns by open circles.

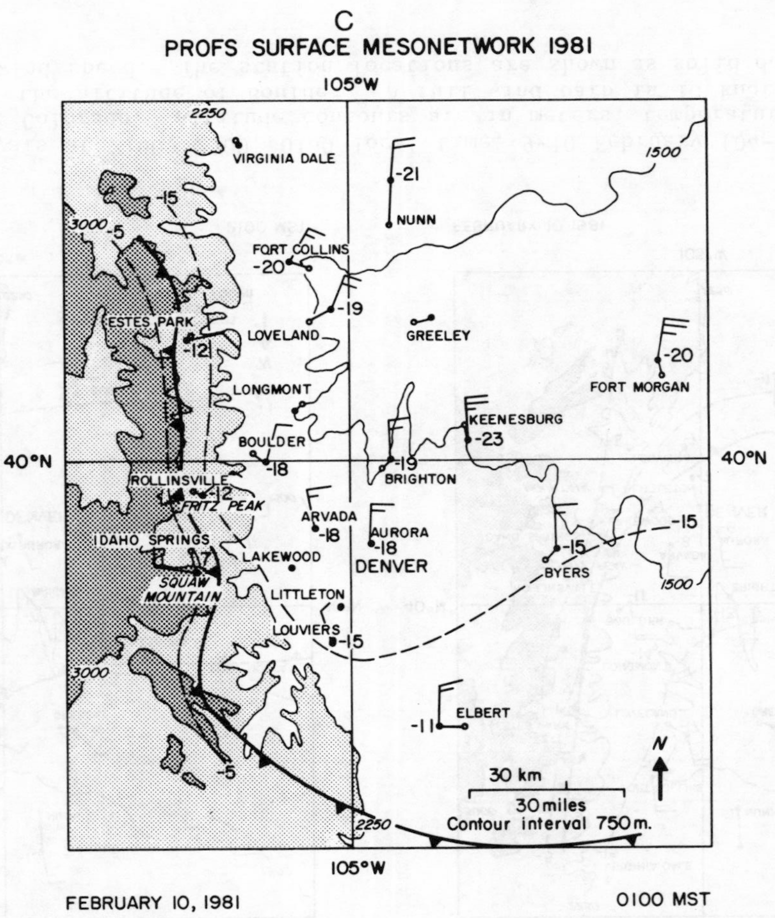

C

PROFS SURFACE MESONETWORK 1981

FEBRUARY 10, 1981

0100 MST

Fig. 19 (Continued)

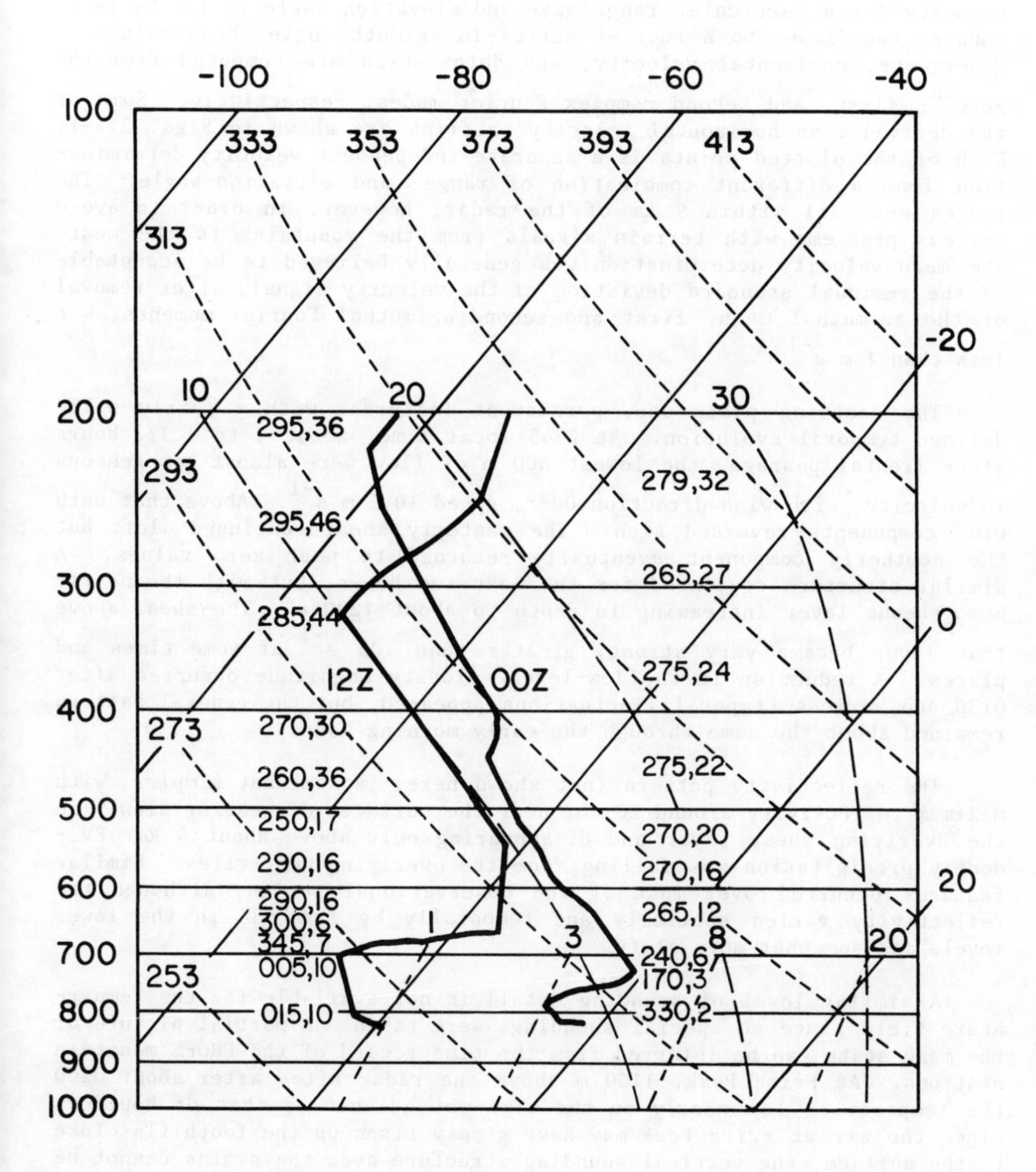

Fig. 20 Temperature and wind sounding profiles at Denver on 10 February 1981. The winds are shown by direction and speed in knots, with the 00Z sounding to the right and the 12Z sounding to the left.

due originally to Browning and Wexler [3], the Doppler-measured radial velocity for a particular range gate and elevation angle is fit by least squares techniques to a Fourier series in azimuth angle. Mean values of divergence, horizontal velocity, and deformation are computed from the zeroth, first, and second complex Fourier modes, respectively. Some of the derived mean horizontal velocity profiles are shown in Figs. 21a-f. Each of the plotted points is a separate independent velocity determination from a different combination of range- and elevation-angle. The ranges were all within 9 km of the radar, however, in order to avoid serious problems with terrain signals from the mountains to the west. The mean velocity determination was generally believed to be acceptable if the residual standard deviation of the velocity signal, after removal of the azimuthal mean, first and second azimuthal Fourier moments, was less than 1 m s^{-1}.

The sounding plots show persistent features, with a fairly well-defined temporal evolution. At 2345 local time, about 1 to 1 1/2 hours after frontal passage, the lowest 600 m of flow were almost homogeneous in velocity, with wind direction 008°, speed 10.5 m s^{-1}. Above that both wind components reversed sign. The westerly shear continued aloft but the southerly component eventually returned to near zero values. A similar structure continued for the next two hours, but with the nearly homogeneous layer increasing in depth to about 1200 m. The shear above that layer became very strong, greater than .05 s^{-1} at some times and places. A reduction in the low-level velocity magnitude occurred after 0130 and various temporal fluctuations appeared, but the general pattern remained about the same through the early morning hours.

The reflectivity pattern (not shown here) is somewhat simpler, with maximum reflectivity around 20 dBZ near the surface, decreasing slowly in the overlying shear layer and disappearing only above about 4 km. Evidently precipitation was falling from the overlying westerlies. Similar features occurred over most of the observational period, although the reflectivity varied spatially and temporally by ± 5 dBZ in the lower levels and somewhat more aloft.

A similar level of sounding detail is not available for the temperature field since no special soundings were taken. A partial picture of the mean state can be inferred from the time record of the PROFS mountain stations. At Fritz Peak, 1100 m above the radar site, after about 0200 the temperature lay nearly on the same wet adiabat as that of Boulder. Since the air at Fritz Peak may have simply risen up the foothills close to the surface, the vertical sounding structure over the plains cannot be unambiguously determined from this data. Because the Arctic air was moving in rapidly over the initially warmer surface, it seems plausible to assume that the cold air mass over Boulder and the foothills was nearly homogeneous in θ_e (defined with respect to ice saturation), and so could be regarded as a well-mixed layer. Between 2300 and 0100 the temperature at Fritz Peak dropped about 18°C, during which time the apparent frontal surface over Boulder lifted about 500 m, indicating a

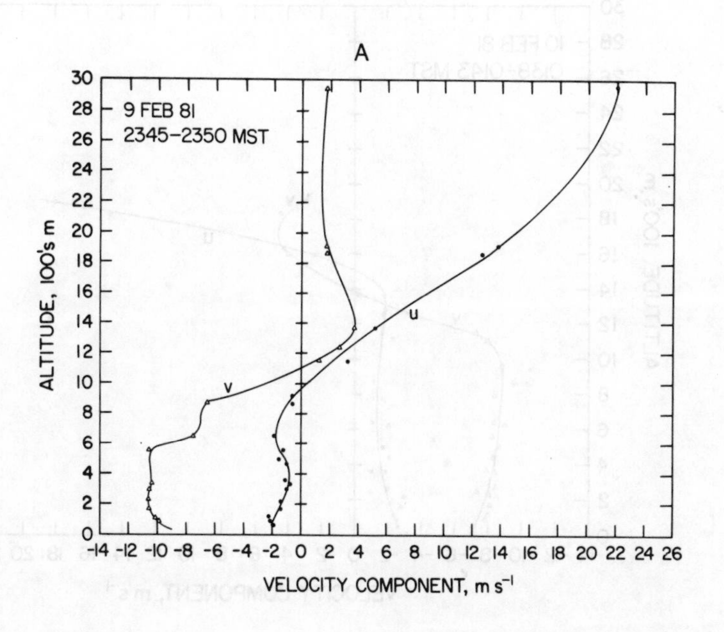

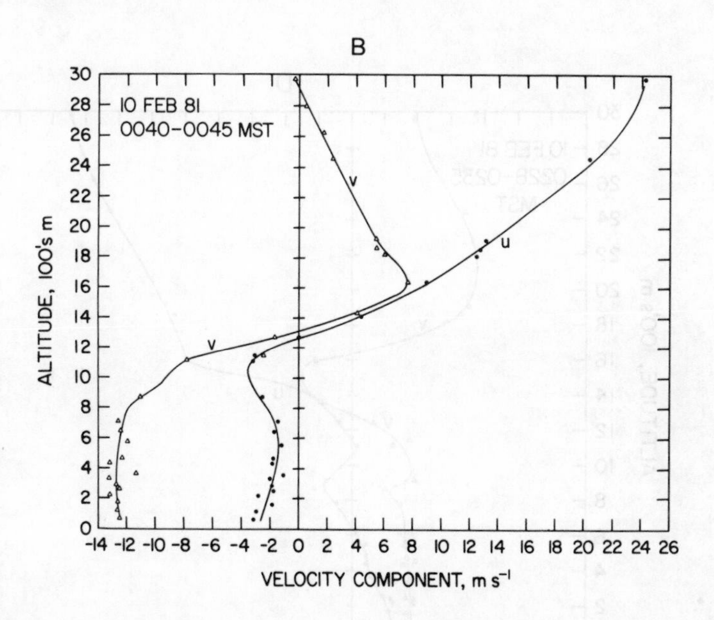

Fig. 21a-f Wind component profiles determined by analysis of Doppler radar data taken near Boulder starting at about 1-1/2 hours after frontal passage. Each point corresponds to a determination from a different elevation angle and range gate but all are within 9 km of the radar site.

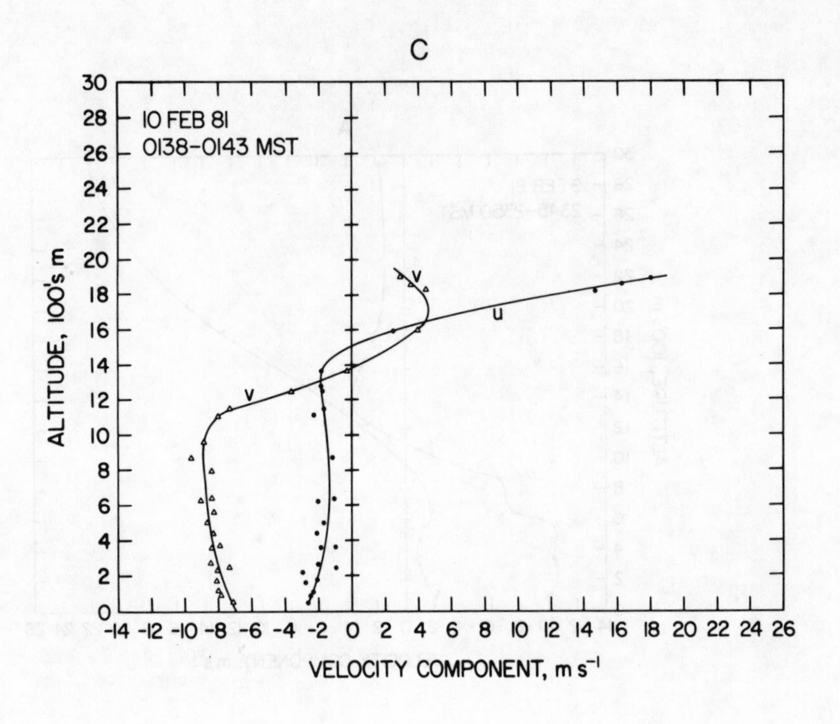

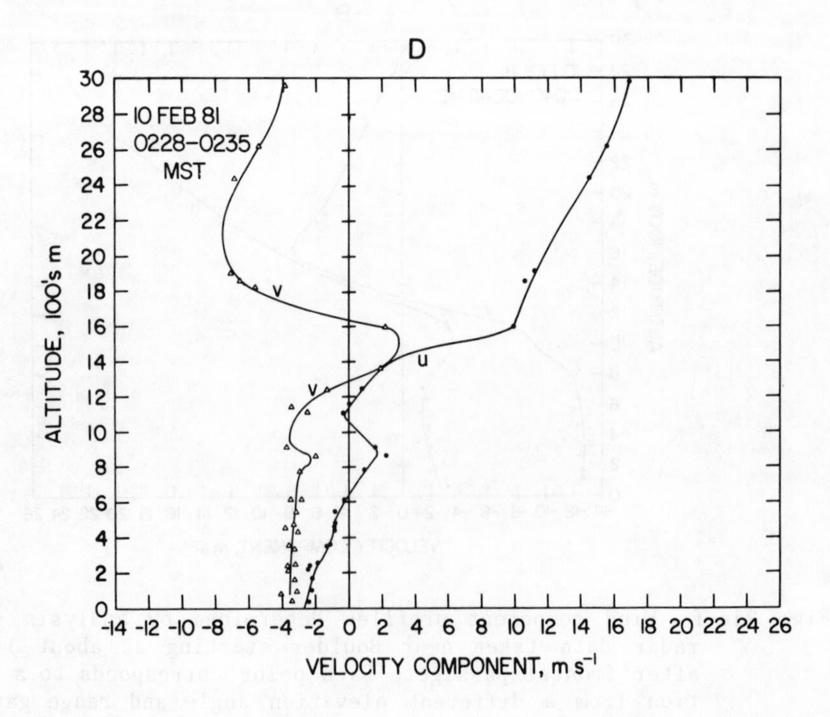

Fig. 21 (Continued)

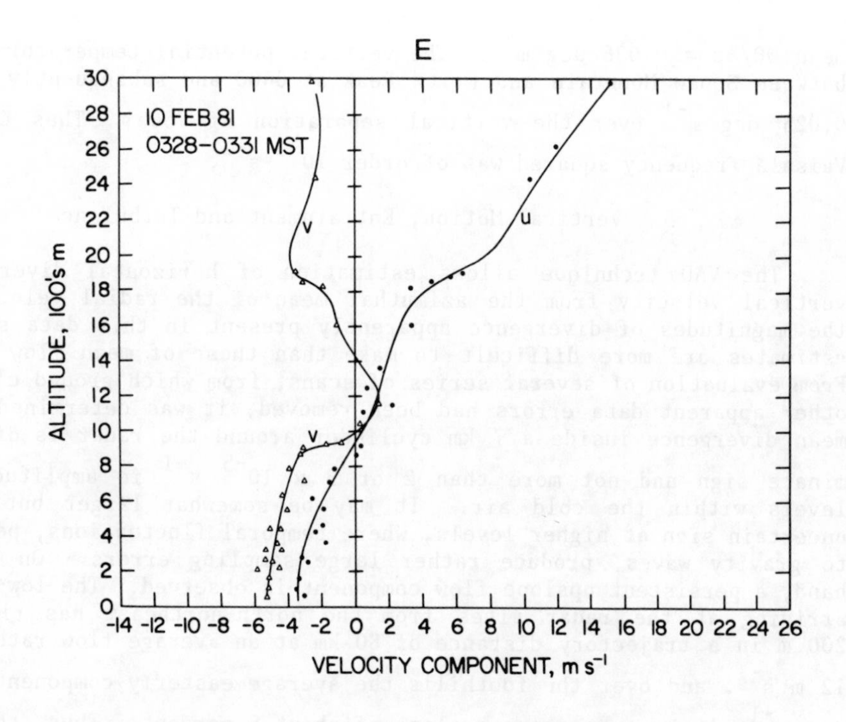

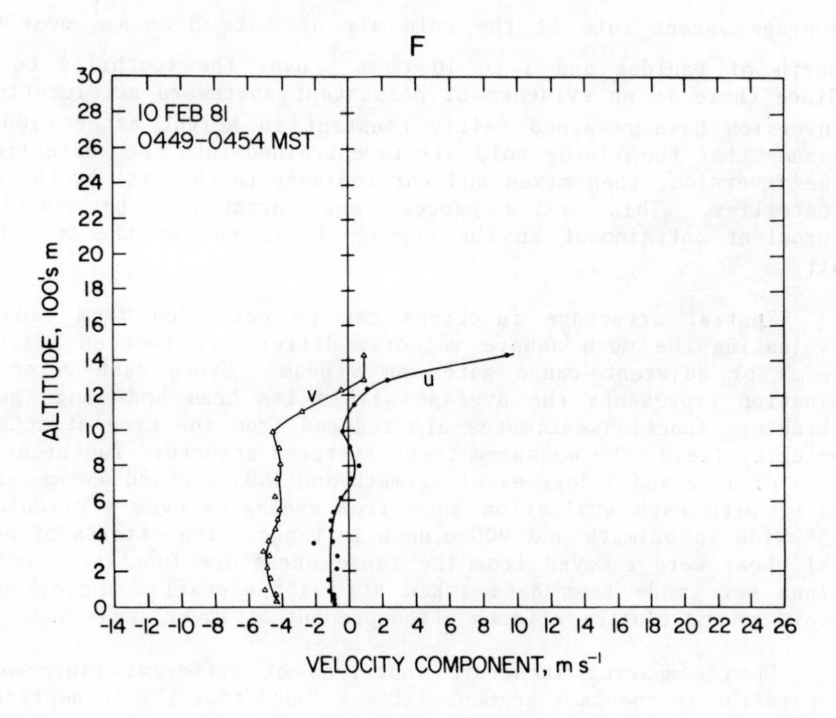

Fig. 21 (Continued)

mean $\partial\theta/\partial z = 0.036$ deg m^{-1}. The vertical potential temperature gradient between Squaw Mountain and Fritz Peak at 0400 and subsequently was about 0.025 deg m^{-1} over the vertical separation of 750 m. Thus the Brunt-Vaisala frequency squared was of order 10^{-3} s^{-2}.

5. Vertical Motion, Entrainment and Turbulence

The VAD technique allows estimation of horizontal divergence and vertical velocity from the azimuthal mean of the radial velocity. For the magnitudes of divergence apparently present in this data set, these estimates are more difficult to make than those of mean flow velocity. From evaluation of several series of scans, from which ground clutter and other apparent data errors had been removed, it was determined that the mean divergence inside a 7 km cylinder around the radar is of indeterminate sign and not more than 2 or 3 x 10^{-5} s^{-1} in amplitude at all levels within the cold air. It may be somewhat larger but still of uncertain sign at higher levels, where temporal fluctuations, perhaps due to gravity waves, produce rather large sampling errors. On the other hand, a persistent upslope flow component is observed. The low-level air arriving at the radar sites from the north-northeast has risen about 200 m in a trajectory distance of 80 km at an average flow rate of 8 to 12 m s^{-1}, and over the foothills the average easterly component of 1 to 2 m s^{-1} rises up an average slope of about 5 percent. Thus, there is an average ascent rate of the cold air of 2 to 3 cm s^{-1} over the plains north of Boulder and 5 to 10 cm s^{-1} over the foothills to the west. Since there is no evidence of persistent southward acceleration and the inversion base remained fairly constant in height after 0100, we must assume that the rising cold air is entrained into the westerlies through the inversion, then mixed and carried away to the east by the overriding westerlies. This mixing process must presumably be associated with turbulent entrainment in the strong shear zone at the top of the cold air.

Spatial structure functions can be estimated from radar data by evaluating the mean square velocity differences between adjacent radar beams or adjacent range gates on a beam. Since each velocity determination represents the average within its beam and range gate, these structure function estimates are reduced from the true structure of the velocity field. We measured these filtered structure functions at intervals of 1, 2 and 3 degrees of azimuth and 150, 300 and 450 m range intervals, with each evaluation made from averaging over a circular segment 45° wide in azimuth and 900 m deep in range. The effects of mean vertical shear were removed from the range structure function. These evaluations were made from data taken at a 15° elevation angle, in order to avoid ground clutter effects often present at lower angle scans.

Upon comparing structure functions of different range and angular separation in the same segment, it was found that the assumption of a 2/3 power law collapsed the data fairly well in most cases. This law would hold if the signals were produced by isotropic turbulence in an inertial range. Since intervals of 300 and 450 m are probably outside the inertial range and the effects of filtering are not removed, the above result should probably be regarded as empirical and somewhat fortuitous. It does, however, allow combining the structure function data to a single vertical profile, which is plotted in Fig. 22. Thus the structure function values shown are to be regarded as the mean square differences of radial velocities at 150 m interval, averaged over all azimuths. The values obtained at other intervals have been included in the averaging after multiplying them by the two-thirds power of the ratio between their interval and 150 m. Figure 22 also shows the mean shear determined from the observed mean velocities. The radar data were all taken at about 0106 local time (0806 Z).

The structure function records show very low values, perhaps in the noise level, at the top of the cold air, rising to a maximum at the level of the strongest shear, decreasing a little above that level, but then increasing again to the top of the region of available data. The reason for the large values at high levels is unknown. Gravity waves are likely involved but perhaps radar noise levels in the regions of decreasing reflectivity may be significant.

6. Interpretation of Observed Mean State

The following is a schematic and somewhat speculative outline of the processes which maintain the observed flow. An outstanding feature of the low-level flow is that it appears to be almost antitriptic, that is, straight down the north-south pressure gradient, with almost no geostrophic turning. In part, this is the result of fictitious sea-level pressure corrections across the sloping terrain of the high plains east of the mountains. An 850-mb map tends to show contours tilted a little to the south of west, so that the surface wind direction is probably about 50 or 60° to the left of geostrophic. The surface easterly geostrophic wind is, however, of 30 to 40 m s^{-1} in magnitude. The strong temperature gradient in the cold air, about 5°/100 km, plus the extreme frontal gradient, produces about a 50 m s^{-1} geostrophic shear through the frontal surface, assuming that the winds above 2000 to 2500 m are approximately geostrophic. The strongly positive $\partial p/\partial y$ in the cold air is overlain by a strongly negative $\partial p/\partial y$ in air near the top of the frontal zone. It is hypothesized that a negative turbulent flux of v-velocity exists near the top of the cold air, the vertical gradient of which tends to balance the geostrophic deficits in both layers.

The weak orographic upslope motion must be balanced by downward heat flux, through turbulent entrainment, in a self-regulating way. If more cold air were moved up the mountain than could be entrained away by the upper westerlies, an adverse pressure gradient would develop which would stop the upslope flow, while excessive entrainment would cause shallowing of the inversion and allow more cold air to be drawn up the slope. Although the structure function data are suggestive of turbulent processes, the details of the entrainment mechanism are somewhat concealed

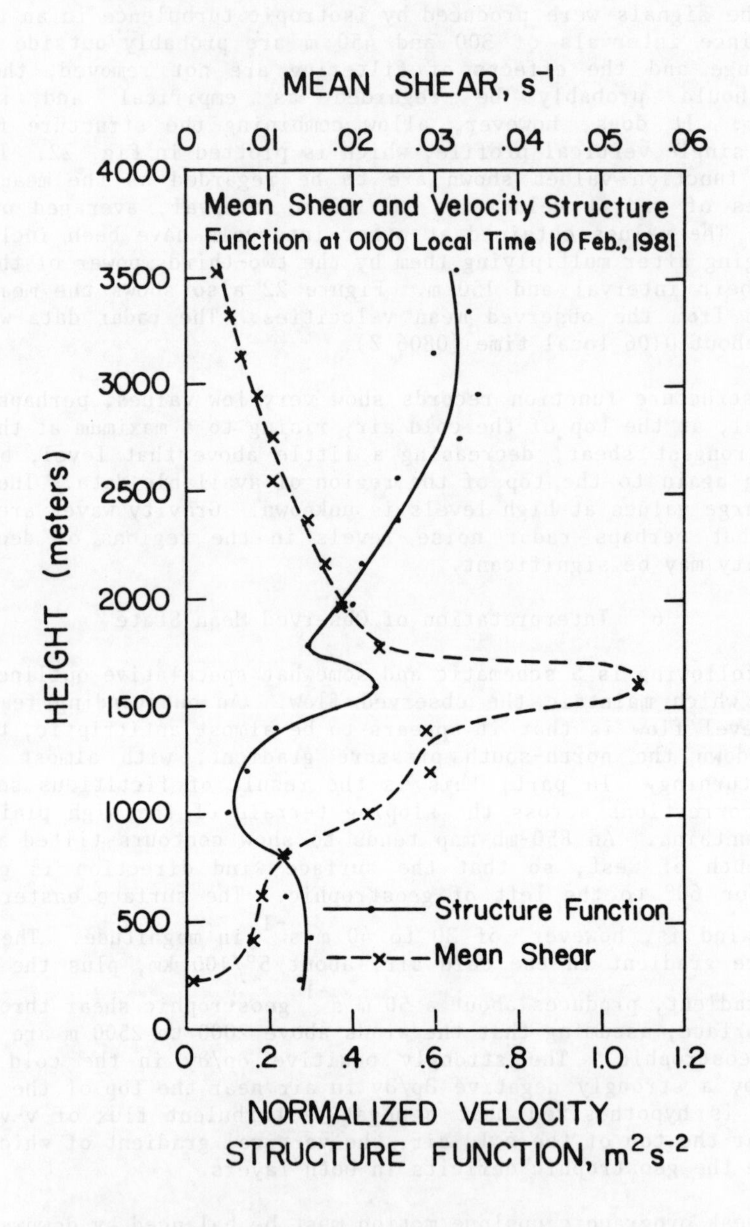

Fig. 22 Radial velocity structure functions (solid curve) as determined by Doppler radar at an elevation angle of 15°. Separation distance is 150 m, with radial and azimuthal structure function averaged together and normalized as indicated in the text. Mean vertical shear amplitude of horizontal wind (dashed curve).

because it is probably taking place more strongly over the foothills where the westerly flow coming over the mountains first comes in contac with the cold upslope air.

REFERENCES

1. Barcilon, A., J.C. Jusem and P.G. Drazin: On the two-dimensional hydrostatic flow of a stream of moist air over a mountain ridge. Geophys. Astrophys. Fluid Dyn., 13, 125-140 (1979).

2. Barcilon, A., J.C. Jusem and S. Blumsack: Pseudo-adiabatic flow over a two-dimensional ridge. Geophys. Astrophys. Fluid Dyn., 16, 19-33 (1980).

3. Browning, K.A. and R. Wexler: The determination of kinematic properties of a wind field using Doppler radar. J. Appl. Meteor., 7, 105-113 (1968).

4. Deardorff, J.W.: On the magnitude of the subgrid scale eddy coefficient. J. Comput. Phys., 7, 120-133 (1971).

5. Deardorff, J.W.: Numerical investigation of neutral and unstable planetary boundary layers. J. Atmos. Sci., 29, 91-115 (1972).

6. Durran, D.R.: The effects of moisture on mountain lee waves. Doctoral Dissertation, Massachusetts Institute of Technology, Cooperative Thesis No. 65, Massachusetts Institute of Technology and National Center for Atmospheric Research, 142 pp. (1981).

7. Durran, D.R. and J.B. Klemp: On the effects of moisture on the Brunt-Vaisala frequency. To be published in J. Atmos. Sci. (1982a).

8. Durran, D.R. and J.B. Klemp: The effects of moisture on trapped mountain lee waves. To be published in J. Atmos. Sci. (1982b).

9. Gal-Chen, T. and R. Somerville: Numerical solutions of the Navier-Stokes equations with topography. J. Comput. Phys., 17, 209-223 (1975).

10. Kessler, E.: On the distribution and continuity of water substance in atmospheric circulation. Meteorological Monographs No. 32, American Meteorological Society, 84 pp. (1969).

11. Klemp, J.B. and D.K. Lilly: Numerical simulations of hydrostatic mountain waves. J. Atmos. Sci., 35, 78-106 (1978).

12. Klemp, J.B. and D.K. Lilly: Mountain waves and momentum flux. Orographic Effects in Planetary Flows, GARP Publication Series 23, 116-142 (1980).

13. Klemp, J.B. and R. Wilhelmson: The simulation of three-dimensional convective storm dynamics. J. Atmos. Sci., 35, 1070-1096 (1978).

14. Lilly, D.K.: On the numerical simulation of buoyant convection. Tellus, 14, 148-172 (1962).

15. Lilly, D.K.: Doppler radar observations of upslope snowstorms. Proc. 20th Conference on Radar Meteorology, 30 November-3 December 1981. American Meteorological Society, Boston, MA, 638-645 (1981).

16. Miles, J.W. and H.E. Huppert: Lee waves in a stratified flow. Part 4. Perturbation approximations. J. Fluid Mech., 35, 497-525 (1969).

17. Richwein, B.A.: The damming effect of the southern Appalachians. Nat. Wea. Dig., 5, 2-12 (1980).

18. Smith, R.B.: The steepening of hydrostatic mountain waves. J. Atmos. Sci., 34, 1634-1654 (1977).

19. Wilson, J., R. Carbone, H. Baynton and R. Serafin: Operational application of Doppler radar. Bull. Amer. Meteor. Soc., 61, 1154-1168 (1980).

5.2 NUMERICAL MODEL FOR A LARGE-SCALE MOUNTAIN-VALLEY BREEZE ON A PLATEAU

Sang Jianguo
Department of Geophysics
Beijing University
The People's Republic of China

and

Elmar R. Reiter
Department of Atmospheric Science
Colorado State University
Fort Collins, CO 80523 USA

ABSTRACT

A three-dimensional numerical model with simplified topography is used to simulate the diurnal variations of the meteorological fields over a plateau. The results reveal that, because of the difference of thermal forcing between the ground surface of the plateau and the surrounding free atmosphere, a warm low forms in the lower levels over the plateau during the daytime and a cold high during the nighttime. A large-scale mountain-valley breeze is induced by horizontal pressure gradients. Reversal between the inward flow at lower levels and the outward flow at higher levels occurs somewhere between 400 and 800 m above the ground. The spatial distribution of the vertical velocity and its diurnal variation are discussed. The results are compared with observations on the Tibetan Plateau.

I. INTRODUCTION

The strong diurnal variations of the wind field in the lower troposphere over the Tibetan Plateau and its vicinity have attracted attention for the last three decades. This so-called large-scale mountain-valley breeze is thought to be induced by the difference of thermal forcing between the surface layer of the plateau and its surrounding free atmosphere. Many related phenomena, such as the diurnal variation of surface temperature, potential heights of pressure levels, the distribution of cumulus clouds, etc., have been summarized in the literatures by Yeh and Gao [7] and by Yang et al. [6]. Because of the scarcity of observations and the complexity of the plateau topography many large-scale phenomena are obscured by local diurnal variations. For example, the large-scale mountain-valley breeze has been noticed for many years, but its strength and its vertical and horizontal extent are still not fully understood. Yang et al. [6] used the components of observed winds normal to the edge of the plateau as an indicator of the strength of the mountain-valley wind system and defined the layer in which the diurnal variations of wind direction were larger than 120° as the layer of the mountain-valley winds. The authors realized that this method could hardly give a complete description of the characteristics of the mountain-valley wind system, especially not in the southeastern part of the plateau where observing stations are located in deep canyons or on peaks of mountains.

Here we outline a numerical modelling approach for the large-scale mountain-valley breeze, using simplified topography. The strength, the vertical and horizontal extent and the diurnal variation of the mountain-valley wind system and its related phenomena are discussed. In order to examine the influence of the smoothness of topography on these wind systems a case with more complicated topography is also tested.

II. BASIC EQUATIONS

The basic equations used in this study are transformed from the Cartesian coordinate system (X, Y, Z) into a terrain-following coordinate system (X, Y, $\bar{Z}$) by the transformation $\bar{Z} = H\,(Z-Z_g)/(H-Z_g)$, where H is the height of the top row of grid points in the model and Z_g(X, Y) is the terrain elevation.

The three-dimensional equations of horizontal momentum, thermal energy, and continuity and the hydrostatic equation in the (X, Y, $\bar{Z}$) coordinate system are similar to those used by Mahrer and Pielke [3], and Pielke [5]

$$\frac{du}{dt} = -\theta\frac{\partial\Pi}{\partial X} + fv + g\,\frac{\bar{Z}-H}{H}\frac{\partial Z_g}{\partial X} + F_u \qquad (1)$$

$$\frac{dv}{dt} = -\theta\frac{\partial\Pi}{\partial Y} - fu + g\,\frac{\bar{Z}-H}{H}\frac{\partial Z_g}{\partial Y} + F_v \qquad (2)$$

$$\frac{d\theta}{dt} = F_\theta \qquad (3)$$

$$\frac{\partial u}{\partial X} + \frac{\partial v}{\partial Y} + \frac{\partial \bar{w}}{\partial \bar{Z}} - \frac{u}{H-Z_g}\frac{\partial Z_g}{\partial X} - \frac{v}{H-Z_g}\frac{\partial Z_g}{\partial Y} = 0 \qquad (4)$$

$$\frac{\partial\Pi}{\partial\bar{Z}} = -\frac{H-Z_g}{H}\frac{g}{\theta} \qquad (5)$$

where

$$\frac{d}{dt} = \frac{\partial}{\partial t} + u\frac{\partial}{\partial X} + v\frac{\partial}{\partial Y} + \bar{w}\frac{\partial}{\partial\bar{Z}}$$

and

$\Pi = C_p\,(p/1000)^{R/C_p}$ is Exner's function.

The turbulent dissipation terms, F_u, F_v and F_θ, can be written as

$$F_\psi = K_H\left(\frac{\partial^2\psi}{\partial X^2} + \frac{\partial^2\psi}{\partial Y^2}\right) + \left(\frac{H}{H-Z_g}\right)^2 \frac{\partial}{\partial\bar{Z}} \left(K_Z\frac{\partial\psi}{\partial\bar{Z}}\right) \tag{6}$$

with $\psi = u, v, \theta$.

The horizontal and vertical exchange coefficients K_H and K_Z will be defined in Section 6.

III. STRUCTURE OF THE MODEL

We use a horizontal grid of 21 x 21 points with a grid interval of 200 km at the center. Near the lateral boundaries the grid intervals are stretched to 250, 312.5, 390.6 and 488.3 km. In vertical the model contains 12 levels with heights of 0, 10, 50, 100, 400, 800, 1500, 2500, 4000, 6000, 9000 and 12000 m.

A simple topography, representing the plateau, is introduced as follows (Fig. 1). It should be pointed out that the grid intervals in the computer plots of Figs. 1, 3, 4 and 5 are not drawn to scale in the outer region, where, as mentioned above, an expanding grid has been used.

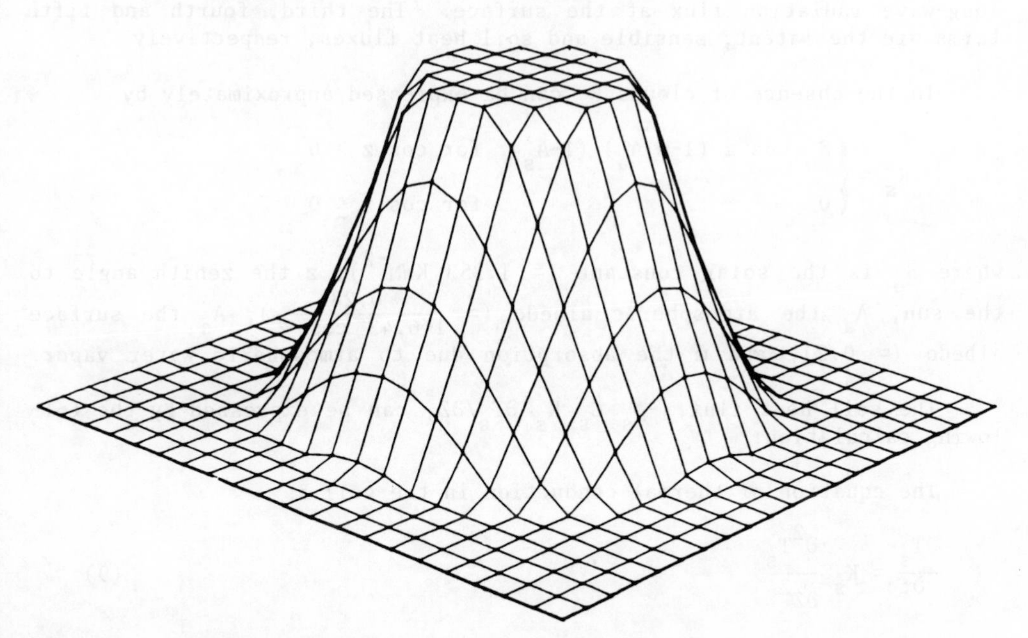

Fig. 1 Topography of the model. The grid intervals in the outer region should be larger than those shown because an expanding grid has been used in the computations (see Section III).

$$Z_g = \begin{cases} 4500 \text{ m} & \text{if } r < 667 \text{ km} \\ A \dfrac{B^2}{(r^2+B^2)} & \text{if } 667 < r < 1500 \text{ km} \\ 2000 \text{ m} & \text{if } r > 1500 \text{ km} \end{cases} \tag{7}$$

where $A = 6500$ m

$B = 1000$ km

$r = \sqrt{(X-X_o)^2 + (Y-Y_o)^2}$ is the distance from the center of the plateau whose coordinates are X_o, Y_o.

IV. SURFACE ENERGY BALANCE

The ground surface temperature, T_s, is computed from

$$R_s - R_L + \rho L u_* q_* + \rho C_p u_* \theta_* - \rho_s C_s K_s \left.\frac{\partial T_s}{\partial Z}\right|_G = 0 \tag{8}$$

where R_s is the net short-wave radiation flux and R_L is the net long-wave radiation flux at the surface. The third, fourth and fifth terms are the latent, sensible and soil heat fluxes, respectively.

In the absence of clouds R_s can be expressed approximately by

$$R_s = \begin{cases} S_o \cos z \,(1-m A_a)(1-A_s) & \text{for } \cos z > 0 \\ 0 & \text{for } \cos z \leq 0 \end{cases}$$

where S_o is the solar constant (= 1.353 KWM^{-2}), z the zenith angle to the sun, A_a the atmospheric albedo (= $\frac{0.28}{1+6.43 \cos z}$), A_s the surface albedo (= 0.2), and m the absorption due to atmospheric water vapor.

The soil heat flux, $-\rho_s C_s K_s \partial T_s/\partial Z$, can be expressed by the following calculation:

The equation of thermal conduction in the soil is

$$\frac{\partial T_s}{\partial t} = K_s \frac{\partial^2 T_s}{\partial Z^2} \tag{9}$$

Assuming that the diurnal variation of soil temperature at depth Z=-d is a sine function of time, the boundary condition for Eq. (9) at that depth is

$$T_s(-d, t) = \bar{T}_s + a \sin(\omega t + \phi) \qquad (10)$$

$\bar{T}_s$ is the mean daily soil temperature.

The solution of Eq. (9) with the boundary condition (Eq. 10) is

$$T_s(Z,t) = \bar{T}_s + a \left[\exp\left(\sqrt{\omega/2K_s}\,(Z+d)\right)\right] \sin\left(\omega t + \sqrt{\omega/2K_s}\,(Z+d) + \phi\right) \qquad (11)$$

Thus,

$$\rho_s C_s K_s \frac{\partial T_s}{\partial Z} = \rho_s C_s K_s a \sqrt{\omega/2K_s} \left[\exp\left(\sqrt{\omega/2K_s}\,(Z+d)\right)\right].$$

$$\{\sin[\omega t + \sqrt{\omega/2K_2}\,(Z+d) + \phi] + \qquad (12)$$

$$\cos[\omega t + \sqrt{\omega/2K_s}\,(Z+d) + \phi]\}$$

From Eq. (11), we arrive at

$$a \exp\left[\sqrt{\omega/2K_s}\,(Z+d)\right] \cos\left[\omega t + \sqrt{\omega/2K_s}\,(Z+d) + \phi\right] = \frac{1}{\omega}\frac{\partial T_s}{\partial t} \qquad (13)$$

$$a \exp\left[\sqrt{\omega/2K_s}\,(Z+d)\right] \sin\left[\omega t + \sqrt{\omega/2K_s}\,(Z+d) + \phi\right] = T_s - \bar{T}_s \qquad (14)$$

Subsitituting (13) and (14) into (12), we obtain

$$\rho_s C_s K_s \frac{\partial T_s}{\partial Z} = \rho_s C_s K_s \sqrt{\omega/2K_s}\left(\frac{1}{\omega}\frac{\partial T_s}{\partial t} + T_s - \bar{T}_s)\right] \qquad (15)$$

At the ground surface

$$\rho_s C_s K_s \left.\frac{\partial T_s}{\partial Z}\right|_G = \rho_s C_s K_s \sqrt{\omega/2K_s}\left(\frac{1}{\omega}\frac{\partial T_s}{\partial t} + T_G - \bar{T}_s)\right] \qquad (16)$$

Substituting Eq. (16) into Eq. (8), we obtain the predictive equation for the ground surface temperature

$$\rho_s C_s \sqrt{\frac{K_s}{2\omega}}\frac{\partial T_G}{\partial t} = R_s - R_L + \rho L u_* q_* + \rho C_p u_* \theta_*$$

$$- \rho_s C_s K_s \sqrt{\omega/2K_s}\,(T_G - \bar{T}_s) \qquad (17)$$

where T_G is the ground surface temperature. The parameters u_*, and θ_* will be discussed in Section 6.

V. INITIAL AND BOUNDARY CONDITIONS

The initial air temperature field is assumed to be uniform horizontally and to have a lapse rate of 6.5°C/km vertically. The surface temperature is taken as 20°C at Z = 2000 m. The initial three-dimentional field of temperature is given by

$$T(x, y, \bar{Z}, 0) = 293 - 0.0065\ (Z - 2000) \quad (18)$$

where $Z = \frac{12000-Z_g}{12000}\ \bar{Z} + Z_g$

is the height above sea level.

The surface pressure at Z = 2000 m is assumed to be 800 mb. The initial pressure field is calculated by using the hydrostatic equation

$$P = P_o\ \{[T_o - \gamma(Z - 2000)]/T_o\}^{g/\gamma R} \quad (19)$$

where $P_o = 800$ mb

$T_o = 293°k$

$\gamma = 6.5°k/km \quad (20)$

The atmosphere is assumed to be initially at rest

$$u = v = \bar{w} = 0 \qquad \text{at } t = 0 \quad (21)$$

The boundary conditions are as follows:

$$u = v = \bar{w} = 0 \qquad \text{at } \bar{Z} = 0 \quad (22)$$

$$\begin{cases} u = v = \bar{w} = 0 \\ \theta = \text{constant} \quad \text{and } \Pi = \text{constant} \end{cases} \qquad \text{at } \bar{Z} = 12000 \text{ m} \quad (23)$$

At the lateral boundaries

$$\frac{\partial\theta}{\partial x} = \frac{\partial\theta}{\partial y} = 0$$
$$\frac{\partial\Pi}{\partial x} = \frac{\partial\Pi}{\partial y} = 0 \quad (24)$$

The mean soil temperature is assumed to equal the initial air temperature

$$\bar{T}(X, Y, t) = 293 - 0.0065(Z_g(X, Y) - 2000) \quad (25)$$

VI. SUBGRID-SCALE PARAMETERIZATION

In the turbulent dissipation terms, Eq. (6), the horizontal eddy exchange coefficient, K_H, is given by

$$K_H = 0.36\ \Delta X \Delta Y[(\frac{\partial v}{\partial X} + \frac{\partial u}{\partial Y})^2 + 0.5((\frac{\partial u}{\partial X})^2 + (\frac{\partial v}{\partial Y})^2)]^{\frac{1}{2}}$$

The vertical eddy exchange coefficient, K_Z, is determined by a Hermite interpolation [4]

$$K_Z = \begin{cases} K_1 = K_2/100 & \text{for } Z > Z_j \\ K_1 + [(Z-Z_j)^2/(Z_j-Z_i)^2]\ \{(K_2-K_1) + (Z-Z_i)\ \cdot \\ \quad \cdot\ [K_3 + 2(K_2-K_1)/(Z_j-Z_i)]\} & \text{for } Z_i \leq Z \leq Z_j \\ (Z/Z_i)\ K_2 & \text{for } Z < Z_i \end{cases}$$

where Z_j and Z_i are the heights of the top of the planetary boundary layer and the top of the surface layer, respectively.

K_1 and K_2 are the values of K_Z at Z_j and Z_i, respectively, and

$$K_3 = \left.\frac{\partial K_Z}{\partial Z}\right|_{Z_i}.$$

In the surface layer K_2 assumes the following forms [1]:

$K_2 = k_o\ u_*\ Z_i/\phi_M$ for momentum

$K_2 = k_o\ u_*\ Z_i/\phi_H$ for heat

where k_o is the von Karman constant (= 0.35).

The dimensionless wind shear, ϕ_M, and dimensionless potential temperature gradient, ϕ_H, can be expressed by the empirical functions

$$\phi_M = \begin{cases} (1-15\zeta)^{-\frac{1}{4}}, & \zeta \leq 0 \\ 1 + 4.7\zeta, & \zeta > 0 \end{cases}$$

$$\phi_H = \begin{cases} 0.74(1-9\zeta)^{-\frac{1}{2}}, & \zeta \leq 0 \\ 0.74 + 4.7\zeta, & \zeta > 0 \end{cases}$$

where $\zeta = Z_i/L_*$

$$L_* = \frac{\bar{\theta}\ u_*}{g\ K_o\ \theta_*}$$

The friction velocity, u_*, and friction potential temperature, θ_*, are given by

$$u_* = k_o \, u\Big|_{Z_i} / [\ln(Z_i/z_o) - \psi_1]$$

$$\theta_* = [\theta\Big|_{Z_i} - \theta_G]/[0.74 \ln(Z_i/z_o) - 0.74 \, \psi_2]$$

where z_o is the roughness length (= 0.1m), $u\Big|_{Z_i}$ is assumed to be 10m/sec, and

$$\psi_1 = \begin{cases} 2\ln[(1+\phi_M^{-1})/2] + \ln[(1+\phi_M^{-2})/2] - \\ \quad -2\tan^{-1}(\phi_M^{-1}) + \Pi/2, & \zeta \leq 0 \\ -4.7\zeta, & \zeta > 0 \end{cases}$$

$$\psi_2 = \begin{cases} \ln[(1+0.74\,\phi_H^{-1})/2], & \zeta \leq 0 \\ -6.35\,\zeta, & \zeta > 0. \end{cases}$$

The diurnal variation of Z_j is expressed by empirical equations suggested by Deardorff [2]

$$\frac{\partial Z_j}{\partial t} = -(u\frac{\partial Z_j}{\partial X} + v\frac{\partial Z_j}{\partial Y}) + \bar{w}\Big|_{Z_j} + \frac{1.8(w_*^3 + 1.1\,u_*^3 - 3.3\,u_* f Z_j)}{g\frac{(Z_j)^2}{\theta}\frac{\partial\theta}{\partial Z}\Big|_{Z_j} + 9\,w_*^2 + 7.2\,u_*^2}$$

for unstable conditions

where

$$w_* = \begin{cases} (-\frac{g}{\theta}\,u_*\,\theta_*\,Z_j)^{-1/3}, & \text{if } \theta_* \leq 0 \\ 0, & \text{if } \theta_* > 0 \end{cases}$$

and

$$\frac{\partial Z_j}{\partial t} = -\left(u\frac{\partial Z_j}{\partial X} + v\frac{\partial Z_j}{\partial Y}\right) + 0.06\,\frac{u_*^2}{Z_j f}\left[1 - \left(\frac{3.3 Z_j f}{u_*}\right)^3\right]$$

for neutral and stable conditions.

The height of Z_i is assumed as

$$Z_i = 0.04\,Z_j.$$

The initial value of Z_j is taken as 200 m.

VII. RESULTS

The integration of the model starts from an atmosphere at rest at 0800 LST (or Beijing time, in the case of the Tibet Plateau). After sunrise, which is assumed to be at 0800 LST, the surface of the plateau and the slopes begin to absorb the solar radiation. The air in contact with the surface warms up more rapidly than the surrounding free atmosphere, rises along the slopes and forms a valley wind system. Figure 2 shows the process of the formation of valley winds at various levels. Since the flow field is nearly axially symmetric, we chose the mean inward flow at the four grid-points 600 km to the north, east, south and west of the plateau center being representative of the valley winds. The inward flow, which we define as negative, begins to build up near the surface and reaches its maximum of 3.9 m/sec at 15 LST. The amplitude of the inward flow decreases with height and turns outward in the upper levels. This outward flow develops at about 600 m above the surface, but its amplitude is much smaller than that of the inward flow. Therefore, in the upper troposphere, where the wind speeds usually are larger than 10 m/sec, the outward flow component induced by surface heating can hardly be detected in data from the real atmosphere.

In the afternoon the solar radiation diminishes. With the decrease of surface heating the inward flow at the lower levels decreases, becomes zero at about 2000 LST, and then changes to outward flow, i.e. to a mountain breeze. The outward flow reaches its maximum at about 0100 LST and then drops gradually to zero at 0800 LST. Thus, a diurnal cycle is established which propagates upward. Its phase lag increases with height. Model results show that in the layer represented by $\bar{Z}$ = 50 m the outward flow is stronger than the inward flow, whereas in the layer above $\bar{Z}$ = 400 m the outward flow is weaker than the inward flow. This result indicates that the mountain breeze layer is shallower than the valley breeze because of the weaker vertical turbulent exchange in the nocturnal boundary layer. However, the horizontal difference of air temperature between the surface layer of the plateau and its vicinity is larger during nighttime than during daytime when strong vertical and horizontal exchange diminishes the horizontal temperature gradient. The large horizontal temperature gradient in the surface layer at night causes the mountain breeze to be stronger.

Figure 2 also shows the diurnal variation of the tangential component of the mountain-valley breeze system. In the lower levels we find anticlockwise tangential flow with a maximum at 1800 LST during the daytime and clockwise flow with a maximum at 0400 LST during the nighttime. At 0800 LST the tangential velocity does not turn back to zero. Because of the inertial oscillation there still remains a tangential component, even in the absence of horizontal pressure gradients. The maximum of the resultant wind of radial and tangential components in the surface layer should occur between 1500 and 1800 LST if the turbulent momentum transport from the upper layers is not taken into consideration.

Figure 3a shows the horizontal flow vectors at 1400 LST and at the level $\bar{Z}$ = 100 m, where the inward flow is fully developed. The strongest wind belt is found along the rim of the plateau. The wind speeds at $\bar{Z}$ =

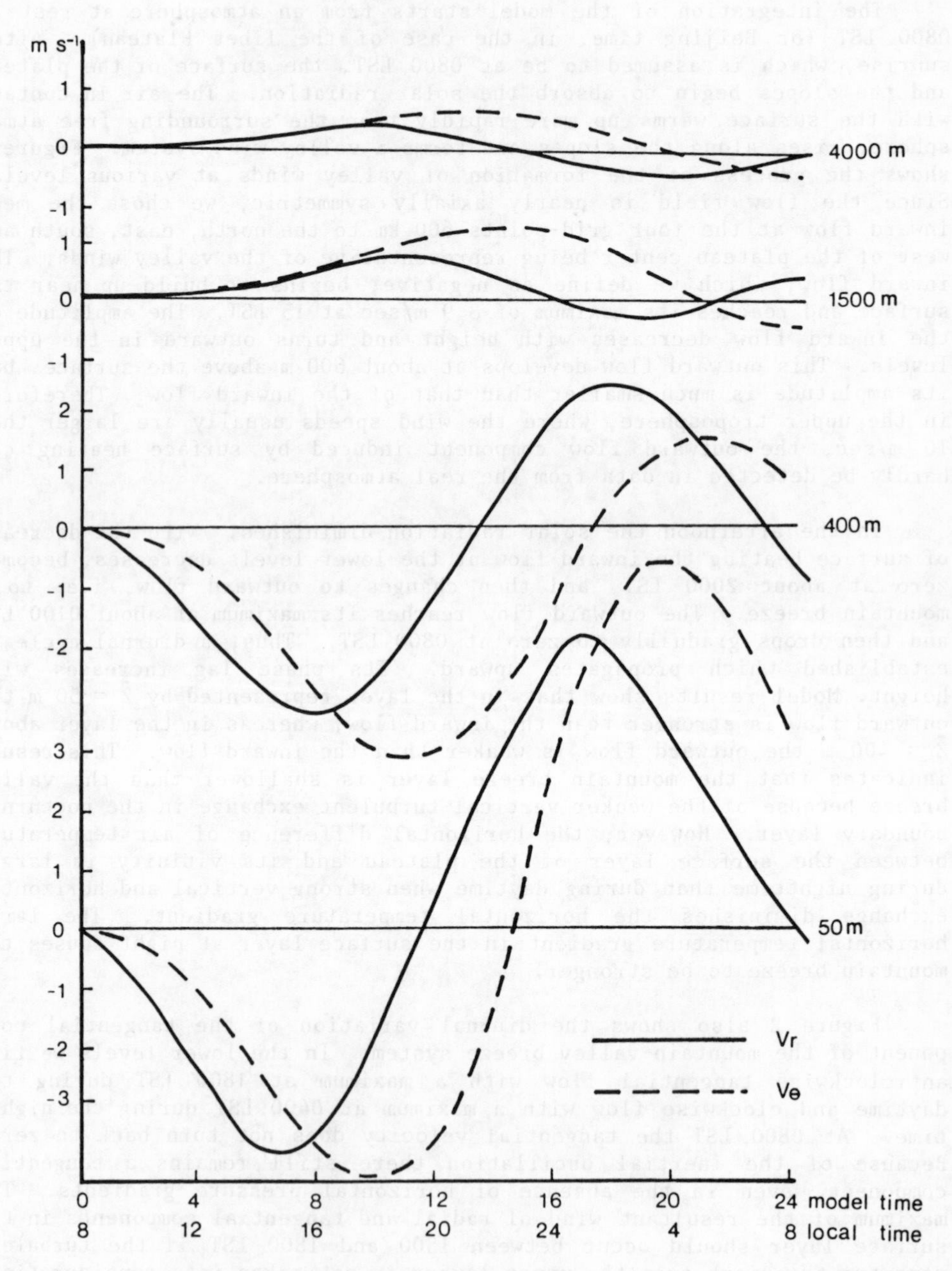

Fig. 2 The diurnal variation of radial velocity, V_r (solid lines outward positive), and tangential velocity, V_θ (dashed lines, clockwise positive), at the levels $\bar{Z}$ = 50 m, $\bar{Z}$ = 400 m, $\bar{Z}$ = 1500 m and $\bar{Z}$ = 4000 m.

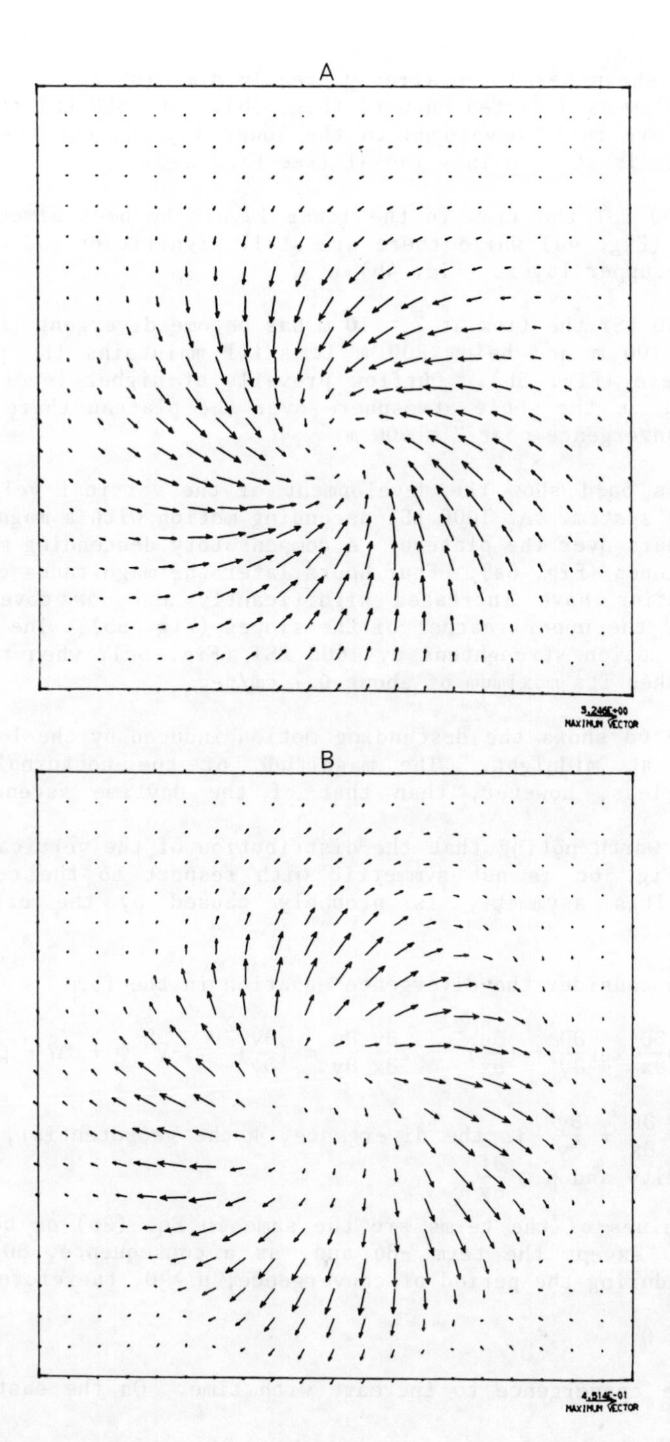

Fig. 3 The horizontal wind vectors (arbitrary scale) at 1400 LST for the levels (a) $\bar{Z}$ = 100 m and (b) $\bar{Z}$ = 800 m. The explanation for the grid intervals in the outer region of this figure is the same as in Fig. 1.

400 m (not shown here) are already greatly diminished. At and above $\bar{Z}$ = 800 m the flow is directed outward (Fig. 3b). At 1400 LST the tangential components are fully developed in the lower layers, but the flow in the upper layers is still mainly radial (see Fig. 2).

At 1800 LST the flow in the lower layers becomes almost completely tangential (Fig. 4a) while there are still significant radial components left in the upper layers (Fig. 4b).

At 2100 LST the flow at $\bar{Z}$ = 10 m has become divergent (Fig. 5a), but above $\bar{Z}$ = 100 m and below 800 m it still maintains the pattern of a valley breeze (Fig. 5b). Outflow prevails at higher levels (e.g. Fig. 5c). Thus, in the whole atmosphere over the plateau there is only one layer of convergence near $\bar{Z}$ = 400 m.

Figures 6a-d show the development of the vertical velocity, $\bar{w}$, in the $(X,Y,\bar{Z})$ system. At 1000 LST ascending motion with a magnitude of 0.1 cm/sec appears over the plateau. A compensatory descending motion occurs over the slopes (Fig. 6a). Four hours later the magnitudes of the vertical velocities have increased significantly and now cover the whole plateau and the upper reaches of the slopes (Fig. 6b). The compensatory descending motion strenghtens by 1600 LST (Fig. 6c), when the ascending motion reaches its maximum of about 0.7 cm/sec.

Figure 6d shows the descending motion induced by the low-level flow divergence at midnight. The magnitude of the nocturnal descending motion is less, however, than that of the daytime ascending motion.

It is worth noting that the distribution of the vertical velocities shown in Fig. 6c is not symmetric with respect to the center of the plateau. This asymmetry is probably caused by the effect of the Coriolis force.

Let us consider the divergence equation in the form

$$\frac{\partial D}{\partial t} + u\frac{\partial D}{\partial x} + v\frac{\partial D}{\partial y} + \left(\frac{\partial u}{\partial x}\right)^2 + 2\frac{\partial v}{\partial x}\frac{\partial u}{\partial y} + \left(\frac{\partial v}{\partial y}\right)^2 = -\nabla^2 \Phi + f\Omega - \beta u \qquad (26)$$

where $D = \frac{\partial u}{\partial x} + \frac{\partial v}{\partial y}$ is the divergence, Φ the geopotential, Ω the relative vorticity and $\beta = \frac{\partial f}{\partial y}$.

The values of the terms are the same in Eq. (26) on both sides of the plateau except the term $-\beta u$ and, as a consequence, $\partial D/\partial t$. On the west side, during the period of convergence, $u > 0$, therefore,

$$-\beta u < 0$$

causing the convergence to increase with time. On the east side $u < 0$

$$-\beta u > 0,$$

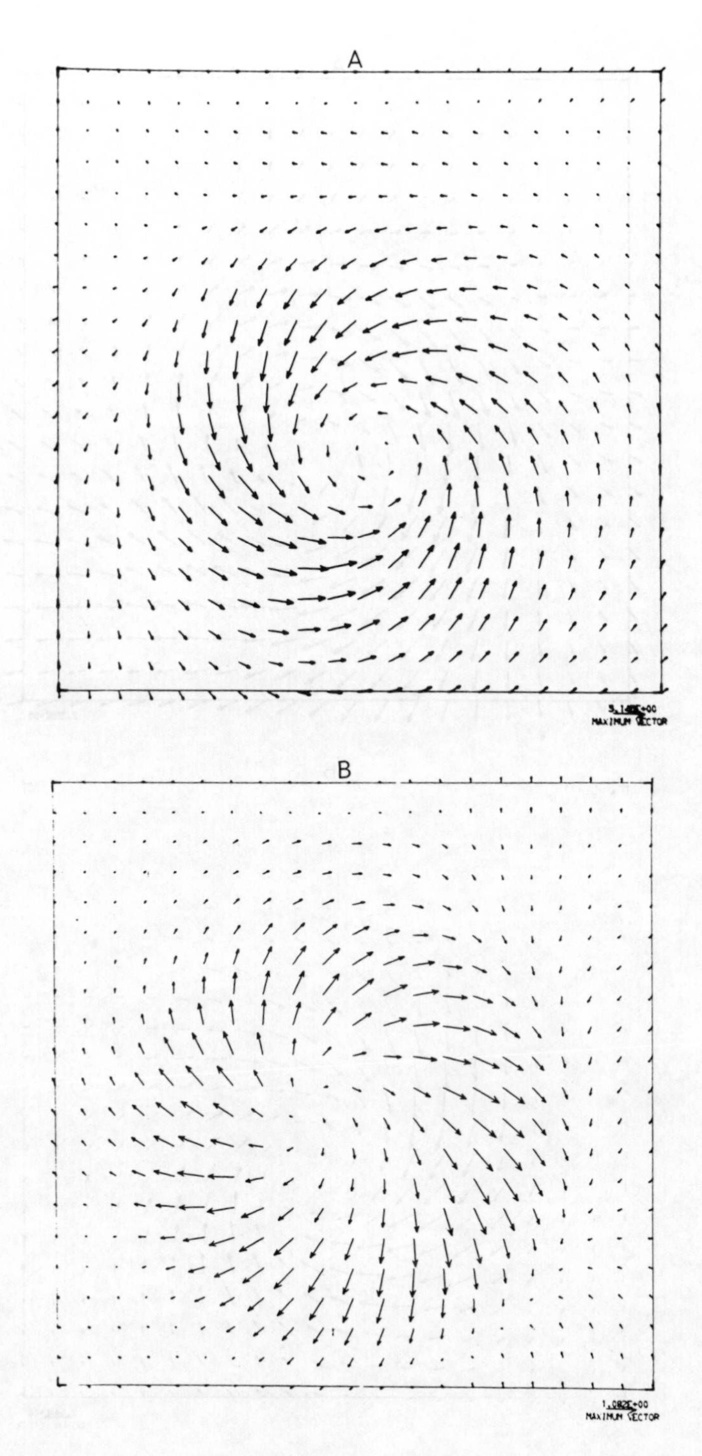

Fig. 4 Similar to Fig. 3, except at 1800 LST for the levels (a) $\bar{Z}$ = 100 m and (b) $\bar{Z}$ = 800 m.

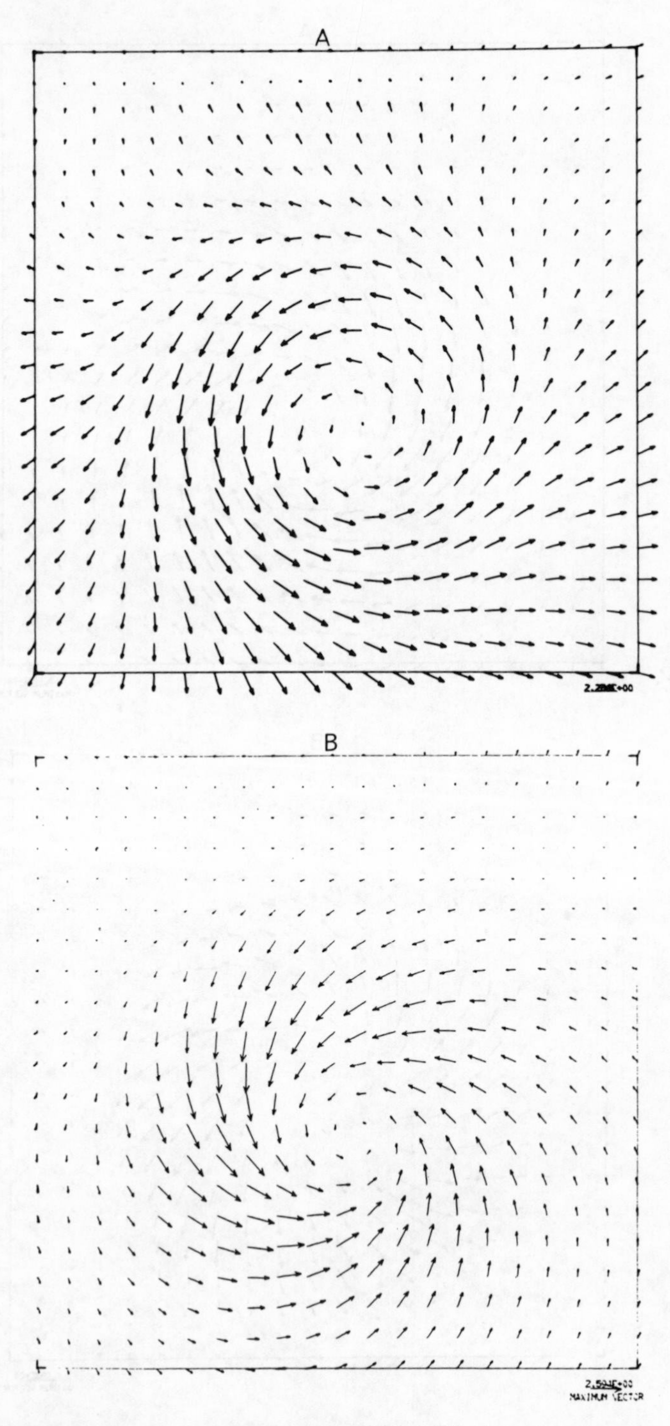

Fig. 5 Similar to Fig. 3, except at 2100 LST for the levels (a) $\bar{Z}$ = 100 m, (b) $\bar{Z}$ = 400 m and (c) $\bar{Z}$ = 800 m.

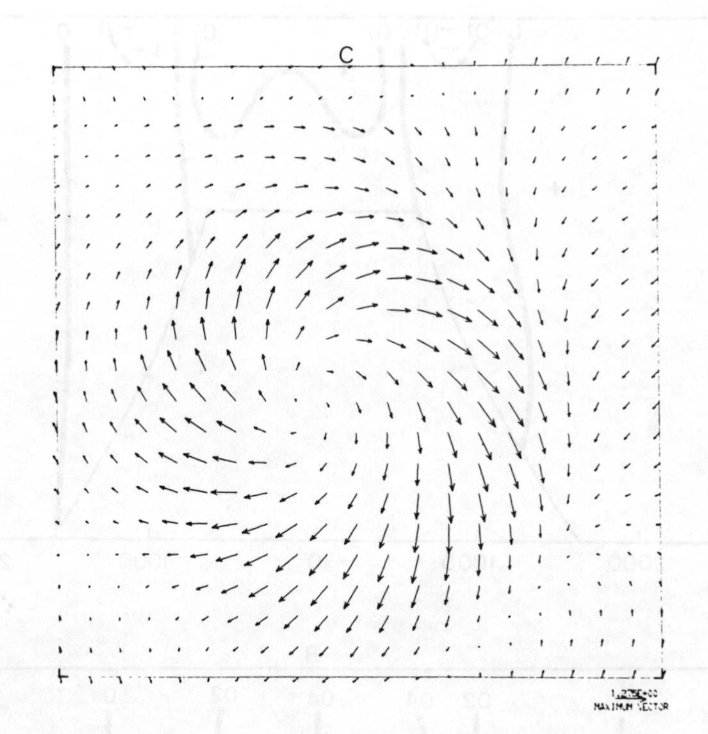

Fig. 5 (Continued)

prompting the convergence to decreases with time. Thus the vertical velocities induced by the convergence are larger on the west side than on the east side.

Inspection of Figs. 3-5 reveals that the centers of convergence at the lower levels and of divergence at the higher levels do not coincide. The axis of the flow system tilts westward with height.

Figure 7 shows the diurnal cycles of the surface potential temperature and of the surface pressure at the center of the plateau. The amplitude of the diurnal variation of pressure calculated by the model is less than that observed over the Tibetan Plateau. The variation there ranges from 1 to 3 mb. The discrepancy is due to the fact that the model cannot simulate the semidiurnal pressure variations induced by the solar tide.

Weather phenomena, such as the convective activity over the Tibetan Plateau and its vicinity, have been thought to be associated with the diurnal thermal forcing. Figure 8 from Yeh and Gao (1979) shows the diurnal variation of the frequency of cumulonimbus clouds observed at stations over and near the Tibetan Plateau. At the stations Ger, Bange

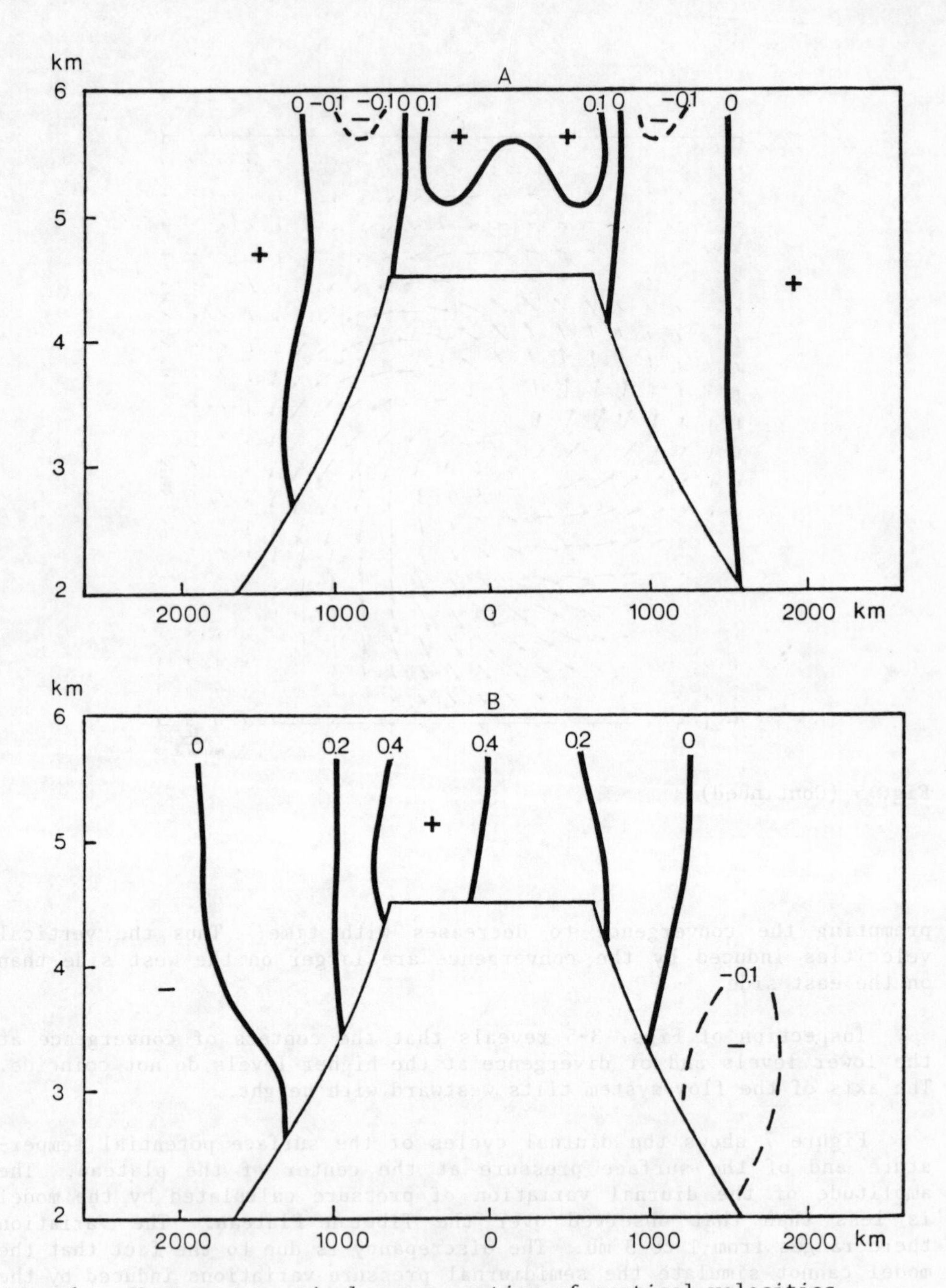

Fig. 6 West-east vertical cross section of vertical velocities, $\bar{w}$, cm sec^{-1}, at times (a) 1000 LST, (b) 1400 LST, (c) 1600 LST and (d) 0200 LST. The abscissa shows the distance from the center of the plateau in km. The ordinate is labelled as the height above sea surface.

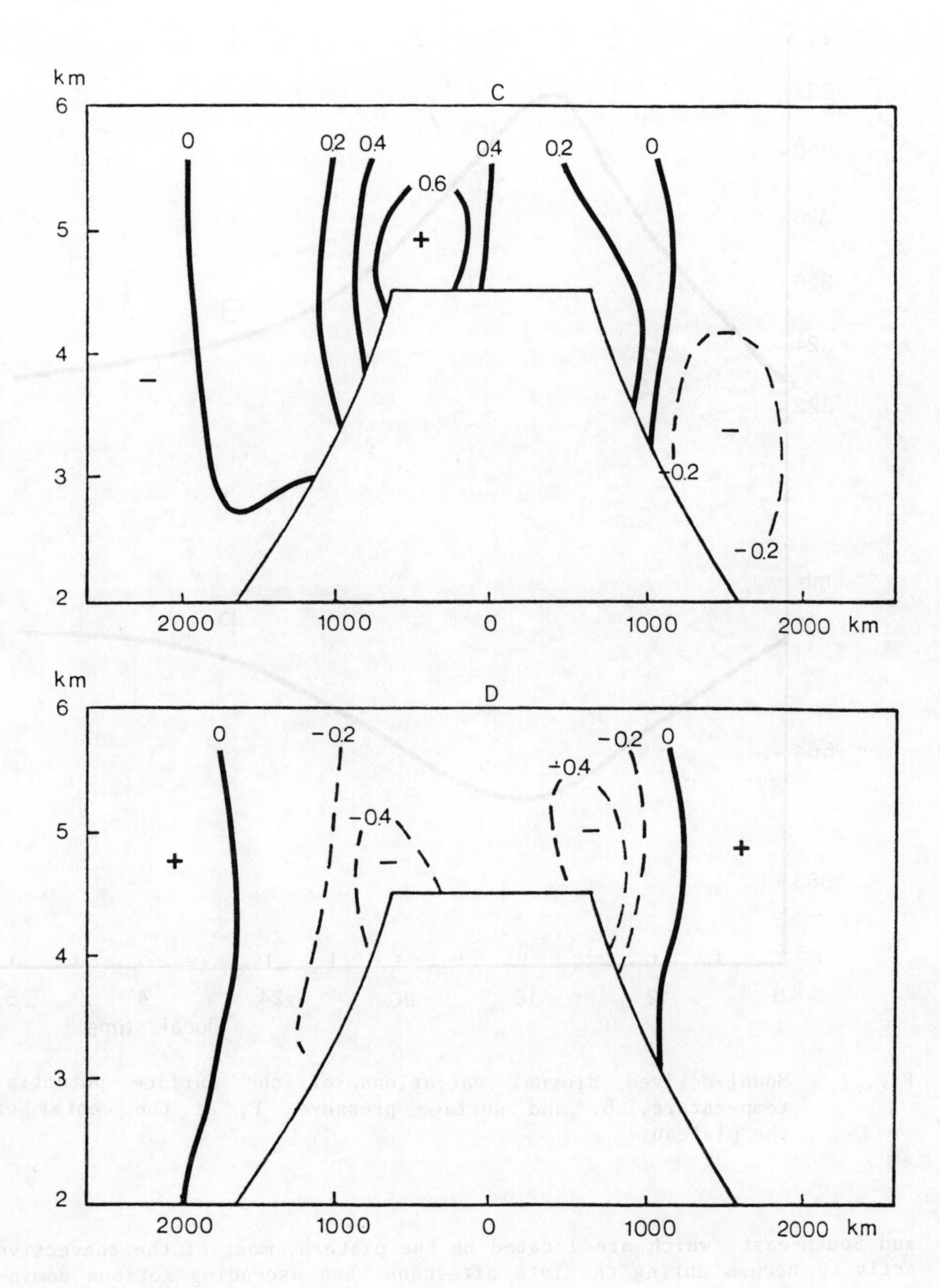

Fig. 6 (Continued)

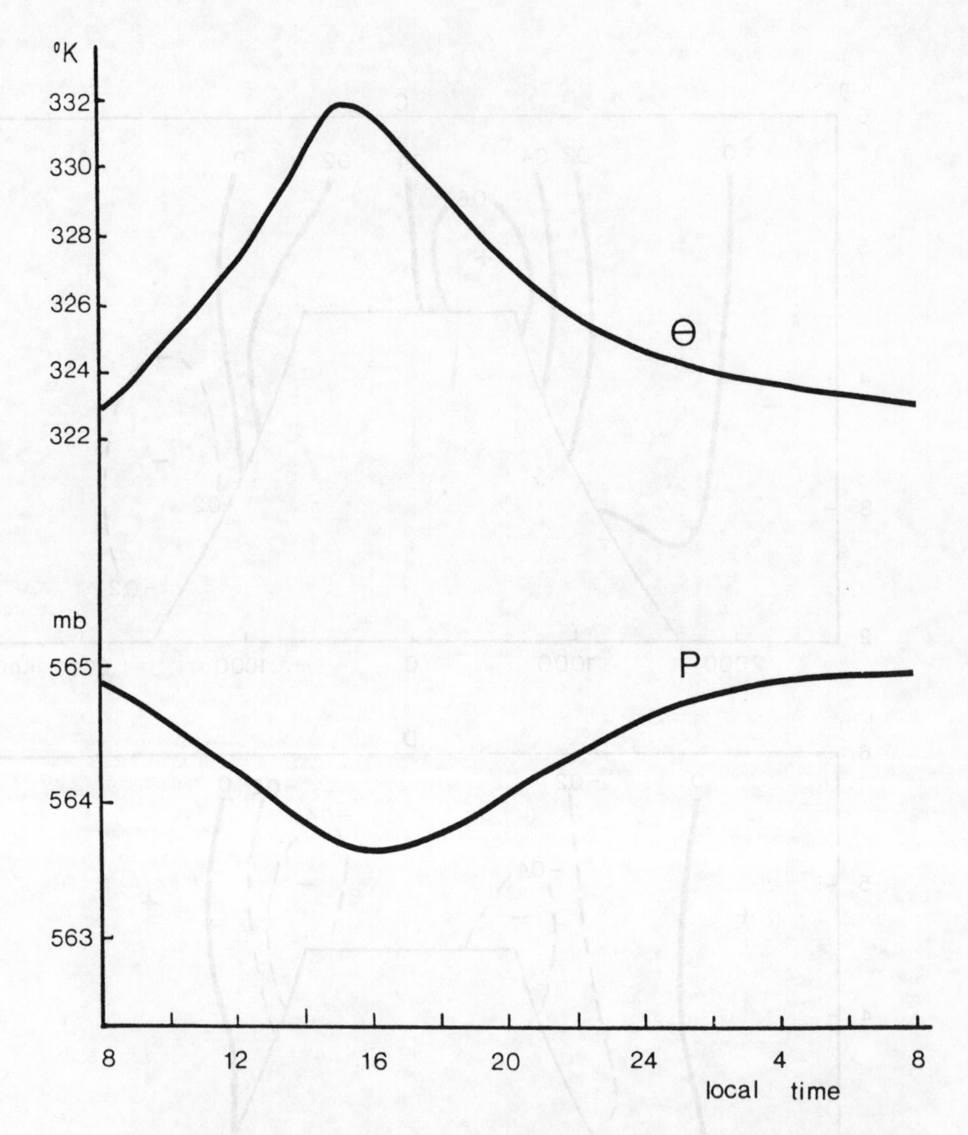

Fig. 7 Model-derived diurnal variations of the surface potential temperature, θ, and surface pressure, P, at the center of the plateau.

and South-east, which are located on the plateau, most of the convective activity occurs during the late afternoon when ascending motions dominate the whole plateau. At night and in the early morning descending motions suppress most of the convective activities and cumulus cloud cover is very low. Similar reasoning can explain the opposite diurnal variations over stations Jielumu, Anbala and Baleili, which are located in the vicinity of the plateau.

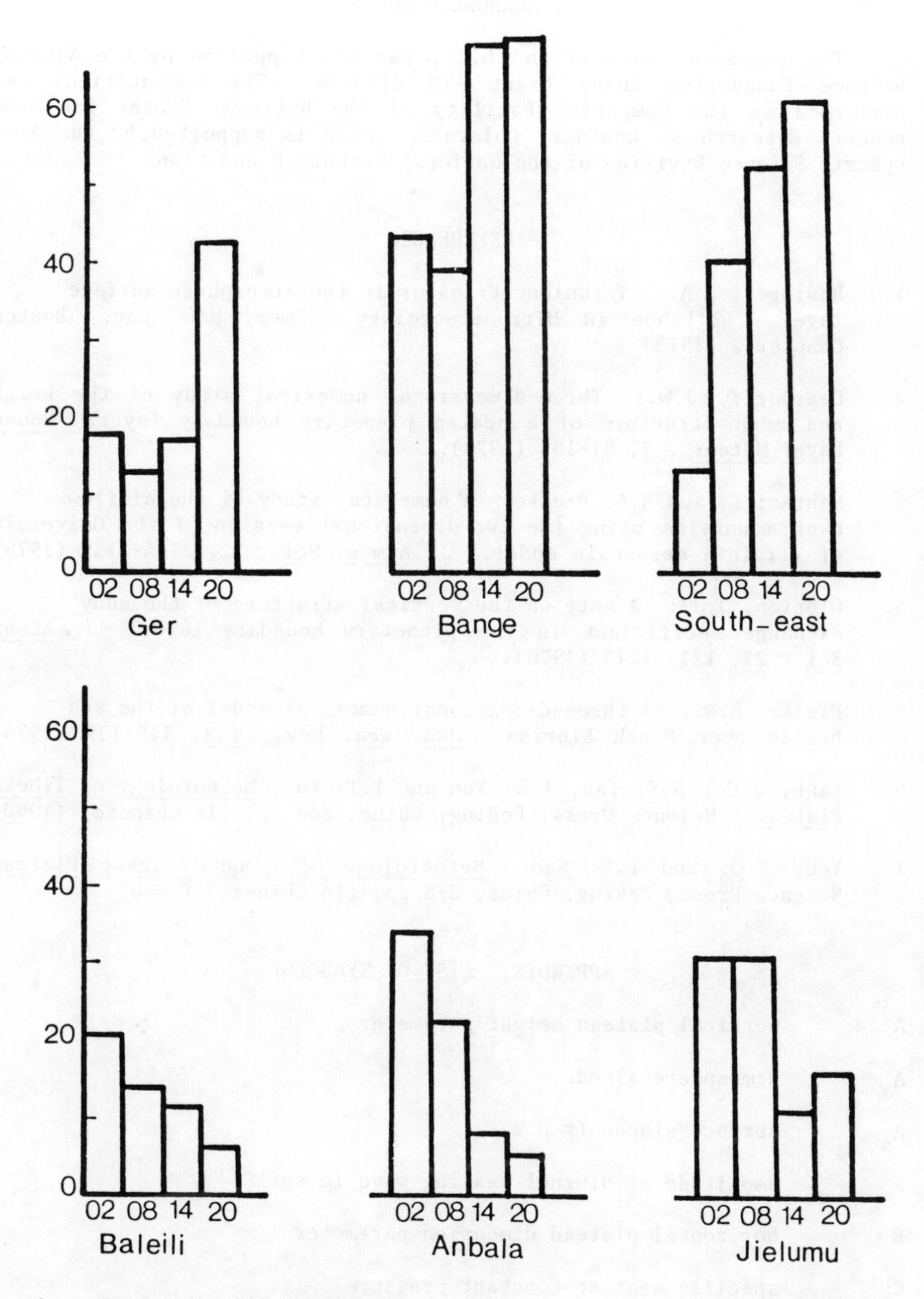

Fig. 8 Frequency of cumulonimbus clouds (ordinate) over stations on the Tibetan Plateau (top) and in its vicinity (bottom). Abscissa is the local time 0200, 0800, 1400 and 2000 LST (after Yeh and Gao [7]).

ACKNOWLEDGMENTS

The research reported in this paper was supported by the National Science Foundation under Grant ATM 8109504. The computations were performed at the Computing Facility of the National Center for Atmospheric Research at Boulder, Colorado. NCAR is supported by the Atmospheric Science Division of the National Science Foundation.

REFERENCES

1. Businger, J.A.: Turbulent transfer in the atmosphere surface layer. Workshop in Micrometeorology. Amer. Met. Soc., Boston, Chapter 2 (1973).

2. Deardorff, J.W.: Three-dimensional numerical study of the height and mean structure of a heated planetary boundary layer. Bound. Layer Meteor., 7, 81-106 (1974).

3. Mahrer, Y. and R.A. Pielke: A numerical study of the airflow over mountains using the two-dimensional version of the University of Virginia mesoscale model. J. Atmos. Sci., 32, 2144-2155 (1975).

4. O'Brien, J.J.: A note on the vertical structure of the eddy exchange coefficient in the planetary boundary layer. J. Atmos. Sci., 27, 1213-1215 (1970).

5. Pielke, R.A.: A three-dimensional numerical model of the sea breeze over South Florida. Mon. Wea. Rev., 102, 115-139 (1974).

6. Yang, J.C., S.Y. Tao, T.C. Yeh and T.T. Ku: Meteorology of Tibetan Plateau. Science Press, Peking, China, 268 pp. (In Chinese) (1960).

7. Yeh, T.C. and Y.X. Gao: Meteorology of Qinghai-Xizang Plateau. Science Press, Peking, China, 278 pp. (In Chinese) (1960).

APPENDIX: LIST OF SYMBOLS

A vertical plateau height parameter

A_a atmosphere albedo

A_s surface albedo (= 0.2)

a amplitude of diurnal heating wave in soil

B horizontal plateau dimension parameter

C_p specific heat at constant pressure

C_s specific heat of soil

D	horizontal divergence
$F_\psi = F_u, F_v, F_\theta$	turbulent dissipation terms
f	Coriolis parameter
H	height of atmosphere in the model
K_H	horizontal exchange coefficient
K_Z	vertical exchange coefficient
K_1	value of K_Z at Z_j
K_2	value of K_Z at Z_i
$K_3 = \left.\frac{\partial K_Z}{\partial Z}\right\vert_{Z_i}$	
K_s	soil heat diffusivity (= $3{:}10^{-6}$ m^2 sec^{-1})
k_o	von Karman constant
L	latent heat of evaporation
L_*	Monin-Obukhov length
m	absorption of short-wave radiation due to water vapor
P	surface pressure
q_*	specific humidity
R_L	net long-wave radiative flux at the surface
R_S	net short-wave radiative flux at the surface
r	radial dimension of plateau
S_o	solar constant
T	air temperature
$\bar{T}$	mean daily soil temperature
T_G	ground surface temperature
T_s	soil temperature
u,v	horizontal wind velocity components
u_*	friction velocity

$\bar{w} = \frac{d\bar{Z}}{dt}$ vertical velocity in system (X, Y, $\bar{Z}$, t)

X,Y Cartesian coordinates

X_o, Y_o coordinates of center of plateau

Z vertical Cartesian coordinate

$\bar{Z}$ vertical terrain-following coordinate

Z_i height of top of the surface layer

Z_j height of top of the planetary boundary layer

Z_g terrain elevation

z zenith angle of sun

z_o roughness length

$\beta = \frac{\partial f}{\partial y}$

$\zeta = Z_i/L_*$

θ potential temperature

θ_* friction potential temperature

Π Exner's function (= $C_p (p/1000)^{R/C_p}$)

ρ_s soil density (= $1.5 \cdot 10^3$ Kg/m^3)

Φ geopotential

ϕ phase lag of soil temperature

ϕ_M dimensionless wind shear

ϕ_H dimensionless potential temperature gradient

Ω relative vorticity

$\omega = 2\pi/(24 \times 3600)$ sec^{-1} frequency of diurnal heating cycle

Subscripts G and s denote ground and soil values, respectively.

5.3 SLOPE WINDS

Fu Baopu
Department of Meteorology
Nanjing University
The People's Republic of China

ABSTRACT

In this paper the general features of slope winds are first stated on the basis of observations made in the field. Then a mathematical model of slope winds is deduced from relatively general conditions, and the principal characteristics of the slope winds are analyzed from the model and compared with the results observed in the field.

I. ANALYSIS OF OBSERVATIONS OF SLOPE WINDS

In February, April and July 1959 we made microclimatic observations at 46 points located on four cross sections of the river valley at Badong, Fengjie, Wanxian and Fuling in Sichuan Province, and obtained numerous slope wind data. Here we will discuss some general features of slope winds in river valleys based on these observations.

Figure 1 shows the diurnal variation of slope winds in the river valleys of Badong, Wanxian and Fuling. The upslope winds begin from one and one-half to three hours after sunrise and increase in strength with the elevation of the sun and the heating of the ground, reaching maximum values of velocity at about midday. From then on they become weaker and weaker, and die down in the evening. The downslope winds usually begin one-half to two hours before sunset and reach their first maximum at one to two hours after sunset. They then weaken slowly for a period of time, then increase again gradually until sunrise of the next day, after which they subside quickly and calm down in one to three hours.

As the diurnal variations of temperature on different slopes differ from each other, the onset times of up- and downslope winds are different also. On clear summer days the beginning and end times of up- and downslope winds on the east slopes and southeast slopes in the Badong and Wanxian valleys are about one hour earlier than those on the west slopes and northwest slopes.

Table 1 gives the occurrence frequencies and mean speeds of up- and downslope winds at two meters above the ground. It shows that upslope winds are stronger than downslope winds in summer, but weaker in winter; furthermore, the strength of upslope winds is greater in summer than in winter, while those of downslope winds behave exactly in the opposite fashion. The slope winds are strongest in Badong, where the valley is deep and the slope is steep. In the summer month of July, the mean maximum speed of the upslope wind is 2.1 m/s, and that of the downslope

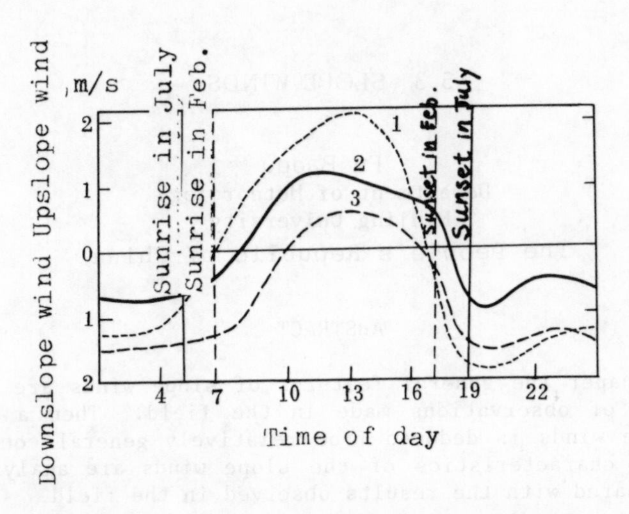

Fig. 1 Daily cycle of slope wind speed at a height of 2 meters above ground. (1) At Badong in July, (2) at Wanxian in July, (3) at Fuling in February.

Table 1 Up- and downslope winds at 2 m.

Site	Depth of Valley (m)	Mean Maximum Speeds (m/s)					
		Downslope Winds			Upslope Winds		
		Feb.	Apr.	July	Feb.	Apr.	July
Badong	510-870	2.6	2.2	1.8	1.2	2.4	2.1
Wanxian	300-530	1.8	1.5	0.9	0.6	1.8	1.1
Fuling	150-220	1.1	0.8	1.0	0.6	/	1.6

Occurence Frequencies of Slope Winds

Site	Depth of Valley (m)	Downslope Winds			Upslope Winds		
		Feb.	Apr.	July	Feb.	Apr.	July
Badong	510-870	47	60	73	13	40	77
Wanxian	300-530	80	87	80	27	27	67
Fuling	150-220	27	53	53	13	7	13

wind is 1.8 m/s. On individual days both peaks of velocity may attain 4 to 5 m/s. In the winter month of February the mean maximum speed of the downslope wind is 2.6 m/s, and that of the upslope wind is 1.2 m/s. On particular days the former may attain 6 m/s, while the latter seldom exceeds 1.5 m/s. The slope winds are weakest in Fuling, where the valley is shallow and the slope gentle, and their peak values not exceeding 2 m/s.

It is also clear from Table 1 that the occurrence frequencies of upslope winds are generally small in winter, because the incoming solar radiation in the daytime is weak and the earth's surface does not heat much. Frequencies of downslope winds are one to two times greater. In summer, as the greater solar altitude and diurnal range of temperature create favorable conditions for the development of local circulation, more than two-thirds of the days show slope winds in the Badong and Wanxian valleys, where the differences of height are large. Even on cloudy days weak slope winds may appear around noon or after midnight for a short time.

The magnitude of temperature fluctuations indicates the rate of heating in the daytime or cooling at night, therefore the velocity of the slope wind is proportional to the diurnal range of temperature (shown in Fig. 2).

The left half of Fig. 3 shows a cross section of the slope in Badong valley, while the right half of the figure expresses the distributions of up- and downslope winds along the slope at 3 meters above the ground. From this figure we can see that the speed of the upslope wind first increases upward along the slope, reaching a peak at the upper part of the slope, then decreases upward, whereas that of the downslope wind increases downward and attains a maximum value in the lower part of the slope, after which it decreases rapidly toward the valley bottom. These results can be interpreted as follows:

Under upslope wind conditions the airstream is subject to buoyancy acceleration so that the wind velocity increases upward. Near the top, however, as the space expands upward and the airstream diverges, the wind velocity decreases upward.

On the other hand, under downslope wind conditions, as the sinking cold air is subjected to gravitational acceleration, the wind velocity increases downward. In the lowest part of the slope the velocity decreases rapidly downward, however, since the airflow is slowed by the valley bottom and by the opposing downslope wind coming from the opposite slope.

According to our observations made in the mountainous regions of Sichuan, Yunnan, Jiangxi and Shanxi, other conditions being similar the speed and the frequency of occurrence of slope winds are related to the vegetation covering the ground. The denser and higher the vegetation, the less the speed and the frequency of occurrence of slope winds.

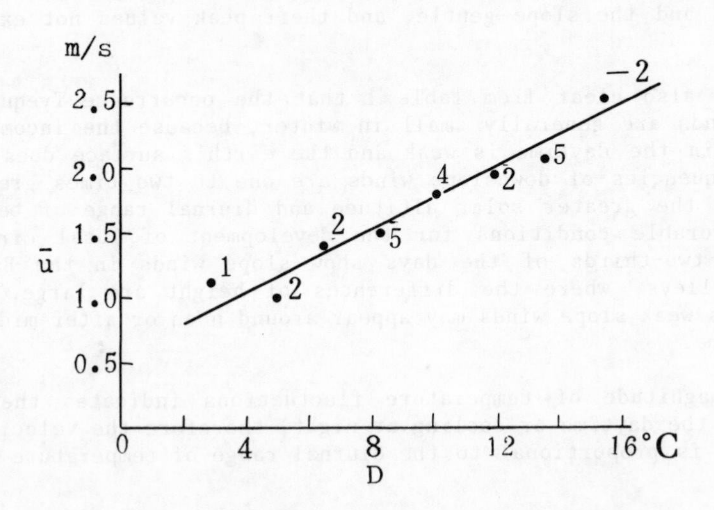

Fig. 2 Dependence of the mean speed of the upslope wind (U) on the diurnal range of air temperature (D) (at Badong).

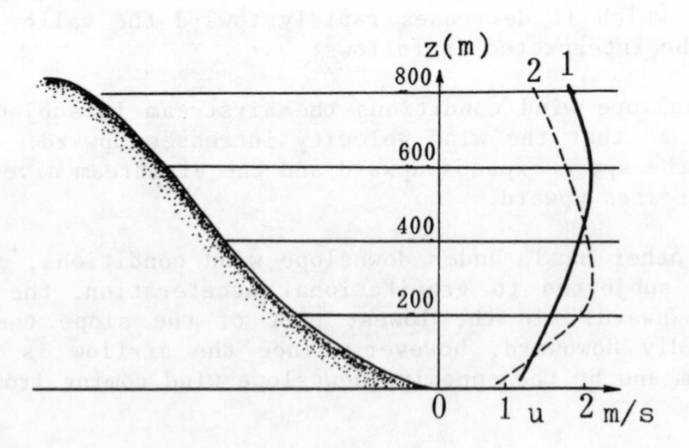

Fig. 3 Distributions of up- and downslope winds at 2 m along the slope at Badong. (1) Upslope wind, (2) downslope wind.

II. MATHEMATICAL ANALYSIS OF SLOPE WINDS

There are a number of theoretical studies of slope winds and various models are given [1-7]. Some of these models are based on assuming a steady state with a constant eddy exchange coefficient K, some assuming an unsteady state with constant K or a steady state with variable K, and some are obtained by numerical simulation. Here we derive a mathematical model of a pure slope wind proceeding from relatively general conditions and compare the observed features of the slope wind with those of the model.

Let OP be a section of a plane inclined at an angle α to the horizontal, and let ξ and ζ be coordinates measured along and normal to the slope, respectively. Since our interest is to find the solution to the problem in the representative middle part of the slope, we suppose that the speed of the slope wind, u, is a function of ζ only, and is always parallel to the slope. In this case, the simplified equation of motion of compressible viscous air and the equation of heat conduction can be written

$$\frac{\partial u}{\partial t} = \frac{\partial}{\partial \zeta}\left(K \frac{\partial u}{\partial \zeta}\right) + \beta g \sin\alpha \cdot \theta' \tag{1}$$

$$\frac{\partial \theta'}{\partial t} = \frac{\partial}{\partial \zeta}\left(K \frac{\partial \theta'}{\partial \zeta}\right) - A \sin\alpha \cdot u \tag{2}$$

where K is the coefficient of turbulent exchange in the ζ direction, g is the gravitational acceleration, and β the coefficient of thermal expansion, θ' is the disturbed potential temperature, specifically the deviation of potential temperature from the initial state. $A = \partial\theta_o/\partial z$, assumed constant, and θ_o is the potential temperature of the undisturbed atmosphere.

The boundary conditions adopted are

$$u = 0, \quad \theta' = \theta_o' = \sum_{(n)} \left[a_n \sin\frac{2n\pi}{\tau}t + b_n \cos\frac{2n\pi}{\tau}t\right] \quad \text{at } \zeta = \zeta_o \tag{3}$$

$$u = 0, \ \theta' = 0 \quad \text{as} \quad \zeta \to \infty \tag{4}$$

where ζ_o is the roughness length for a natural surface.

Supposing $A > 0$ (i.e. the initial atmosphere is considered stable, the most common case in practice), and letting

$$\phi \equiv \sqrt{\frac{A}{\beta g}}\, u + i\theta' , \tag{5}$$

Equations (1) and (2) reduce to

$$\frac{\partial \phi}{\partial t} = \frac{\partial}{\partial \zeta}\left(K \frac{\partial \phi}{\partial \zeta}\right) - ic\phi \tag{6}$$

where

$$c = \sin\alpha \sqrt{A\beta g} \quad . \tag{7}$$

The boundary conditions of Eqs. (3) and (4) thus become

$$\phi = i \sum_{(n)} (a_n \sin \frac{2n\pi}{\tau} t + b_n \cos \frac{2n\pi}{\tau} t) \quad \text{at } \zeta = \zeta_o \tag{8}$$

$$\phi = 0 \quad \text{as} \quad \zeta \rightarrow \infty \; . \tag{9}$$

K is taken as the expression

$$K = K_1 (\frac{\zeta}{\zeta_1})^{1-\varepsilon} \tag{10}$$

where K_1 is the value of K at a fixed reference height, ζ_1, which for convenience may be taken to be unity.

Equation (6) thus reduces to

$$\frac{1}{K_1} \zeta^{1+\varepsilon} \frac{\partial\phi}{\partial t} = \zeta^2 \frac{\partial^2\phi}{\partial\zeta^2} + (1-\varepsilon) \zeta \frac{\partial\phi}{\partial\zeta} - i \frac{c}{K_1} \zeta^{1+\varepsilon} \phi \tag{11}$$

In consideration of Eq. (8), the desired solution of Eq. (11) may be written in the form

$$\phi(\zeta,t) = \sum_{(n)} [Y_n(\zeta)\, e^{i\frac{2n\pi}{\tau}t} + Z_n(\zeta)\, e^{-i\frac{2n\pi}{\tau}t}] \; . \tag{12}$$

Substituting this expression into Eq. (11) gives

$$\begin{aligned} \sum_{(n)} \{ & [\zeta^2 Y_n'' + (1-\varepsilon)\zeta Y_n' - i(\frac{c\tau+2n\pi}{K_1\tau})\zeta^{1+\varepsilon} Y_n] e^{i\frac{2n\pi}{\tau}t} + \\ & + [\zeta^2 Z_n'' + (1-\varepsilon)\zeta Z_n' - i(\frac{c\tau+2n\pi}{K_1\tau})\zeta^{1+\varepsilon} Z_n] e^{-i\frac{2n\pi}{\tau}t} \} = 0 \end{aligned} \tag{13}$$

On equating the coefficients of $e^{i\frac{2n\pi}{\tau}t}$ and $e^{-i\frac{2n\pi}{\tau}t}$ to zero, it follows that

$$\zeta^2 Y_n'' + (1-\varepsilon)\zeta Y_n' - i(\frac{c\tau+2n\pi}{K_1\tau})\zeta^{1+\varepsilon} \, Y_n = 0, \tag{14}$$

$$\zeta^2 Z_n'' + (1-\varepsilon)\zeta Z_n' - i(\frac{c\tau+2n\pi}{K_1\tau})\zeta^{1+\varepsilon} \, Z_n = 0, \tag{15}$$

The solutions of Eqs. (14) and (15), subject to the boundary condition (Eq. 9), are

$$Y_n = A_n \zeta^{\frac{\nu}{2(1-\nu)}} K_\nu(y_n)$$

$$Z_n = B_n \zeta^{\frac{\nu}{2(1-\nu)}} K_\nu(z_n)$$

and hence

$$\phi(\zeta,t) = \sum_{(n)} [A_n \zeta^{\frac{\nu}{2(1-\nu)}} K_\nu(y_n) e^{i\frac{2n\pi}{\tau}t} + B_n \zeta^{\frac{\nu}{2(1-\nu)}} K_\nu(z_n) e^{-i\frac{2n\pi}{\tau}t}] \tag{16}$$

where

$$y_n = 2\delta_n \sqrt{i}\, \zeta^{\frac{1}{2(1-\nu)}}, \quad z_n = 2\mu_n \sqrt{i}\, \zeta^{\frac{1}{2(1-\nu)}}, \quad \nu = \frac{\varepsilon}{1+\varepsilon},$$

$$\delta_n = \sqrt{\frac{c\tau+2n\pi}{K_1\tau}}/(1+\varepsilon), \quad \mu_n = \sqrt{\frac{c\tau-2n\pi}{K_1\tau}}/(1+\varepsilon) \tag{17}$$

$K_\nu(x)$ is the modified Bessel function of the second kind of order ν; A_n and B_n are constants.

Determining the constants A_n and B_n by using the power-series expansion of $K_\nu(x)$ combined with boundary condition Eqs. (8) and (9), we have

$$u = \sqrt{\frac{\beta g}{A}}\, \frac{\sin\nu\pi}{\pi}\, \zeta^{\frac{\nu}{2(1-\nu)}} \sum_{(n)} \{[W^{(1)}(\delta_n,y_n) - W^{(3)}(\mu_n,z_n)] \cos\frac{2n\pi}{\tau}t -$$

$$- [W^{(2)}(\delta_n,y_n) + W^{(4)}(\mu_n,z_n)] \sin\frac{2n\pi}{\tau}t\}, \tag{18}$$

$$\theta' = \frac{\sin\nu\pi}{\pi}\, \zeta^{\frac{\nu}{2(1-\nu)}} \sum_{(n)} \{[W^{(2)}(\delta_n,y_n) - W^{(4)}(\mu_n,z_n)] \cos\frac{2n\pi}{\tau}t +$$

$$+ [W^{(1)}(\delta_n,y_n) + W^{(3)}(\mu_n,z_n)] \sin\frac{2n\pi}{\tau}t\}\,. \tag{19}$$

where

$$W^{(1)}(\delta_n,y_n) = S(\delta_n) R_e K_\nu(y_n) - T(\delta_n) I_m K_\nu(y_n)$$

$$W^{(2)}(\delta_n,y_n) = S(\delta_n)I_mK_\nu(y_n) + T(\delta_n)R_eK_\nu(y_n)$$

$$W^{(3)}(\mu_n,z_n) = U(\mu_n)R_eK_\nu(z_n) + V(\mu_n)I_mK_\nu(z_n)$$

$$W^{(4)}(\mu_n,z_n) = U(\mu_n)I_mK_\nu(z_n) - V(\mu_n)R_eK_\nu(z_n)$$

$$S(\delta_n) = \frac{a_nR(\delta_n,\zeta_o)-b_nH(\delta_n,\zeta_o)}{M(\delta_n,\zeta_o)}$$

$$T(\delta_n) = \frac{b_nR(\mu_n,\zeta_o)+a_nH(\delta_n,\zeta_o)}{M(\delta_n,\zeta_o)}$$

$$U(\mu_n) = \frac{a_nR(\mu_n,\zeta_o)+b_nH(\mu_n,\zeta_o)}{M(\mu_n,\zeta_o)} \tag{20}$$

$$V(\mu_n) = \frac{b_nR(\mu_n,\zeta_o)-a_nH(\mu_n,\zeta_o)}{M(\mu_n,\zeta_o)}$$

$$M(x,y) = R^2(x,y)+H^2(x,y),$$

$$R(x,y) = [\frac{x_n^{-\nu}}{\Gamma(1-\nu)} - \frac{x_n^{\nu}y^{\frac{\nu}{1-\nu}}}{\Gamma(1+\nu)}] \cos\frac{\nu\pi}{4}$$

$$H(x,y) = [\frac{x_n^{-\nu}}{\Gamma(1-\nu)} + \frac{x_n^{\nu}y^{\frac{\nu}{1-\nu}}}{\Gamma(1+\nu)}] \sin\frac{\nu\pi}{4}$$

$R_eK_\nu(z)$ and $I_mK_\nu(z)$ are real and imaginary parts of $K_\nu(z)$, and $\Gamma(x)$ is the gamma function.

In layers of air near the ground, as y_n and z_n are small quantities, $K_\nu(z)$ can be expressed by the approximate formula

$$K_\nu(z) \approx \frac{\pi}{2\sin\nu\pi}[\frac{1}{\Gamma(1-\nu)}(\frac{z}{2})^{-\nu} - \frac{1}{\Gamma(1+\nu)}(\frac{z}{2})^{\nu}] \tag{21}$$

Considering Eq. (17), we have

$$K_\nu(y_n) = \frac{\pi}{2\sin\nu\pi}\zeta^{-\frac{\nu}{2(1-\nu)}}[R(\delta_n,\zeta) - iH(\delta_n,\zeta)],$$

$$K_\nu(z_n) = \frac{\pi}{2\sin\nu\pi}\zeta^{-\frac{\nu}{2(1-\nu)}}[R(\mu_n,\zeta) - iH(\mu_n,\zeta)],$$

and hence

$$u = \frac{1}{\Gamma(1+\nu)}\sqrt{\frac{\beta g}{A}}(\zeta^{\frac{\nu}{1-\nu}} - \zeta_o^{\frac{\nu}{1-\nu}}) \sum_{(n)} [\psi^{(1)}\sin\frac{2n\pi}{\tau}t - \psi^{(2)}\cos\frac{2n\pi}{\tau}t], \quad (22)$$

$$\theta' = \sum_{(n)} \{[a_n - \frac{1}{2\Gamma(1+\nu)}\psi^{(3)}(\zeta^{\frac{\nu}{1-\nu}} - \zeta_o^{\frac{\nu}{1-\nu}})]\sin\frac{2n\pi}{\tau}t +$$

$$+ [b_n - \frac{1}{2\Gamma(1+\nu)}\psi^{(4)}(\zeta^{\frac{\nu}{1-\nu}} - \zeta_o^{\frac{\nu}{1-\nu}})]\cos\frac{2n\pi}{\tau}t\} \quad (23)$$

where

$$\psi^{(1)} = [\delta_n^\nu S(\delta_n) + \mu_n^\nu U(\mu_n)]\sin\frac{\nu\pi}{4} + [\delta_n^\nu T(\delta_n) - \mu_n^\nu V(\mu_n)]\cos\frac{\nu\pi}{4}$$

$$\psi^{(2)} = [\delta_n^\nu S(\delta_n) + \mu_n^\nu U(\mu_n)]\cos\frac{\nu\pi}{4} - [\delta_n^\nu T(\delta_n) + \mu_n^\nu V(\mu_n)]\sin\frac{\nu\pi}{4}$$

$$(24)$$

$$\psi^{(3)} = [\delta_n^\nu S(\delta_n) + \mu_n^\nu U(\mu_n)]\cos\frac{\nu\pi}{4} - [\delta_n^\nu T(\delta_n) - \mu_n^\nu V(\mu_n)]\sin\frac{\nu\pi}{4}$$

$$\psi^{(4)} = [\delta_n^\nu S(\delta_n) - \mu_n^\nu U(\mu_n)]\sin\frac{\nu\pi}{4} + [\delta_n^\nu T(\delta_n) - \mu_n^\nu V(\mu_n)]\cos\frac{\nu\pi}{4}$$

When $\alpha \geq 10°$, assuming $A \geq 0.3°C/100$ m and taking $\beta = \frac{1}{273}$, $g = 9.8$ m/s^2, we have

$$c = \sin\alpha\ \sqrt{A\beta g} > 2 \times 10^{-3}, \quad \frac{2n\pi}{\tau} = 7.3n \times 10^{-5}.$$

Usually it is sufficient to represent the disturbed potential temperature of the surface of the earth by three harmonics (n=3), $\frac{2n\pi}{\tau}$ is one order smaller than c at least, and δ_n and μ_n may be regarded as approximately equal.

Put

$$\delta_n \sim \mu_n = \delta = \frac{\sqrt{\sin\alpha}\ \sqrt{A\beta g/K_1}}{1+\varepsilon}, \quad (25)$$

Equations (22) and (23) then reduce to

$$u = \frac{\sin\frac{\varepsilon\pi}{2(1+\varepsilon)}}{M\Gamma(\frac{1}{1+\varepsilon})\Gamma(\frac{1+2\varepsilon}{1+\varepsilon})}\sqrt{\frac{\beta g}{A}}(\zeta^{\varepsilon}-\zeta_o^{\varepsilon})\sum_{(n)}(a_n \sin\frac{2n\pi}{\tau}t + b_n \cos\frac{2n\pi}{\tau}t) =$$

$$= \frac{\sin\frac{\varepsilon\pi}{2(1+\varepsilon)}}{M\Gamma(\frac{1}{1+\varepsilon})\Gamma(\frac{1+2\varepsilon}{1+\varepsilon})}\sqrt{\frac{\beta g}{A}}\,\theta_o'(\zeta^{\varepsilon}-\zeta_o^{\varepsilon}) \tag{26}$$

$$\theta' = \left\{1 - \frac{\zeta^{\varepsilon}-\zeta_o^{\varepsilon}}{M\Gamma(\frac{1+2\varepsilon}{1+\varepsilon})}\left[\frac{\cos\frac{\varepsilon\pi}{2(1+\varepsilon)}}{\Gamma(\frac{1}{1+\varepsilon})} - \frac{\delta^{\frac{2\varepsilon}{1+\varepsilon}}\zeta_o^{\varepsilon}}{\Gamma(\frac{1+2\varepsilon}{1+\varepsilon})}\right]\right\}\sum_{(n)}(a_n \sin\frac{2n\pi}{\tau}t + b_n \cos\frac{2n\pi}{\tau}t) =$$

$$= \left\{1 - \frac{\zeta^{\varepsilon}-\zeta_o^{\varepsilon}}{M\Gamma(\frac{1+2\varepsilon}{1+\varepsilon})}\left[\frac{\cos\frac{\varepsilon\pi}{2(1+\varepsilon)}}{\Gamma(\frac{1}{1+\varepsilon})} - \frac{\delta^{\frac{2\varepsilon}{1+\varepsilon}}\zeta_o^{\varepsilon}}{\Gamma(\frac{1+2\varepsilon}{1+\varepsilon})}\right]\right\}\theta_o' \tag{27}$$

where

$$M = \left(\frac{\delta^{-\frac{\varepsilon}{1+\varepsilon}}}{\Gamma(\frac{1}{1+\varepsilon})}\right)^2 + \left(\frac{\delta^{\frac{\varepsilon}{1+\varepsilon}}\zeta_o^{\varepsilon}}{\Gamma(\frac{1+2\varepsilon}{1+\varepsilon})}\right)^2 - \frac{2\zeta_o^{\varepsilon}}{\Gamma(\frac{1}{1+\varepsilon})\Gamma(\frac{1+2\varepsilon}{1+\varepsilon})}\cos\frac{\varepsilon\pi}{2(1+\varepsilon)} \tag{28}$$

Taking A = 0.3°C/100 m and ζ_o = 0.03 m, the ratios of the slope wind velocity at two meters above the ground to θ_o', u/θ_o', for various values of ε, K, and α computed according to Eq. (25) are given in Table 2.

From Eqs. (26) and (27), and Table 2 we can see that the speed of the slope wind near the ground is directly proportional to the temperature disturbance θ_o', and it is greater the steeper the slope, the weaker the turbulence, and the smoother the ground surface. These results are consistent with the observations stated in Part I.

Furthermore, Eqs. (26) and (27) show that the variations of slope wind velocity, u, and of the temperature disturbance, θ', with height in the layers of air near the ground conform to a power law. U increases as the ε power of ζ, and θ' decreases as the ε power of ζ.

On clear winter days in areas where the soil is either bare or covered by sparse or low vegetation, the value of ε is about -0.05 in the daytime, but might rise to 0.5 or even larger at night, while the temperature disturbance θ_o', whether in the daytime or at night, amounts to 3

Table 2 Ratios of U (m/s) at ζ = 2m to θ_o (°C), in relation to ε, K_1 and α.

ε	-0.15	-0.05	0.10	0.25	0.50
$K_1(m^2/s)$ / α	0.20	0.15	0.10	0.09	0.07
10°	0.41	0.51	0.55	0.65	0.85
15°	0.51	0.54	0.61	0.73	1.00
25°	0.56	0.60	0.71	0.75	1.30

to 6°C. Consequently, as far as a slope of moderate inclination is concerned, according to Table 2, the maximum speed of the upslope wind at 2 m is about 1 to 3 m/s, and that of the downslope wind 3 to 6 m/s. On clear summer days ε is perhaps less than -0.1 and θ_o' attains 6 to 12°C or more in the afternoon. At night ε usually lies between 0.2 and 0.5 and θ_o' is reduced by half. The maximum velocity of the upslope wind is then about 3 to 6 m/s, and that of the downslope wind 2 to 5 m/s. The results, as with the observation stated above, show that the upslope wind is stronger in summer than in winter and is stronger than the downslope wind in summer but weaker in winter.

For large values of ζ, the asymptotic expansion for $K_\nu(z)$ be used, omitting all but the first term, namely

$$K_\nu(z) = \sqrt{\frac{\pi}{2z}}\, e^{-z} \tag{29}$$

Thus, in consideration of $\delta_n \sim \mu_n = \delta$, the solution of the problem can be reduced to

$$u = \sqrt{\frac{\beta g}{A\pi}}\, \sin\nu\pi \cdot L(\delta)\zeta^{-\frac{1-\varepsilon}{4}} e^{-\sqrt{2}\delta\zeta^m} [f_1 \sin(\frac{\pi}{8} - \frac{\nu\pi}{4} + \sqrt{2}\,\delta\zeta^m) - f_2 \sin(\frac{\pi}{8} + \frac{\nu\pi}{4} + \sqrt{2}\,\delta\zeta^m)] \sum_{(n)} (a_n \sin\frac{2n\pi}{\tau}t + b_n \cos\frac{2n\pi}{\tau}t) \tag{30}$$

$$\theta' = \frac{\sin\nu\pi}{\sqrt{\pi}} L(\delta)\zeta^{-\frac{1-\varepsilon}{4}} e^{-\sqrt{2}\delta\zeta^m} [f_1 \cos(\frac{\pi}{8} - \frac{\nu\pi}{4} + \sqrt{2}\,\delta\zeta^m) - f_2 \cos(\frac{\pi}{8} + \frac{\nu\pi}{4} + \sqrt{2}\,\delta\zeta^m)] \sum_{(n)} (a_n \sin\frac{2n\pi}{\tau}t + b_n \cos\frac{2n\pi}{\tau}t) \tag{31}$$

where

$$f_1 = \delta^{-\nu}/\Gamma(1-\nu), \quad m = \frac{1+\varepsilon}{2},$$

$$f_2 = \delta^{\nu}\zeta_o^{\frac{\nu}{1-\nu}}/\Gamma(1+\nu), \tag{32}$$

$$L(\delta) = \delta^{-\frac{1}{2}}/[R^2(\delta,\zeta_o) + H^2(\delta,\zeta_o)]$$

Setting U = 0 and solving Eq. (30), we obtain the transition height at which the slope wind changes direction as follows:

$$H = \left\{ \frac{1}{\sqrt{2}\delta} \mathrm{arctg} \left[\frac{\Gamma(\frac{1+2\varepsilon}{1+\varepsilon})\sin(\frac{\pi}{8} - \frac{\varepsilon\pi}{4(1+\varepsilon)}) - \Gamma(\frac{1}{1+\varepsilon})\delta^{\frac{2\varepsilon}{1+\varepsilon}}\zeta_o^{\varepsilon}\sin(\frac{\pi}{8} + \frac{\varepsilon\pi}{4(1+\varepsilon)})}{\Gamma(\frac{1+2\varepsilon}{1+\varepsilon})\cos(\frac{\pi}{8} - \frac{\varepsilon\pi}{4(1+\varepsilon)}) - \Gamma(\frac{1}{1+\varepsilon})\delta^{\frac{2\varepsilon}{1+\varepsilon}}\zeta_o^{\varepsilon}\cos(\frac{\pi}{8} - \frac{\varepsilon\pi}{4(1+\varepsilon)})} \right] \right\}^{\frac{2}{1+\varepsilon}} \tag{33}$$

When A = 0.3°C/100 m, the transition heights, H, of slope winds under different conditions, computed according to Eq. (33), are given in Table 3. This table shows that H is dependent on the roughness of the surface (ζ_o), the inclination of the slope (α) and the stability of the atmosphere (ε). The rougher the surface, the gentler the slope and the more unstable the atmosphere, the larger the transition height. Hence, the thickness of the upslope wind during the daytime ($\varepsilon < 0$) is larger than that of the downslope wind at night ($\varepsilon > 0$), and the denser or deeper the vegetation on the ground the thicker the slope wind. Thus, the surface vegetation exerts a lifting effect on the transition height of the slope wind.

Assuming $\alpha = 15°$, $\varepsilon = 0.5$ and $K_1 = 0.07\ m^2/s$, the variations of downslope wind velocity and potential temperature deviation, expressed by the nondimensional quantities

$$\sqrt{\frac{A}{\beta g}}\frac{u}{\theta_o'} \quad \text{and} \quad \frac{\theta'}{\theta_o'},$$

respectively, with nondimensional distance normal to the slope, ζ/H, are computed according to Eqs. (18) and (19) and are plotted in Fig. 4. The distribution of nondimensional temperature $(T-T_o)/\theta_o'$, with A = 0.3°C/100 m and $\theta_o' = 6°C$, is also shown in that figure.

From Fig. 4 we see that the maximum velocity of the downslope wind occurs at an altitude of about one-fifth of the transition height H, and the upper limit of the inversion layer on the slope appears at an altitude of about H/2. These values are in excellent agreement with the of observations made in the valleys of Sichuan and Shensi Provinces.

Table 3 Transition heights of slope winds, m.

ζ_o (m)	0					0.03				
K_1 (m^2/s)	0.20	0.15	0.10	0.09	0.7	0.20	0.15	0.10	0.09	0.07
α \ ε	-0.15	-0.05	0.10	0.25	0.50	-0.15	-0.05	0.10	0.25	0.50
10°	1176	398	162	103	53.1	1293	457	178	110	55.1
15°	583	263	113	75.5	41.7	665	305	125	76.7	42.4
25°	331	150	72.6	52.6	30.0	376	183	80.2	54.5	30.6

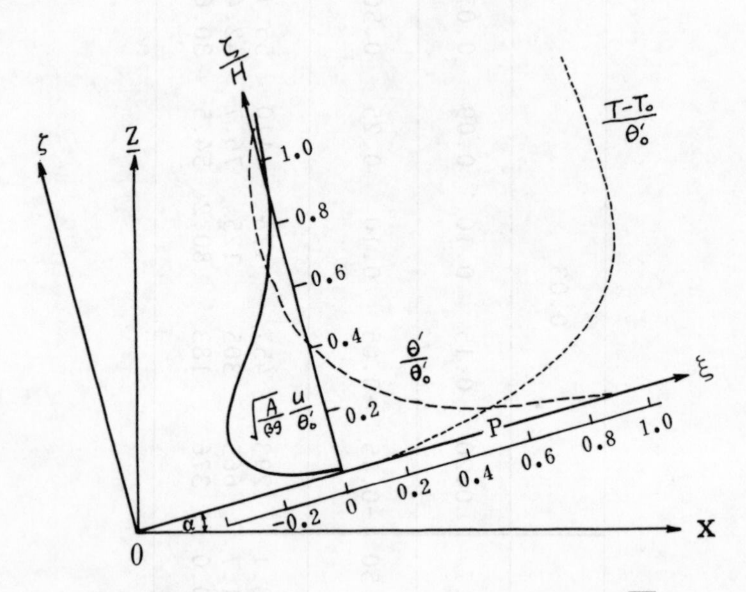

Fig. 4 Theoretical distributions of downslope wind $\sqrt{\frac{A}{\beta g}}\,\frac{u}{\theta_o'}$, potential temperature deviation $\frac{\theta'}{\theta_o'}$, and temperature $\frac{T-T_o}{\theta_o'}$, denoted by nondimensional quantities.

In the particular case, when $\varepsilon = 1$ and thus $K = K_1$ = constant, $\nu = \frac{\varepsilon}{1+\varepsilon} = \frac{1}{2}$, and

$$K_{\frac{1}{2}}(z) = \sqrt{\frac{\pi}{2z}}\, e^{-z}$$

taking $\delta_n = \mu_n = \delta$, Eq. (16) can be reduced to

$$u = \sqrt{\frac{\beta g}{A}}\, e^{-\sqrt{2}\delta(\zeta-\zeta_o)} \sin \sqrt{2}\delta(\zeta-\zeta_o) \sum_{(n)} (a_n \sin\frac{2n\pi}{\tau}t + b_n \cos\frac{2n\pi}{\tau}t),$$

$$\theta' = e^{-\sqrt{2}\delta(\zeta-\zeta_o)} \cos \sqrt{2}\delta(\zeta-\zeta_o) \sum_{(n)} (a_n \sin\frac{2n\pi}{\tau}t + b_n \cos\frac{2n\pi}{\tau}t). \tag{34}$$

Let $b_n = 0$, $\zeta_o = 0$, and $n = 1$. These equations then reduce to Gandin [2] formulas:

$$u = \sqrt{\frac{\beta g}{A}}\, e^{-\sqrt{2}\delta\zeta}\, a_1 \sin \sqrt{2}\delta\zeta \sin\frac{2\pi}{\tau}t, \tag{35}$$

$$\theta' = a_1 e^{-\sqrt{2}\delta\zeta} \cos\sqrt{2}\delta\zeta \sin\frac{2\pi}{\tau}t.$$

If we put $\zeta_o = 0$ and replace

$$\sum_{(n)} (a_n \sin\frac{2n\pi}{\tau}t + b_n \cos\frac{2n\pi}{\tau}t)$$

by θ_o', Eq. (34) gives the Prandtl [6] formulas:

$$u = \sqrt{\frac{\beta g}{A}} \theta_o' e^{-\sqrt{2}\delta\zeta} \sin\sqrt{2}\delta\zeta, \qquad (36)$$

$$\theta' = \theta_o' e^{-\sqrt{2}\delta\zeta} \cos\sqrt{2}\delta\zeta.$$

III. CONCLUSION

The main results of our study can be summarized as follows:

1. The upslope wind in the valley generally begins one and one-half to three hours after sunrise and attains its maximum velocity at about midday, whereas the downslope wind begins one-half to two hours before sunset and attains its first maximum velocity one to two hours after sunset.

2. The upslope wind is stronger than the downslope wind in summer, but weaker in winter. Moreover, the strength of the upslope wind is greater in summer than in winter, while that of the downslope wind behaves in the opposite fashion.

3. The deeper the valley, the steeper the slope, the sparser or lower the vegetation and the larger the daily range of temperature, the stronger the slope wind.

4. The thickness of the upslope wind is generally greater than that of the downslope wind, and the more gentle the slope and the denser or higher the vegetation, the thicker the slope wind.

5. The velocity of the upslope wind increases upward along the slope and attains a maximum value on the upper part of the slope, while that of the downslope wind increases downward and reaches a peak value on the lower part of the slope.

REFERENCES

1. Dobrishman, E.M.: On the slope wind above a nonuniformly heated underlying surface. Trudy Ts.I.P., 43 (70) (In Russian) (1956).

2. Gandin, L.S.: The principles of dynamical meteorology. Gidrometeoizdat, L., (In Russian) (1955).

3. Gutman, L.N.: The theory of mountain-valley wind. Dokl. A.N. SSSR, 15 (2) (In Russian) (1949).

4. Likosov, B.N. and Gutman, L.N.: The turbulent boundary layer above an inclined underlying surface. A.N. SSSR, 8 (8) (In Russian) (1972).

5. Monin, A.S.: On the slope wind. Trudy Ts.I.P., 7 (37) (In Russian) (1948).

6 Prandtl, I., Strömungslehre, Braunschweig (1942).

7. Thyer, N.H.: A theoretical explanation of mountain and valley winds by numerical method. Arch. Met. Geophys. Biokl., 15, Nos. 3 - 4 (1966).

5.4 SELECTED TOPICS ON THE PARAMETERIZATION OF CUMULUS CONVECTION

Bruce A. Albrecht
The Pennsylvania State University
University Park, PA 16802 USA

ABSTRACT

Two topics concerning the parameterization of cumulus convection are discussed. 1) A scheme for predicting the fractional cloud amount associated with nonprecipitating cumulus convection is described. This scheme is applied to oceanic conditions in the subtropics and is used in a one-dimensional boundary layer model to demonstrate the dependence of cloudiness on sea surface temperature and the diurnal variations in solar absorption. The implications of these results to the parameterization of cloudiness over a heated surface are discussed.

2) A simplified approach to the parameterization of the heating and moistening associated with precipitating clouds is described. The convective effects considered are warming and drying due to cumulus induced subsidence, and cooling and moistening due to detrainment and the evaporation of rain. The parameterization is used in a one-dimensional model that represents the vertical structure as polynomials. The model is applied to mean tropical conditions. Sensitivity tests are used to demonstrate the relative importance of the various convective processes and to determine the sensitivity of the convective heating and moistening profiles to variations in cloud model and closure assumptions.

I. INTRODUCTION

Cumulus convection plays an important role in transporting heat, moisture, and momentum in the atmosphere. Furthermore, the latent heat released in precipitating clouds has been shown to be an important source of energy for various atmospheric circulations. In addition to the effect that cumulus clouds have on the heat and moisture budgets of the atmosphere, they can significantly modify the radiation balance of both the atmosphere and the earth's surface.

The growth, maintenance and dissipation of clouds, however, may be regulated by the large-scale fields of temperature, moisture and momentum. The possible interactions between large-scale, convective and radiative processes are shown in Fig. 1.

In models describing the time evolution of the large-scale fields, individual cumulus clouds cannot be resolved. Consequently, it is necessary to parameterize or represent the cumulative effects of these clouds in terms of the large-scale fields.

This paper will consider two particular aspects of the cumulus parameterization problem. The first is the parameterization of the fractional cloud amount associated with shallow nonprecipitating convection. These clouds can significantly alter the radiative budget of the earth's surface which could affect the further development of nonprecipitating and precipitating convection. This type of interaction may be important for the surface heat budget of the Tibetan Plateau and particularly its diurnal variation.

The second topic is on the parameterization of the vertical distribution of heating and moistening associated with deep precipitating cumulus. Kuo and Qian [8] clearly showed the importance of the latent heat release over the Tibetan Plateau in maintaining the large-scale circulation. In this paper particular emphasis is placed on the role of physical processes in determining the vertical distribution of the convective effects.

II. PARAMETERIZATION OF NONPRECIPITATING CLOUD AMOUNTS

In a field of nonprecipitating cumulus clouds, the cloud coverage associated with decaying or dynamically inactive clouds may be a significant fraction of the total visible cloud cover. Albrecht [1] proposed a cloud cover scheme based on the assumption that the fractional coverage of decaying clouds is proportional to the time required for an individual cloud element to decay and mix with its environment. This parameterization gives the cloud cover fraction in terms of the area averaged relative humidity in the cloud layer and the liquid water content of individual convective elements. The cloud fraction given by Albrecht [1] is

$$\sigma = \frac{SR-1}{SR-RH} \tag{1}$$

where RH is the relative humidity in the cloud layer and SR is defined as the ratio of the total water mixing ratio (liquid and vapor) of a cloud element to the saturation mixing ratio of the environment. The convective scale parameter SR may be related to the large-scale temperature and moisture fields using some type of cloud model.

One important implication of Eq. (1) is that for conditions of low liquid water content (SR~1) and high relative humidity (RH~1), the cloud amount is very sensitive to variations in either of these parameters. This behavior is apparent in Fig. 2 where observed cloud amounts for typical trade-wind conditions are compared with the cloud amount calculated using radiosonde data, a simple cloud model, and Eq. (1) applied near the top of the cloud layer. These results are from observations made during the Atlantic Trade-Wind Experiment, 1969 (Augstein et al. [4]). The relative humidity, which is also shown in Fig. 2, is approximately 90 percent for all observed cloud amounts, and higher cloud fractions are associated with slightly higher relative humidities. The cloud cover calculated using Eq. (1), however, shows good correlation with the observed cloud amounts. An important implication of Eq. (1) is that the cloud cover may depend not only on the conditions in the cloud layer, but

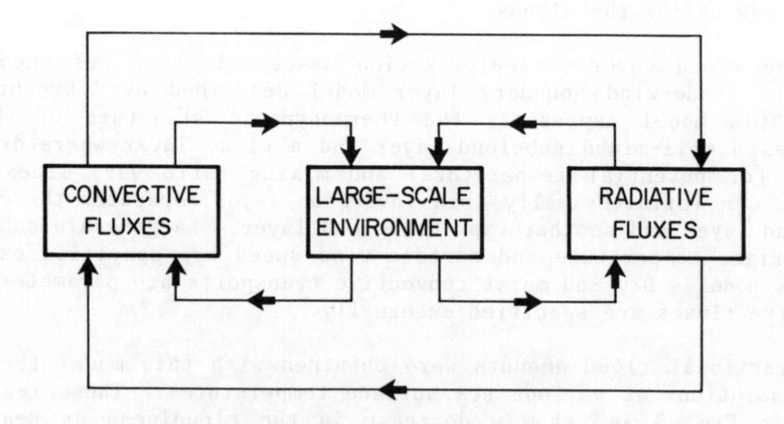

Fig. 1 Schematic diagram showing possible interactions between convective, radiative and large-scale processes.

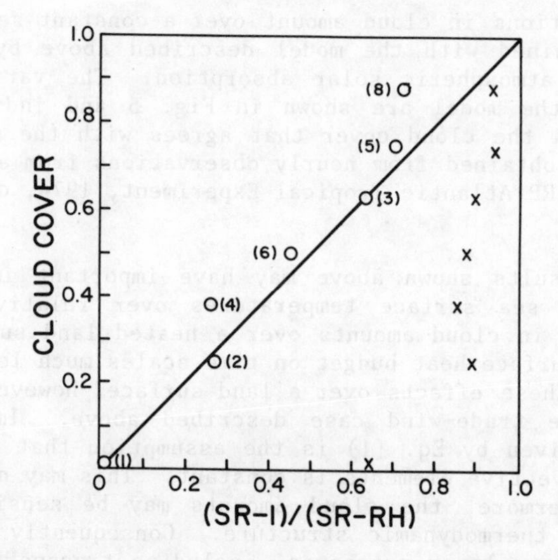

Fig. 2 Observed fractional cloud cover as a function of (SR-1)/(SR-RH) and the relative humidity (x) of the top 20 mb of the cloud layer (after [1]).

also the thermodynamic properties of the subcloud layer since this is the source region for the clouds.

The cloud cover parameterization described above was incorporated into the trade-wind boundary layer model described by Albrecht et al. [2]. This model represents the thermodynamic structure of the trade winds as a well-mixed subcloud layer and a cloud layer where dry static energy (or potential temperature) and mixing ratio vary linearly with height. An infinitesimally thin inversion layer separates the cloud and subcloud layer and another caps the cloud layer. Large-scale subsidence, sea surface temperature and surface wind speed are specified externally in this model. Dry and moist convective transports are parameterized and radiative fluxes are specified externally.

Fractional cloud amounts were obtained with this model for steady-state solutions at various sea surface temperatures. These results are shown in Fig. 3 and show a decrease in the cloudiness as sea surface temperature increases. The surface moisture fluxes obtained for these solutions are shown in Fig. 4 to increase with increasing sea surface temperature and demonstrate the inadequacy of parameterizing the low cloud amounts as a function of the surface moisture fluxes. Changes in the thermodynamic structure of the boundary layer depend on the flux divergence, not the surface fluxes.

Diurnal variations in cloud amount over a constant sea surface temperature were obtained with the model described above by specifying a diurnally varying atmospheric solar absorption. The variations in the cloud cover from the model are shown in Fig. 5 and indicate an early morning maximum in the cloud cover that agrees with the maximum in the low cloud amounts obtained from hourly observations from a ship 15°N and 30°W during the GARP Atlantic Tropical Experiment, 1974, during Phases I and II.

While the results shown above may have important implications to variations in the sea surface temperatures over relatively long time scales, variations in cloud amounts over a heated land surface may have an impact on the surface heat budget on time scales much less than a day. The modelling of these effects over a land surface, however, may be more difficult than the trade-wind case described above. Implicit in the parameterization given by Eq. (1) is the assumption that the production rate of active convective elements is constant. This may not be the case over land. Furthermore, the cloud amounts may be sensitive to small variations in the thermodynamic structure. Consequently a good representation of boundary layer processes, including transports of heat and moisture by the shallow clouds would be needed.

III. PARAMETERIZATION OF CUMULUS HEATING AND MOISTENING PROFILES

The latent heat released in deep precipitating cumulus clouds may be an important source of energy for many atmospheric circulations, including the circulation over Tibet in the summer. If such circulations are to be simulated in numerical models, it is necessary to parameterize the effects of moist convection as a function of the large-scale fields. The

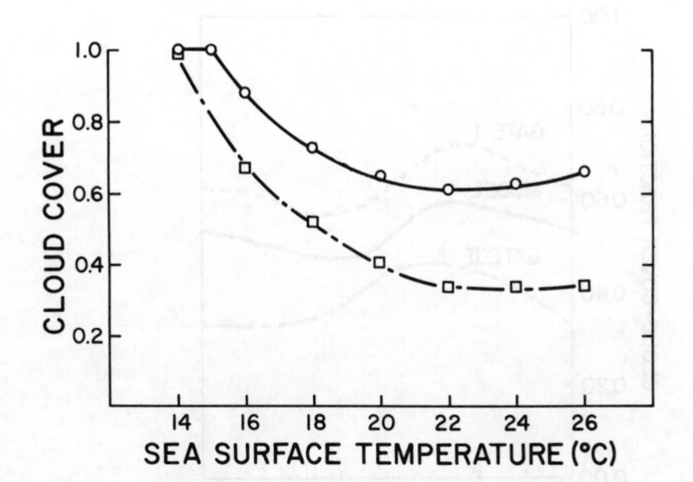

Fig. 3 Model-predicted cloud cover as a function of sea surface temperature for a surface wind speed of 8 ms^{-1} (solid line) and 4 ms^{-1} (long dashes) (after [1]).

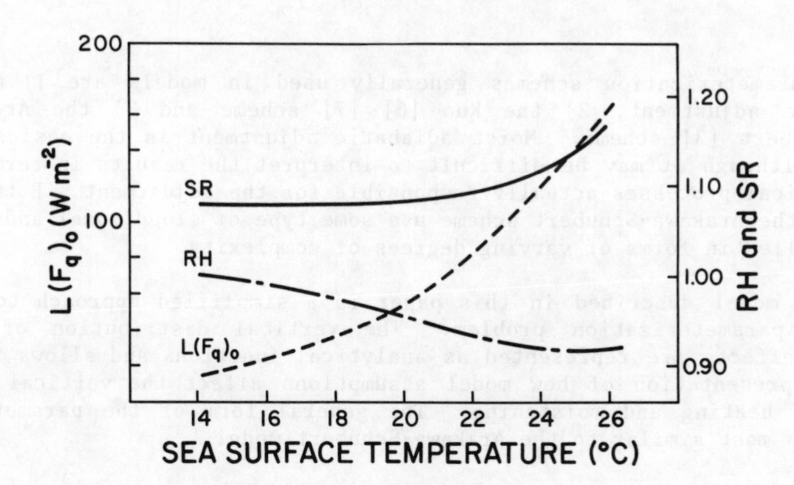

Fig. 4 Surface flux of water vapor, relative humidity and virtual saturation ratio as a function of sea surface temperature for the V_o = 8 ms^{-1} case shown in Fig. 3 (after [1]).

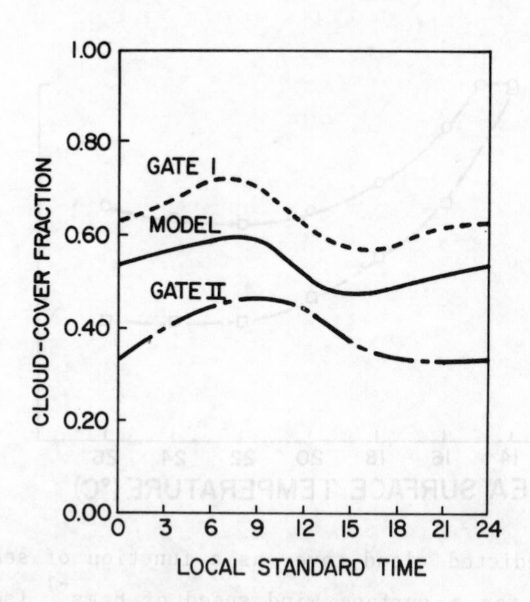

Fig. 5 Dirunal variation in cloud amount obtained from model and that observed from ships at 15°N and 30°W during Phases I and II of GATE.

three parameterization schemes generally used in models are 1) moist adiabatic adjustment, 2) the Kuo [6] [7] scheme and 3) the Arakawa and Schubert [3] scheme. Moist adiabatic adjustment is the easiest to apply, although it may be difficult to interpret the results in terms of the physical processes actually responsible for the adjustment. Both the Kuo and the Arakawa-Schubert scheme use some type of cloud model and have been applied in forms of varying degrees of complexity.

The model described in this paper is a simplified approach to the cumulus parameterization problem. The vertical distribution of the cumulus effects are represented as analytical functions and allows for a clear representation of how model assumptions affect the vertical profiles of heating and moistening. The general form of the parameterization is most similar to the Arakawa-Schubert model.

The thermodynamic structure of the model is represented using dry static energy $s = c_p T + gz$ where c_p is the specific heat of air, T is temperature, g is the acceleration due to gravity and z is the height above the surface ($s \approx c_p \theta$). The moisture is represented by mixing ratio q. It is also convenient to note that moist static energy $h = s + Lq$ ($h \approx c_p \theta_e$ where θ_e is equivalent potential temperature). The vertical distributions of s, q and h for a convectively active region over the western Pacific are shown in Fig. 6. These profiles are from the 100-day average described by Yanai et al. [11] for the Marshall Islands region.

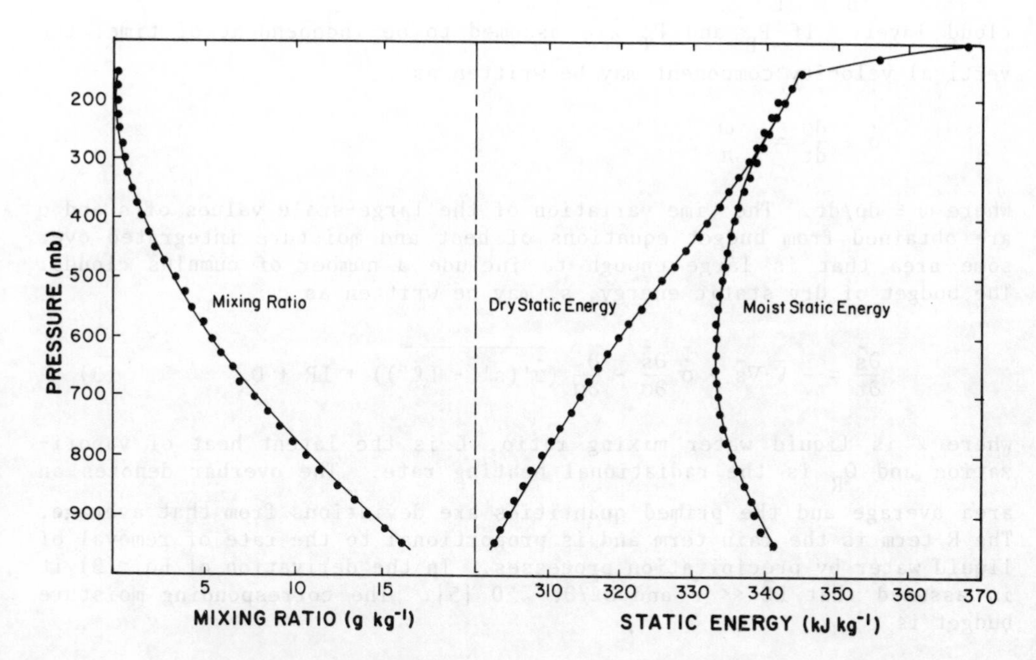

Fig. 6 Thermodynamic structure for 100-day average for Marshall Islands region obtained by Yanai et al. [11]. Cubic fits to the data are shown.

Although details that might be found in an individual sounding have been eliminated by the averaging, it is clear that in a region of active deep convection, sharp gradients between the tropopause and the top of the boundary layer are probably relatively transient features. A cubic fit to the data is shown in Fig. 6 and gives a good representation of the structure.

The cloud layer structure of the model is represented by polynomials. Formally a finite series of shifted Chebyshev polynomials [9] is used for the representation. Although the order of the representation is limited to third order, higher order terms can easily be incorporated into the scheme. Chebyschev polynomials are used since they allow higher order terms that arise in the mathematical formulation to be truncated with relatively little error. The vertical coordinate used in the model is a modified sigma coordinate system. If P_B is the pressure at cloud base and P_T is the pressure at the top of the cloud layer

$$\sigma = \frac{P_B - P}{\pi} \tag{2}$$

where $\pi = P_B - P_T$. Hence $\sigma = 0$ at cloud base and $= 1$ at the top of the cloud layer. If P_B and P_T are assumed to be independent of time, the vertical velocity component may be written as

$$\dot{\sigma} \equiv \frac{d\sigma}{dt} = - \frac{\omega}{\pi}$$

where $\omega \equiv dp/dt$. The time variation of the large-scale values of s and q are obtained from budget equations of heat and moisture integrated over some area that is large enough to include a number of cumulus clouds. The budget of dry static energy, s, may be written as

$$\frac{\partial \bar{s}}{\partial t} = - \bar{V} \cdot \nabla \bar{s} - \bar{\dot{\sigma}} \frac{\partial \bar{s}}{\partial \sigma} - \frac{\partial}{\partial \sigma} \overline{(\dot{\sigma}'(s' - L\ell'))} + LR + Q_R \qquad (3)$$

where ℓ is liquid water mixing ratio, L is the latent heat of vaporization and Q_R is the radiational heating rate. The overbar denotes an area average and the primed quantities are deviations from that average. The R term is the rain term and is proportional to the rate of removal of liquid water by precipitation processes. In the derivation of Eq. (3) it is assumed that $L\bar{\ell} << \bar{s}$ and $\partial\bar{\ell}/\partial t \simeq 0$ [5]. The corresponding moisture budget is

$$\frac{\partial \bar{q}}{\partial t} = - \bar{V} \cdot \nabla \bar{q} - \bar{\dot{\sigma}} \frac{\partial \bar{q}}{\partial \sigma} - \frac{\partial}{\partial \sigma} \overline{(\dot{\sigma}'(q' + \ell'))} - R \qquad (4)$$

The first two terms on the right-hand side of Eqs. (3) and (4) are horizontal and vertical advection terms and are determined by the large-scale fields. The flux divergence terms and the precipitation terms are the terms that need to be parameterized or represented as a function of the large-scale variables.

The parameterization follows the basic scheme used by Ooyama [10], Yanai et al. [11] and Arakawa and Schubert [3]. The flux term in Eq. (3) is then represented as

$$\overline{\dot{\sigma}'(s' - L\ell')} = M_c(s_c - L\ell_c - \bar{s}) \qquad (5)$$

where M_c is an effective mass flux (nondimensional) and the subscript c refers to in-cloud values for a "reference cloud". The effective mass flux is defined as

$$M_c \equiv \Sigma \, \eta_i \, m_i \, \alpha_i \qquad (6)$$

where η_i is the fractional area covered by active updrafts for a cloud type i, m_i is the mass flux for that cloud type and α_i parameterizes differences between the in-cloud values for a particular cloud type and the reference cloud. With this definition α_i will generally vary between

0 and 1. If only a single cloud type is used, the summation in Eq. (6) is no longer needed, m_i may be related directly to the cloud updraft velocity and $\alpha_i = 1$. For a situation where different cloud types are considered, a cloud model could be used to define m_i and α_i, but the fractional coverage η_i is still unknown and must be obtained by some closure assumption [10], [3]. The use of the parameter α_i and combining it with the unknown product $\eta_i m_i$ allows the flux given by Eq. (5) to be written without a summation of the product of the individual mass fluxes that appears in the Arakawa and Schubert [3] formulation.

With some approximation the corresponding flux of total water in Eq. (4) may be written as

$$\overline{\dot{\sigma}'(q' + \ell')} = M_c(q_c + \ell_c - \bar{q}) \; . \tag{7}$$

If the rain water produced per unit height of the reference cloud is r_c, the precipitation term may be approximated as

$$R = M_c \, r_c \; . \tag{8}$$

If the reference cloud is assumed to be a deep cloud, the approximation made in writing Eqs. (7) and (8) using the M_c defined by Eq. (6) is best near the base layer and the top of the cloud layer. For the results presented in this paper the approximation should have no major effect.

A cloud model is needed to obtain the cloud values of q_c, s_c, ℓ_c and r_c. The cloud model relates these parameters to the large-scale parameters. Although a cloud model of any sophistication could be used, entrainment relationships like those used by Yanai et al. [11] are used. These relationships are

$$\frac{\partial(s_c - L\ell_c)}{\partial\sigma} = -\mu_o(s_c - L\ell_c - \bar{s}) + L\, r_c \tag{9}$$

and

$$\frac{\partial(q_c + \ell_c)}{\partial\sigma} = -\mu_o(q_c + \ell_c - \bar{q}) - r_c \tag{10}$$

where μ_o is the entrainment rate for the reference cloud. If the rate of conversion of cloud water to precipitation per unit of σ is assumed to be a specified fraction of the cloud water ℓ_c

$$r_c = c_o \ell_c \tag{11}$$

where c_o is a coefficient that is specified externally. Equations (9) and (10) with (11) may be solved analytically since $\bar{s}$ and $\bar{q}$ are polynomials. The clouds are assumed to have subcloud layer properties at cloud base which defined the boundary conditions for Eqs. (9) and (10). The solutions of Eqs. (9) and (10) are exponential but are approximated using the Chebyschev polynomials.

Equations (9) and (10) may be combined with Eqs. (3), (4), (9) and (10) to write the convective terms in Eqs. (3) and (4) as

$$\left.\frac{\partial \bar{s}}{\partial t}\right|_{conv} = -\left(\frac{\partial M_c}{\partial \sigma} - \mu_o M_c\right)(s_c - L\ell_c - \bar{s}) + M_c \frac{\partial \bar{s}}{\partial \sigma} \tag{12}$$

and

$$\left.\frac{\partial \bar{q}}{\partial t}\right|_{conv} = -\left(\frac{\partial M_c}{\partial \sigma} - \mu_o M_c\right)(q_c + \ell_c - \bar{q}) + M_c \frac{\partial \bar{q}}{\partial \sigma} \tag{13}$$

Physcial meaning may be attached to the terms in these equations. The first term is a detrainment term which generally results in a cooling and moistening as cloud water mixes with and evaporates into with its surroundings. For nonprecipitating clouds this term may be large, while for precipitating clouds ℓ_c decreases and the detrainment effect is diminished. The second term in Eqs. (12) and (13) may be interpreted physically as cumulus-induced subsidence that warms and dries the large-scale.

The cumulus-induced subsidence term dominates for precipitating clouds and is thought to be the principal process by which latent heat released in cumulus clouds is communicated to the large-scale fields. On the cloud-scale, the latent heat released is converted to the kinetic and potential energy of the cloud parcels.

In the derivation of Eq. (12) and (13) it was assumed that cloud water converted to precipitation falls to the surface without being evaporated. The evaporation of rain can lead to a cooling and a moistening in the lower part of the cloud layer. This effect is included in one of the calculations presented below.

The vertical distribution of convective heating and moistening effects may be obtained from the formulation given by Eqs. (12) and (13). Formal closure of the problem requires that M_c be expressed as a function of the large-scale fields or specified externally. It is assumed that M_c has no more than a cubic dependence on σ. Furthermore, no interaction is allowed between convective elements and the tropopause. Thus, $M_c = 0$ at $\sigma = 1$. At cloud base it is assumed that clouds have subcloud thermodynamic properties which in this paper are specified externally using values consistent with the Yanai et al. [11] data. The cloud base mass flux is also prescribed and the resulting sensible heat flux into the cloud layer is approximately zero and the latent heat flux from the subcloud layer into the cloud layer is 190 Wm^{-2}, consistent with the values obtained by Yanai et al. [11]. In a prognostic application of this model, the cloud base mass flux could be obtained by other assumptions (see [2]). Since four coefficients are needed to specify the mass flux as a cubic function, two additional equations are needed to obtain the mass flux distribution. A closure scheme will be discussed below.

The sensitivity of the heating and moistening profiles to specified mass flux profiles was obtained using the two mass flux profiles shown in Fig. 7. The mass flux marked "Yanai" is a cubic approximation to the mass flux diagnostically determined by Yanai et al. [11]. Physically it can be interpreted as a result of shallow and deep clouds. Shallow clouds give the decrease in the flux near cloud base due to detrainment, while deep clouds give the peak in the upper levels since they have a maximum vertical velocity near that level. The decrease in the mass flux at upper levels is due to the detrainment of the deeper clouds. The second profile, designated as $x = 0.25$, is a quadratic function of σ and would be appropriate for a distribution with few shallow or deep clouds, but with many clouds detraining in the middle levels. These mass flux profiles were used in Eqs. (12) and (13) with the in-cloud values obtained from the entrainment relationships. The entrainment rate for the reference cloud was chosen to give a cloud which loses its buoyancy at $\sigma = 1$. The precipitation conversion coefficient is chosen to give a small liquid water content at $\sigma = 1$. The convective heating and moistening profiles for the mass flux profiles shown in Fig. 7 are shown in Figs. 8 and 9. The heating rate values correspond to

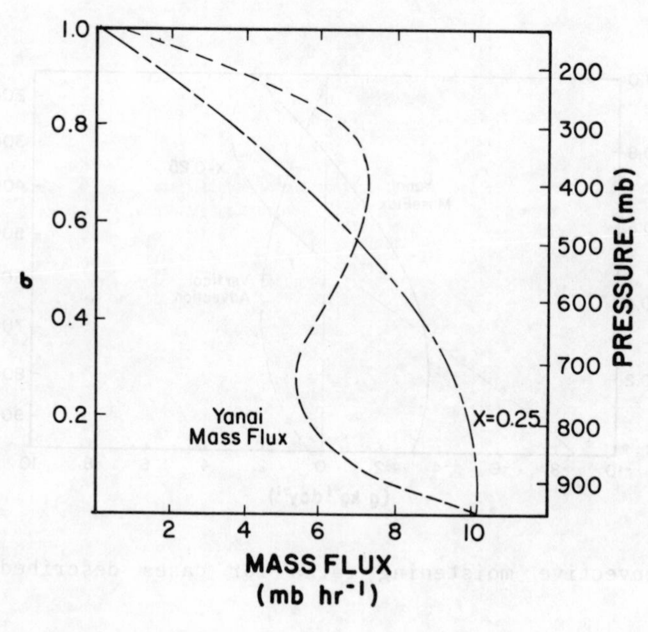

Fig. 7 Mass flux from Yanai et al. [11] and a quadratic mass flux ($x = 0.25$).

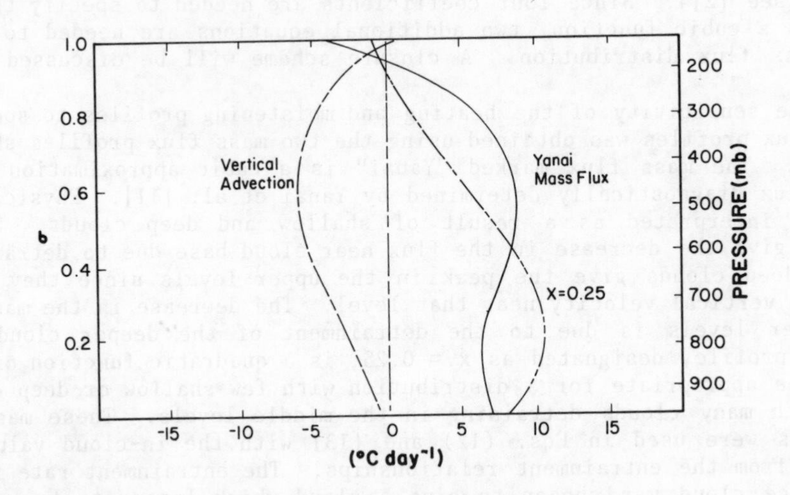

Fig. 8 Heating rates from convective parameterization for the mass flux profiles shown in Fig. 7 and the large-scale vertical advection heating profile.

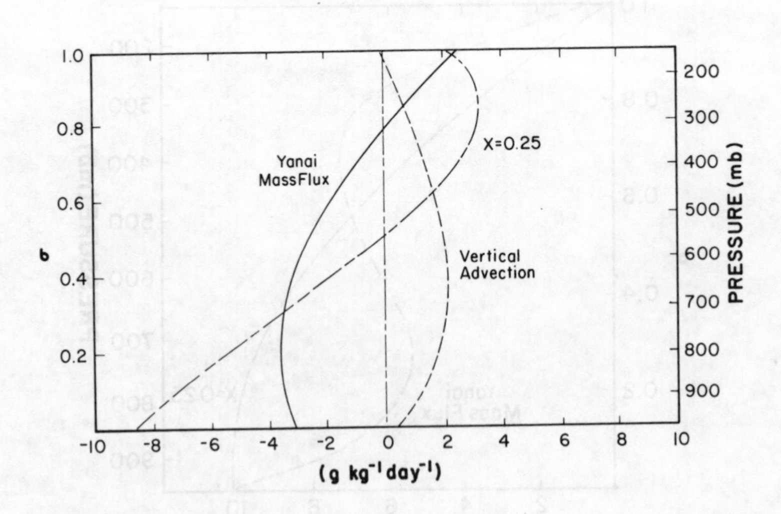

Fig. 9 Convective moistening rates for cases described in Fig. 8.

$$c_p^{-1} \left. \frac{\partial \bar{s}}{\partial t} \right|_{conv}$$

and the vertical distributions of s and q used in the model are the Yanai profiles (Fig. 6). The two mass fluxes give very different heating and moistening profiles. The differences in the heating profiles are principally due to the cumulus-induced subsidence term since the detrainment term is small. The detrainment term is not small, however, in the moisture budget since near cloud base $q_c + \ell_c - \bar{q}$ may be large. As a consequence, shallow clouds (which detrain at lower levels) significantly diminish the strong drying due to the cumulus-induced subsidence. Yanai et al. [11] pointed out the importance of both shallow and deep clouds for the large-scale heat and moisture budgets. The vertical advection terms

$$- \bar{\dot{\sigma}} \frac{\partial \bar{s}}{\partial \sigma} \text{ and } - \bar{\dot{\sigma}} \frac{\partial \bar{q}}{\partial \sigma}$$

from the Yanai study are also shown in Figs. 8 and 9. Not surprisingly, the Yanai mass flux gives a shape to the heating and moistening rates that would most closely balance the vertical advection term. For horizontally homogeneous and steady state conditions, the convective heating would balance the vertical advection and the net radiative cooling; the convective drying would balance the moistening due to large-scale vertical advection. Although these results only confirm the Yanai et al. [11] conclusion that both shallow and deep clouds are needed to maintain the structure of the tropical atmosphere, these conclusions were obtained using a relatively simple parameterization.

The evaporation of falling rain would also help balance the drying in the lower levels. A simple scheme was devised that allowed falling rain to evaporate at a rate proportional to the total rain falling through a level. The simple quadratic mass flux shown in Fig. 7 was used for this calculation and the total heating and moistening and the terms contributing to this total are shown in Figs. 10 and 11. Although the evaporation of rain has a significant effect on the heat budget, it still has a relatively small effect on the moisture budget compared to the subsidence drying. The shape of the moistening profile is not significantly changed from the profile with no evaporation.

The cloud model used in this parameterization is very simple and the conversion of cloud water to rain is a particularly crude aspect of the model. The sensitivity of the heating and moistening profiles to the conversion coefficient c_o is shown in Figs. 12 and 13 for the simple quadratic mass flux profile. This factor strongly modulates heating and moistening rates in the upper levels. The $c_o = 0$ case corresponds to the idealized case of no precipitation. In this case the detrainment cooling average over the whole cloud layer approximately balances the cumulus-induced warming, which is what one would expect with no precipitation. There is, however, some net moistening in the $c_o = 0$ case since there is

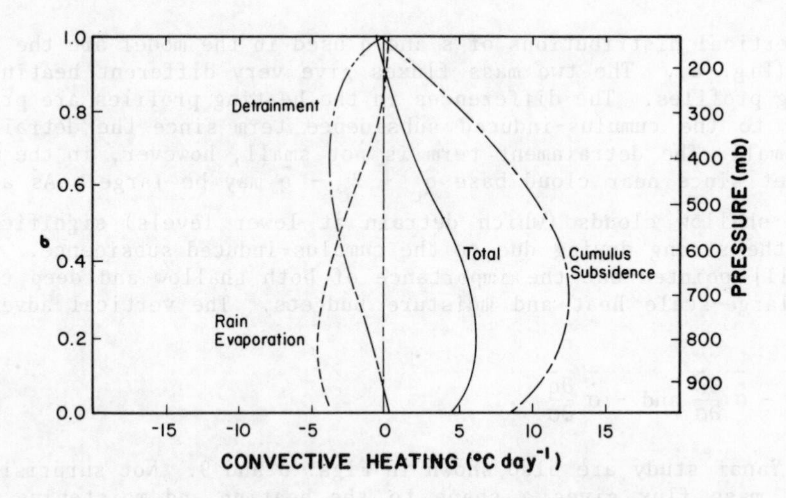

Fig. 10 Convective heating rates for various processes considered by the cumulus parameterization scheme obtained with the quadratic mass flux profile shown in Fig. 7.

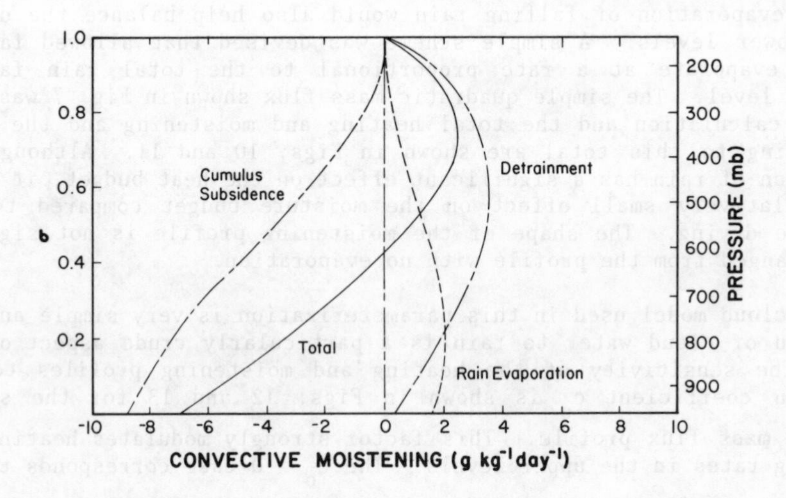

Fig. 11 Convective moistening rates for cases described in Fig. 10.

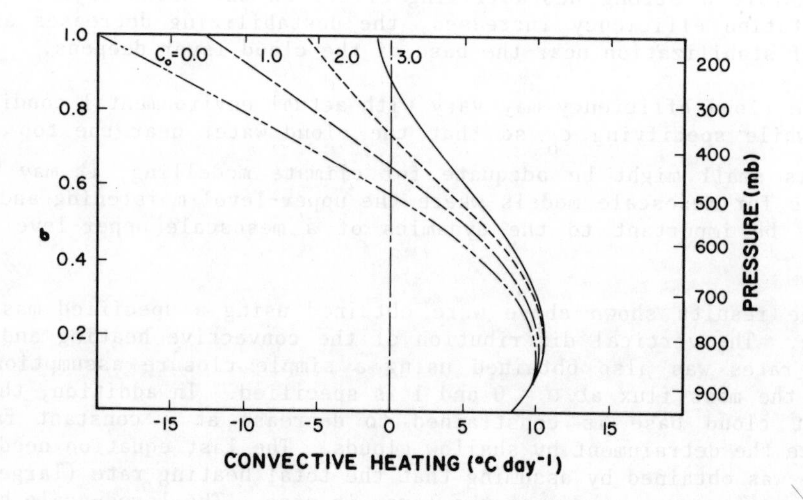

Fig. 12 Convective heating rates for various cloud water to precipitation conversion rates.

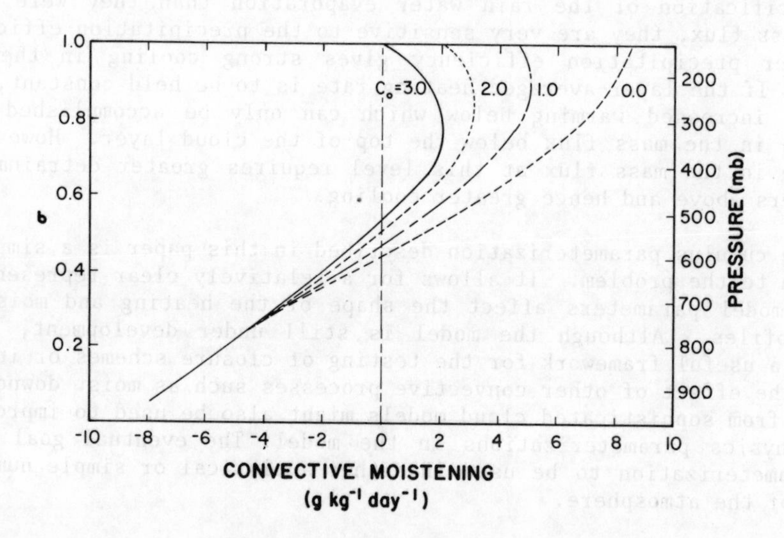

Fig. 13 Convective moistening for cases described in Fig. 12.

a flux of water vapor into the cloud layer. The nonprecipitating case does exhibit a strong destabilizing effect on the cloud layer. As the precipitation efficiency increases, the destabilizing decreases and the layer of stabilization near the base of the cloud layer deepens.

The cloud efficiency may vary with actual environmental conditions. Thus, while specifying c_o so that the cloud water near the top of the layer is small might be adequate for climate modelling, it may be inadequate for mesoscale models where the upper-level moistening and cooling may be important to the dynamics of a mesoscale upper-level cloud system.

The results shown above were obtained using a specified mass flux profile. The vertical distribution of the convective heating and moistening rates was also obtained using a simple closure assumption. As before the mass flux at $\sigma = 0$ and 1 is specified. In addition, the mass flux at cloud base is constrained to decrease at a constant rate to simulate the detrainment by shallow clouds. The last equation needed for closure was obtained by assuming that the total heating rate (large-scale and convective) averaged over the layer is zero. The large-scale heating in this case consisted of the cooling due to vertical advection shown in Fig. 8 and a layer-averaged radiative cooling rate of -1.5C day^{-1}.

Solutions obtained with this closure assumption are shown in Figs. 14 and 15 for $c_o = 3.0$ and $c_o = 1.5$ and with $c_o = 3.0$ and the evaporation of rain. Although the heating and moistening rates are less sensitive to the specification of the rain water evaporation than they were with a fixed mass flux, they are very sensitive to the precipitation efficiency. The lower precipitation efficiency gives strong cooling in the upper levels. If the layer-averaged heating rate is to be held constant, there must be increased warming below which can only be accomplished by an increase in the mass flux below the top of the cloud layer. However, an increase in the mass flux at this level requires greater detrainment in the layers above and hence greater cooling.

The cumulus parameterization described in this paper is a simplified approach to the problem. It allows for a relatively clear representation of how model parameters affect the shape of the heating and moistening rate profiles. Although the model is still under development, it may provide a useful framework for the testing of closure schemes or investigating the effect of other convective processes such as moist downdrafts. Results from sophisticated cloud models might also be used to improve the cloud physics parameterizations in the model. The eventual goal is for the parameterization to be used in either analytical or simple numerical models of the atmosphere.

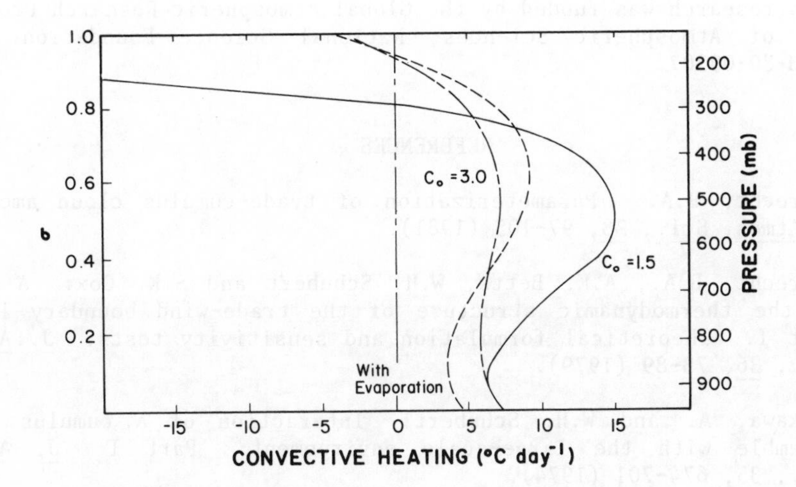

Fig. 14 Convective heating rates for different cloud model assumptions and a mass flux obtained by the closure assumptions described in the text.

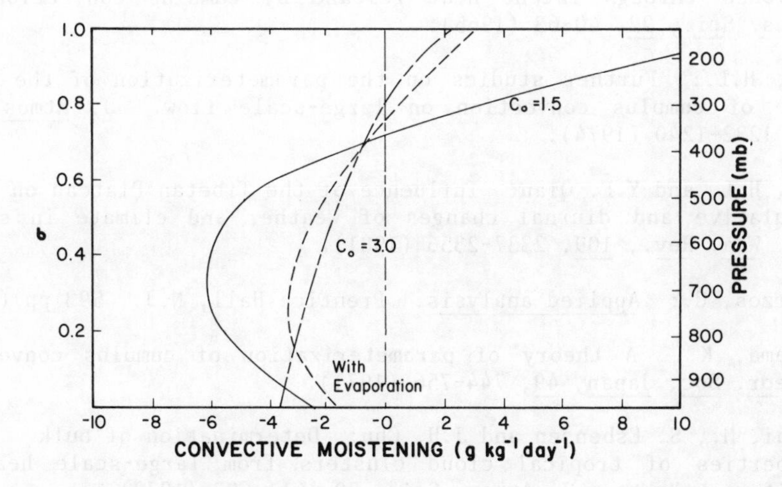

Fig. 15 Convective moistening rates for cases described in Fig. 14.

ACKNOWLEDGMENTS

This research was funded by the Global Atmospheric Research Program, Division of Atmospheric Sciences, National Science Foundation under Grant ATM-80-09307.

REFERENCES

1. Albrecht, B.A.: Parameterization of trade-cumulus cloud amounts. J. Atmos. Sci., 38, 97-105 (1981).

2. Albrecht, B.A., A.K. Betts, W.H. Schubert and S.K. Cox: A model of the thermodynamic structure of the trade-wind boundary layer: Part I. Theoretical formulation and sensitivity tests. J. Atmos. Sci., 36, 73-89 (1979).

3. Arakawa, A. and W.H. Schubert: Interaction of a cumulus cloud ensemble with the large-scale environment: Part I. J. Atmos. Sci., 35, 674-701 (1974).

4. Augstein, A., H. Riehl, F. Ostapoff and V. Wagner: Mass and energy transports in an undisturbed Atlantic trade-wind flow. Mon. Wea. Rev., 101, 101-111 (1973).

5. Betts, A.K.: Parametric interpretation of trade-wind cumulus budget studies. J. Atmos. Sci., 33, 1008-1020 (1975).

6. Kuo, H.L.: On the formation and the intensification of tropical cyclones through latent heat release by cumulus convection. J. Atmos. Sci., 22, 40-63 (1965).

7. Kuo, H.L.: Further studies on the parameterization of the influence of cumulus convection on large-scale flow. J. Atmos. Sci. 31, 1232-1240 (1974).

8. Kuo, H.L. and Y.F. Qian: Influence of the Tibetan Plateau on cumulative and diurnal changes of weather and climate in summer. Mon. Wea. Rev., 109, 2337-2356 (1981).

9. Lanczos, C.: Applied analysis. Prentice Hall, N.J., 593 pp (1964).

10. Ooyama, K.: A theory of parameterization of cumulus convection. Meteor. Soc. Japan, 49, 744-756 (1971).

11. Yanai, M., S. Esbensen and J.H. Chu: Determination of bulk properties of tropical cloud clusters from large-scale heat and mositure budgets. J. Atmos. Sci., 30, 611-627 (1973).

5.5 A STUDY OF THE INFLUENCE OF OROGRAPHY ON DIURNAL VARIATIONS OF TEMPERATURE, PRESSURE, WIND VELOCITY, PRECIPITATION, RELATIVE HUMIDITY AND DURATION OF SUNSHINE

Lin Zhiguang
Academy of Meteorological Science
National Meteorological Bureau
The People's Republic of China

ABSTRACT

Five years of meteorological data are used to study differences between diurnal variations at the tops of mountains and in valleys. The following results are obtained:

1. The diurnal temperature range is small on the summit of high mountains, but it is large in the valleys. Maximum temperature occurs earlier on the summits than in the valleys, i.e. about 1300 LT (local time) on the summits and 1400-1500 in the valleys.

2. The diurnal variation of precipitation on a high mountain is opposite to that in some large valleys: Precipitation on the summits during the day is larger than that at night, and precipitation in the high valleys during the daytime is smaller than at night. In the large valleys of Xizang and Yunnan, night rains follow fair days; in the Sichuan Basin nocturnal rains follow overcast days.

3. The diurnal changes of relative humidity are very small on the summit, but very large in the valley. The maximum occurs in the early morning and the minimum occurs in the afternoon.

4. The diurnal variation of wind speed at the top of large mountains is out of phase with the winds in the valley. In general, the maximum appears at night and the minimum in the daytime on the high mountain, while in the low valley a daytime maximum and nighttime minimum are observed. Over the Xizang Plateau in winter and spring, the diurnal range of wind velocity is about 7-8 m/sec, which is the largest in our country.

5. On the mountain summit, two maxima and one minimum are observed in the diurnal variation of sunshine: The maxima appear at 0900-1000 LT and 1500-1600 LT, the minimum appears at about 1200-1300 LT. In the valley, the diurnal variation has only one maximum that appears at about 0900-1000 LT in the dry season and about 1200 LT in the rainy season.

6. On high mountains the diurnal variation of surface pressure has two peaks which are different from those in the low-lying land. In some high mountains only a single peak is observed. We found that the diurnal range of pressure was proportional to the diurnal range of temperature.

I. INTRODUCTION

This paper deals with the differences in the diurnal variations of meteorological parameters, such as temperature, pressure, relative humidity, wind velocity, precipitation and duration of sunshine at mountaintops (or ridges) and valleys (or basins).

In this study the data collected from 1955-1959 were used to study diurnal variations in pressure, temperature, humidity and duration of sunshine. Data for a slightly longer time period were used to study variations in precipitation. Wind velocity data were obtained for a period from 1973 to 1976. It was found that the results obtained from different sites were fairly consistent and may, to a large extent, represent the general influence of orography on the diurnal variation of meteorological parameters.

It was found that in general the diurnal variations of any of the meteorological parameters, except precipitation and sunshine duration, at the foot of a mountain agree with those in the mountain valley, although the amplitude of the variations of the former are greater. Therefore, in our analysis of such changes, the station at the foot of a mountain was used to represent variations in the valley.

II. THE AIR TEMPERATURE

The influence of orography on diurnal changes of temperature can be represented by the amplitudes and phases of the changes. The curves shown in Fig. 1 represent the annual mean of the diurnal changes of temperature at five summit stations located at Mt. Hua Shan, Tai Shan, Lu Shan, Wutai Shan and Omei Shan, and five valley stations located in the cities of Lhasa, Xian, Xichang, Taian and Taiyuan. As seen from these curves, the diurnal variations of air temperature are relatively small on the summits of the mountains, while they are large at the valley (or shallow valley) stations.

Table 1 gives the daily amplitudes and daily ranges of the diurnal air temperature variations at some mountaintops and valleys and indicates that the diurnal variations of air temperature at mountaintops are smaller than those in the valleys. The mean daily amplitude of temperature, which is defined as the difference between the maximum and minimum temperature of the average diurnal variation, of the five summit stations is only 2.9°C, while that of the five valley and foot stations is 9.6°C. It must be pointed out that the smaller diurnal changes at the summit stations are not due to their greater elevations, but their orographic features. For example, at the station of Lhasa, the elevation of which is 3658 meters, the daily amplitude is 12°C; four times greater than the average of the five summit stations. Similarly, at station Xichang, the elevation of which is 1591 meters, the amplitude is two times the average at the mountaintops.

The air temperature at mountaintop shows relatively little variation due to the fact that the mountaintop experiences persistent mountain-valley breezes. Under these circumstances the diurnal variations of the mountaintops and ridges are similar to those at oceanic stations. In

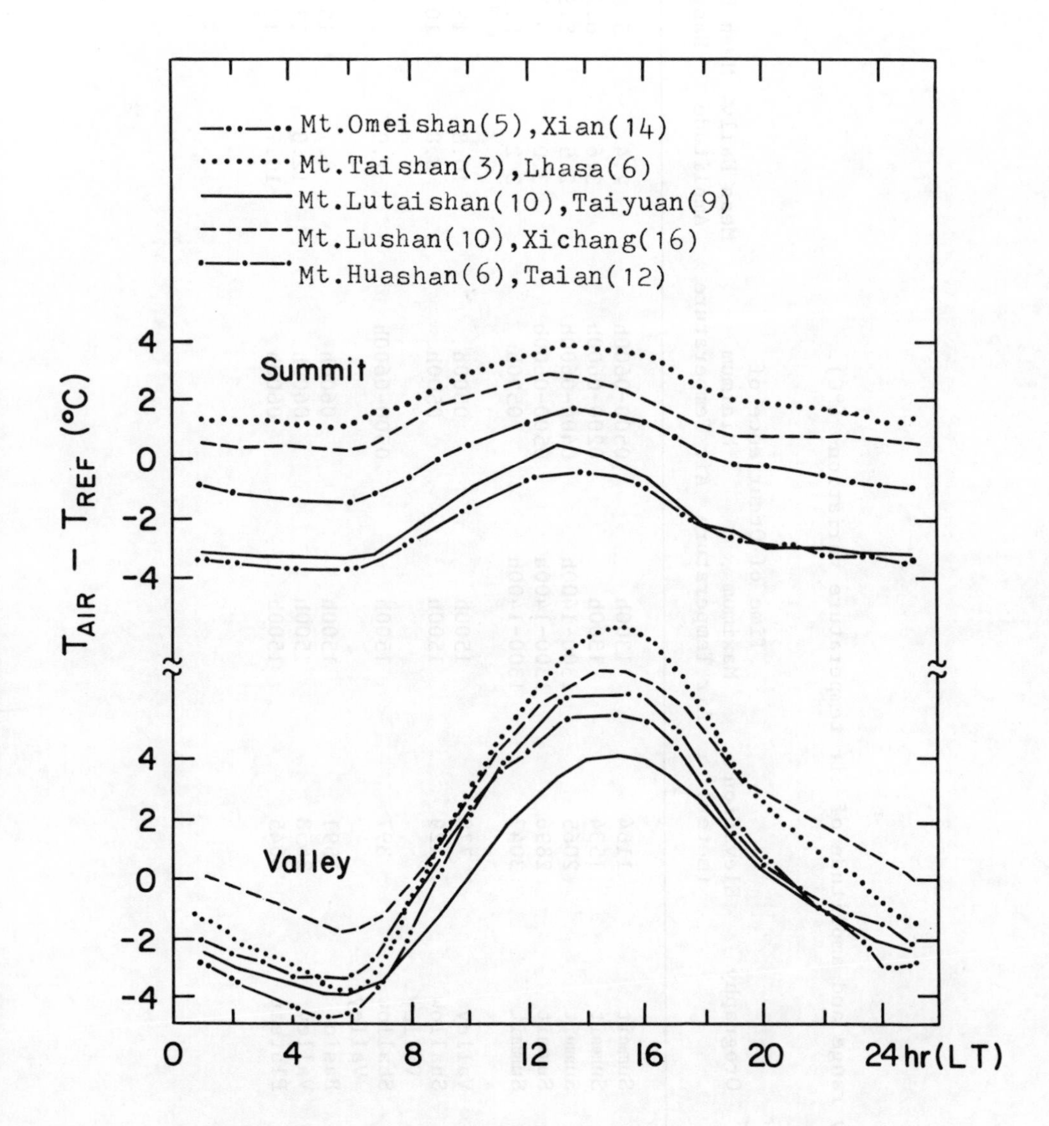

Fig. 1 Diurnal variations of annual mean air temperature at five summit and five valley stations. (For Pinyin spelling refer to text.)

Table 1 Mean daily range and amplitude of air temperature variations (°C).

Obs. Station	Orography	Elevation (meter)	Time of Occurrence of Maximum Air Temperature	Time of Occurrence of Minimum Air Temperature	Mean Daily Amplitude	Mean Daily Range
Mt. Lu Shan	Summit	1164	1300h	0500-0600h	2.4	5.8
Mt. Tai Shan	Summit	1534	1300h	0200-0600h	3.6	6.2
Mt. Hua Shan	Summit	2065	1300-1400h	0400-0600h	2.5	5.8
Mt. Wutai Shan	Summit	2896	1300-1400h	0500-0600h	3.0	6.1
Mt. Omei Shan	Summit	3047	1300-1400h	0500h	3.2	6.3
Taiyuan	Valley	778	1500h	0500h	10.8	13.0
Taian	Shallow Valley	129	1500h	0500h	8.8	10.9
Xian	Shallow Valley	397	1500h	0500-0600h	7.9	9.5
Xichang	Basin	1591	1500h	0600h	8.7	10.3
Lhasa	Valley	3658	1500h	0600h	12.0	13.7
Wudaoliang	Plateau	4645	1500h	0600h	10.7	13.2

contrast, the airflow in a valley is somewhat blocked, and so the diurnal changes of air temperature there are relatively large.

At the plains stations, the mean daily amplitudes of air temperature are intermediate between those at mountaintops and in valleys. Over a plateau area, however, the air density is relatively small so that the daily amplitude of air temperature variations may increase. For example, the station of Wudaoliang has a daily amplitude of 10.7°C and its elevation is 4645 meters (Table 1).

Since the mean amplitude of the air temperature variations is the difference between the maximum and minimum temperatures on the curve of mean diurnal changes of air temperature, it is smaller than the mean range of air temperature (mean of the difference between the daily maxima and minima). The differences between the mean amplitudes and the ranges are greater at a summit station than at a valley station. As seen from Table 1, the average difference for the five summit stations is 3.1°C while for the five valley stations it is 1.8°C. The cause of such differences is that the diurnal temperature variations at the mountaintop are less regular than at the foot of the mountain or in the valley due to the influence of the wind variations at the top of the mountains. This wind variation leads to greater variability in the times of occurrence of maximum and minimum temperatures.

As seen in Table 1, the maximum temperature at both the foot of a mountain and in the valley generally occurs at ∿ 1500 LT, while at a mountaintop it occurs 1 or 2 hours sooner, that is at about 1300-1400 LT. The times at which the minimum temperatures occur are not very different for all of the stations and are generally found at 0500-0600 LT. However, at the stations on high mountains the air temperatures remain nearly constant from 0200 to 0600 LT. That the maximum air temperature over a high mountain occurs at an earlier hour than at other locations is due to the increased cloudiness along the slopes during midday, causing the occurrence of a minimum of sunshine at about 1200-1300 LT. In Japan, for example, the annual mean maximum air temperature at the top of Fujiyama, the elevation of which is 3776 meters, occurs at 1200-1300 LT, while at the foot of the mountain it occurs at 1300-1400 LT [1].

III. PRECIPITATION

Generally speaking, at the summit of a mountain, just as on the plains, more rain falls during the day than at night. But the reverse is true in a large and deep valley or basin.

In this study the influence of orography on precipitation is analyzed not only by amount, but also by total rainfall duration in hours.

1. The Predominance of Daytime Rain at Mountaintops

Figure 2 gives the curves showing the diurnal changes of rainfall amount and rainfall hours from June to August for the four summit stations: Mt. Lu Shan, Jiuxian Shan, Qixia Shan and Heng Shan.

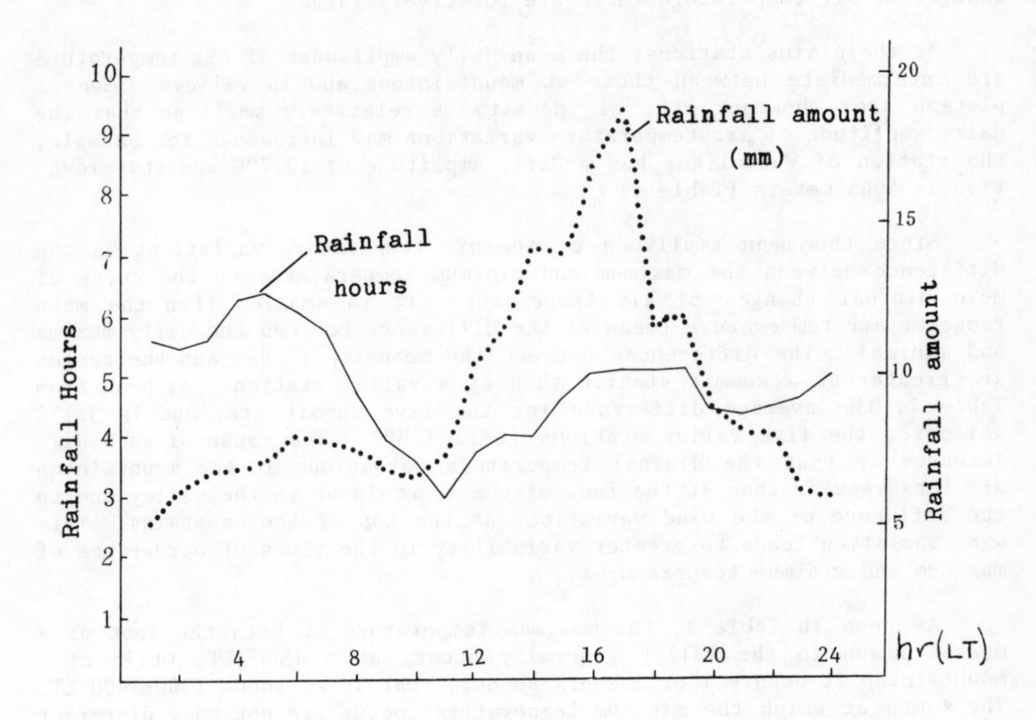

Fig. 2 Diurnal variations of rainfall amount (mm) and rainfall hours for four summit stations (Mt. Lu Shan, Jiuxian Shan, Qixia Shan and Heng Shan) from June to August.

These curves have a single peak with the maximum occurring in the afternoon. This variation is similar to the changes of precipitation generally observed over the plains. The reason for this diurnal variation is that the most intense convection occurs in the afternoon. There is an afternoon peak in the diurnal variation of rainfall, but there is another maximum in the early morning. Furthermore, this maximum is even greater in the rainfall hours than the afternoon maximum. It follows that a relatively large amount of rainfall at mountaintops during the daytime is mainly caused by greater rainfall intensity, but the number of rainfall hours at night still exceeds that of the daytime.

2. The Predominance of Night Rain in High and Large Valleys

By studying the rainfall records at a number of stations we found that the nocturnal rain in many valleys may be classified into two types -- night rain after a sunny day and night rain after an overcast day.

An example for night rain associated with sunny days is given for Lhasa. Figure 3 shows the curves of diurnal changes of rainfall amount

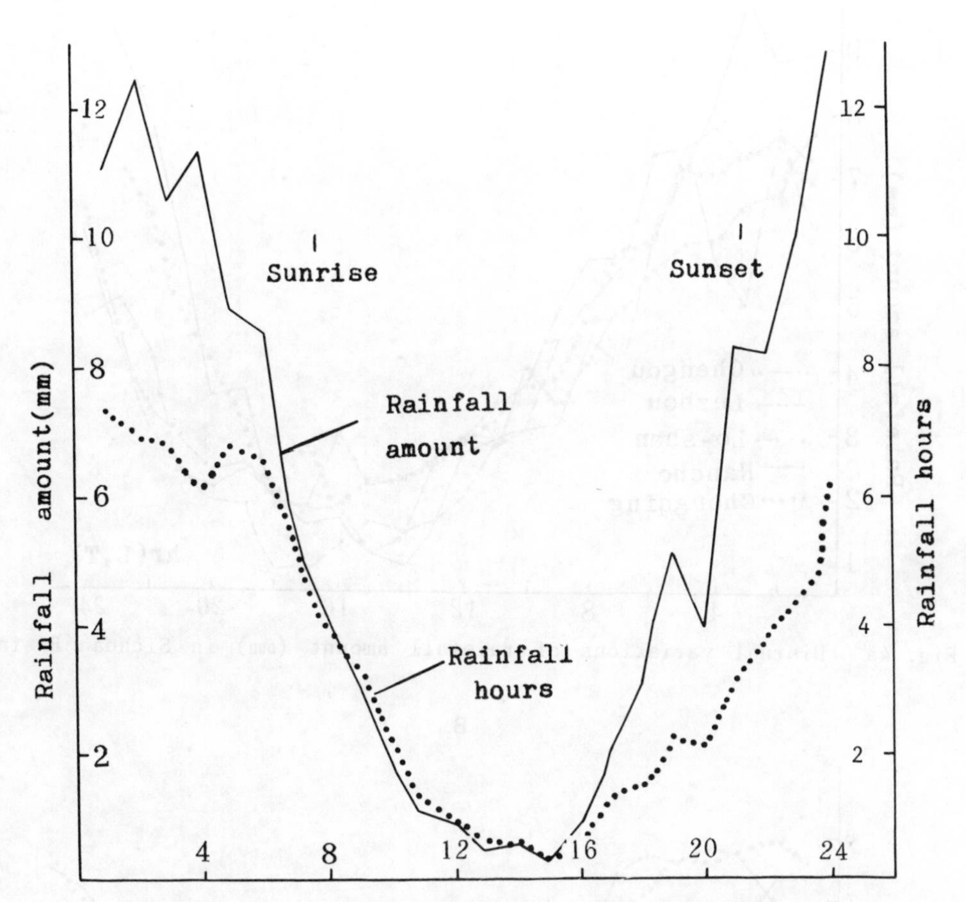

Fig. 3 Diurnal variation of rainfall amount and rainfall hours at Lhasa (valley) July-August.

and rainfall hours from July to August for this station. These curves clearly show that the night rain is much greater than daytime rain. As a matter of fact, there is little rainfall in the afternoon from 1200 to 1600 LT, and the percentage of sunshine is 70-80 percent during the same time period. It follows that the contrast between sunny days and rainy nights is very large.

The nocturnal rain maximum may be mainly due to air being cooled by radiation and flowing downward along the slopes, lifting the warm moist air in the area of the valley.

Rainy nights associated with overcast days occur typically in the Sichuan Basin. Five representative stations are the cities of Chengdu, Luzhou, Loshan, Nanchong and Chongqing. The curves in Fig. 4 represent

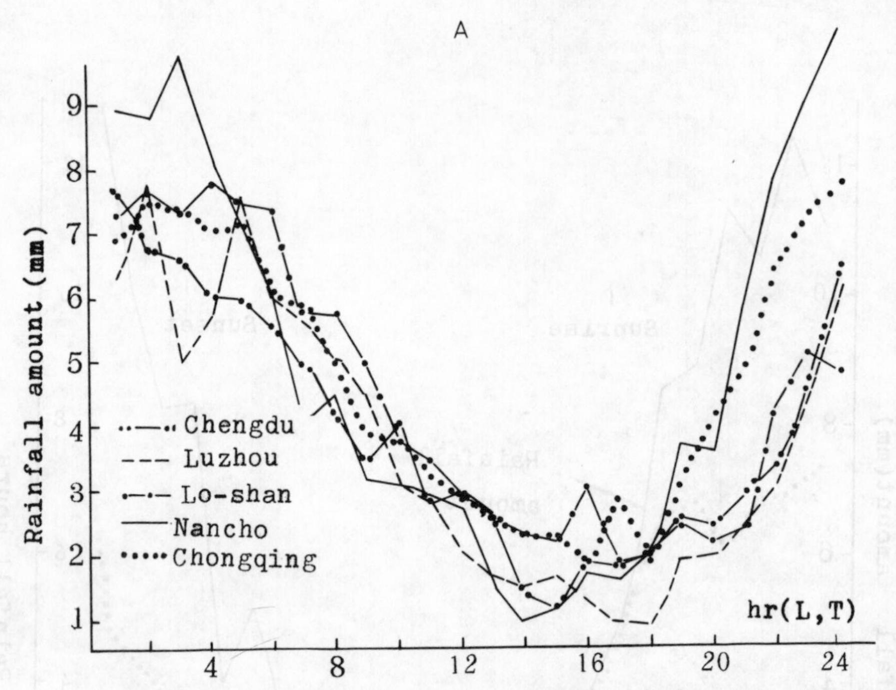

Fig. 4a Diurnal variations of rainfall amount (mm) in Sichuan Basin.

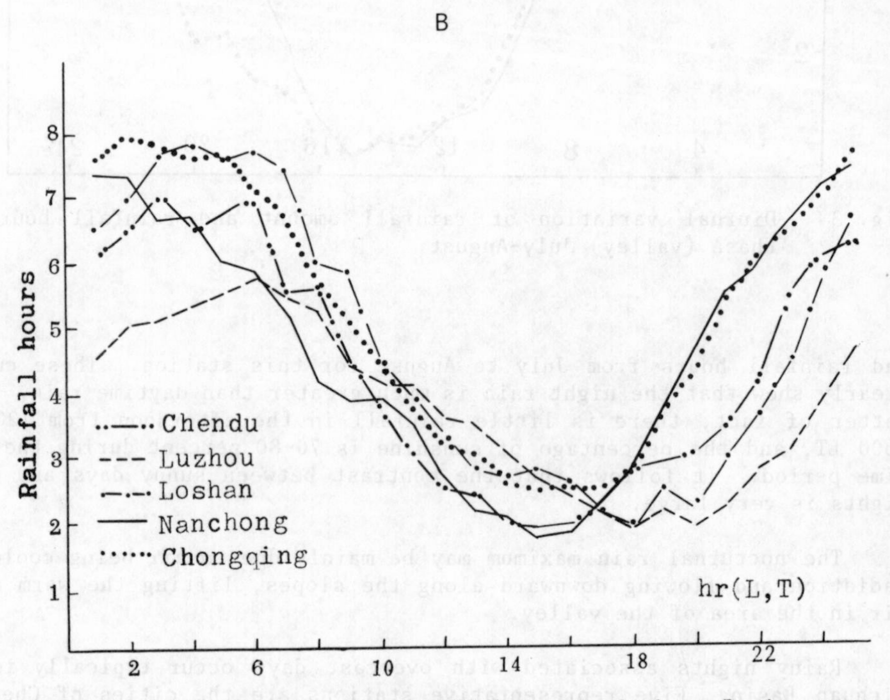

Fig. 4b Diurnal variations of rainfall hours in Sichuan Basin.

the diurnal changes of annual amounts and hours of rainfall at these five stations. From Fig. 4 it is obvious that the night rains in Sichuan Basin are typical for the whole region, and both the curves of rainfall hours and rainfall amounts have a single maximum and minimum.

The Sichuan Basin has significant rainfall at night. It rains very little in the daytime in spite of persistent overcast conditions. During the time period from 1000 to 1500 LT the percentage of sunshine is only 30-40 percent, while that for the sunny-day type shown previously is 70-80 percent.

The amount and hours of night rain vary with season. As a general rule, there is more nocturnal rain during the wet season than the dry season. In the dry season the rainfall hours, however, show a nighttime maximum (Fig. 5). For example, in the city of Chongqing (in Sichuan Province) and Hekou (in Yunnan Province) January and February are the dry season. The diurnal changes of rainfall amount are relatively small, while the variability of rainfall hours is still well defined.

Generally speaking, when all other conditions are equal the amount and hours of nighttime rain increase with the amount of water vapor. For example, at the station of Hekou, the mean water vapor pressure in July and August is 31.3 mb, and the amount of night rainfall during the same period is 632.0 mm, while at the station of Xigaze during the same period, the mean vapor pressure and the amount of night rainfall is only 10.6 mb and 298.3 mm, respectively.

There is no typical type of nighttime rain maximum in many valleys of our country. Most of the have two maxima in rainfall, one in the afternoon due to convection and the other in the early morning, caused by orographic influences. This fact is illustrated by the diurnal variation of rainfall for Kunming (Fig. 6).

IV. WIND VELOCITY

Diurnal variations of wind velocity were obtained for a number of summit and valley (or foot) stations. The diurnal variations of the wind velocity for summit and valley stations are out of phase. For example, at Mt. Tai Shan (Fig. 7) the minimum wind velocity occurs between 1300 and 1600 LT, while a maximum occurs at the same time at the foot of the mountain and in the valley (city of Taian in Fig. 7). This peak may be due mainly to the downward transfer of the momentum from upper levels. Furthermore, the greatest diurnal variation of wind velocity in the valleys at higher elevations also occurs in the afternoon (graphs are omitted here). It follows that the diurnal change of wind velocity, just as that of precipitation, depends mainly upon the orographic conditions and not on the elevation.

On the Qinghai-Xizang Plateau there are areas where the elevation is above 4000 meters and which are wide and open with little surface undulations. The characteristics of the diurnal changes of wind velocity associated with these geographical features are as follows: (The example

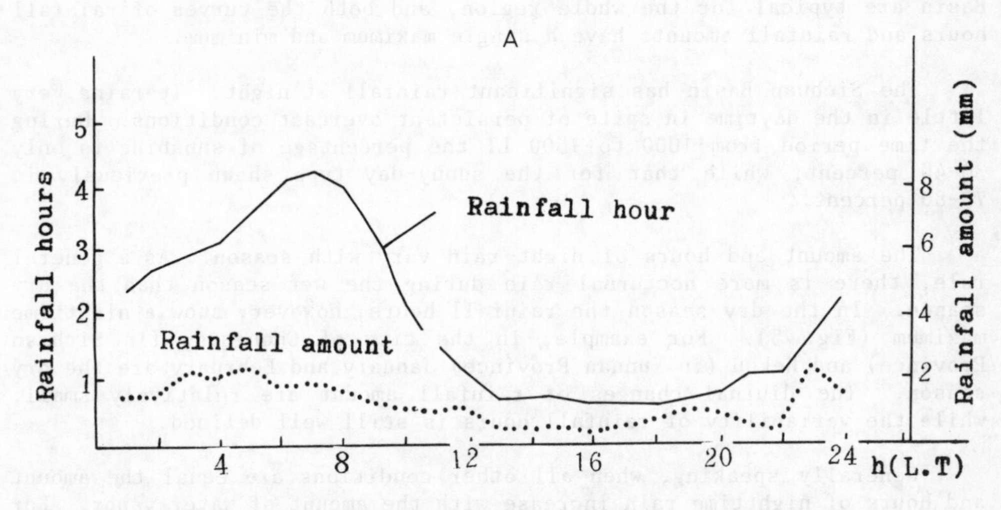

Fig. 5a Diurnal variation of rainfall in Heko in the dry season (January-February)

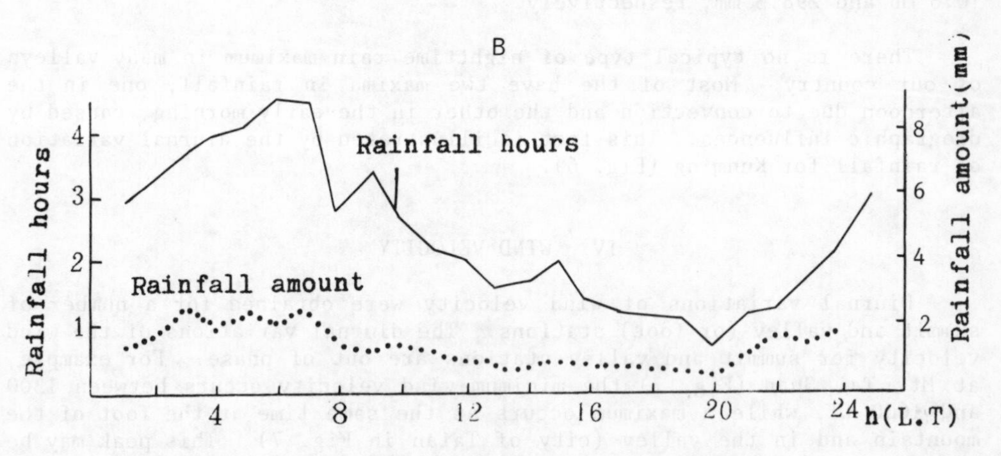

Fig. 5b Diurnal variation of rainfall in Chongqing in the dry season (January-February).

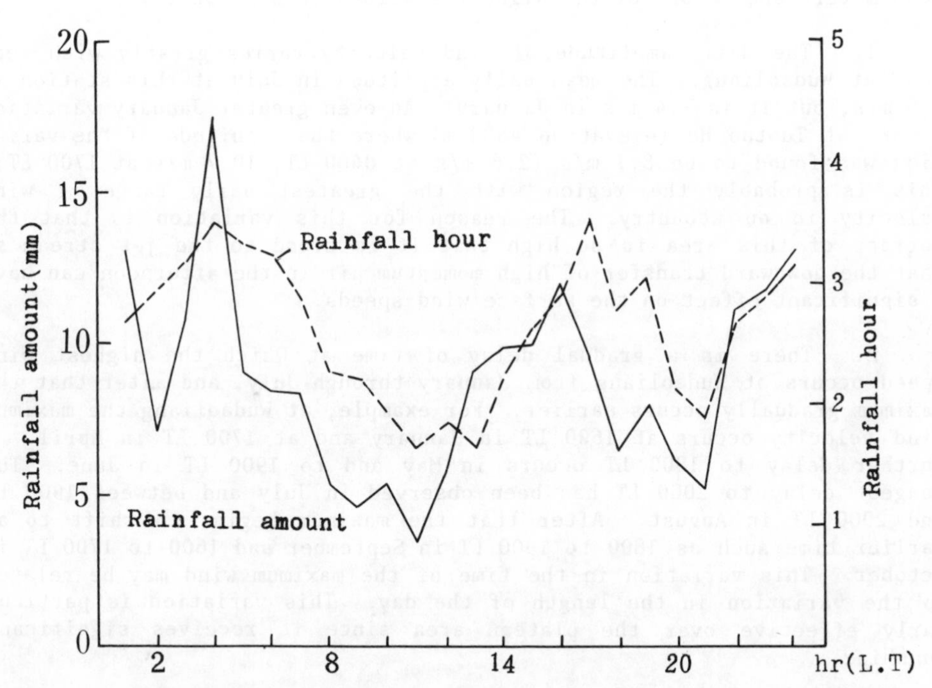

Fig. 6 Diurnal variation of rainfall in the wet season (July-August) in Kunming.

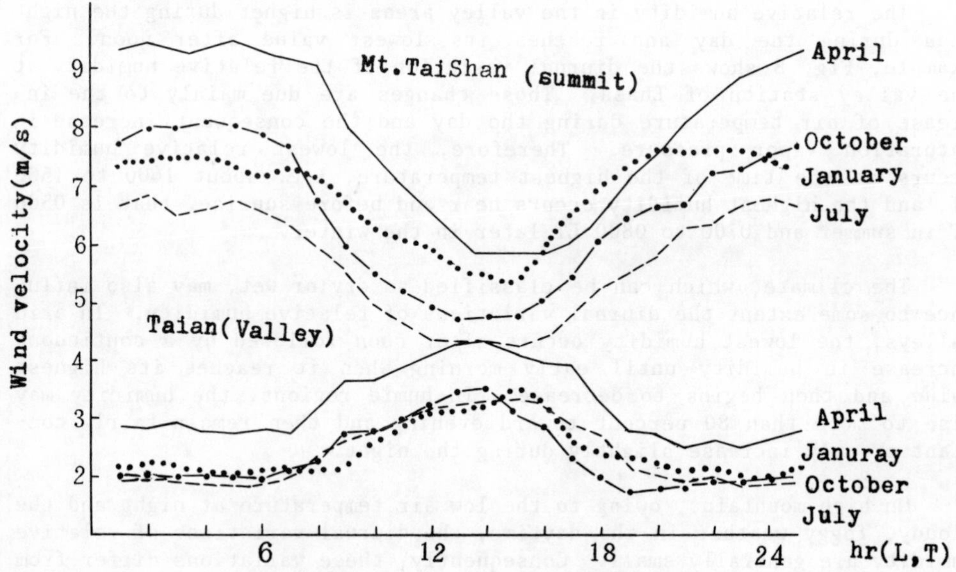

Fig. 7 Diurnal variations of wind velocity at Mt. Tai Shan and the city of Taian.

here is Wudaoliang in the province of Qinghai. It is at an elevation of 4645 meters and is one of the highest stations in our country.)

1. The daily amplitude of wind velocity varies greatly with seasons at Wudaoliang. The mean daily amplitude in July at this station is 3.5 m/s, but it is 7.4 m/s in January. An even greater January variation occurs at Tuotuo He (elevation 4533 m) where the magnitude of the variation was found to be 8.1 m/s (2.6 m/s at 0400 LT, 10.7 m/s at 1700 LT). This is probably the region with the greatest daily range of wind velocity in our country. The reason for this variation is that the surface of this area is so high that it is close to the jet stream so that the downward transfer of high momentum air in the afternoon can have a significant effect on the surface wind speeds.

2. There is a gradual delay of time at which the highest wind speed occurs at Wudaoliang from January through July, and after that the maximum gradually occurs earlier. For example, at Wudaoliang the maximum wind velocity occurs at 1600 LT in January and at 1700 LT in April. A further delay to 1800 LT occurs in May and to 1900 LT in June. The longest delay to 2000 LT has been observed in July and between 1900 LT and 2000 LT in August. After that the maximum begins to shift to an earlier time such as 1800 to 1900 LT in September and 1600 to 1700 LT in October. This variation in the time of the maximum wind may be related to the variation in the length of the day. This variation is particularly effective over the plateau area since it receives significant sunshine.

V. RELATIVE HUMIDITY

The relative humidity in the valley areas is higher during the night than during the day and reaches its lowest value after noon. For example, Fig. 8 shows the diurnal variation of the relative humidity at the valley station of Lhasa. Those changes are due mainly to the increase of air temperature during the day and the consequent increase in saturation vapor pressure. Therefore, the lowest relative humidity occurs at the time of the highest temperature, i.e. about 1400 to 1500 LT, and the highest humidity occurs near and before sunrise, that is 0500 LT in summer and 0700 to 0800 LT later in the winter.

The climate, which can be classified as dry or wet, may also influence to some extent the diurnal variations of relative humidity. In arid valleys, the lowest humidity occurs after noon followed by a continuous increase in humidity until early morning when it reaches its highest value and then begins to decrease. In humid regions, the humidity may rise to more than 80 percent toward evening and then remain fairly constant or only increase slightly during the night.

On high mountains, owing to the low air temperature at night and the cloudy, foggy weather in the daytime, the diurnal variations of relative humidity are generally small. Consequently, these variations differ from those of the valley. In most cases, however, the relative humidity on high mountains is lower in the daytime than at night, but the amplitudes

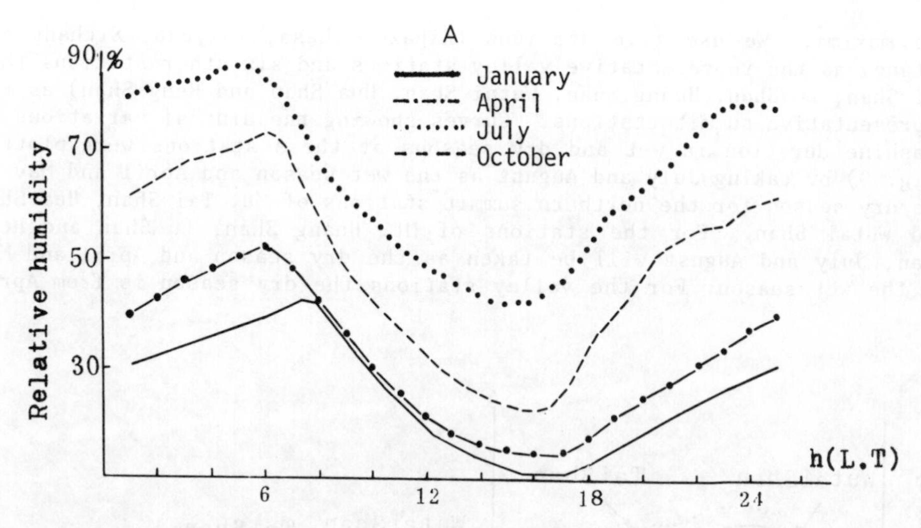

Fig. 8a Diurnal variation of relative humidity (%) at Lhasa (valley).

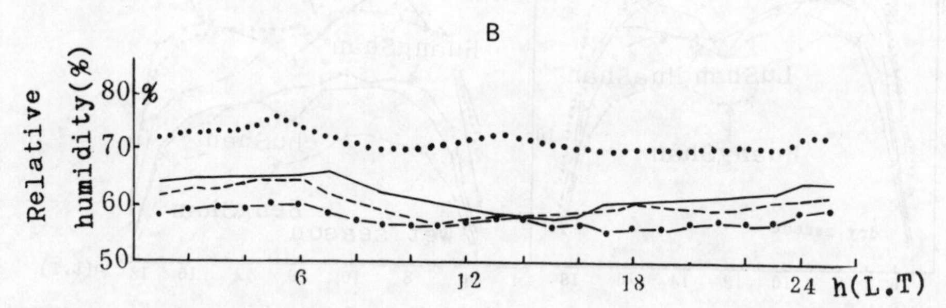

Fig. 8b Diurnal variation of relative humidity (%) at Mt. Hua Shan (summit).

generally do not exceed more than 5-10 percent as shown in Fig. 8 for Mt. Hua Shan. At valley stations, the lowest humidity always occurs after noon, while at summit stations the relative humidity minimum is not pronounced and may occur before noon. This is mainly due to the fact that the cloud amount after noon is larger than that before noon, particularly in the dry season (see next section).

The cause of the small amplitudes of variations in relative humidity at mountaintops may also be related to the small amplitudes of temperature changes. Therefore, at valley stations which are at high elevations the amplitudes of relative humidity variations, just as that of air temperature, may be very large (the graphs are omitted here).

VI. THE DURATION OF SUNSHINE

The diurnal variation of sunshine duration for a valley station is characterized by a single maximum, while that for a summit station has two maxima. We use five stations (Xigaze, Lhasa, Taiyuan, Xichang and Batang) as the representative valley stations and six other stations (Mt. Tai Shan, Lu Shan, Huang Shan, Wutai Shan, Hua Shan and Heng Shan) as the representative summit stations. Curves showing the diurnal variations of sunshine duration in wet and dry seasons at these stations were plotted (Fig. 9) by taking July and August as the wet season and April and May as the dry season for the northern summit stations of Mt. Tai Shan, Hua Shan and Wutai Shan. For the stations of Mt. Huang Shan, Lu Shan and Heng Shan, July and August will be taken as the dry season and April and May as the wet season. For the valley stations the dry season is from April

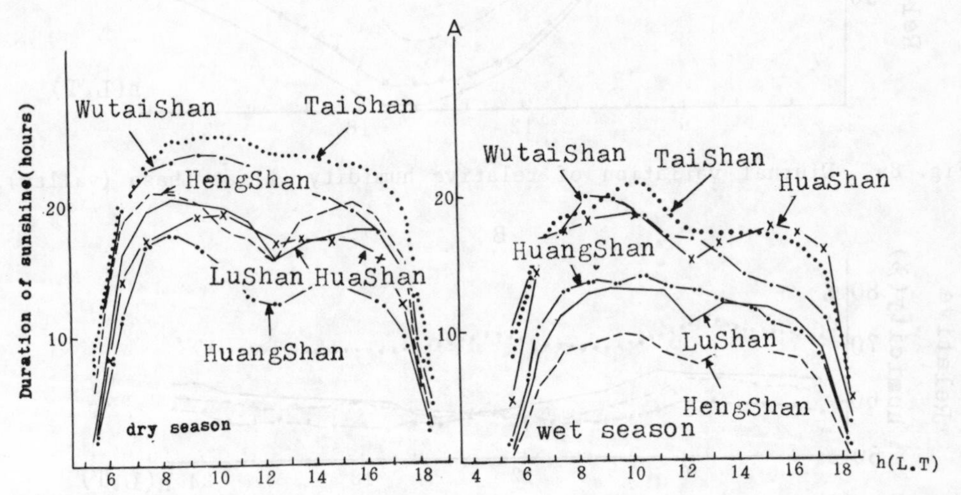

Fig. 9a Diurnal variations of the duration of sunshine at five summit stations.

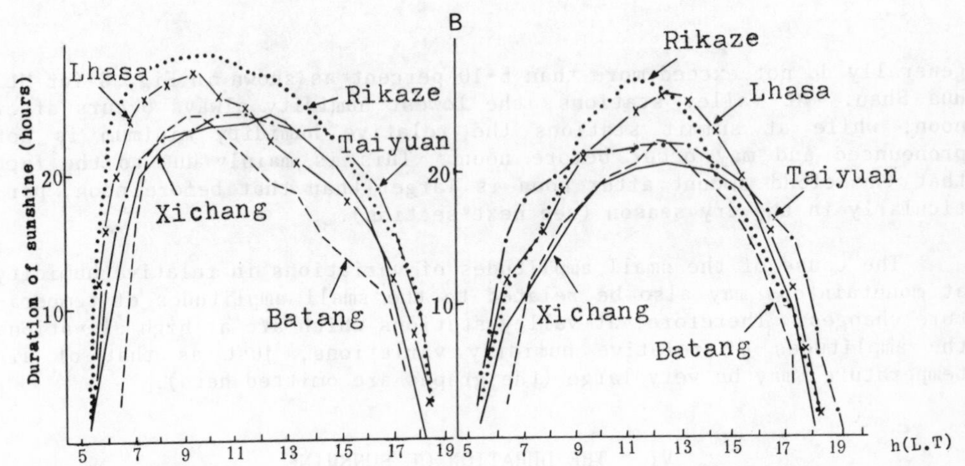

Fig. 9b Diurnal variations of the duration of sunshine at five valley stations.

to May and the wet season is from July to August. The reason for this choice is that the duration of sunshine is more closely related to cloud conditions than to temperature.

A summary of the data shown in Fig. 9 is given below.

1. The curves of sunshine duration for summit stations have two maxima, while valley stations have a single maximum.

The double peaks indicate that on the mountains the amount of sunshine in the morning and afternoon hours is greater than during the rest of the day and reaches its minimum at about 1200 to 1300 LT, the time of greatest cloudiness. On the contrary, the single maximum for valley stations indicates that the largest amount of sunshine occurs only at one time during the day. A relative large amount of cloudiness occurs during morning and afternoon hours and there is a slight minimum in the cloudiness around noon.

2. The differences between the shapes of the curves for dry and wet seasons indicate (Fig. 9) that sunshine duration for the dry season is greater than that for the wet season, and that the double maxima are more distinct in the dry than the wet season. The minima of sunshine in the wet season are not as clearly defined.

3. The times of occurrence of the single peak in the sunshine curves for valley stations are very different between the dry season in which the time is 0900 to 1000 LT and wet season in which the time is 1200 to 1300 LT. The reason for this shift is that there is little cloudiness in the morning in the dry season, and there are relatively small cloud amounts around noon in the wet season.

VII. AIR PRESSURE

At low elevations on the plains there is a double peak in the diurnal variation of air pressure. The primary maximum occurs at 0900 to 1000 LT and the secondary maximum occurs at 2200 to 2300 LT. The primary minimum occurs at 1600 to 1700 LT and a secondary minimum occurs at 0300 to 0400 LT. At some higher elevation stations (> 3000 m), the variation differed from that described above. At these stations there were two maxima in the surface pressure and the time of the principal maximum and minimum and the secondary maximum and minimum are reversed. The diurnal variations at Mt. Wutai Shan have these characteristics. For other stations, such as Omei Shan, this type of variation is only observed during the summer. Moreover, we found that at some high-level stations, such as Lhasa, the curves of diurnal changes of air pressure generally have a single maximum.

The mean daily amplitude of air pressure (Δp) was found to have a positive correlation with the mean daily amplitude of air temperature (Δt). For example at summit stations, Δt is small and Δp is also small, so their points are situated in the lower part of Fig. 10. At valley

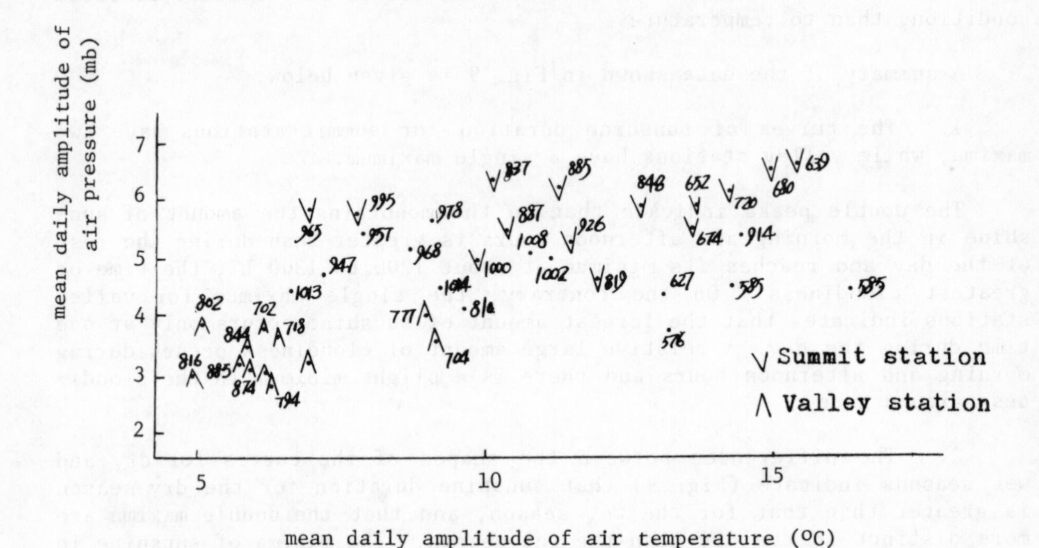

Fig. 10 The relationship between mean amplitude of air temperature (Δt) and air pressure (Δp). Mean annual pressure is 968 mb.

stations, Δt is large and Δp is large, so their points are situated in the upper part of Fig. 10.

Figure 10 shows data for 14 valley stations for which the mean Δt is 11.6°C and Δp is 5.8 mb, and 11 summit stations for which the mean Δt is only 5.3°C and Δp is only 3.3 mb. In general, when Δt = 5°C, Δp is about 3.5 mb; when Δt = 10°C, Δp is about 5 mb; and when Δt = 15°C, Δp is about 6.5 mb.

Furthermore, the results shown in Fig. 10 show that for similar amplitudes of air temperature (Δt), the points for stations of greater elevation are associated with smaller pressure variations. The reason for this relationship may be that the absolute pressure is lower so that diurnal changes of absolute pressure should also be small. For example, Wudiaoliang, Bangehu and Jimai have elevations of 4000-4700 m. While Δt is as large as 13.1-14.3°C, Δp is only 3.7-4.4 mb.

ACKNOWLEDGMENT

The author is much indebted to Professor S.Y. Tao, Institute of Atmospheric Physics, Chinese Academy of Sciences, for making valuable suggestions.

REFERENCES

1. Yoshino, Mosatoshi M.: Climate in a small area. An introduction to local meteorology. University of Tokyo Press (1975).

5.6 THE ATMOSPHERIC TURBIDITY AND ITS EFFECT ON SOLAR RADIATION OVER A CITY IN A VALLEY DURING WINTER

Shen Zhibao, Wang Yaoqi, Ji Guoliang, Wu Jingzhi,
Wang Wenhua and Shui Dengchao
Lanzhou Institute of Plateau Atmospheric Physics
Academia Sinica, The People's Republic of China

ABSTRACT

In winter (December, 1980 to January, 1981), simultaneous observations of atmospheric turbidity and solar radiation were taken at two sites with an elevation difference of 625 m. These sites were located at Lanzhou and at another control site outside of the city. Measurements from these sites were used to obtain the vertical distribution of atmospheric turbidity and the attenuation of solar radiation in the polluted urban atmosphere, especially in the boundary layer where air pollution is rather serious. The attenuation of sunlight in Lanzhou is caused mainly by aerosols. Empirical formulas are established, with which the atmospheric turbidity can be calculated from the direct solar radiation flux density received on the ground or vice versa. The radiation absorption by aerosols and the radiative heating rate in the lower layer are estimated from the observational data, and the effect of air pollution on the energy budget of the ground surface is discussed.

I. INTRODUCTION

Air pollution over all the world due to human activity and industry is changing the chemical components of the atmosphere. Aerosol concentrations increasing continuously in the atmosphere may change the energy budget of the ground surface and the climate of the world by decreasing the solar radiation and altering the incoming long-wave radiation. Air pollution conditions are rather serious over cities. For example, the atmospheric turbidity in Washington, D.C., U.S.A. and Davos, Switzerland has increased by about 57 and 78 percent, respectively, in recent years [1]. Owing to the air pollution, the incoming solar radiation in the central region of London is about 20-30 percent less than that in its suburban districts [2].

The city of Lanzhou is located in a narrow and deep valley on the northeast side of the Tibet (Xizang) Plateau. The wind velocity in the valley is very small in the winter, generally being less than 2 m/s, and there is a persistent strong temperature inversion over the valley and its top is usually a little higher than the mountains on both sides of the valley. The meteorological conditions in the valley prevent the diffusion of pollutants, so the air pollution over Lanzhou is rather serious. In order to study the vertical distribution of the atmospheric turbidity above the city and the effect of air pollution on solar radiation, observations of atmospheric turbidity and solar radiation near the ground were carried out in Lanzhou and its suburban districts during the period from December, 1980 to January, 1981.

This paper is abstracted from papers by Wang Yaoqu et al. [7] and Shen Zhibao et al. [6].

II. EXPERIMENT DESIGN

1. Sites

Simultaneous observations were made in Lanzhou at two sites with an elevation difference of 625 m and with different topographic conditions. One site is at the eastern part of the city (called Lanzhou) and another at the top of the mountain (called Mt. Gaolan). The whole atmsophere over Lanzhou is divided into two layers. An upper layer and lower layer, where the height of the top of Mt. Gaolan is the interface. It is obvious that the lower layer is only 625 m thick (see Fig. 1) The third site, Yuzhong, is used as a control site outside of the city, and is about 40 km to the east of Lanzhou.

2. Instruments

A sunphotometer (model MS-120) was used to measure the atmospheric turbidity. A pyrheliometer with an equatorial mount (model MS-52), a pyranometer (model MS-42), and a net radiometer (model CN-11) were used to measure direct solar radiation, diffusive sky radiation, global radiation and net radiation, respectively. All of these instruments were made in Japan and were calibrated using side-by-side field comparison tests before and after observations. The relative difference in short-wave radiation and atmospheric turbidity obtained with the two instruments was less than 3 percent, which had little effect on the comparative results reported in this paper.

The observations were made once an hour during the day and only data obtained under clear skies were used.

3. Calculating Methods

Using data measured with the sunphotometer, we can calculate the extinction coefficient due to aerosol as

$$\tau_a(\lambda) = \frac{1}{m} \ln \frac{E_o(\lambda)}{E(\lambda) \cdot S^*} - [\tau_R(\lambda) + \tau_{03}(\lambda)] \tag{1}$$

where m is the mass of the air, S* is the multiplying factor due to the mean sun-earth distance, $\tau_R(\lambda)$ and $\tau_{03}(\lambda)$ are the extinction coefficients for Rayleigh scattering and ozone absorption, respectively. The term $E(\lambda)$ is the instrument reading which is directly proportional to $I(\lambda)$, and $E_o(\lambda)$ is the instrument calibration constant which corresponds to the extraterrestrial irradiance $I(\lambda)$. Angström's wavelength exponent α as

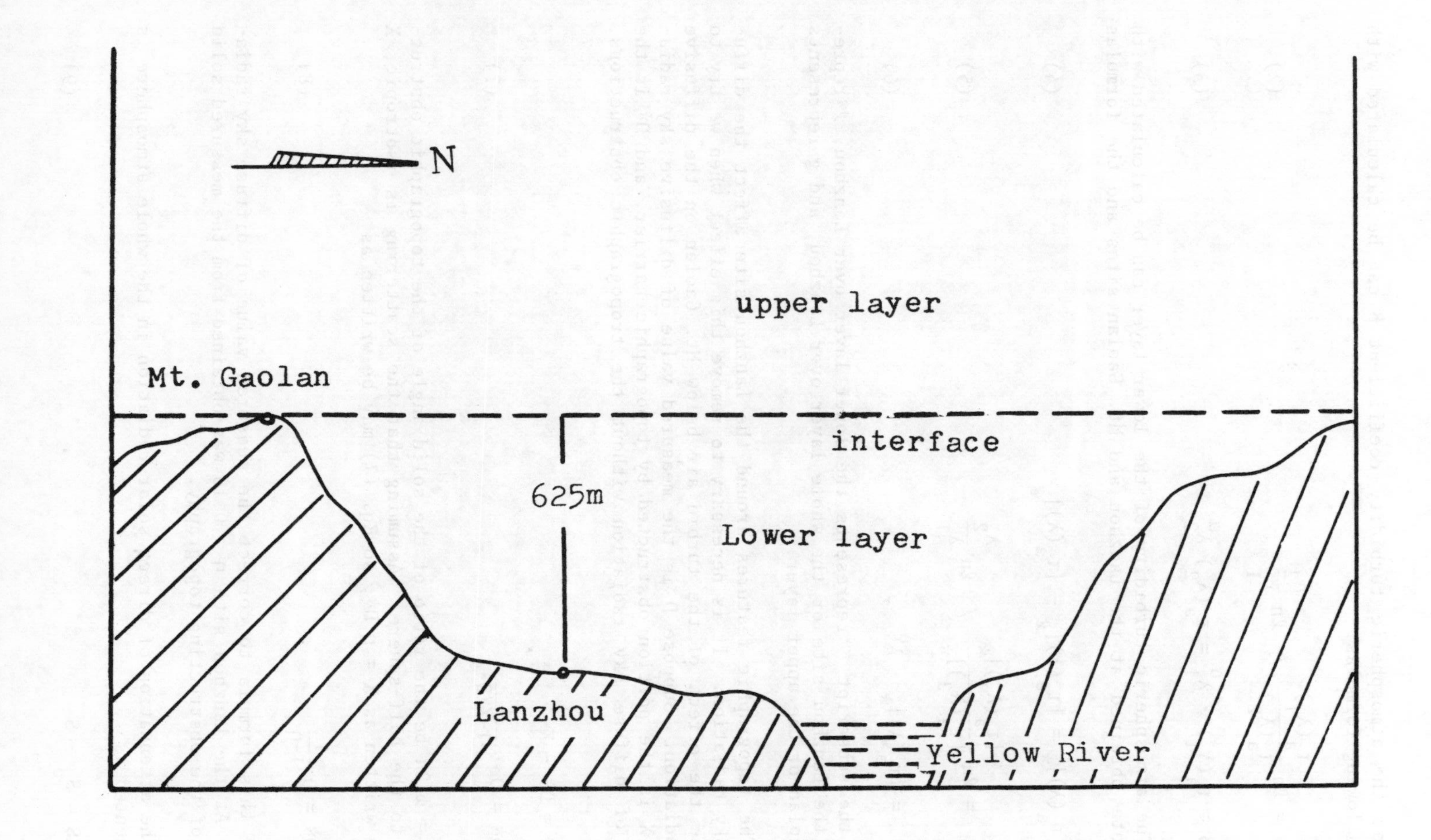

Fig. 1 A sketch of the north-south topographic section of Lanzhou city and the locations of the observational stations.

well as the atmospheric turbidity coefficient β can be calculated with the following formulas:

$$\alpha = \ell n \frac{\tau_a(\lambda_1)}{\tau_a(\lambda_2)} \; / \; \ell n \frac{\lambda_2}{\lambda_1} \tag{2}$$

$$\beta = \tau_a(\lambda_1) \cdot \lambda_1^{\alpha} = \tau_a(\lambda_2)\, \lambda_2^{\alpha} \tag{3}$$

The atmospheric turbidity of the lower layer can be calculated with the data obtained at the Lanzhou and Mt. Gaolan sites and the formulas

$$[\tau_a(\lambda)]_* = [\tau_a(\lambda)]_\ell - [\tau_a(\lambda)]_g \tag{4}$$

$$= \ell n \frac{[\tau_a(\lambda_1)]_*}{[\tau_a(\lambda_2)]_*} \; / \; \ell n \frac{\lambda_2}{\lambda_1} \tag{5}$$

$$= [\tau_a(\lambda)]_* \cdot \lambda^{\alpha} \quad , \tag{6}$$

where the subscript * represents the lower layer over Lanzhou, ℓ represents the Lanzhou site or the whole layer over Lanzhou, and g represents Mt. Gaolan or the upper layer.

The topographic features around the Lanzhou site affect the diffusive sky radiation. It is necessary to remove this effect when we try to isolate the effect of the turbid air below Mt. Gaolan on the diffusive sky radiation. Suppose D is the measured value of diffusive sky radiation, X is the portion obstructed by topographic barrier, and D* is the value of diffuse sky radiation without the tropographic obstructions, then

$$D = D^* - X$$

or $$D^* = D \cdot \frac{1}{1 - X/D^*} \tag{7}$$

Let $\eta = \omega/2\pi$ be the ratio of the solid angle of the topographic obstructions to the half-sphere. Assuming that the scattering is isotropic, X can be written as $X = \eta D^*$, so Eq. (7) may be written as

$$D^* = D \frac{1}{1-\eta} \tag{8}$$

We use this formula to correct the measured value of diffuse sky radiation. At the Lanzhou site $\eta = 0.10$ was obtained from the measured solid angle of the obstructing topography.

The attenuation of direct solar radiation in the whole atmosphere is represented as

$$\Delta S = S_o - S \tag{9}$$

where S_o is the solar constant (1.98 cal cm^{-2} min^{-1}) and S is the direct solar radiation flux density received on a surface perpendicular to the sunbeam. The attenuation of solar radiation in the lower layer over the Lanzhou site is

$$S_* = S_g - S_1 = \Delta S_1 - \Delta S_g \ . \tag{10}$$

Omitting the absorption by CO_2 and O_2, the total attenuation of solar radiation may be written as

$$\Delta S = \Delta S(R) + \Delta S(O_3) + \Delta S(H_2O) + \Delta S(a) \tag{11}$$

where $\Delta S(R)$ is the attenuation due to Rayleigh scattering, $\Delta S(O_3)$ is ozone absorption, $\Delta S(H_2O)$ is water vapor absorption and $\Delta S(a)$ is aerosol attenuation. From Sivkov's [5] empirical formula for the flux density of direct solar radiation in the ideal atmosphere, the first two terms of Eq. (11) can be calculated as

$$\Delta S(R \cdot O_3) = S_o[-0.04 + 0.16 \sqrt{m(0.949 \frac{P}{P_o} + 0.051}\ ] \tag{12}$$

where P is the pressure and P_o is the sea level pressure. The radiation absorption by water vapor is calculated by Möller's empirical formula

$$\Delta S(H_2O) = 0.172(m\ W)^{0.303} \tag{13}$$

where W is the precipitable water content obtained from radiosonde data.

The residual attenuation is taken as $\Delta S(a)$.

III. RESULTS

1. Atmospheric Turbidity

Table 1 gives the mean values of the atmospheric turbidity coefficient, β, and the wavelength exponent, α, at the Lanzhou site, Mt. Gaolan and the lower layer over Lanzhou. The mean value of β in December 1980 was 0.44 for Lanzhou, 0.12 for Mt. Gaolan and 0.32 for the lower layer. This indicates that the turbidity over the city of Lanzhou in winter is very high, although the air above Mt. Gaolan is relatively clean, with 73 percent of the aerosols restricted to the valley. Figure 2 is the cumulative frequency curve of β. The same results can be found in this figure. In the lower layer, in about 76 percent of the cases, β is greater than 0.20, but in only about 2.5 percent of the cases β is higher than 0.20 at the top of Mt. Gaolan.

Table 1 The mean values of the atmospheric turbidity coefficient, β, and wavelength exponent, α, at two sites and in the lower layer over Lanzhou (December 1980). N is the number of cases, σ is the standard error, V% is the standard relative error.

Site	N	β			α		
		$\bar{\beta}$	σ	V%	$\bar{\alpha}$	σ	V%
Lanzhou	115	0.440	0.148	33.6	0.82	0.12	14.6
Mt. Gaolan	119	0.120	0.039	32.5	0.80	0.14	17.5
Lower Layer	114	0.320	0.150	46.9	0.83	0.16	19.5

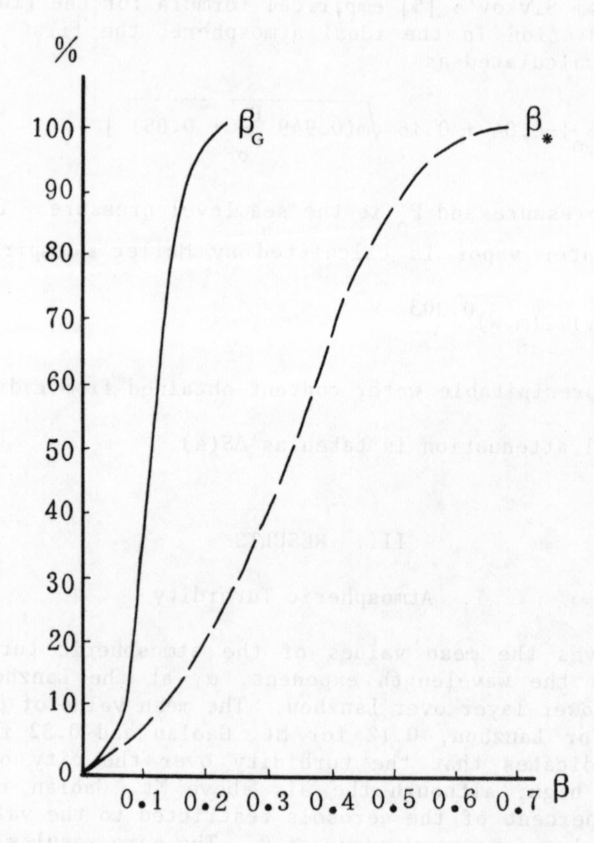

Fig. 2 The accumulated frequency curve for β_g and β_*.

The wavelength exponents at all levels over Lanzhou are almost the same, about 0.80. This result implies that the sizes of the aerosols are uniform in the vertical.

Figure 3 represents the mean diurnal variation curve of β. In the valley, β_* has a maximum value in the early morning, then it starts to decrease when the temperature inversion in the valley breaks down and β_* reaches a minimum in the early evening. On certain days, the temperature inversion in the valley is very strong and persists all morning, so the maximum value of β occurs near noon. The mean diurnal variation of β for Mt. Gaolan is different from that for the lower layer, it reaches a maximum when the most unstable stratification appears and a minimum at daybreak. Therefore, the topography of the valley and its special meteorological conditions are the main natural factors which affect the atmospheric turbidity over Lanzhou in winter.

2. The Attenuation of Solar Radiation

The attenuation of solar radiation over Lanzhou is affected by the atmospheric turbidity mentioned above. Table 2 gives the mean values of

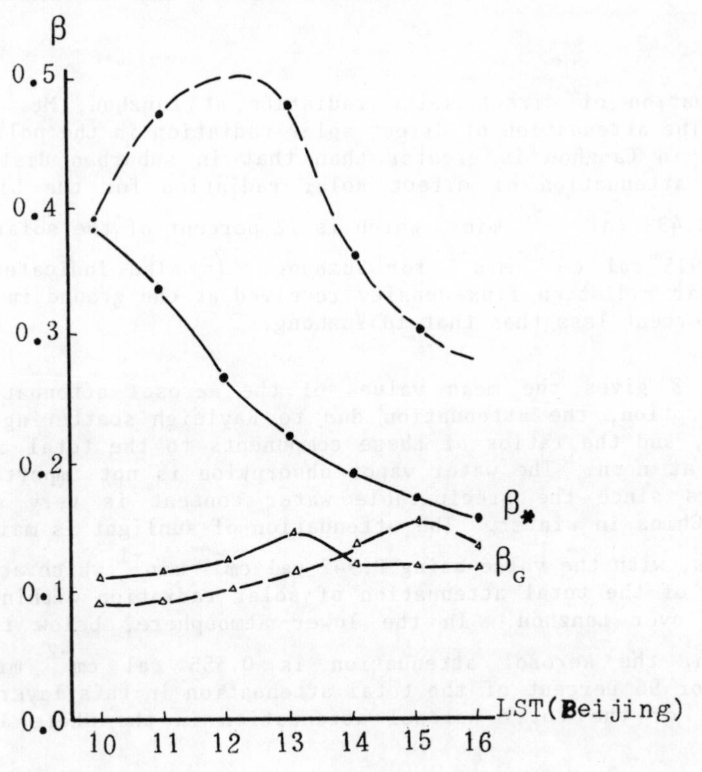

Fig. 3 Mean diurnal variations of β_g and β_*. The dashed lines are for days with very strong inversions.

Table 2 The mean values of the attenuation of direct solar radiation at Lanzhou (ΔS_1), Mt. Gaolan (ΔS_G), in the lower layer (ΔS_*) (December 1980), and at Yuzhong (ΔS_Y) (January 1981). (Unit: cal cm^{-2} min^{-1}.)

Time (Beijing Standard Time)		1000	1100	1200	1300	1400	1500	1600
Lanzhou	ΔS_1	1.765	1.601	1.537	1.432	1.415	1.482	1.596
Mt. Gaolan	ΔS_G	1.062	0.899	0.844	0.851	0.901	0.995	1.167
Lower Layer	ΔS_*	0.703	0.702	0.693	0.581	0.514	0.487	0.429
Yuzhong	ΔS_Y	1.268	1.077	0.969	0.915	0.921	0.949	1.126
$\overline{\Delta S_*}/\overline{\Delta S_\ell}$		0.40	0.44	0.45	0.41	0.36	0.33	0.27

the attenuation of direct solar radiation at Lanzhou, Mt. Gaolan and Yuzhong. The attenuation of direct solar radiation in the polluted urban atmosphere in Lanzhou is greater than that in suburban districts. At noon, the attenuation of direct solar radiation for the Lanzhou site (ΔS_1) is 1.432 cal cm^{-2} min^{-1} which is 72 percent of the solar constant, but is 0.915 cal cm^{-2} min^{-1} for Yuzhong. It also indicates that the direct solar radiation flux density received at the ground in Lanzhou is about 51 percent less than that in Yuzhong.

Table 3 gives the mean values of the aerosol attenuation, water vapor absorption, the attenuation due to Rayleigh scattering and ozone absorption, and the ratios of these components to the total attenuation in winter at noon. The water vapor absorption is not important at the three sites since the precipitable water content is very small over Northwest China in winter. The attenuation of sunlight is mainly caused by aerosols, with the value being 0.942 cal cm^{-2} min^{-1} which accounts for 66 percent of the total attenuation of solar radiation within the whole atmosphere over Lanzhou. In the lower atmosphere, below the top of Mt. Gaolan, the aerosol attenuation is 0.555 cal cm^{-2} min^{-1} which accounts for 96 percent of the total attenuation in this layer and about 59 percent of the total aerosol attenuation in the whole atmosphere.

Table 3 The mean values of the aerosol attenuation $\Delta S(a)$, water vapor absorption $\Delta S(H_2O)$, the attenuation by Rayleigh scattering and ozone absorption $\Delta S(R \cdot O_3)$, (units: cal. cm^{-2} min^{-1}) and the ratios of those components to the total attenuation (%) at Lanzhou, Mt. Gaolan, Yuzhong and in the lower layer in winter at noon.

Site		ΔS	$\Delta S(\cdot O_3)$	$\Delta S(H_2O)$	$\Delta S(a)$
Lanzhou	ΔSi	1.432	0.329	0.161	0.942
	$\Delta Si/\Delta S \cdot 100\%$	100	23	11	66
Mt. Gaolan	ΔSi	0.851	0.315	0.149	0.387
	$\Delta Si/\Delta S \cdot 100\%$	100	37	18	45
Lower Layer	ΔSi	0.581	0.014	0.012	0.555
	$\Delta Si/\Delta S \cdot 100\%$	100	2	2	96
Yuzhong	ΔSi	0.915	0.315	0.153	0.447
	$\Delta Si/\Delta S \cdot 100\%$	100	34	17	49

Because the attenuation of sunlight is mainly caused by aerosols, there is a good correlation between the direct solar radiation flux density reaching the ground surface and the atmopsheric turbidity coefficient. Figure 4 shows a linear correlation between ℓn S and β. The correlation coefficient is as high as -0.99. The empirical formulas obtained from these data are

$$S = 1.454e^{-1.040m\beta} \tag{14}$$

$$\beta = 0.1622 \quad 0.9195 \frac{\ell n\ S}{m} . \tag{15}$$

These formulas can be used to calculate β from S or vice versa in arid areas, such as Northwest China and the Northwest Tibetan (Xizang) Plateau.

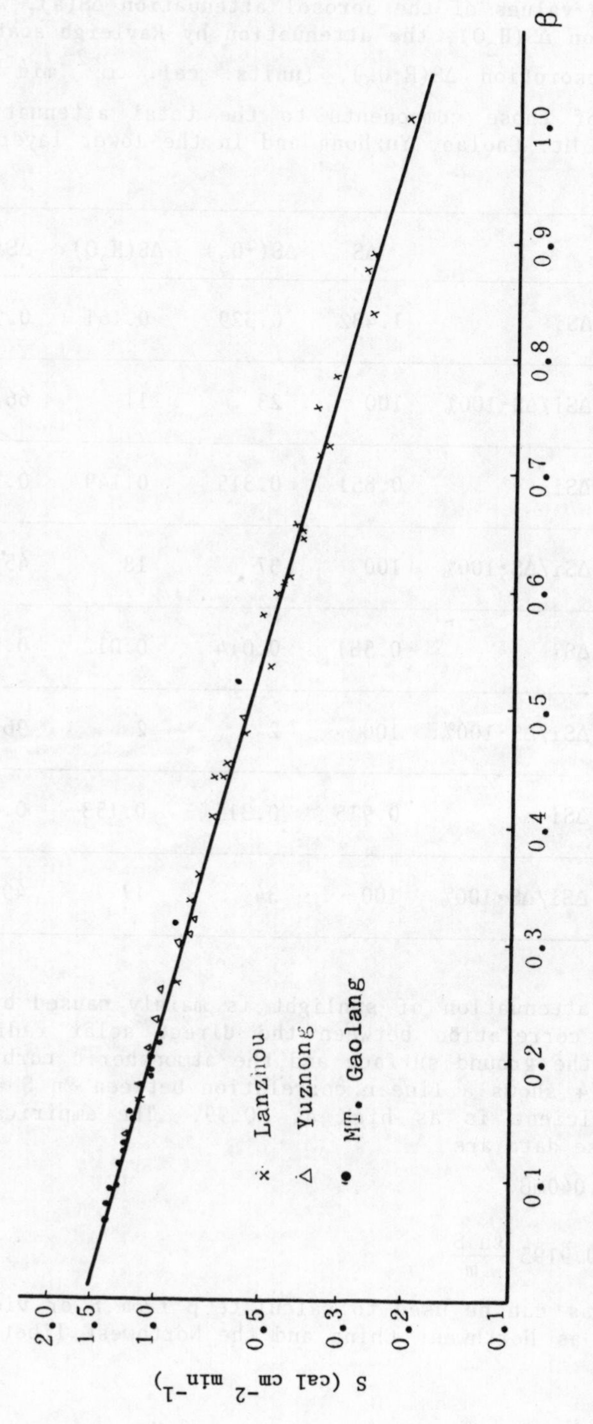

Fig. 4 The correlation curve for β and S at noon.

3. The Radiation Absorption by Aerosol and Its Effect On the Energy Budget of the Surface.

It is well known that aerosols in the atmosphere not only scatter but also absorb solar radiation. Table 4 gives the increase of diffusive sky radiation in the lower layer over Lanzhou where

$$D_* = D_1 - D_G.$$

We can see clearly from Table 4 that D_* is about 35% of the attenuation of direct solar radiation within the same layer in the vertical ($\Delta S_*(a) \cdot \sin h$, where h is the sun's altitude angle). This indicates that a large amount of the solar radiation must be absorbed by aerosol.

Table 4 The mean values of the increase of diffusive sky radiation D_* and the radiation absorption by aerosol $\Delta S_*(a)_{abs}$ (units: cal. cm^{-2} min^{-1}) as well as the radiative heating rate (°C/hr) in the lower layer over Lanzhou.

Time (Beijing Standard Time)	1000	1100	1200	1300	1400	1500	1600	Daily Mean
$D_*=D_1-D_G$	0.047	0.084	0.105	0.101	0.086	0.067	0.033	
$\Delta S_*(a) \cdot \sin h$	0.212	0.292	0.304	0.312	0.250	0.202	0.112	
$\Delta S_*(a)_{abs}$	0.160	0.199	0.223	0.200	0.154	0.128	0.075	0.163
$\partial T/\partial t$	0.72	0.90	1.00	0.90	0.69	0.58	0.34	0.73

In order to estimate the radiation absorption by aerosols in the lower layer over Lanzhou, let ΔD_* represent the forward scattering. According to some observational results [1], we assume that the back-scattering is $\frac{1}{10}$ of the total scattering, then the radiation absorption by aerosols is

$$\Delta S_*(a)_{abs} = \Delta S_*(a) \cdot \sin h - \frac{10}{9} \Delta D_*. \quad (16)$$

The results are also given in Table 4. The mean daily radiation absorption due to aerosols in 0.163 cal cm^{-2} min^{-1}. The thickness of the lower layer (ΔP) is 55 mb and the mean radiative heating rate

$$\frac{\partial T}{\partial t} = \frac{g}{cp} \cdot \frac{\Delta S_*(a)_{abs}}{\Delta P}$$

is 0.73°C/hr. If we consider that the day runs from 9 a.m. to 5 p.m.

(Beijing Standard Time) in winter, the total heating in the daytime reaches 5.8°C which is larger than that reported by Murai et al. [3].

It is certain that the attenuation of solar radiation decreases the net radiation at the surface in Lanzhou, but the radiative heating in the lower layer may increase the incoming long-wave radiation to the surface which may increase the net radiation of the ground. According to the measurements at Hamilton, Ontario, Canada, made by Rouse et al. [4], the incoming long-wave radiation excess largely compensates for the solar radiation deficit. Therefore, the incoming all-wave radiation is about the same for the industrial and control stations both for the daytime and nighttime periods.

Table 5 gives the mean difference of global radiation ($Q=S'+D=S \cdot \sin h+D$) and the net radiation between the Lanzhou site and Mt. Gaolan. In comparison with Mt. Gaolan the global radiation for the Lanzhou site is about 23-44 percent less and the net radiation is about 15-57 percent less.

Table 5 The mean difference of global radiation, $\Delta Q_*=Q_1-Q_g$, and net radiation, $\Delta B_*=B_1-B_g$, between the Lanzhou site and Mt. Gaolan (units: cal cm^{-2} min^{-1}).

	Time (Beijing Standard Time)	1000	1100	1200	1300	1400	1500	1600
Q	$Q_*=Q_1-Q_g$	-0.165	-0.208	-0.235	-0.211	-0.164	-0.134	-0.078
	$Q_*/Q_g \cdot 100\%$	-44	-37	-34	-29	-24	-25	-23
B	$B_*=B_1-B_g$	-0.06	-0.08	-0.09	-0.07	-0.04	-0.03	-0.04
	$B_*/B_g \cdot 100\%$	-50	-35	-31	-24	-15	-17	-57

Figure 5 shows the mean diurnal variation curves of global radiation for Lanzhou, Mt. Gaolan and Yuzhong. The values for Mt. Gaolan and Yuzhong are in close agreement, so we conclude that the net radiation for Lanzhou is also less than that for Yuzhong. This conclusion implies that the heavily polluted atmosphere over Lanzhou decreases the incoming all-wave radiation and cools down the ground surface.

IV. CONCLUSIONS

Based on the results given above, the following conclusions are drawn:

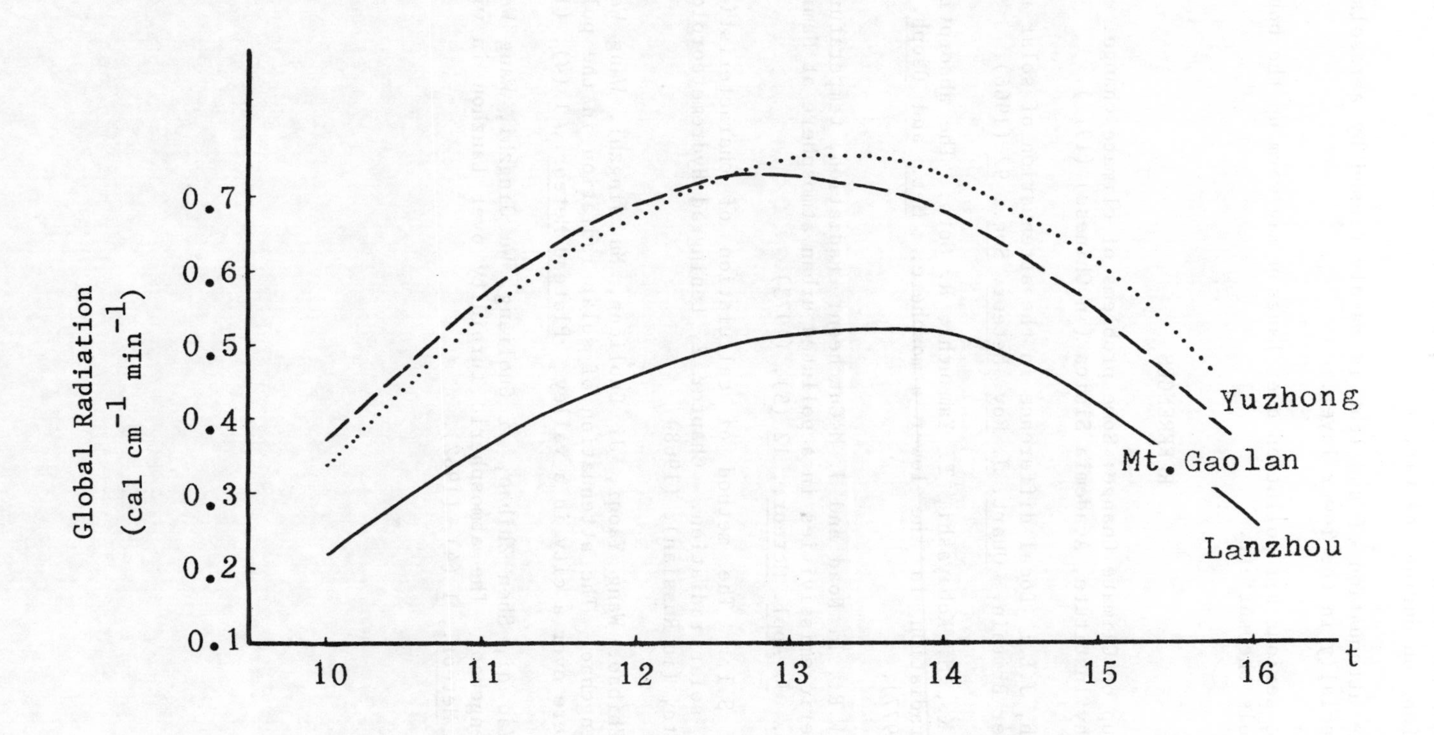

Fig. 5 The mean diurnal variations of global radiation for the Lanzhou site, Mt. Gaolan and Yuzhong.

(1) The topography of the valley and its special meteorological conditions are the main factors which cause serious air pollution over the city of Lanzhou during the winter.

(2) The attenuation of sunlight is mainly caused by aerosols over Lanzhou, especially in the lower layer.

(3) The serious air pollution over Lanzhou may warm up the boundary layer but cools the surface.

REFERENCES

1. The Group on Climate Change: Some problems of climate change. Geography Institute, Academia Sinica (in Chinese) (1977).

2. Monteith, J.L.: Local difference in the attenuation of solar radiation over Britain. Quart. J. Roy. Meteor. Soc., 92 (1966).

3. Murai, K., M. Kobayashi, T. Yamauchi, R. Goto: The absorption of solar radiation in the lower atmosphere. Met. and Geoph., 27, (1), (1977).

4. Rouse, W.R., D. Noad and J. McCutcheon: Radiation, temperature and atmospheric emissivities in a polluted urban atmosphere at Hamilton, Ontario. J. Appl. Meteor., 12 (5), (1973).

5. Sivkov, S.I.: The method of calculation of characteristics of monthly solar radiation. Chapter 2, Leningrad Hydrometeorological Institute, (in Russian), (1968).

6. Shen, Zhibao, Wang Yaoqi, Ji Guoliang, Mu Jingzhi, Wang Wenhua, Shui Dengchao: The attenuation of solar radiation in the polluted atmosphere over a city in a valley. Plateau Meteor., 1 (4), (1982).

7. Wang, Yao qi, Shen Zhibao, Ji Guoliang, Wu Jingzhi, Wang Wenhua, Shui Dengchao: The atmospheric turbidity over Lanzhou in winter. Plateau Meteor., 1 (4), (1982).

SUMMARY AND RECOMMENDATIONS OF DISCUSSION GROUPS

I. LARGE-SCALE EFFECTS

Based on its review of both the synoptic-scale and planetary-scale influences of mountains on the atmospheric general circulation and climate, the discussion group on large-scale effects noted that our knowledge of these influences is rather incomplete. In order to increase our ability to observe, forecast and understand large-scale orographic effects, the group noted that several avenues of observational, modelling and theoretical research need further attention, and recommended that these activities be pursued in a cooperative manner between Chinese and American atmospheric scientists.

1. In the area of observational or diagnostic research, the group recommended that:

   (1) A global year-long data base be identified and made available for cooperative work; these data might consist, for example, of those for the year 1979 (which include the FGGE and its special observing periods during which the large-scale observational network was at a maximum), or the year 1981 (during which the Asian monsoon was especially well developed). Such a data set would permit joint diagnostic studies of the planetary-scale and monsoonal circulation in response to orography in both winter and summer, as well as provide a common verification base for forecast model tests and seasonal simulation experiments. In assembling this data set, efforts should be made to include all available Chinese data as well as the analyses underway in the United States.

   (2) Efforts be made to assemble a regional data base for cooperative diagnostic studies of the heat balance over mountainous terrain. These data should include satellite observations of the free air, surface radiances and cloud properties, as well as surface and upper-air observations during one or more periods of particular meteorological interest over both the Tibetan Plateau and the Rocky Mountain region. From this data base, comparative studies of various methods of determining the atmospheric heat and moisture balances may be performed, and new methods may be developed which are especially appropriate in regions of irregular terrain.

2. In the area of numerical modelling research, the group recommended that:

   (1) A program of cooperative model development and testing be organized in which the effects of various numerical algorithms, computational schemes and model formulations on terrain-induced flow could be examined using American computer systems. Such research might be implemented, for example, between scientists at individual universities, research institutes or operational centers

by means of short- or longer-term visits, and should permit the accumulation of new information on the performance of models for both short-range prediction and extended simulation. Such a research program might also include model studies of the effects of surface albedo and snow cover on the circulation near elevated terrain, especially as it affects the onset and cessation of the Asian monsoon and the flow in the associated transition seasons.

(2) Cooperative research be initiated to develop improved methods of parameterization of those physical processes which are believed to be important in elevated terrain. Particular attention should be given to the parameterization of the mountainous planetary boundary layer, of mountain cumulus clouds, and of surface radiative fluxes in irregular terrain.

3. In the area of theoretical research, the group recommended that:

(1) Attention be given to the possible role of large-scale terrain features in the initiation and maintenance of blocking phenomena, and to the more general question of the possibility of multiple equilibrium circulation states. Such studies might be successfully carried out with simplified dynamical models, and their results examined in the light of the results from more general model studies and observational data.

(2) Consideration be given to the problem of representing the effects of large-scale mountains on the energetics of the atmosphere, such that the locally-important exchanges of potential and kinetic energy can be properly represented in the large-scale energy budget and in the dynamics of the stationary and transient planetary waves.

II. MESOSCALE EFFECTS

Similarities and contrasts in results of studies in our respective countries will add a new dimension of common understanding. Accordingly the subgroup indorsed the concept of another workshop symposium within several years to exchange ideas. The collaboration should be augmented in the interim with close contact among individual investigators to allow exchanges in technical resources ranging from data sets to computational power.

In reviewing papers of the Mountain Meteorology Workshop, the subgroup on mesoscale processes noted that the influence of the Tibetan Plateau is reflected most profoundly in rainfall patterns over China. Not only does the plateau create rather extreme climatic zones of heavy and sparse rainfall, but it exerts strong control on the duration, amount, and frequency of mesoscale precipitation as well. In both regards, the Rocky Mountains play a similar, but less extreme, role in the behavior of United States weather.

Since heavy rainfall plays a critical role in the economy of both nations and in the lives of their peoples, the subgroup felt that joint scientific studies in this area would be of considerable mutual benefit.

Work recently and over the past few decades has suggested that some key fundamental processes link large mountain chains to significant weather events. These include:

1. Surface boundary layer physics
2. Low-level jet stream genesis
3. Cyclogenesis over slopes and lee areas
4. Propagation of synoptic systems over terrain
5. Monsoon interactions with topography.

The convective systems connected with these phenomena deserve more attention than either nation has been able to put forth alone.

An enhanced effort to develop mesoscale climatology in relation to synoptic regimes should be undertaken in both nations, not only to understand better the physical processed involved, but also to aid in practical weather forecasting. Mutual, concerted efforts in 1) diagnosing detailed atmospheric structures, 2) numerical modelling for short-range prediction, and 3) the development of new physical concepts in the fundamental areas listed above should receive high priority.